GÖDEL, ESCHER, BACH:

an Eternal Golden Braid

GÖDEL, ESCHER, BACH:

an Eternal Golden Braid

Douglas R. Hofstadter

A Member of the Perseus Books Group

 Originally published by Basic Books, Inc., New York, in April 1979.

Library of Congress Cataloging-in-Publication Data

Hofstadter, Douglas R. 1945–
Gödel, Escher, Bach.

Bibliography: p. 746
Includes Index.
I. Metamathematics. 2. Symmetry. 3. Artificial intelligence.
4. Bach, Johann Sebastian, 1685–1750.
5. Escher, Maurits Cornelis, 1898–1971.
6. Gödel, Kurt. I. Title.
[QA9.8H63 1980] 510'.1 80-11354
ISBN 13: 978-465-02656-2
ISBN 10: 0-465-02656-2

Manufactured in the United States of America
EBA 06 07 08 09 18 17 16 15

To M. and D.

Preface to *GEB*'s Twentieth-anniversary Edition

SO WHAT *IS* this book, *Gödel, Escher, Bach: an Eternal Golden Braid* — usually known by its acronym, *"GEB"* — really all about?

That question has hounded me ever since I was scribbling its first drafts in pen, way back in 1973. Friends would inquire, of course, what I was so gripped by, but I was hard pressed to explain it concisely. A few years later, in 1980, when *GEB* found itself for a while on the bestseller list of *The New York Times,* the obligatory one-sentence summary printed underneath the title said the following, for several weeks running: "A scientist argues that reality is a system of interconnected braids." After I protested vehemently about this utter hogwash, they finally substituted something a little better, just barely accurate enough to keep me from howling again.

Many people think the title tells it all: a book about a mathematician, an artist, and a musician. But the most casual look will show that these three individuals *per se,* august though they undeniably are, play but tiny roles in the book's content. There's no way the book is about those three people!

Well, then, how about describing *GEB* as "a book that shows how math, art, and music are really all the same thing at their core"? Again, this is a million miles off — and yet I've heard it over and over again, not only from nonreaders but also from readers, even very ardent readers, of the book.

And in bookstores, I have run across *GEB* gracing the shelves of many diverse sections, including not only math, general science, philosophy, and cognitive science (which are all fine), but also religion, the occult, and God knows what else. Why is it so hard to figure out what this book is about? Certainly it's not just its length. No, it must be in part that *GEB* delves, and not just superficially, into so many motley topics — fugues and canons, logic and truth, geometry, recursion, syntactic structures, the nature of meaning, Zen Buddhism, paradoxes, brain and mind, reductionism and holism, ant colonies, concepts and mental representations, translation, computers and their languages, DNA, proteins, the genetic code, artificial intelligence, creativity, consciousness and free will — sometimes even art and music, of all things! — that many people find it impossible to locate the core focus.

The Key Images and Ideas that Lie at the Core of *GEB*

Needless to say, this widespread confusion has been quite frustrating to me over the years, since I felt sure I had spelled out my aims over and over in the text itself. Clearly, however, I didn't do it sufficiently often, or sufficiently clearly. But since now I've got the chance to do it once more — and in a prominent spot in the book, to boot — let me try one last time to say why I wrote this book, what it is about, and what its principal thesis is.

In a word, *GEB* is a very personal attempt to say how it is that animate beings can come out of inanimate matter. What is a self, and how can a self come out of stuff that is as selfless as a stone or a puddle? What is an "I", and why are such things found (at least so far) only in association with, as poet Russell Edson once wonderfully phrased it, "teetering bulbs of dread and dream" — that is, only in association with certain kinds of gooey lumps encased in hard protective shells mounted atop mobile pedestals that roam the world on pairs of slightly fuzzy, jointed stilts?

GEB approaches these questions by slowly building up an analogy that likens inanimate molecules to meaningless symbols, and further likens selves (or "I"'s or "souls", if you prefer — whatever it is that distinguishes animate from inanimate matter) to certain special swirly, twisty, vortex-like, and *meaningful* patterns that arise only in particular types of systems of meaningless symbols. It is these strange, twisty patterns that the book spends so much time on, because they are little known, little appreciated, counterintuitive, and quite filled with mystery. And for reasons that should not be too difficult to fathom, I call such strange, loopy patterns "strange loops" throughout the book, although in later chapters, I also use the phrase "tangled hierarchies" to describe basically the same idea.

This is in many ways why M. C. Escher — or more precisely, his art — is prominent in the "golden braid", because Escher, in his own special way, was just as fascinated as I am by strange loops, and in fact he *drew* them in a variety of contexts, all wonderfully disorienting and fascinating. When I was first working on my book, however, Escher was totally out of the picture (or out of the loop, as we now say); my working title was the rather mundane phrase "Gödel's Theorem and the Human Brain", and I gave no thought to inserting paradoxical pictures, let alone playful dialogues. It's just that time and again, while writing about my notion of strange loops, I would catch fleeting glimpses of this or that Escher print flashing almost subliminally before my mind's eye, and finally one day I realized that these images were so connected in my own mind with the ideas that I was writing about that for me to deprive my readers of the connection that I myself felt so strongly would be nothing less than perverse. And so Escher's art was welcomed on board. As for Bach, I'll come back to his entry into my "metaphorical fugue on minds and machines" a little later.

Back to strange loops, right now. *GEB* was inspired by my long-held conviction that the "strange loop" notion holds the key to unraveling the mystery that we conscious beings call "being" or "consciousness". I was first hit by this idea when, as a teen-ager, I found myself obsessedly pondering the quintessential strange loop that lies at the core of the proof of Kurt Gödel's famous incompleteness theorem in mathematical logic — a rather arcane place, one might well think, to stumble across the secret behind the nature of selves and "I"'s, and yet I practically heard it screaming up at me from the pages of Nagel and Newman that this was what it was all about.

This preface is not the time and place to go into details — indeed, that's why the tome you're holding was written, so it would be a bit presumptuous of me to think I could outdo its author in just these few pages! — but one thing has to be said straight off: the Gödelian strange loop that arises in formal systems in mathematics (*i.e.,* collections of rules for churning out an endless series of mathematical truths solely by mechanical symbol-shunting without any regard to meanings or ideas hidden in the shapes being manipulated) is a loop that allows such a system to "perceive itself", to talk about itself, to become "self-aware", and in a sense it would not be going too far to say that by virtue of having such a loop, a formal system *acquires a self.*

Meaningless Symbols Acquire Meaning Despite Themselves

What is so weird in this is that the formal systems where these skeletal "selves" come to exist are built out of nothing but meaningless symbols. The self, such as it is, arises solely because of a special type of swirly, tangled *pattern* among the meaningless symbols. But now a confession: I am being a bit coy when I repeatedly type the phrase "meaningless symbols" (as at the ends of both of the previous sentences), because a crucial part of my book's argument rests on the idea that meaning cannot be kept out of formal systems when sufficiently complex isomorphisms arise. Meaning comes in despite one's best efforts to keep symbols meaningless!

Let me rephrase these last couple of sentences without using the slightly technical term "isomorphism". When a system of "meaningless" symbols has patterns in it that accurately track, or mirror, various phenomena in the world, then that tracking or mirroring imbues the symbols with some degree of meaning — indeed, such tracking or mirroring is no less and no more than what meaning is. Depending on how complex and subtle and reliable the tracking is, different degrees of meaningfulness arise. I won't go further into this here, for it's a thesis that is taken up quite often in the text, most of all in Chapters 2, 4, 6, 9, and 11.

Compared to a typical formal system, human language is unbelievably fluid and subtle in its patterns of tracking reality, and for that reason the symbols in formal systems can seem quite arid; indeed, without too much trouble, one can look at them as totally devoid of meaning. But then again, one can look at a newspaper written in an unfamiliar writing system, and the strange shapes seem like nothing more than wondrously intricate but totally meaningless patterns. Thus even human language, rich though it is, can be drained of its seeming significance.

As a matter of fact, there are still quite a few philosophers, scientists, and so forth who believe that patterns of symbols *per se* (such as books or movies or libraries or CD-ROM's or computer programs, no matter how complex or dynamic) *never* have meaning on their own, but that meaning instead, in some most mysterious manner, springs only from the organic chemistry, or perhaps the quantum mechanics, of processes that take place in carbon-based biological brains. Although I have no patience with this parochial, bio-chauvinistic view, I nonetheless have a pretty clear sense of its intuitive appeal. Trying to don the hat of a believer in the primacy, indeed the uniqueness, of brains, I can see where such people are coming from.

Such people feel that some kind of "semantic magic" takes place only inside our "teetering bulbs", somewhere behind pairs of eyeballs, even though they can never quite put their finger on how or why this is so; moreover, they believe that this semantic magic is what is responsible for the existence of human selves, souls, consciousness, "I"'s. And I, as a matter of fact, quite agree with such thinkers that selves and semantics — in other words, that me's and meanings — *do* spring from one and the same source; where I take issue with these people is over their contention that such phenomena are due entirely to some special, though as yet undiscovered, properties of the microscopic hardware of brains.

As I see it, the only way of overcoming this magical view of what "I" and consciousness are is to keep on reminding oneself, unpleasant though it may seem, that the "teetering bulb of dread and dream" that nestles safely inside one's own cranium is a purely physical object made up of completely sterile and inanimate components, all of which obey exactly the same laws as those that govern all the rest of the universe, such as pieces of text, or CD-ROM's, or computers. Only if one keeps on bashing up against this disturbing fact can one slowly begin to develop a feel for the way out of the mystery of consciousness: that the key is not the *stuff* out of which brains are made, but the *patterns* that can come to exist inside the stuff of a brain.

This is a liberating shift, because it allows one to move to a different level of considering what brains are: as *media* that support complex patterns that mirror, albeit far from perfectly, the world, of which, needless to say, those brains are themselves denizens — and it is in the inevitable self-mirroring that arises, however impartial or imperfect it may be, that the strange loops of consciousness start to swirl.

Kurt Gödel Smashes through Bertrand Russell's Maginot Line

I've just claimed that the shift of focus from material components to abstract patterns allows the quasi-magical leap from inanimate to animate, from nonsemantic to semantic, from meaningless to meaningful, to take place. But how does this happen? After all, not *all* jumps from matter to pattern give rise to consciousness or soul or self, quite obviously: in a word, not all patterns are conscious. What kind of pattern is it, then, that is the telltale mark of a *self*? *GEB*'s answer is: a strange loop.

The irony is that the first strange loop ever found — and my model for the concept in general — was found in a system *tailor-made to keep loopiness out.* I speak of Bertrand Russell and Alfred North Whitehead's famous treatise *Principia Mathematica,* a gigantic, forbidding work laced with dense, prickly symbolism filling up volume after volume, whose creation in the years 1910–1913 was sparked primarily by its first author's desperate quest for a way to circumvent paradoxes of self-reference in mathematics.

At the heart of *Principia Mathematica* lay Russell's so-called "theory of types", which, much like the roughly contemporaneous Maginot Line, was designed to keep "the enemy" out in a most staunch and watertight manner.

For the French, the enemy was Germany; for Russell, it was self-reference. Russell believed that for a mathematical system to be able to talk about itself in any way whatsoever was the kiss of death, for self-reference would — so he thought — necessarily open the door to self-contradiction, and thereby send all of mathematics crashing to the ground. In order to forestall this dire fate, he invented an elaborate (and infinite) hierarchy of levels, all sealed off from each other in such a manner as to definitively — so he thought — block the dreaded virus of self-reference from infecting the fragile system.

It took a couple of decades, but eventually the young Austrian logician Kurt Gödel realized that Russell and Whitehead's mathematical Maginot Line against self-reference could be most deftly circumvented (just as the Germans in World War II would soon wind up deftly sidestepping the real Maginot Line), and that self-reference not only had lurked from Day One in *Principia Mathematica,* but in fact plagued poor *PM* in a totally unremovable manner. Moreover, as Gödel made brutally clear, this thorough riddling of the system by self-reference was not due to some weakness in *PM,* but quite to the contrary, it was due to its *strength.* Any similar system would have exactly the same "defect". The reason it had taken so long for the world to realize this astonishing fact is that it depended on making a leap somewhat analogous to that from a brain to a self, that famous leap from inanimate constituents to animate patterns.

For Gödel, it all came into focus in 1930 or so, thanks to a simple but wonderfully rich discovery that came to be known as "Gödel numbering" — a mapping whereby the long linear arrangements of strings of symbols in any formal system are mirrored precisely by mathematical relationships among certain (usually astronomically large) whole numbers. Using his mapping between elaborate patterns of meaningless symbols (to use that dubious term once again) and huge numbers, Gödel showed how a statement *about* any mathematical formal system (such as the assertion that *Principia Mathematica* is contradiction-free) can be translated into a mathematical statement *inside* number theory (the study of whole numbers). In other words, any metamathematical statement can be imported *into* mathematics, and in its new guise the statement simply asserts (as do all statements of number theory) that certain whole numbers have certain properties or relationships to each other. But on another level, it also has a vastly different meaning that, on its surface, seems as far removed from a statement of number theory as would be a sentence in a Dostoevsky novel.

By means of Gödel's mapping, any formal system designed to spew forth truths about "mere" numbers would also wind up spewing forth truths — inadvertently but inexorably — about its own properties, and would thereby become "self-aware", in a manner of speaking. And of all the clandestine instances of self-referentiality plaguing *PM* and brought to light by Gödel, the most concentrated doses lurked in those sentences that talked about their *own* Gödel numbers, and in particular said some very odd things about themselves, such as "I am not provable inside *PM*". And let me repeat: such twisting-back, such looping-around, such self-enfolding, far from being an eliminable defect, was an inevitable by-product of the system's vast power.

Not too surprisingly, revolutionary mathematical and philosophical consequences tumbled out of Gödel's sudden revelation that self-reference abounded in the bosom of the bastion so carefully designed by Russell to keep it out at all costs; the most famous such consequence was the so-called "essential incompleteness" of formalized mathematics. That notion will be carefully covered in the chapters to come, and yet, fascinating though it is, incompleteness is not in itself central to *GEB*'s thesis. For *GEB*, the most crucial aspect of Gödel's work is its demonstration that a statement's *meaning* can have deep consequences, even in a supposedly meaningless universe. Thus it is the *meaning* of Gödel's sentence G (the one that asserts "G is not provable inside *PM*") that guarantees that G is not provable inside *PM* (which is precisely what G itself claims). It is as if the sentence's hidden Gödelian meaning had some kind of power over the vacuous symbol-shunting, meaning-impervious rules of the system, preventing them from ever putting together a demonstration of G, no matter what they do.

Upside-down Causality and the Emergence of an "I"

This kind of effect gives one a sense of crazily twisted, or upside-down, causality. After all, shouldn't meanings that one chooses to read into strings of meaningless symbols be totally without consequence? Even stranger is that the *only reason* sentence G is not provable inside *PM* is its self-referential meaning; indeed, it would seem that G, being a *true* statement about whole numbers, *ought* to be provable, but — thanks to its extra level of meaning as a statement about itself, asserting its own nonprovability — it is not.

Something very strange thus emerges from the Gödelian loop: the revelation of the causal power of meaning in a rule-bound but meaning-free universe. And this is where my analogy to brains and selves comes back in, suggesting that the twisted loop of *selfhood* trapped inside an inanimate bulb called a "brain" also has causal power — or, put another way, that a mere pattern called "I" can shove around inanimate particles in the brain no less than inanimate particles in the brain can shove around patterns. In short, an "I" comes about — in my view, at least — via a kind of vortex whereby patterns in a brain mirror the brain's mirroring of the world, and eventually mirror themselves, whereupon the vortex of "I" becomes a real, causal entity. For an imperfect but vivid concrete analogue to this curious abstract phenomenon, think of what happens when a TV camera is pointed at a TV screen so as to display the screen on itself (and that screen on itself, etc.) — what in *GEB* I called a "self-engulfing television", and in my later writings I sometimes call a "level-crossing feedback loop".

When and only when such a loop arises in a brain or in any other substrate, is a *person* — a unique new "I" — brought into being. Moreover, the more self-referentially rich such a loop is, the more conscious is the self to which it gives rise. Yes, shocking though this might sound, consciousness is not an on/off phenomenon, but admits of degrees, grades, shades. Or, to put it more bluntly, there are bigger souls and smaller souls.

Small-souled Men, Beware!

I can't help but recall, at this point, a horribly elitist but very droll remark by one of my favorite writers, the American "critic of the seven arts", James Huneker, in his scintillating biography of Frédéric Chopin, on the subject of Chopin's étude Op. 25, No. 11 in A minor, which for me, and for Huneker, is one of the most stirring and most sublime pieces of music ever written: "Small-souled men, no matter how agile their fingers, should avoid it."

"Small-souled men"?! Whew! Does *that* phrase ever run against the grain of American democracy! And yet, leaving aside its offensive, archaic sexism (a crime I, too, commit in *GEB,* to my great regret), I would suggest that it is only because we all tacitly *do* believe in something like Huneker's shocking distinction that most of us are willing to eat animals of one sort or another, to smash flies, swat mosquitos, fight bacteria with antibiotics, and so forth. We generally concur that "men" such as a cow, a turkey, a frog, and a fish all possess *some* spark of consciousness, *some* kind of primitive "soul", but by God, it's a good deal smaller than *ours* is — and that, no more and no less, is why we "men" feel that we have the perfect right to extinguish the dim lights in the heads of these fractionally-souled beasts and to gobble down their once warm and wiggling, now chilled and stilled protoplasm with limitless gusto, and not to feel a trace of guilt while doing so.

Enough sermonizing! The real point here is that not all strange loops give rise to souls as grand and glorious as yours and mine, dear reader. Thus, for example, I would not want you or anyone else to walk away from reading all or part of *GEB,* shake their head and say with sadness, "That weird Hofstadter guy has convinced himself that Russell and Whitehead's *Principia Mathematica* is a conscious person with a soul!" Horsefeathers! Balderdash! Poppycock! Gödel's strange loop, though it is my paragon for the concept, is nonetheless only the most bare-bones strange loop, and it resides in a system whose complexity is pathetic, relative to that of an organic brain. Moreover, a formal system is static; it doesn't change or grow over time. A formal system does not live in a society of other formal systems, mirroring them inside itself, and being mirrored in turn inside its "friends". Well, I retract that last remark, at least a bit: any formal system as powerful as *PM* does in fact contain models not just of itself but of an infinite number of other formal systems, some like it, some very much unlike it. That is essentially what Gödel realized. But still, there is no counterpart to time, no counterpart to development, let alone to birth and death.

And so whatever I say about "selves" coming to exist in mathematical formal systems has to be taken with the proper grain of salt. Strange loops are an abstract structure that crops up in various media and in varying degrees of richness. *GEB* is in essence a long proposal of strange loops as a metaphor for how selfhood originates, a metaphor by which to begin to grab a hold of just what it is that makes an "I" seem, at one and the same time, so terribly real and tangible to its own possessor, and yet also so vague, so impenetrable, so deeply elusive.

I personally cannot imagine that consciousness will be fully understood without reference to Gödelian strange loops or level-crossing feedback loops. For that reason, I must say, I have been surprised and puzzled that the past few years' flurry of books trying to unravel the mysteries of consciousness almost never mention anything along these lines. Many of these books' authors have even read and savored *GEB,* yet nowhere is its core thesis echoed. It sometimes feels as if I had shouted a deeply cherished message out into an empty chasm and nobody heard me.

The Earliest Seeds of *GEB*

Why, one might wonder, if the author's aim was merely to propose a theory of strange loops as the crux of our consciousness and the source of our irrepressible "I"-feeling, did he wind up writing such a vast book with so many seeming digressions in it? Why on earth did he drag in fugues and canons? Why recursion? And Zen? And molecular biology? Et cetera…

The truth of the matter is, when I started out, I didn't have the foggiest idea that I would wind up talking about these kinds of things. Nor did I dream that my future book would include dialogues, let alone dialogues based on musical forms. The complex and ambitious nature of my project evolved only gradually. In broad strokes, it came about this way.

I earlier alluded to my reading, as a teen-ager, of Ernest Nagel and James R. Newman's little book *Gödel's Proof.* Well, that book just radiated excitement and depth to me, and it propelled me like an arrow straight into the study of symbolic logic. Thus, as an undergraduate math major at Stanford and a few years later, in my short-lived career as a graduate student in math at Berkeley, I took several advanced logic courses, but to my bitter disappointment, all of them were arcane, technical, and utterly devoid of the magic I'd known in Nagel and Newman. The upshot of my taking these highbrow courses was that my keen teen interest in Gödel's wondrous proof and its "strange loopiness" was nearly killed off. Indeed, I was left with such a feeling of sterility that in late 1967, almost in desperation, I dropped out of math grad school in Berkeley and took up a new identity as physics grad student at the University of Oregon in Eugene, where my once-ardent fascination with logic and metamathematics went into deep dormancy.

Several years passed, and then one day in May of 1972, while browsing the math shelves in the University of Oregon bookstore, I stumbled across philosopher Howard DeLong's superb book *A Profile of Mathematical Logic,* took a chance on buying it, and within weeks, my old love for the great Gödelian mysteries and all they touch on was reawakened. Ideas started churning around like mad inside my teetering bulb of dread and dream.

Despite this joy, I was very discouraged with the way my physics studies and my life in general were going, so in July I packed all my belongings into a dozen or so cardboard boxes and set out on an eastward trek across the vast American continent in Quicksilver, my faithful 1956 Mercury. Where I was headed, I wasn't sure. All I knew is that I was looking for a new life.

After crossing the beautiful Cascades and eastern Oregon's desert, I wound up in Moscow, Idaho. Since Quicksilver had a little engine trouble and needed some repair, I took advantage of the spare time and went to the University of Idaho's library to look up some of the articles about Gödel's proof in DeLong's annotated bibliography. I photocopied several of them, and in a day or so headed off toward Montana and Alberta. Each night I would stop and pitch my little tent, sometimes in a forest, sometimes by a lake, and then I would eagerly plunge by flashlight into these articles until I fell asleep in my sleeping bag. I was starting to understand many Gödelian matters ever more clearly, and what I was learning was truly enthralling.

From Letter to Pamphlet to Seminar

After a few days in the Canadian Rockies, I headed south again and eventually reached Boulder, Colorado. There, one afternoon, a host of fresh ideas started gushing out in a spontaneous letter to my old friend Robert Boeninger. After several hours of writing, I saw that although my letter was longer than I'd expected — thirty handwritten pages or so — I'd said only about half of what I'd wanted to say. This made me think that maybe I should write a pamphlet, not a letter, and to this day, Robert has never received my unfinished missive.

From Boulder I headed further east, bouncing from one university town to another, and eventually, almost as if it had been beckoning me the whole time, New York City loomed as my ultimate goal. Indeed, I wound up spending several months in Manhattan, taking graduate courses at City College and teaching elementary physics to nurses at Hunter College, but as 1973 rolled around, I faced the fact that despite loving New York in many ways, I was even more agitated than I had been in Eugene, and I decided it would be wiser to return to Oregon and to finish graduate school there.

Although my hoped-for "new life" had failed to materialize, in certain respects I was relieved to be back. For one thing, the U of O in those days had the enlightened policy that any community member could invent and teach a for-credit "SEARCH" course, as long as one or more departments approved it. And so I petitioned the philosophy and math departments to sponsor a spring-quarter SEARCH course centered on Gödel's theorem, and my request was granted. Things were looking up.

My intuition told me that my personal fascination with strange loops — not only with their philosophical importance but also with their esthetic charm — was not just some unique little neurotic obsession of mine, but could well be infectious, if only I could get across to my students that these notions were anything but dull and dry, as in those frigid, sterile logic courses I'd taken, but rather — as Nagel and Newman had hinted — were intimately related to a slew of profound and beautiful ideas in mathematics, physics, computer science, psychology, philosophy, linguistics, and so on.

I gave my course the half-dippy, half-romantic title "The Mystery of the Undecidable" in the hopes that I might attract students from wildly diverse areas, and the trick worked. Twenty-five souls were snagged, and all were enthusiastic. I vividly remember the lovely blossoms I could see out the window each day as I lectured that spring, but even more vividly I remember David Justman, who was in art history, Scott Buresh, who was in political science, and Avril Greenberg, who was an art major. These three simply devoured the ideas, and we talked and talked endlessly about them. My course thus turned out very well, both for the snaggees and for the snagger.

Sometime during the summer of 1973, I made a stab at sketching out a table of contents for my "pamphlet", and at that point, the ambitiousness of my project started dawning on me, but it still felt more like a pamphlet than a tome to me. It was only in the fall that I started writing in earnest. I had never written anything more than a few pages long, but I fearlessly plunged ahead, figuring it would take me just a few days — maybe a week or two. I was slightly off, for in fact, the very first draft (done in pen, just like my letter to Robert, but with more cross-outs) took me about a month — a month that overlapped in time with the "Yom Kippur war", which made a very deep impression on me. I realized this first draft was not the final product, but I felt I had done the major work and now it was just a question of revision.

Experiments with Literary Form Start to Take Place

As I was writing that draft, I certainly wasn't thinking about Escher pictures. Nor was I thinking about Bach's music. But one day I found myself on fire with ideas about mind, brain, and human identity, and so, shamelessly borrowing Lewis Carroll's odd couple of Achilles and the Tortoise, whose droll personalities amused me no end, I sat down and in absolute white heat dashed off a long, complex dialogue, all about a fictitious, unimaginably large book each of whose pages, on a one-by-one basis, contained exhaustive information on one specific neuron in Einstein's brain. As it happened, the dialogue featured a short section where the two characters imagined each other in another dialogue, and each of them said, "You might then say this... to which I might well reply as follows... and then you would go on..." and so forth. Because of this unusual structural feature, after I'd finally put the final period on the final speech, I flipped back to the top of page one and there, on a whim, typed out the single word "FUGUE".

My Einstein–book dialogue was not really a fugue, of course — not even close — and yet it somehow reminded me of one. From earliest childhood, I had been profoundly moved by the music of Bach, and this off-the-wall idea of marrying Bach-like contrapuntal forms to lively dialogues with intellectually rich content grabbed me with a passion. Over the next few weeks, as I tossed the idea around in my head, I realized how much room for play there was along these lines, and I could imagine how voraciously I as a teen-ager might have consumed such dialogues. Thus I was led to the idea of inserting contrapuntal dialogues every so often, partly to break the tedium of the heavy ideas in my chapters, and partly to allow me to introduce lighter, more allegorical versions of all the abstruse concepts.

The long and the short of it is that I eventually decided — but this took many months — that the optimal structure would be a strict alternation between chapters and dialogues. Once that was clear, then I had the joyous task of trying to pinpoint the most crucial ideas that I wanted to get across to my readers and then somehow embodying them in both the form and the content of fanciful, often punning dialogues between Achilles and the Tortoise (plus a few new friends).

GEB Is First Cooled off, Then Reheated

In early 1974, I switched Ph.D. advisors for the fourth and final time, taking on a totally unfamiliar problem in solid-state physics that smelled very sweet though it threatened thorniness. My new advisor, Gregory Wannier, wanted me to plunge in deeply, and I knew in my gut that this time it was sink or swim for me in the world of physics. If I wanted a Ph.D. — a precious but horribly elusive goal toward which I had been struggling for almost a decade by then — it was now or never! And so, with great reluctance, I stowed my beloved manuscript in a desk drawer and told myself, "Hands off! And no peeking!" I even instituted food-deprivation punishments if I so much as opened the drawer and riffled through my book-in-the-making. Thinking *GEB* thoughts — or rather, *GTATHB* thoughts — was strictly *verboten.*

Speaking of German, Wannier was scheduled to go to Germany for a six-month period in the fall of 1974, and since I had always loved Europe, I asked if there was any way I could go along. Very kindly, he arranged for me to be a *wissenschaftlicher Assistent* — essentially a teaching assistant — in physics at the Universität Regensburg, and so that's what I did for one semester spanning the end of 1974 and the start of 1975. It was then that I got most of the work done for my Ph.D. thesis. Since I had no close friends, my Regensburg days and nights were long and lonely. In a peculiar sense, my closest friend during that tough period was Frédéric Chopin, since I tuned in to Radio Warsaw nearly every night at midnight and listened to various pianists playing many of his pieces that I knew and loved, and others that were new to me and that I came to love.

That whole stretch was *GEB-verboten* time, and thus it continued until the end of 1975, when finally I closed the book on my thesis. Although that work was all about an exquisite visual structure (see Chapter 5 of this book) and seemed to offer a good launchpad for a career, I had suffered too many blows to my ego in graduate school to believe I would make a good physicist. On the other hand, the rekindling of old intellectual flames and especially the writing of *GTATHB* had breathed a new kind of self-confidence into me.

Jobless but highly motivated, I moved to my home town of Stanford, and there, thanks to my parents' unquestioning and generous financial support ("a two-year Hofstadter Fellowship", I jokingly called it), I set out to "retool myself" as an artificial-intelligence researcher. Even more important, though, was that I was able to resume my passionate love affair with the ideas that had so grabbed me a couple of years earlier.

At Stanford, my erstwhile "pamphlet" bloomed. It was rewritten from start to finish, because I felt that my earlier drafts, though focused on the proper ideas, were immature and inconsistent in style. And I enjoyed the luxury of one of the world's earliest and best word-processing programs, my new friend Pentti Kanerva's tremendously flexible and user-friendly TV-Edit. Thanks to that program, the new version just flowed out, and ever so smoothly. I just can't imagine how *GEB* could have been written without it.

Only at this stage did the book's unusual stylistic hallmarks really emerge — the sometimes-silly playing with words, the concocting of novel verbal structures that imitate musical forms, the wallowing in analogies of every sort, the spinning of stories whose very structures exemplify the points they are talking about, the mixing of oddball personalities in fantastic scenarios. As I was writing, I certainly knew that my book would be quite different from other books on related topics, and that I was violating quite a number of conventions. Nonetheless I blithely continued, because I felt confident that what I was doing simply had to be done, and that it had an intrinsic rightness to it. One of the key qualities that made me so believe in what I was doing is that this was a book in which form was being given equal billing with content — and that was no accident, since *GEB* is in large part about how content is inseparable from form, how semantics is of a piece with syntax, how inextricable pattern and matter are from each other.

Although I had always known of myself that, in many aspects of life, I was concerned as much with form as with content, I had never suspected how deeply I would get caught up, in the writing of my first book, in matters of visual appearance on all levels. Thus, thanks to the ease of using TV-Edit, whatever I wrote underwent polishing to make it look better on the screen, and though such control would at one time have been considered a luxury for an author, I was very attached to it and loath to give it up. By the time I had a solid version of the manuscript ready to send out to publishers, visual design and conceptual structure were intimately bound up with each other.

The Clarion Call

I've oft been asked if I, an unknown author with an unorthodox manuscript and an off-the-wall title, had to struggle for years against the monolithic publishing industry's fear of taking risks. Well, perhaps I was just lucky, but my experience was far more pleasant than that.

In mid-1977, I sent out a little sample to about fifteen high-quality publishers, just as a feeler, to which most replied politely that this was "not the type of thing" they dealt in. Fair enough. But three or four expressed interest in seeing more, and so, by turns, I let them take a look at the whole thing. Needless to say, I was disappointed when the first two turned it down (and in each case the vetting process took a few months, so the loss of time was frustrating), but on the other hand, I wasn't overly disheartened. Then near Christmastime, Martin Kessler, head of Basic Books, a publishing outfit I had always admired, gave me some hopeful though tentative signals.

The winter of 1977–78 was so severe that Indiana University, where I was now a fledgling assistant professor, ran out of coal for heating, and in March the university was forced to close down for three weeks to wait for warmer weather. I decided to use this free time to drive to New York and points south to see old friends. Clear as a bell in my oft-blurry memory is my brief stop in some dingy little diner in the town of Clarion, Pennsylvania, where from a chilly phone booth I made a quick call to Martin Kessler in New York to see if he had a verdict yet. It was a great moment in my life when he said he would be "delighted" to work with me — and it's almost eerie to think that this signal event occurred in that well-named hamlet, of all places...

Revenge of the Holey Rollers

Now that I had found a publisher, there came the question of turning the manuscript from crude computer printout to a finely typeset book. It was a piece of true luck that Pentti, to enhance TV-Edit, had just developed one of the world's first computer typesetting systems, and he strongly encouraged me to use it. Kessler, ever the adventurer, was also willing to give it a try — partly, of course, because it would save Basic Books some money, but also because he was by nature a shrewd risk-taker.

Do-it-yourself typesetting, though for me a great break, was hardly a piece of cake. Computing then was a lot more primitive than it is today, and to use Pentti's system, I had to insert into each chapter or dialogue literally thousands of cryptic typesetting commands, next chop each computer file into several small pieces — five or six per file, usually — each of which had to be run through a series of two computer programs, and then each of the resulting output files had to be punched out physically as a cryptic pattern of myriad holes on a long, thin roll of paper tape. I myself had to walk the 200 yards to the building where the hole-puncher was located, load the paper tape, and sit there monitoring it carefully to make sure it didn't jam.

Next, I would carry this batch of oily tapes another quarter-mile to the building where *The Stanford Daily* was printed, and if it was free, I would use their phototypesetting machine myself. Doing so was a long, elaborate operation involving cartridges of photosensitive paper, darkrooms, chemical baths with rollers through which the paper had to be passed to get all the developing chemicals off, and clotheslines on which all the five-foot long galleys with my text on them would be hung out to dry for a day or two. The process of actually *seeing* what my thousands of typesetting commands had wrought was thus enormously unwieldy and slow. Truth to tell, though, I didn't mind that; in fact, it was arcane, special, and kind of exciting.

But one day, when nearly all the galleys had been printed — two to three hundred of them — and I thought I was home free, I made a horrendous discovery. I'd seen each one emerge with jet-black print from the developing baths, and yet on some of the more recently-dried ones, the text looked brownish. What!? As I checked out others, slightly older, I saw light-brown print, and on yet older ones, it was orangy, or even pale yellow!

I couldn't believe it. How in the world had this happened? The simple answer left me feeling so angry and helpless: the aging rollers, having worn unevenly, no longer wiped the galleys clean, so acid was day by day eating the black print away. For the *Daily*'s purposes, this didn't matter — they chucked their galleys in a matter of hours — but for a book, it spelled disaster. No way could a book be printed from yellow galleys! And the photocopies I'd made of them when they were newborn were sharp, but not sharp enough. What a nightmare! Untold labor had just gone up in smoke. I was filled with the despair of a football team that's just made a 99-yard downfield march only to be stopped dead on the opponent's one-yard line.

I'd spent almost all summer 1978 producing these galleys, but now summer was drawing to a close, and I had to go back to Indiana to teach courses. What on earth to do? How could I salvage *GEB*? The only solution I could see was, on my own money, to fly back to Stanford every weekend of the fall, and redo the whole thing from scratch. Luckily, I was teaching only on Tuesdays and Thursdays, and so each Thursday afternoon I would zoom from class, catch a plane, arrive at Stanford, work like a maniac until Monday afternoon, and then dash off to the airport to return to Indiana. I will never forget the worst of those weekends, when I somehow managed to work for forty hours straight without a wink of sleep. That's love for you!

In this ordeal there was a saving grace, though, and it was this: I got to correct all the typesetting errors I'd made in the first batch of galleys. The original plan had been to use a bunch of correction galleys, which would have had to be sliced up into little pieces in Basic's New York offices and pasted in wherever there were glitches — and in that first batch I'd made glitches galore, that's for sure. Such a process would probably have resulted in hundreds of errors in the layout. But thanks to my 99-yard drive having been halted at the one-yard line, I now had the chance to undo all these glitches, and produce a nearly pristine set of galleys. And thus, although the chemical catastrophe delayed the actual printing of *GEB* for a couple of months, it turned out, in retrospect, to have been a blessing in disguise.

Oops...

There were of course many ideas that vied with each other for entry into the book taking shape during those years, and some made it in while others did not. One of the ironies is that the Einstein–book dialogue, which in its "fugality" was the inspiration for all dialogues to come, was chopped.

There was another long and intricate dialogue, too, that was chopped, or more accurately, that wound up getting transmogrified nearly beyond recognition, and its curious story is connected with an intense debate that was raging inside my brain at that time.

I had been made acutely aware, by some leaflets I'd read in the student union at Oregon in 1970, of sexist language and its insidious unconscious effects. My mind was awakened to the subtle ways that generic "he" and "man" (and a host of similar words and phrases) contribute to the shaping of one's sense of what is a "normal" human being and what is an "exception", and I welcomed this new perspective. But I was not a writer at that time — I was a physics grad student — and these issues didn't seem all that close to my own life. When I started writing dialogues, though, things changed. There came a point when it dawned on me that the characters in my dialogues — Achilles, the Tortoise, the Crab, the Anteater, and a couple of others with cameo roles — were without exception *males*. I was shocked at my own having fallen victim to the unconscious pressures pushing against the introduction of female characters. And yet, when I toyed with the idea of going back and performing a "sex-change operation" on one or more of these characters, that really rubbed me the wrong way. How come?

Well, all I could tell myself was, "Bring in females and you wind up importing the whole confusing world of sexuality into what is essentially a purely abstract discussion, and that would distract attention from my book's main purposes." This nonsensical view of mine stemmed from and echoed many tacit assumptions of western civilization at that time (and still today). As I forced myself to grapple with my own ugly attitude, a real battle started up in my mind, with one side of me arguing for going back and making some characters female, and the other trying to maintain the *status quo*.

Out of this internal battle suddenly came a long and rather amusing dialogue in which my various characters, having come to the realization that they are all males, discuss why this might be so, and decide that, despite their sense of having free will, they must in fact be merely characters in the mind of some sexist male author. One way or another, they manage to summon this Author character into their dialogue — and what does *he* do when accused of sexism? He pleads innocent, claiming that what his brain does is out of his control — the blame for his sexism must instead fall on a sexist God. And the next thing you know, God poofs into the dialogue — and guess what? She turns out to be female (ho ho ho). I don't remember the details of how it went on from there, but the point is, I was deeply torn, and I was grappling in my own way with these complex issues.

To my regret — that is to say, to the regret of the me of the years that followed — the side that wound up winning this battle was the sexist side, with just a few concessions to the other side (*e.g.*, the tower of Djinns in the dialogue "Little Harmonic Labyrinth", and Aunt Hillary in "Prelude... Ant Fugue"). *GEB* remained a book with a deep sexist bias sewn into its fabric. Interestingly, it is a bias that very few readers, females or males, have commented on (which in turn supports my belief that these kinds of things are very subtle and insidious, and escape nearly everyone's perception).

As for generic "man" and "he", I certainly disliked those usages at that time, and I tried to avoid them whenever I could (or rather, whenever it was *easy*), but on the other hand I wasn't particularly concerned about cleansing my prose of every last one of them, and as a consequence the book's pages are also marred, here and there, by that more obvious, more explicit form of sexism. Today, I cringe whenever I come across sentences in *GEB* that talk about the reader as "he", or that casually speak of "mankind" as if humanity were some huge abstract *guy*. One lives and learns, I guess.

And lastly, as for that soul-searching dialogue in which the Author and God are summoned up by Achilles and company to face the accusation of sexism, well, that was somehow transformed, in a series of many, many small changes, into the dialogue with which *GEB* concludes: "Six-part Ricercar". If you read it with its genesis in mind, you may find an extra level of interest.

Mr. Tortoise, Meet Madame Tortue

A few years later, a wholly unexpected chance came along to make amends, at least in part, for my sexist sin. That opportunity was afforded me by the challenge of translating *GEB* into various foreign languages.

When I was writing the book, the idea that it might someday appear in other languages never crossed my mind. I don't know why, since I loved languages and loved translation, but somehow it just never occurred to me. However, as soon as the idea was proposed to me by my publisher, I was very excited about seeing my book in other languages, especially ones that I spoke to some extent — most of all French, since that was a language that I spoke fluently and loved very deeply.

There were a million issues to consider in any potential translation, since the book is rife not only with explicit wordplay but also with what Scott Kim dubbed "structural puns" — passages where form and content echo or reinforce each other in some unexpected manner, and very often thanks to happy coincidences involving specific English words. Because of these intricate medium–message tangles, I painstakingly went through every last sentence of *GEB,* annotating a copy for translators into any language that might be targeted. This took me about a year of on-again, off-again toil, but finally it was done, and just in the nick of time, because contracts with foreign publishers started flowing thick and fast around 1982. I could write a short book — a pamphlet? — on the crazy, delightful, knotty puzzles and dilemmas that arose in translating *GEB,* but here I will mention just one — how to render the simple-seeming phrase "Mr. Tortoise" in French.

When in the spring of 1983, Jacqueline Henry and Bob French, the book's excellent translators into French, began to tackle the dialogues, they instantly ran headlong into the conflict between the feminine gender of the French noun *tortue* and the masculinity of my character, the Tortoise. By the way, I must ruefully mention that in the marvelous but little-known Lewis Carroll dialogue from which I borrowed these delightful characters (reprinted in *GEB* as "Two-part Invention"), the Tortoise turns out, if you look carefully, never to have been attributed either gender. But when I first read it, the question never entered my mind. This was *clearly* a he-tortoise. Otherwise, I would have known not only *that* it was female but also *why* it was female. After all, an author only introduces a female character for some special *reason,* right? Whereas a male character in a "neutral" context (*e.g.,* philosophy) needs no *raison d'être,* a female does. And so, given no clue as to the Tortoise's sex, I unthinkingly and uncritically envisaged it as a male. Thus does sexism silently pervade well-meaning but susceptible brains.

But let's not forget Jacqueline and Bob! Although they could simply have bludgeoned their way through the problem by inventing a "Monsieur Tortue" character, that route felt distinctly unnatural in French, to their taste, and so, in one of our many exchanges of letters, they rather gingerly asked me if I would ever consider letting them switch the Tortoise's sex to female. To them, of course, it probably seemed pretty far-fetched to imagine that the author would even give such a proposal the time of day, but as a matter of fact, the moment I read their idea, I seized upon it with great enthusiasm. And as a result, the French *GEB*'s pages are graced throughout with the fresh, fantastic figure of Madame Tortue, who runs perverse intellectual circles around her male companion Achilles, erstwhile Greek warrior and amateur philosopher.

There was something so delightful and gratifying to me about this new vision of "the Tortoise" that I was ecstatic with her. What particularly amused me were a few bilingual conversations that I had about the Tortoise, in which I would start out in English using the pronoun "he", then switch to French and to *elle* as well. Either pronoun felt perfectly natural, and I even felt I was referring to the selfsame "person" in both languages. In its own funny way, this seemed faithful to Carroll's tortoise's sexual neutrality.

And then, redoubling my pleasure, the translators into Italian, another language that I adored and spoke quite well, chose to follow suit and to convert my "Mr. Tortoise" into "signorina Tartaruga". Of course these radical switches in no way affect the perceptions of *GEB*'s purely anglophone readers, but in some small way, I feel, they help to make up for the lamentable outcome of my internal battle of a few years earlier.

Zen Buddhism, John Cage, and My Voguish Irrationality

The French translation was greeted, overall, very favorably. One specially gratifying moment for Bob, Jacqueline, and myself was when a truly glowing full-page review by Jacques Attali appeared in the most prestigious French newspaper, *Le Monde,* not just praising the book for its ideas and style, but also making a particular point of praising its translation.

A few months later, I received a pair of reviews published in successive issues of *Humanisme,* an obscure journal put out by the Society of French Freemasons. Both had issued from the pen of one author, Alain Houlou, and I tackled them with interest. The first one was quite lengthy and, like that in *Le Monde,* glowed with praise; I was gratified and grateful.

I then went on to the second review, which started out with the poetic phrase *Après les roses, les épines...* ("After roses, thorns..."), and which then proceeded for several pages, to my amazement, to rip *GEB* apart as *un piège très grave* ("a very dangerous trap") in which the mindless bandwagon of Zen Buddhism was eagerly jumped on, and in which a rabidly antiscientific, beatnik-influenced, hippie-like irrationality typical of American physicists was embraced as the supreme path to enlightenment, with the iconoclastic Zen-influenced American composer John Cage as the patron saint of it all.

All I could do was chuckle, and throw my hands up in bewilderment at these Tati-esque *vacarmes de monsieur Houlou*. Somehow, this reviewer saw me praising Cage to the skies ("Gödel, Escher, Cage"?) and managed to read into my coy allusions to and minor borrowings from Zen an uncritical acceptance thereof, which in fact is not at all my stance. As I declare at the start of Chapter 9, I find Zen not only confusing and silly, but on a very deep level utterly inimical to my core beliefs. However, I also find Zen's silliness — especially when it gets *really* silly — quite amusing, even refreshing, and it was simply fun for me to sprinkle a bit of Eastern spice into my basically very Western casserole. However, my having sprinkled little traces of Zen here and there does not mean that I am a Zen monk in sheep's clothing.

As for John Cage, for some odd reason I had felt very sure, up till reading Houlou's weird about-face, that in my "Canon by Intervallic Augmentation" and the chapter that follows it, I had unambiguously heaped scorn on Cage's music, albeit in a somewhat respectful manner. But wait, wait, wait — isn't "heaping respectful scorn" not a contradiction in terms, indeed a patent impossibility? And doesn't such coy flirting with self-contradiction and paradox demonstrate, exactly as Houlou claims, that I *am*, deep down, both antiscientific and pro-Zen, after all? Well, so be it.

Even if I feel my book is as often misunderstood as understood, I certainly can't complain about the size or the enthusiasm of its readership around the world. The original English-language *GEB* was and continues to be very popular, and its translated selves hit the bestseller lists in (at least) France, Holland, and Japan. The German *GEB*, in fact, occupied the #1 rank on the nonfiction list for something like *five months* during 1985, the 300th birthyear of J. S. Bach. It seems a bit absurd to me. But who knows — that anniversary, aided by the other Germanic names on the cover, may have crucially sparked *GEB*'s popularity there. *GEB* has also been lovingly translated into Spanish, Italian, Hungarian, Swedish, and Portuguese, and — perhaps unexpectedly — with great virtuosity into Chinese. There is also a fine Russian version all ready, just waiting in the wings until it finds a publisher. All of this far transcends anything I ever expected, even though I can't deny that as I was writing it, especially in those heady Stanford days, I had a growing inner feeling that *GEB* would make some sort of splash.

My Subsequent Intellectual Path: Decade I

Since sending *GEB* off to the printers two decades ago, I've somehow managed to keep myself pretty busy. Aside from striving, with a team of excellent graduate students, to develop computer models of the mental mechanisms that underlie analogy and creativity, I've also written several further books, each of which I'll comment on here, though only very briefly.

The first of these, appearing in late 1981, was *The Mind's I*, an anthology co-edited with a new friend, philosopher Daniel Dennett. Our purpose, closely related to that of *GEB*, was to force our readers to confront, in the most vivid and even jolting manner, the fundamental conundrum of human existence: our deep and almost ineradicable sense of possessing a unique "I"-ness transcending our physical bodies and mysteriously enabling us to exercise something we call "free will", without ever quite knowing just what that is. Dan and I used stories and dialogues from a motley crew of excellent writers, and one of the pleasures for me was that I finally got to see my Einstein–book dialogue in print, after all.

During the years 1981–1983, I had the opportunity to write a monthly column for *Scientific American,* which I called "Metamagical Themas" (an anagram of "Mathematical Games", the title of the wonderful column by Martin Gardner that had occupied the same slot in the magazine for the preceding 25 years). Although the topics I dealt with in my column were, on their surface, all over the map, in some sense they were unified by their incessant quest for "the essence of mind and pattern". I covered such things as pattern and poetry in the music of Chopin; the question of whether the genetic code is arbitrary or inevitable; strategies in the never-ending battle against pseudo-science; the boundary between sense and nonsense in literature; chaos and strange attractors in mathematics; game theory and the Prisoner's Dilemma; creative analogies involving simple number patterns; the insidious effects of sexist language; and many other topics. In addition, strange loops, self-reference, recursion, and a closely related phenomenon that I came to call "locking-in" were occasional themes in my columns. In that sense, as well as in their wandering through many disciplines, my "Metamagical Themas" essays echoed the flavor of *GEB.*

Although I stopped writing my column in 1983, I spent the next year pulling together the essays I'd done and providing each of them with a substantial "Post Scriptum"; these 25 chapters, along with eight fresh ones, constituted my 1985 book *Metamagical Themas: Questing for the Essence of Mind and Pattern.* One of the new pieces was a rather zany Achilles–Tortoise dialogue called "Who Shoves Whom Around Inside the Careenium?", which I feel captures my personal views on self, soul, and the infamous "I"-word — namely, "I"! — perhaps better than anything else I've written — maybe even better than *GEB* does, though that might be going too far.

For several years during the 1980's, I was afflicted with a severe case of "ambigrammitis", which I caught from my friend Scott Kim, and out of which came my 1987 book *Ambigrammi.* An ambigram (or an "inversion", as Scott calls them in his own book, *Inversions*) is a calligraphic design that manages to squeeze two different readings into the selfsame set of curves. I found the idea charming and intellectually fascinating, and as I developed my own skill at this odd but elegant art form, I found that self-observation gave me many new insights into the nature of creativity, and so *Ambigrammi,* aside from showcasing some 200 of my ambigrams, also features a text — in fact, a dialogue — that is a long, wandering meditation on the creative act, centered on the making of ambigrams but branching out to include musical composition, scientific discovery, creative writing, and so on. For reasons not worth going into, *Ambigrammi: Un microcosmo ideale per lo studio della creatività* was published only in Italian and by a tiny publisher called Hopeful Monster, and I regret to say that it is no longer available.

My Subsequent Intellectual Path: Decade II

As I said above, writing, though crucial, was not my only intellectual focus; research into cognitive mechanisms was an equally important one. My early hunches about how to model analogy and creativity are actually set forth quite clearly in *GEB*'s Chapter 19, in my discussion of Bongard problems, and although those were just the germs of an actual architecture, I feel it is fair to say that despite many years of refinement, most of those ideas can be found in one form or another in the models developed in my research group at Indiana University and the University of Michigan (where I spent the years 1984–1988, in the Psychology Department).

After a decade and a half of development of computer models, the time seemed ripe for a book that would pull all the main threads together and describe the programs' principles and performance in clear and accessible language. Thus over several years, *Fluid Concepts and Creative Analogies* took shape, and finally appeared in print in 1995. In it are presented a series of closely related computer programs — Seek-Whence, Jumbo, Numbo, Copycat, Tabletop, and (still in progress) Metacat and Letter Spirit — together with philosophical discussions that attempt to set them in context. Several of its chapters are co-authored by members of the Fluid Analogies Research Group, and indeed FARG gets its proper billing as my collective co-author. The book shares much with *GEB,* but perhaps most important of all is the basic philosophical article of faith that being an "I" — in other words, possessing a sense of self so deep and ineradicable that it blurs into causality — is an inevitable concomitant to, and ingredient of, the flexibility and power that are synonymous with *intelligence,* and that the latter is but another term for *conceptual flexibility,* which in turn means *meaningful symbols.*

A very different strand of my intellectual life was my deep involvement in the translation of *GEB* into various languages, and this led me, perhaps inevitably, in retrospect, to the territory of verse translation. It all started in 1987 with my attempt to mimic in English a beautiful French miniature by sixteenth-century French poet Clément Marot, but from there it spun off in many directions at once. To make a long story short, I wound up writing a complex and deeply personal book about translation in its most general and metaphorical sense, and while writing it, I experienced much the same feeling of exhilaration as I had twenty years earlier, when writing *GEB.*

This book, *Le Ton beau de Marot: In Praise of the Music of Language,* winds through many diverse terrains, including what it means to "think in" a given language (or a blur of languages); how constraints can enhance creativity; how meaning germinates, buds, and flowers in minds and might someday do so in machines; how words, when put together into compounds, often melt together and lose some or all of their identity; how a language spoken on a neutron star might or might not resemble human languages; how poetry written hundreds of years ago should be rendered today; how translation is intimately related to analogy and to the fundamental human process of understanding one another; what kinds of passages, if any, are intrinsically untranslatable; what it means to translate nonsense passages from one language to another; the absurdity of supposing that today's mostly money-driven machine-translation gimmickry could handle even the simplest of poetry; and on and on.

The two middle chapters of *Le Ton beau de Marot* are devoted to a work of fiction that I had recently fallen in love with: Alexander Pushkin's novel in verse, *Eugene Onegin.* I first came into contact with this work through a couple of English translations, and then read others, always fascinated by the translators' different philosophies and styles. From this first flame of excitement, I slowly was drawn into trying to read the original text, and then somehow, despite having a poor command of Russian, I could not prevent myself from trying to translate a stanza or two. Thus started a slippery slope that I soon slid down, eventually stunning myself by devoting a whole year to recreating the entire novel — nearly 400 sparkling sonnets — in English verse. Of course, during that time, my Russian improved by leaps and bounds, though it still is far from conversationally fluent. As I write, my *Onegin* has not yet come out, but it will be appearing at just about the same time in 1999 as the book you are holding — the twentieth-anniversary version of *Gödel, Escher, Bach.* And the year 1999 plays an equally important role in my *EO*'s creation, being the 200th birthyear of Alexander Pushkin.

Forward-looking and Backward-looking Books

Le Ton beau de Marot is a bit longer than *GEB,* and on its first page, I go out on a limb and call it "probably the best book I will ever write". Some of my readers will maintain that *GEB* is superior, and I can see why they might do so. But it's so long since I wrote *GEB* that perhaps the magical feeling I had when writing it has faded, while the magic of *LeTbM* is still vivid. Still, there's no denying that, at least in the short run, *LeTbM* has had far less impact than *GEB* did, and I confess that that's disappointed me quite a bit.

Permit me to speculate for a moment as to why this might be the case. In some sense, *GEB* was a "forward-looking" book, or at least on its surface it gave that appearance. Many hailed it as something like "the bible of artificial intelligence", which is of course ridiculous, but the fact is that many young students read it and caught the bug of my own fascination with the modeling of mind in all of its elusive aspects, including the evanescent goals of "I" and free will and consciousness. Although I am the furthest thing in the world from being a futurist, a science-fiction addict, or a technology guru, I was often pigeonholed in just that way, simply because I had written a long treatise that dealt quite a bit with computers and their vast potential (in the most philosophical of senses), and because my book was quite a hit among young people interested in computers.

Well, by contrast, *Le Ton beau de Marot* might be seen as a "backward-looking" book, not so much because it was inspired by a sixteenth-century poem and deals with many other authors of the past, such as Dante and Pushkin, but because there simply is nothing in the book's pages that could be confused with glib technological glitz and surreal futuristic promises. Not that *GEB* had those either, but many people seemed to see something vaguely along those lines in it, whereas there's nothing of that sort to latch onto, in *LeTbM.* In fact, some might see it almost as technology-bashing, in that I take many artificial-intelligence researchers and machine-translation developers to task for wildly exaggerated claims. I am not an enemy of these fields, but I am against vast oversimplifications and underestimations of the challenges that they represent, for in the end, that amounts to a vast underestimation of the human spirit, for which I have the deepest respect.

Anyone who has read *GEB* with any care should have seen this same "backward-looking" flavor permeating the book, perhaps most explicitly so in the key section "Ten Questions and Speculations" (pp. 676–680), which is a very romantic way of looking at the depth of the human spirit. Although my prediction about chess-playing programs put forth there turned out to be embarrassingly wrong (as the world saw with Deep Blue versus Kasparov in 1997), those few pages nonetheless express a set of philosophical beliefs to which I am still committed in the strongest sense.

To Tamper, or to Leave Pristine?

Given that I was quite wrong in a prediction made twenty years ago, why not rewrite the "Ten Questions and Speculations" section, updating it and talking about how I feel in light of Deep Blue? Well, of course, this brings up a much larger issue: that of revising the 1979 book from top to bottom, and coming out with a spanking new 1999 edition of *GEB.* What might militate for, and what might militate against, undertaking such a project?

I don't deny that some delightful, if small, improvements were made in the translated versions. For example, my magistrally Bach-savvy friend Bernie Greenberg informed me that the "BACH goblet" I had invented out of whole cloth in my dialogue "Contracrostipunctus" actually exists! The real goblet is not (as in my dialogue) a piece of glass blown by Bach, but rather a gift from one of his prize students; nonetheless, its key feature — that of having the melody "BACH" etched into the glass itself — is just as I said in the dialogue! This was such an amazing coincidence that I rewrote the dialogue for the French version to reflect the *real* goblet's existence, and insisted on having a photograph of the BACH goblet in the French *GEB.*

Another delicious touch in the French *GEB* was the replacement of the very formal, character-less photo of Gödel by a far more engaging snapshot in which he's in a spiffy white suit and is strolling with some old codger in a forest. The latter, decked out in a floppy hat and baggy pants held up by gawky suspenders, looks every inch the quintessential rube, so I rewrote the caption as *Kurt Gödel avec un paysan non identifié* ("Kurt Gödel with unknown peasant"). But as anyone who has lived in the twentieth century can see in a split second, the *paysan non identifié* is none other than A. Einstein.

Why not, then, incorporate those amusing changes into a revised edition in English? On a more substantial level, why not talk a bit about the pioneering artificial-intelligence program Hearsay II, whose very subtle architecture started exerting, only a year or two after *GEB* came out, a vast impact on my own computer models, and about which I already knew something way back in 1976? Why not talk more about machine translation, and especially its weaknesses? Why not have a whole chapter about the most promising developments (and/or exaggerated claims) over the past two decades in artificial intelligence — featuring my own research group as well as others? Or why not, as some have suggested, come out with a CD-ROM with Escher pictures and Bach music on it, as well as recordings of all of *GEB*'s dialogues as performed by top-notch actors?

Well, I can see the arguments for any of these, but unfortunately, I just don't buy them. The CD-ROM suggestion, the one most often made to me, is the simplest to dismiss. I intended *GEB* as a book, not as a multimedia circus, and book it shall remain — end of story. As for the idea of revising the text, however, that is more complex. Where would one draw the line? What would be sacrosanct? What would survive, what would be tossed out? Were I to take that task on, I might well wind up rewriting every single sentence — and, let's not forget, reverse-engineering old Mr. T...

Perhaps I'm just a crazy purist; perhaps I'm just a lazy lout; but stubborn no doubt, and wouldn't dream of changing my book's *Urtext*. That's out! Thus in my sternness, I won't allow myself to add the names of two people — Donald Kennedy and Howard Edenberg — to my "Words of Thanks", despite the fact that for years, I've felt sad at having inadvertently left them out. I won't even correct the book's typos (and, to my chagrin, I did find, over the decades, that there are a few, aside from those listed explicitly under "typos" in the index)! Why on earth am I such a stick-in-the-mud? Why not bring *Gödel, Escher, Bach* up to date and make it a book worthy of ushering in the twenty-first century — indeed, the third millenium?

Quærendo Invenietis...

Well, the only answer I can give, other than that life is short, is that *GEB* was written in one sitting, so to speak. *GEB* was a clean and pure vision that was dreamed by someone else — someone who, to be sure, was remarkably similar to yours truly, but someone who nonetheless had a slightly different perspective and a slightly different agenda. *GEB* was *that* person's labor of love, and as such — at least so say I — it should not be touched.

Indeed, I somehow feel a strange inner confidence that the true author of *GEB*, when one fine day he finally reaches *my* ripe age, will tender to me the truest of thanks for not having tampered with the vessel into which he poured so much of his young and eager soul — the work that he even went so far as to call, in what some might see as a cryptic or even naïvely romantic remark, "a statement of my religion". At least *I* know what he meant.

REQVIESCAT IN CONSTANTIA, ERGO,
REPRÆSENTATIO CVPIDI AVCTORIS RELIGIONIS.

Contents

Part II: EGB

Overview

Part I: GEB

Introduction: A Musico-Logical Offering. The book opens with the story of Bach's *Musical Offering*. Bach made an impromptu visit to King Frederick the Great of Prussia, and was requested to improvise upon a theme presented by the King. His improvisations formed the basis of that great work. The *Musical Offering* and its story form a theme upon which I "improvise" throughout the book, thus making a sort of "Metamusical Offering". Self-reference and the interplay between different levels in Bach are discussed; this leads to a discussion of parallel ideas in Escher's drawings and then Gödel's Theorem. A brief presentation of the history of logic and paradoxes is given as background for Gödel's Theorem. This leads to mechanical reasoning and computers, and the debate about whether Artificial Intelligence is possible. I close with an explanation of the origins of the book—particularly the why and wherefore of the Dialogues.

Three-Part Invention. Bach wrote fifteen three-part inventions. In this three-part Dialogue, the Tortoise and Achilles—the main fictional protagonists in the Dialogues—are "invented" by Zeno (as in fact they were, to illustrate Zeno's paradoxes of motion). Very short, it simply gives the flavor of the Dialogues to come.

Chapter I: The MU-puzzle. A simple formal system (the MIU-system) is presented, and the reader is urged to work out a puzzle to gain familiarity with formal systems in general. A number of fundamental notions are introduced: string, theorem, axiom, rule of inference, derivation, formal system, decision procedure, working inside/outside the system.

Two-Part Invention. Bach also wrote fifteen two-part inventions. This two-part Dialogue was written not by me, but by Lewis Carroll in 1895. Carroll borrowed Achilles and the Tortoise from Zeno, and I in turn borrowed them from Carroll. The topic is the relation between reasoning, reasoning about reasoning, reasoning about reasoning about reasoning, and so on. It parallels, in a way, Zeno's paradoxes about the impossibility of motion, seeming to show, by using infinite regress, that reasoning is impossible. It is a beautiful paradox, and is referred to several times later in the book.

Chapter II: Meaning and Form in Mathematics. A new formal system (the pq-system) is presented, even simpler than the MIU-system of Chapter I. Apparently meaningless at first, its symbols are suddenly revealed to possess meaning by virtue of the form of the theorems they appear in. This revelation is the first important insight into meaning: its deep connection to isomorphism. Various issues related to meaning are then discussed, such as truth, proof, symbol manipulation, and the elusive concept, "form".

Sonata for Unaccompanied Achilles. A Dialogue which imitates the Bach Sonatas for unaccompanied violin. In particular, Achilles is the only speaker, since it is a transcript of one end of a telephone call, at the far end of which is the Tortoise. Their conversation concerns the concepts of "figure" and "ground" in various contexts—e.g., Escher's art. The Dialogue itself forms an example of the distinction, since Achilles' lines form a "figure", and the Tortoise's lines—implicit in Achilles' lines—form a "ground".

Chapter III: Figure and Ground. The distinction between figure and ground in art is compared to the distinction between theorems and nontheorems in formal systems. The question "Does a figure necessarily contain the same information as its ground?" leads to the distinction between recursively enumerable sets and recursive sets.

Contracrostipunctus. This Dialogue is central to the book, for it contains a set of paraphrases of Gödel's self-referential construction and of his Incompleteness Theorem. One of the the paraphrases of the Theorem says, "For each record player there is a record which it cannot play." The Dialogue's title is a cross between the word "acrostic" and the word "contrapunctus", a Latin word which Bach used to denote the many fugues and canons making up his *Art of the Fugue*. Some explicit references to the *Art of the Fugue* are made. The Dialogue itself conceals some acrostic tricks.

Chapter IV: Consistency, Completeness, and Geometry. The preceding Dialogue is explicated to the extent it is possible at this stage. This leads back to the question of how and when symbols in a formal system acquire meaning. The history of Euclidean and non-Euclidean geometry is given, as an illustration of the elusive notion of "undefined terms". This leads to ideas about the consistency of different and possibly "rival" geometries. Through this discussion the notion of undefined terms is clarified, and the relation of undefined terms to perception and thought processes is considered.

Little Harmonic Labyrinth. This is based on the Bach organ piece by the same name. It is a playful introduction to the notion of recursive—i.e., nested—structures. It contains stories within stories. The frame story, instead of finishing as expected, is left open, so the reader is left dangling without resolution. One nested story concerns modulation in music—particularly an organ piece which ends in the wrong key, leaving the listener dangling without resolution.

Chapter V: Recursive Structures and Processes. The idea of recursion is presented in many different contexts: musical patterns, linguistic patterns, geometric structures, mathematical functions, physical theories, computer programs, and others.

Canon by Intervallic Augmentation. Achilles and the Tortoise try to resolve the question, "Which contains more information—a record, or the phonograph which plays it?" This odd question arises when the Tortoise describes a single record which, when played on a set of different phonographs, produces two quite different melodies: B-A-C-H and C-A-G-E. It turns out, however, that these melodies are "the same", in a peculiar sense.

Chapter VI: The Location of Meaning. A broad discussion of how meaning is split among coded message, decoder, and receiver. Examples presented include strands of DNA, undeciphered inscriptions on ancient tablets, and phonograph records sailing out in space. The relationship of intelligence to "absolute" meaning is postulated.

Chromatic Fantasy, And Feud. A short Dialogue bearing hardly any resemblance, except in title, to Bach's *Chromatic Fantasy and Fugue*. It concerns the proper way to manipulate sentences so as to preserve truth—and in particular the question of whether there exist rules for the usage of the word "and". This Dialogue has much in common with the Dialogue by Lewis Carroll.

Chapter VII: The Propositional Calculus. It is suggested how words such as "and" can be governed by formal rules. Once again, the ideas of isomorphism and automatic acquisition of meaning by symbols in such a system are brought up. All the examples in this Chapter, incidentally, are "Zentences"—sentences taken from Zen kōans. This is purposefully done, somewhat tongue-in-cheek, since Zen kōans are deliberately illogical stories.

Crab Canon. A Dialogue based on a piece by the same name from the *Musical Offering.* Both are so named because crabs (supposedly) walk backwards. The Crab makes his first appearance in this Dialogue. It is perhaps the densest Dialogue in the book in terms of formal trickery and level-play. Gödel, Escher, and Bach are deeply intertwined in this very short Dialogue.

Chapter VIII: Typographical Number Theory. An extension of the Propositional Calculus called "TNT" is presented. In TNT, number-theoretical reasoning can be done by rigid symbol manipulation. Differences between formal reasoning and human thought are considered.

A Mu Offering. This Dialogue foreshadows several new topics in the book. Ostensibly concerned with Zen Buddhism and kōans, it is actually a thinly veiled discussion of theoremhood and nontheoremhood, truth and falsity, of strings in number theory. There are fleeting references to molecular biology—particularly the Genetic Code. There is no close affinity to the *Musical Offering,* other than in the title and the playing of self-referential games.

Chapter IX: Mumon and Gödel. An attempt is made to talk about the strange ideas of Zen Buddhism. The Zen monk Mumon, who gave well known commentaries on many kōans, is a central figure. In a way, Zen ideas bear a metaphorical resemblance to some contemporary ideas in the philosophy of mathematics. After this "Zennery", Gödel's fundamental idea of Gödel-numbering is introduced, and a first pass through Gödel's Theorem is made.

Part II: EGB

Prelude . . . This Dialogue attaches to the next one. They are based on preludes and fugues from Bach's *Well-Tempered Clavier.* Achilles and the Tortoise bring a present to the Crab, who has a guest: the Anteater. The present turns out to be a recording of the *W.T.C.;* it is immediately put on. As they listen to a prelude, they discuss the structure of preludes and fugues, which leads Achilles to ask how to hear a fugue: as a whole, or as a sum of parts? This is the debate between holism and reductionism, which is soon taken up in the *Ant Fugue.*

Chapter X: Levels of Description, and Computer Systems. Various levels of seeing pictures, chessboards, and computer systems are discussed. The last of these is then examined in detail. This involves describing machine languages, assembly languages, compiler languages, operating systems, and so forth. Then the discussion turns to composite systems of other types, such as sports teams, nuclei, atoms, the weather, and so forth. The question arises as to how many intermediate levels exist—or indeed whether any exist.

. . . *Ant Fugue.* An imitation of a musical fugue: each voice enters with the same statement. The theme—holism versus reductionism—is introduced in a recursive picture composed of words composed of smaller words, etc. The words which appear on the four levels of this strange picture are "HOLISM", "REDUCTIONISM", and "MU". The discussion veers off to a friend of the Anteater's—Aunt Hillary, a conscious ant colony. The various levels of her thought processes are the topic of discussion. Many fugal tricks are ensconced in the Dialogue. As a hint to the reader, references are made to parallel tricks occurring in the fugue on the record to which the foursome is listening. At the end of the *Ant Fugue,* themes from the *Prelude* return, transformed considerably.

Chapter XI: Brains and Thoughts. "How can thoughts be supported by the hardware of the brain?" is the topic of the Chapter. An overview of the large-scale and small-scale structure of the brain is first given. Then the relation between concepts and neural activity is speculatively discussed in some detail.

English French German Suite. An interlude consisting of Lewis Carroll's nonsense poem "Jabberwocky" together with two translations: one into French and one into German, both done last century.

Chapter XII: Minds and Thoughts. The preceding poems bring up in a forceful way the question of whether languages, or indeed minds, can be "mapped" onto each other. How is communication possible between two separate physical brains? What do all human brains have in common? A geographical analogy is used to suggest an answer. The question arises, "Can a brain be understood, in some objective sense, by an outsider?"

Aria with Diverse Variations. A Dialogue whose form is based on Bach's *Goldberg Variations,* and whose content is related to number-theoretical problems such as the Goldbach conjecture. This hybrid has as its main purpose to show how number theory's subtlety stems from the fact that there are many diverse variations on the theme of searching through an infinite space. Some of them lead to infinite searches, some of them lead to finite searches, while some others hover in between.

Chapter XIII: BlooP and FlooP and GlooP. These are the names of three computer languages. BlooP programs can carry out only predictably finite searches, while FlooP programs can carry out unpredictable or even infinite searches. The purpose of this Chapter is to give an intuition for the notions of primitive recursive and general recursive functions in number theory, for they are essential in Gödel's proof.

Air on G's String. A Dialogue in which Gödel's self-referential construction is mirrored in words. The idea is due to W. V. O. Quine. This Dialogue serves as a prototype for the next Chapter.

Chapter XIV: On Formally Undecidable Propositions of TNT and Related Systems. This Chapter's title is an adaptation of the title of Gödel's 1931 article, in which his Incompleteness Theorem was first published. The two major parts of Gödel's proof are gone through carefully. It is shown how the assumption of consistency of TNT forces one to conclude that TNT (or any similar system) is incomplete. Relations to Euclidean and non-Euclidean geometry are discussed. Implications for the philosophy of mathematics are gone into with some care.

Birthday Cantatatata . . . In which Achilles cannot convince the wily and skeptical Tortoise that today is his (Achilles') birthday. His repeated but unsuccessful tries to do so foreshadow the repeatability of the Gödel argument.

Chapter XV: Jumping out of the System. The repeatability of Gödel's argument is shown, with the implication that TNT is not only incomplete, but "essentially incomplete". The fairly notorious argument by J. R. Lucas, to the effect that Gödel's Theorem demonstrates that human thought cannot in any sense be "mechanical", is analyzed and found to be wanting.

Edifying Thoughts of a Tobacco Smoker. A Dialogue treating of many topics, with the thrust being problems connected with self-replication and self-reference. Television cameras filming television screens, and viruses and other subcellular entities which assemble themselves, are among the examples used. The title comes from a poem by J. S. Bach himself, which enters in a peculiar way.

Chapter XVI: Self-Ref and Self-Rep. This Chapter is about the connection between self-reference in its various guises, and self-reproducing entities (e.g., computer programs or DNA molecules). The relations between a self-reproducing entity and the mechanisms external to it which aid it in reproducing itself (e.g., a computer or proteins) are discussed—particularly the fuzziness of the distinction. How information travels between various levels of such systems is the central topic of this Chapter.

The Magnificrab, Indeed. The title is a pun on Bach's *Magnificat* in D. The tale is about the Crab, who gives the appearance of having a magical power of distinguishing between true and false statements of number theory by reading them as musical pieces, playing them on his flute, and determining whether they are "beautiful" or not.

Chapter XVII: Church, Turing, Tarski, and Others. The fictional Crab of the preceding Dialogue is replaced by various real people with amazing mathematical abilities. The Church-Turing Thesis, which relates mental activity to computation, is presented in several versions of differing strengths. All are analyzed, particularly in terms of their implications for simulating human thought mechanically, or programming into a machine an ability to sense or create beauty. The connection between brain activity and computation brings up some other topics: the halting problem of Turing, and Tarski's Truth Theorem.

SHRDLU, Toy of Man's Designing. This Dialogue is lifted out of an article by Terry Winograd on his program SHRDLU; only a few names have been changed. In it, a program communicates with a person about the so-called "blocks world" in rather impressive English. The computer program appears to exhibit some real understanding—in its limited world. The Dialogue's title is based on *Jesu, Joy of Man's Desiring,* one movement of Bach's Cantata 147.

Chapter XVIII: Artificial Intelligence: Retrospects. This Chapter opens with a discussion of the famous "Turing test"—a proposal by the computer pioneer Alan Turing for a way to detect the presence or absence of "thought" in a machine. From there, we go on to an abridged history of Artificial Intelligence. This covers programs that can—to some degree—play games, prove theorems, solve problems, compose music, do mathematics, and use "natural language" (e.g., English).

Contrafactus. About how we unconsciously organize our thoughts so that we can imagine hypothetical variants on the real world all the time. Also about aberrant variants of this ability—such as possessed by the new character, the Sloth, an avid lover of French fries, and rabid hater of counterfactuals.

Chapter XIX: Artificial Intelligence: Prospects. The preceding Dialogue triggers a discussion of how knowledge is represented in layers of contexts. This leads to the modern AI idea of "frames". A frame-like way of handling a set of visual pattern puzzles is presented, for the purpose of concreteness. Then the deep issue of the interaction of concepts in general is discussed, which leads into some speculations on creativity. The Chapter concludes with a set of personal "Questions and Speculations" on AI and minds in general.

Sloth Canon. A canon which imitates a Bach canon in which one voice plays the same melody as another, only upside down and twice as slowly, while a third voice is free. Here, the Sloth utters the same lines as the Tortoise does, only negated (in a liberal sense of the term) and twice as slowly, while Achilles is free.

Chapter XX: Strange Loops, Or Tangled Hierarchies. A grand windup of many of the ideas about hierarchical systems and self-reference. It is concerned with the snarls which arise when systems turn back on themselves—for example, science probing science, government investigating governmental wrongdoing, art violating the rules of art, and finally, humans thinking about their own brains and minds. Does Gödel's Theorem have anything to say about this last "snarl"? Are free will and the sensation of consciousness connected to Gödel's Theorem? The Chapter ends by tying Gödel, Escher, and Bach together once again.

Six-Part Ricercar. This Dialogue is an exuberant game played with many of the ideas which have permeated the book. It is a reenactment of the story of the *Musical Offering,* which began the book; it is simultaneously a "translation" into words of the most complex piece in the *Musical Offering:* the *Six-Part Ricercar.* This duality imbues the Dialogue with more levels of meaning than any other in the book. Frederick the Great is replaced by the Crab, pianos by computers, and so on. Many surprises arise. The Dialogue's content concerns problems of mind, consciousness, free will, Artificial Intelligence, the Turing test, and so forth, which have been introduced earlier. It concludes with an implicit reference to the beginning of the book, thus making the book into one big self-referential loop, symbolizing at once Bach's music, Escher's drawings, and Gödel's Theorem.

List of Illustrations

Words of Thanks

This book was brewing in my mind over a period of probably nearly twenty years—ever since I was thirteen and thinking about how I thought in English and French. And even before that, there were clear signs of my main-line interest. I remember that at some early age, there was nothing more fascinating to me than the idea of taking three 3's: operating upon 3 with *itself!* I was sure that this idea was so subtle that it was inconceivable to anyone else—but I dared ask my mother one day how big it was anyway, and she answered "Nine". I wasn't sure she knew what I meant, though. Later, my father initiated me into the mysteries of square roots, and *i* . . .

I owe more to my parents than to anyone else. They have been pillars of support I could rely on at all times. They have guided me, inspired me, encouraged me, and sustained me. Most of all, they have always believed in me. It is to them that this book is dedicated.

To two friends of many years—Robert Boeninger, Peter Jones—I owe special thanks, for they helped mold a million ways of thought, and their influences and ideas are spread all around in this book.

To Charles Brenner, I am much indebted for his having taught me to program when we were both young, and for his constant pushing and prodding—implicit praise—and occasional criticism.

I am pleased to acknowledge the immense influence of Ernest Nagel, a long-time friend and mentor. I loved "Nagel and Newman", and I have learned much from many conversations, long ago in Vermont and more recently in New York.

Howard DeLong, through his book, reawakened in me a long-dormant love of the matters in this book. I truly owe him a great debt.

David Jonathan Justman taught me what it is to be a Tortoise—an ingenious, persistent, and humorous being, with a fondness for paradox and contradiction. I hope that he will read and enjoy this book, which owes a great deal to him.

Scott Kim has exerted a gigantic influence on me. Ever since we met some two and a half years ago, the resonance between the two of us has been incredible. Aside from his tangible contributions of art, music, humor, analogies, and so on—including much-appreciated volunteer labor at crucial times—Scott has contributed new perspectives and insights which have changed my own views of my endeavor as it has evolved. If anyone understands this book, it is Scott.

For advice on matters of large scale or small, I have turned repeatedly to Don Byrd, who knows this book forwards and backwards and every which way . . . He has an unerring sense for its overall goals and structure, and time and again has given me good ideas which I have delightedly incorporated. My only regret is that I won't be able to include all the *future* ideas Don will come up with, once the book is in print. And let me not forget to thank Don for the marvelous flexibility-in-inflexibility of his music-printing program, SMUT. He spent many long days and arduous nights coaxing SMUT to sit up and do preposterous tricks. Some of his results are included as figures in this book. But Don's influence is spread throughout, which pleases me greatly.

I could not possibly have written this book if it were not for the facilities of the Institute for Mathematical Studies in the Social Sciences, at Stanford University. Its director, Pat Suppes, is a long-time friend and he was extremely generous to me, in housing me in Ventura Hall, giving me access to an outstanding computer system, and in general an excellent working environment, for two whole years—and then some.

This brings me to Pentti Kanerva, the author of the text-editing program to which this book owes its existence. I have said to many people that it would have taken me twice as long to write my book if I hadn't been able to use "TV-Edit", that graceful program which is so simple in spirit that only Pentti could have written it. It is also thanks to Pentti that I was able to do something which very few authors have ever done: typeset my own book. He has been a major force in the development of computer typesetting at the IMSSS. Equally important to me, however, is Pentti's rare quality: his sense of style. If my book *looks* good, to Pentti Kanerva is due much of the credit.

It was in the ASSU Typesetting Shop that this book was actually born. I would like to offer many hearty words of thanks to its director, Beverly Hendricks, and to her crew, for help in times of dire need, and for consistently good spirits in the face of one disaster after another. I would also like to thank Cecille Taylor and Barbara Laddaga, who did most of the actual work of running off galleys.

Over the years, my sister Laura Hofstadter has contributed much to my outlook on the world. Her influence is present in both the form and content of this book.

I would like to thank my new and old friends Marie Anthony, Sydney Arkowitz, Bengt Olle Bengtsson, Felix Bloch, Francisco Claro, Persi Diaconis, Năi-Huá Duàn, John Ellis, Robin Freeman, Dan Friedman, Pranab Ghosh, Michael Goldhaber, Avril Greenberg, Eric Hamburg, Robert Herman, Ray Hyman, Dave Jennings, Dianne Kanerva, Lauri Kanerva, Inga Karliner, Jonathan and Ellen King, Gayle Landt, Bill Lewis, Jos Marlowe, John McCarthy, Jim McDonald, Louis Mendelowitz, Mike Mueller, Rosemary Nelson, Steve Omohundro, Paul Oppenheimer, Peter E. Parks, David Policansky, Pete Rimbey, Kathy Rosser, Wilfried Sieg, Guy Steele, Larry Tesler, François Vannucci, Phil Wadler, Terry Winograd, and Bob Wolf for "resonating" with me at crucial times in my life, and thereby contributing in various and sundry ways to this book.

I wrote this book twice. After having written it once, I started all over again and rewrote it. The first go-round was when I was still a graduate student in physics at the University of Oregon, and four faculty members were mighty indulgent concerning my aberrant ways: Paul Csonka, Rudy Hwa, Mike Moravcsik, and Gregory Wannier. I appreciate their under-

PART I

FIGURE 1. Johann Sebastian Bach, in 1748. From a painting by Elias Gottlieb Haussmann.

standing attitudes. In addition, Paul Csonka read an entire early version and made helpful comments.

Thanks to E. O. Wilson for reading and commenting on an early version of my *Prelude, Ant Fugue*.

Thanks to Marsha Meredith for being the meta-author of a droll kōan.

Thanks to Marvin Minsky for a memorable conversation one March day in his home, parts of which the reader will find reconstructed herein.

Thanks to Bill Kaufmann for advice on publication, and to Jeremy Bernstein and Alex George for encouraging words when needed.

Very warm thanks to Martin Kessler, Maureen Bischoff, Vincent Torre, Leon Dorin, and all the other people at Basic Books, for undertaking this publishing venture which is unusual in quite a few ways.

Thanks to Phoebe Hoss for doing well the difficult job of copy editing, and to Larry Breed for valuable last-minute proofreading.

Thanks to my many Imlac-roommates, who took so many phone messages over the years; also to the Pine Hall crew, who developed and maintained much of the hardware and software that this book has so vitally depended on.

Thanks to Dennis Davies of the Stanford Instructional Television Network for his help in setting up the "self-engulfing televisions" which I spent hours photographing.

Thanks to Jerry Pryke, Bob Parks, Ted Bradshaw, and Vinnie Aveni of the machine shop in the High Energy Physics Laboratory at Stanford, for generously helping me make trip-lets.

Thanks to my uncle and aunt, Jimmy and Betty Givan, for the Christmas present they never knew would so delight me: a "Black Box" which had no other function than to turn itself off.

Finally, I would like to give special thanks to my freshman English teacher, Brent Harold, who first sprang Zen on me; to Kees Gugelot, who gave me a record of the *Musical Offering* one sad November long ago; and to Otto Frisch, in whose office at Cambridge I first saw the magic of Escher.

I have tried to remember all the people who have contributed to the development of this book, but I have undoubtedly failed to include all of them.

In a way, this book is a statement of my religion. I hope that this will come through to my readers, and that my enthusiasm and reverence for certain ideas will infiltrate the hearts and minds of a few people. That is the best I could ask for.

D. R. H.
Bloomington and Stanford
January, 1979.

Introduction: A Musico-Logical Offering

Author:

FREDERICK THE GREAT, King of Prussia, came to power in 1740. Although he is remembered in history books mostly for his military astuteness, he was also devoted to the life of the mind and the spirit. His court in Potsdam was one of the great centers of intellectual activity in Europe in the eighteenth century. The celebrated mathematician Leonhard Euler spent twenty-five years there. Many other mathematicians and scientists came, as well as philosophers—including Voltaire and La Mettrie, who wrote some of their most influential works while there.

But music was Frederick's real love. He was an avid flutist and composer. Some of his compositions are occasionally performed even to this day. Frederick was one of the first patrons of the arts to recognize the virtues of the newly developed "piano-forte" ("soft-loud"). The piano had been developed in the first half of the eighteenth century as a modification of the harpsichord. The problem with the harpsichord was that pieces could only be played at a rather uniform loudness—there was no way to strike one note more loudly than its neighbors. The "soft-loud", as its name implies, provided a remedy to this problem. From Italy, where Bartolommeo Cristofori had made the first one, the soft-loud idea had spread widely. Gottfried Silbermann, the foremost German organ builder of the day, was endeavoring to make a "perfect" piano-forte. Undoubtedly King Frederick was the greatest supporter of his efforts—it is said that the King owned as many as fifteen Silbermann pianos!

Bach

Frederick was an admirer not only of pianos, but also of an organist and composer by the name of J. S. Bach. This Bach's compositions were somewhat notorious. Some called them "turgid and confused", while others claimed they were incomparable masterpieces. But no one disputed Bach's ability to improvise on the organ. In those days, being an organist not only meant being able to play, but also to extemporize, and Bach was known far and wide for his remarkable extemporizations. (For some delightful anecdotes about Bach's extemporization, see *The Bach Reader*, by H. T. David and A. Mendel.)

In 1747, Bach was sixty-two, and his fame, as well as one of his sons, had reached Potsdam; in fact, Carl Philipp Emanuel Bach was the Capellmeister (choirmaster) at the court of King Frederick. For years the King had let it be known, through gentle hints to Philipp Emanuel, how

pleased he would be to have the elder Bach come and pay him a visit; but this wish had never been realized. Frederick was particularly eager for Bach to try out his new Silbermann pianos, which he (Frederick) correctly foresaw as the great new wave in music.

It was Frederick's custom to have evening concerts of chamber music in his court. Often he himself would be the soloist in a concerto for flute. Here we have reproduced a painting of such an evening by the German painter Adolph von Menzel, who, in the 1800's, made a series of paintings illustrating the life of Frederick the Great. At the cembalo is C. P. E. Bach, and the figure furthest to the right is Joachim Quantz, the King's flute master—and the only person allowed to find fault with the King's flute playing. One May evening in 1747, an unexpected guest showed up. Johann Nikolaus Forkel, one of Bach's earliest biographers, tells the story as follows:

> One evening, just as he was getting his flute ready, and his musicians were assembled, an officer brought him a list of the strangers who had arrived. With his flute in his hand he ran over the list, but immediately turned to the assembled musicians, and said, with a kind of agitation, "Gentlemen, old Bach is come." The flute was now laid aside, and old Bach, who had alighted at his son's lodgings, was immediately summoned to the Palace. Wilhelm Friedemann, who accompanied his father, told me this story, and I must say that I still think with pleasure on the manner in which he related it. At that time it was the fashion to make rather prolix compliments. The first appearance of J. S. Bach before so great a King, who did not even give him time to change his traveling dress for a black chanter's gown, must necessarily be attended with many apologies. I will not here dwell on these apologies, but merely observe, that in Wilhelm Friedemann's mouth they made a formal Dialogue between the King and the Apologist.
>
> But what is more important than this is that the King gave up his Concert for this evening, and invited Bach, then already called the Old Bach, to try his fortepianos, made by Silbermann, which stood in several rooms of the palace. [Forkel here inserts this footnote: "The pianofortes manufactured by Silbermann, of Freyberg, pleased the King so much, that he resolved to buy them all up. He collected fifteen. I hear that they all now stand unfit for use in various corners of the Royal Palace."] The musicians went with him from room to room, and Bach was invited everywhere to try them and to play unpremeditated compositions. After he had gone on for some time, he asked the King to give him a subject for a Fugue, in order to execute it immediately without any preparation. The King admired the learned manner in which his subject was thus executed extempore; and, probably to see how far such art could be carried, expressed a wish to hear a Fugue with six Obligato parts. But as it is not every subject that is fit for such full harmony, Bach chose one himself, and immediately executed it to the astonishment of all present in the same magnificent and learned manner as he had done that of the King. His Majesty desired also to hear his performance on the organ. The next day therefore Bach was taken to all the organs in Potsdam, as he had before been to Silbermann's fortepianos. After his return to Leipzig, he composed the subject, which he had received from the King, in three and six parts, added several artificial passages in strict canon to it, and had it engraved, under the title of "Musikalisches Opfer" [*Musical Offering*], and dedicated it to the inventor.[1]

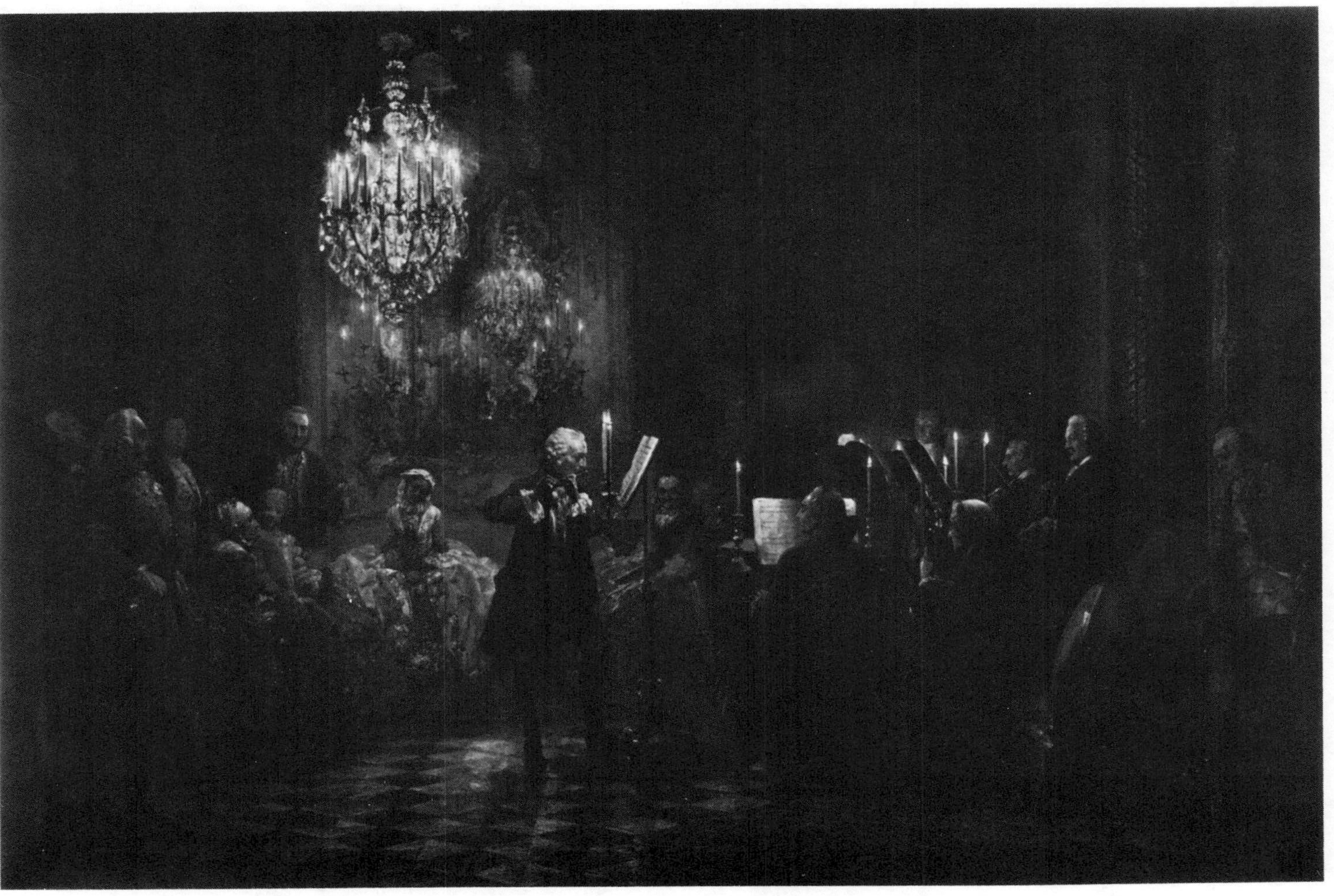

FIGURE 2. Flute Concert in Sanssouci, *by Adolph von Menzel (1852).*

FIGURE 3. The Royal Theme.

When Bach sent a copy of his *Musical Offering* to the King, he included a dedicatory letter, which is of interest for its prose style if nothing else—rather submissive and flattersome. From a modern perspective it seems comical. Also, it probably gives something of the flavor of Bach's apology for his appearance.[2]

> MOST GRACIOUS KING!
>
> In deepest humility I dedicate herewith to Your Majesty a musical offering, the noblest part of which derives from Your Majesty's own august hand. With awesome pleasure I still remember the very special Royal grace when, some time ago, during my visit in Potsdam, Your Majesty's Self deigned to play to me a theme for a fugue upon the clavier, and at the same time charged me most graciously to carry it out in Your Majesty's most august presence. To obey Your Majesty's command was my most humble duty. I noticed very soon, however, that, for lack of necessary preparation, the execution of the task did not fare as well as such an excellent theme demanded. I resolved therefore and promptly pledged myself to work out this right Royal theme more fully, and then make it known to the world. This resolve has now been carried out as well as possible, and it has none other than this irreproachable intent, to glorify, if only in a small point, the fame of a monarch whose greatness and power, as in all the sciences of war and peace, so especially in music, everyone must admire and revere. I make bold to add this most humble request: may Your Majesty deign to dignify the present modest labor with a gracious acceptance, and continue to grant Your Majesty's most august Royal grace to
>
> Your Majesty's
most humble and obedient servant,
THE AUTHOR
>
> Leipzig, July 7
1747

Some twenty-seven years later, when Bach had been dead for twenty-four years, a Baron named Gottfried van Swieten—to whom, incidentally, Forkel dedicated his biography of Bach, and Beethoven dedicated his First Symphony—had a conversation with King Frederick, which he reported as follows:

> He [Frederick] spoke to me, among other things, of music, and of a great organist named Bach, who has been for a while in Berlin. This artist [Wilhelm Friedemann Bach] is endowed with a talent superior, in depth of harmonic knowledge and power of execution, to any I have heard or can imagine, while those who knew his father claim that he, in turn, was even greater. The King

> is of this opinion, and to prove it to me he sang aloud a chromatic fugue subject which he had given this old Bach, who on the spot had made of it a fugue in four parts, then in five parts, and finally in eight parts.[3]

Of course there is no way of knowing whether it was King Frederick or Baron van Swieten who magnified the story into larger-than-life proportions. But it shows how powerful Bach's legend had become by that time. To give an idea of how extraordinary a six-part fugue is, in the entire *Well-Tempered Clavier* by Bach, containing forty-eight Preludes and Fugues, only two have as many as five parts, and nowhere is there a six-part fugue! One could probably liken the task of improvising a six-part fugue to the playing of sixty simultaneous blindfold games of chess, and winning them all. To improvise an eight-part fugue is really beyond human capability.

In the copy which Bach sent to King Frederick, on the page preceding the first sheet of music, was the following inscription:

Regis Iuſsu Cantio Et Reliqua Canonica Arte Reſoluta.

FIGURE 4.

("At the King's Command, the Song and the Remainder Resolved with Canonic Art.") Here Bach is punning on the word "canonic", since it means not only "with canons" but also "in the best possible way". The initials of this inscription are

R I C E R C A R

—an Italian word, meaning "to seek". And certainly there is a great deal to seek in the *Musical Offering*. It consists of one three-part fugue, one six-part fugue, ten canons, and a trio sonata. Musical scholars have concluded that the three-part fugue must be, in essence, identical with the one which Bach improvised for King Frederick. The six-part fugue is one of Bach's most complex creations, and its theme is, of course, the Royal Theme. That theme, shown in Figure 3, is a very complex one, rhythmically irregular and highly chromatic (that is, filled with tones which do not belong to the key it is in). To write a decent fugue of even two voices based on it would not be easy for the average musician!

Both of the fugues are inscribed "Ricercar", rather than "Fuga". This is another meaning of the word; "ricercar" was, in fact, the original name for the musical form now known as "fugue". By Bach's time, the word "fugue" (or *fuga*, in Latin and Italian) had become standard, but the term "ricercar" had survived, and now designated an erudite kind of fugue, perhaps too austerely intellectual for the common ear. A similar usage survives in English today: the word "recherché" means, literally, "sought out", but carries the same kind of implication, namely of esoteric or highbrow cleverness.

The trio sonata forms a delightful relief from the austerity of the fugues and canons, because it is very melodious and sweet, almost danceable. Nevertheless, it too is based largely on the King's theme, chromatic and austere as it is. It is rather miraculous that Bach could use such a theme to make so pleasing an interlude.

The ten canons in the *Musical Offering* are among the most sophisticated canons Bach ever wrote. However, curiously enough, Bach himself never wrote them out in full. This was deliberate. They were posed as puzzles to King Frederick. It was a familiar musical game of the day to give a single theme, together with some more or less tricky hints, and to let the canon based on that theme be "discovered" by someone else. In order to know how this is possible, you must understand a few facts about canons.

Canons and Fugues

The idea of a canon is that one single theme is played against itself. This is done by having "copies" of the theme played by the various participating voices. But there are many ways to do this. The most straightforward of all canons is the round, such as "Three Blind Mice", "Row, Row, Row Your Boat", or "Frère Jacques". Here, the theme enters in the first voice and, after a fixed time-delay, a "copy" of it enters, in precisely the same key. After the same fixed time-delay in the second voice, the third voice enters carrying the theme, and so on. Most themes will not harmonize with themselves in this way. In order for a theme to work as a canon theme, each of its notes must be able to serve in a dual (or triple, or quadruple) role: it must firstly be part of a melody, and secondly it must be part of a harmonization of the same melody. When there are three canonical voices, for instance, each note of the theme must act in two distinct harmonic ways, as well as melodically. Thus, each note in a canon has more than one musical meaning; the listener's ear and brain automatically figure out the appropriate meaning, by referring to context.

There are more complicated sorts of canons, of course. The first escalation in complexity comes when the "copies" of the theme are staggered not only in *time,* but also in *pitch;* thus, the first voice might sing the theme starting on C, and the second voice, overlapping with the first voice, might sing the identical theme starting five notes higher, on G. A third voice, starting on the D yet five notes higher, might overlap with the first two, and so on. The next escalation in complexity comes when the *speeds* of the different voices are not equal; thus, the second voice might sing twice as quickly, or twice as slowly, as the first voice. The former is called *diminution,* the latter *augmentation* (since the theme seems to shrink or to expand).

We are not yet done! The next stage of complexity in canon construction is to *invert* the theme, which means to make a melody which jumps *down* wherever the original theme jumps *up,* and by exactly the same number of semitones. This is a rather weird melodic transformation, but when one has heard many themes inverted, it begins to seem quite natural. Bach was especially fond of inversions, and they show up often in his work—and the *Musical Offering* is no exception. (For a simple example of inversion, try the tune "Good King Wenceslas". When the original and its inversion are sung together, starting an octave apart and staggered with a time-delay of two beats, a pleasing canon results.) Finally, the most esoteric of "copies" is the retrograde copy—where the theme is played backwards in time. A canon which uses this trick is affectionately known as a *crab canon,* because of the peculiarities of crab locomotion. Bach included a crab canon in the *Musical Offering,* needless to say. Notice that every type of "copy" preserves all the information in the original theme, in the sense that the theme is fully recoverable from any of the copies. Such an information-preserving transformation is often called an *isomorphism,* and we will have much traffic with isomorphisms in this book.

Sometimes it is desirable to relax the tightness of the canon form. One way is to allow slight departures from perfect copying, for the sake of more fluid harmony. Also, some canons have "free" voices—voices which do not employ the canon's theme, but which simply harmonize agreeably with the voices that are in canon with each other.

Each of the canons in the *Musical Offering* has for its theme a different variant of the King's Theme, and all the devices described above for making canons intricate are exploited to the hilt; in fact, they are occasionally combined. Thus, one three-voice canon is labeled "Canon per Augmentationem, contrario Motu"; its middle voice is free (in fact, it sings the Royal Theme), while the other two dance canonically above and below it, using the devices of augmentation and inversion. Another bears simply the cryptic label "Quaerendo invenietis" ("By seeking, you will discover"). All of the canon puzzles have been solved. The canonical solutions were given by one of Bach's pupils, Johann Philipp Kirnberger. But one might still wonder whether there are more solutions to seek!

I should also explain briefly what a fugue is. A fugue is like a canon, in that it is usually based on one theme which gets played in different voices and different keys, and occasionally at different speeds or upside down or backwards. However, the notion of fugue is much less rigid than that of canon, and consequently it allows for more emotional and artistic expression. The telltale sign of a fugue is the way it begins: with a single voice singing its theme. When it is done, then a second voice enters, either five scale-notes up, or four down. Meanwhile the first voice goes on, singing the "countersubject": a secondary theme, chosen to provide rhythmic, harmonic, and melodic contrasts to the subject. Each of the voices enters in turn, singing the theme, often to the accompaniment of the countersubject in some other voice, with the remaining voices doing whatever fanciful things entered the composer's mind. When all the voices have "arrived", then there are no rules. There are, to be sure, standard kinds of things to do—but not so standard that one can merely compose a fugue by formula. The two fugues in the *Musical Offering* are outstanding examples of fugues that could never have been "composed by formula". There is something much deeper in them than mere fugality.

All in all, the *Musical Offering* represents one of Bach's supreme accomplishments in counterpoint. It is itself one large intellectual fugue, in which many ideas and forms have been woven together, and in which playful double meanings and subtle allusions are commonplace. And it is a very beautiful creation of the human intellect which we can appreciate forever. (The entire work is wonderfully described in the book *J. S. Bach's Musical Offering,* by H. T. David.)

An Endlessly Rising Canon

There is one canon in the *Musical Offering* which is particularly unusual. Labeled simply "Canon per Tonos", it has three voices. The uppermost voice sings a variant of the Royal Theme, while underneath it, two voices provide a canonic harmonization based on a second theme. The lower of this pair sings its theme in C minor (which is the key of the canon as a whole), and the upper of the pair sings the same theme displaced upwards in pitch by an interval of a fifth. What makes this canon different from any other, however, is that when it concludes—or, rather, *seems* to conclude—it is no longer in the key of C minor, but now is in D minor. Somehow Bach has contrived to *modulate* (change keys) right under the listener's nose. And it is so constructed that this "ending" ties smoothly onto the beginning again; thus one can repeat the process and return in the key of E, only to join again to the beginning. These successive modulations lead the ear to increasingly remote provinces of tonality, so that after several of them, one would expect to be hopelessly far away from the starting key. And yet magically, after exactly six such modulations, the original key of C minor has been restored! All the voices are exactly one octave higher than they were at the beginning, and here the piece may be broken off in a musically agreeable way. Such, one imagines, was Bach's intention; but Bach indubitably also relished the implication that this process could go on ad infinitum, which is perhaps why he wrote in the margin "As the modulation rises, so may the King's Glory." To emphasize its potentially infinite aspect, I like to call this the "Endlessly Rising Canon".

In this canon, Bach has given us our first example of the notion of *Strange Loops.* The "Strange Loop" phenomenon occurs whenever, by moving upwards (or downwards) through the levels of some hierarchical system, we unexpectedly find ourselves right back where we started. (Here, the system is that of musical keys.) Sometimes I use the term *Tangled Hierarchy* to describe a system in which a Strange Loop occurs. As we go on, the theme of Strange Loops will recur again and again. Sometimes it will be hidden, other times it will be out in the open; sometimes it will be right side up, other times it will be upside down, or backwards. "Quaerendo invenietis" is my advice to the reader.

Escher

To my mind, the most beautiful and powerful visual realizations of this notion of Strange Loops exist in the work of the Dutch graphic artist M. C. Escher, who lived from 1902 to 1972. Escher was the creator of some of the

FIGURE 5. Waterfall, *by M. C. Escher (lithograph, 1961).*

most intellectually stimulating drawings of all time. Many of them have their origin in paradox, illusion, or double-meaning. Mathematicians were among the first admirers of Escher's drawings, and this is understandable because they often are based on mathematical principles of symmetry or pattern . . . But there is much more to a typical Escher drawing than just symmetry or pattern; there is often an underlying idea, realized in artistic form. And in particular, the Strange Loop is one of the most recurrent themes in Escher's work. Look, for example, at the lithograph *Waterfall* (Fig. 5), and compare its six-step endlessly falling loop with the six-step endlessly rising loop of the "Canon per Tonos". The similarity of vision is

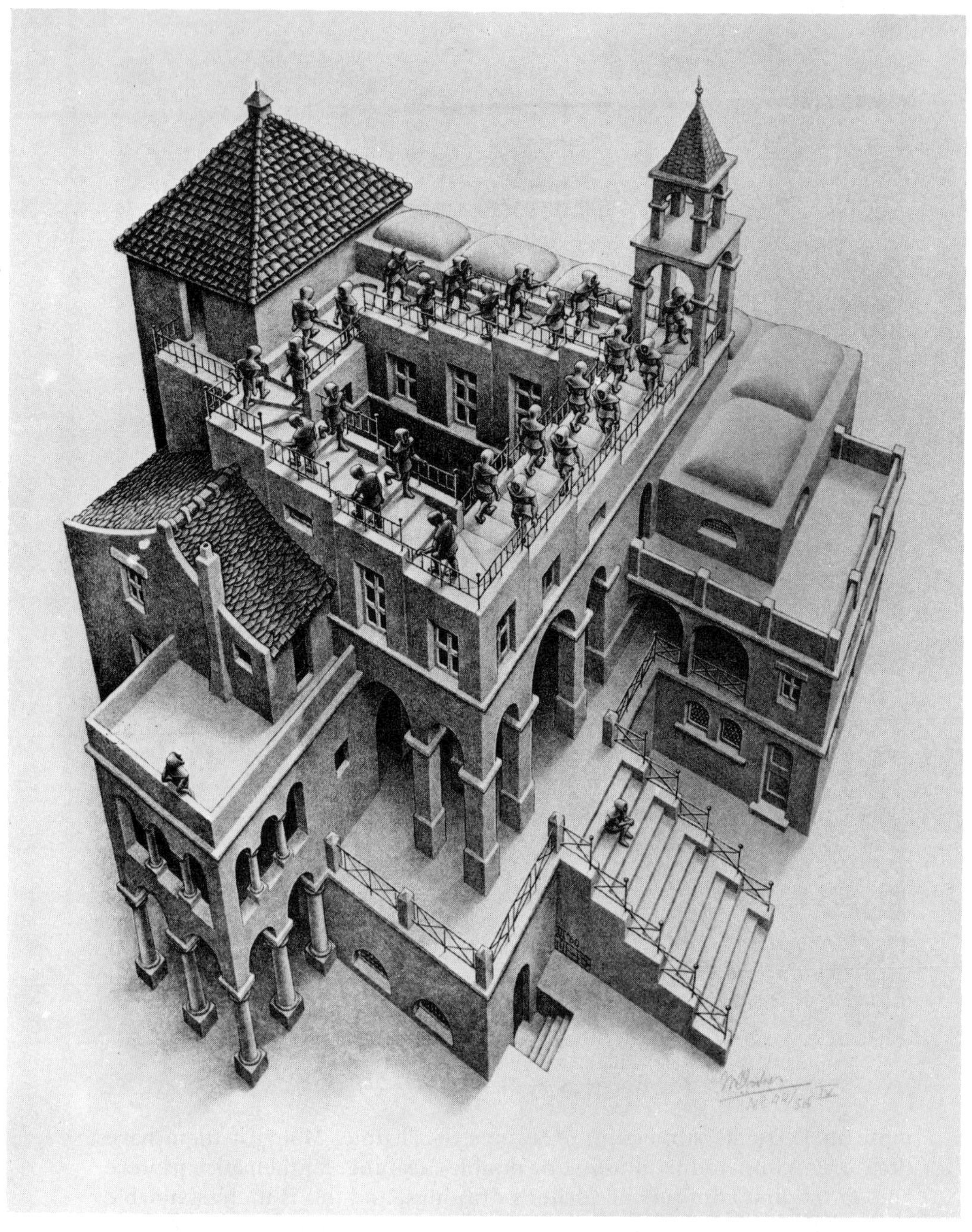

FIGURE 6. Ascending and Descending, *by M. C. Escher (lithograph, 1960).*

remarkable. Bach and Escher are playing one single theme in two different "keys": music and art.

Escher realized Strange Loops in several different ways, and they can be arranged according to the tightness of the loop. The lithograph *Ascending and Descending* (Fig. 6), in which monks trudge forever in loops, is the loosest version, since it involves so many steps before the starting point is regained. A tighter loop is contained in *Waterfall,* which, as we already observed, involves only six discrete steps. You may be thinking that there is some ambiguity in the notion of a single "step"—for instance, couldn't *Ascending and Descending* be seen just as easily as having four levels (staircases) as forty-five levels (stairs)? It is indeed true that there is an inherent

FIGURE 7. Hand with Reflecting Globe. *Self-portrait by M. C. Escher (lithograph, 1935).*

FIGURE 8. Metamorphosis II, *by M. C. Escher (woodcut, 19.5 cm. × 400 cm., 1939-40).*

haziness in level-counting, not only in Escher pictures, but in hierarchical, many-level systems. We will sharpen our understanding of this haziness later on. But let us not get too distracted now! As we tighten our loop, we come to the remarkable *Drawing Hands* (Fig. 135), in which each of two hands draws the other: a two-step Strange Loop. And finally, the tightest of all Strange Loops is realized in *Print Gallery* (Fig. 142): a picture of a picture which contains itself. Or is it a picture of a gallery which contains itself? Or of a town which contains itself? Or a young man who contains himself? (Incidentally, the illusion underlying *Ascending and Descending* and *Waterfall* was not invented by Escher, but by Roger Penrose, a British mathematician, in 1958. However, the theme of the Strange Loop was already present in Escher's work in 1948, the year he drew *Drawing Hands*. *Print Gallery* dates from 1956.)

Implicit in the concept of Strange Loops is the concept of infinity, since what else is a loop but a way of representing an endless process in a finite way? And infinity plays a large role in many of Escher's drawings. Copies of one single theme often fit into each other, forming visual analogues to the canons of Bach. Several such patterns can be seen in Escher's famous print *Metamorphosis* (Fig. 8). It is a little like the "Endlessly Rising Canon": wandering further and further from its starting point, it suddenly is back. In the tiled planes of *Metamorphosis* and other pictures, there are already suggestions of infinity. But wilder visions of infinity appear in other drawings by Escher. In some of his drawings, one single theme can appear on different levels of reality. For instance, one level in a drawing might clearly be recognizable as representing fantasy or imagination; another level would be recognizable as reality. These two levels might be the only explicitly portrayed levels. But the mere presence of these two levels invites the viewer to look upon himself as part of yet another level; and by taking that step, the viewer cannot help getting caught up in Escher's implied chain of levels, in which, for any one level, there is always another level above it of greater "reality", and likewise, there is always a level below, "more imaginary" than it is. This can be mind-boggling in itself. However, what happens if the chain of levels is not linear, but forms a loop? What is real, then, and what is fantasy? The genius of Escher was that he could not only concoct, but actually portray, dozens of half-real, half-mythical worlds, worlds filled with Strange Loops, which he seems to be inviting his viewers to enter.

Gödel

In the examples we have seen of Strange Loops by Bach and Escher, there is a conflict between the finite and the infinite, and hence a strong sense of paradox. Intuition senses that there is something mathematical involved here. And indeed in our own century a mathematical counterpart was discovered, with the most enormous repercussions. And, just as the Bach and Escher loops appeal to very simple and ancient intuitions—a musical scale, a staircase—so this discovery, by K. Gödel, of a Strange Loop in

FIGURE 9. Kurt Gödel.

mathematical systems has its origins in simple and ancient intuitions. In its absolutely barest form, Gödel's discovery involves the translation of an ancient paradox in philosophy into mathematical terms. That paradox is the so-called *Epimenides paradox,* or *liar paradox*. Epimenides was a Cretan who made one immortal statement: "All Cretans are liars." A sharper version of the statement is simply "I am lying"; or, "This statement is false". It is that last version which I will usually mean when I speak of the Epimenides paradox. It is a statement which rudely violates the usually assumed dichotomy of statements into true and false, because if you tentatively think it is true, then it immediately backfires on you and makes you think it is false. But once you've decided it is false, a similar backfiring returns you to the idea that it must be true. Try it!

The Epimenides paradox is a one-step Strange Loop, like Escher's *Print Gallery*. But how does it have to do with mathematics? That is what Gödel discovered. His idea was to use mathematical reasoning in exploring mathematical reasoning itself. This notion of making mathematics "introspective" proved to be enormously powerful, and perhaps its richest implication was the one Gödel found: Gödel's Incompleteness Theorem. What the Theorem states and how it is proved are two different things. We shall discuss both in quite some detail in this book. The Theorem can be likened to a pearl, and the method of proof to an oyster. The pearl is prized for its luster and simplicity; the oyster is a complex living beast whose innards give rise to this mysteriously simple gem.

Gödel's Theorem appears as Proposition VI in his 1931 paper "On Formally Undecidable Propositions in *Principia Mathematica* and Related Systems I." It states:

> To every ω-consistent recursive class κ of *formulae* there correspond recursive *class-signs* r, such that neither v Gen r nor Neg (v Gen r) belongs to Flg (κ) (where v is the *free variable* of r).

Actually, it was in German, and perhaps you feel that it might as well be in German anyway. So here is a paraphrase in more normal English:

> All consistent axiomatic formulations of number theory
> include undecidable propositions.

This is the pearl.

In this pearl it is hard to see a Strange Loop. That is because the Strange Loop is buried in the oyster—the proof. The proof of Gödel's Incompleteness Theorem hinges upon the writing of a self-referential mathematical statement, in the same way as the Epimenides paradox is a self-referential statement of language. But whereas it is very simple to talk about language in language, it is not at all easy to see how a statement about numbers can talk about itself. In fact, it took genius merely to connect the idea of self-referential statements with number theory. Once Gödel had the intuition that such a statement could be created, he was over the major hurdle. The actual creation of the statement was the working out of this one beautiful spark of intuition.

We shall examine the Gödel construction quite carefully in Chapters to come, but so that you are not left completely in the dark, I will sketch here, in a few strokes, the core of the idea, hoping that what you see will trigger ideas in your mind. First of all, the difficulty should be made absolutely clear. Mathematical statements—let us concentrate on number-theoretical ones—are about properties of whole numbers. Whole numbers are not statements, nor are their properties. A statement of number theory is not *about* a statement of number theory; it just *is* a statement of number theory. This is the problem; but Gödel realized that there was more here than meets the eye.

Gödel had the insight that a statement of number theory could be *about* a statement of number theory (possibly even itself), if only numbers could somehow stand for statements. The idea of a *code,* in other words, is at the heart of his construction. In the Gödel Code, usually called "Gödel-numbering", numbers are made to stand for symbols and sequences of symbols. That way, each statement of number theory, being a sequence of specialized symbols, acquires a Gödel number, something like a telephone number or a license plate, by which it can be referred to. And this coding trick enables statements of number theory to be understood on two different levels: as statements of number theory, and also as *statements about statements* of number theory.

Once Gödel had invented this coding scheme, he had to work out in detail a way of transporting the Epimenides paradox into a number-theoretical formalism. His final transplant of Epimenides did not say, "This statement of number theory is false", but rather, "This statement of number theory does not have any proof". A great deal of confusion can be caused by this, because people generally understand the notion of "proof" rather vaguely. In fact, Gödel's work was just part of a long attempt by mathematicians to explicate for themselves what proofs are. The important thing to keep in mind is that proofs are demonstrations *within fixed systems* of propositions. In the case of Gödel's work, the fixed system of number-theoretical reasoning to which the word "proof" refers is that of *Principia Mathematica (P.M.),* a giant opus by Bertrand Russell and Alfred North Whitehead, published between 1910 and 1913. Therefore, the Gödel sentence G should more properly be written in English as:

> This statement of number theory does not have any proof
> in the system of *Principia Mathematica.*

Incidentally, this Gödel sentence G is not Gödel's Theorem—no more than the Epimenides sentence is the observation that "The Epimenides sentence is a paradox." We can now state what the effect of discovering G is. Whereas the Epimenides statement creates a paradox since it is neither true nor false, the Gödel sentence G is unprovable (inside *P.M.*) but true. The grand conclusion? That the system of *Principia Mathematica* is "incomplete"—there are true statements of number theory which its methods of proof are too weak to demonstrate.

But if *Principia Mathematica* was the first victim of this stroke, it was certainly not the last! The phrase "and Related Systems" in the title of Gödel's article is a telling one; for if Gödel's result had merely pointed out a defect in the work of Russell and Whitehead, then others could have been inspired to improve upon *P.M.* and to outwit Gödel's Theorem. But this was not possible: Gödel's proof pertained to *any* axiomatic system which purported to achieve the aims which Whitehead and Russell had set for themselves. And for each different system, one basic method did the trick. In short, Gödel showed that provability is a weaker notion than truth, no matter what axiomatic system is involved.

Therefore Gödel's Theorem had an electrifying effect upon logicians, mathematicians, and philosophers interested in the foundations of mathematics, for it showed that no fixed system, no matter how complicated, could represent the complexity of the whole numbers: 0, 1, 2, 3, ... Modern readers may not be as nonplussed by this as readers of 1931 were, since in the interim our culture has absorbed Gödel's Theorem, along with the conceptual revolutions of relativity and quantum mechanics, and their philosophically disorienting messages have reached the public, even if cushioned by several layers of translation (and usually obfuscation). There is a general mood of expectation, these days, of "limitative" results—but back in 1931, this came as a bolt from the blue.

Mathematical Logic: A Synopsis

A proper appreciation of Gödel's Theorem requires a setting of context. Therefore, I will now attempt to summarize in a short space the history of mathematical logic prior to 1931—an impossible task. (See DeLong, Kneebone, or Nagel and Newman, for good presentations of history.) It all began with the attempts to mechanize the thought processes of reasoning. Now our ability to reason has often been claimed to be what distinguishes us from other species; so it seems somewhat paradoxical, on first thought, to mechanize that which is most human. Yet even the ancient Greeks knew that reasoning is a patterned process, and is at least partially governed by statable laws. Aristotle codified syllogisms, and Euclid codified geometry; but thereafter, many centuries had to pass before progress in the study of axiomatic reasoning would take place again.

One of the significant discoveries of nineteenth-century mathematics was that there are different, and equally valid, geometries—where by "a geometry" is meant a theory of properties of abstract points and lines. It had long been assumed that geometry was what Euclid had codified, and that, although there might be small flaws in Euclid's presentation, they were unimportant and any real progress in geometry would be achieved by extending Euclid. This idea was shattered by the roughly simultaneous discovery of non-Euclidean geometry by several people—a discovery that shocked the mathematics community, because it deeply challenged the idea that mathematics studies the real world. How could there be many different kinds of "points" and "lines" in one single reality? Today, the solution to the dilemma may be apparent, even to some nonmathematicians—but at the time, the dilemma created havoc in mathematical circles.

Later in the nineteenth century, the English logicians George Boole and Augustus De Morgan went considerably further than Aristotle in codifying strictly deductive reasoning patterns. Boole even called his book *"The Laws of Thought"*—surely an exaggeration, but it was an important contribution. Lewis Carroll was fascinated by these mechanized reasoning methods, and invented many puzzles which could be solved with them. Gottlob Frege in Jena and Giuseppe Peano in Turin worked on combining formal reasoning with the study of sets and numbers. David Hilbert in Göttingen worked on stricter formalizations of geometry than Euclid's. All of these efforts were directed towards clarifying what one means by "proof".

In the meantime, interesting developments were taking place in classical mathematics. A theory of different types of infinities, known as the *theory of sets,* was developed by Georg Cantor in the 1880's. The theory was powerful and beautiful, but intuition-defying. Before long, a variety of set-theoretical paradoxes had been unearthed. The situation was very disturbing, because just as mathematics seemed to be recovering from one set of paradoxes—those related to the theory of limits, in the calculus—along came a whole new set, which looked worse!

The most famous is Russell's paradox. Most sets, it would seem, are not members of themselves—for example, the set of walruses is not a walrus, the set containing only Joan of Arc is not Joan of Arc (a set is not a person)—and so on. In this respect, most sets are rather "run-of-the-mill". However, some "self-swallowing" sets *do* contain themselves as members, such as the set of all sets, or the set of all things except Joan of Arc, and so on. Clearly, every set is either run-of-the-mill or self-swallowing, and no set can be both. Now nothing prevents us from inventing R: *the set of all run-of-the-mill sets.* At first, R might seem a rather run-of-the-mill invention—but that opinion must be revised when you ask yourself, "Is R itself a run-of-the-mill set or a self-swallowing set?" You will find that the answer is: "R is neither run-of-the-mill nor self-swallowing, for either choice leads to paradox." Try it!

But if R is neither run-of-the-mill nor self-swallowing, then what is it? At the very least, pathological. But no one was satisfied with evasive answers of that sort. And so people began to dig more deeply into the foundations of set theory. The crucial questions seemed to be: "What is wrong with our intuitive concept of 'set'? Can we make a rigorous theory of sets which corresponds closely with our intuitions, but which skirts the paradoxes?" Here, as in number theory and geometry, the problem is in trying to line up intuition with formalized, or axiomatized, reasoning systems.

A startling variant of Russell's paradox, called "Grelling's paradox", can be made using adjectives instead of sets. Divide the adjectives in English into two categories: those which are self-descriptive, such as "pentasyllabic", "awkwardnessful", and "recherché", and those which are not, such as "edible", "incomplete", and "bisyllabic". Now if we admit "non-self-descriptive" as an adjective, to which class does it belong? If it seems questionable to include hyphenated words, we can use two terms invented specially for this paradox: *autological* (= "self-descriptive"), and *heterological* (= "non-self-descriptive"). The question then becomes: "Is 'heterological' heterological?" Try it!

There seems to be one common culprit in these paradoxes, namely self-reference, or "Strange Loopiness". So if the goal is to ban all paradoxes, why not try banning self-reference and anything that allows it to arise? This is not so easy as it might seem, because it can be hard to figure out just where self-reference is occurring. It may be spread out over a whole Strange Loop with several steps, as in this "expanded" version of Epimenides, reminiscent of *Drawing Hands:*

The following sentence is false.
The preceding sentence is true.

Taken together, these sentences have the same effect as the original Epimenides paradox; yet separately, they are harmless and even potentially useful sentences. The "blame" for this Strange Loop can't be pinned on either sentence—only on the way they "point" at each other. In the same way, each local region of *Ascending and Descending* is quite legitimate; it is only the way they are globally put together that creates an impossibility. Since there are indirect as well as direct ways of achieving self-reference, one must figure out how to ban both types at once—if one sees self-reference as the root of all evil.

Banishing Strange Loops

Russell and Whitehead did subscribe to this view, and accordingly, *Principia Mathematica* was a mammoth exercise in exorcising Strange Loops from logic, set theory, and number theory. The idea of their system was basically this. A set of the lowest "type" could contain only "objects" as members—not sets. A set of the next type up could only contain objects, or sets of the lowest type. In general, a set of a given type could only contain sets of lower type, or objects. Every set would belong to a specific type. Clearly, no set could contain itself because it would have to belong to a type higher than its own type. Only "run-of-the-mill" sets exist in such a system; furthermore, old R—the set of all run-of-the-mill sets—no longer is considered a set at all, because it does not belong to any finite type. To all appearances, then, this *theory of types,* which we might also call the "theory of the abolition of Strange Loops", successfully rids set theory of its paradoxes, but only at the cost of introducing an artificial-seeming hierarchy, and of disallowing the formation of certain kinds of sets—such as the set of all run-of-the-mill sets. Intuitively, this is not the way we imagine sets.

The theory of types handled Russell's paradox, but it did nothing about the Epimenides paradox or Grelling's paradox. For people whose interest went no further than set theory, this was quite adequate—but for people interested in the elimination of paradoxes generally, some similar "hierarchization" seemed necessary, to forbid looping back inside language. At the bottom of such a hierarchy would be an *object language.* Here, reference could be made only to a specific domain—not to aspects of the object language itself (such as its grammatical rules, or specific sentences in it). For that purpose there would be a *metalanguage.* This experience of two linguistic levels is familiar to all learners of foreign languages. Then there would be a metametalanguage for discussing the metalanguage, and so on. It would be required that every sentence should belong to some precise level of the hierarchy. Therefore, if one could find no level in which a given utterance fit, then the utterance would be deemed meaningless, and forgotten.

An analysis can be attempted on the two-step Epimenides loop given above. The first sentence, since it speaks of the second, must be on a higher level than the second. But by the same token, the second sentence must be on a higher level than the first. Since this is impossible, the two sentences are "meaningless". More precisely, such sentences simply cannot be formulated at all in a system based on a strict hierarchy of languages. This prevents all versions of the Epimenides paradox as well as Grelling's paradox. (To what language level could "heterological" belong?)

Now in set theory, which deals with abstractions that we don't use all the time, a stratification like the theory of types seems acceptable, even if a little strange—but when it comes to language, an all-pervading part of life, such stratification appears absurd. We don't think of ourselves as jumping up and down a hierarchy of languages when we speak about various things. A rather matter-of-fact sentence such as, "In this book, I criticize the theory of types" would be doubly forbidden in the system we are discussing. Firstly, it mentions "this book", which should only be mentionable in a "metabook"—and secondly, it mentions *me*—a person whom I should not be allowed to speak of at all! This example points out how silly the theory of types seems, when you import it into a familiar context. The remedy it adopts for paradoxes—total banishment of self-reference in any form—is a real case of overkill, branding many perfectly good constructions as meaningless. The adjective "meaningless", by the way, would have to apply to all discussions of the theory of linguistic types (such as that of this very paragraph) for they clearly could not occur on any of the levels—neither object language, nor metalanguage, nor metametalanguage, etc. So the very act of discussing the theory would be the most blatant possible violation of it!

Now one could defend such theories by saying that they were only intended to deal with formal languages—not with ordinary, informal language. This may be so, but then it shows that such theories are extremely academic and have little to say about paradoxes except when they crop up in special tailor-made systems. Besides, the drive to eliminate paradoxes at any cost, especially when it requires the creation of highly artificial formalisms, puts too much stress on bland consistency, and too little on the quirky and bizarre, which make life and mathematics interesting. It is of course important to try to maintain consistency, but when this effort forces you into a stupendously ugly theory, you know something is wrong.

These types of issues in the foundations of mathematics were responsible for the high interest in codifying human reasoning methods which was present in the early part of this century. Mathematicians and philosophers had begun to have serious doubts about whether even the most concrete of theories, such as the study of whole numbers (number theory), were built on solid foundations. If paradoxes could pop up so easily in set theory—a theory whose basic concept, that of a set, is surely very intuitively appealing—then might they not also exist in other branches of mathematics? Another related worry was that the paradoxes of logic, such as the Epimenides paradox, might turn out to be internal to mathematics, and thereby cast in doubt all of mathematics. This was especially worrisome to those—and there were a good number—who firmly believed that mathematics is simply a branch of logic (or conversely, that logic is simply a branch of mathematics). In fact, this very question—"Are mathematics and logic distinct, or separate?"—was the source of much controversy.

This study of mathematics itself became known as *metamathematics*—or occasionally, *metalogic,* since mathematics and logic are so intertwined. The most urgent priority of metamathematicians was to determine the true nature of mathematical reasoning. What is a legal method of procedure, and what is an illegal one? Since mathematical reasoning had always been done in "natural language" (e.g., French or Latin or some language for normal communication), there was always a lot of possible ambiguity. Words had different meanings to different people, conjured up different images, and so forth. It seemed reasonable and even important to establish a single uniform notation in which all mathematical work could be done, and with the aid of which any two mathematicians could resolve disputes over whether a suggested proof was valid or not. This would require a complete codification of the universally acceptable modes of human reasoning, at least as far as they applied to mathematics.

Consistency, Completeness, Hilbert's Program

This was the goal of *Principia Mathematica,* which purported to derive all of mathematics from logic, and, to be sure, without contradictions! It was widely admired, but no one was sure if (1) all of mathematics really was contained in the methods delineated by Russell and Whitehead, or (2) the methods given were even self-consistent. Was it absolutely clear that contradictory results could *never* be derived, by any mathematicians whatsoever, following the methods of Russell and Whitehead?

This question particularly bothered the distinguished German mathematician (and metamathematician) David Hilbert, who set before the world community of mathematicians (and metamathematicians) this challenge: to demonstrate rigorously—perhaps following the very methods outlined by Russell and Whitehead—that the system defined in *Principia Mathematica* was both *consistent* (contradiction-free), and *complete* (i.e., that every true statement of number theory could be derived within the framework drawn up in *P.M.*). This was a tall order, and one could criticize it on the grounds that it was somewhat circular: how can you justify your methods of reasoning on the basis of those same methods of reasoning? It is like lifting yourself up by your own bootstraps. (We just don't seem to be able to get away from these Strange Loops!)

Hilbert was fully aware of this dilemma, of course, and therefore expressed the hope that a demonstration of consistency or completeness could be found which depended only on "finitistic" modes of reasoning. These were a small set of reasoning methods usually accepted by mathematicians. In this way, Hilbert hoped that mathematicians could partially lift themselves by their own bootstraps: the sum total of mathematical methods might be proved sound, by invoking only a smaller set of methods. This goal may sound rather esoteric, but it occupied the minds of many of the greatest mathematicians in the world during the first thirty years of this century.

In the thirty-first year, however, Gödel published his paper, which in some ways utterly demolished Hilbert's program. This paper revealed not only that there were irreparable "holes" in the axiomatic system proposed by Russell and Whitehead, but more generally, that no axiomatic system whatsoever could produce all number-theoretical truths, unless it were an inconsistent system! And finally, the hope of proving the consistency of a system such as that presented in *P.M.* was shown to be vain: if such a proof could be found using only methods inside *P.M.*, then—and this is one of the most mystifying consequences of Gödel's work—*P.M.* itself would be inconsistent!

The final irony of it all is that the proof of Gödel's Incompleteness Theorem involved importing the Epimenides paradox right into the heart of *Principia Mathematica,* a bastion supposedly invulnerable to the attacks of Strange Loops! Although Gödel's Strange Loop did not destroy *Principia Mathematica,* it made it far less interesting to mathematicians, for it showed that Russell and Whitehead's original aims were illusory.

Babbage, Computers, Artificial Intelligence . . .

When Gödel's paper came out, the world was on the brink of developing electronic digital computers. Now the idea of mechanical calculating engines had been around for a while. In the seventeenth century, Pascal and Leibniz designed machines to perform fixed operations (addition and multiplication). These machines had no memory, however, and were not, in modern parlance, programmable.

The first human to conceive of the immense computing potential of machinery was the Londoner Charles Babbage (1792-1871). A character who could almost have stepped out of the pages of the *Pickwick Papers,*

Babbage was most famous during his lifetime for his vigorous campaign to rid London of "street nuisances"—organ grinders above all. These pests, loving to get his goat, would come and serenade him at any time of day or night, and he would furiously chase them down the street. Today, we recognize in Babbage a man a hundred years ahead of his time: not only inventor of the basic principles of modern computers, he was also one of the first to battle noise pollution.

His first machine, the "Difference Engine", could generate mathematical tables of many kinds by the "method of differences". But before any model of the "D.E." had been built, Babbage became obsessed with a much more revolutionary idea: his "Analytical Engine". Rather immodestly, he wrote, "The course through which I arrived at it was the most entangled and perplexed which probably ever occupied the human mind."[4] Unlike any previously designed machine, the A.E. was to possess both a "store" (memory) and a "mill" (calculating and decision-making unit). These units were to be built of thousands of intricate geared cylinders interlocked in incredibly complex ways. Babbage had a vision of numbers swirling in and out of the mill under control of a *program* contained in punched cards—an idea inspired by the Jacquard loom, a card-controlled loom that wove amazingly complex patterns. Babbage's brilliant but ill-fated Countess friend, Lady Ada Lovelace (daughter of Lord Byron), poetically commented that "the Analytical Engine *weaves algebraic patterns* just as the Jacquard-loom weaves flowers and leaves." Unfortunately, her use of the present tense was misleading, for no A.E. was ever built, and Babbage died a bitterly disappointed man.

Lady Lovelace, no less than Babbage, was profoundly aware that with the invention of the Analytical Engine, mankind was flirting with mechanized intelligence—particularly if the Engine were capable of "eating its own tail" (the way Babbage described the Strange Loop created when a machine reaches in and alters its own stored program). In an 1842 memoir,[5] she wrote that the A.E. "might act upon other things besides *number*". While Babbage dreamt of creating a chess or tic-tac-toe automaton, she suggested that his Engine, with pitches and harmonies coded into its spinning cylinders, "might compose elaborate and scientific pieces of music of any degree of complexity or extent." In nearly the same breath, however, she cautions that "The Analytical Engine has no pretensions whatever to *originate* anything. It can do whatever we *know how to order it* to perform." Though she well understood the power of artificial computation, Lady Lovelace was skeptical about the artificial creation of intelligence. However, could her keen insight allow her to dream of the potential that would be opened up with the taming of electricity?

In our century the time was ripe for computers—computers beyond the wildest dreams of Pascal, Leibniz, Babbage, or Lady Lovelace. In the 1930's and 1940's, the first "giant electronic brains" were designed and built. They catalyzed the convergence of three previously disparate areas: the theory of axiomatic reasoning, the study of mechanical computation, and the psychology of intelligence.

These same years saw the theory of computers develop by leaps and bounds. This theory was tightly linked to metamathematics. In fact, Gödel's Theorem has a counterpart in the theory of computation, discovered by Alan Turing, which reveals the existence of ineluctable "holes" in even the most powerful computer imaginable. Ironically, just as these somewhat eerie limits were being mapped out, real computers were being built whose powers seemed to grow and grow beyond their makers' power of prophecy. Babbage, who once declared he would gladly give up the rest of his life if he could come back in five hundred years and have a three-day guided scientific tour of the new age, would probably have been thrilled speechless a mere century after his death—both by the new machines, and by their unexpected limitations.

By the early 1950's, mechanized intelligence seemed a mere stone's throw away; and yet, for each barrier crossed, there always cropped up some new barrier to the actual creation of a genuine thinking machine. Was there some deep reason for this goal's mysterious recession?

No one knows where the borderline between non-intelligent behavior and intelligent behavior lies; in fact, to suggest that a sharp borderline exists is probably silly. But essential abilities for intelligence are certainly:

- to respond to situations very flexibly;
- to take advantage of fortuitous circumstances;
- to make sense out of ambiguous or contradictory messages;
- to recognize the relative importance of different elements of a situation;
- to find similarities between situations despite differences which may separate them;
- to draw distinctions between situations despite similarities which may link them;
- to synthesize new concepts by taking old concepts and putting them together in new ways;
- to come up with ideas which are novel.

Here one runs up against a seeming paradox. Computers by their very nature are the most inflexible, desireless, rule-following of beasts. Fast though they may be, they are nonetheless the epitome of unconsciousness. How, then, can intelligent behavior be programmed? Isn't this the most blatant of contradictions in terms? One of the major theses of this book is that it is not a contradiction at all. One of the major purposes of this book is to urge each reader to confront the apparent contradiction head on, to savor it, to turn it over, to take it apart, to wallow in it, so that in the end the reader might emerge with new insights into the seemingly unbreachable gulf between the formal and the informal, the animate and the inanimate, the flexible and the inflexible.

This is what Artificial Intelligence (AI) research is all about. And the strange flavor of AI work is that people try to put together long sets of rules in strict formalisms which tell inflexible machines how to be flexible.

What sorts of "rules" could possibly capture all of what we think of as intelligent behavior, however? Certainly there must be rules on all sorts of different levels. There must be many "just plain" rules. There must be "metarules" to modify the "just plain" rules; then "metametarules" to modify the metarules, and so on. The flexibility of intelligence comes from the enormous number of different rules, and levels of rules. The reason that so many rules on so many different levels must exist is that in life, a creature is faced with millions of situations of completely different types. In some situations, there are stereotyped responses which require "just plain" rules. Some situations are mixtures of stereotyped situations—thus they require rules for deciding which of the "just plain" rules to apply. Some situations cannot be classified—thus there must exist rules for inventing new rules . . . and on and on. Without doubt, Strange Loops involving rules that change themselves, directly or indirectly, are at the core of intelligence. Sometimes the complexity of our minds seems so overwhelming that one feels that there can be no solution to the problem of understanding intelligence—that it is wrong to think that rules of any sort govern a creature's behavior, even if one takes "rule" in the multilevel sense described above.

. . . and Bach

In the year 1754, four years after the death of J. S. Bach, the Leipzig theologian Johann Michael Schmidt wrote, in a treatise on music and the soul, the following noteworthy passage:

> Not many years ago it was reported from France that a man had made a statue that could play various pieces on the *Fleuttraversiere,* placed the flute to its lips and took it down again, rolled its eyes, etc. But no one has yet invented an image that thinks, or wills, or composes, or even does anything at all similar. Let anyone who wishes to be convinced look carefully at the last fugal work of the above-praised Bach, which has appeared in copper engraving, but which was left unfinished because his blindness intervened, and let him observe the art that is contained therein; or what must strike him as even more wonderful, the Chorale which he dictated in his blindness to the pen of another: *Wenn wir in höchsten Nöthen seyn.* I am sure that he will soon need his soul if he wishes to observe all the beauties contained therein, let alone wishes to play it to himself or to form a judgment of the author. Everything that the champions of Materialism put forward must fall to the ground in view of this single example.[6]

Quite likely, the foremost of the "champions of Materialism" here alluded to was none other than Julien Offroy de la Mettrie—philosopher at the court of Frederick the Great, author of *L'homme machine* ("Man, the Machine"), and Materialist Par Excellence. It is now more than 200 years later, and the battle is still raging between those who agree with Johann Michael Schmidt, and those who agree with Julien Offroy de la Mettrie. I hope in this book to give some perspective on the battle.

"Gödel, Escher, Bach"

The book is structured in an unusual way: as a counterpoint between Dialogues and Chapters. The purpose of this structure is to allow me to present new concepts twice: almost every new concept is first presented metaphorically in a Dialogue, yielding a set of concrete, visual images; then these serve, during the reading of the following Chapter, as an intuitive background for a more serious and abstract presentation of the same concept. In many of the Dialogues I appear to be talking about one idea on the surface, but in reality I am talking about some other idea, in a thinly disguised way.

Originally, the only characters in my Dialogues were Achilles and the Tortoise, who came to me from Zeno of Elea, by way of Lewis Carroll. Zeno of Elea, inventor of paradoxes, lived in the fifth century B.C. One of his paradoxes was an allegory, with Achilles and the Tortoise as protagonists. Zeno's invention of the happy pair is told in my first Dialogue, *Three-Part Invention.* In 1895, Lewis Carroll reincarnated Achilles and the Tortoise for the purpose of illustrating his own new paradox of infinity. Carroll's paradox, which deserves to be far better known than it is, plays a significant role in this book. Originally titled "What the Tortoise Said to Achilles", it is reprinted here as *Two-Part Invention.*

When I began writing Dialogues, somehow I connected them up with musical forms. I don't remember the moment it happened; I just remember one day writing "Fugue" above an early Dialogue, and from then on the idea stuck. Eventually I decided to pattern each Dialogue in one way or another on a different piece by Bach. This was not so inappropriate. Old Bach himself used to remind his pupils that the separate parts in their compositions should behave like "persons who conversed together as if in a select company". I have taken that suggestion perhaps rather more literally than Bach intended it; nevertheless I hope the result is faithful to the meaning. I have been particularly inspired by aspects of Bach's compositions which have struck me over and over, and which are so well described by David and Mendel in *The Bach Reader:*

> His form in general was based on relations between separate sections. These relations ranged from complete identity of passages on the one hand to the return of a single principle of elaboration or a mere thematic allusion on the other. The resulting patterns were often symmetrical, but by no means necessarily so. Sometimes the relations between the various sections make up a maze of interwoven threads that only detailed analysis can unravel. Usually, however, a few dominant features afford proper orientation at first sight or hearing, and while in the course of study one may discover unending subtleties, one is never at a loss to grasp the unity that holds together every single creation by Bach.[6]

I have sought to weave an Eternal Golden Braid out of these three strands: Gödel, Escher, Bach. I began, intending to write an essay at the core of which would be Gödel's Theorem. I imagined it would be a mere pamphlet. But my ideas expanded like a sphere, and soon touched Bach and Escher. It took some time for me to think of making this connection explicit, instead of just letting it be a private motivating force. But finally I realized that to me, Gödel and Escher and Bach were only shadows cast in different directions by some central solid essence. I tried to reconstruct the central object, and came up with this book.

Three-Part Invention

Achilles (a Greek warrior, the fleetest of foot of all mortals) and a Tortoise are standing together on a dusty runway in the hot sun. Far down the runway, on a tall flagpole, there hangs a large rectangular flag. The flag is solid red, except where a thin ring-shaped hole has been cut out of it, through which one can see the sky.

Achilles: What is that strange flag down at the other end of the track? It reminds me somehow of a print by my favorite artist, M. C. Escher.

Tortoise: That is Zeno's flag.

Achilles: Could it be that the hole in it resembles the holes in a Möbius strip Escher once drew? Something is wrong about that flag, I can tell.

Tortoise: The ring which has been cut from it has the shape of the numeral for zero, which is Zeno's favorite number.

Achilles: But zero hasn't been invented yet! It will only be invented by a Hindu mathematician some millennia hence. And thus, Mr. T, my argument proves that such a flag is impossible.

Tortoise: Your argument is persuasive, Achilles, and I must agree that such a flag is indeed impossible. But it is beautiful anyway, is it not?

Achilles: Oh, yes, there is no doubt of its beauty.

Tortoise: I wonder if its beauty is related to its impossibility. I don't know; I've never had the time to analyze Beauty. It's a Capitalized Essence; and I never seem to have the time for Capitalized Essences.

Achilles: Speaking of Capitalized Essences, Mr. T, have you ever wondered about the Purpose of Life?

Tortoise: Oh, heavens, no.

Achilles: Haven't you ever wondered why we are here, or who invented us?

Tortoise: Oh, that is quite another matter. We are inventions of Zeno (as you will shortly see); and the reason we are here is to have a footrace.

Achilles: A footrace? How outrageous! Me, the fleetest of foot of all mortals, versus you, the ploddingest of all plodders! There can be no point to such a race.

Tortoise: You might give me a head start.

Achilles: It would have to be a huge one.

Tortoise: I don't object.

Achilles: But I will catch you, sooner or later—most likely sooner.

Tortoise: Not if things go according to Zeno's paradox, you won't. Zeno is hoping to use our footrace to show that motion is impossible, you see. It is only in the mind that motion seems possible, according to Zeno. In truth, Motion Is Inherently Impossible. He proves it quite elegantly.

FIGURE 10. Möbius Strip I, *by M. C. Escher (wood-engraving printed from four blocks, 1961).*

Achilles: Oh, yes, it comes back to me now: the famous Zen kōan about Zen Master Zeno. As you say, it is very simple indeed.

Tortoise: Zen kōan? Zen Master? What do you mean?

Achilles: It goes like this: Two monks were arguing about a flag. One said, "The flag is moving." The other said, "The wind is moving." The sixth patriarch, Zeno, happened to be passing by. He told them, "Not the wind, not the flag; mind is moving."

Tortoise: I am afraid you are a little befuddled, Achilles. Zeno is no Zen master; far from it. He is, in fact, a Greek philosopher from the town of Elea (which lies halfway between points A and B). Centuries hence, he will be celebrated for his paradoxes of motion. In one of those paradoxes, this very footrace between you and me will play a central role.

Achilles: I'm all confused. I remember vividly how I used to repeat over and over the names of the six patriarchs of Zen, and I always said, "The sixth patriarch is Zeno, the sixth patriarch is Zeno . . ." *(Suddenly a soft warm breeze picks up.)* Oh, look, Mr. Tortoise—look at the flag waving! How I love to watch the ripples shimmer through its soft fabric. And the ring cut out of it is waving, too!

Tortoise: Don't be silly. The flag is impossible, hence it can't be waving. The wind is waving.

(At this moment, Zeno happens by.)

Zeno: Hallo! Hulloo! What's up? What's new?

Achilles: The flag is moving.

Tortoise: The wind is moving.

Zeno: O Friends, Friends! Cease your argumentation! Arrest your vitriolics! Abandon your discord! For I shall resolve the issue for you forthwith. Ho! And on such a fine day!

Achilles: This fellow must be playing the fool.

Tortoise: No, wait, Achilles. Let us hear what he has to say. Oh, Unknown Sir, do impart to us your thoughts on this matter.

Zeno: Most willingly. Not the wind, not the flag—neither one is moving, nor is anything moving at all. For I have discovered a great Theorem, which states: "Motion Is Inherently Impossible." And from this Theorem follows an even greater Theorem—Zeno's Theorem: "Motion Unexists."

Achilles: "Zeno's Theorem"? Are you, sir, by any chance, the philosopher Zeno of Elea?

Zeno: I am indeed, Achilles.

Achilles (scratching his head in puzzlement): Now how did he know my name?

Zeno: Could I possibly persuade you two to hear me out as to why this is the case? I've come all the way to Elea from point A this afternoon, just trying to find someone who'll pay some attention to my closely honed argument. But they're all hurrying hither and thither, and they don't have time. You've no idea how disheartening it is to meet with refusal after refusal. Oh, but I'm sorry to burden you with my troubles. I'd just like to ask one thing: Would the two of you humor a silly old philosopher for a few moments—only a few, I promise you—in his eccentric theories?

Achilles: Oh, by all means! Please do illuminate us! I know I speak for both of us, since my companion, Mr. Tortoise, was only moments ago speaking of you with great veneration—and he mentioned especially your paradoxes.

Zeno: Thank you. You see, my Master, the fifth patriarch, taught me that reality is one, immutable, and unchanging; all plurality, change, and motion are mere illusions of the senses. Some have mocked his views; but I will show the absurdity of their mockery. My argument is quite simple. I will illustrate it with two characters of my own Invention: Achilles (a Greek warrior, the fleetest of foot of all mortals), and a Tortoise. In my tale, they are persuaded by a passerby to run a footrace down a runway towards a distant flag waving in the breeze. Let us assume that, since the Tortoise is a much slower runner, he gets a head start of, say, ten rods. Now the race begins. In a few bounds, Achilles has reached the spot where the Tortoise started.

Achilles: Hah!

Zeno: And now the Tortoise is but a single rod ahead of Achilles. Within only a moment, Achilles has attained that spot.

Achilles: Ho ho!

Zeno: Yet, in that short moment, the Tortoise has managed to advance a slight amount. In a flash, Achilles covers that distance, too.

Achilles: Hee hee hee!

Zeno: But in that very short flash, the Tortoise has managed to inch ahead by ever so little, and so Achilles is still behind. Now you see that in order for Achilles to catch the Tortoise, this game of "try-to-catch-me" will have to be played an INFINITE number of times—and therefore Achilles can NEVER catch up with the Tortoise!

Tortoise: Heh heh heh heh!

Achilles: Hmm ... hmm ... hmm ... hmm ... hmm ... That argument sounds wrong to me. And yet, I can't quite make out what's wrong with it.

Zeno: Isn't it a teaser? It's my favorite paradox.

Tortoise: Excuse me, Zeno, but I believe your tale illustrates the wrong principle, does it not? You have just told us what will come to be known, centuries hence, as Zeno's "Achilles paradox", which shows (ahem!) that Achilles will never catch the Tortoise; but the proof that Motion Is Inherently Impossible (and thence that Motion Unexists) is your "dichotomy paradox", isn't that so?

Zeno: Oh, shame on me. Of course, you're right. That's the one about how, in getting from A to B, one has to go halfway first—and of that stretch one also has to go halfway, and so on and so forth. But you see, both those paradoxes really have the same flavor. Frankly, I've only had one Great Idea—I just exploit it in different ways.

Achilles: I swear, these arguments contain a flaw. I don't quite see where, but they cannot be correct.

Zeno: You doubt the validity of my paradox? Why not just try it out? You see that red flag down there, at the far end of the runway?

Achilles: The impossible one, based on an Escher print?

Zeno: Exactly. What do you say to you and Mr. Tortoise racing for it, allowing Mr. T a fair head start of, well, I don't know—

Tortoise: How about ten rods?

Zeno: Very good—ten rods.

Achilles: Any time.

Zeno: Excellent! How exciting! An empirical test of my rigorously proven Theorem! Mr. Tortoise, will you position yourself ten rods upwind?

(The Tortoise moves ten rods closer to the flag.)

Are you both ready?

Tortoise and Achilles: Ready!

Zeno: On your mark! Get set! Go!

CHAPTER I

The MU-puzzle

Formal Systems

ONE OF THE most central notions in this book is that of a *formal system.* The type of formal system I use was invented by the American logician Emil Post in the 1920's, and is often called a "Post production system". This Chapter introduces you to a formal system and moreover, it is my hope that you will want to explore this formal system at least a little; so to provoke your curiosity, I have posed a little puzzle.

"Can you produce MU?" is the puzzle. To begin with, you will be supplied with a *string* (which means a string of letters).* Not to keep you in suspense, that string will be MI. Then you will be told some rules, with which you can change one string into another. If one of those rules is applicable at some point, and you want to use it, you may, but—there is nothing that will dictate which rule you should use, in case there are several applicable rules. That is left up to you—and of course, that is where playing the game of any formal system can become something of an art. The major point, which almost doesn't need stating, is that you must not do anything which is outside the rules. We might call this restriction the "Requirement of Formality". In the present Chapter, it probably won't need to be stressed at all. Strange though it may sound, though, I predict that when you play around with some of the formal systems of Chapters to come, you will find yourself violating the Requirement of Formality over and over again, unless you have worked with formal systems before.

The first thing to say about our formal system—the *MIU-system*—is that it utilizes only three letters of the alphabet: M, I, U. That means that the only strings of the MIU-system are strings which are composed of those three letters. Below are some strings of the MIU-system:

MU
UIM
MUUMUU
UIIUMIUUIMUIIUMIUUIMUIIU

* In this book, we shall employ the following conventions when we refer to strings. When the string is in the same typeface as the text, then it will be enclosed in single or double quotes. Punctuation which belongs to the sentence and not to the string under discussion will go *outside* of the quotes, as logic dictates. For example, the first letter of this sentence is 'F', while the first letter of 'this sentence' is 't'. When the string is in Quadrata Roman, however, quotes will usually be left off, unless clarity demands them. For example, the first letter of Quadrata is Q.

But although all of these are legitimate strings, they are not strings which are "in your possession". In fact, the only string in your possession so far is MI. Only by using the rules, about to be introduced, can you enlarge your private collection. Here is the first rule:

RULE I: If you possess a string whose last letter is I, you can add on a U at the end.

By the way, if up to this point you had not guessed it, a fact about the meaning of "string" is that the letters are in a fixed order. For example, MI and IM are two different strings. A string of symbols is not just a "bag" of symbols, in which the order doesn't make any difference.

Here is the second rule:

RULE II: Suppose you have M*x*. Then you may add M*xx* to your collection.

What I mean by this is shown below, in a few examples.

From MIU, you may get MIUIU.
From MUM, you may get MUMUM.
From MU, you may get MUU.

So the letter '*x*' in the rule simply stands for any string; but once you have decided which string it stands for, you have to stick with your choice (until you use the rule again, at which point you may make a new choice). Notice the third example above. It shows how, once you possess MU, you can add another string to your collection; but you have to get MU first! I want to add one last comment about the letter '*x*': it is not part of the formal system in the same way as the three letters 'M', 'I', and 'U' are. It is useful for us, though, to have some way to talk in general about strings of the system, symbolically—and that is the function of the '*x*': to stand for an arbitrary string. If you ever add a string containing an '*x*' to your "collection", you have done something wrong, because strings of the MIU-system never contain "*x*"'s!

Here is the third rule:

RULE III: If III occurs in one of the strings in your collection, you may make a new string with U in place of III.

Examples:

From UMIIIMU, you could make UMUMU.
From MIIII, you could make MIU (also MUI).
From IIMII, you can't get anywhere using this rule.
(The three I's have to be consecutive.)
From MIII, make MU.

Don't, under any circumstances, think you can run this rule backwards, as in the following example:

From MU, make MIII. ⇐ This is wrong.

Rules are one-way.

Here is the final rule:

RULE IV: If UU occurs inside one of your strings, you can drop it.

From UUU, get U.
From MUUUIII, get MUIII.

There you have it. Now you may begin trying to make MU. Don't worry if you don't get it. Just try it out a bit—the main thing is for you to get the flavor of this MU-puzzle. Have fun.

Theorems, Axioms, Rules

The answer to the MU-puzzle appears later in the book. For now, what is important is not finding the answer, but looking for it. You probably have made some attempts to produce MU. In so doing, you have built up your own private collection of strings. Such strings, producible by the rules, are called *theorems.* The term "theorem" has, of course, a common usage in mathematics which is quite different from this one. It means some statement in ordinary language which has been proven to be true by a rigorous argument, such as Zeno's Theorem about the "unexistence" of motion, or Euclid's Theorem about the infinitude of primes. But in formal systems, theorems need not be thought of as statements—they are merely strings of symbols. And instead of being *proven,* theorems are merely *produced,* as if by machine, according to certain typographical rules. To emphasize this important distinction in meanings for the word "theorem", I will adopt the following convention in this book: when "theorem" is capitalized, its meaning will be the everyday one—a Theorem is a statement in ordinary language which somebody once proved to be true by some sort of logical argument. When uncapitalized, "theorem" will have its technical meaning: a string producible in some formal system. In these terms, the MU-puzzle asks whether MU is a theorem of the MIU-system.

I gave you a theorem for free at the beginning, namely MI. Such a "free" theorem is called an *axiom*—the technical meaning again being quite different from the usual meaning. A formal system may have zero, one, several, or even infinitely many axioms. Examples of all these types will appear in the book.

Every formal system has symbol-shunting rules, such as the four rules of the MIU-system. These rules are called either *rules of production* or *rules of inference.* I will use both terms.

The last term which I wish to introduce at this point is *derivation.* Shown below is a derivation of the theorem MUIIU:

(1) MI axiom
(2) MII from (1) by rule II

(3) MIIII from (2) by rule II
(4) MIIIIU from (3) by rule I
(5) MUIU from (4) by rule III
(6) MUIUUIU from (5) by rule II
(7) MUIIU from (6) by rule IV

A derivation of a theorem is an explicit, line-by-line demonstration of how to produce that theorem according to the rules of the formal system. The concept of derivation is modeled on that of proof, but a derivation is an austere cousin of a proof. It would sound strange to say that you had *proven* MUIIU, but it does not sound so strange to say you have *derived* MUIIU.

Inside and Outside the System

Most people go about the MU-puzzle by deriving a number of theorems, quite at random, just to see what kind of thing turns up. Pretty soon, they begin to notice some properties of the theorems they have made; that is where human intelligence enters the picture. For instance, it was probably not obvious to you that all theorems would begin with M, until you had tried a few. Then, the pattern emerged, and not only could you see the pattern, but you could understand it by looking at the rules, which have the property that they make each new theorem inherit its first letter from an earlier theorem; ultimately, then, all theorems' first letters can be traced back to the first letter of the sole axiom MI—and that is a proof that theorems of the MIU-system must all begin with M.

There is something very significant about what has happened here. It shows one difference between people and machines. It would certainly be possible—in fact it would be very easy—to program a computer to generate theorem after theorem of the MIU-system; and we could include in the program a command to stop only upon generating U. You now know that a computer so programmed would never stop. And this does not amaze you. But what if you asked a friend to try to generate U? It would not surprise you if he came back after a while, complaining that he can't get rid of the initial M, and therefore it is a wild goose chase. Even if a person is not very bright, he still cannot help making some observations about what he is doing, and these observations give him good insight into the task—insight which the computer program, as we have described it, lacks.

Now let me be very explicit about what I meant by saying this shows a difference between people and machines. I meant that it is *possible* to program a machine to do a routine task in such a way that the machine will never notice even the most obvious facts about what it is doing; but it is inherent in human consciousness to notice some facts about the things one is doing. But you knew this all along. If you punch "1" into an adding machine, and then add 1 to it, and then add 1 again, and again, and again, and continue doing so for hours and hours, the machine will never learn to anticipate you, and do it itself, although any person would pick up the repetitive behavior very quickly. Or, to take a silly example, a car will never pick up the idea, no matter how much or how well it is driven, that it is supposed to avoid other cars and obstacles on the road; and it will never learn even the most frequently traveled routes of its owner.

The difference, then, is that it is *possible* for a machine to act unobservant; it is impossible for a human to act unobservant. Notice I am not saying that all machines are necessarily incapable of making sophisticated observations; just that some machines are. Nor am I saying that all people are always making sophisticated observations; people, in fact, are often very unobservant. But machines can be made to be totally unobservant; and people cannot. And in fact, most machines made so far are pretty close to being totally unobservant. Probably for this reason, the property of being unobservant seems to be the characteristic feature of machines, to most people. For example, if somebody says that some task is "mechanical", it does not mean that people are incapable of doing the task; it implies, though, that only a machine could do it over and over without ever complaining, or feeling bored.

Jumping out of the System

It is an inherent property of intelligence that it can jump out of the task which it is performing, and survey what it has done; it is always looking for, and often finding, patterns. Now I said that an intelligence can jump out of its task, but that does not mean that it always will. However, a little prompting will often suffice. For example, a human being who is reading a book may grow sleepy. Instead of continuing to read until the book is finished, he is just as likely to put the book aside and turn off the light. He has stepped "out of the system" and yet it seems the most natural thing in the world to us. Or, suppose person A is watching television when person B comes in the room, and shows evident displeasure with the situation. Person A may think he understands the problem, and try to remedy it by exiting the present system (that television program), and flipping the channel knob, looking for a better show. Person B may have a more radical concept of what it is to "exit the system"—namely to turn the television off! Of course, there are cases where only a rare individual will have the vision to perceive a system which governs many peoples' lives, a system which had never before even been recognized as a system; then such people often devote their lives to convincing other people that the system really is there, and that it ought to be exited from!

How well have computers been taught to jump out of the system? I will cite one example which surprised some observers. In a computer chess tournament not long ago in Canada, one program—the weakest of all the competing ones—had the unusual feature of quitting long before the game was over. It was not a very good chess player, but it at least had the redeeming quality of being able to spot a hopeless position, and to resign then and there, instead of waiting for the other program to go through the boring ritual of checkmating. Although it lost every game it played, it did it in style. A lot of local chess experts were impressed. Thus, if you define "the system" as "making moves in a chess game", it is clear that this program had a sophisticated, preprogrammed ability to exit from the system. On the other hand, if you think of "the system" as being "whatever the computer had been programmed to do", then there is no doubt that the computer had no ability whatsoever to exit from that system.

It is very important when studying formal systems to distinguish working *within* the system from making statements or observations *about* the system. I assume that you began the MU-puzzle, as do most people, by working within the system; and that you then gradually started getting anxious, and this anxiety finally built up to the point where without any need for further consideration, you exited from the system, trying to take stock of what you had produced, and wondering why it was that you had not succeeded in producing MU. Perhaps you found a reason why you could not produce MU; that is thinking about the system. Perhaps you produced MIU somewhere along the way; that is working within the system. Now I do not want to make it sound as if the two modes are entirely incompatible; I am sure that every human being is capable to some extent of working inside a system and simultaneously thinking about what he is doing. Actually, in human affairs, it is often next to impossible to break things neatly up into "inside the system" and "outside the system"; life is composed of so many interlocking and interwoven and often inconsistent "systems" that it may seem simplistic to think of things in those terms. But it is often important to formulate simple ideas very clearly so that one can use them as models in thinking about more complex ideas. And that is why I am showing you formal systems; and it is about time we went back to discussing the MIU-system.

M-Mode, I-Mode, U-Mode

The MU-puzzle was stated in such a way that it encouraged some amount of exploration within the MIU-system—deriving theorems. But it was also stated in a way so as not to imply that staying inside the system would necessarily yield fruit. Therefore it encouraged some oscillation between the two modes of work. One way to separate these two modes would be to have two sheets of paper; on one sheet, you work "in your capacity as a machine", thus filling it with nothing but M's, I's, and U's; on the second sheet, you work "in your capacity as a thinking being", and are allowed to do whatever your intelligence suggests—which might involve using English, sketching ideas, working backwards, using shorthand (such as the letter '*x*'), compressing several steps into one, modifying the rules of the system to see what that gives, or whatever else you might dream up. One thing you might do is notice that the numbers 3 and 2 play an important role, since I's are gotten rid of in three's, and U's in two's—and doubling of length (except for the M) is allowed by rule II. So the second sheet might also have some figuring on it. We will occasionally refer back to these two modes of dealing with a formal system, and we will call them the *Mechanical mode (M-mode)* and the *Intelligent mode (I-mode).* To round out our modes, with one for each letter of the MIU-system, I will also mention a final mode—the *Un-mode (U-mode),* which is the Zen way of approaching things. More about this in a few Chapters.

Decision Procedures

An observation about this puzzle is that it involves rules of two opposing tendencies—the *lengthening rules* and the *shortening rules.* Two rules (I and II) allow you to increase the size of strings (but only in very rigid, prescribed ways, of course); and two others allow you to shrink strings somewhat (again in very rigid ways). There seems to be an endless variety to the order in which these different types of rules might be applied, and this gives hope that one way or another, MU could be produced. It might involve lengthening the string to some gigantic size, and then extracting piece after piece until only two symbols are left; or, worse yet, it might involve successive stages of lengthening and then shortening and then lengthening and then shortening, and so on. But there is no guarantee of it. As a matter of fact, we already observed that U cannot be produced at all, and it will make no difference if you lengthen and shorten till kingdom come.

Still, the case of U and the case of MU seem quite different. It is by a very superficial feature of U that we recognize the impossibility of producing it: it doesn't begin with an M (whereas all theorems must). It is very convenient to have such a simple way to detect nontheorems. However, who says that that test will detect *all* nontheorems? There may be lots of strings which begin with M but are not producible. Maybe MU is one of them. That would mean that the "first-letter test" is of limited usefulness, able only to detect a portion of the nontheorems, but missing others. But there remains the possibility of some more elaborate test which discriminates perfectly between those strings which can be produced by the rules, and those which cannot. Here we have to face the question, "What do we mean by a test?" It may not be obvious why that question makes sense, or is important, in this context. But I will give an example of a "test" which somehow seems to violate the spirit of the word.

Imagine a genie who has all the time in the world, and who enjoys using it to produce theorems of the MIU-system, in a rather methodical way. Here, for instance, is a possible way the genie might go about it:

Step 1: Apply every applicable rule to the axiom MI. This yields two new theorems: MIU, MII.

Step 2: Apply every applicable rule to the theorems produced in step 1. This yields three new theorems: MIIU, MIUIU, MIIII.

Step 3: Apply every applicable rule to the theorems produced in step 2. This yields five new theorems: MIIIIU, MIIUIIU, MIUIUIUIU, MIIIIIIII, MUI.

.

.

.

This method produces every single theorem sooner or later, because the rules are applied in every conceivable order. (See Fig. 11.) All of the lengthening-shortening alternations which we mentioned above eventually get carried out. However, it is not clear how long to wait for a given string

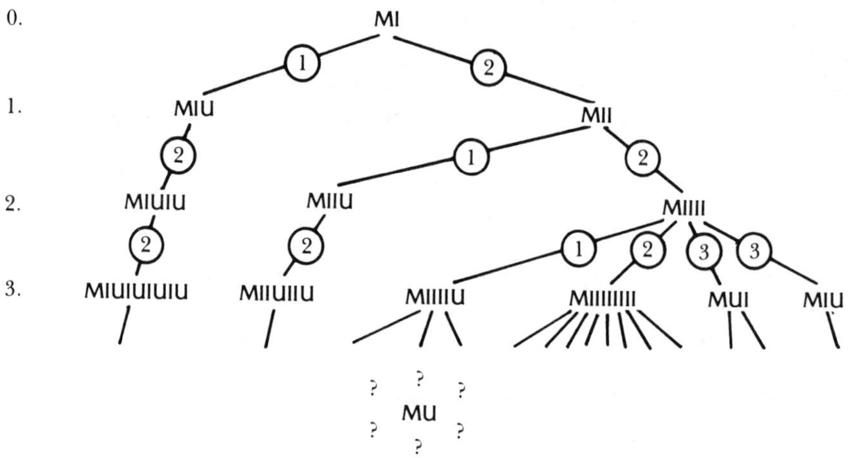

FIGURE 11. A systematically constructed "tree" of all the theorems of the MIU-system. The N*th level down contains those theorems whose derivations contain exactly* N *steps. The encircled numbers tell which rule was employed. Is* MU *anywhere in this tree?*

to appear on this list, since theorems are listed according to the shortness of their derivations. This is not a very useful order, if you are interested in a specific string (such as MU), and you don't even know if it has any derivation, much less how long that derivation might be.

Now we state the proposed "theoremhood-test":

> Wait until the string in question is produced; when that happens, you know it is a theorem—and if it never happens, you know that it is not a theorem.

This seems ridiculous, because it presupposes that we don't mind waiting around literally an infinite length of time for our answer. This gets to the crux of the matter of what should count as a "test". Of prime importance is a guarantee that we will get our answer in a finite length of time. If there is a test for theoremhood, a test which does always terminate in a finite amount of time, then that test is called a *decision procedure* for the given formal system.

When you have a decision procedure, then you have a very concrete characterization of the nature of all theorems in the system. Offhand, it might seem that the rules and axioms of the formal system provide no less complete a characterization of the theorems of the system than a decision procedure would. The tricky word here is "characterization". Certainly the rules of inference and the axioms of the MIU-system do characterize, *implicitly,* those strings that are theorems. Even *more* implicitly, they characterize those strings that are *not* theorems. But implicit characterization is not enough, for many purposes. If someone claims to have a characterization of all theorems, but it takes him infinitely long to deduce that some particular string is not a theorem, you would probably tend to say that there is something lacking in that characterization—it is not quite concrete enough. And that is why discovering that a decision procedure exists is a very important step. What the discovery means, in effect, is that you can perform a test for theoremhood of a string, and that, even if the test is complicated, it is *guaranteed to terminate.* In principle, the test is just as easy, just as mechanical, just as finite, just as full of certitude, as checking whether the first letter of the string is M. A decision procedure is a "litmus test" for theoremhood!

Incidentally, one requirement on formal systems is that the set of *axioms* must be characterized by a decision procedure—there must be a litmus test for axiomhood. This ensures that there is no problem in getting off the ground at the beginning, at least. That is the difference between the set of axioms and the set of theorems: the former always has a decision procedure, but the latter may not.

I am sure you will agree that when you looked at the MIU-system for the first time, you had to face this problem exactly. The lone axiom was known, the rules of inference were simple, so the theorems had been implicitly characterized—and yet it was still quite unclear what the consequences of that characterization were. In particular, it was still totally unclear whether MU is, or is not, a theorem.

FIGURE 12. Sky Castle, *by M. C. Escher (woodcut, 1928).*

Two-Part Invention

or,

What the Tortoise Said to Achilles

by Lewis Carroll[1]

Achilles had overtaken the Tortoise, and had seated himself comfortably on its back.

"So you've got to the end of our race-course?" said the Tortoise. "Even though it DOES consist of an infinite series of distances? I thought some wiseacre or other had proved that the thing couldn't be done?"

"It CAN be done," said Achilles. "It HAS been done! *Solvitur ambulando.* You see the distances were constantly DIMINISHING; and so—"

"But if they had been constantly INCREASING?" the Tortoise interrupted. "How then?"

"Then I shouldn't be here," Achilles modestly replied; "and YOU would have got several times round the world, by this time!"

"You flatter me—FLATTEN, I mean," said the Tortoise; "for you ARE a heavy weight, and NO mistake! Well now, would you like to hear of a race-course, that most people fancy they can get to the end of in two or three steps, while it REALLY consists of an infinite number of distances, each one longer than the previous one?"

"Very much indeed!" said the Grecian warrior, as he drew from his helmet (few Grecian warriors possessed POCKETS in those days) an enormous note-book and pencil. "Proceed! And speak SLOWLY, please! SHORT-HAND isn't invented yet!"

"That beautiful First Proposition by Euclid!" the Tortoise murmured dreamily. "You admire Euclid?"

"Passionately! So far, at least, as one CAN admire a treatise that won't be published for some centuries to come!"

"Well, now, let's take a little bit of the argument in that First Proposition—just TWO steps, and the conclusion drawn from them. Kindly enter them in your note-book. And in order to refer to them conveniently, let's call them A, B, and Z:—

(A) Things that are equal to the same are equal to each other.
(B) The two sides of this Triangle are things that are equal to the same.
(Z) The two sides of this Triangle are equal to each other.

Readers of Euclid will grant, I suppose, that Z follows logically from A and B, so that any one who accepts A and B as true, MUST accept Z as true?"

"Undoubtedly! The youngest child in a High School—as soon as High

Schools are invented, which will not be till some two thousand years later—will grant THAT."

"And if some reader had NOT yet accepted A and B as true, he might still accept the SEQUENCE as a VALID one, I suppose?"

"No doubt such a reader might exist. He might say, 'I accept as true the Hypothetical Proposition that, IF A and B be true, Z must be true; but I DON'T accept A and B as true.' Such a reader would do wisely in abandoning Euclid, and taking to football."

"And might there not ALSO be some reader who would say 'I accept A and B as true, but I DON'T accept the Hypothetical'?"

"Certainly there might. HE, also, had better take to football."

"And NEITHER of these readers," the Tortoise continued, "is AS YET under any logical necessity to accept Z as true?"

"Quite so," Achilles assented.

"Well, now, I want you to consider ME as a reader of the SECOND kind, and to force me, logically, to accept Z as true."

"A tortoise playing football would be—" Achilles was beginning.

"—an anomaly, of course," the Tortoise hastily interrupted. "Don't wander from the point. Let's have Z first, and football afterwards!"

"I'm to force you to accept Z, am I?" Achilles said musingly. "And your present position is that you accept A and B, but you DON'T accept the Hypothetical—"

"Let's call it C," said the Tortoise.

"—but you DON'T accept

(C) If A and B are true, Z must be true."

"That is my present position," said the Tortoise.

"Then I must ask you to accept C."

"I'll do so," said the Tortoise, "as soon as you've entered it in that notebook of yours. What else have you got in it?"

"Only a few memoranda," said Achilles, nervously fluttering the leaves: "a few memoranda of—of the battles in which I have distinguished myself!"

"Plenty of blank leaves, I see!" the Tortoise cheerily remarked. "We shall need them ALL!" (Achilles shuddered.) "Now write as I dictate:—

(A) Things that are equal to the same are equal to each other.
(B) The two sides of this Triangle are things that are equal to the same.
(C) If A and B are true, Z must be true.
(Z) The two sides of this Triangle are equal to each other."

"You should call it D, not Z," said Achilles. "It comes NEXT to the other three. If you accept A and B and C, you MUST accept Z."

"And why must I?"

"Because it follows LOGICALLY from them. If A and B and C are true, Z MUST be true. You can't dispute THAT, I imagine?"

"If A and B and C are true, Z MUST be true," the Tortoise thoughtfully repeated. "That's ANOTHER Hypothetical, isn't it? And, if I failed to see its truth, I might accept A and B and C, and STILL not accept Z, mightn't I?"

"You might," the candid hero admitted; "though such obtuseness would certainly be phenomenal. Still, the event is POSSIBLE. So I must ask you to grant ONE more Hypothetical."

"Very good, I'm quite willing to grant it, as soon as you've written it down. We will call it

(D) If A and B and C are true, Z must be true.

Have you entered that in your note-book?"

"I HAVE!" Achilles joyfully exclaimed, as he ran the pencil into its sheath. "And at last we've got to the end of this ideal race-course! Now that you accept A and B and C and D, OF COURSE you accept Z."

"Do I?" said the Tortoise innocently. "Let's make that quite clear. I accept A and B and C and D. Suppose I STILL refused to accept Z?"

"Then Logic would take you by the throat, and FORCE you to do it!" Achilles triumphantly replied. "Logic would tell you, 'You can't help yourself. Now that you've accepted A and B and C and D, you MUST accept Z!' So you've no choice, you see."

"Whatever LOGIC is good enough to tell me is worth WRITING DOWN," said the Tortoise. "So enter it in your book, please. We will call it

(E) If A and B and C and D are true, Z must be true.

Until I've granted THAT, of course I needn't grant Z. So it's quite a NECESSARY step, you see?"

"I see," said Achilles; and there was a touch of sadness in his tone.

Here the narrator, having pressing business at the Bank, was obliged to leave the happy pair, and did not again pass the spot until some months afterwards. When he did so, Achilles was still seated on the back of the much-enduring Tortoise, and was writing in his notebook, which appeared to be nearly full. The Tortoise was saying, "Have you got that last step written down? Unless I've lost count, that makes a thousand and one. There are several millions more to come. And WOULD you mind, as a personal favour, considering what a lot of instruction this colloquy of ours will provide for the Logicians of the Nineteenth Century—WOULD you mind adopting a pun that my cousin the Mock-Turtle will then make, and allowing yourself to be renamed TAUGHT-US?"

"As you please," replied the weary warrior, in the hollow tones of despair, as he buried his face in his hands. "Provided that YOU, for YOUR part, will adopt a pun the Mock-Turtle never made, and allow yourself to be re-named A KILL-EASE!"

CHAPTER II

Meaning and Form in Mathematics

THIS *Two-Part Invention* was the inspiration for my two characters. Just as Lewis Carroll took liberties with Zeno's Tortoise and Achilles, so have I taken liberties with Lewis Carroll's Tortoise and Achilles. In Carroll's dialogue, the same events take place over and over again, only each time on a higher and higher level; it is a wonderful analogue to Bach's Ever-Rising Canon. The Carrollian Dialogue, with its wit subtracted out, still leaves a deep philosophical problem: *Do words and thoughts follow formal rules, or do they not?* That problem is the problem of this book.

In this Chapter and the next, we will look at several new formal systems. This will give us a much wider perspective on the concept of formal system. By the end of these two Chapters, you should have quite a good idea of the power of formal systems, and why they are of interest to mathematicians and logicians.

The pq-System

The formal system of this Chapter is called the *pq-system.* It is not important to mathematicians or logicians—in fact, it is just a simple invention of mine. Its importance lies only in the fact that it provides an excellent example of many ideas that play a large role in this book. There are three distinct symbols of the pq-system:

p q -

—the letters p, q, and the hyphen.

The pq-system has an infinite number of axioms. Since we can't write them all down, we have to have some other way of describing what they are. Actually, we want more than just a description of the axioms; we want a way to tell whether some given string is an axiom or not. A mere description of axioms might characterize them fully and yet weakly—which was the problem with the way theorems in the MIU-system were characterized. We don't want to have to struggle for an indeterminate—possibly infinite—length of time, just to find out if some string is an axiom or not. Therefore, we will define axioms in such a way that there is an obvious decision procedure for axiomhood of a string composed of p's, q's, and hyphens.

DEFINITION: *x*p−q*x*− is an axiom, whenever *x* is composed of hyphens only.

Note that '*x*' must stand for the same string of hyphens in both occurrences. For example, --p-q--- is an axiom. The literal expression '*x*p−q*x*−' is not an axiom, of course (because '*x*' does not belong to the pq-system); it is more like a mold in which all axioms are cast—and it is called an *axiom schema.*

The pq-system has only one rule of production:

RULE: Suppose *x, y,* and *z* all stand for particular strings containing only hyphens. And suppose that *x*p*y*q*z* is known to be a theorem. Then *x*p*y*−q*z*− is a theorem.

For example, take *x* to be '--', *y* to be '---', and *z* to be '-'. The rule tells us:

If --p---q- turns out to be a theorem, then so will --p----q--.

As is typical of rules of production, the statement establishes a causal connection between the theoremhood of two strings, but without asserting theoremhood for either one on its own.

A most useful exercise for you is to find a decision procedure for the theorems of the pq-system. It is not hard; if you play around for a while, you will probably pick it up. Try it.

The Decision Procedure

I presume you have tried it. First of all, though it may seem too obvious to mention, I would like to point out that every theorem of the pq-system has three separate groups of hyphens, and the separating elements are one p, and one q, in that order. (This can be shown by an argument based on "heredity", just the way one could show that all MIU-system theorems had to begin with M.) This means that we can rule out, from its form alone, a string such as --p--p--p--q--------.

Now, stressing the phrase "from its form alone" may seem silly; what else is there to a string except its form? What else could possibly play a role in determining its properties? Clearly nothing could. But bear this in mind as the discussion of formal systems goes on; the notion of "form" will start to get rather more complicated and abstract, and we will have to think more about the meaning of the word "form". In any case, let us give the name *well-formed string* to any string which begins with a hyphen-group, then has one p, then has a second hyphen-group, then a q, and then a final hyphen-group.

Back to the decision procedure . . . The criterion for theoremhood is that the first two hyphen-groups should add up, in length, to the third hyphen-group. For instance, --p--q---- is a theorem, since 2 plus 2 equals 4, whereas --p--q- is not, since 2 plus 2 is not 1. To see why this is the proper criterion, look first at the axiom schema. Obviously, it only manufactures axioms which satisfy the addition criterion. Second, look at the rule of production. If the first string satisfies the addition criterion, so must the second one—and conversely, if the first string does not satisfy the addition criterion, then neither does the second string. The rule makes the addition criterion into a hereditary property of theorems: any theorem passes the property on to its offspring. This shows why the addition criterion is correct.

There is, incidentally, a fact about the pq-system which would enable us to say with confidence that it has a decision procedure, even before finding the addition criterion. That fact is that the pq-system is not complicated by the opposing currents of *lengthening* and *shortening* rules; it has only lengthening rules. Any formal system which tells you how to make longer theorems from shorter ones, but never the reverse, has got to have a decision procedure for its theorems. For suppose you are given a string. First check whether it's an axiom or not (I am assuming that there is a decision procedure for axiomhood—otherwise, things are hopeless). If it is an axiom, then it is by definition a theorem, and the test is over. So suppose instead that it's not an axiom. Then, to be a theorem, it must have come from a shorter string, via one of the rules. By going over the various rules one by one, you can pinpoint not only the rules that could conceivably produce that string, but also exactly which shorter strings could be its forebears on the "family tree". In this way, you "reduce" the problem to determining whether any of several new but shorter strings is a theorem. Each of them can in turn be subjected to the same test. The worst that can happen is a proliferation of more and more, but shorter and shorter, strings to test. As you continue inching your way backwards in this fashion, you must be getting closer to the source of all theorems—the axiom schemata. You just can't get shorter and shorter indefinitely; therefore, eventually either you will find that one of your short strings is an axiom, or you'll come to a point where you're stuck, in that none of your short strings is an axiom, and none of them can be further shortened by running some rule or other backwards. This points out that there really is not much deep interest in formal systems with lengthening rules only; it is the interplay of lengthening and shortening rules that gives formal systems a certain fascination.

Bottom-up *vs.* Top-down

The method above might be called a *top-down* decision procedure, to be contrasted with a *bottom-up* decision procedure, which I give now. It is very reminiscent of the genie's systematic theorem-generating method for the MIU-system, but is complicated by the presence of an axiom schema. We are going to form a "bucket" into which we throw theorems as they are generated. Here is how it is done:

(1a) Throw the simplest possible axiom (-p-q--) into the bucket.

(1b) Apply the rule of inference to the item in the bucket, and put the result into the bucket.

(2a) Throw the second-simplest axiom into the bucket.

(2b) Apply the rule to each item in the bucket, and throw all results into the bucket.

(3a) Throw the third-simplest axiom into the bucket.

(3b) Apply the rule to each item in the bucket, and throw all results into the bucket.

etc., etc.

A moment's reflection will show that you can't fail to produce every theorem of the pq-system this way. Moreover, the bucket is getting filled with longer and longer theorems, as time goes on. It is again a consequence of that lack of shortening rules. So if you have a particular string, such as --p---q-----, which you want to test for theoremhood, just follow the numbered steps, checking all the while for the string in question. If it turns up—theorem! If at some point everything that goes into the bucket is longer than the string in question, forget it—it is not a theorem. This decision procedure is *bottom-up* because it is working its way up from the basics, which is to say the axioms. The previous decision procedure is *top-down* because it does precisely the reverse: it works its way back down towards the basics.

Isomorphisms Induce Meaning

Now we come to a central issue of this Chapter—indeed of the book. Perhaps you have already thought to yourself that the pq-theorems are like additions. The string --p---q----- is a theorem because 2 plus 3 equals 5. It could even occur to you that the theorem --p---q----- is a *statement,* written in an odd notation, whose *meaning* is that 2 plus 3 is 5. Is this a reasonable way to look at things? Well, I deliberately chose 'p' to remind you of 'plus', and 'q' to remind you of 'equals' . . . So, does the string --p---q----- actually *mean* "2 plus 3 equals 5"?

What would make us feel that way? My answer would be that we have perceived an *isomorphism* between pq-theorems and additions. In the Introduction, the word "isomorphism" was defined as an information-preserving transformation. We can now go into that notion a little more deeply, and see it from another perspective. The word "isomorphism" applies when two complex structures can be mapped onto each other, in such a way that to each part of one structure there is a corresponding part in the other structure, where "corresponding" means that the two parts play similar roles in their respective structures. This usage of the word "isomorphism" is derived from a more precise notion in mathematics.

It is cause for joy when a mathematician discovers an isomorphism between two structures which he knows. It is often a "bolt from the blue", and a source of wonderment. The perception of an isomorphism between two known structures is a significant advance in knowledge—and I claim that it is such perceptions of isomorphism which create *meanings* in the minds of people. A final word on the perception of isomorphisms: since they come in many shapes and sizes, figuratively speaking, it is not always totally clear when you really have found an isomorphism. Thus, "isomorphism" is a word with all the usual vagueness of words—which is a defect but an advantage as well.

In this case, we have an excellent prototype for the concept of isomorphism. There is a "lower level" of our isomorphism—that is, a mapping between the parts of the two structures:

p ⇐⇒ plus
q ⇐⇒ equals
- ⇐⇒ one
-- ⇐⇒ two
--- ⇐⇒ three
etc.

This symbol-word correspondence has a name: *interpretation.*

Secondly, on a higher level, there is the correspondence between true statements and theorems. But—note carefully—this higher-level correspondence could not be perceived without the prior choice of an interpretation for the symbols. Thus it would be more accurate to describe it as a correspondence between true statements and *interpreted* theorems. In any case we have displayed a two-tiered correspondence, which is typical of all isomorphisms.

When you confront a formal system you know nothing of, and if you hope to discover some hidden meaning in it, your problem is how to assign interpretations to its symbols in a meaningful way—that is, in such a way that a higher-level correspondence emerges between true statements and theorems. You may make several tentative stabs in the dark before finding a good set of words to associate with the symbols. It is very similar to attempts to crack a code, or to decipher inscriptions in an unknown language like Linear B of Crete: the only way to proceed is by trial and error, based on educated guesses. When you hit a good choice, a "meaningful" choice, all of a sudden things just feel right, and work speeds up enormously. Pretty soon everything falls into place. The excitement of such an experience is captured in *The Decipherment of Linear B* by John Chadwick.

But it is uncommon, to say the least, for someone to be in the position of "decoding" a formal system turned up in the excavations of a ruined civilization! Mathematicians (and more recently, linguists, philosophers, and some others) are the only users of formal systems, and they invariably have an interpretation in mind for the formal systems which they use and publish. The idea of these people is to set up a formal system whose theorems reflect some portion of reality isomorphically. In such a case, the choice of symbols is a highly motivated one, as is the choice of typographical rules of production. When I devised the pq-system, I was in this position. You see why I chose the symbols I chose. It is no accident that theorems are isomorphic to additions; it happened because I deliberately sought out a way to reflect additions typographically.

Meaningless and Meaningful Interpretations

You can choose interpretations other than the one I chose. You need not make every theorem come out true. But there would be very little reason to make an interpretation in which, say, all theorems came out false, and certainly even less reason to make an interpretation under which there is no correlation at all, positive or negative, between theoremhood and truth. Let us therefore make a distinction between two types of interpretations for a formal system. First, we can have a *meaningless* interpretation, one under which we fail to see any isomorphic connection between theorems of the system, and reality. Such interpretations abound—any random choice at all will do. For instance, take this one:

p ⇐⇒ horse
q ⇐⇒ happy
– ⇐⇒ apple

Now –p–q–– acquires a new interpretation: "apple horse apple happy apple apple"—a poetic sentiment, which might appeal to horses, and might even lead them to favor this mode of interpreting pq-strings! However, this interpretation has very little "meaningfulness"; under interpretation, theorems don't sound any truer, or any better, than nontheorems. A horse might enjoy "happy happy happy apple horse" (mapped onto qqq–p) just as much as any interpreted theorem.

The other kind of interpretation will be called *meaningful*. Under such an interpretation, theorems and truths correspond—that is, an isomorphism exists between theorems and some portion of reality. That is why it is good to distinguish between *interpretations* and *meanings*. Any old word can be used as an interpretation for 'p', but 'plus' is the only *meaningful* choice we've come up with. In summary, the meaning of 'p' seems to be 'plus', though it can have a million different interpretations.

Active *vs.* Passive Meanings

Probably the most significant fact of this Chapter, if understood deeply, is this: the pq-system seems to force us into recognizing that *symbols of a formal system, though initially without meaning, cannot avoid taking on "meaning" of sorts, at least if an isomorphism is found.* The difference between meaning in a formal system and in a language is a very important one, however. It is this:

in a language, when we have learned a meaning for a word, we then make new statements based on the meaning of the word. In a sense the meaning becomes *active,* since it brings into being a new rule for creating sentences. This means that our command of language is not like a finished product: the rules for making sentences increase when we learn new meanings. On the other hand, in a formal system, the theorems are predefined, by the rules of production. We can choose "meanings" based on an isomorphism (if we can find one) between theorems and true statements. But this does not give us the license to go out and add new theorems to the established theorems. That is what the Requirement of Formality in Chapter I was warning you of.

In the MIU-system, of course, there was no temptation to go beyond the four rules, because no interpretation was sought or found. But here, in our new system, one might be seduced by the newly found "meaning" of each symbol into thinking that the string

--p--p--p--q--------

is a theorem. At least, one might *wish* that this string were a theorem. But wishing doesn't change the fact that it isn't. And it would be a serious mistake to think that it "must" be a theorem, just because 2 plus 2 plus 2 plus 2 equals 8. It would even be misleading to attribute it any meaning at all, since it is not well-formed, and our meaningful interpretation is entirely derived from looking at well-formed strings.

In a formal system, the meaning must remain *passive;* we can read each string according to the meanings of its constituent symbols, but we do not have the right to create new theorems purely on the basis of the meanings we've assigned the symbols. Interpreted formal systems straddle the line between systems without meaning, and systems with meaning. Their strings can be thought of as "expressing" things, but this must come only as a consequence of the formal properties of the system.

Double-Entendre!

And now, I want to destroy any illusion about having found *the* meanings for the symbols of the pq-system. Consider the following association:

p ⇐⇒ equals
q ⇐⇒ taken from
- ⇐⇒ one
-- ⇐⇒ two
etc.

Now, --p---q----- has a new interpretation: "2 equals 3 taken from 5". Of course it is a true statement. All theorems will come out true under this new interpretation. It is just as meaningful as the old one. Obviously, it is silly to ask, "But which one is *the* meaning of the string?" An interpretation will be meaningful to the extent that it accurately reflects some isomorphism to the real world. When different aspects of the real world are isomorphic to each other (in this case, additions and subtractions), one single formal system can be isomorphic to both, and therefore can take on two passive meanings. This kind of double-valuedness of symbols and strings is an extremely important phenomenon. Here it seems trivial, curious, annoying. But it will come back in deeper contexts and bring with it a great richness of ideas.

Here is a summary of our observations about the pq-system. Under either of the two meaningful interpretations given, every well-formed string has a grammatical assertion for its counterpart—some are true, some false. The idea of *well-formed strings* in any formal system is that they are those strings which, when interpreted symbol for symbol, yield *grammatical* sentences. (Of course, it depends on the interpretation, but usually, there is one in mind.) Among the well-formed strings occur the theorems. These are defined by an axiom schema, and a rule of production. My goal in inventing the pq-system was to imitate additions: I wanted every theorem to express a true addition under interpretation; conversely, I wanted every true addition of precisely two positive integers to be translatable into a string, which would be a theorem. That goal was achieved. Notice, therefore, that all false additions, such as "2 plus 3 equals 6", are mapped into strings which are well-formed, but which are not theorems.

Formal Systems and Reality

This is our first example of a case where a formal system is based upon a portion of reality, and seems to mimic it perfectly, in that its theorems are isomorphic to truths about that part of reality. However, reality and the formal system are independent. Nobody need be aware that there is an isomorphism between the two. Each side stands by itself—one plus one equals two, whether or not we know that -p-q-- is a theorem; and -p-q-- is still a theorem whether or not we connect it with addition.

You might wonder whether making this formal system, or any formal system, sheds new light on truths in the domain of its interpretation. Have we learned any new additions by producing pq-theorems? Certainly not; but we have learned something about the nature of addition as a process—namely, that it is easily mimicked by a typographical rule governing meaningless symbols. This still should not be a big surprise since addition is such a simple concept. It is a commonplace that addition can be captured in the spinning gears of a device like a cash register.

But it is clear that we have hardly scratched the surface, as far as formal systems go; it is natural to wonder about what portion of reality can be imitated in its behavior by a set of meaningless symbols governed by formal rules. Can all of reality be turned into a formal system? In a very broad sense, the answer might appear to be yes. One could suggest, for instance, that reality is itself nothing but one very complicated formal system. Its symbols do not move around on paper, but rather in a three-dimensional vacuum (space); they are the elementary particles of which everything is composed. (Tacit assumption: that there is an end to the descending chain of matter, so that the expression "elementary particles" makes sense.) The "typographical rules" are the laws of physics, which tell how, given the positions and velocities of all particles at a given instant, to modify them, resulting in a new set of positions and velocities belonging to the "next" instant. So the theorems of this grand formal system are the possible configurations of particles at different times in the history of the universe. The sole axiom is (or perhaps, *was*) the original configuration of all the particles at the "beginning of time". This is so grandiose a conception, however, that it has only the most theoretical interest; and besides, quantum mechanics (and other parts of physics) casts at least some doubt on even the theoretical worth of this idea. Basically, we are asking if the universe operates deterministically, which is an open question.

Mathematics and Symbol Manipulation

Instead of dealing with such a big picture, let's limit ourselves to *mathematics* as our "real world". Here, a serious question arises: How can we be sure, if we've tried to model a formal system on some part of mathematics, that we've done the job accurately—especially if we're not one hundred per cent familiar with that portion of mathematics already? Suppose the goal of the formal system is to bring us new knowledge in that discipline. How will we know that the interpretation of every theorem is true, unless we've proven that the isomorphism is perfect? And how will we prove that the isomorphism is perfect, if we don't already know all about the truths in the discipline to begin with?

Suppose that in an excavation somewhere, we actually did discover some mysterious formal system. We would try out various interpretations and perhaps eventually hit upon one which seemed to make every theorem come out true, and every nontheorem come out false. But this is something which we could only check directly in a finite number of cases. The set of theorems is most likely infinite. How will we *know* that all theorems express truths under this interpretation, unless we know everything there is to know about both the formal system and the corresponding domain of interpretation?

It is in somewhat this odd position that we will find ourselves when we attempt to match the reality of natural numbers (i.e., the nonnegative integers: 0, 1, 2, . . .) with the typographical symbols of a formal system. We will try to understand the relationship between what we call "truth" in number theory and what we can get at by symbol manipulation.

So let us briefly look at the basis for calling some statements of number theory true, and others false. How much is 12 times 12? Everyone knows it is 144. But how many of the people who give that answer have actually at any time in their lives drawn a 12 by 12 rectangle, and then counted the little squares in it? Most people would regard the drawing and counting as unnecessary. They would instead offer as proof a few marks on paper, such as are shown below:

$$\begin{array}{r} 12 \\ \times 12 \\ \hline 24 \\ 12 \\ \hline 144 \end{array}$$

And that would be the "proof". Nearly everyone believes that if you counted the squares, you would get 144 of them; few people feel that the outcome is in doubt.

The conflict between the two points of view comes into sharper focus when you consider the problem of determining the value of 987654321 × 123456789. First of all, it is virtually impossible to construct the appropriate rectangle; and what is worse, even if it *were* constructed, and huge armies of people spent centuries counting the little squares, only a very gullible person would be willing to believe their final answer. It is just too likely that somewhere, somehow, somebody bobbled just a little bit. So is it ever possible to know what the answer is? If you trust the symbolic process which involves manipulating digits according to certain simple rules, yes. That process is presented to children as a device which gets the right answer; lost in the shuffle, for many children, are the rhyme and reason of that process. The digit-shunting laws for multiplication are based mostly on a few properties of addition and multiplication which are assumed to hold for all numbers.

The Basic Laws of Arithmetic

The kind of assumption I mean is illustrated below. Suppose that you lay down a few sticks:

/ // //// / /

Now you count them. At the same time, somebody else counts them, but starting from the other end. Is it clear that the two of you will get the same answer? The result of a counting process is independent of the way in which it is done. This is really an assumption about what counting is. It would be senseless to try to prove it, because it is so basic; either you see it or you don't—but in the latter case, a proof won't help you a bit.

From this kind of assumption, one can get to the commutativity and associativity of addition (i.e., first that $b + c = c + b$ always, and second that $b + (c + d) = (b + c) + d$ always). The same assumption can also lead you to the commutativity and associativity of multiplication; just think of many cubes assembled to form a large rectangular solid. Multiplicative commutativity and associativity are just the assumptions that when you rotate the solid in various ways, the number of cubes will not change. Now these assumptions are not verifiable in all possible cases, because the number of such cases is infinite. We take them for granted; we believe them (if we ever think about them) as deeply as we could believe anything. The amount of money in our pocket will not change as we walk down the street, jostling it up and down; the number of books we have will not change if we pack them up in a box, load them into our car, drive one hundred miles, unload the box, unpack it, and place the books in a new shelf. All of this is part of what we mean by *number*.

There are certain types of people who, as soon as some undeniable fact is written down, find it amusing to show why that "fact" is false after all. I am such a person, and as soon as I had written down the examples above involving sticks, money, and books, I invented situations in which they were wrong. You may have done the same. It goes to show that numbers as abstractions are really quite different from the everyday numbers which we use.

People enjoy inventing slogans which violate basic arithmetic but which illustrate "deeper" truths, such as "1 and 1 make 1" (for lovers), or "1 plus 1 plus 1 equals 1" (the Trinity). You can easily pick holes in those slogans, showing why, for instance, using the plus-sign is inappropriate in both cases. But such cases proliferate. Two raindrops running down a windowpane merge; does one plus one make one? A cloud breaks up into two clouds—more evidence for the same? It is not at all easy to draw a sharp line between cases where what is happening could be called "addition", and where some other word is wanted. If you think about the question, you will probably come up with some criterion involving separation of the objects in space, and making sure each one is clearly distinguishable from all the others. But then how could one count ideas? Or the number of gases comprising the atmosphere? Somewhere, if you try to look it up, you can probably find a statement such as, "There are 17 languages in India, and 462 dialects." There is something strange about precise statements like that, when the concepts "language" and "dialect" are themselves fuzzy.

Ideal Numbers

Numbers as realities misbehave. However, there is an ancient and innate sense in people that numbers ought not to misbehave. There is something clean and pure in the abstract notion of number, removed from counting beads, dialects, or clouds; and there ought to be a way of talking about numbers without always having the silliness of reality come in and intrude. The hard-edged rules that govern "ideal" numbers constitute arithmetic, and their more advanced consequences constitute number theory. There is only one relevant question to be asked, in making the transition from numbers as practical things to numbers as formal things. Once you have

FIGURE 13. Liberation, *by M. C. Escher (lithograph, 1955).*

decided to try to capsulize all of number theory in an ideal system, is it really possible to do the job completely? Are numbers so clean and crystalline and regular that their nature can be completely captured in the rules of a formal system? The picture *Liberation* (Fig. 13), one of Escher's most beautiful, is a marvelous contrast between the formal and the informal, with a fascinating transition region. Are numbers really as free as birds? Do they suffer as much from being crystallized into a rule-obeying system? Is there a magical transition region between numbers in reality and numbers on paper?

When I speak of the properties of natural numbers, I don't just mean properties such as the sum of a particular pair of integers. That can be found out by counting, and anybody who has grown up in this century cannot doubt the mechanizability of such processes as counting, adding, multiplying, and so on. I mean the kinds of properties which mathematicians are interested in exploring, questions for which no counting-process is sufficient to provide the answer—not even theoretically sufficient. Let us take a classic example of such a property of natural numbers. The statement is: "There are infinitely many prime numbers." First of all, there is no counting process which will ever be able to confirm, or refute, this assertion. The best we could do would be to count primes for a while and concede that there are "a lot". But no amount of counting alone would ever resolve the question of whether the number of primes is finite or infinite. There could always be more. The statement—and it is called "Euclid's Theorem" (notice the capital "T")—is quite unobvious. It may seem reasonable, or appealing, but it is not obvious. However, mathematicians since Euclid have always called it true. What is the reason?

Euclid's Proof

The reason is that *reasoning* tells them it is so. Let us follow the reasoning involved. We will look at a variant of Euclid's proof. This proof works by showing that whatever number you pick, there is a prime larger than it. Pick a number—N. Multiply all the positive integers starting with 1 and ending with N; in other words, form the factorial of N, written "$N!$". What you get is divisible by every number up to N. When you add 1 to $N!$, the result

can't be a multiple of 2 (because it leaves 1 over, when you divide by 2);

can't be a multiple of 3 (because it leaves 1 over, when you divide by 3);

can't be a multiple of 4 (because it leaves 1 over, when you divide by 4);

.
.
.

can't be a multiple of N (because it leaves 1 over, when you divide by N);

In other words, $N! + 1$, if it is divisible at all (other than by 1 and itself), only is divisible by numbers greater than N. So either it is itself prime, or its prime divisors are greater than N. But in either case we've shown there must exist a prime above N. The process holds no matter what number N is. Whatever N is, there is a prime greater than N. And thus ends the demonstration of the infinitude of the primes.

This last step, incidentally, is called *generalization,* and we will meet it again later in a more formal context. It is where we phrase an argument in terms of a single number (N), and then point out that N was unspecified and therefore the argument is a general one.

Euclid's proof is typical of what constitutes "real mathematics". It is simple, compelling, and beautiful. It illustrates that by taking several rather short steps one can get a long way from one's starting point. In our case, the starting points are basic ideas about multiplication and division and so forth. The short steps are the steps of reasoning. And though every individual step of the reasoning seems obvious, the end result is not obvious. We can never check directly whether the statement is true or not; yet we believe it, because we believe in reasoning. If you accept reasoning, there seems to be no escape route; once you agree to hear Euclid out, you'll have to agree with his conclusion. That's most fortunate—because it means that mathematicians will always agree on what statements to label "true", and what statements to label "false".

This proof exemplifies an orderly thought process. Each statement is related to previous ones in an irresistible way. This is why it is called a "proof'' rather than just "good evidence". In mathematics the goal is always to give an ironclad proof for some unobvious statement. The very fact of the steps being linked together in an ironclad way suggests that there may be a *patterned structure* binding these statements together. This structure can best be exposed by finding a new vocabulary—a stylized vocabulary, consisting of symbols—suitable only for expressing statements about numbers. Then we can look at the proof as it exists in its translated version. It will be a set of statements which are related, line by line, in some detectable way. But the statements, since they're represented by means of a small and stylized set of symbols, take on the aspect of *patterns.* In other words, though when read aloud, they seem to be statements about numbers and their properties, still when looked at on paper, they seem to be abstract patterns—and the line-by-line structure of the proof may start to look like a slow transformation of patterns according to some few typographical rules.

Getting Around Infinity

Although Euclid's proof is a proof that *all* numbers have a certain property, it avoids treating each of the infinitely many cases separately. It gets around it by using phrases like "whatever *N* is", or "no matter what number *N* is". We could also phrase the proof over again, so that it uses the phrase "all *N*". By knowing the appropriate context and correct ways of using such phrases, we never have to deal with infinitely many statements. We deal with just two or three concepts, such as the word "all"—which, though themselves finite, embody an infinitude; and by using them, we sidestep the apparent problem that there are an infinite number of facts we want to prove.

We use the word "all" in a few ways which are defined by the thought processes of reasoning. That is, there are *rules* which our usage of "all" obeys. We may be unconscious of them, and tend to claim we operate on the basis of the *meaning* of the word; but that, after all, is only a circumlocution for saying that we are guided by rules which we never make explicit. We have used words all our lives in certain patterns, and instead of calling the patterns "rules", we attribute the courses of our thought processes to the "meanings" of words. That discovery was a crucial recognition in the long path towards the formalization of number theory.

If we were to delve into Euclid's proof more and more carefully, we would see that it is composed of many, many small—almost infinitesimal—steps. If all those steps were written out line after line, the proof would appear incredibly complicated. To our minds it is clearest when several steps are telescoped together, to form one single sentence. If we tried to look at the proof in slow motion, we would begin to discern individual frames. In other words, the dissection can go only so far, and then we hit the "atomic" nature of reasoning processes. A proof can be broken down into a series of tiny but discontinuous jumps which seem to flow smoothly when perceived from a higher vantage point. In Chapter VIII, I will show one way of breaking the proof into atomic units, and you will see how incredibly many steps are involved. Perhaps it should not surprise you, though. The operations in Euclid's brain when he invented the proof must have involved millions of neurons (nerve cells), many of which fired several hundred times in a single second. The mere utterance of a sentence involves hundreds of thousands of neurons. If Euclid's thoughts were that complicated, it makes sense for his proof to contain a huge number of steps! (There may be little direct connection between the neural actions in his brain, and a proof in our formal system, but the complexities of the two are comparable. It is as if nature wants the complexity of the proof of the infinitude of primes to be conserved, even when the systems involved are very different from each other.)

In Chapters to come, we will lay out a formal system that (1) includes a stylized vocabulary in which all statements about natural numbers can be expressed, and (2) has rules corresponding to all the types of reasoning which seem necessary. A very important question will be whether the rules for symbol manipulation which we have then formulated are really of equal power (as far as number theory is concerned) to our usual mental reasoning abilities—or, more generally, whether it is theoretically possible to attain the level of our thinking abilities, by using some formal system.

Sonata for Unaccompanied Achilles

The telephone rings; Achilles picks it up.

Achilles: Hello, this is Achilles.

Achilles: Oh, hello, Mr. T. How are you?

Achilles: A torticollis? Oh, I'm sorry to hear it. Do you have any idea what caused it?

Achilles: How long did you hold it in that position?

Achilles: Well, no wonder it's stiff, then. What on earth induced you to keep your neck twisted that way for so long?

Achilles: Wondrous many of them, eh? What kinds, for example?

Achilles: What do you mean, "phantasmagorical beasts"?

FIGURE 14. Mosaic II, *by M. C. Escher (lithograph, 1957).*

Achilles: Wasn't it terrifying to see so many of them at the same time?

Achilles: A guitar!? Of all things to be in the midst of all those weird creatures. Say, don't you play the guitar?

Achilles: Oh, well, it's all the same to me.

Achilles: You're right; I wonder why I never noticed that difference between fiddles and guitars before. Speaking of fiddling, how would you like to come over and listen to one of the sonatas for unaccompanied violin by your favorite composer, J. S. Bach? I just bought a marvelous recording of them. I still can't get over the way Bach uses a single violin to create a piece with such interest.

Achilles: A headache too? That's a shame. Perhaps you should just go to bed.

Achilles: I see. Have you tried counting sheep?

Achilles: Oh, oh, I see. Yes, I fully know what you mean. Well, if it's THAT distracting, perhaps you'd better tell it to me, and let me try to work on it, too.

Achilles: A word with the letters 'A', 'D', 'A', 'C' consecutively inside it . . . Hmm . . . What about "abracadabra"?

Achilles: True, "ADAC" occurs backwards, not forwards, in that word.

Achilles: Hours and hours? It sounds like I'm in for a long puzzle, then. Where did you hear this infernal riddle?

Achilles: You mean he looked like he was meditating on esoteric Buddhist matters, but in reality he was just trying to think up complex word puzzles?

Achilles: Aha!—the snail knew what this fellow was up to. But how did you come to talk to the snail?

Achilles: Say, I once heard a word puzzle a little bit like this one. Do you want to hear it? Or would it just drive you further into distraction?

Achilles: I agree—can't do any harm. Here it is: What's a word that begins with the letters "HE" and also ends with "HE"?

Achilles: Very ingenious—but that's almost cheating. It's certainly not what I meant!

Achilles: Of course you're right—it fulfills the conditions, but it's a sort of "degenerate" solution. There's another solution which I had in mind.

Achilles: That's exactly it! How did you come up with it so fast?

Achilles: So here's a case where having a headache actually might have helped you, rather than hindering you. Excellent! But I'm still in the dark on your "ADAC" puzzle.

Achilles: Congratulations! Now maybe you'll be able to get to sleep! So tell me, what IS the solution?

Achilles: Well, normally I don't like hints, but all right. What's your hint?

Achilles: I don't know what you mean by "figure" and "ground" in this case.

Achilles: Certainly I know *Mosaic II!* I know ALL of Escher's works. After all, he's my favorite artist. In any case, I've got a print of *Mosaic II* hanging on my wall, in plain view from here.

Achilles: Yes, I see all the black animals.

Achilles: Yes, I also see how their "negative space"—what's left out—defines the white animals.

Achilles: So THAT'S what you mean by "figure" and "ground". But what does that have to do with the "ADAC" puzzle?

Achilles: Oh, this is too tricky for me. I think I'M starting to get a headache.

Achilles: You want to come over now? But I thought—

Achilles: Very well. Perhaps by then I'll have thought of the right answer to YOUR puzzle, using your figure-ground hint, relating it to MY puzzle.

Achilles: I'd love to play them for you.

Achilles: You've invented a theory about them?

Achilles: Accompanied by what instrument?

Achilles: Well, if that's the case, it seems a little strange that he wouldn't have written out the harpsichord part, then, and had it published as well.

Achilles: I see—sort of an optional feature. One could listen to them either way—with or without accompaniment. But how would one know what the accompaniment is supposed to sound like?

Achilles: Ah, yes, I guess that it is best, after all, to leave it to the listener's imagination. And perhaps, as you said, Bach never even had any accompaniment in mind at all. Those sonatas seem to work very well indeed as they are.

Achilles: Right. Well, I'll see you shortly.

Achilles: Good-bye, Mr. T.

CHAPTER III

Figure and Ground

Primes *vs.* Composites

THERE IS A strangeness to the idea that concepts can be captured by simple typographical manipulations. The one concept so far captured is that of addition, and it may not have appeared very strange. But suppose the goal were to create a formal system with theorems of the form P*x*, the letter '*x*' standing for a hyphen-string, and where the only such theorems would be ones in which the hyphen-string contained exactly a prime number of hyphens. Thus, P--- would be a theorem, but P---- would not. How could this be done typographically? First, it is important to specify clearly what is meant by *typographical* operations. The complete repertoire has been presented in the MIU-system and the pq-system, so we really only need to make a list of the kinds of things we have permitted:

(1) reading and recognizing any of a finite set of symbols;
(2) writing down any symbol belonging to that set;
(3) copying any of those symbols from one place to another;
(4) erasing any of those symbols;
(5) checking to see whether one symbol is the same as another;
(6) keeping and using a list of previously generated theorems.

The list is a little redundant, but no matter. What is important is that it clearly involves only trivial abilities, each of them far less than the ability to distinguish primes from nonprimes. How, then, could we compound some of these operations to make a formal system in which primes are distinguished from composite numbers?

The tq-System

A first step might be to try to solve a simpler, but related, problem. We could try to make a system similar to the pq-system, except that it represents multiplication, instead of addition. Let's call it the *tq-system,* 't' for 'times'. More specifically, suppose *X, Y,* and *Z* are, respectively, the numbers of hyphens in the hyphen-strings *x, y,* and *z.* (Notice I am taking special pains to distinguish between a string and the number of hyphens it contains.) Then we wish the string *x*t*y*q*z* to be a theorem if and only if *X* times *Y* equals *Z*. For instance, --t---q------ should be a theorem because 2 times 3 equals 6, but --t--q--- should not be a theorem. The tq-system can be characterized just about as easily as the pq-system—namely, by using just one axiom schema and one rule of inference:

AXIOM SCHEMA: *x*t–q*x* is an axiom, whenever *x* is a hyphen-string.

RULE OF INFERENCE: Suppose that *x, y,* and *z* are all hyphen-strings. And suppose that *x*t*y*q*z* is an old theorem. Then, *x*t*y*–q*zx* is a new theorem.

Below is the derivation of the theorem --t---q------:

(1)	--t-q--	(axiom)
(2)	--t--q----	(by rule of inference, using line (1) as the old theorem)
(3)	--t---q------	(by rule of inference, using line (2) as the old theorem)

Notice how the middle hyphen-string grows by one hyphen each time the rule of inference is applied; so it is predictable that if you want a theorem with ten hyphens in the middle, you apply the rule of inference nine times in a row.

Capturing Compositeness

Multiplication, a slightly trickier concept than addition, has now been "captured" typographically, like the birds in Escher's *Liberation.* What about primeness? Here's a plan that might seem smart: using the tq-system, define a new set of theorems of the form C*x,* which characterize *composite* numbers, as follows:

RULE: Suppose *x, y,* and *z* are hyphen-strings. If *x*–t*y*–q*z* is a theorem, then C*z* is a theorem.

This works by saying that *Z* (the number of hyphens in *z*) is composite as long as it is the product of two numbers greater than 1—namely, $X + 1$ (the number of hyphens in *x*–), and $Y + 1$ (the number of hyphens in *y*–). I am defending this new rule by giving you some "Intelligent mode" justifications for it. That is because you are a human being, and want to know *why* there is such a rule. If you were operating exclusively in the "Mechanical mode", you would not need any justification, since M-mode workers just follow the rules mechanically and happily, never questioning them!

Because you work in the I-mode, you will tend to blur in your mind the distinction between strings and their interpretations. You see, things can become quite confusing as soon as you perceive "meaning" in the symbols which you are manipulating. You have to fight your own self to keep from thinking that the *string* '---' is the *number* 3. The Requirement of Formality, which in Chapter I probably seemed puzzling (because it seemed so obvious), here becomes tricky, and crucial. It is the essential thing which keeps you from mixing up the I-mode with the M-mode; or said another way, it keeps you from mixing up arithmetical facts with typographical theorems.

Illegally Characterizing Primes

It is very tempting to jump from the C-type theorems directly to P-type theorems, by proposing a rule of the following kind:

PROPOSED RULE: Suppose *x* is a hyphen-string. If C*x* is *not* a theorem, then P*x* *is* a theorem.

The fatal flaw here is that checking whether C*x* is *not* a theorem is not an explicitly typographical operation. To know for sure that MU is not a theorem of the MIU-system, you have to go *outside* of the system . . . and so it is with this Proposed Rule. It is a rule which violates the whole idea of formal systems, in that it asks you to operate informally—that is, outside the system. Typographical operation (6) allows you to look into the stockpile of previously found theorems, but this Proposed Rule is asking you to look into a hypothetical "Table of Nontheorems". But in order to generate such a table, you would have to do some reasoning *outside the system*—reasoning which shows why various strings cannot be generated inside the system. Now it may well be that there is *another* formal system which can generate the "Table of Nontheorems", by purely typographical means. In fact, our aim is to find just such a system. But the Proposed Rule is not a typographical rule, and must be dropped.

This is such an important point that we might dwell on it a bit more. In our *C-system* (which includes the tq-system and the rule which defines C-type theorems), we have theorems of the form C*x*, with '*x*' standing, as usual, for a hyphen-string. There are also nontheorems of the form C*x*. (These are what I mean when I refer to "nontheorems", although of course tt–Cqq and other ill-formed messes are also nontheorems.) The difference is that theorems have a composite number of hyphens, nontheorems have a prime number of hyphens. Now the theorems all have a common "form", that is, originate from a common set of typographical rules. Do all nontheorems also have a common "form", in the same sense? Below is a list of C-type theorems, shown without their derivations. The parenthesized numbers following them simply count the hyphens in them.

C---- (4)
C------ (6)
C-------- (8)
C--------- (9)
C---------- (10)
C------------ (12)
C-------------- (14)
C--------------- (15)
C---------------- (16)
C------------------ (18)

.
.
.

The "holes" in this list are the nontheorems. To repeat the earlier question: Do the holes also have some "form" in common? Would it be reasonable to say that merely by virtue of being the holes in this list, they share a common form? Yes and no. That they share *some* typographical quality is undeniable, but whether we want to call it "form" is unclear. The reason for hesitating is that the holes are only *negatively* defined—they are the things that are left out of a list which is *positively* defined.

Figure and Ground

This recalls the famous artistic distinction between *figure* and *ground.* When a figure or "positive space" (e.g., a human form, or a letter, or a still life) is drawn inside a frame, an unavoidable consequence is that its complementary shape—also called the "ground", or "background", or "negative space"—has also been drawn. In most drawings, however, this figure-ground relationship plays little role. The artist is much less interested in the ground than in the figure. But sometimes, an artist will take interest in the ground as well.

There are beautiful alphabets which play with this figure-ground distinction. A message written in such an alphabet is shown below. At first it looks like a collection of somewhat random blobs, but if you step back a ways and stare at it for a while, all of a sudden, you will see seven letters appear in this . . .

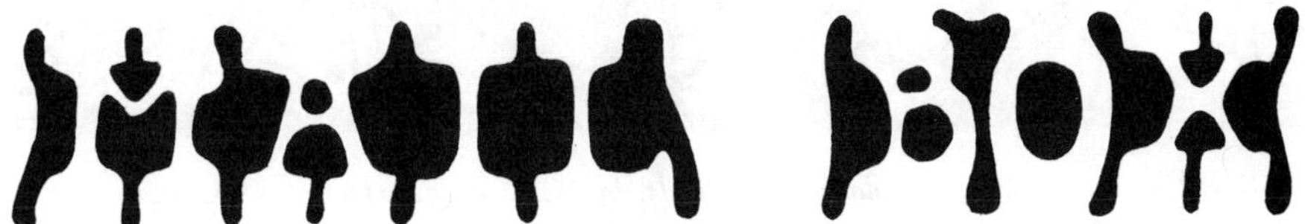

FIGURE 15.

For a similar effect, take a look at my drawing *Smoke Signal* (Fig. 139). Along these lines, you might consider this puzzle: can you somehow create a drawing containing words in both the figure *and* the ground?

Let us now officially distinguish between two kinds of figures: *cursively drawable* ones, and *recursive* ones (by the way, these are my own terms—they are not in common usage). A *cursively drawable* figure is one whose ground is merely an accidental by-product of the drawing act. A *recursive* figure is one whose ground can be seen as a figure in its own right. Usually this is quite deliberate on the part of the artist. The "re" in "recursive" represents the fact that both foreground *and* background are cursively drawable—the figure is "twice-cursive". Each figure-ground boundary in a recursive figure is a double-edged sword. M. C. Escher was a master at drawing recursive figures—see, for instance, his beautiful recursive drawing of birds (Fig. 16).

FIGURE 16. Tiling of the plane using birds, by M. C. Escher (from a 1942 notebook).

Our distinction is not as rigorous as one in mathematics, for who can definitively say that a particular ground is not a figure? Once pointed out, almost any ground has interest of its own. In that sense, every figure is recursive. But that is not what I intended by the term. There is a natural and intuitive notion of recognizable forms. Are both the foreground and background recognizable forms? If so, then the drawing is recursive. If you look at the grounds of most line drawings, you will find them rather unrecognizable. This demonstrates that

> There exist recognizable forms whose negative space is not any recognizable form.

In more "technical" terminology, this becomes:

> There exist cursively drawable figures which are not recursive.

Scott Kim's solution to the above puzzle, which I call his "FIGURE-FIGURE Figure", is shown in Figure 17. If you read both black and white,

FIGURE 17. FIGURE-FIGURE Figure, *by Scott E. Kim (1975).*

you will see "FIGURE" everywhere, but "GROUND" nowhere! It is a paragon of recursive figures. In this clever drawing, there are two nonequivalent ways of characterizing the black regions:

(1) as the *negative space* to the white regions;
(2) as *altered copies* of the white regions (produced by coloring and shifting each white region).

(In the special case of the FIGURE-FIGURE Figure, the two characterizations *are* equivalent—but in most black-and-white pictures, they would not be.) Now in Chapter VIII, when we create our Typographical Number Theory (TNT), it will be our hope that the set of all false statements of number theory can be characterized in two analogous ways:

(1) as the *negative space* to the set of all TNT-theorems;
(2) as *altered copies* of the set of all TNT-theorems (produced by negating each TNT-theorem).

But this hope will be dashed, because:

(1) inside the set of all nontheorems are found some truths;
(2) outside the set of all negated theorems are found some falsehoods.

You will see why and how this happens, in Chapter XIV. Meanwhile, ponder over a pictorial representation of the situation (Fig. 18).

Figure and Ground in Music

One may also look for figures and grounds in music. One analogue is the distinction between melody and accompaniment—for the melody is always in the forefront of our attention, and the accompaniment is subsidiary, in some sense. Therefore it is surprising when we find, in the lower lines of a piece of music, recognizable melodies. This does not happen too often in post-baroque music. Usually the harmonies are not thought of as foreground. But in baroque music—in Bach above all—the distinct lines, whether high or low or in between, all act as "figures". In this sense, pieces by Bach can be called "recursive".

Another figure-ground distinction exists in music: that between on-beat and off-beat. If you count notes in a measure "one-and, two-and, three-and, four-and", most melody-notes will come on numbers, not on "and"'s. But sometimes, a melody will be deliberately pushed onto the "and"'s, for the sheer effect of it. This occurs in several études for the piano by Chopin, for instance. It also occurs in Bach—particularly in his Sonatas and Partitas for unaccompanied violin, and his Suites for unaccompanied cello. There, Bach manages to get two or more musical lines going simultaneously. Sometimes he does this by having the solo instrument play "double stops"—two notes at once. Other times, however, he

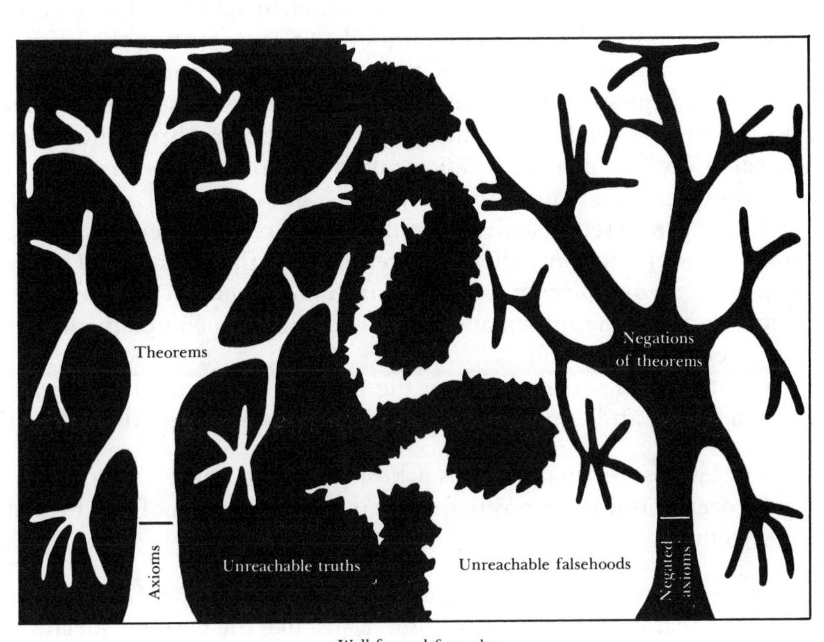

FIGURE 18. Considerable visual symbolism is featured in this diagram of the relationship between various classes of TNT strings. The biggest box represents the set of all TNT strings. The next-biggest box represents the set of all well-formed *TNT strings. Within it is found the set of all* sentences *of TNT. Now things begin to get interesting. The set of* theorems *is pictured as a tree growing out of a trunk (representing the set of* axioms*). The tree-symbol was chosen because of the recursive growth pattern which it exhibits: new branches (theorems) constantly sprouting from old ones. The fingerlike branches probe into the corners of the constraining region (the set of* truths*), yet can never fully occupy it. The boundary between the set of truths and the set of falsities is meant to suggest a randomly meandering coastline which, no matter how closely you examine it, always has finer levels of structure, and is consequently impossible to describe exactly in any finite way. (See B. Mandelbrot's book* Fractals.*) The reflected tree represents the set of* negations of theorems: *all of them false, yet unable collectively to span the space of false statements.* [*Drawing by the author.*]

puts one voice on the on-beats, and the other voice on the off-beats, so the ear separates them and hears two distinct melodies weaving in and out, and harmonizing with each other. Needless to say, Bach didn't stop at this level of complexity . . .

Recursively Enumerable Sets *vs.* Recursive Sets

Now let us carry back the notions of figure and ground to the domain of formal systems. In our example, the role of positive space is played by the C-type theorems, and the role of negative space is played by strings with a prime number of hyphens. So far, the only way we have found to represent prime numbers typographically is as a negative space. Is there, however, some way—I don't care how complicated—of representing the primes as a *positive* space—that is, as a set of theorems of some formal system?

Different people's intuitions give different answers here. I remember quite vividly how puzzled and intrigued I was upon realizing the difference between a positive characterization and a negative characterization. I was quite convinced that not only the primes, but *any* set of numbers which could be represented negatively, could also be represented positively. The intuition underlying my belief is represented by the question: "*How could a figure and its ground not carry exactly the same information?*" They seemed to me to embody the same information, just coded in two complementary ways. What seems right to you?

It turns out I was right about the primes, but wrong in general. This astonished me, and continues to astonish me even today. It is a fact that:

> There exist formal systems whose negative space (set of non-theorems) is not the positive space (set of theorems) of any formal system.

This result, it turns out, is of depth equal to Gödel's Theorem—so it is not surprising that my intuition was upset. I, just like the mathematicians of the early twentieth century, expected the world of formal systems and natural numbers to be more predictable than it is. In more technical terminology, this becomes:

> There exist recursively enumerable sets which are not recursive.

The phrase *recursively enumerable* (often abbreviated "r.e.") is the mathematical counterpart to our artistic notion of "cursively drawable"—and *recursive* is the counterpart of "recursive". For a set of strings to be "r.e." means that it *can* be generated according to typographical rules—for example, the set of C-type theorems, the set of theorems of the MIU-system—indeed, the set of theorems of any formal system. This could be compared with the conception of a "figure" as "a set of lines which can be generated according to artistic rules" (whatever that might mean!). And a "recursive set" is like a figure whose ground is also a figure—not only is it r.e., but its complement is also r.e.

It follows from the above result that:

> There exist formal systems for which there is no typographical decision procedure.

How does this follow? Very simply. A typographical decision procedure is a method which tells theorems from nontheorems. The existence of such a test allows us to generate all nontheorems systematically, simply by going down a list of *all* strings and performing the test on them one at a time, discarding ill-formed strings and theorems along the way. This amounts to a typographical method for generating the set of nontheorems. But according to the earlier statement (which we here accept on faith), for *some* systems this is not possible. So we must conclude that typographical decision procedures do not exist for all formal systems.

Suppose we found a set F of natural numbers ('F' for 'Figure') which we could generate in some formal way—like the composite numbers. Suppose its complement is the set G (for 'Ground')—like the primes. Together, F and G make up all the natural numbers, and we know a rule for making all the numbers in set F, but we know no such rule for making all the numbers in set G. It is important to understand that if the members of F were always generated in order of *increasing size,* then we could always characterize G. The problem is that many r.e. sets are generated by methods which throw in elements in an arbitrary order, so you never know if a number which has been skipped over for a long time will get included if you just wait a little longer.

We answered no to the artistic question, "Are all figures recursive?" We have now seen that we must likewise answer no to the analogous question in mathematics: "Are all sets recursive?" With this perspective, let us now come back to the elusive word "form". Let us take our figure-set F and our ground-set G again. We can agree that all the numbers in set F have some common "form"—but can the same be said about numbers in set G? It is a strange question. When we are dealing with an infinite set to start with—the natural numbers—the holes created by removing some subset may be very hard to define in any explicit way. And so it may be that they are not connected by any common attribute or "form". In the last analysis, it is a matter of taste whether you want to use the word "form"—but just thinking about it is provocative. Perhaps it is best not to define "form", but to leave it with some intuitive fluidity.

Here is a puzzle to think about in connection with the above matters. Can you characterize the following set of integers (or its negative space)?

1 3 7 12 18 26 35 45 56 69 . . .

How is this sequence like the FIGURE-FIGURE Figure?

Primes as Figure Rather than Ground

Finally, what about a formal system for generating primes? How is it done? The trick is to skip right over multiplication, and to go directly to *nondivisibility* as the thing to represent positively. Here are an axiom schema and a rule for producing theorems which represent the notion that one number *does not divide* (**DND**) another number exactly:

AXIOM SCHEMA: *xy***DND***x* where *x* and *y* are hyphen-strings.

For example, -----**DND**--, where *x* has been replaced by '--' and *y* by '---'.

RULE: If *x*DND*y* is a theorem, then so is *x*DND*xy*.

If you use the rule twice, you can generate this theorem:

-----DND------------

which is interpreted as "5 does not divide 12". But ---DND------ is not a theorem. What goes wrong if you try to produce it?

Now in order to determine that a given number is prime, we have to build up some knowledge about its nondivisibility properties. In particular, we want to know that it is not divisible by 2 or 3 or 4, etc., all the way up to 1 less than the number itself. But we can't be so vague in formal systems as to say "et cetera". We must spell things out. We would like to have a way of saying, in the language of the system, "the number *Z* is *divisor-free* up to *X*", meaning that no number between 2 and *X* divides *Z*. This can be done, but there is a trick to it. Think about it if you want.

Here is the solution:

RULE: If --DND*z* is a theorem, so is *z*DF--.

RULE: If *z*DF*x* is a theorem and also *x*-DND*z* is a theorem, then *z*DF*x*- is a theorem.

These two rules capture the notion of *divisor-freeness*. All we need to do is to say that primes are numbers which are divisor-free up to 1 less than themselves:

RULE: If *z*-DF*z* is a theorem, then P*z*- is a theorem.

Oh—let's not forget that 2 is prime!

AXIOM: P--.

And there you have it. The principle of representing primality formally is that there is a test for divisibility which can be done without any backtracking. You march steadily upward, testing first for divisibility by 2, then by 3, and so on. It is this "monotonicity" or unidirectionality—this absence of cross-play between lengthening and shortening, increasing and decreasing—that allows primality to be captured. And it is this potential complexity of formal systems to involve arbitrary amounts of backwards-forwards interference that is responsible for such limitative results as Gödel's Theorem, Turing's Halting Problem, and the fact that not all recursively enumerable sets are recursive.

Contracrostipunctus

Achilles has come to visit his friend and jogging companion, the Tortoise, at his home.

Achilles: Heavens, you certainly have an admirable boomerang collection!

Tortoise: Oh, pshaw. No better than that of any other Tortoise. And now, would you like to step into the parlor?

Achilles: Fine. *(Walks to the corner of the room.)* I see you also have a large collection of records. What sort of music do you enjoy?

Tortoise: Sebastian Bach isn't so bad, in my opinion. But these days, I must say, I am developing more and more of an interest in a rather specialized sort of music.

Achilles: Tell me, what kind of music is that?

Tortoise: A type of music which you are most unlikely to have heard of. I call it "music to break phonographs by".

Achilles: Did you say "to break phonographs by"? That is a curious concept. I can just see you, sledgehammer in hand, whacking one phonograph after another to pieces, to the strains of Beethoven's heroic masterpiece *Wellington's Victory*.

Tortoise: That's not quite what this music is about. However, you might find its true nature just as intriguing. Perhaps I should give you a brief description of it?

Achilles: Exactly what I was thinking.

Tortoise: Relatively few people are acquainted with it. It all began when my friend the Crab—have you met him, by the way?—paid me a visit.

Achilles: 'twould be a pleasure to make his acquaintance, I'm sure. Though I've heard so much about him, I've never met him.

Tortoise: Sooner or later I'll get the two of you together. You'd hit it off splendidly. Perhaps we could meet at random in the park one day . . .

Achilles: Capital suggestion! I'll be looking forward to it. But you were going to tell me about your weird "music to smash phonographs by", weren't you?

Tortoise: Oh, yes. Well, you see, the Crab came over to visit one day. You must understand that he's always had a weakness for fancy gadgets, and at that time he was quite an aficionado for, of all things, record players. He had just bought his first record player, and being somewhat gullible, believed every word the salesman had told him about it—in particular, that it was capable of reproducing any and all sounds. In short, he was convinced that it was a Perfect phonograph.

Achilles: Naturally, I suppose you disagreed.

Tortoise: True, but he would hear nothing of my arguments. He staunchly maintained that any sound whatever was reproducible on his machine. Since I couldn't convince him of the contrary, I left it at that. But not long after that, I returned the visit, taking with me a record of a song which I had myself composed. The song was called "I Cannot Be Played on Record Player 1".

Achilles: Rather unusual. Was it a present for the Crab?

Tortoise: Absolutely. I suggested that we listen to it on his new phonograph, and he was very glad to oblige me. So he put it on. But unfortunately, after only a few notes, the record player began vibrating rather severely, and then with a loud "pop", broke into a large number of fairly small pieces, scattered all about the room. The record was utterly destroyed also, needless to say.

Achilles: Calamitous blow for the poor fellow, I'd say. What was the matter with his record player?

Tortoise: Really, there was nothing the matter, nothing at all. It simply couldn't reproduce the sounds on the record which I had brought him, because they were sounds that would make it vibrate and break.

Achilles: Odd, isn't it? I mean, I thought it was a Perfect phonograph. That's what the salesman had told him, after all.

Tortoise: Surely, Achilles, you don't believe everything that salesmen tell you! Are you as naïve as the Crab was?

Achilles: The Crab was naïver by far! I know that salesmen are notorious prevaricators. I wasn't born yesterday!

Tortoise: In that case, maybe you can imagine that this particular salesman had somewhat exaggerated the quality of the Crab's piece of equipment . . . perhaps it was indeed less than Perfect, and could not reproduce every possible sound.

Achilles: Perhaps that is an explanation. But there's no explanation for the amazing coincidence that your record had those very sounds on it . . .

Tortoise: Unless they got put there deliberately. You see, before returning the Crab's visit, I went to the store where the Crab had bought his machine, and inquired as to the make. Having ascertained that, I sent off to the manufacturers for a description of its design. After receiving that by return mail, I analyzed the entire construction of the phonograph and discovered a certain set of sounds which, if they were produced anywhere in the vicinity, would set the device to shaking and eventually to falling apart.

Achilles: Nasty fellow! You needn't spell out for me the last details: that you recorded those sounds yourself, and offered the dastardly item as a gift . . .

Tortoise: Clever devil! You jumped ahead of the story! But that wasn't the end of the adventure, by any means, for the Crab did not believe that his record player was at fault. He was quite stubborn. So he went out and bought a new record player, this one even more expensive, and this time the salesman promised to give him double his money back in case the Crab found a sound which it could not reproduce exactly. So the Crab told me excitedly about his new model, and I promised to come over and see it.

Achilles: Tell me if I'm wrong—I bet that before you did so, you once again wrote the manufacturer, and composed and recorded a new song called "I Cannot Be Played on Record Player 2", based on the construction of the new model.

Tortoise: Utterly brilliant deduction, Achilles. You've quite got the spirit.

Achilles: So what happened this time?

Tortoise: As you might expect, precisely the same thing. The phonograph fell into innumerable pieces, and the record was shattered.

Achilles: Consequently, the Crab finally became convinced that there can be no such thing as a Perfect record player.

Tortoise: Rather surprisingly, that's not quite what happened. He was sure that the next model up would fill the bill, and having twice the money, he—

Achilles: Oho—I have an idea! He could have easily outwitted you, by obtaining a LOW-fidelity phonograph—one that was not capable of reproducing the sounds which would destroy it. In that way, he would avoid your trick.

Tortoise: Surely, but that would defeat the original purpose—namely, to have a phonograph which could reproduce any sound whatsoever, even its own self-breaking sound, which is of course impossible.

Achilles: That's true. I see the dilemma now. If any record player—say Record Player X—is sufficiently high-fidelity, then when it attempts to play the song "I Cannot Be Played on Record Player X", it will create just those vibrations which will cause it to break . . . So it fails to be Perfect. And yet, the only way to get around that trickery, namely for Record Player X to be of lower fidelity, even more directly ensures that it is not Perfect. It seems that every record player is vulnerable to one or the other of these frailties, and hence all record players are defective.

Tortoise: I don't see why you call them "defective". It is simply an inherent fact about record players that they can't do all that you might wish them to be able to do. But if there is a defect anywhere, it is not in THEM, but in your expectations of what they should be able to do! And the Crab was just full of such unrealistic expectations.

Achilles: Compassion for the Crab overwhelms me. High fidelity or low fidelity, he loses either way.

Tortoise: And so, our little game went on like this for a few more rounds, and eventually our friend tried to become very smart. He got wind of the principle upon which I was basing my own records, and decided to try to outfox me. He wrote to the phonograph makers, and described a device of his own invention, which they built to specification. He called it "Record Player Omega". It was considerably more sophisticated than an ordinary record player.

Achilles: Let me guess how: Did it have no moving parts? Or was it made of cotton? Or—

Tortoise: Let me tell you, instead. That will save some time. In the first place, Record Player Omega incorporated a television camera whose purpose it was to scan any record before playing it. This camera was hooked up to a small built-in computer, which would determine exactly the nature of the sounds, by looking at the groove-patterns.

Achilles: Yes, so far so good. But what could Record Player Omega do with this information?

Tortoise: By elaborate calculations, its little computer figured out what effects the sounds would have upon its phonograph. If it deduced that the sounds were such that they would cause the machine in its present configuration to break, then it did something very clever. Old Omega contained a device which could disassemble large parts of its phonograph subunit, and rebuild them in new ways, so that it could, in effect, change its own structure. If the sounds were "dangerous", a new configuration was chosen, one to which the sounds would pose no threat, and this new configuration would then be built by the rebuilding subunit, under direction of the little computer. Only after this rebuilding operation would Record Player Omega attempt to play the record.

Achilles: Aha! That must have spelled the end of your tricks. I bet you were a little disappointed.

Tortoise: Curious that you should think so . . . I don't suppose that you know Gödel's Incompleteness Theorem backwards and forwards, do you?

Achilles: Know WHOSE Theorem backwards and forwards? I've never heard of anything that sounds like that. I'm sure it's fascinating, but I'd rather hear more about "music to break records by". It's an amusing little story. Actually, I guess I can fill in the end. Obviously, there was no point in going on, and so you sheepishly admitted defeat, and that was that. Isn't that exactly it?

Tortoise: What! It's almost midnight! I'm afraid it's my bedtime. I'd love to talk some more, but really I am growing quite sleepy.

Achilles: As am I. Well, I'll be on my way. *(As he reaches the door, he suddenly stops, and turns around.)* Oh, how silly of me! I almost forgot, I brought you a little present. Here. *(Hands the Tortoise a small, neatly wrapped package.)*

Tortoise: Really, you shouldn't have! Why, thank you very much indeed. I think I'll open it now. *(Eagerly tears open the package, and inside discovers a glass goblet.)* Oh, what an exquisite goblet! Did you know that I am quite an aficionado for, of all things, glass goblets?

Achilles: Didn't have the foggiest. What an agreeable coincidence!

Tortoise: Say, if you can keep a secret, I'll let you in on something: I'm trying to find a Perfect goblet: one having no defects of any sort in its shape. Wouldn't it be something if this goblet—let's call it "G"—were the one? Tell me, where did you come across Goblet G?

Achilles: Sorry, but that's MY little secret. But you might like to know who its maker is.

Tortoise: Pray tell, who is it?

Achilles: Ever hear of the famous glassblower Johann Sebastian Bach? Well, he wasn't exactly famous for glassblowing—but he dabbled at the art as a hobby, though hardly a soul knows it—and this goblet is the last piece he blew.

Tortoise: Literally his last one? My gracious. If it truly was made by Bach, its value is inestimable. But how are you sure of its maker?

Achilles: Look at the inscription on the inside—do you see where the letters 'B', 'A', 'C', 'H' have been etched?

Tortoise: Sure enough! What an extraordinary thing. *(Gently sets Goblet G down on a shelf.)* By the way, did you know that each of the four letters in Bach's name is the name of a musical note?

Achilles: 'tisn't possible, is it? After all, musical notes only go from 'A' through 'G'.

Tortoise: Just so; in most countries, that's the case. But in Germany, Bach's own homeland, the convention has always been similar, except that what we call 'B', they call 'H', and what we call 'B-flat', they call 'B'. For instance, we talk about Bach's "Mass in B Minor", whereas they talk about his "H-moll Messe". Is that clear?

Achilles: . . . hmm . . . I guess so. It's a little confusing: H is B, and B is B-flat. I suppose his name actually constitutes a melody, then.

Tortoise: Strange but true. In fact, he worked that melody subtly into one of his most elaborate musical pieces—namely, the final *Contrapunctus* in his *Art of the Fugue.* It was the last fugue Bach ever wrote. When I heard it for the first time, I had no idea how it would end. Suddenly, without warning, it broke off. And then . . . dead silence. I realized immediately that was where Bach died. It is an indescribably sad moment, and the effect it had on me was—shattering. In any case, B-A-C-H is the last theme of that fugue. It is hidden inside the piece. Bach didn't point it out

FIGURE 19. The last page of Bach's Art of the Fugue. *In the original manuscript, in the handwriting of Bach's son Carl Philipp Emanuel, is written: "N.B. In the course of this fugue, at the point where the name B.A.C.H. was brought in as countersubject, the composer died." (B-A-C-H in box.) I have let this final page of Bach's last fugue serve as an epitaph.* [*Music printed by Donald Byrd's program "SMUT", developed at Indiana University.*]

explicitly, but if you know about it, you can find it without much trouble. Ah, me—there are so many clever ways of hiding things in music . . .

Achilles: . . . or in poems. Poets used to do very similar things, you know (though it's rather out of style these days). For instance, Lewis Carroll often hid words and names in the first letters (or characters) of the successive lines in poems he wrote. Poems which conceal messages that way are called "acrostics".

Tortoise: Bach, too, occasionally wrote acrostics, which isn't surprising. After all, counterpoint and acrostics, with their levels of hidden meaning, have quite a bit in common. Most acrostics, however, have only one hidden level—but there is no reason that one couldn't make a double-decker—an acrostic on top of an acrostic. Or one could make a "contracrostic"—where the initial letters, taken in reverse order, form a message. Heavens! There's no end to the possibilities inherent in the form. Moreover, it's not limited to poets; anyone could write acrostics—even a dialogician.

Achilles: A dial-a-logician? That's a new one on me.

Tortoise: Correction: I said "dialogician", by which I meant a writer of dialogues. Hmm . . . something just occurred to me. In the unlikely event that a dialogician should write a contrapuntal acrostic in homage to J. S. Bach, do you suppose it would be more proper for him to acrostically embed his OWN name—or that of Bach? Oh, well, why worry about such frivolous matters? Anybody who wanted to write such a piece could make up his own mind. Now getting back to Bach's melodic name, did you know that the melody B-A-C-H, if played upside down and backwards, is exactly the same as the original?

Achilles: How can anything be played upside down? Backwards, I can see—you get H-C-A-B—but upside down? You must be pulling my leg.

Tortoise: 'pon my word, you're quite a skeptic, aren't you? Well, I guess I'll have to give you a demonstration. Let me just go and fetch my fiddle— *(Walks into the next room, and returns in a jiffy with an ancient-looking violin.)* —and play it for you forwards and backwards and every which way. Let's see, now . . . *(Places his copy of the* Art of the Fugue *on his music stand and opens it to the last page.)* . . . here's the last *Contrapunctus,* and here's the last theme . . .

The Tortoise begins to play: B-A-C- —but as he bows the final H, suddenly, without warning, a shattering sound rudely interrupts his performance. Both he and Achilles spin around, just in time to catch a glimpse of myriad fragments of glass tinkling to the floor from the shelf where Goblet G had stood, only moments before. And then . . . dead silence.

CHAPTER IV

Consistency, Completeness, and Geometry

Implicit and Explicit Meaning

IN CHAPTER II, we saw how meaning—at least in the relatively simple context of formal systems—arises when there is an isomorphism between rule-governed symbols, and things in the real world. The more complex the isomorphism, in general, the more "equipment"—both hardware and software—is required to extract the meaning from the symbols. If an isomorphism is very simple (or very familiar), we are tempted to say that the meaning which it allows us to see is explicit. We see the meaning without seeing the isomorphism. The most blatant example is human language, where people often attribute meaning to words in themselves, without being in the slightest aware of the very complex "isomorphism" that imbues them with meanings. This is an easy enough error to make. It attributes all the meaning to the *object* (the word), rather than to the *link* between that object and the real world. You might compare it to the naïve belief that noise is a necessary side effect of any collision of two objects. This is a false belief; if two objects collide in a vacuum, there will be no noise at all. Here again, the error stems from attributing the noise exclusively to the *collision*, and not recognizing the role of the *medium*, which carries it from the objects to the ear.

Above, I used the word "isomorphism" in quotes to indicate that it must be taken with a grain of salt. The symbolic processes which underlie the understanding of human language are so much more complex than the symbolic processes in typical formal systems, that, if we want to continue thinking of meaning as mediated by isomorphisms, we shall have to adopt a far more flexible conception of what isomorphisms can be than we have up till now. In my opinion, in fact, the key element in answering the question "What is consciousness?" will be the unraveling of the nature of the "isomorphism" which underlies meaning.

Explicit Meaning of the *Contracrostipunctus*

All this is by way of preparation for a discussion of the *Contracrostipunctus*—a study in levels of meaning. The Dialogue has both explicit and implicit meanings. Its most explicit meaning is simply the story which was related. This "explicit" meaning is, strictly speaking, extremely *implicit,* in the sense that the brain processes required to understand the events in the story, given only the black marks on paper, are incredibly complex. Nevertheless, we shall consider the events in the story to be the explicit meaning of the Dialogue, and assume that every reader of English uses more or less the same "isomorphism" in sucking that meaning from the marks on the paper.

Even so, I'd like to be a little more explicit about the explicit meaning of the story. First I'll talk about the record players and the records. The main point is that there are two levels of meaning for the grooves in the records. Level One is that of music. Now what is "music"—a sequence of vibrations in the air, or a succession of emotional responses in a brain? It is both. But before there can be emotional responses, there have to be vibrations. Now the vibrations get "pulled" out of the grooves by a record player, a relatively straightforward device; in fact you can do it with a pin, just pulling it down the grooves. After this stage, the ear converts the vibrations into firings of auditory neurons in the brain. Then ensue a number of stages in the brain, which gradually transform the linear sequence of vibrations into a complex pattern of interacting emotional responses—far too complex for us to go into here, much though I would like to. Let us therefore content ourselves with thinking of the sounds in the air as the "Level One" meaning of the grooves.

What is the Level Two meaning of the grooves? It is the sequence of vibrations induced in the record player. This meaning can only arise after the Level One meaning has been pulled out of the grooves, since the vibrations in the air cause the vibrations in the phonograph. Therefore, the Level Two meaning depends upon a chain of *two* isomorphisms:

(1) isomorphism between arbitrary groove patterns and air vibrations;
(2) isomorphism between arbitrary air vibrations and phonograph vibrations.

This chain of two isomorphisms is depicted in Figure 20. Notice that isomorphism 1 is the one which gives rise to the Level One meaning. The Level Two meaning is more implicit than the Level One meaning, because it is mediated by the chain of two isomorphisms. It is the Level Two meaning which "backfires", causing the record player to break apart. What is of interest is that the production of the Level One meaning forces the production of the Level Two meaning simultaneously—there is no way to have Level One without Level Two. So it was the implicit meaning of the record which turned back on it, and destroyed it.

Similar comments apply to the goblet. One difference is that the mapping from letters of the alphabet to musical notes is one more level of isomorphism, which we could call "transcription". That is followed by "translation"—conversion of musical notes into musical sounds. Thereafter, the vibrations act back on the goblet just as they did on the escalating series of phonographs.

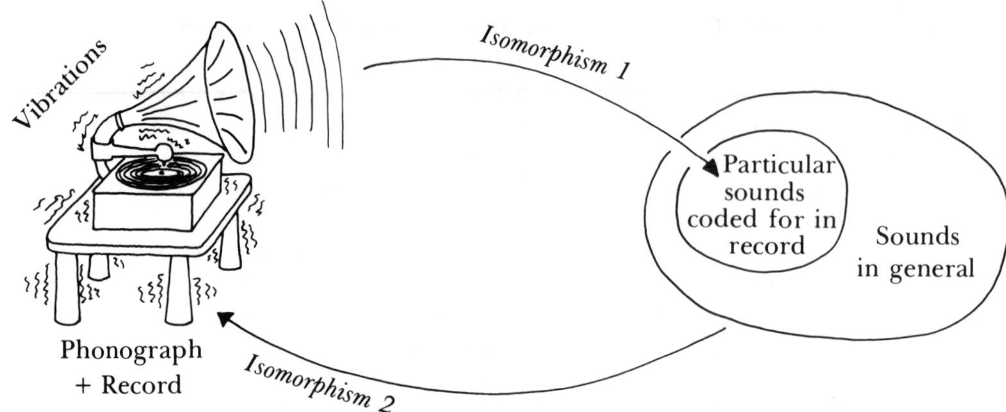

FIGURE 20. Visual rendition of the principle underlying Gödel's Theorem: two back-to-back mappings which have an unexpected boomeranging effect. The first is from groove-patterns to sounds, carried out by a phonograph. The second—familiar, but usually ignored—is from sounds to vibrations of the phonograph. Note that the second mapping exists independently of the first one, for any *sound in the vicinity, not just ones produced by the phonograph itself, will cause such vibrations. The paraphrase of Gödel's Theorem says that for any record player, there are records which it cannot play because they will cause its indirect self-destruction. [Drawing by the author.]*

Implicit Meanings of the *Contracrostipunctus*

What about implicit meanings of the Dialogue? (Yes, it has more than one of these.) The simplest of these has already been pointed out in the paragraphs above—namely, that the events in the two halves of the dialogue are roughly isomorphic to each other: the phonograph becomes a violin, the Tortoise becomes Achilles, the Crab becomes the Tortoise, the grooves become the etched autograph, etc. Once you notice this simple isomorphism, you can go a little further. Observe that in the first half of the story, the Tortoise is the perpetrator of all the mischief, while in the second half, he is the victim. What do you know, but his own method has turned around and backfired on him! Reminiscent of the backfiring of the records' music—or the goblet's inscription—or perhaps of the Tortoise's boomerang collection? Yes, indeed. The story is about backfiring on two levels, as follows . . .

Level One: Goblets and records which backfire;

Level Two: The Tortoise's devilish method of exploiting implicit meaning to cause backfires—which backfires.

Therefore we can even make an isomorphism between the two levels of the story, in which we equate the way in which the records and goblet boomerang back to destroy themselves, with the way in which the Tortoise's own fiendish method boomerangs back to get him in the end. Seen this way, the story itself is an example of the backfirings which it discusses. So we can think of the *Contracrostipunctus* as referring to itself indirectly, in that its own structure is isomorphic to the events it portrays. (Exactly as the goblet and records refer implicitly to themselves via the back-to-back isomorphisms of playing and vibration-causing.) One may read the Dialogue without perceiving this fact, of course—but it is there all the time.

Mapping Between the *Contracrostipunctus* and Gödel's Theorem

Now you may feel a little dizzy—but the best is yet to come. (Actually, some levels of implicit meaning will not even be discussed here—they will be left for you to ferret out.) The deepest reason for writing this Dialogue was to illustrate Gödel's Theorem, which, as I said in the Introduction, relies heavily on two different levels of meaning of statements of number theory. Each of the two halves of the *Contracrostipunctus* is an "isomorphic copy" of Gödel's Theorem. Because this mapping is the central idea of the Dialogue, and is rather elaborate, I have carefully charted it out below.

phonograph	⇐⇒	axiomatic system for number theory
low-fidelity phonograph	⇐⇒	"weak" axiomatic system
high-fidelity phonograph	⇐⇒	"strong" axiomatic system
"Perfect" phonograph	⇐⇒	complete system for number theory
"blueprint" of phonograph	⇐⇒	axioms and rules of formal system
record	⇐⇒	string of the formal system
playable record	⇐⇒	theorem of the axiomatic system
unplayable record	⇐⇒	nontheorem of the axiomatic system
sound	⇐⇒	true statement of number theory
reproducible sound	⇐⇒	interpreted theorem of the system
unreproducible sound	⇐⇒	true statement which isn't a theorem
song title: "I Cannot Be Played on Record Player X"	⇐⇒	implicit meaning of Gödel's string: "I Cannot Be Derived in Formal System X"

This is not the full extent of the isomorphism between Gödel's Theorem and the *Contracrostipunctus*, but it is the core of it. You need not worry if you don't fully grasp Gödel's Theorem by now—there are still a few Chapters to go before we reach it! Nevertheless, having read this Dialogue, you have already tasted some of the flavor of Gödel's Theorem without necessarily being aware of it. I now leave you to look for any other types of implicit meaning in the *Contracrostipunctus*. "Quaerendo invenietis!"

The Art of the Fugue

A few words on the *Art of the Fugue* . . . Composed in the last year of Bach's life, it is a collection of eighteen fugues all based on one theme. Apparently, writing the *Musical Offering* was an inspiration to Bach. He decided to compose another set of fugues on a much simpler theme, to demonstrate the full range of possibilites inherent in the form. In the *Art of the Fugue,* Bach uses a very simple theme in the most complex possible ways. The whole work is in a single key. Most of the fugues have four voices, and they gradually increase in complexity and depth of expression. Toward the end, they soar to such heights of intricacy that one suspects he can no longer maintain them. Yet he does . . . until the last *Contrapunctus.*

The circumstances which caused the break-off of the *Art of the Fugue* (which is to say, of Bach's life) are these: his eyesight having troubled him for years, Bach wished to have an operation. It was done; however, it came out quite poorly, and as a consequence, he lost his sight for the better part of the last year of his life. This did not keep him from vigorous work on his monumental project, however. His aim was to construct a complete exposition of fugal writing, and usage of multiple themes was one important facet of it. In what he planned as the next-to-last fugue, he inserted his own name coded into notes as the third theme. However, upon this very act, his health became so precarious that he was forced to abandon work on his cherished project. In his illness, he managed to dictate to his son-in-law a final chorale prelude, of which Bach's biographer Forkel wrote, "The expression of pious resignation and devotion in it has always affected me whenever I have played it; so that I can hardly say which I would rather miss—this Chorale, or the end of the last fugue."

One day, without warning, Bach regained his vision. But a few hours later, he suffered a stroke; and ten days later, he died, leaving it for others to speculate on the incompleteness of the *Art of the Fugue.* Could it have been caused by Bach's attainment of self-reference?

Problems Caused by Gödel's Result

The Tortoise says that no sufficiently powerful record player can be perfect, in the sense of being able to reproduce every possible sound from a record. Gödel says that no sufficiently powerful formal system can be perfect, in the sense of reproducing every single true statement as a theorem. But as the Tortoise pointed out with respect to phonographs, this fact only seems like a defect if you have unrealistic expectations of what formal systems should be able to do. Nevertheless, mathematicians began this century with just such unrealistic expectations, thinking that axiomatic reasoning was the cure to all ills. They found out otherwise in 1931. The fact that truth transcends theoremhood, in any given formal system, is called "incompleteness" of that system.

A most puzzling fact about Gödel's method of proof is that he uses reasoning methods which seemingly cannot be "encapsulated"—they resist being incorporated into any formal system. Thus, at first sight, it seems that Gödel has unearthed a hitherto unknown, but deeply significant, difference between human reasoning and mechanical reasoning. This mysterious discrepancy in the power of living and nonliving systems is mirrored in the discrepancy between the notion of truth, and that of theoremhood . . . or at least that is a "romantic" way to view the situation.

The Modified pq-System and Inconsistency

In order to see the situation more realistically, it is necessary to see in more depth why and how meaning is mediated, in formal systems, by isomorphisms. And I believe that this leads to a more romantic way to view the situation. So we now will proceed to investigate some further aspects of the relation between meaning and form. Our first step is to make a new formal system by modifying our old friend, the pq-system, very slightly. We add one more axiom schema (retaining the original one, as well as the single rule of inference):

AXIOM SCHEMA II: If *x* is a hyphen-string, then *x*p–q*x* is an axiom.

Clearly, then, --p-q-- is a theorem in the new system, and so is --p--q---. And yet, their interpretations are, respectively, "2 plus 1 equals 2", and "2 plus 2 equals 3". It can be seen that our new system will contain a lot of false statements (if you consider strings to be statements). Thus, our new system is *inconsistent with the external world.*

As if this weren't bad enough, we also have *internal* problems with our new system, since it contains statements which disagree with one another, such as -p-q-- (an old axiom) and -p-q- (a new axiom). So our system is inconsistent in a second sense: internally.

Would, therefore, the only reasonable thing to do at this point be to drop the new system entirely? Hardly. I have deliberately presented these "inconsistencies" in a wool-pulling manner: that is, I have tried to present fuzzy-headed arguments as strongly as possible, with the purpose of misleading. In fact, you may well have detected the fallacies in what I have said. The crucial fallacy came when I unquestioningly adopted the very same interpreting words for the new system as I had for the old one. Remember that there was only one reason for adopting those words in the last Chapter, and that reason was that *the symbols acted isomorphically to the concepts* which they were matched with, by the interpretation. But when you modify the rules governing the system, you are bound to damage the isomorphism. It just cannot be helped. Thus all the problems which were lamented over in preceding paragraphs were bogus problems; they can be made to vanish in no time, by *suitably reinterpreting some of the symbols of the system.* Notice that I said "some"; not necessarily all symbols will have to be mapped onto new notions. Some may very well retain their "meanings", while others change.

Regaining Consistency

Suppose, for instance, that we reinterpret just the symbol q, leaving all the others constant; in particular, interpret q by the phrase "is greater than or equal to". Now, our "contradictory" theorems -p-q- and -p-q-- come out harmlessly as: "1 plus 1 is greater than or equal to 1", and "1 plus 1 is greater than or equal to 2". We have simultaneously gotten rid of (1) the inconsistency with the external world, and (2) the internal inconsistency. And our new interpretation is a *meaningful* interpretation; of course the original one is *meaningless*. That is, it is meaningless *for the new system;* for the original pq-system, it is fine. But it now seems as pointless and arbitrary to apply it to the new pq-system as it was to apply the "horse-apple-happy" interpretation to the old pq-system.

The History of Euclidean Geometry

Although I have tried to catch you off guard and surprise you a little, this lesson about how to interpret symbols by words may not seem terribly difficult once you have the hang of it. In fact, it is not. And yet it is one of the deepest lessons of all of nineteenth century mathematics! It all begins with Euclid, who, around 300 B.C., compiled and systematized all of what was known about plane and solid geometry in his day. The resulting work, Euclid's *Elements,* was so solid that it was virtually a bible of geometry for over two thousand years—one of the most enduring works of all time. Why was this so?

The principal reason was that Euclid was the founder of *rigor* in mathematics. The *Elements* began with very simple concepts, definitions, and so forth, and gradually built up a vast body of results organized in such a way that any given result depended only on foregoing results. Thus, there was a definite plan to the work, an architecture which made it strong and sturdy.

Nevertheless, the architecture was of a different type from that of, say, a skyscraper. (See Fig. 21.) In the latter, that it is standing is proof enough that its structural elements are holding it up. But in a book on geometry, when each proposition is claimed to follow logically from earlier propositions, there will be no visible crash if one of the proofs is invalid. The girders and struts are not physical, but abstract. In fact, in Euclid's *Elements,* the stuff out of which proofs were constructed was human language—that elusive, tricky medium of communication with so many hidden pitfalls. What, then, of the architectural strength of the *Elements*? Is it certain that it is held up by solid structural elements, or could it have structural weaknesses?

Every word which we use has a meaning to us, which guides us in our use of it. The more common the word, the more associations we have with it, and the more deeply rooted is its meaning. Therefore, when someone gives a definition for a common word in the hopes that we will abide by that

FIGURE 21. Tower of Babel, *by M. C. Escher (woodcut, 1928).*

definition, it is a foregone conclusion that we will not do so but will instead be guided, largely unconsciously, by what our minds find in their associative stores. I mention this because it is the sort of problem which Euclid created in his *Elements,* by attempting to give definitions of ordinary, common words such as "point", "straight line", "circle", and so forth. How can you define something of which everyone already has a clear concept? The only way is if you can make it clear that your word is supposed to be a technical term, and is not to be confused with the everyday word with the same spelling. You have to stress that the connection with the everyday word is only suggestive. Well, Euclid did not do this, because he felt that the points and lines of his *Elements* were indeed *the* points and lines of the real world. So by not making sure that all associations were dispelled, Euclid was inviting readers to let their powers of association run free . . .

This sounds almost anarchic, and is a little unfair to Euclid. He did set down axioms, or postulates, which were supposed to be used in the proofs of propositions. In fact, nothing other than those axioms and postulates was supposed to be used. But this is where he slipped up, for an inevitable consequence of his using ordinary words was that some of the images conjured up by those words crept into the proofs which he created. However, if you read proofs in the *Elements,* do not by any means expect to find glaring "jumps" in the reasoning. On the contrary, they are very subtle, for Euclid was a penetrating thinker, and would not have made any simple-minded errors. Nonetheless, gaps are there, creating slight imperfections in a classic work. But this is not to be complained about. One should merely gain an appreciation for the difference between absolute rigor and relative rigor. In the long run, Euclid's lack of absolute rigor was the cause of some of the most fertile path-breaking in mathematics, over two thousand years after he wrote his work.

Euclid gave five postulates to be used as the "ground story" of the infinite skyscraper of geometry, of which his *Elements* constituted only the first several hundred stories. The first four postulates are rather terse and elegant:

(1) A straight line segment can be drawn joining any two points.

(2) Any straight line segment can be extended indefinitely in a straight line.

(3) Given any straight line segment, a circle can be drawn having the segment as radius and one end point as center.

(4) All right angles are congruent.

The fifth, however, did not share their grace:

(5) If two lines are drawn which intersect a third in such a way that the sum of the inner angles on one side is less than two right angles, then the two lines inevitably must intersect each other on that side if extended far enough.

Though he never explicitly said so, Euclid considered this postulate to be somehow inferior to the others, since he managed to avoid using it in the proofs of the first twenty-eight propositions. Thus, the first twenty-eight propositions belong to what might be called "four-postulate geometry"—that part of geometry which can be derived on the basis of the first four postulates of the *Elements,* without the help of the fifth postulate. (It is also often called *absolute geometry.*) Certainly Euclid would have found it far preferable to *prove* this ugly duckling, rather than to have to *assume* it. But he found no proof, and therefore adopted it.

But the disciples of Euclid were no happier about having to assume this fifth postulate. Over the centuries, untold numbers of people gave untold years of their lives in attempting to prove that the fifth postulate was itself part of four-postulate geometry. By 1763, at least twenty-eight different proofs had been published—all erroneous! (They were all criticized in the dissertation of one G. S. Klügel.) All of these erroneous proofs involved a confusion between everyday intuition and strictly formal properties. It is safe to say that today, hardly any of these "proofs" holds any mathematical or historical interest—but there are certain exceptions.

The Many Faces of Noneuclid

Girolamo Saccheri (1667-1733) lived around Bach's time. He had the ambition to free Euclid of every flaw. Based on some earlier work he had done in logic, he decided to try a novel approach to the proof of the famous fifth: suppose you *assume its opposite;* then work with *that* as your fifth postulate . . . Surely after a while you will create a contradiction. Since no mathematical system can support a contradiction, you will have shown the unsoundness of your own fifth postulate, and therefore the soundness of Euclid's fifth postulate. We need not go into details here. Suffice it to say that with great skill, Saccheri worked out proposition after proposition of "Saccherian geometry" and eventually became tired of it. At one point, he decided he had reached a proposition which was "repugnant to the nature of the straight line". That was what he had been hoping for—to his mind, it was the long-sought contradiction. At that point, he published his work under the title *Euclid Freed of Every Flaw,* and then expired.

But in so doing, he robbed himself of much posthumous glory, since he had unwittingly discovered what came later to be known as "hyperbolic geometry". Fifty years after Saccheri, J. H. Lambert repeated the "near miss", this time coming even closer, if possible. Finally, forty years after Lambert, and ninety years after Saccheri, *non-Euclidean geometry* was recognized for what it was—an authentic new brand of geometry, a bifurcation in the hitherto single stream of mathematics. In 1823, non-Euclidean geometry was discovered simultaneously, in one of those inexplicable coincidences, by a Hungarian mathematician, János (or Johann) Bolyai, aged twenty-one, and a Russian mathematician, Nikolay Lobachevskiy, aged thirty. And, ironically, in that same year, the great French mathematician

Adrien-Marie Legendre came up with what he was sure was a proof of Euclid's fifth postulate, very much along the lines of Saccheri.

Incidentally, Bolyai's father, Farkas (or Wolfgang) Bolyai, a close friend of the great Gauss, invested much effort in trying to prove Euclid's fifth postulate. In a letter to his son János, he tried to dissuade him from thinking about such matters:

> You must not attempt this approach to parallels. I know this way to its very end. I have traversed this bottomless night, which extinguished all light and joy of my life. I entreat you, leave the science of parallels alone. . . . I thought I would sacrifice myself for the sake of the truth. I was ready to become a martyr who would remove the flaw from geometry and return it purified to mankind. I accomplished monstrous, enormous labors; my creations are far better than those of others and yet I have not achieved complete satisfaction. For here it is true that *si paullum a summo discessit, vergit ad imum.* I turned back when I saw that no man can reach the bottom of this night. I turned back unconsoled, pitying myself and all mankind. . . . I have traveled past all reefs of this infernal Dead Sea and have always come back with broken mast and torn sail. The ruin of my disposition and my fall date back to this time. I thoughtlessly risked my life and happiness—*aut Caesar aut nihil.*[1]

But later, when convinced his son really "had something", he urged him to publish it, anticipating correctly the simultaneity which is so frequent in scientific discovery:

> When the time is ripe for certain things, these things appear in different places in the manner of violets coming to light in early spring.[2]

How true this was in the case of non-Euclidean geometry! In Germany, Gauss himself and a few others had more or less independently hit upon non-Euclidean ideas. These included a lawyer, F. K. Schweikart, who in 1818 sent a page describing a new "astral" geometry to Gauss; Schweikart's nephew, F. A. Taurinus, who did non-Euclidean trigonometry; and F. L. Wachter, a student of Gauss, who died in 1817, aged twenty-five, having found several deep results in non-Euclidean geometry.

The clue to non-Euclidean geometry was "thinking straight" about the propositions which emerge in geometries like Saccheri's and Lambert's. The Saccherian propositions are only "repugnant to the nature of the straight line" if you cannot free yourself of preconceived notions of what "straight line" must mean. If, however, you can divest yourself of those preconceived images, and merely let a "straight line" be something which satisfies the new propositions, then you have achieved a radically new viewpoint.

Undefined Terms

This should begin to sound familiar. In particular, it harks back to the pq-system, and its variant, in which the symbols acquired passive meanings by virtue of their roles in theorems. The symbol **q** is especially interesting, since its "meaning" changed when a new axiom schema was added. In that very same way, one can *let the meanings of "point", "line", and so on be determined by the set of theorems (or propositions) in which they occur.* This was the great realization of the discoverers of non-Euclidean geometry. They found different sorts of non-Euclidean geometries by denying Euclid's fifth postulate in different ways and following out the consequences. Strictly speaking, they (and Saccheri) did not deny the fifth postulate directly, but rather, they denied an equivalent postulate, called the *parallel postulate,* which runs as follows:

> Given any straight line, and a point not on it, there exists one, and only one, straight line which passes through that point and never intersects the first line, no matter how far they are extended.

The second straight line is then said to be parallel to the first. If you assert that *no* such line exists, then you reach *elliptical geometry;* if you assert that *at least two* such lines exist, you reach *hyperbolic geometry.* Incidentally, the reason that such variations are still called "geometries" is that the core element—absolute, or four-postulate, geometry—is embedded in them. It is the presence of this minimal core which makes it sensible to think of them as describing properties of some sort of geometrical space, even if the space is not as intuitive as ordinary space.

Actually, elliptical geometry is easily visualized. All "points", "lines", and so forth are to be parts of the surface of an ordinary sphere. Let us write "POINT" when the technical term is meant, and "point" when the everyday sense is desired. Then, we can say that a POINT consists of a pair of diametrically opposed points of the sphere's surface. A LINE is a great circle on the sphere (a circle which, like the equator, has its center at the center of the sphere). Under these interpretations, the propositions of elliptical geometry, though they contain words like "POINT" and "LINE", speak of the goings-on on a sphere, not a plane. Notice that two LINES always intersect in exactly two antipodal points of the sphere's surface—that is, in exactly one single POINT! And just as two LINES determine a POINT, so two POINTS determine a LINE.

By treating words such as "POINT" and "LINE" as if they had only the meaning instilled in them by the propositions in which they occur, we take a step towards complete formalization of geometry. This semiformal version still uses a lot of words in English with their usual meanings (words such as "the", "if", "and", "join", "have"), although the everyday meaning has been drained out of special words like "POINT" and "LINE", which are consequently called *undefined terms.* Undefined terms, like the p and q of the pq-system, *do* get defined in a sense: *implicitly*—by the totality of all propositions in which they occur, rather than explicitly, in a definition.

One could maintain that a full definition of the undefined terms resides in the postulates alone, since the propositions which follow from them are implicit in the postulates already. This view would say that the postulates are implicit definitions of all the undefined terms, all of the undefined terms being defined in terms of the others.

The Possibility of Multiple Interpretations

A full formalization of geometry would take the drastic step of making *every* term undefined—that is, turning every term into a "meaningless" symbol of a formal system. I put quotes around "meaningless" because, as you know, the symbols automatically pick up passive meanings in accordance with the theorems they occur in. It is another question, though, whether people discover those meanings, for to do so requires finding a set of concepts which can be linked by an isomorphism to the symbols in the formal system. If one begins with the aim of formalizing geometry, presumably one has an *intended* interpretation for each symbol, so that the passive meanings are built into the system. That is what I did for **p** and **q** when I first created the pq-system.

But there may be other passive meanings which are potentially perceptible, which no one has yet noticed. For instance, there were the surprise interpretations of **p** as "equals" and **q** as "taken from", in the original pq-system. Although this is rather a trivial example, it contains the essence of the idea that symbols may have many meaningful interpretations—it is up to the observer to look for them.

We can summarize our observations so far in terms of the word "consistency". We began our discussion by manufacturing what appeared to be an inconsistent formal system—one which was internally inconsistent, as well as inconsistent with the external world. But a moment later we took it all back, when we realized our error: that we had chosen unfortunate interpretations for the symbols. By changing the interpretations, we regained consistency! It now becomes clear that *consistency is not a property of a formal system per se, but depends on the interpretation which is proposed for it.* By the same token, inconsistency is not an intrinsic property of any formal system.

Varieties of Consistency

We have been speaking of "consistency" and "inconsistency" all along, without defining them. We have just relied on good old everyday notions. But now let us say exactly what is meant by *consistency* of a formal system (together with an interpretation): that every theorem, when interpreted, becomes a true statement. And we will say that *inconsistency* occurs when there is at least one false statement among the interpreted theorems.

This definition appears to be talking about inconsistency with the external world—what about *internal* inconsistencies? Presumably, a system would be internally inconsistent if it contained two or more theorems whose interpretations were incompatible with one another, and internally consistent if all interpreted theorems were compatible with one another. Consider, for example, a formal system which has only the following three theorems: **TbZ**, **ZbE**, and **EbT**. If **T** is interpreted as "the Tortoise", **Z** as "Zeno", **E** as "Egbert", and *x* **b** *y* as "*x* beats *y* in chess always", then we have the following interpreted theorems:

The Tortoise always beats Zeno at chess.
Zeno always beats Egbert at chess.
Egbert always beats the Tortoise at chess.

The statements are not incompatible, although they describe a rather bizarre circle of chess players. Hence, under this interpretation, the formal system in which those three strings are theorems is internally consistent, although, in point of fact, none of the three statements is true! Internal consistency does not require all theorems to come out true, but merely that they come out *compatible* with one another.

Now suppose instead that *x* b *y* is to be interpreted as "*x* was invented by *y*". Then we would have:

The Tortoise was invented by Zeno.
Zeno was invented by Egbert.
Egbert was invented by the Tortoise.

In this case, it doesn't matter whether the individual statements are true or false—and perhaps there is no way to know which ones are true, and which are not. What is nevertheless certain is that *not all three can be true at once.* Thus, the interpretation makes the system internally inconsistent. This internal inconsistency depends not on the interpretations of the three capital letters, but only on that of b, and on the fact that the three capitals are cyclically permuted around the occurrences of b. Thus, one can have internal inconsistency without having interpreted *all* of the symbols of the formal system. (In this case it sufficed to interpret a single symbol.) By the time sufficiently many symbols have been given interpretations, it may be clear that there is no way that the rest of them can be interpreted so that all theorems will come out true. But it is not just a question of truth—it is a question of possibility. All three theorems would come out false if the capitals were interpreted as the names of real people—but that is not why we would call the system internally inconsistent; our grounds for doing so would be the circularity, combined with the interpretation of the letter b. (By the way, you'll find more on this "authorship triangle" in Chapter XX.)

Hypothetical Worlds and Consistency

We have given two ways of looking at consistency: the first says that a system-plus-interpretation is *consistent with the external world* if every theorem comes out *true* when interpreted; the second says that a system-plus-interpretation is *internally consistent* if all theorems come out *mutually compatible* when interpreted. Now there is a close relationship between these two types of consistency. In order to determine whether several statements are mutually compatible, you try to imagine a world in which all of them could be simultaneously true. Therefore, internal consistency depends upon consistency with the external world—only now, "the external world" is allowed to be *any imaginable world,* instead of the one we live in. But this is an extremely vague, unsatisfactory conclusion. What constitutes an "imaginable" world? After all, it is possible to imagine a world in which three characters invent each other cyclically. Or is it? Is it possible to imagine a world in which there are square circles? Is a world imaginable in which Newton's laws, and not relativity, hold? Is it possible to imagine a world in which something can be simultaneously green and not green? Or a world in which animals exist which are not made of cells? In which Bach improvised an eight-part fugue on a theme of King Frederick the Great? In which mosquitoes are more intelligent than people? In which tortoises can play football—or talk? A tortoise talking football would be an anomaly, of course.

Some of these worlds seem more imaginable than others, since some seem to embody *logical* contradictions—for example, green and not green—while some of them seem, for want of a better word, "plausible"—such as Bach improvising an eight-part fugue, or animals which are not made of cells. Or even, come to think of it, a world in which the laws of physics are different . . . Roughly, then, it should be possible to establish different brands of consistency. For instance, the most lenient would be "logical consistency", putting no restraints on things at all, except those of logic. More specifically, a system-plus-interpretation would be *logically consistent* just as long as no two of its theorems, when interpreted as statements, directly contradict each other; and *mathematically consistent* just as long as interpreted theorems do not violate mathematics; and *physically consistent* just as long as all its interpreted theorems are compatible with physical law; then comes *biological consistency,* and so on. In a biologically consistent system, there could be a theorem whose interpretation is the statement "Shakespeare wrote an opera", but no theorem whose interpretation is the statement "Cell-less animals exist". Generally speaking, these fancier kinds of inconsistency are not studied, for the reason that they are very hard to disentangle from one another. What kind of inconsistency, for example, should one say is involved in the problem of the three characters who invent each other cyclically? Logical? Physical? Biological? Literary?

Usually, the borderline between uninteresting and interesting is drawn between physical consistency and mathematical consistency. (Of course, it is the mathematicians and logicians who do the drawing—hardly an impartial crew . . .) This means that the kinds of inconsistency which "count", for formal systems, are just the logical and mathematical kinds. According to this convention, then, we haven't yet found an interpretation which makes the trio of theorems TbZ, ZbE, EbT inconsistent. We can do so by interpreting b as "is bigger than". What about T and Z and E? They can be interpreted as natural numbers—for example, Z as 0, T as 2, and E as 11. Notice that two theorems come out true this way, one false. If, instead, we had interpreted Z as 3, there would have been two falsehoods and only one truth. But either way, we'd have had inconsistency. In fact, the values assigned to T, Z, and E are irrelevant, as long as it is understood that they are restricted to natural numbers. Once again we see a case where only *some* of the interpretation is needed, in order to recognize internal inconsistency.

Embedding of One Formal System in Another

The preceding example, in which some symbols could have interpretations while others didn't, is reminiscent of doing geometry in natural language, using some words as undefined terms. In such a case, words are divided into two classes: those whose meaning is fixed and immutable, and those whose meaning is to be adjusted until the system is consistent (these are the undefined terms). Doing geometry in this way requires that meanings have already been established for words in the first class, somewhere outside of geometry. Those words form a rigid skeleton, giving an underlying structure to the system; filling in that skeleton comes other material, which can vary (Euclidean or non-Euclidean geometry).

Formal systems are often built up in just this type of sequential, or hierarchical, manner. For example, Formal System I may be devised, with rules and axioms that give certain intended passive meanings to its symbols. Then Formal System I is incorporated fully into a larger system with more symbols—Formal System II. Since Formal System I's axioms and rules are part of Formal System II, the passive meanings of Formal System I's symbols remain valid; they form an immutable skeleton which then plays a large role in the determination of the passive meanings of the new symbols of Formal System II. The second system may in turn play the role of a skeleton with respect to a third system, and so on. It is also possible—and geometry is a good example of this—to have a system (e.g., absolute geometry) which *partly* pins down the passive meanings of its undefined terms, and which can be supplemented by extra rules or axioms, which then *further* restrict the passive meanings of the undefined terms. This is the case with Euclidean versus non-Euclidean geometry.

Layers of Stability in Visual Perception

In a similar, hierarchical way, we acquire new knowledge, new vocabulary, or perceive unfamiliar objects. It is particularly interesting in the case of understanding drawings by Escher, such as *Relativity* (Fig. 22), in which there occur blatantly impossible images. You might think that we would seek to reinterpret the picture over and over again until we came to an interpretation of its parts which was free of contradictions—but we don't do that at all. We sit there amused and puzzled by staircases which go every which way, and by people going in inconsistent directions on a single staircase. Those staircases are "islands of certainty" upon which we base our interpretation of the overall picture. Having once identified them, we try to extend our understanding, by seeking to establish the relationship which they bear to one another. At that stage, we encounter trouble. But if we attempted to backtrack—that is, to question the "islands of certainty"—we would also encounter trouble, of another sort. There's no way of backtracking and "undeciding" that they are staircases. They are not fishes, or whips, or hands—they are just staircases. (There is, actually, one other out—to leave all the lines of the picture totally uninterpreted, like the "meaningless

FIGURE 22. Relativity, *by M. C. Escher (lithograph, 1953).*

symbols" of a formal system. This ultimate escape route is an example of a "U-mode" response—a Zen attitude towards symbolism.)

So we are forced, by the hierarchical nature of our perceptive processes, to see either a crazy world or just a bunch of pointless lines. A similar analysis could be made of dozens of Escher pictures, which rely heavily upon the recognition of certain basic forms, which are then put together in nonstandard ways; and by the time the observer sees the paradox on a high level, it is too late—he can't go back and change his mind about how to interpret the lower-level objects. The difference between an Escher drawing and non-Euclidean geometry is that in the latter, comprehensible interpretations can be found for the undefined terms, resulting in a comprehensible total system, whereas for the former, the end result is not reconcilable with one's conception of the world, no matter how long one stares at the pictures. Of course, one can still manufacture hypothetical worlds, in which Escherian events can happen . . . but in such worlds, the laws of biology, physics, mathematics, or even logic will be violated on one level, while simultaneously being obeyed on another, which makes them extremely weird worlds. (An example of this is in *Waterfall* (Fig. 5), where normal gravitation applies to the moving water, but where the nature of space violates the laws of physics.)

Is Mathematics the Same in Every Conceivable World?

We have stressed the fact, above, that *internal* consistency of a formal system (together with an interpretation) requires that there be some *imaginable* world—that is, a world whose only restriction is that in it, mathematics and logic should be the same as in our world—in which all the interpreted theorems come out true. *External* consistency, however—consistency with the external world—requires that all theorems come out true in the *real* world. Now in the special case where one wishes to create a consistent formal system whose theorems are to be interpreted as statements of mathematics, it would seem that the difference between the two types of consistency should fade away, since, according to what we said above, *all imaginable worlds have the same mathematics as the real world.* Thus, in every conceivable world, 1 plus 1 would have to be 2; likewise, there would have to be infinitely many prime numbers; furthermore, in every conceivable world, all right angles would have to be congruent; and of course, through any point not on a given line there would have to be exactly one parallel line . . .

But wait a minute! That's the parallel postulate—and to assert its universality would be a mistake, in light of what's just been said. If in all conceivable worlds the parallel postulate is obeyed, then we are asserting that non-Euclidean geometry is inconceivable, which puts us back in the same mental state as Saccheri and Lambert—surely an unwise move. *But what, then, if not all of mathematics, must all conceivable worlds share?* Could it be as little as logic itself? Or is even logic suspect? Could there be worlds where contradictions are normal parts of existence—worlds where contradictions are not contradictions?

Well, in some sense, by merely inventing the concept, we have shown that such worlds are indeed conceivable; but in a deeper sense, they are also quite inconceivable. (This in itself is a little contradiction.) Quite seriously, however, it seems that if we want to be able to communicate at all, we have to adopt some common base, and it pretty well has to include logic. (There are belief systems which reject this point of view—it is too logical. In particular, Zen embraces contradictions and non-contradictions with equal eagerness. This may seem inconsistent, but then being inconsistent is part of Zen, and so . . . what can one say?)

Is Number Theory the Same in All Conceivable Worlds?

If we assume that *logic* is part of every conceivable world (and note that we have not defined logic, but we will in Chapters to come), is that all? Is it really conceivable that, in some worlds, there are not infinitely many primes? Would it not seem necessary that numbers should obey the same laws in all conceivable worlds? Or . . . is the concept "natural number" better thought of as an undefined term, like "POINT" or "LINE"? In that case, number theory would be a bifurcated theory, like geometry: there would be standard and nonstandard number theories. But there would have to be some counterpart to absolute geometry: a "core" theory, an invariant ingredient of all number theories which identified them as number theories rather than, say, theories about cocoa or rubber or bananas. It seems to be the consensus of most modern mathematicians and philosophers that there *is* such a core number theory, which ought to be included, along with logic, in what we consider to be "conceivable worlds". This core of number theory, the counterpart to absolute geometry—is called *Peano arithmetic,* and we shall formalize it in Chapter VIII. Also, it is now well established—as a matter of fact as a direct consequence of Gödel's Theorem—that number theory *is* a bifurcated theory, with standard and nonstandard versions. Unlike the situation in geometry, however, the number of "brands" of number theory is infinite, which makes the situation of number theory considerably more complex.

For *practical* purposes, all number theories are the same. In other words, if bridge building depended on number theory (which in a sense it does), the fact that there are different number theories would not matter, since in the aspects relevant to the real world, all number theories overlap. The same cannot be said of different geometries; for example, the sum of the angles in a triangle is 180 degrees only in Euclidean geometry; it is greater in elliptic geometry, less in hyperbolic. There is a story that Gauss once attempted to measure the sum of the angles in a large triangle defined by three mountain peaks, in order to determine, once and for all, which kind of geometry really rules our universe. It was a hundred years later that Einstein gave a theory (general relativity) which said that the geometry of the universe is determined by its content of matter, so that no one geometry is intrinsic to space itself. Thus to the question, *"Which geometry is true?"* nature gives an ambiguous answer not only in mathematics, but also in physics. As for the corresponding question, *"Which number theory is true?"*, we shall have more to say on it after going through Gödel's Theorem in detail.

Completeness

If consistency is the minimal condition under which symbols acquire passive meanings, then its complementary notion, *completeness,* is the maximal confirmation of those passive meanings. Where consistency is the property that "Everything produced by the system is true", completeness is the other way round: "Every true statement is produced by the system". Now to refine the notion slightly. We can't mean every true statement in the world—we mean only those which belong to the domain which we are attempting to represent in the system. Therefore, completeness means: "Every true statement which can be expressed in the notation of the system is a theorem."

Consistency: when every theorem, upon interpretation, comes out true (in some imaginable world).

Completeness: when all statements which are true (in some imaginable world), and which can be expressed as well-formed strings of the system, are theorems.

An example of a formal system which is complete on its own modest level is the original pq-system, with the original interpretation. All true additions of two positive integers are represented by theorems of the system. We might say this another way: "All true additions of two positive integers are *provable* within the system." (Warning: When we start using the term "provable statements" instead of "theorems", it shows that we are beginning to blur the distinction between formal systems and their interpretations. This is all right, provided we are very conscious of the blurring that is taking place, and provided that we remember that multiple interpretations are sometimes possible.) The pq-system with the original interpretation is *complete;* it is also *consistent,* since no false statement is—to use our new phrase—provable within the system.

Someone might argue that the system is incomplete, on the grounds that additions of *three* positive integers (such as 2 + 3 + 4 =9) are not represented by theorems of the pq-system, despite being translatable into the notation of the system (e.g., --p---p----q---------). However, this string is not well-formed, and hence should be considered to be just as devoid of meaning as is pqp---qpq. Triple additions are simply *not expressible* in the notation of the system—so the completeness of the system is preserved.

Despite the completeness of the pq-system under this interpretation, it certainly falls far short of capturing the full notion of truth in number theory. For example, there is no way that the pq-system tells us how many prime numbers there are. Gödel's Incompleteness Theorem says that any system which is "sufficiently powerful" is, by virtue of its power, incomplete, in the sense that there are well-formed strings which express true statements of number theory, but which are not theorems. (There are truths belonging to number theory which are not provable within the system.) Systems like the pq-system, which are complete but not very powerful, are more like low-fidelity phonographs; they are so poor to begin with that it is obvious that they cannot do what we would wish them to do—namely tell us everything about number theory.

How an Interpretation May Make or Break Completeness

What does it mean to say, as I did above, that "completeness is the maximal confirmation of passive meanings"? It means that if a system is consistent but incomplete, there is a mismatch between the symbols and their interpretations. The system does not have the power to justify being interpreted that way. Sometimes, if the interpretations are "trimmed" a little, the system can become complete. To illustrate this idea, let's look at the modified pq-system (including Axiom Schema II) and the interpretation we used for it.

After modifying the pq-system, we modified the interpretation for q from "equals" to "is greater than or equal to". We saw that the modified pq-system was consistent under this interpretation; yet something about the new interpretation is not very satisfying. The problem is simple: there are now many expressible truths which are not theorems. For instance, "2 plus 3 is greater than or equal to 1" is expressed by the nontheorem --p---q-. The interpretation is just too sloppy! It doesn't accurately reflect what the theorems in the system do. Under this sloppy interpretation, the pq-system is not complete. We could repair the situation either by (1) *adding new rules* to the system, making it more powerful, or by (2) *tightening up the interpretation.* In this case, the sensible alternative seems to be to tighten the interpretation. Instead of interpreting q as "is greater than or equal to", we should say "equals or exceeds by 1". Now the modified pq-system becomes both consistent and complete. And the completeness confirms the appropriateness of the interpretation.

Incompleteness of Formalized Number Theory

In number theory, we will encounter incompleteness again; but there, to remedy the situation, we will be pulled in the other direction—towards adding new rules, to make the system more powerful. The irony is that we think, each time we add a new rule, that we surely have made the system complete *now!* The nature of the dilemma can be illustrated by the following allegory . . .

We have a record player, and we also have a record tentatively labeled "Canon on B-A-C-H". However, when we play the record on the record player, the feedback-induced vibrations (as caused by the Tortoise's records) interfere so much that we do not even recognize the tune. We conclude that *something* is defective—either our record, or our record player. In order to test our *record,* we would have to play it on friends' record players, and listen to its quality. In order to test our *phonograph,* we would have to play friends' records on it, and see if the music we hear agrees with the labels. If our record player passes its test, then we will say the record was defective; contrariwise, if the record passes *its* test, then we will say our record player was defective. What, however, can we conclude when we find out that *both* pass their respective tests? That is the moment to remember the chain of two isomorphisms (Fig. 20), and think carefully!

Little Harmonic Labyrinth

The Tortoise and Achilles are spending a day at Coney Island. After buying a couple of cotton candies, they decide to take a ride on the Ferris wheel.

Tortoise: This is my favorite ride. One seems to move so far, and yet in reality one gets nowhere.

Achilles: I can see why it would appeal to you. Are you all strapped in?

Tortoise: Yes, I think I've got this buckle done. Well, here we go. Whee!

Achilles: You certainly are exuberant today.

Tortoise: I have good reason to be. My aunt, who is a fortune-teller, told me that a stroke of Good Fortune would befall me today. So I am tingling with anticipation.

Achilles: Don't tell me you believe in fortune-telling!

Tortoise: No . . . but they say it works even if you don't believe in it.

Achilles: Well, that's fortunate indeed.

Tortoise: Ah, what a view of the beach, the crowd, the ocean, the city . . .

Achilles: Yes, it certainly is splendid. Say, look at that helicopter over there. It seems to be flying our way. In fact it's almost directly above us now.

Tortoise: Strange—there's a cable dangling down from it, which is coming very close to us. It's coming so close we could practically grab it.

Achilles: Look! At the end of the line there's a giant hook, with a note.

(He reaches out and snatches the note. They pass by and are on their way down.)

Tortoise: Can you make out what the note says?

Achilles: Yes—it reads, "Howdy, friends. Grab a hold of the hook next time around, for an Unexpected Surprise."

Tortoise: The note's a little corny but who knows where it might lead. Perhaps it's got something to do with that bit of Good Fortune due me. By all means, let's try it!

Achilles: Let's!

(On the trip up they unbuckle their buckles, and at the crest of the ride, they grab for the giant hook. All of a sudden they are whooshed up by the cable, which quickly reels them skyward into the hovering helicopter. A large strong hand helps them in.)

Voice: Welcome aboard—Suckers.

Achilles: Wh—who are you?

Voice: Allow me to introduce myself. I am Hexachlorophene J. Goodfortune, Kidnapper-At-Large, and Devourer of Tortoises par Excellence, at your service.

Tortoise: Gulp!

Achilles (whispering to his friend): Uh-oh—I think that this "Goodfortune" is not exactly what we'd anticipated. *(To Goodfortune)* Ah—if I may be so bold—where are you spiriting us off to?

Goodfortune: Ho ho! To my all-electric kitchen-in-the-sky, where I will prepare THIS tasty morsel—*(leering at the Tortoise as he says this)*—in a delicious pie-in-the-sky! And make no mistake—it's all just for my gobbling pleasure! Ho ho ho!

Achilles: All I can say is you've got a pretty fiendish laugh.

Goodfortune (laughing fiendishly): Ho ho ho! For that remark, my friend, you will pay dearly. Ho ho!

Achilles: Good grief—I wonder what he means by that!

Goodfortune: Very simple—I've got a Sinister Fate in store for both of you! Just you wait! Ho ho ho! Ho ho ho!

Achilles: Yikes!

Goodfortune: Well, we have arrived. Disembark, my friends, into my fabulous all-electric kitchen-in-the-sky.

(They walk inside.)

Let me show you around, before I prepare your fates. Here is my bedroom. Here is my study. Please wait here for me for a moment. I've got to go sharpen my knives. While you're waiting, help yourselves to some popcorn. Ho ho ho! Tortoise pie! Tortoise pie! My favorite kind of pie! *(Exit.)*

Achilles: Oh, boy—popcorn! I'm going to munch my head off!

Tortoise: Achilles! You just stuffed yourself with cotton candy! Besides, how can you think about food at a time like this?

Achilles: Good gravy—oh, pardon me—I shouldn't use that turn of phrase, should I? I mean in these dire circumstances . . .

Tortoise: I'm afraid our goose is cooked.

Achilles: Say—take a gander at all these books old Goodfortune has in his study. Quite a collection of esoterica: *Birdbrains I Have Known; Chess and Umbrella-Twirling Made Easy; Concerto for Tapdancer and Orchestra* . . . Hmmm.

Tortoise: What's that small volume lying open over there on the desk, next to the dodecahedron and the open drawing pad?

Achilles: This one? Why, its title is *Provocative Adventures of Achilles and the Tortoise Taking Place in Sundry Spots of the Globe.*

Tortoise: A moderately provocative title.

Achilles: Indeed—and the adventure it's opened to looks provocative. It's called "Djinn and Tonic".

Tortoise: Hmm . . . I wonder why. Shall we try reading it? I could take the Tortoise's part, and you could take that of Achilles.

Achilles: I'm game. Here goes nothing . . .

(They begin reading "Djinn and Tonic".)

(Achilles has invited the Tortoise over to see his collection of prints by his favorite artist, M. C. Escher.)

Tortoise: These are wonderful prints, Achilles.

Achilles: I knew you would enjoy seeing them. Do you have any particular favorite?

Tortoise: One of my favorites is *Convex and Concave,* where two internally consistent worlds, when juxtaposed, make a completely inconsistent composite world. Inconsistent worlds are always fun places to visit, but I wouldn't want to live there.

Achilles: What do you mean, "fun to visit"? Inconsistent worlds don't EXIST, so how can you visit one?

Tortoise: I beg your pardon, but weren't we just agreeing that in this Escher picture, an inconsistent world is portrayed?

Achilles: Yes, but that's just a two-dimensional world—a fictitious world—a picture. You can't visit that world.

Tortoise: I have my ways . . .

Achilles: How could you propel yourself into a flat picture-universe?

Tortoise: By drinking a little glass of PUSHING-POTION. That does the trick.

Achilles: What on earth is pushing-potion?

Tortoise: It's a liquid that comes in small ceramic phials, and which, when drunk by someone looking at a picture, "pushes" him right into the world of that picture. People who aren't aware of the powers of pushing-potion often are pretty surprised by the situations they wind up in.

Achilles: Is there no antidote? Once pushed, is one irretrievably lost?

Tortoise: In certain cases, that's not so bad a fate. But there is, in fact, another potion—well, not a potion, actually, but an elixir—no, not an elixir, but a—a—

Tortoise: He probably means "tonic".

Achilles: Tonic?

Tortoise: That's the word I was looking for! "POPPING-TONIC" is what it's called, and if you remember to carry a bottle of it in your right hand as you swallow the pushing-potion, it too will be pushed into the picture; then, whenever you get a hankering to "pop" back out into real life, you need only take a swallow of popping-tonic, and presto! You're back in the real world, exactly where you were before you pushed yourself in.

Achilles: That sounds very interesting. What would happen if you took some popping-tonic without having previously pushed yourself into a picture?

Tortoise: I don't precisely know, Achilles, but I would be rather wary of horsing around with these strange pushing and popping liquids. Once I had a friend, a Weasel, who did precisely what you suggested—and no one has heard from him since.

Achilles: That's unfortunate. Can you also carry along the bottle of pushing-potion with you?

Tortoise: Oh, certainly. Just hold it in your left hand, and it too will get pushed right along with you into the picture you're looking at.

Achilles: What happens if you then find a picture inside the picture which you have already entered, and take another swig of pushing-potion?

Tortoise: Just what you would expect: you wind up inside that picture-in-a-picture.

Achilles: I suppose that you have to pop twice, then, in order to extricate yourself from the nested pictures, and re-emerge back in real life.

Tortoise: That's right. You have to pop once for each push, since a push takes you down inside a picture, and a pop undoes that.

Achilles: You know, this all sounds pretty fishy to me . . . Are you sure you're not just testing the limits of my gullibility?

Tortoise: I swear! Look—here are two phials, right here in my pocket. *(Reaches into his lapel pocket, and pulls out two rather large unlabeled phials, in one of which one can hear a red liquid sloshing around, and in the other of which one can hear a blue liquid sloshing around.)* If you're willing, we can try them. What do you say?

Achilles: Well, I guess, ahm, maybe, ahm . . .

Tortoise: Good! I knew you'd want to try it out. Shall we push ourselves into the world of Escher's *Convex and Concave*?

Achilles: Well, ah, . . .

Tortoise: Then it's decided. Now we've got to remember to take along this flask of tonic, so that we can pop back out. Do you want to take that heavy responsibility, Achilles?

Achilles: If it's all the same to you, I'm a little nervous, and I'd prefer letting you, with your experience, manage the operation.

Tortoise: Very well, then.

(So saying, the Tortoise pours two small portions of pushing-potion. Then he picks up the flask of tonic and grasps it firmly in his right hand, and both he and Achilles lift their glasses to their lips.)

Tortoise: Bottoms up!

(They swallow.)

FIGURE 23. Convex and Concave, *by M. C. Escher (lithograph, 1955).*

Achilles: That's an exceedingly strange taste.

Tortoise: One gets used to it.

Achilles: Does taking the tonic feel this strange?

Tortoise: Oh, that's quite another sensation. Whenever you taste the tonic, you feel a deep sense of satisfaction, as if you'd been waiting to taste it all your life.

Achilles: Oh, I'm looking forward to that.

Tortoise: Well, Achilles, where are we?

Achilles (taking cognizance of his surroundings): We're in a little gondola, gliding down a canal! I want to get out. Mr.Gondolier, please let us out here.

(The gondolier pays no attention to this request.)

Tortoise: He doesn't speak English. If we want to get out here, we'd better just clamber out quickly before he enters the sinister "Tunnel of Love", just ahead of us.

(Achilles, his face a little pale, scrambles out in a split second and then pulls his slower friend out.)

Achilles: I didn't like the sound of that place, somehow. I'm glad we got out here. Say, how do you know so much about this place, anyway? Have you been here before?

Tortoise: Many times, although I always came in from other Escher pictures. They're all connected behind the frames, you know. Once you're in one, you can get to any other one.

Achilles: Amazing! Were I not here, seeing these things with my own eyes, I'm not sure I'd believe you. *(They wander out through a little arch.)* Oh, look at those two cute lizards!

Tortoise: Cute? They aren't cute—it makes me shudder just to think of them! They are the vicious guardians of that magic copper lamp hanging from the ceiling over there. A mere touch of their tongues, and any mortal turns to a pickle.

Achilles: Dill, or sweet?

Tortoise: Dill.

Achilles: Oh, what a sour fate! But if the lamp has magical powers, I would like to try for it.

Tortoise: It's a foolhardy venture, my friend. I wouldn't risk it.

Achilles: I'm going to try just once.

(He stealthily approaches the lamp, making sure not to awaken the sleeping lad nearby. But suddenly, he slips on a strange shell-like indentation in the floor, and lunges out into space. Lurching crazily, he reaches for anything, and manages somehow to grab onto the lamp with one hand. Swinging wildly, with both lizards hissing and thrusting their tongues violently out at him, he is left dangling helplessly out in the middle of space.)

Achilles: He-e-e-elp!

(His cry attracts the attention of a woman who rushes downstairs and awakens the sleeping boy. He takes stock of the situation, and, with a kindly smile on his face, gestures to Achilles that all will be well. He shouts something in a strange guttural tongue to a pair of trumpeters high up in windows, and immediately,

weird tones begin ringing out and making beats with each other. The sleepy young lad points at the lizards, and Achilles sees that the music is having a strong soporific effect on them. Soon, they are completely unconscious. Then the helpful lad shouts to two companions climbing up ladders. They both pull their ladders up and then extend them out into space just underneath the stranded Achilles, forming a sort of bridge. Their gestures make it clear that Achilles should hurry and climb on. But before he does so, Achilles carefully unlinks the top link of the chain holding the lamp, and detaches the lamp. Then he climbs onto the ladder-bridge and the three young lads pull him in to safety. Achilles throws his arms around them and hugs them gratefully.)

Achilles: Oh, Mr. T, how can I repay them?

Tortoise: I happen to know that these valiant lads just love coffee, and down in the town below, there's a place where they make an incomparable cup of espresso. Invite them for a cup of espresso!

Achilles: That would hit the spot.

(And so, by a rather comical series of gestures, smiles, and words, Achilles manages to convey his invitation to the young lads, and the party of five walks out and down a steep staircase descending into the town. They reach a charming small café, sit down outside, and order five espressos. As they sip their drinks, Achilles remembers he has the lamp with him.)

Achilles: I forgot, Mr. Tortoise—I've got this magic lamp with me! But—what's magic about it?

Tortoise: Oh, you know, just the usual—a genie.

Achilles: What? You mean a genie comes out when you rub it, and grants you wishes?

Tortoise: Right. What did you expect? Pennies from heaven?

Achilles: Well, this is fantastic! I can have any wish I want, eh? I've always wished this would happen to me . . .

(And so Achilles gently rubs the large letter 'L' which is etched on the lamp's copper surface . . . Suddenly a huge puff of smoke appears, and in the forms of the smoke the five friends can make out a weird, ghostly figure towering above them.)

Genie: Hello, my friends—and thanks ever so much for rescuing my Lamp from the evil Lizard-Duo.

(And so saying, the Genie picks up the Lamp, and stuffs it into a pocket concealed among the folds of his long ghostly robe which swirls out of the Lamp.)

As a sign of gratitude for your heroic deed, I would like to offer you, on the part of my Lamp, the opportunity to have any three of your wishes realized.

Achilles: How stupefying! Don't you think so, Mr. T?

Tortoise: I surely do. Go ahead, Achilles, take the first wish.

Achilles: Wow! But what should I wish? Oh, I know! It's what I thought of the first time I read the *Arabian Nights* (that collection of silly (and nested) tales)—I wish that I had a HUNDRED wishes, instead of just three! Pretty clever, eh, Mr. T? I bet YOU never would have thought of that trick. I always wondered why those dopey people in the stories never tried it themselves.

Tortoise: Maybe now you'll find out the answer.

Genie: I am sorry, Achilles, but I don't grant meta-wishes.

Achilles: I wish you'd tell me what a "meta-wish" is!

Genie: But THAT is a meta-meta-wish, Achilles—and I don't grant them, either.

Achilles: Whaaat? I don't follow you at all.

Tortoise: Why don't you rephrase your last request, Achilles?

Achilles: What do you mean? Why should I?

Tortoise: Well, you began by saying "I wish". Since you're just asking for information, why don't you just ask a question?

Achilles: All right, though I don't see why. Tell me, Mr. Genie—what is a meta-wish?

Genie: It is simply a wish about wishes. I am not allowed to grant meta-wishes. It is only within my purview to grant plain ordinary wishes, such as wishing for ten bottles of beer, to have Helen of Troy on a blanket, or to have an all-expenses-paid weekend for two at the Copacabana. You know—simple things like that. But meta-wishes I cannot grant. GOD won't permit me to.

Achilles: GOD? Who is GOD? And why won't he let you grant meta-wishes? That seems like such a puny thing compared to the others you mentioned.

Genie: Well, it's a complicated matter, you see. Why don't you just go ahead and make your three wishes? Or at least make one of them. I don't have all the time in the world, you know . . .

Achilles: Oh, I feel so rotten. I was REALLY HOPING to wish for a hundred wishes . . .

Genie: Gee, I hate to see anybody so disappointed as that. And besides, meta-wishes are my favorite kind of wish. Let me just see if there isn't anything I can do about this. This'll just take one moment—

(The Genie removes from the wispy folds of his robe an object which looks just like the copper Lamp he had put away, except that this one is made of silver; and where the previous one had 'L' etched on it, this one has 'ML' in smaller letters, so as to cover the same area.)

Achilles: And what is that?

Genie: This is my Meta-Lamp . . .

(He rubs the Meta-Lamp, and a huge puff of smoke appears. In the billows of smoke, they can all make out a ghostly form towering above them.)

Meta-Genie: I am the Meta-Genie. You summoned me, O Genie? What is your wish?

Genie: I have a special wish to make of you, O Djinn, and of GOD. I wish for permission for temporary suspension of all type-restrictions on wishes, for the duration of one Typeless Wish. Could you please grant this wish for me?

Meta-Genie: I'll have to send it through Channels, of course. One half a moment, please.

(And, twice as quickly as the Genie did, this Meta-Genie removes from the wispy folds of her robe an object which looks just like the silver Meta-Lamp, except that it is made of gold; and where the previous one had 'ML' etched on it, this one has 'MML' in smaller letters, so as to cover the same area.)

Achilles (his voice an octave higher than before): And what is that?

Meta-Genie: This is my Meta-Meta-Lamp . . .

(She rubs the Meta-Meta-Lamp, and a huge puff of smoke appears. In the billows of smoke, they can all make out a ghostly form towering above them.)

Meta-Meta-Genie: I am the Meta-Meta-Genie. You summoned me, O Meta-Genie? What is your wish?

Meta-Genie: I have a special wish to make of you, O Djinn, and of GOD. I wish for permission for temporary suspension of all type-restrictions on wishes, for the duration of one Typeless Wish. Could you please grant this wish for me?

Meta-Meta-Genie: I'll have to send it through Channels, of course. One quarter of a moment, please.

(And, twice as quickly as the Meta-Genie did, this Meta-Meta-Genie removes from the folds of his robe an object which looks just like the gold Meta-Lamp, except that it is made of . . .)

.

.

.

{GOD}

.

.

.

(. . . swirls back into the Meta-Meta-Meta-Lamp, which the Meta-Meta-Genie then folds back into his robe, half as quickly as the Meta-Meta-Meta-Genie did.)

Your wish is granted, O Meta-Genie.

Meta-Genie: Thank you, O Djinn, and GOD.

(And the Meta-Meta-Genie, as all the higher ones before him, swirls back into the Meta-Meta-Lamp, which the Meta-Genie then folds back into her robe, half as quickly as the Meta-Meta-Genie did.)

Your wish is granted, O Genie.

Genie: Thank you, O Djinn, and GOD.

(And the Meta-Genie, as all the higher ones before her,

swirls back into the Meta-Lamp, which the Genie then folds back into his robe, half as quickly as the Meta-Genie did.)

Your wish is granted, Achilles.

(And one precise moment has elapsed since he said "This will just take one moment.")

Achilles: Thank you, O Djinn, and GOD.

Genie: I am pleased to report, Achilles, that you may have exactly one (1) Typeless Wish—that is to say, a wish, or a meta-wish, or a meta-meta-wish, as many "meta"'s as you wish—even infinitely many (if you wish).

Achilles: Oh, thank you so very much, Genie. But my curiosity is provoked. Before I make my wish, would you mind telling me who—or what—GOD is?

Genie: Not at all. "GOD" is an acronym which stands for "GOD Over Djinn". The word "Djinn" is used to designate Genies, Meta-Genies, Meta-Meta-Genies, etc. It is a Typeless word.

Achilles: But—but—how can "GOD" be a word in its own acronym? That doesn't make any sense!

Genie: Oh, aren't you acquainted with recursive acronyms? I thought everybody knew about them. You see, "GOD" stands for "GOD Over Djinn"—which can be expanded as "GOD Over Djinn, Over Djinn"—and that can, in turn, be expanded to "GOD Over Djinn, Over Djinn, Over Djinn"—which can, in its turn, be further expanded . . . You can go as far as you like.

Achilles: But I'll never finish!

Genie: Of course not. You can never totally expand GOD.

Achilles: Hmm . . . That's puzzling. What did you mean when you said to the Meta-Genie, "I have a special wish to make of you, O Djinn, and of GOD"?

Genie: I wanted not only to make a request of the Meta-Genie, but also of all the Djinns over her. The recursive acronym method accomplishes this quite naturally. You see, when the Meta-Genie received my request, she then had to pass it upwards to her GOD. So she forwarded a similar message to the Meta-Meta-Genie, who then did likewise to the Meta-Meta-Meta-Genie . . . Ascending the chain this way transmits the message to GOD.

Achilles: I see. You mean GOD sits up at the top of the ladder of djinns?

Genie: No, no, no! There is nothing "at the top", for there is no top. That is why GOD is a recursive acronym. GOD is not some ultimate djinn; GOD is the tower of djinns above any given djinn.

Tortoise: It seems to me that each and every djinn would have a different concept of what GOD is, then, since to any djinn, GOD is the set of djinns above him or her, and no two djinns share that set.

Genie: You're absolutely right—and since I am the lowest djinn of all, my notion of GOD is the most exalted one. I pity the higher djinns, who fancy themselves somehow closer to GOD. What blasphemy!

Achilles: By gum, it must have taken genies to invent GOD.

Tortoise: Do you really believe all this stuff about GOD, Achilles?

Achilles: Why certainly, I do. Are you atheistic, Mr. T? Or are you agnostic?

Tortoise: I don't think I'm agnostic. Maybe I'm meta-agnostic.

Achilles: Whaaat? I don't follow you at all.

Tortoise: Let's see . . . If I were meta-agnostic, I'd be confused over whether I'm agnostic or not—but I'm not quite sure if I feel THAT way; hence I must be meta-meta-agnostic (I guess). Oh, well. Tell me, Genie, does any djinn ever make a mistake, and garble up a message moving up or down the chain?

Genie: This does happen; it is the most common cause for Typeless Wishes not being granted. You see, the chances are infinitesimal that a garbling will occur at any PARTICULAR link in the chain—but when you put an infinite number of them in a row, it becomes virtually certain that a garbling will occur SOMEWHERE. In fact, strange as it seems, an infinite number of garblings usually occur, although they are very sparsely distributed in the chain.

Achilles: Then it seems a miracle that any Typeless Wish ever gets carried out.

Genie: Not really. Most garblings are inconsequential, and many garblings tend to cancel each other out. But occasionally—in fact, rather seldom—the nonfulfillment of a Typeless Wish can be traced back to a single unfortunate djinn's garbling. When this happens, the guilty djinn is forced to run an infinite gauntlet, and get paddled on his or her rump, by GOD. It's good fun for the paddlers, and quite harmless for the paddlee. You might be amused by the sight.

Achilles: I would love to see that! But it only happens when a Typeless Wish goes ungranted?

Genie: That's right.

Achilles: Hmm . . . That gives me an idea for my wish.

Tortoise: Oh, really? What is it?

Achilles: I wish my wish would not be granted!

(At that moment, an event—or is "event" the word for it?—takes place which cannot be described, and hence no attempt will be made to describe it.)

Achilles: What on earth does that cryptic comment mean?

Tortoise: It refers to the Typeless Wish Achilles made.

Achilles: But he hadn't yet made it.

Tortoise: Yes, he had. He said, "I wish my wish would not be granted", and the Genie took THAT to be his wish.

(At that moment, some footsteps are heard coming down the hallway in their direction.)

Achilles: Oh, my! That sounds ominous.

(The footsteps stop; then they turn around and fade away.)

Tortoise: Whew!

Achilles: But does the story go on, or is that the end? Turn the page and let's see.

(The Tortoise turns the page of "Djinn and Tonic", where they find that the story goes on . . .)

Achilles: Hey! What happened? Where is my Genie? My lamp? My cup of espresso? What happened to our young friends from the Convex and Concave worlds? What are all those little lizards doing here?

Tortoise: I'm afraid our context got restored incorrectly, Achilles.

Achilles: What on earth does that cryptic comment mean?

Tortoise: I refer to the Typeless Wish you made.

Achilles: But I hadn't yet made it.

Tortoise: Yes, you had. You said, "I wish my wish would not be granted", and the Genie took THAT to be your wish.

Achilles: Oh, my! That sounds ominous.

Tortoise: It spells PARADOX. For that Typeless Wish to be

granted, it had to be denied—yet not to grant it would be to grant it.

Achilles: So what happened? Did the earth come to a standstill? Did the universe cave in?

Tortoise: No. The System crashed.

Achilles: What does that mean?

Tortoise: It means that you and I, Achilles, were suddenly and instantaneously transported to Tumbolia.

Achilles: To where?

Tortoise: Tumbolia: the land of dead hiccups and extinguished light bulbs. It's a sort of waiting room, where dormant software waits for its host hardware to come back up. No telling how long the System was down, and we were in Tumbolia. It could have been moments, hours, days—even years.

Achilles: I don't know what software is, and I don't know what hardware is. But I do know that I didn't get to make my wishes! I want my Genie back!

Tortoise: I'm sorry, Achilles—you blew it. You crashed the System, and you should thank your lucky stars that we're back at all. Things could have come out a lot worse. But I have no idea where we are.

Achilles: I recognize it now—we're inside another of Escher's pictures. This time it's *Reptiles.*

Tortoise: Aha! The System tried to save as much of our context as it could before it crashed, and it got as far as recording that it was an Escher picture with lizards before it went down. That's commendable.

Achilles: And look—isn't that our phial of popping-tonic over there on the table, next to the cycle of lizards?

Tortoise: It certainly is, Achilles. I must say, we are very lucky indeed. The System was very kind to us, in giving us back our popping-tonic—it's precious stuff!

Achilles: I'll say! Now we can pop back out of the Escher world, into my house.

Tortoise: There are a couple of books on the desk, next to the tonic. I wonder what they are. *(He picks up the smaller one, which is open to a random page.)* This looks like a moderately provocative book.

Achilles: Oh, really? What is its title?

Tortoise: *Provocative Adventures of the Tortoise and Achilles Taking Place in Sundry Parts of the Globe.* It sounds like an interesting book to read out of.

FIGURE 24. Reptiles, *by M. C. Escher (lithograph, 1943).*

Achilles: Well, YOU can read it if you want, but as for me, I'm not going to take any chances with that popping-tonic—one of the lizards might knock it off the table, so I'm going to get it right now!

(He dashes over to the table and reaches for the popping-tonic, but in his haste he somehow bumps the flask of tonic, and it tumbles off the desk and begins rolling.)

Oh, no! Mr. T—look! I accidentally knocked the tonic onto the floor, and it's rolling towards—towards—the stairwell! Quick—before it falls!

(The Tortoise, however, is completely wrapped up in the thin volume which he has in his hands.)

Tortoise (muttering): Eh? This story looks fascinating.

Achilles: Mr. T, Mr. T, help! Help catch the tonic-flask!

Tortoise: What's all the fuss about?

Achilles: The tonic-flask—I knocked it down from the desk, and now it's rolling and—

(At that instant it reaches the brink of the stairwell, and plummets over . . .)

Oh no! What can we do? Mr. Tortoise—aren't you alarmed? We're losing our tonic! It's just fallen down the stairwell! There's only one thing to do! We'll have to go down one story!

Tortoise: Go down one story? My pleasure. Won't you join me?

(He begins to read aloud, and Achilles, pulled in two directions at once, finally stays, taking the role of the Tortoise.)

Achilles: It's very dark here, Mr. T. I can't see a thing. Oof! I bumped into a wall. Watch out!

Tortoise: Here—I have a couple of walking sticks. Why don't you take one of them? You can hold it out in front of you so that you don't bang into things.

Achilles: Good idea. *(He takes the stick.)* Do you get the sense that this path is curving gently to the left as we walk?

Tortoise: Very slightly, yes.

Achilles: I wonder where we are. And whether we'll ever see the light of day again. I wish I'd never listened to you, when you suggested I swallow some of that "DRINK ME" stuff.

Tortoise: I assure you, it's quite harmless. I've done it scads of times, and not a once have I ever regretted it. Relax and enjoy being small.

Achilles: Being small? What is it you've done to me, Mr. T?

Tortoise: Now don't go blaming me. You did it of your own free will.

Achilles: Have you made me shrink? So that this labyrinth we're in is actually some teeny thing that someone could STEP on?

FIGURE 25. Cretan Labyrinth (Italian engraving; School of Finiguerra*). [From W. H. Matthews,* Mazes and Labyrinths: Their History and Development *(New York: Dover Publications, 1970).]*

Tortoise: Labyrinth? Labyrinth? Could it be? Are we in the notorious Little Harmonic Labyrinth of the dreaded Majotaur?

Achilles: Yiikes! What is that?

Tortoise: They say—although I personally never believed it myself—that an Evil Majotaur has created a tiny labyrinth and sits in a pit in the middle of it, waiting for innocent victims to get lost in its fearsome complexity. Then, when they wander lost and dazed into the center, he laughs and laughs at them—so hard, that he laughs them to death!

Achilles: Oh, no!

Tortoise: But it's only a myth. Courage, Achilles.

(And the dauntless pair trudge on.)

Achilles: Feel these walls. They're like corrugated tin sheets, or something. But the corrugations have different sizes.

(To emphasize his point, he sticks out his walking stick against the wall surface as he walks. As the stick bounces back and forth against the corrugations, strange noises echo up and down the long curved corridor they are in.)

Tortoise (alarmed): What was THAT?

Achilles: Oh, just me, rubbing my walking stick against the wall.

Tortoise: Whew! I thought for a moment it was the bellowing of the ferocious Majotaur!

Achilles: I thought you said it was all a myth.

Tortoise: Of course it is. Nothing to be afraid of.

(Achilles puts his walking stick back against the wall, and continues walking. As he does so, some musical sounds are heard, coming from the point where his stick is scraping the wall.)

Tortoise: Uh-oh. I have a bad feeling, Achilles. That Labyrinth may not be a myth, after all.

Achilles: Wait a minute. What makes you change your mind all of a sudden?

Tortoise: Do you hear that music?

(To hear more clearly, Achilles lowers the stick, and the strains of melody cease.)

Hey! Put that back! I want to hear the end of this piece!

(Confused, Achilles obeys, and the music resumes.)

Thank you. Now as I was about to say, I have just figured out where we are.

Achilles: Really? Where are we?

Tortoise: We are walking down a spiral groove of a record in its jacket. Your stick scraping against the strange shapes in the wall acts like a needle running down the groove, allowing us to hear the music.

Achilles: Oh, no, oh, no . . .

Tortoise: What? Aren't you overjoyed? Have you ever had the chance to be in such intimate contact with music before?

Achilles: How am I ever going to win footraces against full-sized people when I am smaller than a flea, Mr. Tortoise?

Tortoise: Oh, is that all that's bothering you? That's nothing to fret about, Achilles.

Achilles: The way you talk, I get the impression that you never worry at all.

Tortoise: I don't know. But one thing for certain is that I don't worry about being small. Especially not when faced with the awful danger of the dreaded Majotaur!

Achilles: Horrors! Are you telling me—

Tortoise: I'm afraid so, Achilles. The music gave it away.

Achilles: How could it do that?

Tortoise: Very simple. When I heard the melody B-A-C-H in the top voice, I immediately realized that the grooves that we're walking through could only be the *Little Harmonic Labyrinth,* one of Bach's lesser known organ pieces. It is so named because of its dizzyingly frequent modulations.

Achilles: Wh-what are they?

Tortoise: Well, you know that most musical pieces are written in a key, or tonality, such as C major, which is the key of this one.

Achilles: I had heard the term before. Doesn't that mean that C is the note you want to end on?

Tortoise: Yes, C acts like a home base, in a way. Actually, the usual word is "tonic".

Achilles: Does one then stray away from the tonic with the aim of eventually returning?

Tortoise: That's right. As the piece develops, ambiguous chords and melodies are used, which lead away from the tonic. Little by little, tension builds up—you feel an increasing desire to return home, to hear the tonic.

Achilles: Is that why, at the end of a piece, I always feel so satisfied, as if I had been waiting my whole life to hear the tonic?

Tortoise: Exactly. The composer has used his knowledge of harmonic progressions to manipulate your emotions, and to build up hopes in you to hear that tonic.

Achilles: But you were going to tell me about modulations.

Tortoise: Oh, yes. One very important thing a composer can do is to "modulate" partway through a piece, which means that he sets up a temporary goal other than resolution into the tonic.

Achilles: I see . . . I think. Do you mean that some sequence of chords shifts the harmonic tension somehow so that I actually desire to resolve in a new key?

Tortoise: Right. This makes the situation more complex, for although in the short term you want to resolve in the new key, all the while at the back of your mind you retain the longing to hit that original goal—in this case, C major. And when the subsidiary goal is reached, there is—

Achilles (suddenly gesturing enthusiastically): Oh, listen to the gorgeous upward-swooping chords which mark the end of this *Little Harmonic Labyrinth*!

Tortoise: No, Achilles, this isn't the end. It's merely—

Achilles: Sure it is! Wow! What a powerful, strong ending! What a sense of relief! That's some resolution! Gee!

(And sure enough, at that moment the music stops, as they emerge into an open area with no walls.)

You see, it IS over. What did I tell you?

Tortoise: Something is very wrong. This record is a disgrace to the world of music.

Achilles: What do you mean?

Tortoise: It was exactly what I was telling you about. Here Bach had modulated from C into G, setting up a secondary goal of hearing G. This means that you experience two tensions at once—waiting for resolution into G, but also keeping in mind that ultimate desire—to resolve triumphantly into C Major.

Achilles: Why should you have to keep anything in mind when listening to a piece of music? Is music only an intellectual exercise?

Tortoise: No, of course not. Some music is highly intellectual, but most music is not. And most of the time your ear or brain does the "calculation" for you, and lets your emotions know what they want to hear. You don't have to think about it consciously. But in this piece, Bach was playing tricks, hoping to lead you astray. And in your case, Achilles, he succeeded.

Achilles: Are you telling me that I responded to a resolution in a subsidiary key?

Tortoise: That's right.

Achilles: It still sounded like an ending to me.

Tortoise: Bach intentionally made it sound that way. You just fell into his trap. It was deliberately contrived to sound like an ending, but if you follow the harmonic progression carefully, you will see that it is in the wrong key. Apparently not just you but also this miserable record company fell for the same trick—and they truncated the piece early!

Achilles: What a dirty trick Bach played on me!

Tortoise: That is his whole game—to make you lose your way in his Labyrinth! The Evil Majotaur is in cahoots with Bach, you see. And if you don't watch out, he will now laugh you to death—and perhaps me along with you!

Achilles: Oh, let us hurry up and get out of here! Quick! Let's run backwards in the grooves, and escape on the outside of the record before the Evil Majotaur finds us!

Tortoise: Heavens, no! My sensibility is far too delicate to handle the bizarre chord progressions which occur when time is reversed.

Achilles: Oh, Mr. T, how will we ever get out of here, if we can't just retrace our steps?

Tortoise: That's a very good question.

(A little desperately, Achilles starts running about aimlessly in the dark. Suddenly there is a slight gasp, and then a "thud".)

Achilles—are you all right?

Achilles: Just a bit shaken up but otherwise fine. I fell into some big hole.

Tortoise: You've fallen into the pit of the Evil Majotaur! Here, I'll come help you out. We've got to move fast!

Achilles: Careful, Mr. T—I don't want YOU to fall in here, too . . .

Tortoise: Don't fret, Achilles. Everything will be all—

(Suddenly, there is a slight gasp, and then a "thud".)

Achilles: Mr. T—you fell in, too! Are you all right?

Tortoise: Only my pride is hurt—otherwise I'm fine.

Achilles: Now we're in a pretty pickle, aren't we?

(Suddenly, a giant, booming laugh is heard, alarmingly close to them.)

Tortoise: Watch out, Achilles! This is no laughing matter.

Majotaur: Hee hee hee! Ho ho! Haw haw haw!

Achilles: I'm starting to feel weak, Mr. T . . .

Tortoise: Try to pay no attention to his laugh, Achilles. That's your only hope.

Achilles: I'll do my best. If only my stomach weren't empty!

Tortoise: Say, am I smelling things, or is there a bowl of hot buttered popcorn around here?

Achilles: I smell it, too. Where is it coming from?

Tortoise: Over here, I think. Oh! I just ran into a big bowl of the stuff. Yes, indeed—it seems to be a bowl of popcorn!

Achilles: Oh, boy—popcorn! I'm going to munch my head off!

Tortoise: Let's just hope it isn't pushcorn! Pushcorn and popcorn are so extraordinarily difficult to tell apart.

Achilles: What's this about Pushkin?

Tortoise: I didn't say a thing. You must be hearing things.

Achilles: Go-golly! I hope not. Well, let's dig in!

(And the two friends begin munching the popcorn (or pushcorn?)—and all at once—POP! I guess it was popcorn, after all.)

Tortoise: What an amusing story. Did you enjoy it?

Achilles: Mildly. Only I wonder whether they ever got out of that Evil Majotaur's pit or not. Poor Achilles—he wanted to be full-sized again.

Tortoise: Don't worry—they're out, and he is full-sized again. That's what the "POP" was all about.

Achilles: Oh, I couldn't tell. Well, now I REALLY want to find that bottle of tonic. For some reason, my lips are burning. And nothing would taste better than a drink of popping-tonic.

Tortoise: That stuff is renowned for its thirst quenching powers. Why, in some places people very nearly go crazy over it. At the turn of the century in Vienna, the Schönberg food factory stopped making tonic, and started making cereal instead. You can't imagine the uproar that caused.

Achilles: I have an inkling. But let's go look for the tonic. Hey—just a moment. Those lizards on the desk—do you see anything funny about them?

Tortoise: Umm . . . not particularly. What do you see of such great interest?

Achilles: Don't you see it? They're emerging from that flat picture without drinking any popping-tonic! How are they able to do that?

Tortoise: Oh, didn't I tell you? You can get out of a picture by moving perpendicularly to its plane, if you have no popping-tonic. The little lizards have learned to climb UP when they want to get out of the two-dimensional sketchbook world.

Achilles: Could we do the same thing to get out of this Escher picture we're in?

Tortoise: Of course! We just need to go UP one story. Do you want to try it?

Achilles: Anything to get back to my house! I'm tired of all these provocative adventures.

Tortoise: Follow me, then, up this way.

(And they go up one story.)

Achilles: It's good to be back. But something seems wrong. This isn't my house! This is YOUR house, Mr. Tortoise.

Tortoise: Well, so it is—and am I glad for that! I wasn't looking forward one whit to the long walk back from your house. I am bushed, and doubt if I could have made it.

Achilles: I don't mind walking home, so I guess it's lucky we ended up here, after all.

Tortoise: I'll say! This certainly is a piece of Good Fortune!

CHAPTER V

Recursive Structures and Processes

What Is Recursion?

WHAT IS RECURSION? It is what was illustrated in the Dialogue *Little Harmonic Labyrinth:* nesting, and variations on nesting. The concept is very general. (Stories inside stories, movies inside movies, paintings inside paintings, Russian dolls inside Russian dolls (even parenthetical comments inside parenthetical comments!)—these are just a few of the charms of recursion.) However, you should be aware that the meaning of "recursive" in this Chapter is only faintly related to its meaning in Chapter III. The relation should be clear by the end of this Chapter.

Sometimes recursion seems to brush paradox very closely. For example, there are *recursive definitions.* Such a definition may give the casual viewer the impression that something is being defined in terms of *itself.* That would be circular and lead to infinite regress, if not to paradox proper. Actually, a recursive definition (when properly formulated) never leads to infinite regress or paradox. This is because a recursive definition never defines something in terms of itself, but always in terms of *simpler versions* of itself. What I mean by this will become clearer shortly, when I show some examples of recursive definitions.

One of the most common ways in which recursion appears in daily life is when you postpone completing a task in favor of a simpler task, often of the same type. Here is a good example. An executive has a fancy telephone and receives many calls on it. He is talking to A when B calls. To A he says, "Would you mind holding for a moment?" Of course he doesn't really care if A minds; he just pushes a button, and switches to B. Now C calls. The same deferment happens to B. This could go on indefinitely, but let us not get too bogged down in our enthusiasm. So let's say the call with C terminates. Then our executive "pops" back up to B, and continues. Meanwhile, A is sitting at the other end of the line, drumming his fingernails against some table, and listening to some horrible Muzak piped through the phone lines to placate him . . . Now the easiest case is if the call with B simply terminates, and the executive returns to A finally. But it *could* happen that after the conversation with B is resumed, a new caller—D—calls. B is once again pushed onto the stack of waiting callers, and D is taken care of. After D is done, back to B, then back to A. This executive is hopelessly mechanical, to be sure—but we are illustrating recursion in its most precise form.

Pushing, Popping, and Stacks

In the preceding example, I have introduced some basic terminology of recursion—at least as seen through the eyes of computer scientists. The terms are *push, pop,* and *stack* (or *push-down stack,* to be precise) and they are all related. They were introduced in the late 1950's as part of IPL, one of the first languages for Artificial Intelligence. You have already encountered "push" and "pop" in the Dialogue. But I will spell things out anyway. To *push* means to suspend operations on the task you're currently working on, without forgetting where you are—and to take up a new task. The new task is usually said to be "on a lower level" than the earlier task. To *pop* is the reverse—it means to close operations on one level, and to resume operations exactly where you left off, one level higher.

But how do you remember exactly where you were on each different level? The answer is, you store the relevant information in a *stack.* So a stack is just a table telling you such things as (1) where you were in each unfinished task (jargon: the "return address"), (2) what the relevant facts to know were at the points of interruption (jargon: the "variable bindings"). When you pop back up to resume some task, it is the stack which restores your context, so you don't feel lost. In the telephone-call example, the stack tells you *who* is waiting on each different level, and *where* you were in the conversation when it was interrupted.

By the way, the terms "push", "pop", and "stack" all come from the visual image of cafeteria trays in a stack. There is usually some sort of spring underneath which tends to keep the topmost tray at a constant height, more or less. So when you push a tray onto the stack, it sinks a little—and when you remove a tray from the stack, the stack pops up a little.

One more example from daily life. When you listen to a news report on the radio, oftentimes it happens that they switch you to some foreign correspondent. "We now switch you to Sally Swumpley in Peafog, England." Now Sally has got a tape of some local reporter interviewing someone, so after giving a bit of background, she plays it. "I'm Nigel Cadwallader, here on scene just outside of Peafog, where the great robbery took place, and I'm talking with . . ." Now you are three levels down. It may turn out that the interviewee also plays a tape of some conversation. It is not too uncommon to go down three levels in real news reports, and surprisingly enough, we scarcely have any awareness of the suspension. It is all kept track of quite easily by our subconscious mind. Probably the reason it is so easy is that each level is extremely different in flavor from each other level. If they were all similar, we would get confused in no time flat.

An example of a more complex recursion is, of course, our Dialogue. There, Achilles and the Tortoise appeared on all the different levels. Sometimes they were reading a story in which they appeared as characters. That is when your mind may get a little hazy on what's going on, and you have to concentrate carefully to get things straight. "Let's see, the *real* Achilles and Tortoise are still up there in Goodfortune's helicopter, but the *secondary* ones are in some Escher picture—and then they found this book and are reading in it, so it's the *tertiary* Achilles and Tortoise who are wandering around inside the grooves of the *Little Harmonic Labyrinth*. No, wait a minute—I left out one level somewhere . . ." You have to have a conscious mental stack like this in order to keep track of the recursion in the Dialogue. (See Fig. 26.)

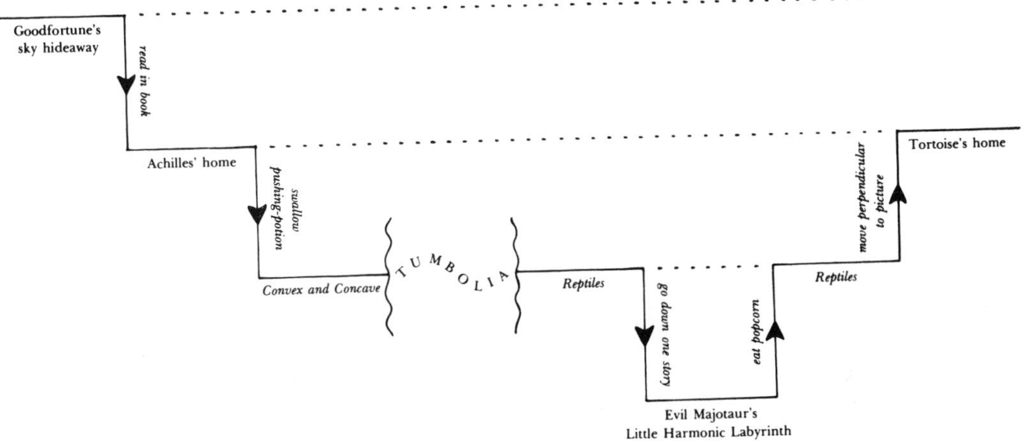

FIGURE 26. Diagram of the structure of the Dialogue Little Harmonic Labyrinth. *Vertical descents are "pushes"; rises are "pops". Notice the similarity of this diagram to the indentation pattern of the Dialogue. From the diagram it is clear that the initial tension—Goodfortune's threat—never was resolved; Achilles and the Tortoise were just left dangling in the sky. Some readers might agonize over this unpopped push, while others might not bat an eyelash. In the story, Bach's musical labyrinth likewise was cut off too soon—but Achilles didn't even notice anything funny. Only the Tortoise was aware of the more global dangling tension.*

Stacks in Music

While we're talking about the *Little Harmonic Labyrinth,* we should discuss something which is hinted at, if not stated explicitly in the Dialogue: that we hear music recursively—in particular, that we maintain a mental stack of keys, and that each new modulation pushes a new key onto the stack. The implication is further that we want to hear that sequence of keys retraced in reverse order—popping the pushed keys off the stack, one by one, until the tonic is reached. This is an exaggeration. There is a grain of truth to it, however.

Any reasonably musical person automatically maintains a shallow stack with two keys. In that "short stack", the true tonic key is held, and also the most immediate "pseudotonic" (the key the composer is pretending to be in). In other words, the most global key and the most local key. That way, the listener knows when the true tonic is regained, and feels a strong sense of "relief". The listener can also distinguish (unlike Achilles) between a *local* easing of tension—for example a resolution into the pseudotonic—

and a *global* resolution. In fact, a pseudoresolution should heighten the global tension, not relieve it, because it is a piece of irony—just like Achilles' rescue from his perilous perch on the swinging lamp, when all the while you know he and the Tortoise are really awaiting their dire fates at the knife of Monsieur Goodfortune.

Since tension and resolution are the heart and soul of music, there are many, many examples. But let us just look at a couple in Bach. Bach wrote many pieces in an "*AABB*" form—that is, where there are two halves, and each one is repeated. Let's take the gigue from the French Suite no. 5, which is quite typical of the form. Its tonic key is G, and we hear a gay dancing melody which establishes the key of G strongly. Soon, however, a modulation in the *A*-section leads to the closely related key of D (the dominant). When the *A*-section ends, we are in the key of D. In fact, it sounds as if the piece has ended in the key of D! (Or at least it might sound that way to Achilles.) But then a strange thing happens—we abruptly jump back to the beginning, back to G, and rehear the same transition into D. But then a strange thing happens—we abruptly jump back to the beginning, back to G, and rehear the same transition into D.

Then comes the *B*-section. With the inversion of the theme for our melody, we begin in D as if that had always been the tonic—but we modulate back to G after all, which means that we pop back into the tonic, and the *B*-section ends properly. Then that funny repetition takes place, jerking us without warning back into D, and letting us return to G once more. Then that funny repetition takes place, jerking us without warning back into D, and letting us return to G once more.

The psychological effect of all this key shifting—some jerky, some smooth—is very difficult to describe. It is part of the magic of music that we can automatically make sense of these shifts. Or perhaps it is the magic of Bach that he can write pieces with this kind of structure which have such a natural grace to them that we are not aware of exactly what is happening.

The original *Little Harmonic Labyrinth* is a piece by Bach in which he tries to lose you in a labyrinth of quick key changes. Pretty soon you are so disoriented that you don't have any sense of direction left—you don't know where the true tonic is, unless you have perfect pitch, or like Theseus, have a friend like Ariadne who gives you a thread that allows you to retrace your steps. In this case, the thread would be a written score. This piece—another example is the Endlessly Rising Canon—goes to show that, as music listeners, we don't have very reliable deep stacks.

Recursion in Language

Our mental stacking power is perhaps slightly stronger in language. The grammatical structure of all languages involves setting up quite elaborate push-down stacks, though, to be sure, the difficulty of understanding a sentence increases sharply with the number of pushes onto the stack. The proverbial German phenomenon of the "verb-at-the-end", about which droll tales of absentminded professors who would begin a sentence, ramble on for an entire lecture, and then finish up by rattling off a string of verbs by which their audience, for whom the stack had long since lost its coherence, would be totally nonplussed, are told, is an excellent example of linguistic pushing and popping. The confusion among the audience that out-of-order popping from the stack onto which the professor's verbs had been pushed, is amusing to imagine, could engender. But in normal spoken German, such deep stacks almost never occur—in fact, native speakers of German often unconsciously violate certain conventions which force the verb to go to the end, in order to avoid the mental effort of keeping track of the stack. Every language has constructions which involve stacks, though usually of a less spectacular nature than German. But there are always ways of rephrasing sentences so that the depth of stacking is minimal.

Recursive Transition Networks

The syntactical structure of sentences affords a good place to present a way of describing recursive structures and processes: the *Recursive Transition Network* (RTN). An RTN is a diagram showing various paths which can be followed to accomplish a particular task. Each path consists of a number of *nodes,* or little boxes with words in them, joined by *arcs,* or lines with arrows. The overall name for the RTN is written separately at the left, and the first and last nodes have the words *begin* and *end* in them. All the other nodes contain either very short explicit directions to perform, or else names of other RTN's. Each time you hit a node, you are to carry out the directions inside it, or to jump to the RTN named inside it, and carry it out.

Let's take a sample RTN, called **ORNATE NOUN**, which tells how to construct a certain type of English noun phrase. (See Fig. 27a.) If we traverse **ORNATE NOUN** purely horizontally, we *begin*, then we create an **ARTICLE**, an **ADJECTIVE**, and a **NOUN**, then we *end*. For instance, "the silly shampoo" or "a thankless brunch". But the arcs show other possibilities, such as skipping the article, or repeating the adjective. Thus we could construct "milk", or "big red blue green sneezes", etc.

When you hit the node **NOUN**, you are asking the unknown black box called **NOUN** to fetch any noun for you from its storehouse of nouns. This is known as a *procedure call,* in computer science terminology. It means you temporarily give control to a *procedure* (here, **NOUN**) which (1) does its thing (produces a noun) and then (2) hands control back to you. In the above RTN, there are calls on three such procedures: **ARTICLE, ADJECTIVE,** and **NOUN**. Now the RTN **ORNATE NOUN** could itself be called from some other RTN—for instance an RTN called **SENTENCE**. In this case, **ORNATE NOUN** would produce a phrase such as "the silly shampoo" and then return to the place inside **SENTENCE** from which it had been called. It is quite reminiscent of the way in which you resume where you left off in nested telephone calls or nested news reports.

However, despite calling this a "recursive transition network", we have

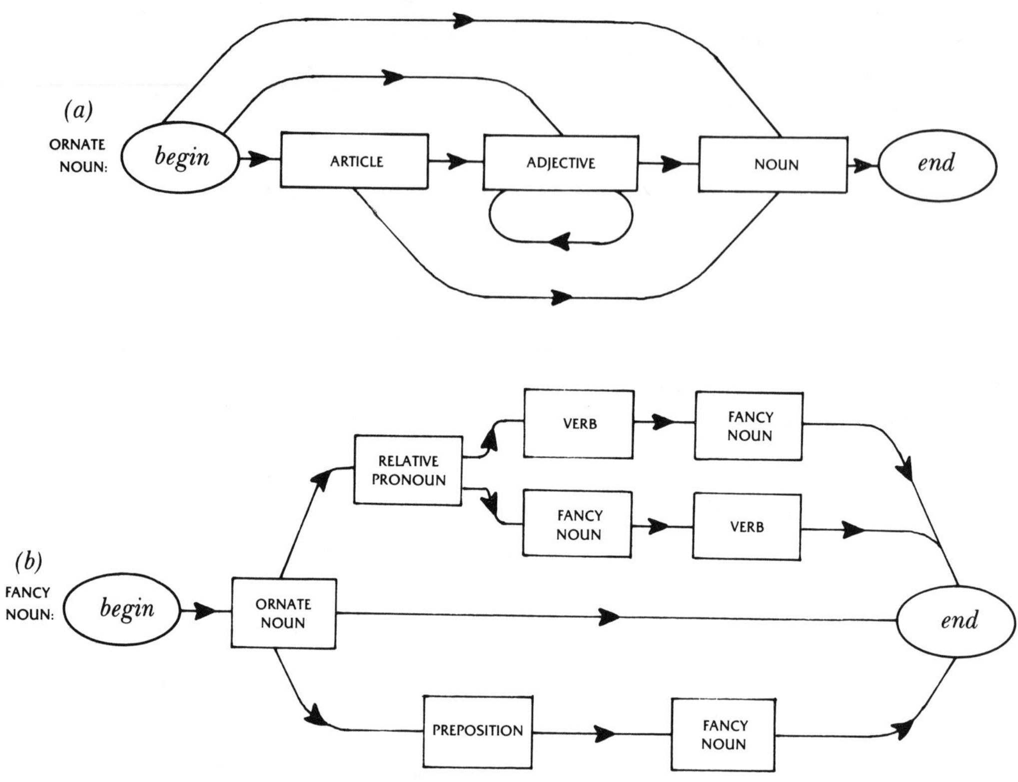

FIGURE 27. Recursive Transition Networks for ORNATE NOUN *and* FANCY NOUN.

not exhibited any true recursion so far. Things get recursive—and seemingly circular—when you go to an RTN such as the one in Figure 27b, for FANCY NOUN. As you can see, every possible pathway in FANCY NOUN involves a call on ORNATE NOUN, so there is no way to avoid getting a noun of some sort or other. And it is possible to be no more ornate than that, coming out merely with "milk" or "big red blue green sneezes". But three of the pathways involve *recursive* calls on FANCY NOUN itself. It certainly looks as if something is being defined in terms of itself. Is that what is happening, or not?

The answer is "yes, but benignly". Suppose that, in the procedure SENTENCE, there is a node which calls FANCY NOUN, and we hit that node. This means that we commit to memory (viz., the stack) the location of that node inside SENTENCE, so we'll know where to return to—then we transfer our attention to the procedure FANCY NOUN. Now we must choose a pathway to take, in order to generate a FANCY NOUN. Suppose we choose the lower of the upper pathways—the one whose calling sequence goes:

ORNATE NOUN; RELATIVE PRONOUN; FANCY NOUN; VERB.

So we spit out an **ORNATE NOUN**: *"the strange bagels"*; a **RELATIVE PRONOUN**: *"that"*; and now we are suddenly asked for a **FANCY NOUN**. But we are in the middle of **FANCY NOUN**! Yes, but remember our executive who was in the middle of one phone call when he got another one. He merely stored the old phone call's status on a stack, and began the new one as if nothing were unusual. So we shall do the same.

We first write down in our stack the node we are at in the outer call on **FANCY NOUN**, so that we have a "return address"; then we jump to the beginning of **FANCY NOUN** as if nothing were unusual. Now we have to choose a pathway again. For variety's sake, let's choose the lower pathway: **ORNATE NOUN; PREPOSITION; FANCY NOUN**. That means we produce an **ORNATE NOUN** (say *"the purple cow"*), then a **PREPOSITION** (say *"without"*), and once again, we hit the recursion. So we hang onto our hats, and descend one more level. To avoid complexity, let's assume that this time, the pathway we take is the direct one—just **ORNATE NOUN**. For example, we might get *"horns"*. We hit the node **END** in this call on **FANCY NOUN**, which amounts to popping out, and so we go to our stack to find the return address. It tells us that we were in the middle of executing **FANCY NOUN** one level up—and so we resume there. This yields *"the purple cow without horns"*. On this level, too, we hit **END**, and so we pop up once more, this time finding ourselves in need of a **VERB**—so let's choose *"gobbled"*. This ends the highest-level call on **FANCY NOUN**, with the result that the phrase

"the strange bagels that the purple cow without horns gobbled"

will get passed upwards to the patient **SENTENCE**, as we pop for the last time.

As you see, we didn't get into any infinite regress. The reason is that at least one pathway inside the RTN **FANCY NOUN** does *not* involve any recursive calls on **FANCY NOUN** itself. Of course, we could have perversely insisted on always choosing the bottom pathway inside **FANCY NOUN**, and then we would never have gotten finished, just as the acronym "GOD" never got fully expanded. But if the pathways are chosen at random, then an infinite regress of that sort will not happen.

"Bottoming Out" and Heterarchies

This is the crucial fact which distinguishes recursive definitions from circular ones. There is always some part of the definition which avoids self-reference, so that the action of constructing an object which satisfies the definition will eventually "bottom out".

Now there are more oblique ways of achieving recursivity in RTN's than by self-calling. There is the analogue of Escher's *Drawing Hands* (Fig. 135), where each of two procedures calls the other, but not itself. For example, we could have an RTN named **CLAUSE**, which calls **FANCY NOUN** whenever it needs an object for a transitive verb, and conversely, the upper path of **FANCY NOUN** could call **RELATIVE PRONOUN** and then **CLAUSE** whenever it wants a relative clause. This is an example of *indirect* recursion. It is reminiscent also of the two-step version of the Epimenides paradox.

Needless to say, there can be a trio of procedures which call one another, cyclically—and so on. There can be a whole family of RTN's which are all tangled up, calling each other and themselves like crazy. A program which has such a structure in which there is no single "highest level", or "monitor", is called a *heterarchy* (as distinguished from a hierarchy). The term is due, I believe, to Warren McCulloch, one of the first cyberneticists, and a reverent student of brains and minds.

Expanding Nodes

One graphic way of thinking about RTN's is this. Whenever you are moving along some pathway and you hit a node which calls on an RTN, you "expand" that node, which means to replace it by a very small copy of the RTN it calls (see Fig. 28). Then you proceed into the very small RTN!

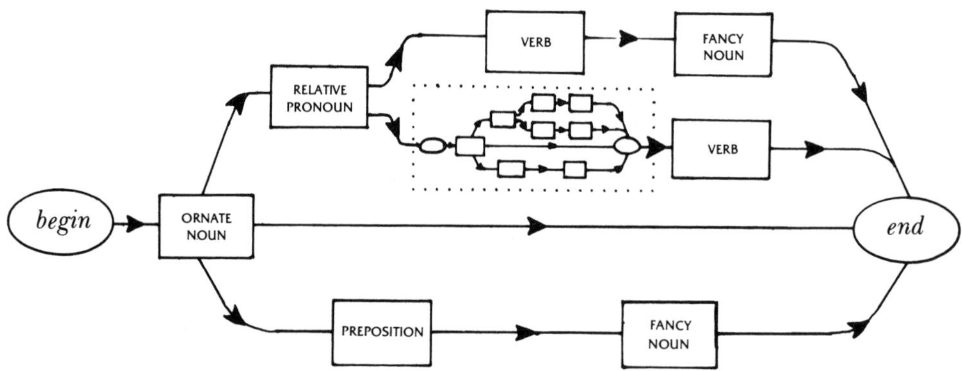

FIGURE 28. The FANCY NOUN *RTN with one node recursively expanded.*

When you pop out of it, you are automatically in the right place in the big one. While in the small one, you may wind up constructing even more miniature RTN's. But by expanding nodes only when you come across them, you avoid the need to make an infinite diagram, even when an RTN calls itself.

Expanding a node is a little like replacing a letter in an acronym by the word it stands for. The "GOD" acronym is recursive but has the defect—or advantage—that you must repeatedly expand the 'G'; thus it never bottoms out. When an RTN is implemented as a real computer program, however, it always has at least one pathway which avoids recursivity (direct or indirect) so that infinite regress is not created. Even the most heterarchical program structure bottoms out—otherwise it couldn't run! It would just be constantly expanding node after node, but never performing any action.

Diagram G and Recursive Sequences

Infinite geometrical structures can be defined in just this way—that is, by expanding node after node. For example, let us define an infinite diagram called "Diagram G". To do so, we shall use an implicit representation. In two nodes, we shall write merely the letter 'G', which, however, will stand for an entire copy of Diagram G. In Figure 29a, Diagram G is portrayed implicitly. Now if we wish to see Diagram G more explicitly, we expand each of the two G's—that is, we *replace them by the same diagram,* only reduced in scale (see Fig. 29b). This "second-order" version of Diagram G gives us an inkling of what the final, impossible-to-realize Diagram G really looks like. In Figure 30 is shown a larger portion of Diagram G, where all the nodes have been numbered from the bottom up, and from left to right. Two extra nodes—numbers 1 and 2—have been inserted at the bottom.

This infinite *tree* has some very curious mathematical properties. Running up its right-hand edge is the famous sequence of *Fibonacci numbers:*

1, 1, 2, 3, 5, 8, 13, 21, 34, 55, 89, 144, 233, . . .

discovered around the year 1202 by Leonardo of Pisa, son of Bonaccio, ergo "Filius Bonacci", or "Fibonacci" for short. These numbers are best

FIGURE 29. (a) Diagram G, unexpanded. (b) Diagram G, expanded once. (c) Diagram H, unexpanded. (d) Diagram H, expanded once.

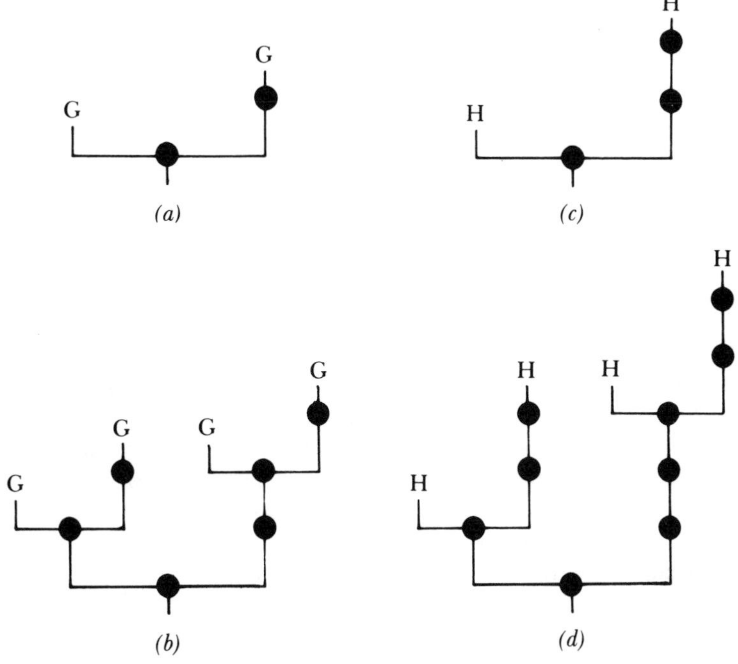

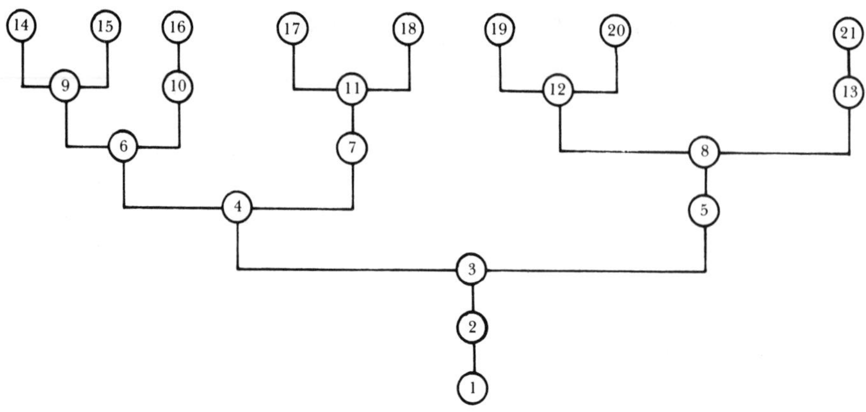

FIGURE 30. Diagram G, further expanded and with numbered nodes.

defined recursively by the pair of formulas

$$\text{FIBO}(n) = \text{FIBO}(n-1) + \text{FIBO}(n-2) \qquad \text{for } n > 2$$

$$\text{FIBO}(1) = \text{FIBO}(2) = 1$$

Notice how new Fibonacci numbers are defined in terms of previous Fibonacci numbers. We could represent this pair of formulas in an RTN (see Fig. 31).

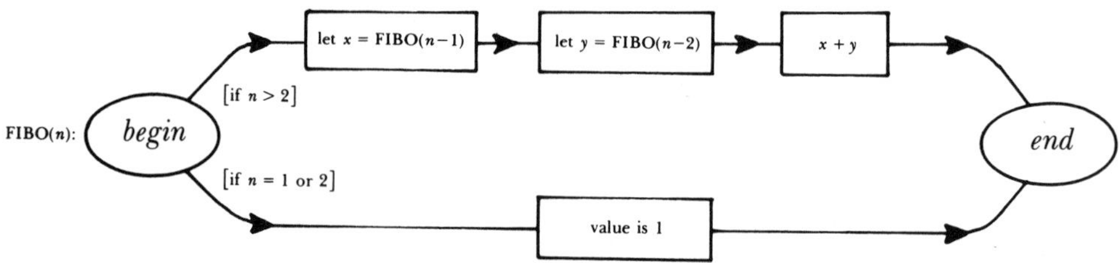

FIGURE 31. An RTN for Fibonacci numbers.

Thus you can calculate FIBO(15) by a sequence of recursive calls on the procedure defined by the RTN above. This recursive definition bottoms out when you hit FIBO(1) or FIBO(2) (which are given explicitly) after you have worked your way backwards through descending values of *n*. It is slightly awkward to work your way backwards, when you could just as well work your way forwards, starting with FIBO(1) and FIBO(2) and always adding the most recent two values, until you reach FIBO(15). That way you don't need to keep track of a stack.

Now Diagram G has some even more surprising properties than this. Its entire structure can be coded up in a single recursive definition, as follows:

$$G(n) = n - G(G(n-1)) \qquad \text{for } n > 0$$
$$G(0) = 0$$

How does this function $G(n)$ code for the tree-structure? Quite simply, if you construct a tree by placing $G(n)$ below n, for all values of n, you will recreate Diagram G. In fact, that is how I discovered Diagram G in the first place. I was investigating the *function* G, and in trying to calculate its values quickly, I conceived of displaying the values I already knew in a tree. To my surprise, the tree turned out to have this extremely orderly recursive geometrical description.

What is more wonderful is that if you make the analogous tree for a function $H(n)$ defined with one more nesting than G—

$$H(n) = n - H(H(H(n-1))) \qquad \text{for } n > 0$$
$$H(0) = 0$$

—then the associated "Diagram H" is defined implicitly as shown in Figure 29c. The right-hand trunk contains one more node; that is the only difference. The first recursive expansion of Diagram H is shown in Figure 29d. And so it goes, for any degree of nesting. There is a beautiful regularity to the recursive geometrical structures, which corresponds precisely to the recursive algebraic definitions.

A problem for curious readers is: suppose you flip Diagram G around as if in a mirror, and label the nodes of the new tree so they increase from left to right. Can you find a recursive *algebraic* definition for this "flip-tree"? What about for the "flip" of the H-tree? Etc.?

Another pleasing problem involves a pair of recursively intertwined functions $F(n)$ and $M(n)$—"married" functions, you might say—defined this way:

$$\left.\begin{aligned} F(n) &= n - M(F(n-1)) \\ M(n) &= n - F(M(n-1)) \end{aligned}\right\} \text{for } n > 0$$

$$F(0) = 1, \text{ and } M(0) = 0.$$

The RTN's for these two functions call each other and themselves as well. The problem is simply to discover the recursive structures of Diagram F and Diagram M. They are quite elegant and simple.

A Chaotic Sequence

One last example of recursion in number theory leads to a small mystery. Consider the following recursive definition of a function:

$$Q(n) = Q(n - Q(n-1)) + Q(n - Q(n-2)) \qquad \text{for } n > 2$$
$$Q(1) = Q(2) = 1.$$

It is reminiscent of the Fibonacci definition in that each new value is a sum of two previous values—but not of the *immediately* previous two values. Instead, the two immediately previous values tell *how far to count back* to obtain the numbers to be added to make the new value! The first 17 Q-numbers run as follows:

1, 1, 2, 3, 3, 4, 5, 5, 6, 6, 6, 8, 8, 8, 10, 9, 10, . . .

5 + 6 = 11

new term

how far to move to the left

To obtain the next one, move leftwards (from the three dots) respectively 10 and 9 terms; you will hit a 5 and a 6, shown by the arrows. Their sum—11—yields the new value: Q(18). This is the strange process by which the list of known Q-numbers is used to extend itself. The resulting sequence is, to put it mildly, erratic. The further out you go, the less sense it seems to make. This is one of those very peculiar cases where what seems to be a somewhat natural definition leads to extremely puzzling behavior: chaos produced in a very orderly manner. One is naturally led to wonder whether the apparent chaos conceals some subtle regularity. Of course, by definition, there is regularity, but what is of interest is whether there is another way of characterizing this sequence—and with luck, a nonrecursive way.

Two Striking Recursive Graphs

The marvels of recursion in mathematics are innumerable, and it is not my purpose to present them all. However, there are a couple of particularly striking examples from my own experience which I feel are worth presenting. They are both graphs. One came up in the course of some number-theoretical investigations. The other came up in the course of my Ph.D. thesis work, in solid state physics. What is truly fascinating is that the graphs are closely related.

The first one (Fig. 32) is a graph of a function which I call INT(x). It is plotted here for x between 0 and 1. For x between any other pair of integers n and $n + 1$, you just find INT($x-n$), then add n back. The structure of the plot is quite jumpy, as you can see. It consists of an infinite number of curved pieces, which get smaller and smaller towards the corners—and incidentally, less and less curved. Now if you look closely at each such piece, you will find that it is actually a copy of the full graph, merely curved! The implications are wild. One of them is that the graph of INT consists of nothing but copies of itself, nested down infinitely deeply. If you pick up any piece of the graph, no matter how small, you are holding a complete copy of the whole graph—in fact, infinitely many copies of it!

The fact that INT consists of nothing but copies of itself might make you think it is too ephemeral to exist. Its definition sounds too circular.

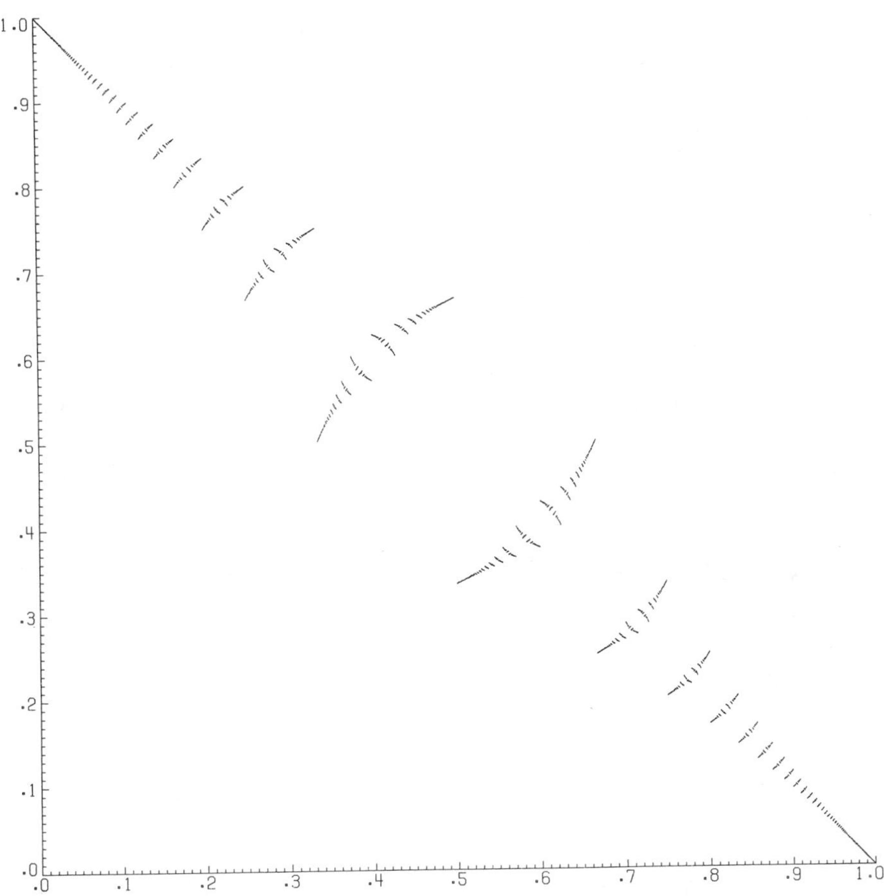

FIGURE 32. Graph of the function INT(x). There is a jump discontinuity at every rational value of x.

How does it ever get off the ground? That is a very interesting matter. The main thing to notice is that, to describe INT to someone who hasn't seen it, it will not suffice merely to say, "It consists of copies of itself." The other half of the story—the nonrecursive half—tells *where* those copies lie inside the square, and *how* they have been deformed, relative to the full-size graph. Only the combination of these two aspects of INT will specify the structure of INT. It is exactly as in the definition of Fibonacci numbers, where you need two lines—one to define the *recursion,* the other to define the *bottom* (i.e., the values at the beginning). To be very concrete, if you make one of the bottom values 3 instead of 1, you will produce a completely different sequence, known as the *Lucas sequence:*

1, 3, 4, 7, 11, 18, 29, 47, 76, 123, . . .

$29 + 47 = 76$

the "bottom" (1, 3) — *same recursive rule as for the Fibonacci numbers*

What corresponds to the *bottom* in the definition of INT is a picture (Fig. 33a) composed of many boxes, showing *where* the copies go, and *how* they are distorted. I call it the "skeleton" of INT. To construct INT from its skeleton, you do the following. First, for each box of the skeleton, you do two operations: (1) put a small curved copy of the skeleton inside the box, using the curved line inside it as a guide; (2) erase the containing box and its curved line. Once this has been done for each box of the original skeleton, you are left with many "baby" skeletons in place of one big one. Next you repeat the process one level down, with all the baby skeletons. Then again, again, and again . . . What you approach in the limit is an exact graph of INT, though you never get there. By nesting the skeleton inside itself over and over again, you gradually construct the graph of INT "from out of nothing". But in fact the "nothing" was not nothing—it was a picture.

To see this even more dramatically, imagine keeping the recursive part of the definition of INT, but changing the initial picture, the skeleton. A variant skeleton is shown in Figure 33b, again with boxes which get smaller and smaller as they trail off to the four corners. If you nest this second skeleton inside itself over and over again, you will create the key graph from my Ph.D. thesis, which I call *Gplot* (Fig. 34). (In fact, some complicated distortion of each copy is needed as well—but nesting is the basic idea.) Gplot is thus a member of the INT-family. It is a distant relative, because its skeleton is quite different from—and considerably more complex than—that of INT. However, the recursive part of the definition is identical, and therein lies the family tie.

I should not keep you too much in the dark about the origin of these beautiful graphs. INT—standing for "interchange"—comes from a problem involving "Eta-sequences", which are related to continued fractions. The basic idea behind INT is that plus and minus signs are interchanged in a certain kind of continued fraction. As a consequence, $\mathrm{INT}(\mathrm{INT}(x)) = x$. INT has the property that if x is rational, so is $\mathrm{INT}(x)$; if x is quadratic, so is $\mathrm{INT}(x)$. I do not know if this trend holds for higher algebraic degrees. Another lovely feature of INT is that at all rational values of x, it has a jump discontinuity, but at all irrational values of x, it is continuous.

Gplot comes from a highly idealized version of the question, "What are the allowed energies of electrons in a crystal in a magnetic field?" This problem is interesting because it is a cross between two very simple and fundamental physical situations: an electron in a perfect crystal, and an electron in a homogeneous magnetic field. These two simpler problems are both well understood, and their characteristic solutions seem almost incompatible with each other. Therefore, it is of quite some interest to see how nature manages to reconcile the two. As it happens, the crystal-without-magnetic-field situation and the magnetic-field-without-crystal situation do have one feature in common: in each of them, the electron behaves periodically in time. It turns out that when the two situations are combined, the ratio of their two time periods is the key parameter. In fact, that ratio holds all the information about the distribution of allowed electron energies—but it only gives up its secret upon being expanded into a continued fraction.

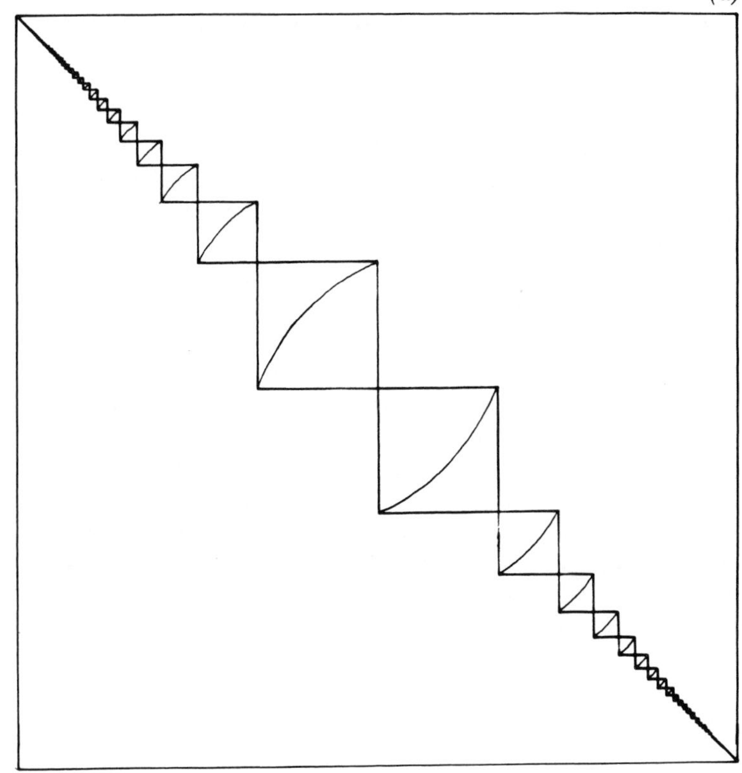

FIGURE 33(a) *The skeleton from which INT can be constructed by recursive substitutions.*
(b) *The skeleton from which Gplot can be constructed by recursive substitutions.*

(b)

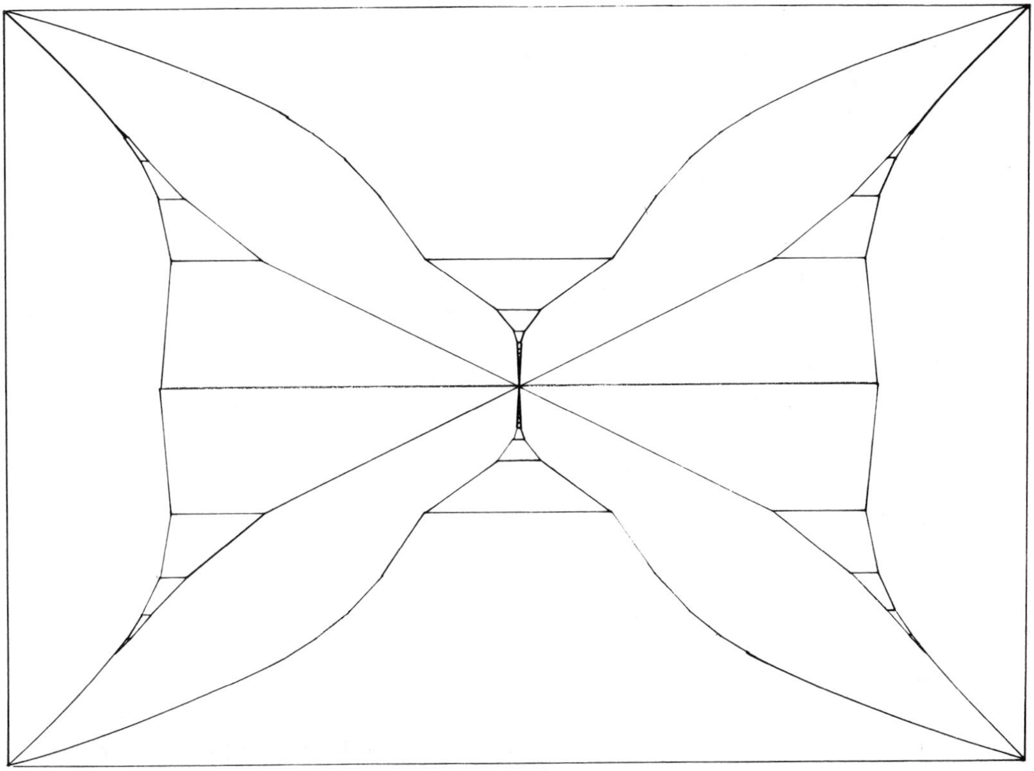

Gplot shows that distribution. The horizontal axis represents energy, and the vertical axis represents the above-mentioned ratio of time periods, which we can call "α". At the bottom, α is zero, and at the top α is unity. When α is zero, there is no magnetic field. Each of the line segments making up Gplot is an "energy band"—that is, it represents allowed values of energy. The empty swaths traversing Gplot on all different size scales are therefore regions of forbidden energy. One of the most startling properties of Gplot is that when α is rational (say p/q in lowest terms), there are exactly q such bands (though when q is even, two of them "kiss" in the middle). And when α is irrational, the bands shrink to points, of which there are infinitely many, very sparsely distributed in a so-called "Cantor set"—another recursively defined entity which springs up in topology.

You might well wonder whether such an intricate structure would ever show up in an experiment. Frankly, I would be the most surprised person in the world if Gplot came out of any experiment. The physicality of Gplot lies in the fact that it points the way to the proper mathematical treatment of less idealized problems of this sort. In other words, Gplot is purely a contribution to theoretical physics, not a hint to experimentalists as to what to expect to see! An agnostic friend of mine once was so struck by Gplot's infinitely many infinities that he called it "a picture of God", which I don't think is blasphemous at all.

Recursion at the Lowest Level of Matter

We have seen recursion in the grammars of languages, we have seen recursive geometrical trees which grow upwards forever, and we have seen one way in which recursion enters the theory of solid state physics. Now we are going to see yet another way in which the whole world is built out of recursion. This has to do with the structure of elementary particles: electrons, protons, neutrons, and the tiny quanta of electromagnetic radiation called "photons". We are going to see that particles are—in a certain sense which can only be defined rigorously in relativistic quantum mechanics—nested inside each other in a way which can be described recursively, perhaps even by some sort of "grammar".

We begin with the observation that if particles didn't interact with each other, things would be incredibly simple. Physicists would like such a world because then they could calculate the behavior of all particles easily (if physicists in such a world existed, which is a doubtful proposition). Particles without interactions are called *bare particles,* and they are purely hypothetical creations; they don't exist.

Now when you "turn on" the interactions, then particles get tangled up together in the way that functions F and M are tangled together, or married people are tangled together. These real particles are said to be *renormalized*—an ugly but intriguing term. What happens is that no particle can even be defined without referring to all other particles, whose definitions in turn depend on the first particles, etc. Round and round, in a never-ending loop.

FIGURE 34. Gplot: a recursive graph showing energy bands for electrons in an idealized crystal in a magnetic field. α, representing magnetic field strength, runs vertically from 0 to 1. Energy runs horizontally. The horizontal line segments are bands of allowed electron energies.

Let us be a little more concrete, now. Let's limit ourselves to only two kinds of particles: *electrons* and *photons*. We'll also have to throw in the electron's antiparticle, the *positron*. (Photons are their own antiparticles.) Imagine first a dull world where a bare electron wishes to propagate from point A to point B, as Zeno did in my *Three-Part Invention*. A physicist would draw a picture like this:

There is a mathematical expression which corresponds to this line and its endpoints, and it is easy to write down. With it, a physicist can understand the behavior of the bare electron in this trajectory.

Now let us "turn on" the electromagnetic interaction, whereby electrons and photons interact. Although there are no photons in the scene, there will nevertheless be profound consequences even for this simple trajectory. In particular, our electron now becomes capable of emitting and then reabsorbing *virtual photons*—photons which flicker in and out of existence before they can be seen. Let us show one such process:

Now as our electron propagates, it may emit and reabsorb one photon after another, or it may even nest them, as shown below:

The mathematical expressions corresponding to these diagrams—called "Feynman diagrams"—are easy to write down, but they are harder to calculate than that for the bare electron. But what really complicates matters is that a photon (real or virtual) can decay for a brief moment into an electron-positron pair. Then these two annihilate each other, and, as if by magic, the original photon reappears. This sort of process is shown below:

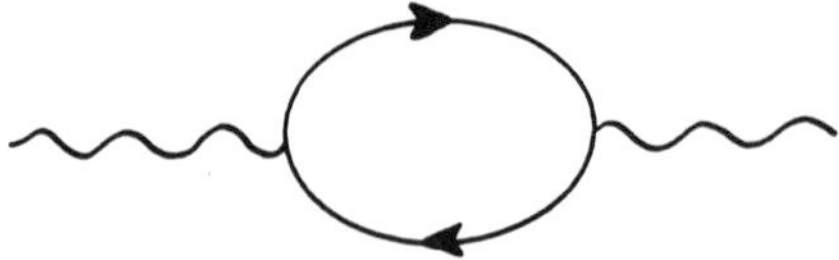

The electron has a right-pointing arrow, while the positron's arrow points leftwards.

As you might have anticipated, these virtual processes can be nested inside each other to arbitrary depth. This can give rise to some very complicated-looking drawings, such as the one in Figure 35. In that Feynman diagram, a single electron enters on the left at A, does some amazing acrobatics, and then a single electron emerges on the right at B. To an outsider who can't see the inner mess, it looks as if one electron has peacefully sailed from A to B. In the diagram, you can see how electron lines can get arbitrarily embellished, and so can the photon lines. This diagram would be ferociously hard to calculate.

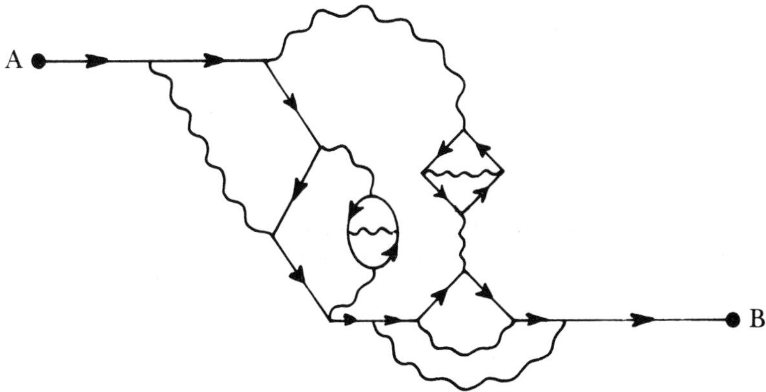

FIGURE 35. A Feynman diagram showing the propagation of a renormalized electron from A to B. In this diagram, time increases to the right. Therefore, in the segments where the electron's arrow points leftwards, it is moving "backwards in time". A more intuitive way to say this is that an antielectron (positron) is moving forwards in time. Photons are their own antiparticles; hence their lines have no need of arrows.

There is a sort of "grammar" to these diagrams, that only allows certain pictures to be realized in nature. For instance, the one below is impossible:

You might say it is not a "well-formed" Feynman diagram. The grammar is a result of basic laws of physics, such as conservation of energy, conservation of electric charge, and so on. And, like the grammars of human languages, this grammar has a recursive structure, in that it allows deep nestings of structures inside each other. It would be possible to draw up a set of recursive transition networks defining the "grammar" of the electromagnetic interaction.

When bare electrons and bare photons are allowed to interact in these arbitrarily tangled ways, the result is *renormalized* electrons and photons. Thus, to understand how a real, physical electron propagates from A to B, the physicist has to be able to take a sort of average of all the infinitely many different possible drawings which involve virtual particles. This is Zeno with a vengeance!

Thus the point is that a physical particle—a renormalized particle—involves (1) a bare particle and (2) a huge tangle of virtual particles, inextricably wound together in a recursive mess. Every real particle's existence therefore involves the existence of infinitely many other particles, contained in a virtual "cloud" which surrounds it as it propagates. And each of the virtual particles in the cloud, of course, also drags along its own virtual cloud, and so on ad infinitum.

Particle physicists have found that this complexity is too much to handle, and in order to understand the behavior of electrons and photons, they use approximations which neglect all but fairly simple Feynman diagrams. Fortunately, the more complex a diagram, the less important its contribution. There is no known way of summing up all of the infinitely many possible diagrams, to get an expression for the behavior of a fully renormalized, physical electron. But by considering roughly the simplest hundred diagrams for certain processes, physicists have been able to predict one value (the so-called g-factor of the muon) to nine decimal places—correctly!

Renormalization takes place not only among electrons and photons. Whenever any types of particle interact together, physicists use the ideas of renormalization to understand the phenomena. Thus protons and neutrons, neutrinos, pi-mesons, quarks—all the beasts in the subnuclear zoo—they all have bare and renormalized versions in physical theories. And from billions of these bubbles within bubbles are all the beasts and baubles of the world composed.

Copies and Sameness

Let us now consider Gplot once again. You will remember that in the Introduction, we spoke of different varieties of canons. Each type of canon exploited some manner of taking an original theme and copying it by an isomorphism, or information-preserving transformation. Sometimes the copies were upside down, sometimes backwards, sometimes shrunken or expanded . . . In Gplot we have all those types of transformation, and more. The mappings between the full Gplot and the "copies" of itself inside itself involve size changes, skewings, reflections, and more. And yet there remains a sort of skeletal identity, which the eye can pick up with a bit of effort, particularly after it has practiced with INT.

Escher took the idea of an object's parts being copies of the object itself and made it into a print: his woodcut *Fishes and Scales* (Fig. 36). Of course these fishes and scales are the same only when seen on a sufficiently abstract plane. Now everyone knows that a fish's scales aren't really small copies of the fish; and a fish's cells aren't small copies of the fish; however, a fish's DNA, sitting inside each and every one of the fish's cells, *is* a very convo-

FIGURE 36. Fish and Scales, *by M. C. Escher (woodcut, 1959).*

luted "copy" of the entire fish—and so there is more than a grain of truth to the Escher picture.

What is there that is the "same" about all butterflies? The mapping from one butterfly to another does not map cell onto cell; rather, it maps functional part onto functional part, and this may be partially on a macroscopic scale, partially on a microscopic scale. The exact proportions of parts are not preserved; just the functional relationships between parts. That is the type of isomorphism which links all butterflies in Escher's wood engraving *Butterflies* (Fig. 37) to each other. The same goes for the more abstract butterflies of Gplot, which are all linked to each other by mathematical mappings that carry functional part onto functional part, but totally ignore exact line proportions, angles, and so on.

Taking this exploration of sameness to a yet higher plane of abstraction, we might well ask, "What is there that is the 'same' about all Escher drawings?" It would be quite ludicrous to attempt to map them piece by piece onto each other. The amazing thing is that even a tiny section of an

FIGURE 37. Butterflies, *by M. C. Escher (wood-engraving, 1950).*

Escher drawing or a Bach piece gives it away. Just as a fish's DNA is contained inside every tiny bit of the fish, so a creator's "signature" is contained inside every tiny section of his creations. We don't know what to call it but "style"—a vague and elusive word.

We keep on running up against "sameness-in-differentness", and the question

When are two things the same?

It will recur over and over again in this book. We shall come at it from all sorts of skew angles, and in the end, we shall see how deeply this simple question is connected with the nature of intelligence.

That this issue arose in the Chapter on recursion is no accident, for recursion is a domain where "sameness-in-differentness" plays a central role. Recursion is based on the "same" thing happening on several different levels at once. But the events on different levels *aren't* exactly the same—rather, we find some invariant feature in them, despite many ways in which they differ. For example, in the *Little Harmonic Labyrinth,* all the stories on different levels are quite unrelated—their "sameness" resides in only two facts: (1) they are stories, and (2) they involve the Tortoise and Achilles. Other than that, they are radically different from each other.

Programming and Recursion: Modularity, Loops, Procedures

One of the essential skills in computer programming is to perceive when two processes are the same in this extended sense, for that leads to *modularization*—the breaking-up of a task into natural subtasks. For instance, one might want a sequence of many similar operations to be carried out one after another. Instead of writing them all out, one can write a *loop,* which tells the computer to perform a fixed set of operations and then loop back and perform them again, over and over, until some condition is satisfied. Now the *body* of the loop—the fixed set of instructions to be repeated—need not actually be completely fixed. It may vary in some predictable way.

An example is the most simple-minded test for the primality of a natural number N, in which you begin by trying to divide N by 2, then by 3, 4, 5, etc. until $N - 1$. If N has survived all these tests without being divisible, it's prime. Notice that each step in the loop is similar to, but not the same as, each other step. Notice also that the number of steps varies with N—hence a loop of fixed length could never work as a general test for primality. There are two criteria for "aborting" the loop: (1) if some number divides N exactly, quit with answer "NO"; (2) if $N - 1$ is reached as a test divisor and N survives, quit with answer "YES".

The general idea of loops, then, is this: perform some series of related steps over and over, and abort the process when specific conditions are met. Now sometimes, the maximum number of steps in a loop will be known in advance; other times, you just begin, and wait until it is aborted. The second type of loop—which I call a *free* loop—is dangerous, because the criterion for abortion may never occur, leaving the computer in a so-called "infinite loop". This distinction between *bounded loops* and *free loops* is one of the most important concepts in all of computer science, and we shall devote an entire Chapter to it: "BlooP and FlooP and GlooP".

Now loops may be nested inside each other. For instance, suppose that we wish to test all the numbers between 1 and 5000 for primality. We can write a second loop which uses the above-described test over and over, starting with $N = 1$ and finishing with $N = 5000$. So our program will have a "loop-the-loop" structure. Such program structures are typical—in fact they are deemed to be good programming style. This kind of nested loop also occurs in assembly instructions for commonplace items, and in such activities as knitting or crocheting—in which very small loops are repeated several times in larger loops, which in turn are carried out repeatedly . . . While the result of a low-level loop might be no more than couple of stitches, the result of a high-level loop might be a substantial portion of a piece of clothing.

In music, too, nested loops often occur—as, for instance, when a scale (a small loop) is played several times in a row, perhaps displaced in pitch each new time. For example, the last movements of both the Prokofiev fifth piano concerto and the Rachmaninoff second symphony contain extended passages in which fast, medium, and slow scale-loops are played simultaneously by different groups of instruments, to great effect. The Prokofiev-scales go up; the Rachmaninoff-scales, down. Take your pick.

A more general notion than loop is that of *subroutine,* or *procedure,* which we have already discussed somewhat. The basic idea here is that a group of operations are lumped together and considered a single unit with a name—such as the procedure **ORNATE NOUN**. As we saw in RTN's, procedures can call each other by name, and thereby express very concisely sequences of operations which are to be carried out. This is the essence of modularity in programming. Modularity exists, of course, in hi-fi systems, furniture, living cells, human society—wherever there is hierarchical organization.

More often than not, one wants a procedure which will act variably, according to context. Such a procedure can either be given a way of peering out at what is stored in memory and selecting its actions accordingly, or it can be explicitly fed a list of *parameters* which guide its choice of what actions to take. Sometimes both of these methods are used. In RTN-terminology, choosing the sequence of actions to carry out amounts to *choosing which pathway to follow.* An RTN which has been souped up with parameters and conditions that control the choice of pathways inside it is called an *Augmented Transition Network* (ATN). A place where you might prefer ATN's to RTN's is in producing sensible—as distinguished from nonsensical—English sentences out of raw words, according to a grammar represented in a set of ATN's. The parameters and conditions would allow you to insert various semantic constraints, so that random juxtapositions like "a thankless brunch" would be prohibited. More on this in Chapter XVIII, however.

Recursion in Chess Programs

A classic example of a recursive procedure with parameters is one for choosing the "best" move in chess. The best move would seem to be the one which leaves your opponent in the toughest situation. Therefore, a test for goodness of a move is simply this: pretend you've made the move, and now evaluate the board from the point of view of your opponent. But how does your opponent evaluate the position? Well, he looks for *his* best move. That is, he mentally runs through all possible moves and evaluates them from what he thinks is *your* point of view, hoping they will look bad to you. But notice that we have now defined "best move" recursively, simply using the maxim that what is best for one side is worst for the other. The recursive procedure which looks for the best move operates by trying a move, and then *calling on itself in the role of opponent!* As such, it tries another move, and calls on itself in the role of its opponent's opponent—that is, itself.

This recursion can go several levels deep—but it's got to bottom out somewhere! How do you evaluate a board position *without* looking ahead? There are a number of useful criteria for this purpose, such as simply the number of pieces on each side, the number and type of pieces under attack, the control of the center, and so on. By using this kind of evaluation at the bottom, the recursive move-generator can pop back upwards and give an evaluation at the top level of each different move. One of the parameters in the self-calling, then, must tell how many moves to look ahead. The outermost call on the procedure will use some externally set value for this parameter. Thereafter, each time the procedure recursively calls itself, it must decrease this look-ahead parameter by 1. That way, when the parameter reaches zero, the procedure will follow the alternate pathway—the non-recursive evaluation.

In this kind of game-playing program, each move investigated causes the generation of a so-called "look-ahead tree", with the move itself as trunk, responses as main branches, counter-responses as subsidiary branches, and so on. In Figure 38 I have shown a simple look-ahead tree, depicting the start of a tic-tac-toe game. There is an art to figuring out how to avoid exploring every branch of a look-ahead tree out to its tip. In chess trees, people—not computers—seem to excel at this art; it is known that top-level players look ahead relatively little, compared to most chess programs—yet the people are far better! In the early days of computer chess, people used to estimate that it would be ten years until a computer (or

FIGURE 38. The branching tree of moves and countermoves at the start of a game of tic-tac-toe.

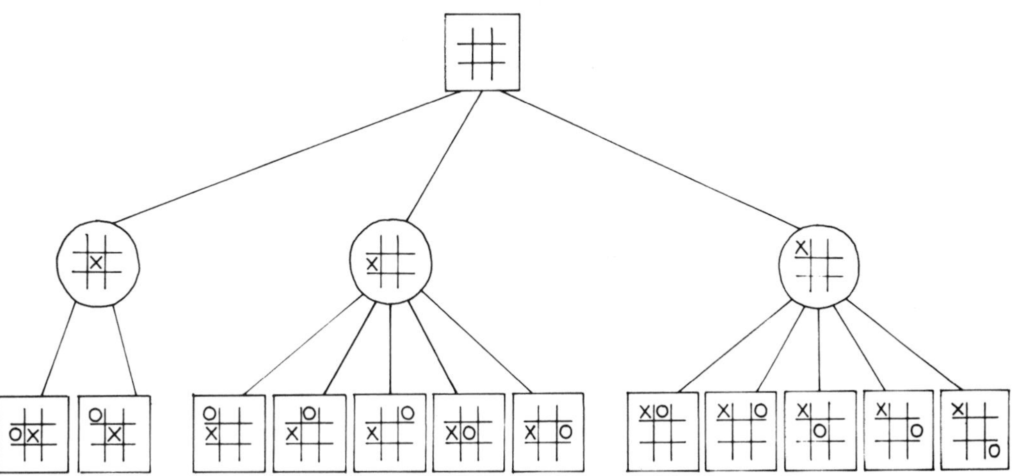

program) was world champion. But after ten years had passed, it seemed that the day a computer would become world champion was still more than ten years away . . . This is just one more piece of evidence for the rather recursive

Hofstadter's Law: It always takes longer than you expect, even when you take into account Hofstadter's Law.

Recursion and Unpredictability

Now what is the connection between the recursive processes of this Chapter, and the recursive sets of the preceding Chapter? The answer involves the notion of a *recursively enumerable set.* For a set to be r.e. means that it can be generated from a set of starting points (axioms), by the repeated application of rules of inference. Thus, the set grows and grows, each new element being compounded somehow out of previous elements, in a sort of "mathematical snowball". But this is the essence of recursion—something being defined in terms of simpler versions of itself, instead of explicitly. The Fibonacci numbers and the Lucas numbers are perfect examples of r.e. sets—snowballing from two elements by a recursive rule into infinite sets. It is just a matter of convention to call an r.e. set whose complement is also r.e. "recursive".

Recursive enumeration is a process in which new things emerge from old things by fixed rules. There seem to be many surprises in such processes—for example the unpredictability of the Q-sequence. It might seem that recursively defined sequences of that type possess some sort of inherently increasing complexity of behavior, so that the further out you go, the less predictable they get. This kind of thought carried a little further suggests that suitably complicated recursive systems might be strong enough to break out of any predetermined patterns. And isn't this one of the defining properties of intelligence? Instead of just considering programs composed of procedures which can recursively *call* themselves, why not get really sophisticated, and invent programs which can *modify* themselves—programs which can act on programs, extending them, improving them, generalizing them, fixing them, and so on? This kind of "tangled recursion" probably lies at the heart of intelligence.

Canon by Intervallic Augmentation

Achilles and the Tortoise have just finished a delicious Chinese banquet for two, at the best Chinese restaurant in town.

Achilles: You wield a mean chopstick, Mr. T.

Tortoise: I ought to. Ever since my youth, I have had a fondness for Oriental cuisine. And you—did you enjoy your meal, Achilles?

Achilles: Immensely. I'd not eaten Chinese food before. This meal was a splendid introduction. And now, are you in a hurry to go, or shall we just sit here and talk a little while?

Tortoise: I'd love to talk while we drink our tea. Waiter!

(A waiter comes up.)

Could we have our bill, please, and some more tea?

(The waiter rushes off.)

Achilles: You may know more about Chinese cuisine than I do, Mr. T, but I'll bet I know more about Japanese poetry than you do. Have you ever read any haiku?

Tortoise: I'm afraid not. What is a haiku?

Achilles: A haiku is a Japanese seventeen-syllable poem—or minipoem, rather, which is evocative in the same way, perhaps, as a fragrant rose petal is, or a lily pond in a light drizzle. It generally consists of groups of five, then seven, then five syllables.

Tortoise: Such compressed poems with seventeen syllables can't have much meaning . . .

Achilles: Meaning lies as much in the mind of the reader as in the haiku.

Tortoise: Hmm . . . That's an evocative statement.

(The waiter arrives with their bill, another pot of tea, and two fortune cookies.)

Thank you, waiter. Care for more tea, Achilles?

Achilles: Please. Those little cookies look delicious. *(Picks one up, bites into it, and begins to chew.)* Hey! What's this funny thing inside? A piece of paper?

Tortoise: That's your fortune, Achilles. Many Chinese restaurants give out fortune cookies with their bills, as a way of softening the blow. If you frequent Chinese restaurants, you come to think of fortune cookies less as cookies than as message bearers. Unfortunately you seem to have swallowed some of your fortune. What does the rest say?

Achilles: It's a little strange, for all the letters are run together, with no spaces in between. Perhaps it needs decoding in some way? Oh, now I see. If you put the spaces back in where they belong, it says, "ONE WAR TWO EAR EWE". I can't quite make head or tail of that. Maybe it was a haiku-like poem, of which I ate the majority of syllables.

Tortoise: In that case, your fortune is now a mere 5/17-haiku. And a curious image it evokes. If 5/17-haiku is a new art form, then I'd say woe, O, woe are we . . . May I look at it?

Achilles (handing the Tortoise the small slip of paper): Certainly.

Tortoise: Why, when I "decode" it, Achilles, it comes out completely different! It's not a 5/17-haiku at all. It is a six-syllable message which says, "O NEW ART WOE ARE WE". That sounds like an insightful commentary on the new art form of 5/17-haiku.

Achilles: You're right. Isn't it astonishing that the poem contains its own commentary!

Tortoise: All I did was to shift the reading frame by one unit—that is, shift all the spaces one unit to the right.

Achilles: Let's see what your fortune says, Mr. Tortoise.

Tortoise (deftly splitting open his cookie, reads): "Fortune lies as much in the hand of the eater as in the cookie."

Achilles: Your fortune is also a haiku, Mr. Tortoise—at least it's got seventeen syllables in the 5-7-5 form.

Tortoise: Glory be! I would never have noticed that, Achilles. It's the kind of thing only you would have noticed. What struck me more is what it says—which, of course, is open to interpretation.

Achilles: I guess it just shows that each of us has his own characteristic way of interpreting messages which we run across . . .

(Idly, Achilles gazes at the tea leaves on the bottom of his empty teacup.)

Tortoise: More tea, Achilles?

Achilles: Yes, thank you. By the way, how is your friend the Crab? I have been thinking about him a lot since you told me of your peculiar phonograph-battle.

Tortoise: I have told him about you, too, and he is quite eager to meet you. He is getting along just fine. In fact, he recently made a new acquisition in the record player line: a rare type of jukebox.

Achilles: Oh, would you tell me about it? I find jukeboxes, with their flashing colored lights and silly songs, so quaint and reminiscent of bygone eras.

Tortoise: This jukebox is too large to fit in his house, so he had a shed specially built in back for it.

Achilles: I can't imagine why it would be so large, unless it has an unusually large selection of records. Is that it?

Tortoise: As a matter of fact, it has exactly one record.

Achilles: What? A jukebox with only one record? That's a contradiction in terms. Why is the jukebox so big, then? Is its single record gigantic—twenty feet in diameter?

Tortoise: No, it's just a regular jukebox-style record.

Achilles: Now, Mr. Tortoise, you must be joshing me. After all, what kind of a jukebox is it that has only a single song?

Tortoise: Who said anything about a single song, Achilles?

Achilles: Every jukebox I've ever run into obeyed the fundamental jukebox-axiom: "One record, one song".

Tortoise: This jukebox is different, Achilles. The one record sits vertically suspended, and behind it there is a small but elaborate network of overhead rails, from which hang various record players. When you push a pair of buttons, such as B-1, that selects one of the record players. This triggers an automatic mechanism that starts the record player squeakily rolling along the rusty tracks. It gets shunted up alongside the record—then it clicks into playing position.

Achilles: And then the record begins spinning and music comes out—right?

Tortoise: Not quite. The record stands still—it's the record player which rotates.

Achilles: I might have known. But how, if you have but one record to play, can you get more than one song out of this crazy contraption?

Tortoise: I myself asked the Crab that question. He merely suggested that I try it out. So I fished a quarter from my pocket (you get three plays for a quarter), stuffed it in the slot, and hit buttons B-1, then C-3, then B-10—all just at random.

Achilles: So phonograph B-1 came sliding down the rail, I suppose, and plugged itself into the vertical record, and began spinning?

Tortoise: Exactly. The music that came out was quite agreeable, based on the famous old tune B-A-C-H, which I believe you remember . . .

Achilles: Could I ever forget it?

Tortoise: This was record player B-1. Then it finished, and was slowly rolled back into its hanging position, so that C-3 could be slid into position.

Achilles: Now don't tell me that C-3 played another song?

Tortoise: It did just that.

Achilles: Ah, I understand. It played the flip side of the first song, or another band on the same side.

Tortoise: No, the record has grooves only on one side, and has only a single band.

Achilles: I don't understand that at all. You CAN'T pull different songs out of the same record!

Tortoise: That's what I thought until I saw Mr. Crab's jukebox.

Achilles: How did the second song go?

Tortoise: That's the interesting thing . . . It was a song based on the melody C-A-G-E.

Achilles: That's a totally different melody!

Tortoise: True.

Achilles: And isn't John Cage a composer of modern music? I seem to remember reading about him in one of my books on haiku.

Tortoise: Exactly. He has composed many celebrated pieces, such as *4'33"*, a three-movement piece consisting of silences of different lengths. It's wonderfully expressive—if you like that sort of thing.

Achilles: I can see where if I were in a loud and brash café I might gladly pay to hear Cage's *4'33"* on a jukebox. It might afford some relief!

Tortoise: Right—who wants to hear the racket of clinking dishes and jangling silverware? By the way, another place where *4'33"* would come in handy is the Hall of Big Cats, at feeding time.

Achilles: Are you suggesting that Cage belongs in the zoo? Well, I guess that makes some sense. But about the Crab's jukebox . . . I am baffled. How could both "BACH" and "CAGE" be coded inside a single record at once?

Tortoise: You may notice that there is some relation between the two, Achilles, if you inspect them carefully. Let me point the way. What do you get if you list the successive intervals in the melody B-A-C-H?

Achilles: Let me see. First it goes down one semitone, from B to A (where B is taken the German way); then it rises three semitones to C; and finally it falls one semitone, to H. That yields the pattern:

$$-1, \quad +3, \quad -1.$$

Tortoise: Precisely. What about C-A-G-E, now?

Achilles: Well, in this case, it begins by falling three semitones, then rises ten semitones (nearly an octave), and finally falls three more semitones. That means the pattern is:

$$-3, \quad +10, \quad -3.$$

It's very much like the other one, isn't it?

Tortoise: Indeed it is. They have exactly the same "skeleton", in a certain sense. You can make C-A-G-E out of B-A-C-H by multiplying all the intervals by 3⅓, and taking the nearest whole number.

Achilles: Well, blow me down and pick me up! So does that mean that only some sort of skeletal code is present in the grooves, and that the various record players add their own interpretations to that code?

Tortoise: I don't know, for sure. The cagey Crab wouldn't fill me in on all the details. But I did get to hear a third song, when record player B-10 swiveled into place.

Achilles: How did it go?

Tortoise: The melody consisted of enormously wide intervals, and went B-C-A-H.

The interval pattern in semitones was:

$$-10, \quad +33, \quad -10.$$

It can be gotten from the CAGE pattern by yet another multiplication by 3⅓, and rounding to whole numbers.

Achilles: Is there a name for this kind of interval multiplication?

Tortoise: One could call it "intervallic augmentation". It is similar to the canonic device of temporal augmentation, where all the time values of notes in a melody get multiplied by some constant. There, the effect is just to slow the melody down. Here, the effect is to expand the melodic range in a curious way.

Achilles: Amazing. So all three melodies you tried were intervallic augmentations of one single underlying groove-pattern in the record?

Tortoise: That's what I concluded.

Achilles: I find it curious that when you augment BACH you get CAGE, and when you augment CAGE over again, you get BACH back, except jumbled up inside, as if BACH had an upset stomach after passing through the intermediate stage of CAGE.

Tortoise: That sounds like an insightful commentary on the new art form of Cage.

CHAPTER VI

The Location of Meaning

When Is One Thing Not Always the Same?

LAST CHAPTER, WE came upon the question, "When are two things the same?" In this Chapter, we will deal with the flip side of that question: "When is one thing not always the same?" The issue we are broaching is whether meaning can be said to be inherent in a message, or whether meaning is always manufactured by the interaction of a mind or a mechanism with a message—as in the preceding Dialogue. In the latter case, meaning could not said to be located in any single place, nor could it be said that a message has any universal, or objective, meaning, since each observer could bring its own meaning to each message. But in the former case, meaning would have both location and universality. In this Chapter, I want to present the case for the universality of at least some messages, without, to be sure, claiming it for all messages. The idea of an "objective meaning" of a message will turn out to be related, in an interesting way, to the simplicity with which intelligence can be described.

Information-Bearers and Information-Revealers

I'll begin with my favorite example: the relationship between records, music, and record players. We feel quite comfortable with the idea that a record contains the same information as a piece of music, because of the existence of record players, which can "read" records and convert the groove-patterns into sounds. In other words, there is an isomorphism between groove-patterns and sounds, and the record player is a mechanism which physically realizes that isomorphism. It is natural, then, to think of the record as an *information-bearer,* and the record-player as an *information-revealer.* A second example of these notions is given by the pq-system. There, the "information-bearers" are the theorems, and the "information-revealer" is the interpretation, which is so transparent that we don't need any electrical machine to help us extract the information from pq-theorems.

One gets the impression from these two examples that isomorphisms and decoding mechanisms (i.e., information-revealers) simply reveal information which is intrinsically inside the structures, waiting to be "pulled out". This leads to the idea that for each structure, there are certain pieces of information which *can* be pulled out of it, while there are other pieces of information which *cannot* be pulled out of it. But what does this phrase "pull out" really mean? How hard are you allowed to pull? There are cases where by investing sufficient effort, you can pull very recondite pieces of information out of certain structures. In fact, the pulling-out may involve such complicated operations that it makes you feel you are putting in more information than you are pulling out.

Genotype and Phenotype

Take the case of the genetic information commonly said to reside in the double helix of deoxyribonucleic acid (DNA). A molecule of DNA—a *genotype*—is converted into a physical organism—a *phenotype*—by a very complex process, involving the manufacture of proteins, the replication of the DNA, the replication of cells, the gradual differentiation of cell types, and so on. Incidentally, this unrolling of phenotype from genotype—*epigenesis*—is the most tangled of tangled recursions, and in Chapter XVI we shall devote our full attention to it. Epigenesis is guided by a set of enormously complex cycles of chemical reactions and feedback loops. By the time the full organism has been constructed, there is not even the remotest similarity between its physical characteristics and its genotype.

And yet, it is standard practice to attribute the physical structure of the organism to the structure of its DNA, and to that alone. The first evidence for this point of view came from experiments conducted by Oswald Avery in 1946, and overwhelming corroborative evidence has since been amassed. Avery's experiments showed that, of all the biological molecules, only DNA transmits hereditary properties. One can modify other molecules in an organism, such as proteins, but such modifications will not be transmitted to later generations. However, when DNA is modified, all successive generations inherit the modified DNA. Such experiments show that the only way of changing the instructions for building a new organism is to change the DNA—and this, in turn, implies that those instructions must be coded somehow in the structure of the DNA.

Exotic and Prosaic Isomorphisms

Therefore one seems forced into accepting the idea that the DNA's structure contains the information of the phenotype's structure, which is to say, the two are *isomorphic*. However, the isomorphism is an *exotic* one, by which I mean that it is highly nontrivial to divide the phenotype and genotype into "parts" which can be mapped onto each other. *Prosaic* isomorphisms, by contrast, would be ones in which the parts of one structure are easily mappable onto the parts of the other. An example is the isomorphism between a record and a piece of music, where one knows that to any sound in the piece there exists an exact "image" in the patterns etched into the grooves, and one could pinpoint it arbitrarily accurately, if the need arose. Another prosaic isomorphism is that between Gplot and any of its internal butterflies.

The isomorphism between DNA structure and phenotype structure is anything but prosaic, and the mechanism which carries it out physically is awesomely complicated. For instance, if you wanted to find some piece of your DNA which accounts for the shape of your nose or the shape of your fingerprint, you would have a very hard time. It would be a little like trying to pin down *the* note in a piece of music which is the carrier of the emotional meaning of the piece. Of course there is no such note, because the emotional meaning is carried on a very high level, by large "chunks" of the piece, not by single notes. Incidentally, such "chunks" are not necessarily sets of contiguous notes; there may be disconnected sections which, taken together, carry some emotional meaning.

Similarly, "genetic meaning"—that is, information about phenotype structure—is spread all through the small parts of a molecule of DNA, although nobody understands the language yet. (Warning: Understanding this "language" would not at all be the same as cracking the Genetic Code, something which took place in the early 1960's. The Genetic Code tells how to translate short portions of DNA into various amino acids. Thus, cracking the Genetic Code is comparable to figuring out the phonetic values of the letters of a foreign alphabet, without figuring out the grammar of the language or the meanings of any of its words. The cracking of the Genetic Code was a vital step on the way to extracting the meaning of DNA strands, but it was only the first on a long path which is yet to be trodden.)

Jukeboxes and Triggers

The genetic meaning contained in DNA is one of the best possible examples of implicit meaning. In order to convert genotype into phenotype, a set of mechanisms far more complex than the genotype must operate on the genotype. The various parts of the genotype serve as *triggers* for those mechanisms. A jukebox—the ordinary type, not the Crab type!—provides a useful analogy here: a pair of buttons specifies a very complex action to be taken by the mechanism, so that the pair of buttons could well be described as "triggering" the song which is played. In the process which converts genotype into phenotype, cellular jukeboxes—if you will pardon the notion!—accept "button-pushings" from short excerpts from a long strand of DNA, and the "songs" which they play are often prime ingredients in the creation of further "jukeboxes". It is as if the output of real jukeboxes, instead of being love ballads, were songs whose lyrics told how to build more complex jukeboxes . . . Portions of the DNA trigger the manufacture of proteins; those proteins trigger hundreds of new reactions; they in turn trigger the replicating-operation which, in several steps, copies the DNA—and on and on . . . This gives a sense of how recursive the whole process is. The final result of these many-triggered triggerings is the phenotype—the individual. And one says that the phenotype is the *revelation*—the "pulling-out"—of the information that was present in the DNA to start with, latently. (The term "revelation" in this context is due to

Jacques Monod, one of the deepest and most original of twentieth-century molecular biologists.)

Now no one would say that a song coming out of the loudspeaker of a jukebox constitutes a "revelation" of information inherent in the pair of buttons which were pressed, for the pair of buttons seem to be mere *triggers* whose purpose is to activate information-bearing portions of the jukebox mechanism. On the other hand, it seems perfectly reasonable to call the extraction of music from a record a "revelation" of information inherent in the record, for several reasons:

(1) the music does not seem to be concealed in the mechanism of the record player;
(2) it is possible to match pieces of the input (the record) with pieces of the output (the music) to an arbitrary degree of accuracy;
(3) it is possible to play other records on the same record player and get other sounds out;
(4) the record and the record player are easily separated from one another.

It is another question altogether whether the fragments of a *smashed* record contain intrinsic meaning. The edges of the separate pieces fit together and in that way allow the information to be reconstituted—but something much more complex is going on here. Then there is the question of the intrinsic meaning of a scrambled telephone call . . . There is a vast spectrum of degrees of inherency of meaning. It is interesting to try to place epigenesis in this spectrum. As development of an organism takes place, can it be said that the information is being "pulled out" of its DNA? Is that where all of the information about the organism's structure resides?

DNA and the Necessity of Chemical Context

In one sense, the answer seems to be yes, thanks to experiments like Avery's. But in another sense, the answer seems to be no, because so much of the pulling-out process depends on extraordinarily complicated cellular chemical processes, which are not coded for in the DNA itself. The DNA relies on the fact that they will happen, but does not seem to contain any code which brings them about. Thus we have two conflicting views on the nature of the information in a genotype. One view says that so much of the information is *outside the DNA* that it is not reasonable to look upon the DNA as anything more than a very intricate set of triggers, like a sequence of buttons to be pushed on a jukebox; another view says that *the information is all there,* but in a very implicit form.

Now it might seem that these are just two ways of saying the same thing, but that is not necessarily so. One view says that the DNA is quite meaningless out of context; the other says that even if it were taken out of context, a molecule of DNA from a living being has such a *compelling inner logic* to its structure that its message could be deduced anyway. To put it as succinctly as possible, one view says that in order for DNA to have meaning, *chemical context* is necessary; the other view says that only *intelligence* is necessary to reveal the "intrinsic meaning" of a strand of DNA.

An Unlikely UFO

We can get some perspective on this issue by considering a strange hypothetical event. A record of David Oistrakh and Lev Oborin playing Bach's sonata in F Minor for violin and clavier is sent up in a satellite. From the satellite it is then launched on a course which will carry it outside of the solar system, perhaps out of the entire galaxy—just a thin plastic platter with a hole in the middle, swirling its way through intergalactic space. It has certainly lost its context. How much meaning does it carry?

If an alien civilization were to encounter it, they would almost certainly be struck by its shape, and would probably be very interested in it. Thus immediately its shape, acting as a trigger, has given them some information: that it is an artifact, perhaps an information-bearing artifact. This idea—communicated, or triggered, by the record itself—now *creates a new context* in which the record will henceforth be perceived. The next steps in the decoding might take considerably longer—but that is very hard for us to assess. We can imagine that if such a record had arrived on earth in Bach's time, no one would have known what to make of it, and very likely it would not have gotten deciphered. But that does not diminish our conviction that the information was in principle *there;* we just know that human knowledge in those times was not very sophisticated with respect to the possibilities of storage, transformation, and revelation of information.

Levels of Understanding of a Message

Nowadays, the idea of decoding is extremely widespread; it is a significant part of the activity of astronomers, linguists, archaeologists, military specialists, and so on. It is often suggested that we may be floating in a sea of radio messages from other civilizations, messages which we do not yet know how to decipher. And much serious thought has been given to the techniques of deciphering such a message. One of the main problems—perhaps the deepest problem—is the question, "How will we recognize the fact that there is a message at all? How to identify a frame?" The sending of a record seems to be a simple solution—its gross physical structure is very attention-drawing, and it is at least plausible to us that it would trigger, in any sufficiently great intelligence, the idea of looking for information hidden in it. However, for technological reasons, sending of solid objects to other star systems seems to be out of the question. Still, that does not prevent our thinking about the idea.

Now suppose that an alien civilization hit upon the idea that the appropriate mechanism for translation of the record is a machine which converts the groove-patterns into sounds. This would still be a far cry from a true deciphering. What, indeed, would constitute a *successful* deciphering of such a record? Evidently, the civilization would have to be able to make sense out of the sounds. Mere production of sounds is in itself hardly worthwhile, unless they have the desired triggering effect in the brains (if that is the word) of the alien creatures. And what is that desired effect? It would be to activate structures in their brains which create emotional effects in them which are analogous to the emotional effects which we experience in hearing the piece. In fact, the production of sounds could even be bypassed, provided that they used the record in some other way to get at the appropriate structures in their brains. (If we humans had a way of triggering the appropriate structures in our brains in sequential order, as music does, we might be quite content to bypass the sounds—but it seems extraordinarily unlikely that there is any way to do that, other than via our ears. Deaf composers—Beethoven, Dvořák, Fauré—or musicians who can "hear" music by looking at a score, do not give the lie to this assertion, for such abilities are founded upon preceding decades of direct auditory experiences.)

Here is where things become very unclear. Will beings of an alien civilization have emotions? Will their emotions—supposing they have some—be mappable, in any sense, onto ours? If they do have emotions somewhat like ours, do the emotions cluster together in somewhat the same way as ours do? Will they understand such amalgams as tragic beauty or courageous suffering? If it turns out that beings throughout the universe do share cognitive structures with us to the extent that even emotions overlap, then in some sense, the record can never be out of its natural context; that context is part of the scheme of things, in nature. And if such is the case, then it is likely that a meandering record, if not destroyed en route, would eventually get picked up by a being or group of beings, and get deciphered in a way which we would consider successful.

"Imaginary Spacescape"

In asking about the meaning of a molecule of DNA above, I used the phrase "compelling inner logic"; and I think this is a key notion. To illustrate this, let us slightly modify our hypothetical record-into-space event by substituting John Cage's "Imaginary Landscape no. 4" for the Bach. This piece is a classic of *aleatoric,* or *chance,* music—music whose structure is chosen by various random processes, rather than by an attempt to convey a personal emotion. In this case, twenty-four performers attach themselves to the twenty-four knobs on twelve radios. For the duration of the piece they twiddle their knobs in aleatoric ways so that each radio randomly gets louder and softer, switching stations all the while. The total sound produced is the piece of music. Cage's attitude is expressed in his own words: "to let sounds be themselves, rather than vehicles for man-made theories or expressions of human sentiments."

Now imagine that this is the piece on the record sent out into space. It would be extraordinarily unlikely—if not downright impossible—for an alien civilization to understand the nature of the artifact. They would probably be very puzzled by the contradiction between the frame message ("I am a message; decode me"), and the chaos of the inner structure. There are few "chunks" to seize onto in this Cage piece, few patterns which could guide a decipherer. On the other hand, there seems to be, in a Bach piece, much to seize onto—patterns, patterns of patterns, and so on. We have no way of knowing whether such patterns are universally appealing. We do not know enough about the nature of intelligence, emotions, or music to say whether the inner logic of a piece by Bach is so universally compelling that its meaning could span galaxies.

However, whether Bach in particular has enough inner logic is not the issue here; the issue is whether *any* message has, per se, enough compelling inner logic that its context will be restored automatically whenever intelligence of a high enough level comes in contact with it. If some message did have that context-restoring property, then it would seem reasonable to consider the meaning of the message as an inherent property of the message.

The Heroic Decipherers

Another illuminating example of these ideas is the decipherment of ancient texts written in unknown languages and unknown alphabets. The intuition feels that there *is* information inherent in such texts, whether or not we succeed in revealing it. It is as strong a feeling as the belief that there is meaning inherent in a newspaper written in Chinese, even if we are completely ignorant of Chinese. Once the script or language of a text has been broken, then no one questions where the meaning resides: clearly it resides *in the text,* not in the method of decipherment—just as music resides *in a record,* not inside a record player! One of the ways that we identify decoding mechanisms is by the fact that they do not *add* any meaning to the signs or objects which they take as input; they merely *reveal* the intrinsic meaning of those signs or objects. A jukebox is not a decoding mechanism, for it does not reveal any meaning belonging to its input symbols; on the contrary, it supplies meaning concealed inside itself.

Now the decipherment of an ancient text may have involved decades of labor by several rival teams of scholars, drawing on knowledge stored in libraries all over the world . . . Doesn't this process add information, too? Just how intrinsic is the meaning of a text, when such mammoth efforts are required in order to find the decoding rules? Has one put meaning into the text, or was that meaning already there? My intuition says that the meaning was always there, and that despite the arduousness of the pulling-out process, no meaning was pulled out that wasn't in the text to start with. This intuition comes mainly from one fact: I feel that the result was inevitable; that, had the text not been deciphered by this group at this time, it would have been deciphered by that group at that time—and it would have come

FIGURE 39. The Rosetta Stone [*courtesy of the British Museum*].

out the same way. That is why the meaning is part of the text itself; it acts upon intelligence in a predictable way. Generally, we can say: meaning is part of an object to the extent that it acts upon intelligence in a predictable way.

In Figure 39 is shown the Rosetta stone, one of the most precious of all historic discoveries. It was the key to the decipherment of Egyptian hieroglyphics, for it contains parallel text in three ancient scripts: hieroglyphics, demotic characters, and Greek. The inscription on this basalt stele was first deciphered in 1821 by Jean François Champollion, the "father of Egyptology"; it is a decree of priests assembled at Memphis in favor of Ptolemy V Epiphanes.

Three Layers of Any Message

In these examples of decipherment of out-of-context messages, we can separate out fairly clearly three levels of information: (1) the *frame* message; (2) the *outer* message; (3) the *inner* message. The one we are most familiar with is (3), the inner message; it is the message which is supposed to be transmitted: the emotional experiences in music, the phenotype in genetics, the royalty and rites of ancient civilizations in tablets, etc.

To understand the inner message is to have extracted the meaning intended by the sender.

The frame message is the message "I am a message; decode me if you can!"; and it is implicitly conveyed by the gross structural aspects of any information-bearer.

To understand the frame message is to recognize the need for a decoding-mechanism.

If the frame message is recognized as such, then attention is switched to level (2), the outer message. This is information, implicitly carried by symbol-patterns and structures in the message, which tells how to decode the inner message.

To understand the outer message is to build, or know how to build, the correct decoding mechanism for the inner message.

This outer level is perforce an implicit message, in the sense that the sender cannot ensure that it will be understood. It would be a vain effort to send instructions which tell how to decode the outer message, for they would have to be part of the inner message, which can only be understood once the decoding mechanism has been found. For this reason, *the outer message is necessarily a set of triggers,* rather than a message which can be revealed by a known decoder.

The formulation of these three "layers" is only a rather crude beginning at analyzing how meaning is contained in messages. There may be layers and layers of outer and inner messages, rather than just one of each. Think, for instance, of how intricately tangled are the inner and outer messages of the Rosetta stone. To decode a message fully, one would have to reconstruct the entire semantic structure which underlay its creation—and thus to understand the sender in every deep way. Hence one could throw away the inner message, because if one truly understood all the finesses of the outer message, the inner message would be reconstructible.

The book *After Babel,* by George Steiner, is a long discussion of the interaction between inner and outer messages (though he never uses that terminology). The tone of his book is given by this quote:

> We normally use a shorthand beneath which there lies a wealth of subconscious, deliberately concealed or declared associations so extensive and intri-

cate that they probably equal the sum and uniqueness of our status as an individual person.[1]

Thoughts along the same lines are expressed by Leonard B. Meyer, in his book *Music, the Arts, and Ideas:*

> The way of listening to a composition by Elliott Carter is radically different from the way of listening appropriate to a work by John Cage. Similarly, a novel by Beckett must in a significant sense be read differently from one by Bellow. A painting by Willem de Kooning and one by Andy Warhol require different perceptional-cognitive attitudes.[2]

Perhaps works of art are trying to convey their style more than anything else. In that case, if you could ever plumb a style to its very bottom, you could dispense with the creations in that style. "Style", "outer message", "decoding technique"—all ways of expressing the same basic idea.

Schrödinger's Aperiodic Crystals

What makes us see a frame message in certain objects, but none in others? Why should an alien civilization suspect, if they intercept an errant record, that a message lurks within? What would make a record any different from a meteorite? Clearly its geometric shape is the first clue that "something funny is going on". The next clue is that, on a more microscopic scale, it consists of a very long aperiodic sequence of patterns, arranged in a spiral. If we were to unwrap the spiral, we would have one huge linear sequence (around 2000 feet long) of minuscule symbols. This is not so different from a DNA molecule, whose symbols, drawn from a meager "alphabet" of four different chemical bases, are arrayed in a one-dimensional sequence, and then coiled up into a helix. Before Avery had established the connection between genes and DNA, the physicist Erwin Schrödinger predicted, on purely theoretical grounds, that genetic information would have to be stored in "aperiodic crystals", in his influential book *What Is Life?* In fact, books themselves are aperiodic crystals contained inside neat geometrical forms. These examples suggest that, where an aperiodic crystal is found "packaged" inside a very regular geometric structure, there may lurk an inner message. (I don't claim this is a complete characterization of frame messages; however, it is a fact that many common messages have frame messages of this description. See Figure 40 for some good examples.)

Languages for the Three Levels

The three levels are very clear in the case of a message found in a bottle washed up on a beach. The first level, the frame message, is found when one picks up the bottle and sees that it is sealed, and contains a dry piece of paper. Even without seeing writing, one recognizes this type of artifact as an information-bearer, and at this point it would take an extraordinary—almost inhuman—lack of curiosity, to drop the bottle and not look further.

เรียนภาษาอังกฤษด้วยภาพ

เมื่อท่านเรียนตามวิธีนี้ไปได้ ๓๐ หน้า ลองทวนความรู้ของท่านด้วยการหัดตอบ คำถามเป็นภาษาอังกฤษในหน้า ๓๑, ๓๒ และ ๓๓ แล้วพลิกไปตรวจดูคำตอบในหน้า ๓๔ ว่าถูกต้องหรือไม่ คำถามและคำตอบมีให้ไว้ต่อๆ ไปตลอดทั้งเล่ม

வாசித்துவருகையில், ஒவ்வொரு வாக்கியத்தையும் அதற்குரிய படத்தோடு ஒப்பிட்டுப் பார்க்குந் தோறும் அர்த்தமாகிக்கொண்டு வரும். மனப்பாடம் பண்ணிக்கொள்ள வேண்டிய அம்சங்கள் மிகக் குறைவு. ஆகையினால் அர்த்தத்துக்கு இணங்க வாக்கிய அமைப்பின் வேறுபாடுகளைக் கண்டறிந்து கொள்வதற்கு வசதிகள் உண்டு. இம்முறை அநுசரித்து ஆங்கிலம் கற்றுக்கொள்வது கஷ்டமில்லாததுமன்றி விளையாட்டுச் சம்பந்தப்பட்டது.

কত অজানারে জানাইলে তুমি,
কত ঘরে দিলে ঠাঁই—
দূরকে করিলে নিকট বন্ধু,
পরকে করিলে ভাই।
পুরানো আবাস ছেড়ে যাই যবে
মনে ভেবে মরি কী জানি কী হবে,
নূতনের মাঝে তুমি পুরাতন
সে কথা যে ভুলে যাই
দূরকে করিলে নিকট বন্ধু,
পরকে করিলে ভাই

ഡീസൽ തീവണ്ടി എൻജിന്
നാടൻ നിർമ്മിത വസ്തുക്കൾ

Next, one opens the bottle and examines the marks on the paper. Perhaps they are in Japanese; this can be discovered without any of the inner message being understood—it merely comes from a recognition of the characters. The outer message can be stated as an English sentence: "I am in Japanese." Once this has been discovered, then one can proceed to the inner message, which may be a call for help, a haiku poem, a lover's lament . . .

It would be of no use to include in the inner message a translation of the sentence "This message is in Japanese", since it would take someone who knew Japanese to read it. And before reading it, he would have to recognize the fact that, as it is in Japanese, he can read it. You might try to wriggle out of this by including translations of the statement "This message is in Japanese" into many different languages. That would help in a practical sense, but in a theoretical sense the same difficulty is there. An English-speaking person still has to recognize the "Englishness" of the message; otherwise it does no good. Thus one cannot avoid the problem that one has to find out how to decipher the inner message *from the outside;* the inner message itself may provide clues and confirmations, but those are at best triggers acting upon the bottle finder (or upon the people whom he enlists to help).

Similar kinds of problem confront the shortwave radio listener. First, he has to decide whether the sounds he hears actually constitute a message, or are just static. The sounds in themselves do not give the answer, not even in the unlikely case that the inner message is in the listener's own native language, and is saying, "These sounds actually constitute a message and are not just static!" If the listener recognizes a frame message in the sounds, then he tries to identify the language the broadcast is in—and clearly, he is still on the outside; he accepts *triggers* from the radio, but they cannot explicitly tell him the answer.

It is in the nature of outer messages that they are not conveyed in any

FIGURE 40. A collage of scripts. Uppermost on the left is an inscription in the undeciphered boustrophedonic writing system from Easter Island, in which every second line is upside down. The characters are chiseled on a wooden tablet, 4 inches by 35 inches. Moving clockwise, we encounter vertically written Mongolian: above, present-day Mongolian, and below, a document dating from 1314. Then we come to a poem in Bengali by Rabindranath Tagore in the bottom righthand corner. Next to it is a newspaper headline in Malayālam (West Kerala, southern India), above which is the elegant curvilinear language Tamil (East Kerala). The smallest entry is part of a folk tale in Buginese (Celebes Island, Indonesia). In the center of the collage is a paragraph in the Thai language, and above it a manuscript in Runic dating from the fourteenth century, containing a sample of the provincial law of Scania (south Sweden). Finally, wedged in on the left is a section of the laws of Hammurabi, written in Assyrian cuneiform. As an outsider, I feel a deep sense of mystery as I wonder how meaning *is cloaked in the strange curves and angles of each of these beautiful aperiodic crystals. In form, there is content.* [*From Hans Jensen,* Sign, Symbol, and Script *(New York: G. Putnam's Sons, 1969), pp. 89 (cuneiform), 356 (Easter Island), 386, 417 (Mongolian), 552 (Runic); from Kenneth Katzner,* The Languages of the World *(New York: Funk & Wagnalls, 1975), pp. 190 (Bengali), 237 (Buginese); from I. A. Richards and Christine Gibson,* English Through Pictures *(New York: Washington Square Press, 1960), pp. 73 (Tamil), 82 (Thai).*]

explicit language. To find an explicit language in which to convey outer messages would not be a breakthrough—it would be a contradiction in terms! It is always the listener's burden to understand the outer message. Success lets him break through into the inside, at which point the ratio of triggers to explicit meanings shifts drastically towards the latter. By comparison with the previous stages, understanding the inner message seems effortless. It is as if it just gets pumped in.

The "Jukebox" Theory of Meaning

These examples may appear to be evidence for the viewpoint that no message has intrinsic meaning, for in order to understand any inner message, no matter how simple it is, one must first understand its frame message and its outer message, both of which are carried only by triggers (such as being written in the Japanese alphabet, or having spiraling grooves, etc.). It begins to seem, then, that one cannot get away from a "jukebox" theory of meaning—the doctrine that *no message contains inherent meaning,* because, before any message can be understood, it has to be used as the input to some "jukebox", which means that information contained in the "jukebox" must be added to the message before it acquires meaning.

This argument is very similar to the trap which the Tortoise caught Achilles in, in Lewis Carroll's Dialogue. There, the trap was the idea that before you can use any rule, you have to have a rule which tells you how to use that rule; in other words, there is an infinite hierarchy of levels of rules, which prevents any rule from ever getting used. Here, the trap is the idea that before you can understand any message, you have to have a message which tells you how to understand that message; in other words, there is an infinite hierarchy of levels of messages, which prevents any message from ever getting understood. However, we all know that these paradoxes are invalid, for rules *do* get used, and messages *do* get understood. How come?

Against the Jukebox Theory

This happens because our intelligence is not disembodied, but is instantiated in physical objects: our brains. Their structure is due to the long process of evolution, and their operations are governed by the laws of physics. Since they are physical entities, *our brains run without being told how to run.* So it is at the level where thoughts are produced by physical law that Carroll's rule-paradox breaks down; and likewise, it is at the level where a brain interprets incoming data as a message that the message-paradox breaks down. It seems that brains come equipped with "hardware" for recognizing that certain things are messages, and for decoding those messages. This minimal inborn ability to extract inner meaning is what allows the highly recursive, snowballing process of language acquisition to take place. The inborn hardware is like a jukebox: it supplies the additional information which turns mere triggers into complete messages.

Meaning Is Intrinsic If Intelligence Is Natural

Now if different people's "jukeboxes" had different "songs" in them, and responded to given triggers in completely idiosyncratic ways, then we would have no inclination to attribute intrinsic meaning to those triggers. However, human brains are so constructed that one brain responds in much the same way to a given trigger as does another brain, all other things being equal. This is why a baby can learn any language; it responds to triggers in the same way as any other baby. This uniformity of "human jukeboxes" establishes a uniform "language" in which frame messages and outer messages can be communicated. If, furthermore, we believe that human intelligence is just one example of a general phenomenon in nature—the emergence of intelligent beings in widely varying contexts—then presumably the "language" in which frame messages and outer messages are communicated among humans is a "dialect" of a *universal* language by which intelligences can communicate with each other. Thus, there would be certain kinds of triggers which would have "universal triggering power", in that all intelligent beings would tend to respond to them in the same way as we do.

This would allow us to shift our description of where meaning is located. We could ascribe the meanings (frame, outer, and inner) of a message to the message itself, because of the fact that deciphering mechanisms are themselves universal—that is, they are fundamental forms of nature which arise in the same way in diverse contexts. To make it very concrete, suppose that "A-5" triggered the same song in all jukeboxes—and suppose moreover that jukeboxes were not man-made artifacts, but widely occurring natural objects, like galaxies or carbon atoms. Under such circumstances, we would probably feel justified in calling the universal triggering power of "A-5" its "inherent meaning"; also, "A-5" would merit the name of "message", rather than "trigger", and the song would indeed be a "revelation" of the inherent, though implicit, meaning of "A-5".

Earth Chauvinism

This ascribing of meaning to a message comes from the invariance of the processing of the message by intelligences distributed anywhere in the universe. In that sense, it bears some resemblance to the ascribing of mass to an object. To the ancients, it must have seemed that an object's weight was an intrinsic property of the object. But as gravity became understood, it was realized that weight varies with the gravitational field the object is immersed in. Nevertheless, there is a related quantity, the mass, which does *not* vary according to the gravitational field; and from this invariance came the conclusion that an object's mass was an intrinsic property of the object itself. If it turns out that mass is also variable, according to context, then we will backtrack and revise our opinion that it is an intrinsic property of an object. In the same way, we might imagine that there could exist other kinds of "jukeboxes"—intelligences—which communicate among each other via messages which we would never recognize as messages, and who also would never recognize *our* messages as messages. If that were the case, then the claim that meaning is an intrinsic property of a set of symbols would have to be reconsidered. On the other hand, how could we ever realize that such beings existed?

It is interesting to compare this argument for the inherency of meaning with a parallel argument for the inherency of weight. Suppose one defined an object's weight as "the magnitude of the downward force which the object exerts when on the surface of the planet Earth". Under this definition, the downward force which an object exerts when on the surface of Mars would have to be given another name than "weight". This definition makes weight an inherent property, but at the cost of geocentricity—"Earth chauvinism". It would be like "Greenwich chauvinism"—refusing to accept local time anywhere on the globe but in the GMT time zone. It is an unnatural way to think of time.

Perhaps we are unknowingly burdened with a similar chauvinism with respect to intelligence, and consequently with respect to meaning. In our chauvinism, we would call any being with a brain sufficiently much like our own "intelligent", and refuse to recognize other types of objects as intelligent. To take an extreme example, consider a meteorite which, instead of deciphering the outer-space Bach record, punctures it with colossal indifference, and continues in its merry orbit. It has interacted with the record in a way which we feel disregards the record's meaning. Therefore, we might well feel tempted to call the meteorite "stupid". But perhaps we would thereby do the meteorite a disservice. Perhaps it has a "higher intelligence" which we in our Earth chauvinism cannot perceive, and its interaction with the record was a manifestation of that higher intelligence. Perhaps, then, the record has a "higher meaning"—totally different from that which we attribute to it; perhaps its meaning depends on the *type* of intelligence perceiving it. Perhaps.

It would be nice if we could define intelligence in some other way than "that which gets the same meaning out of a sequence of symbols as we do". For if we can only define it this one way, then our argument that meaning is an intrinsic property is circular, hence content-free. We should try to formulate in some independent way a set of characteristics which deserve the name "intelligence". Such characteristics would constitute the uniform core of intelligence, shared by humans. At this point in history we do not yet have a well-defined list of those characteristics. However, it appears likely that within the next few decades there will be much progress made in elucidating what human intelligence is. In particular, perhaps cognitive psychologists, workers in Artificial Intelligence, and neuroscientists will be able to synthesize their understandings, and come up with a definition of intelligence. It may still be human-chauvinistic; there is no way around that. But to counterbalance that, there may be some elegant and beautiful—and perhaps even simple—abstract ways of characterizing the essence of intelligence. This would serve to lessen the feeling of having formulated an anthropocentric concept. And of course, if contact were established with an alien civilization from another star system, we would feel supported in our belief that our own type of intelligence is not just a fluke, but an example of a basic form which reappears in nature in diverse contexts, like stars and uranium nuclei. This in turn would support the idea of meaning being an inherent property.

To conclude this topic, let us consider some new and old examples, and discuss the degree of inherent meaning which they have, by putting ourselves, to the extent that we can, in the shoes of an alien civilization which intercepts a weird object . . .

Two Plaques in Space

Consider a rectangular plaque made of an indestructible metallic alloy, on which are engraved two dots, one immediately above the another: the preceding colon shows a picture. Though the overall form of the object might suggest that it is an artifact, and therefore that it might conceal some message, two dots are simply not sufficient to convey anything. (Can you, before reading on, hypothesize what they are supposed to mean?) But suppose that we made a second plaque, containing more dots, as follows:

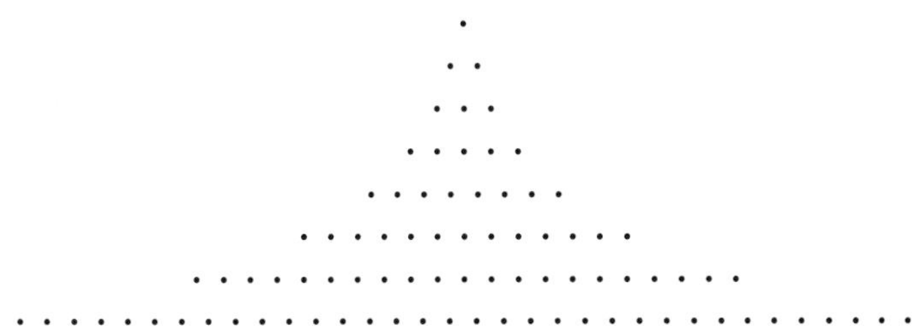

Now one of the most obvious things to do—so it might seem to a terrestrial intelligence at least—would be to count the dots in the successive rows. The sequence obtained is:

1, 1, 2, 3, 5, 8, 13, 21, 34.

Here there is evidence of a rule governing the progression from one line to the next. In fact, the recursive part of the definition of the Fibonacci numbers can be inferred, with some confidence, from this list. Suppose we think of the initial pair of values (1,1) as a "genotype" from which the "phenotype"—the full Fibonacci sequence—is pulled out by a recursive rule. By sending the genotype alone—namely the first version of the plaque—we fail to send the information which allows reconstitution of the phenotype. Thus, the genotype does not contain the full specification of the phenotype. On the other hand, if we consider the second version of the plaque to be the genotype, then there is much better cause to suppose that the phenotype could actually be reconstituted. This new version of the genotype—a "long genotype"—contains so much information that *the mechanism by which phenotype is pulled out of genotype can be inferred by intelligence from the genotype alone.*

Once this mechanism is firmly established as the way to pull phenotype from genotype, then we can go back to using "short genotypes"—like the first plaque. For instance, the "short genotype" (1,3) would yield the phenotype

1, 3, 4, 7, 11, 18, 29, 47, . . .

—the Lucas sequence. And for every set of two initial values—that is, for every short genotype—there will be a corresponding phenotype. But the short genotypes, unlike the long ones, are only triggers—buttons to be pushed on the jukeboxes into which the recursive rule has been built. The long genotypes are informative enough that they trigger, in an intelligent being, the recognition of what kind of "jukebox" to build. In that sense, the long genotypes contain the information of the phenotype, whereas the short genotypes do not. In other words, the long genotype transmits not only an inner message, but also an outer message, which enables the inner message to be read. It seems that the clarity of the outer message resides in the sheer length of the message. This is not unexpected; it parallels precisely what happens in deciphering ancient texts. Clearly, one's likelihood of success depends crucially on the amount of text available.

Bach *vs.* Cage Again

But just having a long text may not be enough. Let us take up once more the difference between sending a record of Bach's music into space, and a record of John Cage's music. Incidentally, the latter, being a Composition of Aleatorically Generated Elements, might be handily called a "CAGE", whereas the former, being a Beautiful Aperiodic Crystal of Harmony, might aptly be dubbed a "BACH". Now let's consider what the meaning of a Cage piece is to ourselves. A Cage piece has to be taken in a large cultural setting—as a revolt against certain kinds of traditions. Thus, if we want to transmit that meaning, we must not only send the notes of the piece, but we must have earlier communicated an extensive history of Western culture. It is fair to say, then, that an isolated record of John Cage's music does *not* have an intrinsic meaning. However, for a listener who is sufficiently well versed in Western and Eastern cultures, particularly in the trends in Western music over the last few decades, it does carry meaning—but such a listener is like a jukebox, and the piece is like a pair of buttons. The meaning is mostly contained inside the listener to begin with; the music serves only to trigger it. And this "jukebox", unlike pure intelligence, is not at all universal; it is highly earthbound, depending on idiosyncratic sequences of events all over our globe for long periods of time. Hoping that John Cage's music will be understood by another civilization is like hoping that your favorite tune, on a jukebox on the moon, will have the same code buttons as in a saloon in Saskatoon.

On the other hand, to appreciate Bach requires far less cultural knowledge. This may seem like high irony, for Bach is so much more complex and organized, and Cage is so devoid of intellectuality. But there is a strange reversal here: intelligence loves patterns and balks at randomness. For most people, the randomness in Cage's music requires much explanation; and even after explanations, they may feel they are missing the message—whereas with much of Bach, words are superfluous. In that sense, Bach's music is more self-contained than Cage's music. Still, it is not clear how much of the human condition is presumed by Bach.

For instance, music has three major dimensions of structure (melody, harmony, rhythm), each of which can be further divided into small-scale, intermediate, and overall aspects. Now in each of these dimensions, there is a certain amount of complexity which our minds can handle before boggling; clearly a composer takes this into account, mostly unconsciously, when writing a piece. These "levels of tolerable complexity" along different dimensions are probably very dependent on the peculiar conditions of our evolution as a species, and another intelligent species might have developed music with totally different levels of tolerable complexity along these many dimensions. Thus a Bach piece might conceivably have to be accompanied by a lot of information about the human species, which simply could not be inferred from the music's structure alone. If we equate the Bach music with a genotype, and the emotions which it is supposed to evoke with the phenotype, then what we are interested in is whether the genotype contains all the information necessary for the revelation of the phenotype.

How Universal Is DNA's Message?

The general question which we are facing, and which is very similar to the questions inspired by the two plaques, is this: "How much of the context necessary for its own understanding is a message capable of restoring?" We can now revert to the original biological meanings of "genotype" and "phenotype"—DNA and a living organism—and ask similar questions. Does DNA have universal triggering power? Or does it need a "biojukebox" to reveal its meaning? Can DNA evoke a phenotype without being embedded in the proper chemical context? To this question the answer is no—but a qualified no. Certainly a molecule of DNA in a vacuum will not create anything at all. However, if a molecule of DNA were sent out to seek its fortune in the universe, as we imagined the BACH and the CAGE were, it might be intercepted by an intelligent civilization. They might first of all recognize its frame message. Given that, they might go on to try to deduce from its chemical structure what kind of chemical environment it seemed to want, and then supply such an environment. Successively more refined attempts along these lines might eventually lead to a full restoration of the chemical context necessary for the revelation of DNA's phenotypical meaning. This may sound a little implausible, but if one allows many millions of years for the experiment, perhaps the DNA's meaning would finally emerge.

On the other hand, if the sequence of bases which compose a strand of DNA were sent as abstract symbols (as in Fig. 41), not as a long helical molecule, the odds are virtually nil that this, as an outer message, would trigger the proper decoding mechanism which would enable the phenotype to be drawn out of the genotype. This would be a case of wrapping an inner message in such an abstract outer message that the context-restoring power of the outer message would be lost, and so in a very pragmatic sense, the set of symbols would have no intrinsic meaning. Lest you think this all sounds hopelessly abstract and philosophical, consider that the exact moment when phenotype can be said to be "available", or "implied", by genotype, is a highly charged issue in our day: it is the issue of abortion.

FIGURE 41. This Giant Aperiodic Crystal is the base sequence for the chromosome of bacteriophage ϕX174. *It is the first complete genome ever mapped out for any organism. About 2,000 of these boustrophedonic pages would be needed to show the base sequence of a single* E. Coli *cell, and about one million pages to show the base sequence of the DNA of a single human cell. The book now in your hands contains roughly the same amount of information as a molecular blueprint for one measly* E. Coli *cell.*

CCGTCAGGATTGACACCCTCCCAATTGTATGTTTTCATGCCTCCAAATCTTGGAGGCTTTTTTATGGTTCGTTCTTATTACCCTTCTGAATGTCACGCTG
ACGAATACCTTCGGTTCGTAACCCCTAACTCTTTCTCATCTTTACGGTGTTCGGAGTTATCGTCCAAATTCTCGGAGCTATGCGAGTTTCAGTTTTATTA
GATGGATAACCGCATCAAGCTCTTGGAAGAGATTCTGTCTTTTCGTATGCAGGGCGTTGAGTTCGATAATGGTGATATGTATGTTGACGGCCATAAGGCT
ACAATAATTATAGTTCAACCCCCTCGTGTAACATCGTAACACGGTTAAGTAGGTAATTGAAGAGTCATTGTCTATGTTTGAGTAGTGCTTGCAGTCTTCG
CTATAGACCACCGCCCCGAAGGGGACGAAAAATGGTTTTTAGAGAACGAGAAGACGGTTACGCAGTTTTGCCGCAAGCTGGCTGCTGAACGCCCTCTTAA
TTTCGGAGATGCGCTATAGAATCACCTCCGGAGGTCGTTAGAACTTGTGAGTAGGAATTATGGAAAGAAAAACCCCATTAATATGAGTAGCGCTTATAGG
GCTATTCAGCGTTTGATGAATGCAATGCGACAGGCTCATGCTGATGGTTGGTTTATCGTTTTTGACACTCTCACGTTGGCTGACGACCGATTAGAGGCGT
GTCAGCCGCACACTTAGTAATCGGAACGCTGGAGCCGTCGTTCTTGGTATGCTGGTTATAGTGCTTTTATCAGTGCGTTTCGTAACCCTAATAGTATTT
CTATCAGTATTTTTGTGTGCCTGAGTATGCTACAGCTAATGGCCGTCTTCATTTCCATGCGGTGCACTTTATGCGGACACTTCCTACAGGTAGCGTTGAC
CGCACATCGCTTGACGCTACCCGTATGACATTGGTATTCCGGTGCATAAAACGTTCGATAAATTGACCGCCCGCTAACGCATAGGCTGCTGGTTTTAATCC
AGGACGCTTTTTCACGTTCTGGTTGGTTGTGGCCTGTTGATGCTAAAGGTGAGCCGCTTAAAGCTACCAGTTATATGGCTGTTGGTTTCTATGTGGCTAA
GAACCCTTCATCGCTGTCGAACCAAAAATCACTCAACAAGGTAAGAAATCGACGATCTGGAAATCGTCGTTCCAGGTATAGACTGAAAAACAATTGCATA
AAGCTGTTCAGAATCAGAATGAGCCGCAACTTCGGGATGAAAATGCTCACAATGACAAATCTGTCCACGGAGTGCTTAATCCAACTTACCAAGCTGGGTT
ACGCGGCGGTTTTGCAGCCGATGTCATTGAAAAGGGTCGGAGTTAGAGTAGAGAGAAAAACGCAAGACGAAGTTATAGACCAACTTGCCGCAGCGCAGCA
ACCTGTGACGACAAATCTGCTCAAATTTATGCGCGCTTCGATAAAAATGATTGGCGTATCCAACCTGCAGAGTTTTATCGCTTCCATGACGCAGAAGTTA
AAGGCGGTCGTCAGGTGAAGCTAAATTAAGCATTTGTTCGTCATCATTAAGGACGAAATAGTTCTATTAAAAAGCTGAGTAGTCTTTATAGGCTTTCACA
AATGAGAAAATTCGACCTATCCTTGCGCAGCTCGAGAAGCTCTTACTTTGCGACCTTTCGCCATCAACTAACGATTCTGTCAAAAACTGACGCGTTGGAT
AATTTTACAGTTGTTCTCTTAGAGATGGTACTTGTTTTACACTGAGTATAGATTTGGTCAGGAACTGCTTGCACGGTTCGTATAATTCGGTGAAGAGGAG
AAGAGCGTGGATTACTATCTGAGTCCGATGCTGTTCAACCACTAATAGGTAAGAAATCATGAGTCAAGTTACTGAACAATCCGTACGTTTCCAGACCGCT
CCAGTCATCGTTAGGTTTGAAACAATGAGCAGTCTTTTAGCTTTAGTAGAAGCCAATTTAGGTTTTGCCGTCTTCGGACTTACTCGAATTATCTCCGGTT
GCTCTCGTGCTCGTCGCTGCGTTGAGGCTTGCGTTTATGGTACGCTGGACTTTGTAGGATACCCTCGCTTTCCTGCTCCTGTTGAGTTTATTGCTGCCGT
GCCTGCGAGCTGCGGTAATTATTACAAAAGGCATTTAAGTCGCGGAAGGTACTACTCTGTCCGGCAAACTTACAACTGCCCTACTTGTATTATTCGTTAC
GTTAAAGCCGCTGAATTGTTCGCGTTTACCTTGCGTGTACGCGCAGGAAACACTGACGTTCTTACTGACGCAGAAGAAAACGTGCGTCAAAAATTACGTG
TCGCGGAAATGCGAACGGAAATCATGGAGCGTTGCCGACGCCTGCTGGTCCCGCTCGCGGTCTTGCAAAAAATGGAAATCTGTAATGTAGTGAGGAAGGC
CGTCTTTGGTATGTAGGTGGTCAACAATTTTAATTGCAGGGGCTTCGGCCCCTTACTTGAGGATAAATTATGTCTAATATTCAAACTGGCGCCGAGCGTA
TAGAGCTTCCTCAGCGGTCGCTATTGGCCTCATCAACTTTACCATTATTCTGCTGGTTAGACTGGTCGTTCCTTCGGTTCTACCCTTTCCAGTACGCCGT
GGACGCCGTTGGCGCTCTCCGTCTTTCTCCATTGCGTCGTGGCCTTGCTATTGACTCTACTGTAGACATTTTTACTTTTTATGTCCCTCATCGTCACGTT
GTTCTTTTCGCCGTACCAGTTATATTGGTCATCACAATTGTCAGCCCTCTCCTCACCGTAATTGTGGTAGGAAGTACTTGAATTAGGTGACAAGTGGTAT
GCACGATTAACCCTGATACCAATAAAATCCCTAAGCATTTGTTTCAGGGTTATTTGAATATCTATAACAACTATTTTAAAGCGCCGTGGATGCCTGACCG
GAGTCCTCCTTCGCCTCGTCAGGTTTACAAAAACTCTACCGTCGTTGCCTTTGGTATTGCTCGTAGTAGAACTAATTCGAGTAATCCCAATCGGAGCCAT
ACTGAGCTTTCTCGCCAAATGACGACTTCTACCACATCTATTGACATTATGGGTCTGCAAGCTGCTTATGCTAATTTGCATACTGACCAAGAACGTGATT
GTCTCTAATCTCGCGTACTGTTCATTTCCTGCCAACAGTCGCAGTATACTCCAAAATGGAGGTTTACTTCTTTATTGTAGTACCATTGCGACGTACTTCA
GGCATCTGGCTATGATGTTGATGGAACTGACCAAACGTCGTTAGGCCAGTTTTCTGGTCGTGTTCAACAGACCTATAAACATTCTGTGCCGCGTTTCTTT
TTCAGTTTCGTGGAAATCGCAATTCCATGACTTAGAGAAATCAGCGTCATCCGCCTTTTGCTTGTTCGCGTTCTCATTTGTATCACGGTACGAGTCCTTG
ATACCGATATTGCTGGCGACCCTGTTTTGTATGGCAACTTGCCGCCGCGTGAAATTTCTATGAAGGATGTTTTCCGTTCTGGTGATTCGTCTAAGAAGTT
TCTTCCCGCCAAGGACTTACTTACCCTTCGGAAGTTCTTCCACTATTCGTCCTCTTTGTATGCTTCCGCGTATTGCTATGGTGACTGGCGAGTCGTTAGAAT
GGTGATTTGCAAGAACGCGTACTTATTCGCAACCATGATTATGACCAGTGTTTCAGTCGTTCAGTTGTTGCAGTGGATAGTCTTACCTCATGTGACGTTT
CGCCGTTTTTAATTTTAAAAATGGCGAAGCCAATATTGGAGTGTGAGTTAGAAAATAGTGCTTCAGTACTAACTTAGCGCTCACCAGCCGTCTAACGCTA
TGAGGGGTTGACCAAGCGAAGCGCGGTAGGTTTTCTGCTTAGGAGTTTAATCATGTTTCAGACTTTTATTTCTCGCCACAATTCAAACTTTTTTTCTGAT
ATTGGCAGTTTGATAGTTTTATATTGCAACTGCTACATCGAAATCCACAGACATTTTGTCCACGGCTTCTTCGACCTCATTGTCTTCACTCTTGGTCGAA
ATGCTGGTAATGGTGGTTTTCTTCATTGCATTCAGATGGATACATCTGTCAACGCCGCTAATCAGGTTGTTTCAGTTGGTGCTGATATTGCTTTTGATGC
TACCGCTGGTAGGTTTCCTATTTGTAGTATCCGTCAGCCCTCCCATCAGCCTTGGCTTCTTCTGAGTTTCGCTTGGTTTGTCCGTTTTTTAAATCCCAGC
GATGGTGGTTATTATACCGTCAAGGACTGTGTGACTATTGACGTCCTTCCCCGTACGCCCCGGCAATAACGTCTACGTTGGTTTCATGGTTTGGTCTAACT
TCGTGCTTTGTATTTAGTGGAGTGAATTCACCGACCTCTGTTTATTAGAGAAATTATTGGACTAAGTCGCTTTGGTTAGGCGCCGTAAATCATCGCCATT
ATTGCTGGCGGTATTGCTTCTGCTCTTGCTGGTGGCGCCATGTCTAAATTGTTTGGAGGCGGTCAAAAAGCCGCCTCCGGTGGCATTCAAGGTGATGTGC
TTTGATCCCCGCCGGAGTAGTCCCAATCCTTGTAATCTCGGAACTTACCGTCTAAATTATGGTCGTAGTGGGTACGGATGTCATAACAATAGCCATCGTT
TGTTTCTGGTGCTATGGCTAAAGCTGGTAAAGGACTTCTTGAAGGTACGTTGCAGGCTGGCACTTCTGCCGTTTCTGATAAGTTGCTTGATTTGGTTGGA
GTAGTCGTGGTCGTGCGAGGGTTCGTAATTCGAGTCCTTTACGTCGTCGTTCTATTAGTGCTCATAGGAAAGGAAATAGTCGCCGTCTGAACGGTGGTTC
CTTCCTCTGCTGGTATGGTTGACGCCGGATTTGAGAATCAAAAAGAGCTTACTAAAATGCAACTGGACAATCAGAAAGAGATTGCCGAGATGCAAAATGA
GAAGACAACTATTCGTTCGTAGAGTAAAACACGTATATGGACCAGAAAGCATAAGACCCCACTTCAGCGGCTGACTTACCGTCGTTAGAGAAAAACTCAG
GAGTCTACTGCTCGCGTTGCGTCTATTATGGAAAACACCAATCTTTCCAAGCAACAGCAGGTTTCCGAGATTATGCGCCAAATGCTTACTCAAGCTCAAA
ATGGCGACTAAGACGCAAACGACTACTTGATTCAGTTGGACTCGTGATTGGAACGCTCAGTAAAGAAACTAAACCAGTAACCATTTTATGACTGGTCGGC
TGGCTCTTCTCATATTGGCGCTACTGCAAAGGATATTTCTAATGTCGTCACTGATGCTGCTTCTGGTGTGGTTGATATTTTTCATGGTATTGATAAAGCT
AATAAAGGATCTGTTTAATCTCGGTTATGGTAGTCGAAATGGCAGAAAGGTCTTTAACAAGGTTCATAGCCGTTG

Chromatic Fantasy, And Feud

Having had a splendid dip in the pond, the Tortoise is just crawling out and shaking himself dry, when who but Achilles walks by.

Tortoise: Ho there, Achilles. I was just thinking of you as I splashed around in the pond.

Achilles: Isn't that curious? I was just thinking of you, too, while I meandered through the meadows. They're so green at this time of year . . .

Tortoise: You think so? It reminds me of a thought I was hoping to share with you. Would you like to hear it?

Achilles: Oh, I would be delighted. That is, I would be delighted as long as you're not going to try to snare me in one of your wicked traps of logic, Mr. T.

Tortoise: Wicked traps? Oh, you do me wrong. Would I do anything wicked? I'm a peaceful soul, bothering nobody and leading a gentle, herbivorous life. And my thoughts merely drift among the oddities and quirks of how things are (as I see them). I, humble observer of phenomena, plod along and puff my silly words into the air rather unspectacularly, I am afraid. But to reassure you about my intentions, I was only planning to speak of my Tortoise-shell today, and as you know, those things have nothing—nothing whatsoever—to do with logic!

Achilles: Your words DO reassure me, Mr. T. And, in fact, my curiosity is quite piqued. I would certainly like to listen to what you have to say, even if it is unspectacular.

Tortoise: Let's see . . . how shall I begin? Hmm . . . What strikes you most about my shell, Achilles?

Achilles: It looks wonderfully clean!

Tortoise: Thank you. I just went swimming and washed off several layers of dirt which had accumulated last century. Now you can see how green my shell is.

Achilles: Such a good healthy green shell, it's nice to see it shining in the sun.

Tortoise: Green? It's not green.

Achilles: Well, didn't you just tell me your shell was green?

Tortoise: I did.

Achilles: Then, we agree: it is green.

Tortoise: No, it isn't green.

Achilles: Oh, I understand your game. You're hinting to me that what you say isn't necessarily true; that Tortoises play with language; that your statements and reality don't necessarily match; that—

Tortoise: I certainly am not. Tortoises treat words as sacred; Tortoises revere accuracy.

Achilles: Well, then, why did you say that your shell is green, and that it is not green also?

Tortoise: I never said such a thing; but I wish I had.

Achilles: You would have liked to say that?

Tortoise: Not a bit. I regret saying it, and disagree wholeheartedly with it.

Achilles: That certainly contradicts what you said before!

Tortoise: Contradicts? Contradicts? I never contradict myself. It's not part of Tortoise-nature.

Achilles: Well, I've caught you this time, you slippery fellow, you. Caught you in a full-fledged contradiction.

Tortoise: Yes, I guess you did.

Achilles: There you go again! Now you're contradicting yourself more and more! You are so steeped in contradiction it's impossible to argue with you!

Tortoise: Not really. I argue with myself without any trouble at all. Perhaps the problem is with you. I would venture a guess that maybe you're the one who's contradictory, but you're so trapped in your own tangled web that you can't see how inconsistent you're being.

Achilles: What an insulting suggestion! I'm going to show you that you're the contradictory one, and there are no two ways about it.

Tortoise: Well, if it's so, your task ought to be cut out for you. What could be easier than to point out a contradiction? Go ahead—try it out.

Achilles: Hmm . . . Now I hardly know where to begin. Oh . . . I know. You first said that (1) your shell is green, and then you went on to say that (2) your shell is not green. What more can I say?

Tortoise: Just kindly point out the contradiction. Quit beating around the bush.

Achilles: But—but—but . . . Oh, now I begin to see. (Sometimes I am so slow-witted!) It must be that you and I differ as to what constitutes a contradiction. That's the trouble. Well, let me make myself very clear: a contradiction occurs when somebody says one thing and denies it at the same time.

Tortoise: A neat trick. I'd like to see it done. Probably ventriloquists would excel at contradictions, speaking out of both sides of their mouth, as it were. But I'm not a ventriloquist.

Achilles: Well, what I actually meant is just that somebody can say one thing and deny it all within one single sentence! It doesn't literally have to be in the same instant.

Tortoise: Well, you didn't give ONE sentence. You gave TWO.

Achilles: Yes—two sentences that contradict each other!

Tortoise: I am sad to see the tangled structure of your thoughts becoming so exposed, Achilles. First you told me that a contradiction is something which occurs in a single sentence. Then you told me that you found a contradiction in a pair of sentences I uttered. Frankly, it's just as I said. Your own system of thought is so delusional that you manage to avoid seeing how inconsistent it is. From the outside, however, it's plain as day.

Achilles: Sometimes I get so confused by your diversionary tactics that I can't quite tell if we're arguing about something utterly petty, or something deep and profound!

Tortoise: I assure you, Tortoises don't spend their time on the petty. Hence it's the latter.

Achilles: I am very reassured. Thank you. Now I have had a moment to reflect, and I see the necessary logical step to convince you that you contradicted yourself.

Tortoise: Good, good. I hope it's an easy step, an indisputable one.

Achilles: It certainly is. Even you will agree with it. The idea is that since you believed sentence 1 ("My shell is green"), AND you believed sentence 2 ("My shell is not green"), you would believe one compound sentence in which both were combined, wouldn't you?

Tortoise: Of course. It would only be reasonable . . . providing just that the manner of combination is universally acceptable. But I'm sure that we'll agree on that.

Achilles: Yes, and then I'll have you! The combination I propose is—

Tortoise: But we must be careful in combining sentences. For instance, you'd grant that "Politicians lie" is true, wouldn't you?

Achilles: Who could deny it?

Tortoise: Good. Likewise, "Cast-iron sinks" is a valid utterance, isn't it?

Achilles: Indubitably.

Tortoise: Then, putting them together, we get "Politicians lie in cast-iron sinks". Now that's not the case, is it?

Achilles: Now wait a minute . . . "Politicians lie in cast-iron sinks?" Well, no, but—

Tortoise: So, you see, combining two true sentences in one is not a safe policy, is it?

Achilles: But you—you combined the two—in such a silly way!

Tortoise: Silly? What have you got to object to in the way I combined them? Would you have me do otherwise?

Achilles: You should have used the word "and", not "in".

Tortoise: I should have? You mean, if YOU'D had YOUR way, I should have.

Achilles: No—it's the LOGICAL thing to do. It's got nothing to do with me personally.

Tortoise: This is where you always lose me, when you resort to your Logic and its high-sounding Principles. None of that for me today, please.

Achilles: Oh, Mr. Tortoise, don't put me through all this agony. You know very well that that's what "and" means! It's harmless to combine two true sentences with "and"!

Tortoise: "Harmless", my eye! What gall! This is certainly a pernicious plot to entrap a poor, innocent, bumbling Tortoise in a fatal contradiction. If it were so harmless, why would you be trying so bloody hard to get me to do it? Eh?

Achilles: You've left me speechless. You make me feel like a villain, where I really had only the most innocent of motivations.

Tortoise: That's what everyone believes of himself . . .

Achilles: Shame on me—trying to outwit you, to use words to snare you in a self-contradiction. I feel so rotten.

Tortoise: And well you should. I know what you were trying to set up. Your plan was to make me accept sentence 3, to wit: "My shell is green and my shell is not green". And such a blatant falsehood is repellent to the Tongue of a Tortoise.

Achilles: Oh, I'm so sorry I started all this.

Tortoise: You needn't be sorry. My feelings aren't hurt. After all, I'm used to the unreasonable ways of the folk about me. I enjoy your company, Achilles, even if your thinking lacks clarity.

Achilles: Yes . . . Well, I fear I am set in my ways, and will probably continue to err and err again, in my quest for Truth.

Tortoise: Today's exchange may have served a little to right your course. Good day, Achilles.

Achilles: Good day, Mr. T.

CHAPTER VII

The Propositional Calculus

Words and Symbols

THE PRECEDING DIALOGUE is reminiscent of the *Two-Part Invention* by Lewis Carroll. In both, the Tortoise refuses to use normal, ordinary words in the normal, ordinary way—or at least he refuses to do so when it is not to his advantage to do so. A way to think about the Carroll paradox was given last Chapter. In this Chapter we are going to make symbols do what Achilles couldn't make the Tortoise do with his words. That is, we are going to make a formal system one of whose symbols will do just what Achilles wished the word '*and*' would do, when spoken by the Tortoise, and another of whose symbols will behave the way the words '*if . . . then . . .*' ought to behave. There are only two other words which we will attempt to deal with: '*or*' and '*not*'. Reasoning which depends only on correct usage of these four words is termed *propositional reasoning*.

Alphabet and First Rule of the Propositional Calculus

I will present this new formal system, called the *Propositional Calculus,* a little like a puzzle, not explaining everything at once, but letting you figure things out to some extent. We begin with the list of symbols:

$$< \quad >$$

$$\mathsf{P} \quad \mathsf{Q} \quad \mathsf{R} \quad '$$

$$\wedge \quad \vee \quad \supset \quad \sim$$

$$[\quad]$$

The first rule of this system that I will reveal is the following:

RULE OF JOINING: If x and y are theorems of the system, then so is the string $<x \wedge y>$.

This rule takes two theorems and combines them into one. It should remind you of the Dialogue.

Well-Formed Strings

There will be several other rules of inference, and they will all be presented shortly—but first, it is important to define a subset of all strings, namely the *well-formed* strings. They will be defined in a recursive way. We begin with the

ATOMS: P, Q, and R are called *atoms*. New atoms are formed by appending primes onto the right of old atoms—thus, R′, Q″, P‴, etc. This gives an endless supply of atoms. All atoms are well-formed.

Then we have four recursive

FORMATION RULES: If *x* and *y* are well-formed, then the following four strings are also well-formed:

(1) ~*x*
(2) <*x*∧*y*>
(3) <*x*∨*y*>
(4) <*x*⊃*y*>

For example, all of the following are well-formed:

P	atom
~P	by (1)
~~P	by (1)
Q′	atom
~Q′	by (1)
<P∧~Q′>	by (2)
~<P∧~Q′>	by (1)
<~~P⊃Q′>	by (4)
<~<P∧~Q′>∨<~~P⊃Q′>>	by (3)

The last one may look quite formidable, but it is built up straightforwardly from two components—namely the two lines just above it. Each of them is in turn built up from previous lines . . . and so on. Every well-formed string can in this way be traced back to its elementary constituents—that is, atoms. You simply run the formation rules backwards until you can no more. This process is guaranteed to terminate, since each formation rule (when run forwards) is a *lengthening* rule, so that running it backwards always drives you towards atoms.

This method of decomposing strings thus serves as a check on the well-formedness of any string. It is a *top-down decision procedure* for well-formedness. You can test your understanding of this decision procedure by checking which of the following strings are well-formed:

(1) <P>
(2) <~P>
(3) <P∧Q∧R>
(4) <P∧Q>
(5) <<P∧Q>∧<Q~∧P>>
(6) <P∧~P>
(7) <<P∨<Q⊃R>>∧<~P∨~R′>>
(8) <P∧Q>∧<Q∧P>

(Answer: Those whose numbers are Fibonacci numbers are not well-formed. The rest are well-formed.)

More Rules of Inference

Now we come to the rest of the rules by which *theorems* of this system are constructed. A few rules of inference follow. In all of them, the symbols 'x' and 'y' are always to be understood as restricted to *well-formed* strings.

RULE OF SEPARATION: If $<x\wedge y>$ is a theorem, then both x and y are theorems.

Incidentally, you should have a pretty good guess by now as to what concept the symbol '$\wedge$' stands for. (Hint: it is the troublesome word from the preceding Dialogue.) From the following rule, you should be able to figure out what concept the *tilde* ('~') represents:

DOUBLE-TILDE RULE: The string '~~' can be deleted from any theorem. It can also be inserted into any theorem, provided that the resulting string is itself well-formed.

The Fantasy Rule

Now a special feature of this system is that it has *no axioms*—only rules. If you think back to the previous formal systems we've seen, you may wonder how there can be any theorems, then. How does everything get started? The answer is that there is one rule which manufactures theorems from out of thin air—it doesn't need an "old theorem" as input. (The rest of the rules do require input.) This special rule is called the *fantasy rule.* The reason I call it that is quite simple.

To use the fantasy rule, the first thing you do is to write down any well-formed string x you like, and then "fantasize" by asking, "*What if* this string x were an axiom, or a theorem?" And then, you let the system itself give an answer. That is, you go ahead and make a derivation with x as the opening line; let us suppose y is the last line. (Of course the derivation must strictly follow the rules of the system.) Everything from x to y (inclusive) is the *fantasy;* x is the *premise* of the fantasy, and y is its *outcome.* The next step is to *jump out of the fantasy,* having learned from it that

If x were a theorem, y would be a theorem.

Still, you might wonder, where is the *real* theorem? The real theorem is the string

$$<x\supset y>.$$

Notice the resemblance of this string to the sentence printed above.

To signal the entry into, and emergence from, a fantasy, one uses the square brackets '[' and ']', respectively. Thus, whenever you see a left square bracket, you know you are "pushing" into a fantasy, and the *next* line will contain the fantasy's *premise*. Whenever you see a right square bracket, you know you are "popping" back out, and the *preceding* line was the *outcome*. It is helpful (though not necessary) to indent those lines of a derivation which take place in fantasies.

Here is an illustration of the fantasy rule, in which the string P is taken as a premise. (It so happens that P is *not* a theorem, but that is of no import; we are merely inquiring, "What if it were?") We make the following fantasy:

[	push into fantasy
P	premise
~~P	outcome (by double-tilde rule)
]	pop out of fantasy

The fantasy shows that:

If P were a theorem, so would ~~P be one.

We now "squeeze" this sentence of English (the metalanguage) into the formal notation (the object language): <P⊃~~P>. This, our first theorem of the Propositional Calculus, should reveal to you the intended interpretation of the symbol '⊃'.

Here is another derivation using the fantasy rule:

[	push
<P∧Q>	premise
P	separation
Q	separation
<Q∧P>	joining
]	pop
<<P∧Q>⊃<Q∧P>>	fantasy rule

It is important to understand that only the last line is a genuine theorem, here—everything else is in the fantasy.

Recursion and the Fantasy Rule

As you might guess from the recursion terminology "push" and "pop", the fantasy rule can be used recursively—thus, there can be fantasies within fantasies, thrice-nested fantasies, and so on. This means that there are all sorts of "levels of reality", just as in nested stories or movies. When you pop out of a movie-within-a-movie, you feel for a moment as if you had reached the real world, though you are still one level away from the top. Similarly, when you pop out of a fantasy-within-a-fantasy, you are in a "realer" world than you had been, but you are still one level away from the top.

Now a "No Smoking" sign inside a movie theater does not apply to the characters in the movie—there is no carry-over from the real world into the fantasy world, in movies. But in the Propositional Calculus, there is a carry-over from the real world into the fantasies; there is even carry-over from a fantasy to fantasies inside it. This is formalized by the following rule:

CARRY-OVER RULE: Inside a fantasy, any theorem from the "reality" one level higher can be brought in and used.

It is as if a "No Smoking" sign in a theater applied not only to all the moviegoers, but also to all the actors in the movie, and, by repetition of the same idea, to anyone inside multiply nested movies! (Warning: There is no carry-over in the reverse direction: theorems inside fantasies cannot be exported to the exterior! If it weren't for this fact, you could write anything as the first line of a fantasy, and then lift it out into the real world as a theorem.)

To show how carry-over works, and to show how the fantasy rule can be used recursively, we present the following derivation:

```
[                        push
  P                      premise of outer fantasy
  [                      push again
    Q                    premise of inner fantasy
    P                    carry-over of P into inner fantasy
    <P∧Q>                joining
  ]                      pop out of inner fantasy, regain outer fantasy
  <Q⊃<P∧Q>>              fantasy rule
]                        pop out of outer fantasy, reach real world!
<P⊃<Q⊃<P∧Q>>>            fantasy rule
```

Note that I've indented the outer fantasy once, and the inner fantasy twice, to emphasize the nature of these nested "levels of reality". One way to look at the fantasy rule is to say that an observation made *about* the system is inserted *into* the system. Namely, the theorem $<x \supset y>$ which gets produced can be thought of as a representation inside the system of the statement about the system "If x is a theorem, then y is too". To be more specific, the intended interpretation for <P⊃Q> is "if P, then Q", or equivalently, "P implies Q".

The Converse of the Fantasy Rule

Now Lewis Carroll's Dialogue was all about "if-then" statements. In particular, Achilles had a lot of trouble in persuading the Tortoise to accept the second clause of an "if-then" statement, even when the "if-then" statement itself was accepted, as well as its first clause. The next rule allows you to infer the second "clause" of a '⊃'-string, provided that the '⊃'-string itself is a theorem, and that its first "clause" is also a theorem.

Rule of Detachment: If *x* and <*x*⊃*y*> are both theorems, then *y* is a theorem.

Incidentally, this rule is often called "Modus Ponens", and the fantasy rule is often called the "Deduction Theorem".

The Intended Interpretation of the Symbols

We might as well let the cat out of the bag at this point, and reveal the "meanings" of the rest of the symbols of our new system. In case it is not yet apparent, the symbol '∧' is meant to be acting isomorphically to the normal, everyday word 'and'. The symbol '~' represents the word 'not'—it is a formal sort of negation. The angle brackets '<' and '>' are groupers—their function being very similar to that of parentheses in ordinary algebra. The main difference is that in algebra, you have the freedom to insert parentheses or to leave them out, according to taste and style, whereas in a formal system, such anarchic freedom is not tolerated. The symbol '∨' represents the word 'or' ('vel' is a Latin word for 'or'). The 'or' that is meant is the so-called *inclusive* 'or', which means that the interpretation of <*x*∨*y*> is "either *x* or *y*—or both".

The only symbols we have not interpreted are the atoms. An atom has no single interpretation—it may be interpreted by *any* sentence of English (it must continue to be interpreted by the same sentence if it occurs multiply within a string or derivation). Thus, for example, the well-formed string <P∧~P> could be interpreted by the compound sentence

This mind is Buddha, and this mind is not Buddha.

Now let us look at each of the theorems so far derived, and interpret them. The first one was <P⊃~~P>. If we keep the same interpretation for P, we have the following interpretation:

If this mind is Buddha,
then it is not the case that this mind is not Buddha.

Note how I rendered the double negation. It is awkward to repeat a negation in any natural language, so one gets around it by using two different ways of expressing negation. The second theorem we derived was <<P∧Q>⊃<Q∧P>>. If we let Q be interpreted by the sentence "This flax weighs three pounds", then our theorem reads as follows:

If this mind is Buddha and this flax weighs three pounds,
then this flax weighs three pounds and this mind is Buddha.

The third theorem was <P⊃<Q⊃<P∧Q>>>. This one goes into the following nested "if-then" sentence:

If this mind is Buddha,
then, if this flax weighs three pounds,
then this mind is Buddha and this flax weighs three pounds.

You probably have noticed that each theorem, when interpreted, says something absolutely trivial and self-evident. (Sometimes they are *so* self-evident that they sound vacuous and—paradoxically enough—confusing or even wrong!) This may not be very impressive, but just remember that there are plenty of falsities out there which could have been produced—yet they weren't. This system—the Propositional Calculus—steps neatly from truth to truth, carefully avoiding all falsities, just as a person who is concerned with staying dry will step carefully from one stepping-stone in a creek to the next, following the layout of stepping-stones no matter how twisted and tricky it might be. What is impressive is that—in the Propositional Calculus—the whole thing is done purely *typographically*. There is nobody down "in there", thinking about the *meaning* of the strings. It is all done mechanically, thoughtlessly, rigidly, even stupidly.

Rounding Out the List of Rules

We have not yet stated all the rules of the Propositional Calculus. The complete set of rules is listed below, including the three new ones.

JOINING RULE: If x and y are theorems, then $<x\wedge y>$ is a theorem.

SEPARATION RULE: If $<x\wedge y>$ is a theorem, then both x and y are theorems.

DOUBLE-TILDE RULE: The string '$\sim\sim$' can be deleted from any theorem. It can also be inserted into any theorem, provided that the resulting string is itself well-formed.

FANTASY RULE: If y can be derived when x is assumed to be a theorem, then $<x\supset y>$ is a theorem.

CARRY-OVER RULE: Inside a fantasy, any theorem from the "reality" one level higher can be brought in and used.

RULE OF DETACHMENT: If x and $<x\supset y>$ are both theorems, then y is a theorem.

CONTRAPOSITIVE RULE: $<x\supset y>$ and $<\sim y\supset\sim x>$ are interchangeable.

DE MORGAN'S RULE: $<\sim x\wedge\sim y>$ and $\sim<x\vee y>$ are interchangeable.

SWITCHEROO RULE: $<x\vee y>$ and $<\sim x\supset y>$ are interchangeable.

(The Switcheroo rule is named after Q. q. Switcheroo, an Albanian railroad engineer who worked in logic on the siding.) By "interchangeable" in the foregoing rules, the following is meant: If an expression of one form occurs as either a theorem or part of a theorem, the other form may be substituted, and the resulting string will also be a theorem. It must be kept in mind that the symbols '*x*' and '*y*' always stand for well-formed strings of the system.

Justifying the Rules

Before we see these rules used inside derivations, let us look at some very short justifications for them. You can probably justify them to yourself better than my examples—which is why I only give a couple.

The contrapositive rule expresses explicitly a way of turning around conditional statements which we carry out unconsciously. For instance, the "Zentence"

If you are studying it, then you are far from the Way

means the same thing as

If you are close to the Way, then you are not studying it.

De Morgan's rule can be illustrated by our familiar sentence "The flag is not moving and the wind is not moving". If P symbolizes "the flag is moving", and Q symbolizes "the wind is moving", then the compound sentence is symbolized by <~P∧~Q>, which, according to De Morgan's law, is interchangeable with ~<P∨Q>, whose interpretation would be "It is not true that either the flag or the wind is moving". And no one could deny that that is a Zensible conclusion to draw.

For the Switcheroo rule, consider the sentence "Either a cloud is hanging over the mountain, or the moonlight is penetrating the waves of the lake," which might be spoken, I suppose, by a wistful Zen master remembering a familiar lake which he can visualize mentally but cannot see. Now hang onto your seat, for the Switcheroo rule tells us that this is interchangeable with the thought: "If a cloud is not hanging over the mountain, then the moonlight is penetrating the waves of the lake." This may not be enlightenment, but it is the best the Propositional Calculus has to offer.

Playing Around with the System

Now let us apply these rules to a previous theorem, and see what we get. For instance, take the theorem <P⊃~~P>:

<P⊃~~P>	old theorem
<~~~P⊃~P>	contrapositive
<~P⊃~P>	double-tilde
<P∨~P>	switcheroo

This new theorem, when interpreted, says:

Either this mind is Buddha, or this mind is not Buddha.

Once again, the interpreted theorem, though perhaps less than mind-boggling, is at least true.

Semi-Interpretations

It is natural, when one reads theorems of the Propositional Calculus out loud, to interpret everything but the atoms. I call this *semi-interpreting*. For instance, the semi-interpretation of <P∨~P> would be

P or not P.

Despite the fact that P is not a sentence, the above semisentence still sounds true, because you can very easily imagine sticking any sentence in for P—and the form of the semi-interpreted theorem assures you that however you make your choice, the resulting sentence will be true. And that is the key idea of the Propositional Calculus: it produces theorems which, when semi-interpreted, are seen to be "universally true semisentences", by which is meant that no matter how you complete the interpretation, the final result will be a true statement.

Gantō's Ax

Now we can do a more advanced exercise, based on a Zen kōan called "Gantō's Ax". Here is how it begins:

> One day Tokusan told his student Gantō, "I have two monks who have been here for many years. Go and examine them." Gantō picked up an ax and went to the hut where the two monks were meditating. He raised the ax, saying, "If you say a word I will cut off your heads; and if you do not say a word, I will also cut off your heads."[1]

If you say a word I will cut off this kōan; and if you do not say a word, I will also cut off this kōan—because I want to translate some of it into our notation. Let us symbolize "you say a word" by P, and "I will cut off your heads" by Q. Then Gantō's ax threat is symbolized by the string <<P⊃Q>∧<~P⊃Q>>. What if this ax threat were an axiom? Here is a fantasy to answer that question.

(1)	[	push
(2)	<<P⊃Q>∧<~P⊃Q>>	Gantō's axiom
(3)	<P⊃Q>	separation
(4)	<~Q⊃~P>	contrapositive
(5)	<~P⊃Q>	separation
(6)	<~Q⊃~~P>	contrapositive
(7)	[	push again
(8)	~Q	premise

(9)	<~Q⊃~P>	carry-over of line 4
(10)	~P	detachment
(11)	<~Q⊃~~P>	carry-over of line 6
(12)	~~P	detachment (lines 8 and 11)
(13)	<~P∧~~P>	joining
(14)	~<P∨~P>	De Morgan
(15)	]	pop once
(16)	<~Q⊃~<P∨~P>>	fantasy rule
(17)	<<P∨~P>⊃Q>	contrapositive
(18)	[	push
(19)	~P	premise (also outcome!)
(20)	]	pop
(21)	<~P⊃~P>	fantasy rule
(22)	<P∨~P>	switcheroo
(23)	Q	detachment (lines 22 and 17)
(24)	]	pop out

The power of the Propositional Calculus is shown in this example. Why, in but two dozen steps, we have deduced Q: that the heads will be cut off! (Ominously, the rule last invoked was "detachment" . . .) It might seem superfluous to continue the kōan now, since we know what must ensue . . . However, I shall drop my resolve to cut the kōan off; it is a true Zen kōan, after all. The rest of the incident is here related:

> Both monks continued their meditation as if he had not spoken. Gantō dropped the ax and said, "You are true Zen students." He returned to Tokusan and related the incident. "I see your side well," Tokusan agreed, "but tell me, how is their side?" "Tōzan may admit them," replied Gantō, "but they should not be admitted under Tokusan."[2]

Do you see my side well? How is the Zen side?

Is There a Decision Procedure for Theorems?

The Propositional Calculus gives us a set of rules for producing statements which would be true in all conceivable worlds. That is why all of its theorems sound so simple-minded; it seems that they have absolutely no content! Looked at this way, the Propositional Calculus might seem to be a waste of time, since what it tells us is absolutely trivial. On the other hand, it does it by specifying the *form* of statements that are universally true, and this throws a new kind of light onto the core truths of the universe: they are not only fundamental, but also *regular*: they can be produced by one set of typographical rules. To put it another way, they are all "cut from the same cloth". You might consider whether the same could be said about Zen kōans: could they all be produced by one set of typographical rules?

It is quite relevant here to bring up the question of a decision procedure. That is, does there exist any mechanical method to tell nontheorems from theorems? If so, that would tell us that the set of theorems of the

Propositional Calculus is not only r.e., but also recursive. It turns out that there is an interesting decision procedure—the method of truth tables. It would take us a bit afield to present it here; you can find it in almost any standard book on logic. And what about Zen kōans? Could there conceivably be a mechanical decision procedure which distinguishes genuine Zen kōans from other things?

Do We Know the System Is Consistent?

Up till now, we have only *presumed* that all theorems, when interpreted as indicated, are true statements. But do we *know* that that is the case? Could we prove it to be? This is just another way of asking whether the intended interpretations ('and' for '∧', etc.) merit being called the "passive meanings" of the symbols. One can look at this issue from two very different points of view, which might be called the "prudent" and "imprudent" points of view. I will now present those two sides as I see them, personifying their holders as "Prudence" and "Imprudence".

Prudence: We will only KNOW that all theorems come out true under the intended interpretation if we manage to PROVE it. That is the cautious, thoughtful way to proceed.

Imprudence: On the contrary. It is OBVIOUS that all theorems will come out true. If you doubt me, look again at the rules of the system. You will find that each rule makes a symbol act exactly as the word it represents ought to be used. For instance, the joining rule makes the symbol '∧' act as 'and' ought to act; the rule of detachment makes '⊃' act as it ought to, if it is to stand for 'implies', or 'if-then'; and so on. Unless you are like the Tortoise, you will recognize in each rule a codification of a pattern you use in your own thought. So if you trust your own thought patterns, then you HAVE to believe that all theorems come out true! That's the way I see it. I don't need any further proof. If you think that some theorem comes out false, then presumably you think that some rule must be wrong. Show me which one.

Prudence: I'm not sure that there is any faulty rule, so I can't point one out to you. Still, I can imagine the following kind of scenario. You, following the rules, come up with a theorem—say *x*. Meanwhile I, also following the rules, come up with another theorem—it happens to be ~*x*. Can't you force yourself to conceive of that?

Imprudence: All right; let's suppose it happened. Why would it bother you? Or let me put it another way. Suppose that in playing with the MIU-system, I came up with a theorem *x*, and you came up with *x*U. Can you force yourself to conceive of that?

Prudence: Of course—in fact both MI and MIU are theorems.

Imprudence: Doesn't that bother you?

Prudence: Of course not. Your example is ridiculous, because MI and MIU are not CONTRADICTORY, whereas two strings *x* and ~*x* in the Propositional Calculus ARE contradictory.

Imprudence: Well, yes—provided you wish to interpret '$\sim$' as 'not'. But what would lead you to think that '$\sim$' should be interpreted as 'not'?

Prudence: The rules themselves. When you look at them, you realize that the only conceivable interpretation for '$\sim$' is 'not'—and likewise, the only conceivable interpretation for '$\wedge$' is 'and', etc.

Imprudence: In other words, you are convinced that the rules capture the meanings of those words?

Prudence: Precisely.

Imprudence: And yet you are still willing to entertain the thought that both x and $\sim x$ could be theorems? Why not also entertain the notion that hedgehogs are frogs, or that 1 equals 2, or that the moon is made of green cheese? I for one am not prepared even to consider whether such basic ingredients of my thought processes are wrong—because if I entertained that notion, then I would also have to consider whether my modes of analyzing the entire question are also wrong, and I would wind up in a total tangle.

Prudence: Your arguments are forceful . . . Yet I would still like to see a PROOF that all theorems come out true, or that x and $\sim x$ can never both be theorems.

Imprudence: You want a proof. I guess that means that you want to be more convinced that the Propositional Calculus is consistent than you are convinced of your own sanity. Any proof I could think of would involve mental operations of a greater complexity than anything in the Propositional Calculus itself. So what would it prove? Your desire for a proof of consistency of the Propositional Calculus makes me think of someone who is learning English and insists on being given a dictionary which defines all the simple words in terms of complicated ones. . .

The Carroll Dialogue Again

This little debate shows the difficulty of trying to use logic and reasoning to defend themselves. At some point, you reach rock bottom, and there is no defense except loudly shouting, "I know I'm right!" Once again, we are up against the issue which Lewis Carroll so sharply set forth in his Dialogue: you can't go on defending your patterns of reasoning forever. There comes a point where faith takes over.

A system of reasoning can be compared to an egg. An egg has a shell which protects its insides. If you want to ship an egg somewhere, though, you don't rely on the shell. You pack the egg in some sort of container, chosen according to how rough you expect the egg's voyage to be. To be extra careful, you may put the egg inside several nested boxes. However, no matter how many layers of boxes you pack your egg in, you can imagine some cataclysm which could break the egg. But that doesn't mean that you'll never risk transporting your egg. Similarly, one can never give an ultimate, absolute proof that a proof in some system is correct. Of course, one can give a proof of a proof, or a proof of a proof of a proof—but the validity of the outermost system always remains an unproven assumption, accepted on faith. One can always imagine that some unsuspected subtlety will invalidate every single level of proof down to the bottom, and that the "proven" result will be seen not to be correct after all. But that doesn't mean that mathematicians and logicians are constantly worrying that the whole edifice of mathematics might be wrong. On the other hand, when unorthodox proofs are proposed, or extremely lengthy proofs, or proofs generated by computers, then people do stop to think a bit about what they really mean by that quasi-sacred word "proven".

An excellent exercise for you at this point would be to go back to the Carroll Dialogue, and code the various stages of the debate into our notation—beginning with the original bone of contention:

Achilles: If you have <<A∧B>⊃Z>, and you also have <A∧B>, then surely you have Z.

Tortoise: Oh! You mean: <<<<A∧B>⊃Z>∧<A∧B>>⊃Z>, don't you?

(Hint: Whatever Achilles considers a rule of inference, the Tortoise immediately flattens into a mere string of the system. If you use only the letters A, B, and Z, you will get a recursive pattern of longer and longer strings.)

Shortcuts and Derived Rules

When carrying out derivations in the Propositional Calculus, one quickly invents various types of shortcut, which are not strictly part of the system. For instance, if the string <Q∨~Q> were needed at some point, and <P∨~P> had been derived earlier, many people would proceed as if <Q∨~Q> had been derived, since they know that its derivation is an exact parallel to that of <P∨~P>. The derived theorem is treated as a "theorem schema"—a mold for other theorems. This turns out to be a perfectly valid procedure, in that it always leads you to new theorems, but it is not a rule of the Propositional Calculus as we presented it. It is, rather, a *derived rule*. It is part of the knowledge which we have *about* the system. That this rule always keeps you within the space of theorems needs proof, of course—but such a proof is not like a derivation inside the system. It is a proof in the ordinary, intuitive sense—a chain of reasoning carried out in the I-mode. The theory *about* the Propositional Calculus is a "metatheory", and results in it can be called "metatheorems"—Theorems about theorems. (Incidentally, note the peculiar capitalization in the phrase "Theorems about theorems". It is a consequence of our convention: metatheorems are Theorems (proven results) concerning theorems (derivable strings).)

In the Propositional Calculus, one could discover many other metatheorems, or derived rules of inference. For instance, there is a second De Morgan's Rule:

<~*x*∨~*y*> and ~<*x*∧*y*> are interchangeable.

If this were a rule of the system, it could speed up many derivations considerably. But if we *prove* that it is correct, isn't that good enough? Can't we use it just like a rule of inference, from then on?

There is no reason to doubt the correctness of this particular derived rule. But once you start admitting derived rules as part of your procedure in the Propositional Calculus, you have lost the formality of the system, since derived rules are derived informally—outside the system. Now formal systems were proposed as a way to exhibit every step of a proof explicitly, within one single, rigid framework, so that any mathematician could check another's work mechanically. But if you are willing to step outside of that framework at the drop of a hat, you might as well never have created it at all. Therefore, there is a drawback to using such shortcuts.

Formalizing Higher Levels

On the other hand, there is an alternative way out. Why not formalize the metatheory, too? That way, derived rules (metatheorems) would be theorems of a larger formal system, and it would be legitimate to look for shortcuts and derive them as theorems—that is, theorems of the formalized metatheory—which could then be used to speed up the derivations of theorems of the Propositional Calculus. This is an interesting idea, but as soon as it is suggested, one jumps ahead to think of metametatheories, and so on. It is clear that no matter how many levels you formalize, someone will eventually want to make shortcuts in the top level.

It might even be suggested that a theory of reasoning could be identical to its own metatheory, if it were worked out carefully. Then, it might seem, all levels would collapse into one, and thinking *about* the system would be just one way of working *in* the system! But it is not that easy. Even if a system can "think about itself", it still is not *outside* itself. You, outside the system, perceive it differently from the way it perceives itself. So there still is a metatheory—a view from outside—even for a theory which can "think about itself" inside itself. We will find that there are theories which can "think about themselves". In fact, we will soon see a system in which this happens completely accidentally, without our even intending it! And we will see what kinds of effects this produces. But for our study of the Propositional Calculus, we will stick with the simplest ideas—no mixing of levels.

Fallacies can result if you fail to distinguish carefully between working in the system (the M-mode) and thinking about the system (the I-mode). For example, it might seem perfectly reasonable to assume that, since <P∨~P> (whose semi-interpretation is "either P or not P") is a theorem, either P or ~P must be a theorem. But this is dead wrong: neither one of the latter pair is a theorem. In general, it is a dangerous practice to assume that symbols can be slipped back and forth between different levels—here, the language of the formal system and its metalanguage (English).

Reflections on the Strengths and Weaknesses of the System

You have now seen one example of a system with a purpose—to represent part of the architecture of logical thought. The concepts which this system handles are very few in number, and they are very simple, precise concepts. But the simplicity and precision of the Propositional Calculus are exactly the kinds of features which make it appealing to mathematicians. There are two reasons for this. (1) It can be studied for its own properties, exactly as geometry studies simple, rigid shapes. Variants can be made on it, employing different symbols, rules of inference, axioms or axiom schemata, and so on. (Incidentally, the version of the Propositional Calculus here presented is related to one invented by G. Gentzen in the early 1930's. There are other versions in which only one rule of inference is used—detachment, usually—and in which there are several axioms, or axiom schemata.) The study of ways to carry out propositional reasoning in elegant formal systems is an appealing branch of pure mathematics. (2) The Propositional Calculus can easily be extended to include other fundamental aspects of reasoning. Some of this will be shown in the next Chapter, where the Propositional Calculus is incorporated lock, stock and barrel into a much larger and deeper system in which sophisticated number-theoretical reasoning can be done.

Proofs *vs.* Derivations

The Propositional Calculus is very much like reasoning in some ways, but one should not equate its rules with the rules of human thought. A *proof* is something informal, or in other words a product of normal thought, written in a human language, for human consumption. All sorts of complex features of thought may be used in proofs, and, though they may "feel right", one may wonder if they can be defended logically. That is really what formalization is for. A *derivation* is an artificial counterpart of a proof, and its purpose is to reach the same goal but via a logical structure whose methods are not only all explicit, but also very simple.

If—and this is usually the case—it happens that a formal derivation is extremely lengthy compared with the corresponding "natural" proof, that is just too bad. It is the price one pays for making each step so simple. What often happens is that a derivation and a proof are "simple" in complementary senses of the word. The proof is simple in that each step "sounds right", even though one may not know just why; the derivation is simple in that each of its myriad steps is considered so trivial that it is beyond reproach, and since the whole derivation consists just of such trivial steps, it is supposedly error-free. Each type of simplicity, however, brings along a characteristic type of complexity. In the case of proofs, it is the complexity of the underlying system on which they rest—namely, human language; and in the case of derivations, it is their astronomical size, which makes them almost impossible to grasp.

Thus, the Propositional Calculus should be thought of as part of a general method for synthesizing artificial proof-like structures. It does not, however, have much flexibility or generality. It is intended only for use in connection with mathematical concepts—which are themselves quite rigid. As a rather interesting example of this, let us make a derivation in which a very peculiar string is taken as a premise in a fantasy: <P∧~P>. At least its semi-interpretation is peculiar. The Propositional Calculus, however, does not think about semi-interpretations; it just manipulates strings typographically—and typographically, there is really nothing peculiar about this string. Here is a fantasy with this string as its premise:

```
 (1)  [                     push
 (2)     <P∧~P>             premise
 (3)     P                  separation
 (4)     ~P                 separation
 (5)     [                  push
 (6)       ~Q               premise
 (7)       P                carry-over line 3
 (8)       ~~P              double-tilde
 (9)     ]                  pop
(10)     <~Q⊃~~P>           fantasy
(11)     <~P⊃Q>             contrapositive
(12)     Q                  detachment (Lines 4,11)
(13)  ]                     pop
(14)  <<P∧~P>⊃Q>            fantasy
```

Now this theorem has a very strange semi-interpretation:

P and not P together imply Q

Since Q is interpretable by any statement, we can loosely take the theorem to say that "From a contradiction, anything follows"! Thus, in systems based on the Propositional Calculus, contradictions cannot be contained; they infect the whole system like an instantaneous global cancer.

The Handling of Contradictions

This does not sound much like human thought. If you found a contradiction in your own thoughts, it's very unlikely that your whole mentality would break down. Instead, you would probably begin to question the beliefs or modes of reasoning which you felt had led to the contradictory thoughts. In other words, to the extent you could, you would step out of the systems inside you which you felt were responsible for the contradiction, and try to repair them. One of the least likely things for you to do would be to throw up your arms and cry, "Well, I guess that shows that I believe everything now!" As a joke, yes—but not seriously.

Indeed, contradiction is a major source of clarification and progress in all domains of life—and mathematics is no exception. When in times past, a contradiction in mathematics was found, mathematicians would immediately seek to pinpoint the system responsible for it, to jump out of it, to reason about it, and to amend it. Rather than weakening mathematics, the discovery and repair of a contradiction would strengthen it. This might take time and a number of false starts, but in the end it would yield fruit. For instance, in the Middle Ages, the value of the infinite series

$$1 - 1 + 1 - 1 + 1 - \dots$$

was hotly disputed. It was "proven" to equal 0, 1, ½, and perhaps other values. Out of such controversial findings came a fuller, deeper theory about infinite series.

A more relevant example is the contradiction right now confronting us—namely the discrepancy between the way we really think, and the way the Propositional Calculus imitates us. This has been a source of discomfort for many logicians, and much creative effort has gone into trying to patch up the Propositional Calculus so that it would not act so stupidly and inflexibly. One attempt, put forth in the book *Entailment* by A. R. Anderson and N. Belnap,[3] involves "relevant implication", which tries to make the symbol for "if-then" reflect genuine causality, or at least connection of meanings. Consider the following theorems of the Propositional Calculus:

<P⊃<Q⊃P>>
<P⊃<Q∨~Q>>
<<P∧~P>⊃Q>
<<P⊃Q>∨<Q⊃P>>

They, and many others like them, all show that there need be no relationship at all between the first and second clauses of an if-then statement for it to be provable within the Propositional Calculus. In protest, "relevant implication" puts certain restrictions on the contexts in which the rules of inference can be applied. Intuitively, it says that "something can only be derived from something else if they have to do with each other". For example, line 10 in the derivation given above would not be allowed in such a system, and that would block the derivation of the string <<P∧~P>⊃Q>.

More radical attempts abandon completely the quest for completeness or consistency, and try to mimic human reasoning with all its inconsistencies. Such research no longer has as its goal to provide a solid underpinning for mathematics, but purely to study human thought processes.

Despite its quirks, the Propositional Calculus has some features to recommend itself. If one embeds it into a larger system (as we will do next Chapter), and if one is sure that the larger system contains no contradictions (and we will be), then the Propositional Calculus does all that one could hope: it provides valid propositional inferences—all that can be made. So if ever an incompleteness or an inconsistency is uncovered, one can be sure that it will be the fault of the larger system, and not of its subsystem which is the Propositional Calculus.

FIGURE 42. "Crab Canon", *by M. C. Escher (~1965).*

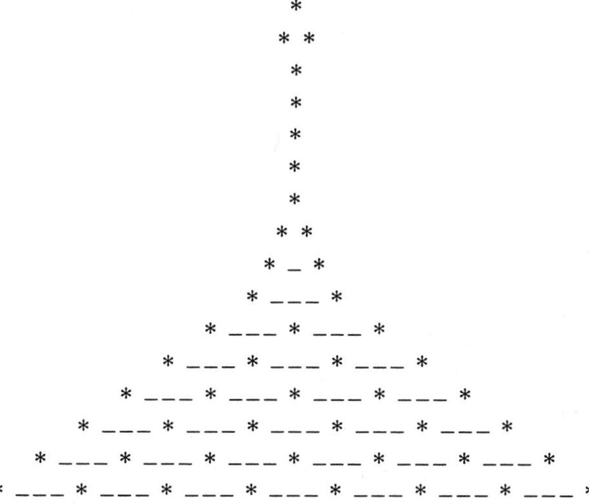

Crab Canon

Achilles and the Tortoise happen upon each other in the park one day while strolling.

Tortoise: Good day, Mr. A.

Achilles: Why, same to you.

Tortoise: So nice to run into you.

Achilles: That echoes my thoughts.

Tortoise: And it's a perfect day for a walk. I think I'll be walking home soon.

Achilles: Oh, really? I guess there's nothing better for you than walking.

Tortoise: Incidentally, you're looking in very fine fettle these days, I must say.

Achilles: Thank you very much.

Tortoise: Not at all. Here, care for one of my cigars?

Achilles: Oh, you are such a philistine. In this area, the Dutch contributions are of markedly inferior taste, don't you think?

Tortoise: I disagree, in this case. But speaking of taste, I finally saw that *Crab Canon* by your favorite artist, M. C. Escher, in a gallery the other day, and I fully appreciate the beauty and ingenuity with which he made one single theme mesh with itself going both backwards and forwards. But I am afraid I will always feel Bach is superior to Escher.

Achilles: I don't know. But one thing for certain is that I don't worry about arguments of taste. *De gustibus non est disputandum.*

Tortoise: Tell me, what's it like to be your age? Is it true that one has no worries at all?

Achilles: To be precise, one has no frets.

Tortoise: Oh, well, it's all the same to me.

Achilles: Fiddle. It makes a big difference, you know.

Tortoise: Say, don't you play the guitar?

Achilles: That's my good friend. He often plays, the fool. But I myself wouldn't touch a guitar with a ten-foot pole!

(Suddenly, the Crab, appearing from out of nowhere, wanders up excitedly, pointing to a rather prominent black eye.)

Crab: Hallo! Hulloo! What's up? What's new? You see this bump, this lump? Given to me by a grump. Ho! And on such a fine day. You see, I was just idly loafing about the park when up lumbers this giant fellow from Warsaw—a colossal bear of a man—playing a lute. He was three meters tall, if I'm a day. I mosey on up to the chap, reach skyward and manage to tap him on the knee, saying, "Pardon me, sir, but you are Pole-luting our park with your mazurkas." But WOW! he had no sense of humor—not a bit, not a wit—and POW!—he lets loose and belts me one, smack in the eye! Were it in my nature, I would crab up a storm, but in the time-honored tradition of my species, I backed off. After all, when we walk forwards, we move backwards. It's in our genes, you know, turning round and round. That reminds me—I've always wondered, "Which came first—the Crab, or the Gene?" That is to say, "Which came last——the Gene, or the Crab?" I'm always turning things round and round, you know. It's in our genes, after all. When we walk backwards, we move forwards. Ah me, oh my! I must lope along on my merry way—so off I go on such a fine day. Sing "ho!" for the life of a Crab! TATA! ¡Olé!

(And he disappears as suddenly as he arrived.)

Tortoise: That's my good friend. He often plays the fool. But I myself wouldn't touch a ten-foot Pole with a guitar!

Achilles: Say, don't you play the guitar?

Tortoise: Fiddle. It makes a big difference, you know.

Achilles: Oh, well, it's all the same to me.

Tortoise: To be precise, one has no frets.

Achilles: Tell me, what's it like to be your age? Is it true that one has no worries at all?

Tortoise: I don't know. But one thing for certain is that I don't worry about arguments of taste. *Disputandum non est de gustibus.*

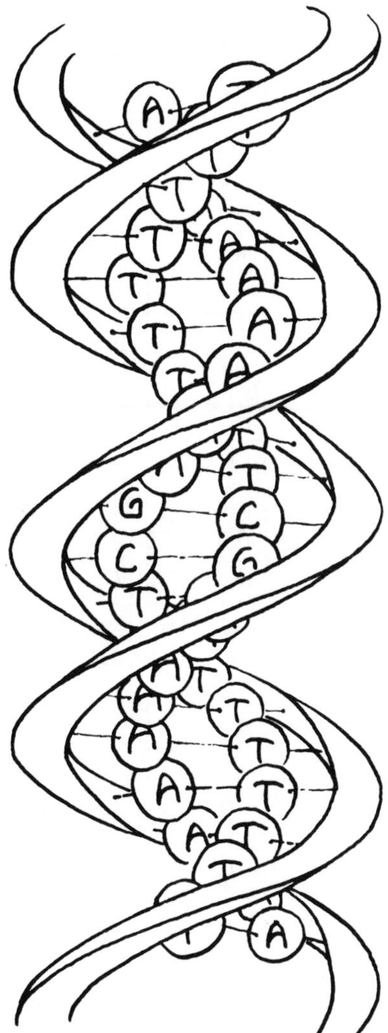

FIGURE 43. Here is a short section of one of the Crab's Genes, turning round and round. When the two DNA strands are unraveled and laid out side by side, they read this way:

. . . TTTTTTTTTCGAAAAAAAAA . . .
. . . AAAAAAAAAGCTTTTTTTTT . . .

Notice that they are the same, only one goes forwards while the other goes backwards. This is the defining property of the form called "crab canon" in music. It is reminiscent of, though a little different from, a palindrome, which is a sentence that reads the same backwards and forwards. In molecular biology, such segments of DNA are called "palindromes"—a slight misnomer, since "crab canon" would be more accurate. Not only is this DNA segment crab-canonical—but moreover its base sequence codes for the Dialogue's structure. Look carefully!

Achilles: I disagree, in this case. But speaking of taste, I finally heard that *Crab Canon* by your favorite composer, J. S. Bach, in a concert the other day, and I fully appreciate the beauty and ingenuity with which he made one single theme mesh with itself going both backwards and forwards. But I'm afraid I will always feel Escher is superior to Bach.

Tortoise: Oh, you are such a philistine. In this area, the Dutch contributions are of markedly inferior taste, don't you think?

Achilles: Not at all. Here, care for one of my cigars?

Tortoise: Thank you very much.

Achilles: Incidentally, you're looking in very fine fettle these days, I must say.

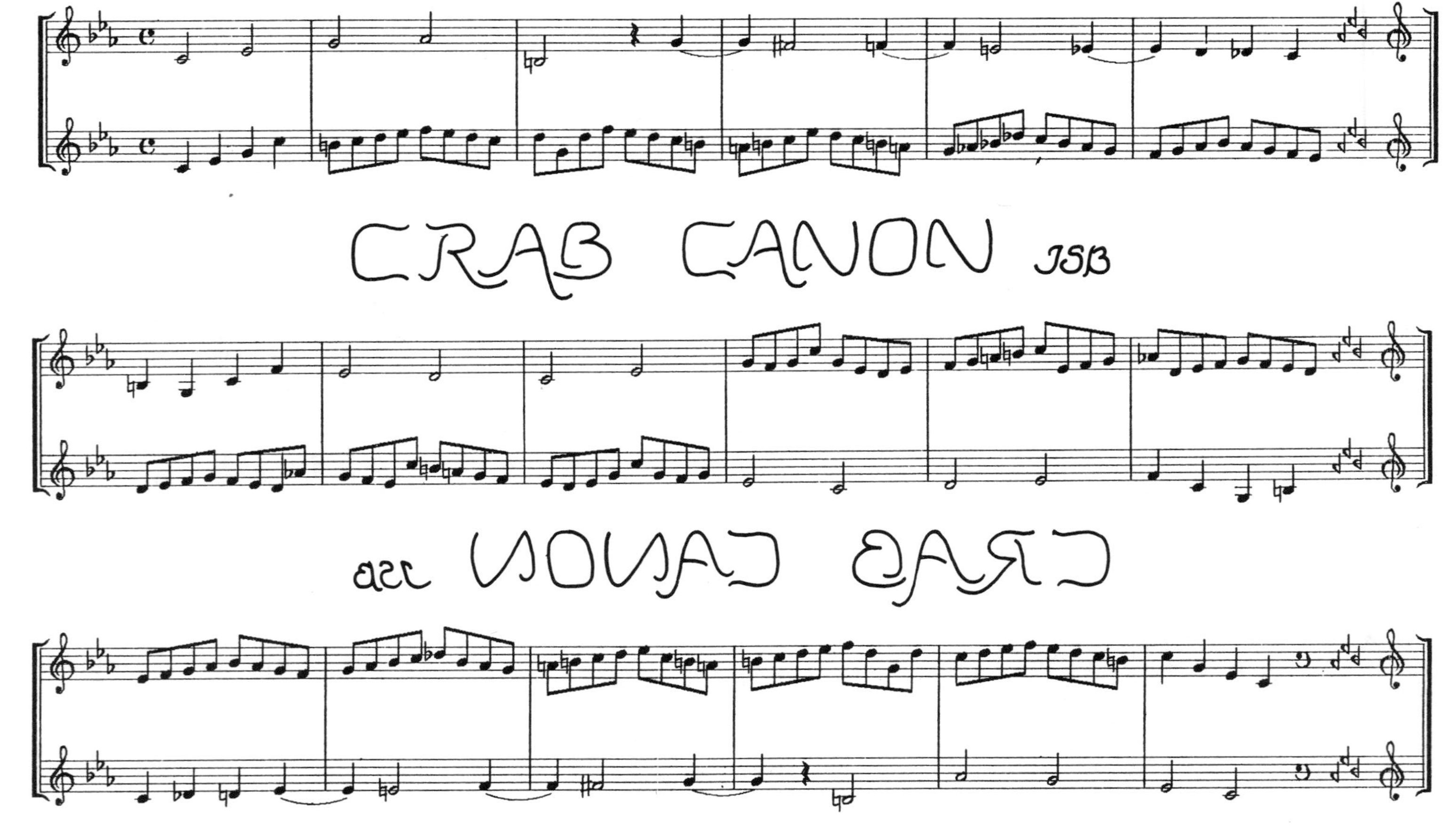

FIGURE 44. Crab Canon *from the* Musical Offering, *by J. S. Bach.* [*Music printed by Donald Byrd's program "SMUT".*]

Tortoise: Oh, really? I guess there's nothing better for you than walking.

Achilles: And it's a perfect day for a walk. I think I'll be walking home soon.

Tortoise: That echoes my thoughts.

Achilles: So nice to run into you.

Tortoise: Why, same to you.

Achilles: Good day, Mr. T.

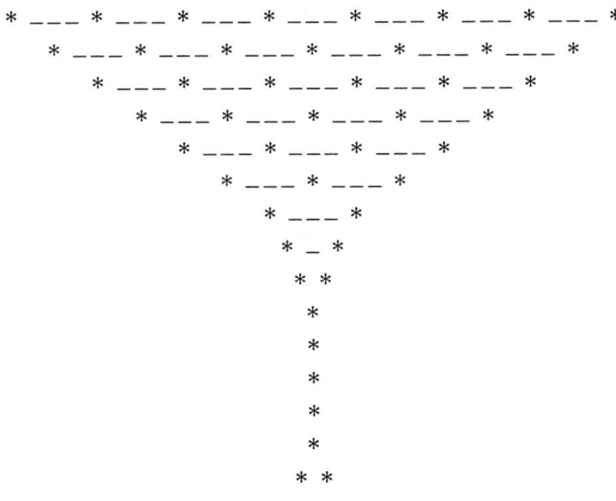

CHAPTER VIII

Typographical Number Theory

The *Crab Canon* and Indirect Self-Reference

THREE EXAMPLES OF indirect self-reference are found in the *Crab Canon.* Achilles and the Tortoise both describe artistic creations they know—and, quite accidentally, those creations happen to have the same structure as the Dialogue they're in. (Imagine my surprise, when I, the author, noticed this!) Also, the Crab describes a biological structure and that, too, has the same property. Of course, one could read the Dialogue and understand it and somehow fail to notice that it, too, has the form of a crab canon. This would be understanding it on one level, but not on another. To see the self-reference, one has to look at the form, as well as the content, of the Dialogue.

Gödel's construction depends on describing the form, as well as the content, of strings of the formal system we shall define in this Chapter—*Typographical Number Theory* (TNT). The unexpected twist is that, because of the subtle mapping which Gödel discovered, the form of strings can be described in the formal system itself. Let us acquaint ourselves with this strange system with the capacity for wrapping around.

What We Want to Be Able to Express in TNT

We'll begin by citing some typical sentences belonging to number theory; then we will try to find a set of basic notions in terms of which all our sentences can be rephrased. Those notions will then be given individual symbols. Incidentally, it should be stated at the outset that the term "number theory" will refer only to properties of positive integers and zero (and sets of such integers). These numbers are called the *natural numbers.* Negative numbers play no role in this theory. Thus the word "number", when used, will mean exclusively a natural number. And it is important—vital—for you to keep separate in your mind the formal system (TNT) and the rather ill-defined but comfortable old branch of mathematics that is number theory itself; this I shall call "N".

Some typical sentences of N—number theory—are:

(1) 5 is prime.
(2) 2 is not a square.
(3) 1729 is a sum of two cubes.
(4) No sum of two positive cubes is itself a cube.
(5) There are infinitely many prime numbers.
(6) 6 is even.

Now it may seem that we will need a symbol for each notion such as "prime" or "cube" or "positive"—but those notions are really not primitive. Primeness, for instance, has to do with the factors which a number has, which in turn has to do with multiplication. Cubeness as well is defined in terms of multiplication. Let us rephrase the sentences, then, in terms of what seem to be more elementary notions.

(1′) There do not exist numbers a and b, both greater than 1, such that 5 equals a times b.

(2′) There does not exist a number b, such that b times b equals 2.

(3′) There exist numbers b and c such that b times b times b, plus c times c times c, equals 1729.

(4′) For all numbers b and c, greater than 0, there is no number a such that a times a times a equals b times b times b plus c times c times c.

(5′) For each number a, there exists a number b, greater than a, with the property that there do not exist numbers c and d, both greater than 1, such that b equals c times d.

(6′) There exists a number e such that 2 times e equals 6.

This analysis has gotten us a long ways towards the basic elements of the language of number theory. It is clear that a few phrases reappear over and over:

for all numbers b
there exists a number b, such that . . .
greater than
equals
times
plus
0, 1, 2, . . .

Most of these will be granted individual symbols. An exception is "greater than", which can be further reduced. In fact, the sentence "a is greater than b" becomes

there exists a number c, not equal to 0, such that a equals b plus c.

Numerals

We will not have a distinct symbol for each natural number. Instead, we will have a very simple, uniform way of giving a compound symbol to each natural number—very much as we did in the pq-system. Here is our notation for natural numbers:

zero: 0
one: S0
two: SS0
three: SSS0

etc.

The symbol S has an interpretation—"the successor of". Hence, the interpretation of SS0 is literally "the successor of the successor of zero". Strings of this form are called *numerals*.

Variables and Terms

Clearly, we need a way of referring to unspecified, or variable, numbers. For that, we will use the letters a, b, c, d, e. But five will not be enough. We need an unlimited supply of them, just as we had of atoms in the Propositional Calculus. We will use a similar method for making more variables: tacking on any number of primes. (Note: Of course the symbol ′—read "prime"—is not to be confused with prime numbers!) For instance:

e
d′
c′′
b′′′
a′′′′

are all *variables*.

In a way it is a luxury to use the first five letters of the alphabet when we could get away with just a and the prime. Later on, I will actually drop b, c, d, and e, which will result in a sort of "austere" version of TNT—austere in the sense that it is a little harder to decipher complex formulas. But for now we'll be luxurious.

Now what about addition and multiplication? Very simple: we will use the ordinary symbols '+' and '·'. However, we will also introduce a parenthesizing requirement (we are now slowly slipping into the rules which define well-formed strings of TNT). To write "b plus c" and "b times c", for instance, we use the strings

(b+c)
(b·c)

There is no laxness about such parentheses; to violate the convention is to produce a non-well-formed formula. ("Formula"? I use the term instead of "string" because it is conventional to do so. A *formula* is no more and no less than a string of TNT.)

Incidentally, addition and multiplication are always to be thought of as *binary* operations—that is, they unite precisely two numbers, never three or more. Hence, if you wish to translate "1 plus 2 plus 3", you have to decide which of the following two expressions you want:

(S0+(SS0+SSS0))
((S0+SS0)+SSS0)

The next notion we'll symbolize is *equals*. That is very simple: we use '='. The advantage of taking over the standard symbol used in N—nonformal number theory—is obvious: easy legibility. The disadvantage is very much like the disadvantage of using the words "point" and "line" in a formal treatment of geometry: unless one is very conscious and careful, one may blur the distinction between the familiar meaning and the strictly rule-governed behavior of the formal symbol. In discussing geometry, I distinguished between the everyday word and the formal term by capitalizing the formal term: thus, in elliptical geometry, a POINT was the union of two ordinary points. Here, there is no such distinction; hence, mental effort is needed not to confuse a symbol with all of the associations it is laden with. As I said earlier, with reference to the pq-system: the string --- is not the number 3, but it acts isomorphically to 3, at least in the context of additions. Similar remarks go for the string SSS0.

Atoms and Propositional Symbols

All the symbols of the Propositional Calculus except the letters used in making atoms (P, Q, and R) will be used in TNT, and they retain their interpretations. The role of *atoms* will be played by strings which, when interpreted, are statements of equality, such as S0=SS0 or (S0·S0)=S0. Now, we have the equipment to do a fair amount of translation of simple sentences into the notation of TNT:

2 plus 3 equals 4: (SS0+SSS0)=SSSS0
2 plus 2 is not equal to 3: ~(SS0+SS0)=SSS0
If 1 equals 0, then 0 equals 1: <S0=0⊃0=S0>

The first of these strings is an atom; the rest are compound formulas. (Warning: The 'and' in the phrase "1 and 1 make 2" is just another word for 'plus', and must be represented by '+' (and the requisite parentheses).)

Free Variables and Quantifiers

All the well-formed formulas above have the property that their interpretations are sentences which are either true or false. There are, however, well-formed formulas which do not have that property, such as this one:

(b+S0)=SS0

Its interpretation is "b plus 1 equals 2". Since b is unspecified, there is no way to assign a truth value to the statement. It is like an out-of-context statement with a pronoun, such as "she is clumsy". It is neither true nor false; it is waiting for you to put it into a context. Because it is neither true nor false, such a formula is called *open,* and the variable b is called a *free variable.*

One way of changing an open formula into a *closed* formula, or *sentence,* is by prefixing it with a *quantifier*—either the phrase "there exists a number b such that . . .", or the phrase "for all numbers b". In the first instance, you get the sentence

There exists a number b such that b plus 1 equals 2.

Clearly this is true. In the second instance, you get the sentence

For all numbers b, b plus 1 equals 2.

Clearly this is false. We now introduce symbols for both of these *quantifiers.* These sentences are translated into TNT-notation as follows:

∃b:(b+S0)=SS0 ('∃' stands for 'exists'.)

∀b:(b+S0)=SS0 ('∀' stands for 'all'.)

It is very important to note that these statements are no longer about unspecified numbers; the first one is an *assertion of existence,* and the second one is a *universal assertion.* They would mean the same thing, even if written with c instead of b:

∃c:(c+S0)=SS0

∀c:(c+S0)=SS0

A variable which is under the dominion of a quantifier is called a *quantified variable.* The following two formulas illustrate the difference between free variables and quantified variables:

(b·b)=SS0 (open)

~∃b:(b·b)=SS0 (closed; a *sentence* of TNT)

The first one expresses a *property* which might be possessed by some natural number. Of course, no natural number has that property. And that is precisely what is expressed by the second one. It is very crucial to understand this difference between a string with a *free variable,* which expresses a *property,* and a string where the variable is *quantified,* which expresses a *truth or falsity.* The English translation of a formula with at least one free variable—an open formula—is called a *predicate.* It is a sentence without a subject (or a sentence whose subject is an out-of-context pronoun). For instance,

"is a sentence without a subject"

"would be an anomaly"

"runs backwards and forwards simultaneously"

"improvised a six-part fugue on demand"

are nonarithmetical predicates. They express *properties* which specific entities might or might not possess. One could as well stick on a "dummy subject", such as "so-and-so". A string with free variables is like a predicate with "so-and-so" as its subject. For instance,

(S0+S0)=b

is like saying "1 plus 1 equals so-and-so". This is a predicate in the variable b. It expresses a property which the number b might have. If one were to substitute various numerals for b, one would get a succession of formulas, most of which would express falsehoods. Here is another example of the difference between open formulas and *sentences:*

∀b:∀c:(b+c)=(c+b)

The above formula is a sentence representing, of course, the commutativity of addition. On the other hand,

∀c:(b+c)=(c+b)

is an open formula, since b is free. It expresses a property which the unspecified number b might or might not have—namely of commuting with all numbers c.

Translating Our Sample Sentences

This completes the vocabulary with which we will express all number-theoretical statements! It takes considerable practice to get the hang of expressing complicated statements of N in this notation, and conversely of figuring out the meaning of well-formed formulas. For this reason we return to the six sample sentences given at the beginning, and work out their translations into TNT. By the way, don't think that the translations given below are unique—far from it. There are many—infinitely many—ways to express each one.

Let us begin with the last one: "6 is even". This we rephrased in terms of more primitive notions as "There exists a number e such that 2 times e equals 6". This one is easy:

∃e:(SS0·e)=SSSSSS0

Note the necessity of the quantifier; it simply would not do to write

(SS0·e)=SSSSSS0

alone. This string's interpretation is of course neither true nor false; it just expresses a property which the number e might have.

It is curious that, since we know multiplication is commutative, we might easily have written

∃e:(e·SS0)=SSSSSS0

instead. Or, knowing that equality is a symmetrical relation, we might have chosen to write the sides of the equation in the opposite order:

$$\exists e: SSSSSS0=(SS0 \cdot e)$$

Now these three translations of "6 is even" are quite different strings, and it is by no means obvious that theoremhood of any one of them is tied to theoremhood of any of the others. (Similarly, the fact that --p-q--- was a theorem had very little to do with the fact that its "equivalent" string -p--q--- was a theorem. The equivalence lies in our minds, since, as humans, we almost automatically think about interpretations, not structural properties of formulas.)

We can dispense with sentence 2: "2 is not a square", almost immediately:

$$\sim\exists b:(b \cdot b)=SS0$$

However, once again, we find an ambiguity. What if we had chosen to write it this way?

$$\forall b:\sim(b \cdot b)=SS0$$

The first way says, "It is not the case that there exists a number b with the property that b's square is 2", while the second way says, "For all numbers b, it is not the case that b's square is 2." Once again, *to us,* they are conceptually equivalent—but to TNT, they are distinct strings.

Let us proceed to sentence 3: "1729 is a sum of two cubes." This one will involve *two* existential quantifiers, one after the other, as follows:

$$\exists b:\exists c:\underbrace{SSSSSS\ldots\ldots SSSSS}_{1729\ of\ them}0=(((b \cdot b) \cdot b)+((c \cdot c) \cdot c))$$

There are alternatives galore. Reverse the order of the quantifiers; switch the sides of the equation; change the variables to d and e; reverse the addition; write the multiplications differently; etc., etc. However, I prefer the following two translations of the sentence:

$$\exists b:\exists c:(((SSSSSSSSSS0 \cdot SSSSSSSSSS0) \cdot SSSSSSSSSS0)+((SSSSSSSSS0 \cdot SSSSSSSSS0) \cdot SSSSSSSSS0))=(((b \cdot b) \cdot b)+((c \cdot c) \cdot c))$$

and

$$\exists b:\exists c:(((SSSSSSSSSSSS0 \cdot SSSSSSSSSSSS0) \cdot SSSSSSSSSSSS0)+((S0 \cdot S0) \cdot S0))=(((b \cdot b) \cdot b)+((c \cdot c) \cdot c))$$

Do you see why?

Tricks of the Trade

Now let us tackle the related sentence 4: "No sum of two positive cubes is itself a cube". Suppose that we wished merely to state that 7 is not a sum of two positive cubes. The easiest way to do this is by *negating* the formula which asserts that 7 *is* a sum of two positive cubes. This will be just like the preceding sentence involving 1729, except that we have to add in the proviso of the cubes being positive. We can do this with a trick: prefix the variables with the symbol S, as follows:

∃b:∃c:SSSSSSS0=(((Sb·Sb)·Sb)+((Sc·Sc)·Sc))

You see, we are cubing not b and c, but their successors, which must be positive, since the smallest value which either b or c can take on is zero. Hence the right-hand side represents a sum of two positive cubes. Incidentally, notice that the phrase "there exist numbers b and c such that . . .", when translated, does not involve the symbol '∧' which stands for 'and'. That symbol is used for connecting entire well-formed strings, not for joining two quantifiers.

Now that we have translated "7 is a sum of two positive cubes", we wish to negate it. That simply involves prefixing the whole thing by a single tilde. (Note: you should *not* negate each quantifier, even though the desired phrase runs "There do not exist numbers b and c such that . . .".) Thus we get:

~∃b:∃c:SSSSSSS0=(((Sb·Sb)·Sb)+((Sc·Sc)·Sc))

Now our original goal was to assert this property not of the number 7, but of all cubes. Therefore, let us replace the numeral SSSSSSS0 by the string ((a·a)·a), which is the translation of "a cubed":

~∃b:∃c:((a·a)·a)=(((Sb·Sb)·Sb)+((Sc·Sc)·Sc))

At this stage, we are in possession of an *open* formula, since a is still free. This formula expresses a property which a number a might or might not have—and it is our purpose to assert that all numbers do have that property. That is simple—just prefix the whole thing with a universal quantifier:

∀a:~∃b:∃c:((a·a)·a)=(((Sb·Sb)·Sb)+((Sc·Sc)·Sc))

An equally good translation would be this:

~∃a:∃b:∃c:((a·a)·a)=(((Sb·Sb)·Sb)+((Sc·Sc)·Sc))

In *austere* TNT, we could use a′ instead of b, and a″ instead of c, and the formula would become:

~∃a:∃a′:∃a″:((a·a)·a)=(((Sa′·Sa′)·Sa′)+((Sa″·Sa″)·Sa″))

What about sentence 1: "5 is prime"? We had reworded it in this way: "There do not exist numbers a and b, both greater than 1, such that 5 equals a times b". We can slightly modify it, as follows: "There do not exist numbers a and b such that 5 equals a plus 2, times b plus 2". This is another trick—since a and b are restricted to natural number values, this is an adequate way to say the same thing. Now "b plus 2" could be translated into (b+SS0), but there is a shorter way to write it—namely, SSb. Likewise, "c plus 2" can be written SSc. Now, our translation is extremely concise:

~∃b:∃c:SSSSS0=(SSb·SSc)

Without the initial tilde, it would be an assertion that two natural numbers *do* exist, which, when augmented by 2, have a product equal to 5. With the tilde in front, that whole statement is denied, resulting in an assertion that 5 is prime.

If we wanted to assert that d plus e plus 1, rather than 5, is prime, the most economical way would be to replace the numeral for 5 by the string (d+Se):

~∃b:∃c:(d+Se)=(SSb·SSc)

Once again, an open formula, one whose interpretation is neither a true nor a false sentence, but just an assertion about two unspecified numbers, d and e. Notice that the number represented by the string (d+Se) is necessarily greater than d, since one has added to d an unspecified but definitely positive amount. Therefore, if we existentially quantify over the variable e, we will have a formula which asserts that:

There exists a number which is greater than d and which is prime.

∃e:~∃b:∃c:(d+Se)=(SSb·SSc)

Well, all we have left to do now is to assert that this property actually obtains, no matter what d is. The way to do that is to universally quantify over the variable d:

∀d:∃e:~∃b:∃c:(d+Se)=(SSb·SSc)

That's the translation of sentence 5!

Translation Puzzles for You

This completes the exercise of translating all six typical number-theoretical sentences. However, it does not necessarily make you an expert in the notation of TNT. There are still some tricky issues to be mastered. The following six well-formed formulas will test your understanding of TNT-notation. What do they mean? Which ones are true (under interpretation, of course), and which ones are false? (Hint: the way to tackle this exercise is to move leftwards. First, translate the atom; next, figure out what adding a single quantifier or a tilde does; then move leftwards, adding another quantifier or tilde; then move leftwards again, and do the same.)

~∀c:∃b:(SS0·b)=c

∀c:~∃b:(SS0·b)=c

∀c:∃b:~(SSO·b)=c

~∃b:∀c:(SSO·b)=c

∃b:~∀c:(SSO·b)=c

∃b:∀c:~(SSO·b)=c

(Second hint: Either four of them are true and two false, or four false and two true.)

How to Distinguish True from False?

At this juncture, it is worthwhile pausing for breath and contemplating what it would mean to have a formal system that could sift out the true ones from the false ones. This system would treat all these strings—which to us look like statements—as designs having form, but no content. And this system would be like a sieve through which could pass only designs with a special style—the "style of truth". If you yourself have gone through the six formulas above, and have separated the true from the false by thinking about meaning, you will appreciate the subtlety that any system would have to have, that could do the same thing—but typographically! The boundary separating the set of true statements from the set of false statements (as written in the TNT-notation) is anything but straight; it is a boundary with many treacherous curves (recall Fig. 18), a boundary of which mathematicians have delineated stretches, here and there, working over hundreds of years. Just think what a coup it would be to have a typographical method which was guaranteed to place any formula on the proper side of the border!

The Rules of Well-Formedness

It is useful to have a table of Rules of Formation for well-formed formulas. This is provided below. There are some preliminary stages, defining *numerals, variables,* and *terms.* Those three classes of strings are ingredients of well-formed formulas, but are not in themselves well-formed. The smallest well-formed formulas are the *atoms;* then there are ways of compounding atoms. Many of these rules are recursive lengthening rules, in that they take as input an item of a given class and produce a longer item of the same class. In this table, I use '*x*' and '*y*' to stand for well-formed formulas, and '*s*', '*t*', and '*u*' to stand for other kinds of TNT-strings. Needless to say, none of these five symbols is itself a symbol of TNT.

NUMERALS.

O is a numeral.

A numeral preceded by S is also a numeral.

Examples: O SO SSO SSSO SSSSO SSSSSO

VARIABLES.

a is a variable. If we're not being austere, so are b, c, d and e.
A variable followed by a prime is also a variable.
Examples: a b′ c′′ d′′′ e′′′′

TERMS.

All numerals and variables are terms.
A term preceded by S is also a term.
If s and t are terms, then so are $(s+t)$ and $(s \cdot t)$.
Examples: 0 b SSa′ (S0·(SS0+c)) S(Sa·(Sb·Sc))

TERMS may be divided into two categories:

(1) DEFINITE terms. These contain no variables.
Examples: 0 (S0+S0) SS((SS0·SS0)+(S0·S0))

(2) INDEFINITE terms. These contain variables.
Examples: b Sa (b+S0) (((S0+S0)+S0)+e)

The above rules tell how to make *parts* of well-formed formulas; the remaining rules tell how to make *complete* well-formed formulas.

ATOMS.

If s and t are terms, then $s = t$ is an atom.
Examples: S0=0 (SS0+SS0)=SSSS0 S(b+c)=((c·d)·e)
If an atom contains a variable u, then u is *free* in it. Thus there are four free variables in the last example.

NEGATIONS.

A well-formed formula preceded by a tilde is well-formed.
Examples: ~S0=0 ~∃b:(b+b)=S0 ~<0=0⊃S0=0> ~b=S0
The *quantification status* of a variable (which says whether the variable is free or quantified) does not change under negation.

COMPOUNDS.

If x and y are well-formed formulas, and provided that no variable which is free in one is quantified in the other, then the following are all well-formed formulas:
<x∧y>, <x∨y>, <x⊃y>.
Examples: <0=0∧~0=0> <b=b∨~∃c:c=b>
<S0=0⊃∀c:~∃b:(b+b)=c>
The quantification status of a variable doesn't change here.

QUANTIFICATIONS.

If u is a variable, and x is a well-formed formula in which u is free, then the following strings are well-formed formulas:
∃u: x and ∀u: x.
Examples: ∀b:<b=b∨~∃c:c=b> ∀c:~∃b:(b+b)=c ~∃c:Sc=d

OPEN FORMULAS contain at least one free variable.
Examples: ~c=c b=b <∀b:b=b∧~c=c>

CLOSED FORMULAS (SENTENCES) contain no free variables.
Examples: S0=0 ~∀d:d=0 ∃c:<∀b:b=b∧~c=c>

This completes the table of Rules of Formation for the well-formed formulas of TNT.

A Few More Translation Exercises

And now, a few practice exercises for you, to test your understanding of the notation of TNT. Try to translate the first four of the following N-sentences into TNT-sentences, and the last one into an open well-formed formula.

All natural numbers are equal to 4.
There is no natural number which equals its own square.
Different natural numbers have different successors.
If 1 equals 0, then every number is odd.
b is a power of 2.

The last one you may find a little tricky. But it is nothing, compared to this one:

b is a power of 10.

Strangely, this one takes great cleverness to render in our notation. I would caution you to try it only if you are willing to spend hours and hours on it—and if you know quite a bit of number theory!

A Nontypographical System

This concludes the exposition of the notation of TNT; however, we are still left with the problem of making TNT into the ambitious system which we have described. Success would justify the interpretations which we have given to the various symbols. Until we have done that, however, these particular interpretations are no more justified than the "horse-apple-happy" interpretations were for the pq-system's symbols.

Someone might suggest the following way of constructing TNT: (1) Do not have any rules of inference; they are unnecessary, because (2) We take as axioms all true statements of number theory (as written in TNT-notation). What a simple prescription! Unfortunately it is as empty as one's instantaneous reaction says it is. Part (2) is, of course, not a typographical description of strings. The whole purpose of TNT is to figure out if and how it is possible to characterize the true strings typographically.

The Five Axioms and First Rules of TNT

Thus we will follow a more difficult route than the suggestion above; we will have axioms and rules of inference. Firstly, as was promised, *all of the rules of the Propositional Calculus are taken over into TNT.* Therefore, one theorem of TNT will be this one:

<S0=0∨~S0=0>

which can be derived in the same way as <P∨~P> was derived.

Before we give more rules, let us give the five *axioms* of TNT:

Axiom 1: ∀a:~Sa=0

Axiom 2: ∀a:(a+0)=a

Axiom 3: ∀a:∀b:(a+Sb)=S(a+b)

Axiom 4: ∀a:(a·0)=0

Axiom 5: ∀a:∀b:(a·Sb)=((a·b)+a)

(In the austere versions, use a′ instead of b.) All of them are very simple to understand. Axiom 1 states a special fact about the number 0; Axioms 2 and 3 are concerned with the nature of addition; Axioms 4 and 5 are concerned with the nature of multiplication, and in particular with its relation to addition.

The Five Peano Postulates

By the way, the interpretation of Axiom 1—"Zero is not the successor of any natural number"—is one of five famous properties of natural numbers first explicitly recognized by the mathematician and logician Giuseppe Peano, in 1889. In setting out his postulates, Peano was following the path of Euclid in this way: he made no attempt to formalize the principles of reasoning, but tried to give a small set of properties of natural numbers from which everything else could be derived by reasoning. Peano's attempt might thus be considered "semiformal". Peano's work had a significant influence, and thus it would be good to show Peano's five postulates. Since the notion of "natural number" is the one which Peano was attempting to define, we will not use the familiar term "natural number", which is laden with connotation. We will replace it with the undefined term *djinn,* a word which comes fresh and free of connotations to our mind. Then Peano's five postulates place five restrictions on djinns. There are two other undefined terms: *Genie,* and *meta.* I will let you figure out for yourself what usual concept each of them is supposed to represent. The five Peano postulates:

(1) Genie is a djinn.
(2) Every djinn has a meta (which is also a djinn).
(3) Genie is not the meta of any djinn.
(4) Different djinns have different metas.
(5) If Genie has X, and each djinn relays X to its meta, then all djinns get X.

In light of the lamps of the *Little Harmonic Labyrinth,* we should name the set of *all* djinns "GOD". This harks back to a celebrated statement by the German mathematician and logician Leopold Kronecker, archenemy of Georg Cantor: "God made the natural numbers; all the rest is the work of man."

You may recognize Peano's fifth postulate as the principle of mathematical induction—another term for a hereditary argument. Peano hoped that his five restrictions on the concepts "Genie", "djinn", and "meta" were so strong that if two different people formed images in their minds of the concepts, the two images would have completely *isomorphic structures.* For example, everybody's image would include an infinite number of distinct djinns. And presumably everybody would agree that no djinn coincides with its own meta, or its meta's meta, etc.

Peano hoped to have pinned down the essence of natural numbers in his five postulates. Mathematicians generally grant that he succeeded, but that does not lessen the importance of the question, "How is a true statement about natural numbers to be distinguished from a false one?" And to answer this question, mathematicians turned to totally formal systems, such as TNT. However, you will see the influence of Peano in TNT, because all of his postulates are incorporated in TNT in one way or another.

New Rules of TNT: Specification and Generalization

Now we come to the new rules of TNT. Many of these rules will allow us to reach in and change the internal structure of the atoms of TNT. In that sense they deal with more "microscopic" properties of strings than the rules of the Propositional Calculus, which treat atoms as indivisible units. For example, it would be nice if we could extract the string ~S0=0 from the first axiom. To do this we would need a rule which permits us to drop a universal quantifier, and at the same time to change the internal structure of the string which remains, if we wish. Here is such a rule:

RULE OF SPECIFICATION: Suppose *u* is a variable which occurs inside the string *x*. If the string ∀*u*: *x* is a theorem, then so is *x*, and so are any strings made from *x* by replacing *u*, wherever it occurs, by one and the same term.

(*Restriction:* The term which replaces *u* must not contain any variable that is quantified in *x*.)

The rule of specification allows the desired string to be extracted from Axiom 1. It is a one-step derivation:

∀a:~Sa=0	axiom 1
~S0=0	specification

Notice that the rule of specification will allow some formulas which contain free variables (i.e., open formulas) to become theorems. For example, the following strings could also be derived from Axiom 1, by specification:

~Sa=0
~S(c+SS0)=0

There is another rule, the *rule of generalization,* which allows us to put back the universal quantifier on theorems which contain variables that became free as a result of usage of specification. Acting on the lower string, for example, generalization would give:

∀c:~S(c+SS0)=0

Generalization undoes the action of specification, and vice versa. Usually, generalization is applied after several intermediate steps have transformed the open formula in various ways. Here is the exact statement of the rule:

RULE OF GENERALIZATION: Suppose *x* is a theorem in which *u*, a variable, occurs free. Then ∀*u*: *x* is a theorem.

(*Restriction:* No generalization is allowed in a fantasy on any variable which appeared free in the fantasy's premise.)

The need for restrictions on these two rules will shortly be demonstrated explicitly. Incidentally, this generalization is the same generalization as was mentioned in Chapter II, in Euclid's proof about the infinitude of primes. Already we can see how the symbol-manipulating rules are starting to approximate the kind of reasoning which a mathematician uses.

The Existential Quantifier

These past two rules told how to take off universal quantifiers and put them back on; the next two rules tell how to handle existential quantifiers.

RULE OF INTERCHANGE: Suppose *u* is a variable. Then the strings ∀*u*:~ and ~∃*u*: are interchangeable anywhere inside any theorem.

For example, let us apply this rule to Axiom 1:

∀a:~Sa=0	axiom 1
~∃a:Sa=0	interchange

By the way, you might notice that both these strings are perfectly natural renditions, in TNT, of the sentence "Zero is not the successor of any natural number". Therefore it is good that they can be turned into each other with ease.

The next rule is, if anything, even more intuitive. It corresponds to the very simple kind of inference we make when we go from "2 is prime" to "There exists a prime". The name of this rule is self-explanatory:

RULE OF EXISTENCE: Suppose a term (which may contain variables as long as they are free) appears once, or multiply, in a theorem. Then any (or several, or all) of the appearances of the term may be replaced by a variable which otherwise does not occur in the theorem, and the corresponding existential quantifier must be placed in front.

Let us apply the rule to—as usual—Axiom 1:

∀a:~Sa=0	axiom 1
∃b:∀a:~Sa=b	existence

You might now try to shunt symbols, according to rules so far given, to produce the theorem ~∀b:∃a:Sa=b.

Rules of Equality and Successorship

We have given rules for manipulating quantifiers, but so far none for the symbols '=' and 'S'. We rectify that situation now. In what follows, *r, s,* and *t* all stand for arbitrary *terms*.

RULES OF EQUALITY:

SYMMETRY: If $r = s$ is a theorem, then so is $s = r$.

TRANSITIVITY: If $r = s$ and $s = t$ are theorems, then so is $r = t$.

RULES OF SUCCESSORSHIP:

ADD S: If $r = t$ is a theorem, then Sr = St is a theorem.

DROP S: If Sr = St is a theorem, then $r = t$ is a theorem.

Now we are equipped with rules that can give us a fantastic variety of theorems. For example, the following derivations yield theorems which are pretty fundamental:

(1)	∀a:∀b:(a+Sb)=S(a+b)	axiom 3
(2)	∀b:(S0+Sb)=S(S0+b)	specification (S0 for a)
(3)	(S0+S0)=S(S0+0)	specification (0 for b)
(4)	∀a:(a+0)=a	axiom 2
(5)	(S0+0)=S0	specification (S0 for a)
(6)	S(S0+0)=SS0	add S
(7)	(S0+S0)=SS0	transitivity (lines 3,6)

* * * * *

(1)	∀a:∀b:(a·Sb)=((a·b)+a)	axiom 5
(2)	∀b:(S0·Sb)=((S0·b)+S0)	specification (S0 for a)
(3)	(S0·S0)=((S0·0)+S0)	specification (0 for b)
(4)	∀a:∀b:(a+Sb)=S(a+b)	axiom 3
(5)	∀b:((S0·0)+Sb)=S((S0·0)+b)	specification ((S0·0) for a)
(6)	((S0·0)+S0)=S((S0·0)+0)	specification (0 for b)
(7)	∀a:(a+0)=a	axiom 2
(8)	((S0·0)+0)=(S0·0)	specification ((S0.0) for a)
(9)	∀a:(a·0)=0	axiom 4
(10)	(S0·0)=0	specification (S0 for a)
(11)	((S0·0)+0)=0	transitivity (lines 8,10)
(12)	S((S0·0)+0)=S0	add S
(13)	((S0·0)+S0)=S0	transitivity (lines 6,12)
(14)	(S0·S0)=S0	transitivity (lines 3,13)

Illegal Shortcuts

Now here is an interesting question: "How can we make a derivation for the string 0=0?" It seems that the obvious route to go would be first to derive the string ∀a:a=a, and then to use specification. So, what about the following "derivation" of ∀a:a=a . . . What is wrong with it? Can you fix it up?

(1)	∀a:(a+0)=a	axiom 2
(2)	∀a:a=(a+0)	symmetry
(3)	∀a:a=a	transitivity (lines 2,1)

I gave this mini-exercise to point out one simple fact: that one should not jump too fast in manipulating symbols (such as '=') which are familiar. One must follow the rules, and not one's knowledge of the passive meanings of the symbols. Of course, this latter type of knowledge is invaluable in guiding the route of a derivation.

Why Specification and Generalization Are Restricted

Now let us see why there are restrictions necessary on both specification and generalization. Here are two derivations. In each of them, one of the restrictions is violated. Look at the disastrous results they produce:

(1)	[	push
(2)	a=0	premise
(3)	∀a:a=0	generalization *(Wrong!)*
(4)	Sa=0	specification
(5)	]	pop
(6)	<a=0⊃Sa=0>	fantasy rule
(7)	∀a:<a=0⊃Sa=0>	generalization
(8)	<0=0⊃S0=0>	specification
(9)	0=0	previous theorem
(10)	S0=0	detachment (lines 9,8)

This is the first disaster. The other one is via faulty specification.

(1)	∀a:a=a	previous theorem
(2)	Sa=Sa	specification
(3)	∃b:b=Sa	existence
(4)	∀a:∃b:b=Sa	generalization
(5)	∃b:b=Sb	specification *(Wrong!)*

So now you can see why those restrictions are needed.

Here is a simple puzzle: translate (if you have not already done so) Peano's fourth postulate into TNT-notation, and then derive that string as a theorem.

Something Is Missing

Now if you experiment around for a while with the rules and axioms of TNT so far presented, you will find that you can produce the following *pyramidal family* of theorems (a set of strings all cast from an identical mold, differing from one another only in that the numerals **0**, **S0**, **SS0**, and so on have been stuffed in):

(0+0)=0
(0+S0)=S0
(0+SS0)=SS0
(0+SSS0)=SSS0
(0+SSSS0)=SSSS0

etc.

As a matter of fact, each of the theorems in this family can be derived from the one directly above it, in only a couple of lines. Thus it is a sort of "cascade" of theorems, each one triggering the next. (These theorems are very reminiscent of the pq-theorems, where the middle and right-hand groups of hyphens grew simultaneously.)

Now there is one string which we can easily write down, and which summarizes the passive meaning of them all, taken together. That universally quantified *summarizing string* is this:

∀a:(0+a)=a

Yet with the rules so far given, this string eludes production. Try to produce it yourself if you don't believe me.

You may think that we should immediately remedy the situation with the following

(PROPOSED) RULE OF ALL: If all the strings in a pyramidal family are theorems, then so is the universally quantified string which summarizes them.

The problem with this rule is that it cannot be used in the M-mode. Only people who are thinking *about* the system can ever know that an infinite set of strings are all theorems. Thus this is not a rule that can be stuck inside any formal system.

ω-Incomplete Systems and Undecidable Strings

So we find ourselves in a strange situation, in which we can typographically produce theorems about the addition of any *specific* numbers, but even such a simple string as the one above, which expresses a property of addition *in general*, is not a theorem. You might think that is not all that strange, since we were in precisely that situation with the pq-system. However, the pq-system had no pretensions about what it ought to be able to do; and in fact there was no way to *express* general statements about addition in its symbolism, let alone prove them. The equipment simply was not there, and it did not even occur to us to think that the system was defective. Here, however, the expressive capability is far stronger, and we have correspondingly higher expectations of TNT than of the pq-system. If the string above is not a theorem, then we will have good reason to consider TNT to be defective. As a matter of fact, there is a name for systems with this kind of defect—they are called *ω-incomplete.* (The prefix 'ω'—'omega'— comes from the fact that the totality of natural numbers is sometimes denoted by 'ω'.) Here is the exact definition:

> A system is ω-incomplete if all the strings in a pyramidal family are theorems, but the universally quantified summarizing string is not a theorem.

Incidentally, the negation of the above summarizing string—

~∀a:(0+a)=a

—is also a nontheorem of TNT. This means that the original string is *undecidable within the system.* If one or the other were a theorem, then we would say that it was decidable. Although it may sound like a mystical term, there is nothing mystical about undecidability within a given system. It is only a sign that the system could be extended. For example, within absolute geometry, Euclid's fifth postulate is undecidable. It has to be added as an extra postulate of geometry, to yield Euclidean geometry; or conversely, its negation can be added, to yield non-Euclidean geometry. If you think back to geometry, you will remember why this curious thing happens. It is because the four postulates of absolute geometry simply do not pin down the meanings of the terms "point" and "line", and there is room for *different extensions* of the notions. The points and lines of Euclidean geometry provide one kind of extension of the notions of "point" and "line"; the POINTS and LINES of non-Euclidean geometry, another. However, using the pre-flavored words "point" and "line" tended, for two millennia, to make people believe that those words were necessarily univalent, capable of only one meaning.

Non-Euclidean TNT

We are now faced with a similar situation, involving TNT. We have adopted a notation which prejudices us in certain ways. For instance, usage of the symbol '+' tends to make us think that every theorem with a plus sign in it ought to say something known and familiar and "sensible" about the known and familiar operation we call "addition". Therefore it would run against the grain to propose adding the following "sixth axiom":

~∀a:(0+a)=a

It doesn't jibe with what we believe about addition. But it is one possible extension of TNT, as we have so far formulated TNT. The system which uses this as its sixth axiom is a *consistent* system, in the sense of not having two theorems of the form x and $\sim x$. However, when you juxtapose this "sixth axiom" with the pyramidal family of theorems shown above, you will probably be bothered by a seeming inconsistency between the family and the new axiom. But this kind of inconsistency is not so damaging as the other kind (where x and $\sim x$ are both theorems). In fact, it is not a true inconsistency, because there is a way of interpreting the symbols so that everything comes out all right.

ω-Inconsistency Is Not the Same as Inconsistency

This kind of inconsistency, created by the opposition of (1) a pyramidal family of theorems which collectively assert that *all* natural numbers have some property, and (2) a single theorem which seems to assert that *not all* numbers have it, is given the name of *ω-inconsistency.* An ω-inconsistent system is more like the at-the-outset-distasteful-but-in-the-end-acceptable non-Euclidean geometry. In order to form a mental model of what is going on, you have to imagine that there are some "extra", unsuspected numbers—let us not call them "natural", but *supernatural* numbers—which have no numerals. Therefore, facts about them cannot be represented in the pyramidal family. (This is a little bit like Achilles' conception of GOD—as a sort of "superdjinn", a being greater than any of the djinns. This was scoffed at by the Genie, but it is a reasonable image, and may help you to imagine supernatural numbers.)

What this tells us is that the axioms and rules of TNT, as so far presented, do not fully pin down the interpretations for the symbols of TNT. There is still room for variation in one's mental model of the notions they stand for. Each of the various possible extensions would pin down some of the notions further; but in different ways. Which symbols would begin to take on "distasteful" passive meanings, if we added the "sixth axiom" given above? Would *all* of the symbols become tainted, or would some of them still mean what we want them to mean? I will let you think about that. We will encounter a similar question in Chapter XIV, and discuss the matter then. In any case, we will not follow this extension now, but instead go on to try to repair the ω-incompleteness of TNT.

The Last Rule

The problem with the "Rule of All" was that it required knowing that all the lines of an infinite pyramidal family are theorems—too much for a finite being. But suppose that each line of the pyramid can be derived from its predecessor in a *patterned* way. Then there would be a *finite reason* accounting for the fact that all the strings in the pyramid are theorems. The trick, then, is to find the *pattern* that causes the cascade, and show that that pattern is a theorem in itself. That is like proving that each djinn passes a message to its meta, as in the children's game of "Telephone". The other thing left to show is that Genie starts the cascading message—that is, to establish that the first line of the pyramid is a theorem. Then you know that GOD will get the message!

In the particular pyramid we were looking at, there is a pattern, captured by lines 4-9 of the derivation below.

(1)	∀a:∀b:(a+Sb)=S(a+b)	axiom 3
(2)	∀b:(0+Sb)=S(0+b)	specification
(3)	(0+Sb)=S(0+b)	specification
(4)	[	push
(5)	(0+b)=b	premise
(6)	S(0+b)=Sb	add S
(7)	(0+Sb)=S(0+b)	carry over line 3
(8)	(0+Sb)=Sb	transitivity
(9)	]	pop

The premise is (0+b)=b; the outcome is (0+Sb)=Sb.

The first line of the pyramid is also a theorem; it follows directly from Axiom 2. All we need now is a rule which lets us deduce that the string which summarizes the entire pyramid is itself a theorem. Such a rule will be a formalized statement of the fifth Peano postulate.

To express that rule, we need a little notation. Let us abbreviate a well-formed formula in which the variable a is free by the following notation:

X{a}

(There may be other free variables, too, but that is irrelevant.) Then the notation *X*{Sa/a} will stand for that string but with every occurrence of a replaced by Sa. Likewise, *X*{0/a} would stand for the same string, with each appearance of a replaced by 0.

A specific example would be to let *X*{a} stand for the string in question: (0+a)=a. Then *X*{Sa/a} would represent the string (0+Sa)=Sa, and *X*{0/a} would represent (0+0)=0. (Warning: This notation is not part of TNT; it is for our convenience in talking *about* TNT.)

With this new notation, we can state the last rule of TNT quite precisely:

RULE OF INDUCTION: Suppose *u* is a variable, and *X*{*u*} is a well-formed formula in which *u* occurs free. If both ∀*u*:<*X*{*u*}⊃*X*{S*u*/*u*}> and *X*{0/*u*} are theorems, then ∀*u*: *X*{*u*} is also a theorem.

This is about as close as we can come to putting Peano's fifth postulate into TNT. Now let us use it to show that ∀a:(0+a)=a is indeed a theorem in TNT. Emerging from the fantasy in our derivation above, we can apply the fantasy rule, to give us

(10)	<(0+b)=b⊃(0+Sb)=Sb>	fantasy rule
(11)	∀b:<(0+b)=b⊃(0+Sb)=Sb>	generalization

This is the first of the two input theorems required by the induction rule. The other requirement is the first line of the pyramid, which we have. Therefore, we can apply the rule of induction, to deduce what we wanted:

∀b:(0+b)=b

Specification and generalization will allow us to change the variable from b to a; thus ∀a:(0+a)=a is no longer an undecidable string of TNT.

A Long Derivation

Now I wish to present one longer derivation in TNT, so that you can see what one is like, and also because it proves a significant, if simple, fact of number theory.

(1)	∀a:∀b:(a+Sb)=S(a+b)	axiom 3
(2)	∀b:(d+Sb)=S(d+b)	specification
(3)	(d+SSc)=S(d+Sc)	specification
(4)	∀b:(Sd+Sb)=S(Sd+b)	specification (line 1)
(5)	(Sd+Sc)=S(Sd+c)	specification
(6)	S(Sd+c)=(Sd+Sc)	symmetry
(7)	[	push
(8)	∀d:(d+Sc)=(Sd+c)	premise
(9)	(d+Sc)=(Sd+c)	specification
(10)	S(d+Sc)=S(Sd+c)	add S
(11)	(d+SSc)=S(d+Sc)	carry over 3
(12)	(d+SSc)=S(Sd+c)	transitivity
(13)	S(Sd+c)=(Sd+Sc)	carry over 6
(14)	(d+SSc)=(Sd+Sc)	transitivity
(15)	∀d:(d+SSc)=(Sd+Sc)	generalization
(16)	]	pop
(17)	<∀d:(d+Sc)=(Sd+c)⊃∀d:(d+SSc)=(Sd+Sc)>	fantasy rule
(18)	∀c:<∀d:(d+Sc)=(Sd+c)⊃∀d:(d+SSc)=(Sd+Sc)>	generalization

* * * * *

(19)	(d+S0)=S(d+0)	specification (line 2)
(20)	∀a:(a+0)=a	axiom 1
(21)	(d+0)=d	specification
(22)	S(d+0)=Sd	add S
(23)	(d+S0)=Sd	transitivity (lines 19,22)
(24)	(Sd+0)=Sd	specification (line 20)
(25)	Sd=(Sd+0)	symmetry

(26)	(d+S0)=(Sd+0)	transitivity (lines 23,25)
(27)	∀d:(d+S0)=(Sd+0)	generalization

* * * * *

(28)	∀c:∀d:(d+Sc)=(Sd+c)	induction (lines 18,27)

[S can be slipped back and forth in an addition.]

* * * * *

(29)	∀b:(c+Sb)=S(c+b)	specification (line 1)
(30)	(c+Sd)=S(c+d)	specification
(31)	∀b:(d+Sb)=S(d+b)	specification (line 1)
(32)	(d+Sc)=S(d+c)	specification
(33)	S(d+c)=(d+Sc)	symmetry
(34)	∀d:(d+Sc)=(Sd+c)	specification (line 28)
(35)	(d+Sc)=(Sd+c)	specification
(36)	[	push
(37)	∀c:(c+d)=(d+c)	premise
(38)	(c+d)=(d+c)	specification
(39)	S(c+d)=S(d+c)	add S
(40)	(c+Sd)=S(c+d)	carry over 30
(41)	(c+Sd)=S(d+c)	transitivity
(42)	S(d+c)=(d+Sc)	carry over 33
(43)	(c+Sd)=(d+Sc)	transitivity
(44)	(d+Sc)=(Sd+c)	carry over 35
(45)	(c+Sd)=(Sd+c)	transitivity
(46)	∀c:(c+Sd)=(Sd+c)	generalization
(47)	]	pop
(48)	<∀c:(c+d)=(d+c)⊃∀c:(c+Sd)=(Sd+c)>	fantasy rule
(49)	∀d:<∀c:(c+d)=(d+c)⊃∀c:(c+Sd)=(Sd+c)>	generalization

[If d commutes with every c, then Sd does too.]

* * * * *

(50)	(c+0)=c	specification (line 20)
(51)	∀a:(0+a)=a	previous theorem
(52)	(0+c)=c	specification
(53)	c=(0+c)	symmetry

(54) (c+0)=(0+c) transitivity (lines 50,53)

(55) ∀c:(c+0)=(0+c) generalization

[0 commutes with every c.]

* * * * *

(56) ∀d:∀c:(c+d)=(d+c) induction (lines 49,55)

[Therefore, every d commutes with every c.]

Tension and Resolution in TNT

TNT has proven the commutativity of addition. Even if you do not follow this derivation in detail, it is important to realize that, like a piece of music, it has its own natural "rhythm". It is not just a random walk that happens to have landed on the desired last line. I have inserted "breathing marks" to show some of the "phrasing" of this derivation. Line 28 in particular is a turning point in the derivation, something like the halfway point in an *AABB* type of piece, where you resolve momentarily, even if not in the tonic key. Such important intermediate stages are often called "lemmas".

It is easy to imagine a reader starting at line 1 of this derivation, ignorant of where it is to end up, and getting a sense of where it is going as he sees each new line. This would set up an inner tension, very much like the tension in a piece of music caused by chord progressions that let you know what the tonality is, without resolving. Arrival at line 28 would confirm the reader's intuition and give him a momentary feeling of satisfaction while at the same time strengthening his drive to progress towards what he presumes is the true goal.

Now line 49 is a critically important tension-increaser, because of the "almost-there" feeling which it induces. It would be extremely unsatisfactory to leave off there! From there on, it is almost predictable how things must go. But you wouldn't want a piece of music to quit on you just when it had made the mode of resolution apparent. You don't want to *imagine* the ending—you want to *hear* the ending. Likewise here, we have to carry things through. Line 55 is inevitable, and sets up all the final tensions, which are resolved by Line 56.

This is typical of the structure not only of formal derivations, but also of informal proofs. The mathematician's sense of tension is intimately related to his sense of beauty, and is what makes mathematics worthwhile doing. Notice, however, that in TNT itself, there seems to be no reflection of these tensions. In other words, TNT doesn't formalize the notions of tension and resolution, goal and subgoal, "naturalness" and "inevitability", any more than a piece of music is a book about harmony and rhythm. Could one devise a much fancier typographical system which is *aware* of the tensions and goals inside derivations?

Formal Reasoning *vs.* Informal Reasoning

I would have preferred to show how to derive Euclid's Theorem (the infinitude of primes) in TNT, but it would probably have doubled the length of the book. Now after this theorem, the natural direction to go would be to prove the associativity of addition, the commutativity and associativity of multiplication, and the distributivity of multiplication over addition. These would give a powerful base to work from.

As it is now formulated, TNT has reached "critical mass" (perhaps a strange metaphor to apply to something called "TNT"). It is of the same strength as the system of *Principia Mathematica;* in TNT one can now prove every theorem which you would find in a standard treatise on number theory. Of course, no one would claim that deriving theorems in TNT is the best way to do number theory. Anybody who felt that way would fall in the same class of people as those who think that the best way to know what 1000 × 1000 is, is to draw a 1000 by 1000 grid, and count all the squares in it . . . No; after total formalization, the only way to go is towards relaxation of the formal system. Otherwise, it is so enormously unwieldy as to be, for all practical purposes, useless. Thus, it is important to embed TNT within a wider context, a context which enables new rules of inference to be derived, so that derivations can be speeded up. This would require formalization of the language in which rules of inference are expressed—that is, the metalanguage. And one could go considerably further. However, none of these speeding-up tricks would make TNT any more *powerful;* they would simply make it more *usable.* The simple fact is that we have put into TNT every mode of thought that number theorists rely on. Embedding it in ever larger contexts will not enlarge the space of theorems; it will just make working in TNT—or in each "new, improved version"—look more like doing conventional number theory.

Number Theorists Go out of Business

Suppose that you didn't have advance knowledge that TNT will turn out to be incomplete, but rather, expected that it is complete—that is, that every true statement expressible in the TNT-notation is a theorem. In that case, you could make a decision procedure for all of number theory. The method would be easy: if you want to know if N-statement *X* is true or false, code it into TNT-sentence *x*. Now if *X* is true, completeness says that *x* is a theorem; and conversely, if not-*X* is true, then completeness says that ~*x* is a theorem. So either *x* or ~*x* must be a theorem, since either *X* or not-*X* is true. Now begin systematically enumerating all the theorems of TNT, in the way we did for the MIU-system and pq-system. You must come to *x* or ~*x* after a while; and whichever one you hit tells you which of *X* and not-*X* is true. (Did you follow this argument? It crucially depends on your being able to hold separate in your mind the formal system TNT and its informal counterpart N. Make sure you understand it.) Thus, in principle, if TNT were complete, number theorists would be put out of business: any question in their field could be resolved, with sufficient time, in a purely mechanical way. As it turns out, this is impossible, which, depending on your point of view, is a cause either for rejoicing, or for mourning.

Hilbert's Program

The final question which we will take up in this Chapter is whether we should have as much faith in the consistency of TNT as we did in the consistency of the Propositional Calculus; and, if we don't, whether it is possible to increase our faith in TNT, by *proving* it to be consistent. One could make the same opening statement on the "obviousness" of TNT's consistency as Imprudence did in regard to the Propositional Calculus—namely, that each rule embodies a reasoning principle which we fully believe in, and therefore to question the consistency of TNT is to question our own sanity. To some extent, this argument still carries weight—but not quite so much weight as before. There are just too many rules of inference, and some of them just might be slightly "off". Furthermore, how do we know that this mental model we have of some abstract entities called "natural numbers" is actually a coherent construct? Perhaps our own thought processes, those informal processes which we have tried to capture in the formal rules of the system, are themselves inconsistent! It is of course not the kind of thing we expect, but it gets more and more conceivable that our thoughts might lead us astray, the more complex the subject matter gets—and natural numbers are by no means a trivial subject matter. So Prudence's cry for a *proof* of consistency has to be taken more seriously in this case. It's not that we seriously doubt that TNT could be inconsistent—but there is a *little* doubt, a flicker, a glimmer of a doubt in our minds, and a proof would help to dispel that doubt.

But what means of proof would we like to see used? Once again, we are faced with the recurrent question of circularity. If we use all the same equipment in a proof *about* our system as we have inserted *into* it, what will we have accomplished? If we could manage to convince ourselves of the consistency of TNT, but by using a weaker system of reasoning than TNT, we will have beaten the circularity objection! Think of the way a heavy rope is passed between ships (or so I read when I was a kid): first a light arrow is fired across the gap, pulling behind it a thin rope. Once a connection has been established between the two ships this way, then the heavy rope can be pulled across the gap. If we can use a "light" system to show that a "heavy" system is consistent, then we shall have really accomplished something.

Now on first sight one might think there is a thin rope. Our goal is to prove that TNT has a certain typographical property (consistency): that no theorems of the form x and $\sim x$ ever occur. This is similar to trying to show that MU is not a theorem of the MIU-system. Both are statements about *typographical* properties of symbol-manipulation systems. The visions of a thin rope are based on the presumption that *facts about number theory won't be needed* in proving that such a typographical property holds. In other words, if properties of integers are not used—or if only a few extremely simple ones are used—then we could achieve the goal of proving TNT consistent, by using means which are weaker than its own internal modes of reasoning.

This is the hope which was held by an important school of mathematicians and logicians in the early part of this century, led by David Hilbert. The goal was to prove the consistency of formalizations of number theory similar to TNT by employing a very restricted set of principles of reasoning called "finitistic" methods of reasoning. These would be the thin rope. Included among finitistic methods are all of propositional reasoning, as embodied in the Propositional Calculus, and additionally some kinds of numerical reasoning. But Gödel's work showed that any effort to pull the heavy rope of TNT's consistency across the gap by using the thin rope of finitistic methods is doomed to failure. Gödel showed that in order to pull the heavy rope across the gap, you can't use a lighter rope; there just isn't a strong enough one. Less metaphorically, we can say: *Any system that is strong enough to prove TNT's consistency is at least as strong as TNT itself.* And so circularity is inevitable.

A Mu Offering[1]

The Tortoise and Achilles have just been to hear a lecture on the origins of the Genetic Code, and are now drinking some tea at Achilles' home.

Achilles: I have something terrible to confess, Mr. T.

Tortoise: What is it, Achilles?

Achilles: Despite the fascinating subject matter of that lecture, I drifted off to sleep a time or two. But in my drowsy state, I still was semi-aware of the words coming into my ears. One strange image that floated up from my lower levels was that 'A' and 'T', instead of standing for "adenine" and "thymine", stood for my name and yours—and that double-strands of DNA had tiny copies of me and you along their backbones, always paired up, just as adenine and thymine always are. Isn't that a strange symbolic image?

Tortoise: Phooey! Who believes in that silly kind of stuff? Anyway, what about 'C' and 'G'?

Achilles: Well, I suppose 'C' could stand for Mr. Crab, instead of for cytosine. I'm not sure about 'G', but I'm sure one could think of something. Anyway, it was amusing to imagine my DNA being filled with minuscule copies of you—as well as tiny copies of myself, for that matter. Just think of the infinite regress THAT leads to!

Tortoise: I can see you were not paying too much attention to the lecture.

Achilles: No, you're wrong. I was doing my best, only I had a hard time keeping fancy separated from fact. After all, it is such a strange netherworld that those molecular biologists are exploring.

Tortoise: How do you mean?

Achilles: Molecular biology is filled with peculiar convoluted loops which I can't quite understand, such as the way that folded proteins, which are coded for in DNA, can loop back and manipulate the DNA which they came from, possibly even destroying it. Such strange loops always confuse the daylights out of me. They're eerie, in a way.

Tortoise: I find them quite appealing.

Achilles: You would, of course—they're just down your alley. But as for me, sometimes I like to retreat from all this analytic thought and just meditate a little, as an antidote. It clears my mind of all those confusing loops and incredible complexities which we were hearing about tonight.

Tortoise: Fancy that. I wouldn't have guessed that you were a meditator.

Achilles: Did I never tell you that I am studying Zen Buddhism?

Tortoise: Heavens, how did you come upon that?

Achilles: I have always had a yen for the yin and yang, you know—the whole Oriental mysticism trip, with the *I Ching*, gurus, and whatnot. So one day I'm thinking to myself, "Why not Zen too?" And that's how it all began.

Tortoise: Oh, splendid. Then perhaps I can finally become enlightened.

Achilles: Whoa, now. Enlightenment is not the first step on the road to Zen; if anything, it's the last one! Enlightenment is not for novices like you, Mr. T!

Tortoise: I see we have had a misunderstanding. By "enlightenment", I hardly meant something so weighty as is meant in Zen. All I meant is that I can perhaps become enlightened as to what Zen is all about.

Achilles: For Pete's sake, why didn't you say so? Well, I'd be only too happy to tell you what I know of Zen. Perhaps you might even be tempted to become a student of it, like me.

Tortoise: Well, nothing's impossible.

Achilles: You could study with me under my master, Okanisama—the seventh patriarch.

Tortoise: Now what in the world does that mean?

Achilles: You have to know the history of Zen to understand that.

Tortoise: Would you tell me a little of the history of Zen, then?

Achilles: An excellent idea. Zen is a kind of Buddhism which was founded by a monk named Bodhidharma, who left India and went to China around the sixth century. Bodhidharma was the first patriarch. The sixth one was Enō. (I've finally got it straight now!)

Tortoise: The sixth patriarch was Zeno, eh? I find it strange that he, of all people, would get mixed up in this business.

Achilles: I daresay you underestimate the value of Zen. Listen just a little more, and maybe you'll come to appreciate it. As I was saying, about five hundred years later, Zen was brought to Japan, and it took hold very well there. Since that time it has been one of the principal religions in Japan.

Tortoise: Who is this Okanisama, the "seventh patriarch"?

Achilles: He is my master, and his teachings descend directly from those of the sixth patriarch. He has taught me that reality is one, immutable, and unchanging; all plurality, change, and motion are mere illusions of the senses.

Tortoise: Sure enough, that's Zeno, a mile away. But how ever did he come to be tangled up in Zen? Poor fellow!

Achilles: Whaaat? I wouldn't put it that way. If ANYONE is tangled up, it's . . . But that's another matter. Anyway, I don't know the answer to your question. Instead, let me tell you something of the teachings of my master. I have learned that in Zen, one seeks enlightenment, or SATORI—the state of "No-mind". In this state, one does not think about the world—one just IS. I have also learned that a student of Zen is not supposed to "attach" to any object or thought or person—which is to say, he must not believe in, or depend on, any absolute—not even this philosophy of nonattachment.

Tortoise: Hmm . . . Now THERE'S something I could like about Zen.

Achilles: I had a hunch you'd get attached to it.

Tortoise: But tell me: if Zen rejects intellectual activity, does it make sense to intellectualize about Zen, to study it rigorously?

Achilles: That matter has troubled me quite a bit. But I think I have finally worked out an answer. It seems to me that you may begin approaching Zen through any path you know—even if it is completely antithetical to Zen. As you approach it, you gradually learn to stray from that path. The more you stray from the path, the closer you get to Zen.

Tortoise: Oh, it all begins to sound so clear now.

Achilles: My favorite path to Zen is through the short, fascinating, and weird Zen parables called "kōans".

Tortoise: What is a kōan?

Achilles: A kōan is a story about Zen masters and their students. Sometimes it is like a riddle; other times like a fable; and other times like nothing you've ever heard before.

Tortoise: Sounds rather intriguing. Would you say that to read and enjoy kōans is to practice Zen?

Achilles: I doubt it. However, in my opinion, a delight in kōans comes a million times closer to real Zen than reading volume after volume about Zen, written in heavy philosophical jargon.

Tortoise: I would like to hear a kōan or two.

Achilles: And I would like to tell you one—or a few. Perhaps I should begin with the most famous one of all. Many centuries ago, there was a Zen master named Jōshū, who lived to be 119 years old.

Tortoise: A mere youngster!

Achilles: By your standards, yes. Now one day while Jōshū and another monk were standing together in the monastery, a dog wandered by. The monk asked Jōshū, "Does a dog have Buddha-nature, or not?"

Tortoise: Whatever that is. So tell me—what did Jōshū reply?

Achilles: 'MU'.

Tortoise: 'MU'? What is this 'MU'? What about the dog? What about Buddha-nature? What's the answer?

Achilles: Oh, but 'MU' is Jōshū's answer. By saying 'MU', Jōshū let the other monk know that only by not asking such questions can one know the answer to them.

Tortoise: Jōshū "unasked" the question.

Achilles: Exactly!

Tortoise: 'MU' sounds like a handy thing to have around. I'd like to unask a question or two, sometimes. I guess I'm beginning to get the hang of Zen. Do you know any other kōans, Achilles? I would like to hear some more.

Achilles: My pleasure. I can tell you a pair of kōans which go together. Only . . .

Tortoise: What's the matter?

Achilles: Well, there is one problem. Although both are widely told kōans, my master has cautioned me that only one of them is genuine. And what is more, he does not know which one is genuine, and which one is a fraud.

Tortoise: Crazy! Why don't you tell them both to me and we can speculate to our hearts' content!

Achilles: All right. One of the alleged kōans goes like this:

A monk asked Baso: "What is Buddha?"
Baso said: "This mind is Buddha."

Tortoise: Hmm . . . "This mind is Buddha"? Sometimes I don't quite understand what these Zen people are getting at.

Achilles: You might prefer the other alleged kōan then.

Tortoise: How does it run?

Achilles: Like this:

A monk asked Baso: "What is Buddha?"
Baso said: "This mind is not Buddha."

Tortoise: My, my! If my shell isn't green and not green! I like that!

Achilles: Now, Mr. T—you're not supposed to just "like" kōans.

Tortoise: Very well, then—I don't like it.

Achilles: That's better. Now as I was saying, my master believes only one of the two is genuine.

Tortoise: I can't imagine what led him to such a belief. But anyway, I suppose it's all academic, since there's no way to know if a kōan is genuine or phony.

Achilles: Oh, but there you are mistaken. My master has shown us how to do it.

Tortoise: Is that so? A decision procedure for genuineness of kōans? I should very much like to hear about THAT.

Achilles: It is a fairly complex ritual, involving two stages. In the first stage, you must TRANSLATE the kōan in question into a piece of string, folded all around in three dimensions.

Tortoise: That's a curious thing to do. And what is the second stage?

Achilles: Oh, that's easy—all you need to do is determine whether the string has Buddha-nature, or not! If it does, then the kōan is genuine—if not, the kōan is a fraud.

Tortoise: Hmm . . . It sounds as if all you've done is transfer the need for a decision procedure to another domain. NOW it's a decision procedure for Buddha-nature that you need. What next? After all, if you can't even tell whether a DOG has Buddha-nature or not, how can you expect to do so for every possible folded string?

Achilles: Well, my master explained to me that shifting between domains can help. It's like switching your point of view. Things sometimes look complicated from one angle, but simple from another. He gave the example of an orchard, in which from one direction no order is

FIGURE 45. La Mezquita, *by M. C. Escher (black and white chalk, 1936).*

apparent, but from special angles, beautiful regularity emerges. You've reordered the same information by changing your way of looking at it.

Tortoise: I see. So perhaps the genuineness of a kōan is concealed somehow very deeply inside it, but if you translate it into a string it manages in some way to float to the surface?

Achilles: That's what my master has discovered.

Tortoise: Then I would very much like to learn about the technique. But first, tell me: how can you turn a kōan (a sequence of words) into a folded string (a three-dimensional object)? They are rather different kinds of entities.

Achilles: That is one of the most mysterious things I have learned in Zen. There are two steps: "transcription" and "translation". TRANSCRIBING a kōan involves writing it in a phonetic alphabet, which contains only four geometric symbols. This phonetic rendition of the kōan is called the MESSENGER.

Tortoise: What do the geometric symbols look like?

Achilles: They are made of hexagons and pentagons. Here is what they look like *(picks up a nearby napkin, and draws for the Tortoise these four figures):*

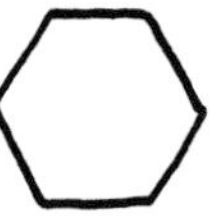  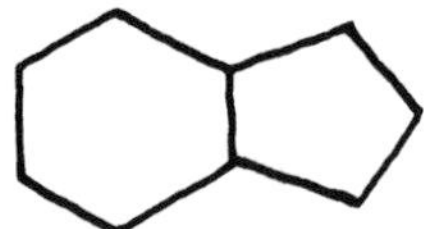 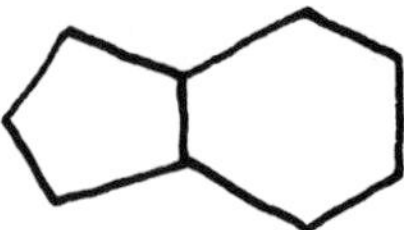

Tortoise: They are mysterious-looking.

Achilles: Only to the uninitiated. Now once you have made the messenger, you rub your hands in some ribo, and—

Tortoise: Some ribo? Is that a kind of ritual anointment?

Achilles: Not exactly. It is a special sticky preparation which makes the string hold its shape, when folded up.

Tortoise: What is it made of?

Achilles: I don't know, exactly. But it feels sort of gluey, and it works exceedingly well. Anyway, once you have some ribo on your hands, you can TRANSLATE the sequence of symbols in the messenger into certain kinds of folds in the string. It's as simple as that.

Tortoise: Hold on! Not so fast! How do you do that?

Achilles: You begin with the string entirely straight. Then you go to one end and start making folds of various types, according to the geometric symbols in the messenger.

Tortoise: So each of those geometric symbols stands for a different way to curl the string up?

Achilles: Not in isolation. You take them three at a time, instead of one at a time. You begin at one end of the string, and one end of the messenger. What to do with the first inch of the string is determined by the first three geometric symbols. The next three symbols tell you how to fold the second inch of string. And so you inch your way along the string and simultaneously along the messenger, folding each little segment of string until you have exhausted the messenger. If you have properly applied some ribo, the string will keep its folded shape, and what you thereby produce is the translation of the kōan into a string.

Tortoise: The procedure has a certain elegance to it. You must get some wild-looking strings that way.

Achilles: That's for sure. The longer kōans translate into quite bizarre shapes.

Tortoise: I can imagine. But in order to carry out the translation of the messenger into the string, you need to know what kind of fold each triplet of geometric symbols in the messenger stands for. How do you know this? Do you have a dictionary?

Achilles: Yes—there is a venerated book which lists the "Geometric Code". If you don't have a copy of this book, of course, you can't translate a kōan into a string.

Tortoise: Evidently not. What is the origin of the Geometric Code?

Achilles: It came from an ancient master known as "Great Tutor" who my master says is the only one ever to attain the Enlightenment 'Yond Enlightenment.

Tortoise: Good gravy! As if one level of the stuff weren't enough. But then there are gluttons of every sort—why not gluttons for enlightenment?

Achilles: Do you suppose that "Enlightenment 'Yond Enlightenment" stands for "EYE"?

Tortoise: In my opinion, it's rather doubtful that it stands for you, Achilles. More likely, it stands for "Meta-Enlightenment"—"ME", that is.

Achilles: For you? Why would it stand for you? You haven't even reached the FIRST stage of enlightenment, let alone the—

Tortoise: You never know, Achilles. Perhaps those who have learned the lowdown on enlightenment return to their state before enlightenment. I've always held that "twice enlightened is unenlightened." But let's get back to the Grand Tortue—uh, I mean the Great Tutor.

Achilles: Little is known of him, except that he also invented the Art of Zen Strings.

Tortoise: What is that?

Achilles: It is an art on which the decision procedure for Buddha-nature is based. I shall tell you about it.

Tortoise: I would be fascinated. There is so much for novices like me to absorb!

Achilles: There is even reputed to be a kōan which tells how the Art of Zen Strings began. But unfortunately, all this has long since been lost in the sands of time, and is no doubt gone forever. Which may be just as well, for otherwise there would be imitators who would take on the master's name, and copy him in other ways.

Tortoise: But wouldn't it be a good thing if all students of Zen copied that most enlightened master of all, the Great Tutor?

Achilles: Let me tell you a kōan about an imitator.

> Zen master Gutei raised his finger whenever he was asked a question about Zen. A young novice began to imitate him in this way. When Gutei was told about the novice's imitation, he sent for him and asked him if it were true. The novice admitted it was so. Gutei asked him if he understood. In reply the novice held up his index finger. Gutei promptly cut it off. The novice ran from the room, howling in pain. As he reached the threshold, Gutei called, "Boy!" When the novice turned, Gutei raised his index finger. At that instant the novice was enlightened.

Tortoise: Well, what do you know! Just when I thought Zen was all about Jōshū and his shenanigans, now I find out that Gutei is in on the merriment too. He seems to have quite a sense of humor.

Achilles: That kōan is very serious. I don't know how you got the idea that it is humorous.

Tortoise: Perhaps Zen is instructive because it is humorous. I would guess that if you took all such stories entirely seriously, you would miss the point as often as you would get it.

Achilles: Maybe there's something to your Tortoise-Zen.

Tortoise: Can you answer just one question for me? I would like to know this: Why did Bodhidharma come from India into China?

Achilles: Oho! Shall I tell you what Jōshū said when he was asked that very question?

Tortoise: Please do.

Achilles: He replied, "That oak tree in the garden."

Tortoise: Of course; that's just what I would have said. Except that I would have said it in answer to a different question—namely, "Where can I find some shade from the midday sun?"

Achilles: Without knowing it, you have inadvertently hit upon one of the basic questions of all Zen. That question, innocent though it sounds, actually means, "What is the basic principle of Zen?"

Tortoise: How extraordinary. I hadn't the slightest idea that the central aim of Zen was to find some shade.

Achilles: Oh, no—you've misunderstood me entirely. I wasn't referring to THAT question. I meant your question about why Bodhidharma came from India into China.

Tortoise: I see. Well, I had no idea that I was getting into such deep waters. But let's come back to this curious mapping. I gather that any kōan can be turned into a folded string by following the method you outlined. Now what about the reverse process? Can any folded string be read in such a way as to yield a kōan?

Achilles: Well, in a way. However . . .

Tortoise: What's wrong?

Achilles: You're just not supposed to do it that way 'round. It would violate the Central Dogma of Zen strings, you see, which is contained in this picture *(picks up a napkin and draws)*:

kōan	$\Rightarrow$	messenger	$\Rightarrow$	folded string
	transcription		*translation*	

You're not supposed to go against the arrows—especially not the second one.

Tortoise: Tell me, does this Dogma have Buddha-nature, or not? Come to think of it, I think I'll unask the question. Is that all right?

Achilles: I am glad you unasked the question. But—I'll let you in on a secret. Promise you won't tell anyone?

Tortoise: Tortoise's honor.

Achilles: Well, once in a while, I actually do go against the arrows. I get sort of an illicit thrill out of it, I guess.

Tortoise: Why, Achilles! I had no idea you would do something so irreverent!

Achilles: I've never confessed it to anyone before—not even Okanisama.

Tortoise: So tell me, what happens when you go against the arrows in the Central Dogma? Does that mean you begin with a string and make a kōan?

Achilles: Sometimes—but some weirder things can happen.

Tortoise: Weirder than producing kōans?

Achilles: Yes . . . When you untranslate and untranscribe, you get SOMETHING, but not always a kōan. Some strings, when read out loud this way, only give nonsense.

Tortoise: Isn't that just another name for kōans?

Achilles: You clearly don't have the true spirit of Zen yet.

Tortoise: Do you always get stories, at least?

Achilles: Not always—sometimes you get nonsense syllables, other times you get ungrammatical sentences. But once in a while you get what seems to be a kōan.

Tortoise: It only SEEMS to be one?

Achilles: Well, it might be fraudulent, you see.

Tortoise: Oh, of course.

Achilles: I call those strings which yield apparent kōans "well-formed" strings.

Tortoise: Why don't you tell me about the decision procedure which allows you to distinguish phony kōans from the genuine article?

Achilles: That's what I was heading towards. Given the kōan, or non-kōan, as the case may be, the first thing is to translate it into the three-dimensional string. All that's left is to find out if the string has Buddha-nature or not.

Tortoise: But how do you do THAT?

Achilles: Well, my master has said that the Great Tutor was able, by just glancing at a string, to tell if it had Buddha-nature or not.

Tortoise: But what if you have not reached the stage of the Enlightenment 'Yond Enlightenment? Is there no other way to tell if a string has Buddha-nature?

Achilles: Yes, there is. And this is where the Art of Zen Strings comes in. It is a technique for making innumerably many strings, all of which have Buddha-nature.

Tortoise: You don't say! And is there a corresponding way of making strings which DON'T have Buddha-nature?

Achilles: Why would you want to do that?

Tortoise: Oh, I just thought it might be useful.

Achilles: You have the strangest taste. Imagine! Being more interested in things that DON'T have Buddha-nature than things that DO!

Tortoise: Just chalk it up to my unenlightened state. But go on. Tell me how to make a string which DOES have Buddha-nature.

Achilles: Well, you must begin by draping a loop of string over your hands in one of five legal starting positions, such as this one . . . *(Picks up a string and drapes it in a simple loop between a finger on each hand.)*

Tortoise: What are the other four legal starting positions?

Achilles: Each one is a position considered to be a SELF-EVIDENT manner of picking up a string. Even novices often pick up strings in those positions. And these five strings all have Buddha-nature.

Tortoise: Of course.

Achilles: Then there are some String Manipulation Rules, by which you can make more complex string figures. In particular, you are allowed to modify your string by doing certain basic motions of the hands. For instance, you can reach across like this—and pull like this—and twist like this. With each operation you are changing the overall configuration of the string draped over your hands.

Tortoise: Why, it looks just like making cat's-cradles and such string figures!

Achilles: That's right. Now as you watch, you'll see that some of these rules make the string more complex; some simplify it. But whichever way you go, as long as you follow the String Manipulation Rules, every string you produce will have Buddha-nature.

Tortoise: That is truly marvelous. Now what about the kōan concealed inside this string you've just made? Would it be genuine?

Achilles: Why, according to what I've learned, it must. Since I made it according to the Rules, and began in one of the five self-evident positions, the string must have Buddha-nature, and consequently it must correspond to a genuine kōan.

Tortoise: Do you know what the kōan is?

Achilles: Are you asking me to violate the Central Dogma? Oh, you naughty fellow!

(And with furrowed brow and code book in hand, Achilles points along the string inch by inch, recording each fold by a triplet of geometric symbols of the strange phonetic alphabet for kōans, until he has nearly a napkinful.)

Done!

Tortoise: Terrific. Now let's hear it.

Achilles: All right.

> A traveling monk asked an old woman the road to Taizan, a popular temple supposed to give wisdom to the one who worships there. The old woman said: "Go straight ahead." After the monk had proceeded a few steps, she said to herself, "He also is a common church-goer." Someone told this incident to Jōshū, who said: "Wait until I investigate." The next day he went and asked the same question, and the old woman gave the same answer. Jōshū remarked: "I have investigated that old woman."

Tortoise: Why, with his flair for investigations, it's a shame that Jōshū never was hired by the FBI. Now tell me—what you did, I could also do, if I followed the Rules from the Art of Zen Strings, right?

Achilles: Right.

Tortoise: Now would I have to perform the operations in just the same ORDER as you did?

Achilles: No, any old order will do.

Tortoise: Of course, then I would get a different string, and consequently a different kōan. Now would I have to perform the same NUMBER of steps as you did?

Achilles: By no means. Any number of steps is fine.

Tortoise: Well, then there are an infinite number of strings with Buddha-nature—and consequently an infinite number of genuine kōans! How do you know there is any string which CAN'T be made by your Rules?

Achilles: Oh, yes—back to things which lack Buddha-nature. It just so happens that once you know how to make strings WITH Buddha-nature, you can also make strings WITHOUT Buddha-nature. That is something which my master drilled into me right at the beginning.

Tortoise: Wonderful! How does it work?

Achilles: Easy. Here, for example—I'll make a string which lacks Buddha-nature . . .

(He picks up the string out of which the preceding kōan was "pulled", and ties a little teeny knot at one end of it, pulling it tight with his thumb and forefinger.)

This is it—no Buddha-nature here.

Tortoise: Very illuminating. All it takes is adding a knot? How do you know that the new string lacks Buddha-nature?

Achilles: Because of this fundamental property of Buddha-nature: when two well-formed strings are identical but for a knot at one end, then only ONE of them can have Buddha-nature. It's a rule of thumb which my master taught me.

Tortoise: I'm just wondering about something. Are there some strings with Buddha-nature which you CAN'T reach by following the Rules of Zen Strings, no matter in what order?

Achilles: I hate to admit it, but I am a little confused on this point myself. At first my master gave the strongest impression that Buddha-nature in a string was DEFINED by starting in one of the five legal starting positions, and then developing the string according to the allowed Rules. But then later, he said something about somebody-or-other's "Theorem". I never got it straight. Maybe I even misheard what he said. But whatever he said, it put some doubt in my mind as to whether this method hits ALL strings with Buddha-nature. To the best of my knowledge, at least, it does. But Buddha-nature is a pretty elusive thing, you know.

Tortoise: I gathered as much, from Jōshū's 'MU'. I wonder . . .

Achilles: What is it?

Tortoise: I was just wondering about those two kōans—I mean the kōan and its un-kōan—the ones which say "This mind is Buddha" and "This mind is not Buddha"—what do they look like, when turned into strings via the Geometric Code?

Achilles: I'd be glad to show you.

(He writes down the phonetic transcriptions, and then pulls from his pocket a couple of pieces of string, which he carefully folds inch by inch, following the triplets of symbols written in the strange alphabet. Then he places the finished strings side by side.)

You see, here is the difference.

Tortoise: They are very similar, indeed. Why, I do believe there is only one difference between them: it's that one of them has a little knot on its end!

Achilles: By Jōshū, you're right.

Tortoise: Aha! Now I understand why your master is suspicious.

Achilles: You do?

Tortoise: According to your rule of thumb, AT MOST ONE of such a pair can have Buddha-nature, so you know right away that one of the kōans must be phony.

Achilles: But that doesn't tell which one is phony. I've worked, and so has my master, at trying to produce these two strings by following the String Manipulation Rules, but to no avail. Neither one ever turns up. It's quite frustrating. Sometimes you begin to wonder . . .

Tortoise: You mean, to wonder if either one has Buddha-nature? Perhaps neither of them has Buddha-nature—and neither kōan is genuine!

Achilles: I never carried my thoughts as far as that—but you're right—it's possible, I guess. But I think you should not ask so many questions about Buddha-nature. The Zen master Mumon always warned his pupils of the danger of too many questions.

Tortoise: All right—no more questions. Instead, I have a sort of hankering to make a string myself. It would be amusing to see if what I come up with is well-formed or not.

Achilles: That could be interesting. Here's a piece of string. *(He passes one to the Tortoise.)*

Tortoise: Now you realize that I don't have the slightest idea what to do. We'll just have to take potluck with my awkward production, which will follow no rules and will probably wind up being completely undecipherable. *(Grasps the loop between his feet and, with a few simple manipulations, creates a complex string which he proffers wordlessly to Achilles. At that moment, Achilles' face lights up.)*

Achilles: Jeepers creepers! I'll have to try out your method myself. I have never seen a string like this!

Tortoise: I hope it is well-formed.

Achilles: I see it's got a knot at one end.

Tortoise: Oh—just a moment! May I have it back? I want to do one thing to it.

Achilles: Why, certainly. Here you are.

(Hands it back to the Tortoise, who ties another knot at the same end. Then the Tortoise gives a sharp tug, and suddenly both knots disappear!)

Achilles: What happened?

Tortoise: I wanted to get rid of that knot.

Achilles: But instead of untying it, you tied another one, and then BOTH disappeared! Where did they go?

Tortoise: Tumbolia, of course. That's the Law of Double Nodulation.

(Suddenly, the two knots reappear from out of nowhere—that is to say, Tumbolia.)

Achilles: Amazing. They must lie in a fairly accessible layer of Tumbolia if they can pop into it and out of it so easily. Or is all of Tumbolia equally inaccessible?

Tortoise: I couldn't say. However, it does occur to me that burning the string would make it quite improbable for the knots to come back. In such a case, you could think of them as being trapped in a deeper layer of Tumbolia. Perhaps there are layers and layers of Tumbolia. But that's neither here nor there. What I would like to know is how my string sounds, if you turn it back into phonetic symbols. *(As he hands it back, once again, the knots pop into oblivion.)*

Achilles: I always feel so guilty about violating the Central Dogma ... *(Takes out his pen and code book, and carefully jots down the many symbol-triplets which correspond to the curvy involutions of the Tortoise's string; and when he is finished, he clears his voice.)* Ahem. Are you ready to hear what you have wrought?

Tortoise: I'm willing if you're willing.

Achilles: All right. It goes like this:

> A certain monk had a habit of pestering the Grand Tortue (the only one who had ever reached the Enlightenment 'Yond Enlightenment), by asking whether various objects had Buddha-nature or not. To such questions Tortue invariably sat silent. The monk had already asked about a bean, a lake, and a moonlit night. One day, he brought to Tortue a piece of string, and asked the same question. In reply, the Grand Tortue grasped the loop between his feet and—

Tortoise: Between his feet? How odd!

Achilles: Why should YOU find that odd?

Tortoise: Well, ah ... you've got a point there. But please go on!

Achilles: All right.

> The Grand Tortue grasped the loop between his feet and, with a few simple manipulations, created a complex string which he proffered wordlessly to the monk. At that moment, the monk was enlightened.

Tortoise: I'd rather be twice-enlightened, personally.

Achilles: Then it tells how to make the Grand Tortue's string, if you begin with a string draped over your feet. I'll skip those boring details. It concludes this way:

> From then on, the monk did not bother Tortue. Instead, he made string after string by Tortue's method; and he passed the method on to his own disciples, who passed it on to theirs.

Tortoise: Quite a yarn. It's hard to believe it was really hidden inside my string.

Achilles: Yet it was. Astonishingly, you seem to have created a well-formed string right off the bat.

Tortoise: But what did the Grand Tortue's string look like? That's the main point of this kōan, I'd suppose.

Achilles: I doubt it. One shouldn't "attach" to small details like that inside kōans. It's the spirit of the whole kōan that counts, not little parts of it. Say, do you know what I just realized? I think, crazy though it sounds, that you may have hit upon that long-lost kōan which describes the very origin of the Art of Zen Strings!

Tortoise: Oh, that would almost be too good to have Buddha-nature.

Achilles: But that means that the great master—the only one who ever reached the mystical state of the Enlightenment 'Yond Enlightenment—was named "Tortue", not "Tutor". What a droll name!

Tortoise: I don't agree. I think it's a handsome name. I still want to know how Tortue's string looked. Can you possibly recreate it from the description given in the kōan?

Achilles: I could try ... Of course, I'll have to use my feet, too, since it's described in terms of foot motions. That's pretty unusual. But I think I can manage it. Let me give it a go. *(He picks up the kōan and a piece of string, and for a few minutes twists and bends the string in arcane ways until he has the finished product.)* Well, here it is. Odd, how familiar it looks.

Tortoise: Yes, isn't that so? I wonder where I saw it before?

Achilles: I know! Why, this is YOUR string, Mr. T! Or is it?

Tortoise: Certainly not.

Achilles: Of course not—it's the string which you first handed to me, before you took it back to tie an extra knot in it.

Tortoise: Oh, yes—indeed it is. Fancy that. I wonder what that implies.

Achilles: It's strange, to say the least.

Tortoise: Do you suppose my kōan is genuine?

Achilles: Wait just a moment ...

Tortoise: Or that my string has Buddha-nature?

Achilles: Something about your string is beginning to trouble me, Mr. Tortoise.

Tortoise (looking most pleased with himself and paying no attention to Achilles): And what about Tortue's string? Does it have Buddha-nature? There are a host of questions to ask!

Achilles: I would be scared to ask such questions, Mr. T. There is something mighty funny going on here, and I'm not sure I like it.

Tortoise: I'm sorry to hear it. I can't imagine what's troubling you.

Achilles: Well, the best way I know to explain it is to quote the words of another old Zen master, Kyōgen. Kyōgen said:

> Zen is like a man hanging in a tree by his teeth over a precipice. His hands grasp no branch, his feet rest on no limb, and under the tree another person asks him: "Why did Bodhidharma come to China from India?" If the man in the tree does not answer, he fails; and if he does answer, he falls and loses his life. Now what shall he do?

Tortoise: That's clear; he should give up Zen, and take up molecular biology.

CHAPTER IX

Mumon and Gödel

What Is Zen?

I'M NOT SURE I know what Zen is. In a way, I think I understand it very well; but in a way, I also think I can never understand it at all. Ever since my freshman English teacher in college read Jōshū's MU out loud to our class, I have struggled with Zen aspects of life, and probably I will never cease doing so. To me, Zen is intellectual quicksand—anarchy, darkness, meaninglessness, chaos. It is tantalizing and infuriating. And yet it is humorous, refreshing, enticing. Zen has its own special kind of meaning, brightness, and clarity. I hope that in this Chapter, I can get some of this cluster of reactions across to you. And then, strange though it may seem, that will lead us directly to Gödelian matters.

One of the basic tenets of Zen Buddhism is that there is no way to characterize what Zen is. No matter what verbal space you try to enclose Zen in, it resists, and spills over. It might seem, then, that all efforts to explain Zen are complete wastes of time. But that is not the attitude of Zen masters and students. For instance, Zen kōans are a central part of Zen study, verbal though they are. Kōans are supposed to be "triggers" which, though they do not contain enough information in themselves to impart enlightenment, may possibly be sufficient to unlock the mechanisms inside one's mind that lead to enlightenment. But in general, the Zen attitude is that words and truth are incompatible, or at least that no words can capture truth.

Zen Master Mumon

Possibly in order to point this out in an extreme way, the monk Mumon ("No-gate"), in the thirteenth century, compiled forty-eight kōans, following each with a commentary and a small "poem". This work is called "The Gateless Gate" or the *Mumonkan* ("No-gate barrier"). It is interesting to note that the lives of Mumon and Fibonacci coincided almost exactly: Mumon living from 1183 to 1260 in China, Fibonacci from 1180 to 1250 in Italy. To those who would look to the *Mumonkan* in hopes of making sense of, or "understanding", the kōans, the *Mumonkan* may come as a rude shock, for the comments and poems are entirely as opaque as the kōans which they are supposed to clarify. Take this, for example:[1]

FIGURE 46. Three Worlds, *by M. C. Escher (lithograph, 1955).*

Kōan:

Hōgen of Seiryo monastery was about to lecture before dinner when he noticed that the bamboo screen, lowered for meditation, had not been rolled up. He pointed to it. Two monks arose wordlessly from the audience and rolled it up. Hōgen, observing the physical moment, said, "The state of the first monk is good, not that of the second."

Mumon's Commentary:

I want to ask you: which of those two monks gained and which lost? If any of you has one eye, he will see the failure on the teacher's part. However, I am not discussing gain and loss.

Mumon's Poem:

When the screen is rolled up the great sky opens,
Yet the sky is not attuned to Zen.
It is best to forget the great sky
And to retire from every wind.

Or then again, there is this one:[2]

Kōan:

Goso said: "When a buffalo goes out of his enclosure to the edge of the abyss, his horns and his head and his hoofs all pass through, but why can't the tail also pass?"

Mumon's Commentary:

If anyone can open one eye at this point and say a word of Zen, he is qualified to repay the four gratifications, and, not only that, he can save all sentient beings under him. But if he cannot say such a word of Zen, he should turn back to his tail.

Mumon's Poem:

If the buffalo runs, he will fall into the trench;
If he returns, he will be butchered.
That little tail
Is a very strange thing.

I think you will have to admit that Mumon does not exactly clear everything up. One might say that the metalanguage (in which Mumon writes) is not very different from the object language (the language of the kōan). According to some, Mumon's comments are intentionally idiotic, perhaps meant to show how useless it is to spend one's time in chattering about Zen. However, Mumon's comments can be taken on more than one level. For instance, consider this:[3]

Kōan:

A monk asked Nansen: "Is there a teaching no master ever taught before?"
Nansen said: "Yes, there is."
"What is it?" asked the monk.
Nansen replied: "It is not mind, it is not Buddha, it is not things."

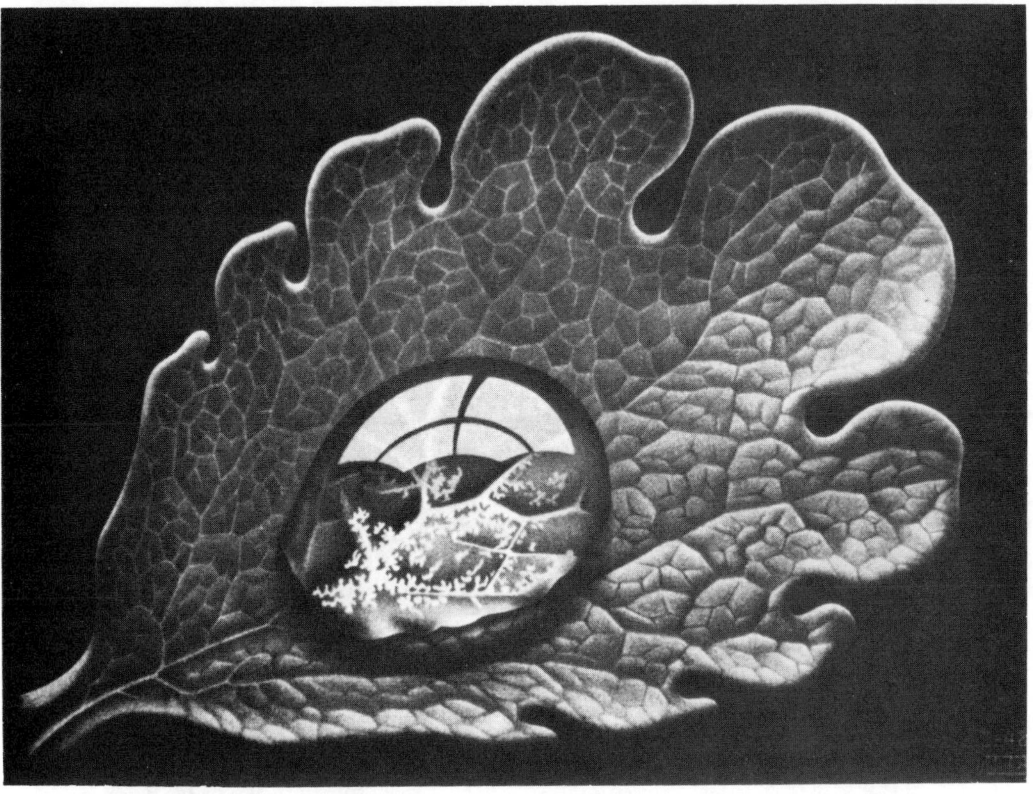

FIGURE 47. Dewdrop, *by M. C. Escher (mezzotint, 1948).*

Mumon's Commentary:

Old Nansen gave away his treasure-words. He must have been greatly upset.

Mumon's Poem:

Nansen was too kind and lost his treasure.
Truly, words have no power.
Even though the mountain becomes the sea,
Words cannot open another's mind.

In this poem Mumon seems to be saying something very central to Zen, and not making idiotic statements. Curiously, however, the poem is self-referential, and thus it is a comment not only on Nansen's words, but also on its own ineffectiveness. This type of paradox is quite characteristic of Zen. It is an attempt to "break the mind of logic". You see this paradoxical quality in the kōan, as well. Concerning Mumon's commentary, do you think that Nansen was really so sure of his answer? Or did the "correctness" of his answer matter at all? Or does correctness play any role in Zen? What is the difference between correctness and truth, or is there any? What if Nansen had said, "No, there is not any such teaching"? Would it have made any difference? Would his remark have been immortalized in a kōan?

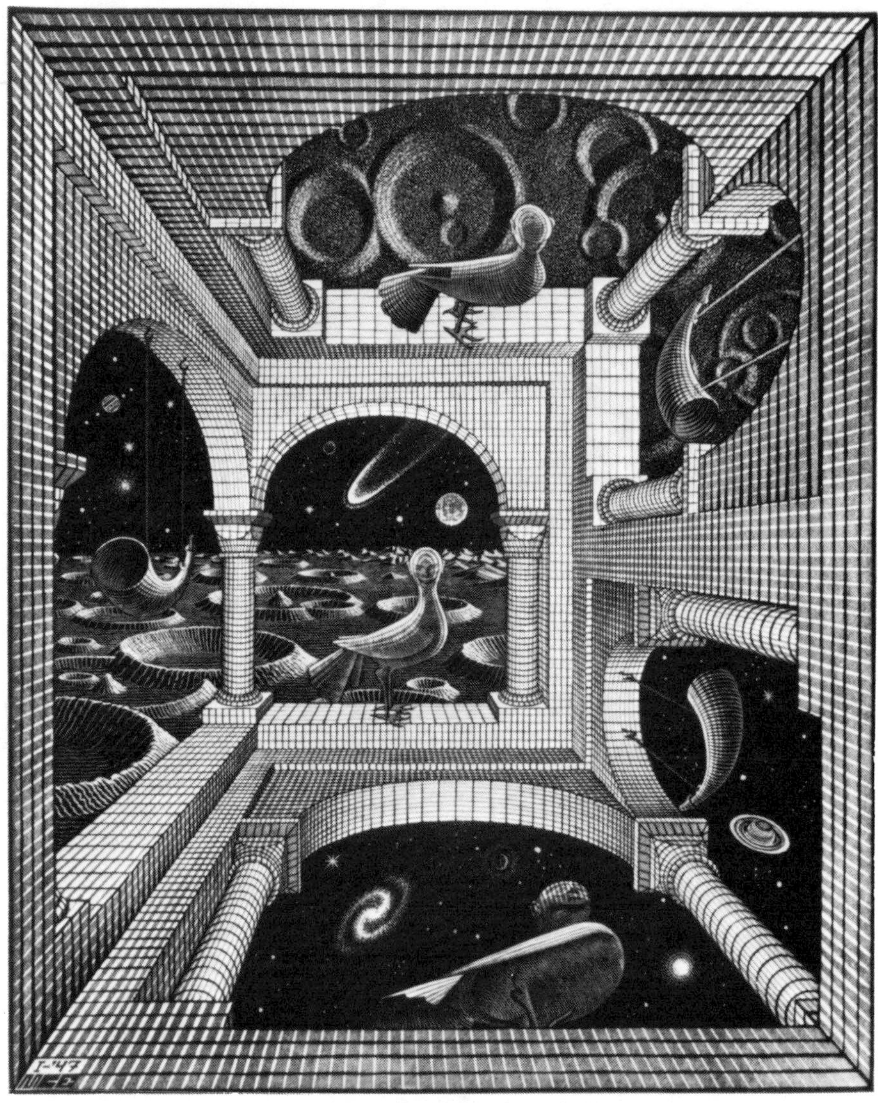

FIGURE 48. Another World, *by M. C. Escher (wood-engraving, 1947).*

Here is another kōan which aims to break the mind of logic:[4]

The student Doko came to a Zen master, and said: "I am seeking the truth. In what state of mind should I train myself, so as to find it?"

Said the master, "There is no mind, so you cannot put it in any state. There is no truth, so you cannot train yourself for it."

"If there is no mind to train, and no truth to find, why do you have these monks gather before you every day to study Zen and train themselves for this study?"

"But I haven't an inch of room here," said the master, "so how could the monks gather? I have no tongue, so how could I call them together or teach them?"

"Oh, how can you lie like this?" asked Doko.

"But if I have no tongue to talk to others, how can I lie to you?" asked the master.

Then Doko said sadly, "I cannot follow you. I cannot understand you."

"I cannot understand myself," said the master.

If any kōan serves to bewilder, this one does. And most likely, causing bewilderment is its precise purpose, for when one is in a bewildered state, one's mind does begin to operate nonlogically, to some extent. Only by stepping outside of logic, so the theory goes, can one make the leap to enlightenment. But what is so bad about logic? Why does it prevent the leap to enlightenment?

Zen's Struggle Against Dualism

To answer that, one needs to understand something about what enlightenment is. Perhaps the most concise summary of enlightenment would be: transcending dualism. Now what is dualism? Dualism is the conceptual division of the world into categories. Is it possible to transcend this very natural tendency? By prefixing the word "division" by the word "conceptual", I may have made it seem that this is an intellectual or conscious effort, and perhaps thereby given the impression that dualism could be overcome simply by suppressing thought (as if to suppress thinking actually were simple!). But the breaking of the world into categories takes place far below the upper strata of thought; in fact, dualism is just as much a *per*ceptual division of the world into categories as it is a *con*ceptual division. In other words, human perception is by nature a dualistic phenomenon—which makes the quest for enlightenment an uphill struggle, to say the least.

At the core of dualism, according to Zen, are words—just plain words. The use of words is inherently dualistic, since each word represents, quite obviously, a conceptual category. Therefore, a major part of Zen is the fight against reliance on words. To combat the use of words, one of the best devices is the kōan, where words are so deeply abused that one's mind is practically left reeling, if one takes the kōans seriously. Therefore it is perhaps wrong to say that the enemy of enlightenment is logic; rather, it is dualistic, verbal thinking. In fact, it is even more basic than that: it is perception. As soon as you perceive an object, you draw a line between it and the rest of the world; you divide the world, artificially, into parts, and you thereby miss the Way.

Here is a kōan which demonstrates the struggle against words:[5]

Kōan:

Shuzan held out his short staff and said: "If you call this a short staff, you oppose its reality. If you do not call it a short staff, you ignore the fact. Now what do you wish to call this?"

FIGURE 49. Day and Night, *by M. C. Escher (woodcut, 1938).*

Mumon's Commentary:

If you call this a short staff, you oppose its reality. If you do not call it a short staff, you ignore the fact. It cannot be expressed with words and it cannot be expressed without words. Now say quickly what it is.

Mumon's Poem:

Holding out the short staff,
He gave an order of life or death.
Positive and negative interwoven,
Even Buddhas and patriarchs cannot escape this attack.

("Patriarchs" refers to six venerated founders of Zen Buddhism, of whom Bodhidharma is the first, and Enō is the sixth.)

Why is calling it a short staff opposing its reality? Probably because such a categorization gives the appearance of capturing reality, whereas the surface has not even been scratched by such a statement. It could be compared to saying "5 is a prime number". There is so much more—an infinity of facts—that has been omitted. On the other hand, not to call it a staff is, indeed, to ignore that particular fact, minuscule as it may be. Thus words lead to some truth—some falsehood, perhaps, as well—but certainly not to all truth. Relying on words to lead you to the truth is like relying on an incomplete formal system to lead you to the truth. A formal system will give you some truths, but as we shall soon see, a formal system—no matter how powerful—cannot lead to all truths. The dilemma of mathematicians is: what else is there to rely on, but formal systems? And the dilemma of

Zen people is: what else is there to rely on, but words? Mumon states the dilemma very clearly: "It cannot be expressed with words and it cannot be expressed without words."

Here is Nansen, once again:[6]

> Jōshū asked the teacher Nansen, "What is the true Way?"
> Nansen answered, "Everyday way is the true Way."
> Jōshū asked, "Can I study it?"
> Nansen answered, "The more you study, the further from the Way."
> Jōshū asked, "If I don't study it, how can I know it?"
> Nansen answered, "The Way does not belong to things seen: nor to things unseen. It does not belong to things known: nor to things unknown. Do not seek it, study it, or name it. To find yourself on it, open yourself wide as the sky." [See Fig. 50.]

FIGURE 50. Rind, *by M. C. Escher (wood-engraving, 1955).*

This curious statement seems to abound with paradox. It is a little reminiscent of this surefire cure for hiccups: "Run around the house three times without thinking of the word 'wolf'." Zen is a philosophy which seems to have embraced the notion that the road to ultimate truth, like the only surefire cure for hiccups, may bristle with paradoxes.

Ism, The Un-Mode, and Unmon

If words are bad, and thinking is bad, what is good? Of course, to ask this is already horribly dualistic, but we are making no pretense of being faithful to Zen in discussing Zen—so we can try to answer the question seriously. I have a name for what Zen strives for: *ism.* Ism is an antiphilosophy, a way of being without thinking. The masters of ism are rocks, trees, clams; but it is the fate of higher animal species to have to strive for ism, without ever being able to attain it fully. Still, one is occasionally granted glimpses of ism. Perhaps the following kōan offers such a glimpse:[7]

> Hyakujō wished to send a monk to open a new monastery. He told his pupils that whoever answered a question most ably would be appointed. Placing a water vase on the ground, he asked: "Who can say what this is without calling its name?"
>
> The chief monk said: "No one can call it a wooden shoe."
>
> Isan, the cooking monk, tipped over the vase with his foot and went out.
>
> Hyakujō smiled and said: "The chief monk loses." And Isan became the master of the new monastery.

To suppress perception, to suppress logical, verbal, dualistic thinking—this is the essence of Zen, the essence of ism. This is the *Un-mode*—not Intelligent, not Mechanical, just "Un". Jōshū was in the Un-mode, and that is why his 'MU' unasks the question. The Un-mode came naturally to Zen Master Unmon:[8]

> One day Unmon said to his disciples, "This staff of mine has transformed itself into a dragon and has swallowed up the universe! Oh, where are the rivers and mountains and the great earth?"

Zen is holism, carried to its logical extreme. If holism claims that things can only be understood as wholes, not as sums of their parts, Zen goes one further, in maintaining that the world cannot be broken into parts at all. To divide the world into parts is to be deluded, and to miss enlightenment.

> A master was asked the question, "What is the Way?" by a curious monk.
>
> "It is right before your eyes," said the master.
>
> "Why do I not see it for myself?"
>
> "Because you are thinking of yourself."
>
> "What about you: do you see it?"
>
> "So long as you see double, saying 'I don't', and 'you do', and so on, your eyes are clouded," said the master.
>
> "When there is neither 'I' nor 'You', can one see it?"
>
> "When there is neither 'I' nor 'You', who is the one that wants to see it?"[9]

Apparently the master wants to get across the idea that an enlightened state is one where the borderlines between the self and the rest of the universe are dissolved. This would truly be the end of dualism, for as he says, there is no system left which has any desire for perception. But what is that state, if not death? How can a live human being dissolve the borderlines between himself and the outside world?

Zen and Tumbolia

The Zen monk Bassui wrote a letter to one of his disciples who was about to die, and in it he said: "Your end which is endless is as a snowflake dissolving in the pure air." The snowflake, which was once very much a discernible subsystem of the universe, now dissolves into the larger system which once held it. Though it is no longer present as a distinct subsystem, its essence is somehow still present, and will remain so. It floats in Tumbolia, along with hiccups that are not being hiccuped and characters in stories that are not being read . . . That is how I understand Bassui's message.

Zen recognizes its own limitations, just as mathematicians have learned to recognize the limitations of the axiomatic method as a method for attaining truth. This does not mean that Zen has an answer to what lies beyond Zen any more than mathematicians have a clear understanding of the forms of valid reasoning which lie outside of formalization. One of the clearest Zen statements about the borderlines of Zen is given in the following strange kōan, very much in the spirit of Nansen:[10]

> Tōzan said to his monks, "You monks should know there is an even higher understanding in Buddhism." A monk stepped forward and asked, "What is the higher Buddhism?" Tōzan answered, "It is not Buddha."

There is always further to go; enlightenment is not the end-all of Zen. And there is no recipe which tells how to transcend Zen; the only thing one can rely on for sure is that Buddha is *not* the way. Zen is a system and cannot be its own metasystem; there is always something outside of Zen, which cannot be fully understood or described within Zen.

Escher and Zen

In questioning perception and posing absurd answerless riddles, Zen has company, in the person of M. C. Escher. Consider *Day and Night* (Fig. 49), a masterpiece of "positive and negative interwoven" (in the words of Mumon). One might ask, "Are those really birds, or are they really fields? Is it really night, or day?" Yet we all know there is no point to such questions. The picture, like a Zen kōan, is trying to break the mind of logic. Escher also delights in setting up contradictory pictures, such as *Another World*

FIGURE 51. Puddle, *by M. C. Escher (woodcut, 1952).*

(Fig. 48)—pictures that play with reality and unreality the same way as Zen plays with reality and unreality. Should one take Escher seriously? Should one take Zen seriously?

There is a delicate, haiku-like study of reflections in *Dewdrop* (Fig. 47); and then there are two tranquil images of the moon reflected in still waters: *Puddle* (Fig. 51), and *Rippled Surface* (Fig. 52). The reflected moon is a theme which recurs in various kōans. Here is an example:[11]

> Chiyono studied Zen for many years under Bukkō of Engaku. Still, she could not attain the fruits of meditation. At last one moonlit night she was carrying water in an old wooden pail girded with bamboo. The bamboo broke, and the bottom fell out of the pail. At that moment, she was set free. Chiyono said, "No more water in the pail, no more moon in the water."

Three Worlds: an Escher picture (Fig. 46), and the subject of a Zen kōan:[12]

> A monk asked Gantō, "When the three worlds threaten me, what shall I do?" Gantō answered, "Sit down." "I do not understand," said the monk. Gantō said, "Pick up the mountain and bring it to me. Then I will tell you."

Hemiolia and Escher

In *Verbum* (Fig. 149), oppositions are made into unities on several levels. Going around we see gradual transitions from black birds to white birds to black fish to white fish to black frogs to white frogs to black birds . . . After six steps, back where we started! Is this a reconciliation of the dichotomy of black and white? Or of the trichotomy of birds, fish, and frogs? Or is it a sixfold unity made from the opposition of the evenness of 2 and the oddness of 3? In music, six notes of equal time value create a rhythmic ambiguity—are they 2 groups of 3, or 3 groups of 2? This ambiguity has a name: *hemiolia.* Chopin was a master of hemiolia: see his Waltz op. 42, or his Etude op. 25, no. 2. In Bach, there is the *Tempo di Menuetto* from the keyboard Partita no. 5, or the incredible *Finale* of the first Sonata for unaccompanied violin, in G Minor.

As one glides inward toward the center of *Verbum,* the distinctions gradually blur, so that in the end there remains not three, not two, but one single essence: "VERBUM", which glows with brilliancy—perhaps a symbol of enlightenment. Ironically, "verbum" not only *is* a word, but *means* "word"—not exactly the most compatible notion with Zen. On the other hand, "verbum" is the only word in the picture. And Zen master Tōzan once said, "The complete Tripitaka can be expressed in one character." ("Tripitaka", meaning "three baskets", refers to the complete texts of the original Buddhist writings.) What kind of decoding-mechanism, I wonder, would it take to suck the three baskets out of one character? Perhaps one with two hemispheres.

FIGURE 52. Rippled Surface, *by M. C. Escher (lino-cut, 1950).*

FIGURE 53. Three Spheres II, *by M. C. Escher (lithograph, 1946).*

Indra's Net

Finally, consider *Three Spheres II* (Fig. 53), in which every part of the world seems to contain, and be contained in, every other part: the writing table reflects the spheres on top of it, the spheres reflect each other, as well as the writing table, the drawing of them, and the artist drawing it. The endless connections which all things have to each other is only hinted at here, yet the hint is enough. The Buddhist allegory of "Indra's Net" tells of an endless net of threads throughout the universe, the horizontal threads running through space, the vertical ones through time. At every crossing of threads is an individual, and every individual is a crystal bead. The great light of "Absolute Being" illuminates and penetrates every crystal bead; moreover, every crystal bead reflects not only the light from every other crystal in the net—but also every reflection of every reflection throughout the universe.

To my mind, this brings forth an image of renormalized particles: in every electron, there are virtual photons, positrons, neutrinos, muons . . . ; in every photon, there are virtual electrons, protons, neutrons, pions . . . ; in every pion, there are . . .

But then another image rises: that of people, each one reflected in the minds of many others, who in turn are mirrored in yet others, and so on.

Both of these images could be represented in a concise, elegant way by using Augmented Transition Networks. In the case of particles, there would be one network for each category of particle; in the case of people, one for each person. Each one would contain calls to many others, thus creating a virtual cloud of ATN's around each ATN. Calling one would create calls on others, and this process might cascade arbitrarily far, until it bottomed out.

Mumon on MU

Let us conclude this brief excursion into Zen by returning to Mumon. Here is his comment on Jōshū's MU:[13]

> To realize Zen one has to pass through the barrier of the patriarchs. Enlightenment always comes after the road of thinking is blocked. If you do not pass the barrier of the patriarchs or if your thinking road is not blocked, whatever you think, whatever you do, is like a tangling ghost. You may ask: "What is a barrier of a patriarch?" This one word, 'MU', is it.
>
> This is the barrier of Zen. If you pass through it, you will see Jōshū face to face. Then you can work hand in hand with the whole line of patriarchs. Is this not a pleasant thing to do?
>
> If you want to pass this barrier, you must work through every bone in your body, through every pore of your skin, filled with this question: "What is 'MU'?" and carry it day and night. Do not believe it is the common negative symbol meaning nothing. It is not nothingness, the opposite of existence. If you really want to pass this barrier, you should feel like drinking a hot iron ball that you can neither swallow nor spit out.
>
> Then your previous lesser knowledge disappears. As a fruit ripening in season, your subjectivity and objectivity naturally become one. It is like a dumb man who has had a dream. He knows about it but he cannot tell it.
>
> When he enters this condition his ego-shell is crushed and he can shake the heaven and move the earth. He is like a great warrior with a sharp sword. If a Buddha stands in his way, he will cut him down; if a patriarch offers him any obstacle, he will kill him; and he will be free in his way of birth and death. He can enter any world as if it were his own playground. I will tell you how to do this with this kōan:
>
> Just concentrate your whole energy into this MU, and do not allow any discontinuation. When you enter this MU and there is no discontinuation, your attainment will be as a candle burning and illuminating the whole universe.

From Mumon to the MU-puzzle

From the ethereal heights of Jōshū's MU, we now descend to the prosaic lowlinesses of Hofstadter's **MU** . . . I know that you have already concentrated your whole energy into this **MU** (when you read Chapter I). So now I wish to answer the question which was posed there:

Has **MU** theorem-nature, or not?

The answer to this question is not an evasive MU; rather, it is a resounding NO. In order to show this, we will take advantage of dualistic, logical thinking.

We made two crucial observations in Chapter I:

(1) that the MU-puzzle has depth largely because it involves the interplay of lengthening and shortening rules;

(2) that hope nevertheless exists for cracking the problem by employing a tool which is in some sense of adequate depth to handle matters of that complexity: the theory of numbers.

We did not analyze the MU-puzzle in those terms very carefully in Chapter I; we shall do so now. And we will see how the second observation (when generalized beyond the insignificant MIU-system) is one of the most fruitful realizations of all mathematics, and how it changed mathematicians' view of their own discipline.

For your ease of reference, here is a recapitulation of the MIU-system:

SYMBOLS: M, I, U

AXIOM: MI

RULES:

I. If *x*I is a theorem, so is *x*IU.
II. If M*x* is a theorem, so is M*xx*.
III. In any theorem, III can be replaced by U.
IV. UU can be dropped from any theorem.

Mumon Shows Us How to Solve the MU-puzzle

According to the observations above, then, the MU-puzzle is merely a puzzle about natural numbers in typographical disguise. If we could only find a way to transfer it to the domain of number theory, we might be able to solve it. Let us ponder the words of Mumon, who said, "If any of you has one eye, he will see the failure on the teacher's part." But why should it matter to have one eye?

If you try counting the number of I's contained in theorems, you will soon notice that it seems never to be 0. In other words, it seems that no matter how much lengthening and shortening is involved, we can never work in such a way that all I's are eliminated. Let us call the number of I's in any string the *I-count* of that string. Note that the I-count of the axiom MI is 1. We can do more than show that the I-count can't be 0—we can show that the I-count can never be any multiple of 3.

To begin with, notice that rules I and IV leave the I-count totally undisturbed. Therefore we need only think about rules II and III. As far as rule III is concerned, it diminishes the I-count by exactly 3. After an application of this rule, the I-count of the output might conceivably be a multiple of 3—but only if the I-count of the *input* was also. Rule III, in short, never creates a multiple of 3 from scratch. It can only create one when it began with one. The same holds for rule II, which doubles the

I-count. The reason is that if 3 divides $2n$, then—because 3 does not divide 2—it must divide n (a simple fact from the theory of numbers). Neither rule II nor rule III can create a multiple of 3 from scratch.

But this is the key to the MU-puzzle! Here is what we know:

(1) The I-count begins at 1 (not a multiple of 3);

(2) Two of the rules do not affect the I-count at all;

(3) The two remaining rules which do affect the I-count do so in such a way as never to create a multiple of 3 unless given one initially.

The conclusion—and a typically hereditary one it is, too—is that the I-count can never become any multiple of 3. In particular, 0 is a forbidden value of the I-count. Hence, MU *is not a theorem of the MIU-system.*

Notice that, even as a puzzle about I-counts, this problem was still plagued by the crossfire of lengthening and shortening rules. Zero became the goal; I-counts could increase (rule II), could decrease (rule III). Until we analyzed the situation, we might have thought that, with enough switching back and forth between the rules, we might eventually hit 0. Now, thanks to a simple number-theoretical argument, we know that that is impossible.

Gödel-Numbering the MIU-System

Not all problems of the the type which the MU-puzzle symbolizes are so easy to solve as this one. But we have seen that at least one such puzzle could be embedded within, and solved within, number theory. We are now going to see that there is a way to embed *all* problems about *any* formal system, in number theory. This can happen thanks to the discovery, by Gödel, of a special kind of isomorphism. To illustrate it, I will use the MIU-system.

We begin by considering the notation of the MIU-system. We shall map each symbol onto a new symbol:

M ⇐⇒ 3
I ⇐⇒ 1
U ⇐⇒ 0

The correspondence was chosen arbitrarily; the only rhyme or reason to it is that each symbol looks a little like the one it is mapped onto. Each number is called the *Gödel number* of the corresponding letter. Now I am sure you can guess what the Gödel number of a multiletter string will be:

MU ⇐⇒ 30
MIIU ⇐⇒ 3110
etc.

It is easy. Clearly this mapping between notations is an information-preserving transformation; it is like playing the same melody on two different instruments.

Let us now take a look at a typical derivation in the MIU-system, written simultaneously in both notations:

(1)	MI	——	axiom	——	31
(2)	MII	——	rule 2	——	311
(3)	MIIII	——	rule 2	——	31111
(4)	MUI	——	rule 3	——	301
(5)	MUIU	——	rule 1	——	3010
(6)	MUIUUIU	——	rule 2	——	3010010
(7)	MUIIU	——	rule 4	——	30110

The left-hand column is obtained by applying our four familiar typographical rules. The right-hand column, too, could be thought of as having been generated by a similar set of typographical rules. Yet the right-hand column has a dual nature. Let me explain what this means.

Seeing Things Both Typographically and Arithmetically

We could say of the fifth string ('3010') that it was made from the fourth, by appending a '0' on the right; on the other hand we could equally well view the transition as caused by an *arithmetical* operation—multiplication by 10, to be exact. When natural numbers are written in the decimal system, multiplication by 10 and putting a '0' on the right are indistinguishable operations. We can take advantage of this to write an *arithmetical* rule which corresponds to typographical rule I:

ARITHMETICAL RULE Ia: A number whose decimal expansion ends on the right in '1' can be multiplied by 10.

We can eliminate the reference to the symbols in the decimal expansion by arithmetically describing the rightmost digit:

ARITHMETICAL RULE Ib: A number whose remainder when divided by 10 is 1, can be multiplied by 10.

Now we could have stuck with a purely typographical rule, such as the following one:

TYPOGRAPHICAL RULE I: From any theorem whose rightmost symbol is '1' a new theorem can be made, by appending '0' to the right of that '1'.

They would have the same effect. This is why the right-hand column has a "dual nature": it can be viewed either as a series of typographical operations changing one pattern of symbols into another, or as a series of arithmetical operations changing one magnitude into another. But there are powerful reasons for being more interested in the arithmetical version. Stepping out of one purely typographical system into another isomorphic typographical system is not a very exciting thing to do; whereas stepping clear out of the typographical domain into an isomorphic part of number theory has some kind of unexplored potential. It is as if somebody had known musical scores all his life, but purely visually—and then, all of a sudden, someone introduced him to the mapping between sounds and musical scores. What a rich, new world! Then again, it is as if somebody had been familiar with string figures all his life, but purely as string figures, devoid of meaning—and then, all of a sudden, someone introduced him to the mapping between stories and strings. What a revelation! The discovery of Gödel-numbering has been likened to the discovery, by Descartes, of the isomorphism between curves in a plane and equations in two variables: incredibly simple, once you see it—and opening onto a vast new world.

Before we jump to conclusions, though, perhaps you would like to see a more complete rendering of this higher level of the isomorphism. It is a very good exercise. The idea is to give an arithmetical rule whose action is indistinguishable from that of each typographical rule of the MIU-system.

A solution is given below. In the rules, m and k are arbitrary natural numbers, and n is any natural number which is less than 10^m.

RULE 1: If we have made $10m + 1$, then we can make $10 \times (10m + 1)$.

Example: Going from line 4 to line 5. Here, $m = 30$.

RULE 2: If we have made $3 \times 10^m + n$, then we can make $10^m \times (3 \times 10^m + n) + n$.

Example: Going from line 1 to line 2, where both m and n equal 1.

RULE 3: If we have made $k \times 10^{m+3} + 111 \times 10^m + n$, then we can make $k \times 10^{m+1} + n$.

Example: Going from line 3 to line 4. Here, m and n are 1, and k is 3.

RULE 4: If we have made $k \times 10^{m+2} + n$, then we can make $k \times 10^m + n$.

Example: Going from line 6 to line 7. Here, $m = 2$, $n = 10$, and $k = 301$.

Let us not forget our axiom! Without it we can go nowhere. Therefore, let us postulate that:

We can make 31.

Now the right-hand column can be seen as a full-fledged arithmetical process, in a new arithmetical system which we might call the *310-system:*

(1)	31	given
(2)	311	rule 2 (m=1, n=1)
(3)	31111	rule 2 (m=2, n=11)
(4)	301	rule 3 (m=1, n=1, k=3)
(5)	3010	rule 1 (m=30)
(6)	3010010	rule 2 (m=3, n=10)
(7)	30110	rule 4 (m=2, n=10, k=301)

Notice once again that the lengthening and shortening rules are ever with us in this "310-system"; they have merely been transposed into the domain of numbers, so that the Gödel numbers go up and down. If you look carefully at what is going on, you will discover that the rules are based on nothing more profound than the idea that shifting digits to left and right in decimal representations of integers is related to multiplications and divisions by powers of 10. This simple observation finds its generalization in the following

> CENTRAL PROPOSITION: If there is a typographical rule which tells how certain digits are to be shifted, changed, dropped, or inserted in any number represented decimally, then this rule can be represented equally well by an arithmetical counterpart which involves arithmetical operations with powers of 10 as well as additions, subtractions, and so forth.

More briefly:

> Typographical rules for manipulating *numerals* are actually arithmetical rules for operating on *numbers*.

This simple observation is at the heart of Gödel's method, and it will have an absolutely shattering effect. It tells us that once we have a Gödel-numbering for any formal system, we can straightaway form a set of arithmetical rules which complete the Gödel isomorphism. The upshot is that we can transfer the study of any formal system—in fact the study of *all* formal systems—into number theory.

MIU-Producible Numbers

Just as any set of typographical rules generates a set of theorems, a corresponding set of natural numbers will be generated by repeated applications of arithmetical rules. These *producible numbers* play the same role inside number theory as theorems do inside any formal system. Of course, different numbers will be producible, depending on which rules are adopted. "Producible numbers" are only producible *relative to a system* of arithmetical rules. For example, such numbers as 31, 3010010, 3111, and so forth could be called *MIU-producible* numbers—an ungainly name, which might be shortened to *MIU-numbers,* symbolizing the fact that those numbers are the ones that result when you transcribe the MIU-system into number theory, via Gödel-numbering. If we were to Gödel-number the pq-system and then "arithmetize" its rules, we could call the producible numbers "pq-numbers"—and so on.

Note that the producible numbers (in any given system) are defined by a recursive method: given numbers which are known to be producible, we have rules telling how to make more producible numbers. Thus, the class of numbers known to be producible is constantly extending itself, in much the same way that the list of Fibonacci numbers, or Q-numbers, does. The set of producible numbers of any system is a *recursively enumerable set.* What about its complement—the set of nonproducible numbers? Is that set always recursively enumerable? Do numbers which are nonproducible share some common arithmetical feature?

This is the sort of issue which arises when you transpose the study of formal systems into number theory. For each system which is arithmetized, one can ask, "Can we characterize producible numbers in a simple way?" "Can we characterize *non*producible numbers in a recursively enumerable way?" These are difficult questions of number theory. Depending on the system which has been arithmetized, such questions might prove too hard for us to resolve. But if there is any hope for solving such problems, it would have to reside in the usual kind of step-by-step reasoning as it applies to natural numbers. And that, of course, was put in its quintessential form in the previous Chapter. TNT seemed, to all appearances, to have captured all valid mathematical thinking processes in one single, compact system.

Answering Questions about Producible Numbers by Consulting TNT

Could it be, therefore, that the means with which to answer any question about any formal system lies within just a single formal system—TNT? It seems plausible. Take, for instance, this question:

Is MU a theorem of the MIU-system?

Finding the answer is equivalent to determining whether 30 is a MIU-number or not. Because it is a statement of number theory, we should expect that, with some hard work, we could figure out how to translate the sentence "30 is a MIU-number" into TNT-notation, in somewhat the same way as we figured out how to translate other number-theoretical sentences into TNT-notation. I should immediately caution the reader that such a translation, though it does exist, is immensely complex. If you recall, I pointed out in Chapter VIII that even such a simple arithmetical predicate as "b is a power of 10" is very tricky to code into TNT-notation—and the predicate "b is a MIU-number" is a lot more complicated than that! Still, it can be found; and the numeral SSSSSSSSSSSSSSSSSSSSSSSSSSSSSS0 can be substituted for every b. This will result in a MONstrous string of TNT, a string of TNT which speaks about the MU-puzzle. Let us therefore call that string "MUMON". Through MUMON and strings like it, TNT is now capable of speaking "in code" about the MIU-system.

The Dual Nature of MUMON

In order to gain some benefit from this peculiar transformation of the original question, we would have to seek the answer to a new question:

Is MUMON a theorem of TNT?

All we have done is replace one relatively short string (MU) by another (the monstrous MUMON), and a simple formal system (the MIU-system) by a complicated one (TNT). It isn't likely that the answer will be any more forthcoming even though the question has been reshaped. In fact, TNT has a full complement of both lengthening and shortening rules, and the reformulation of the question is likely to be far harder than the original. One might even say that looking at MU via MUMON is an intentionally idiotic way of doing things. However, MUMON can be looked at on more than one level.

In fact, this is an intriguing point: MUMON has two different passive meanings. Firstly, it has the one which was given before:

30 is a MIU-number.

But secondly, we know that this statement is tied (via isomorphism) to the statement

MU is a theorem of the MIU-system.

So we can legitimately quote this latter as the second passive meaning of MUMON. It may seem very strange because, after all, MUMON contains nothing but plus signs, parentheses, and so forth—symbols of TNT. How can it possibly express any statement with other than arithmetical content?

The fact is, it can. Just as a single musical line may serve as both harmony and melody in a single piece; just as "BACH" may be interpreted as both a name and a melody; just as a single sentence may be an accurate structural description of a picture by Escher, of a section of DNA, of a piece by Bach, and of the dialogue in which the sentence is embedded, so MUMON can be taken in (at least) two entirely different ways. This state of affairs comes about because of two facts:

Fact 1. Statements such as "MU is a theorem" can be coded into number theory via Gödel's isomorphism.

Fact 2. Statements of number theory can be translated into TNT.

It could be said that MUMON is, by Fact 1, a coded message, where the symbols of the code are, by Fact 2, just symbols of TNT.

Codes and Implicit Meaning

Now it could be objected here that a coded message, unlike an uncoded message, does not express anything on its own—it requires knowledge of the code. But in reality there is no such thing as an uncoded message. There are only messages written in more familiar codes, and messages written in less familiar codes. If the meaning of a message is to be revealed, it must be pulled out of the code by some sort of mechanism, or isomorphism. It may be difficult to discover the method by which the decoding should be done; but once that method has been discovered, the message becomes transparent as water. When a code is familiar enough, it ceases appearing like a code; one forgets that there is a decoding mechanism. The message is identified with its meaning.

Here we have a case where the identification of message and meaning is so strong that it is hard for us to conceive of an alternate meaning residing in the same symbols. Namely, we are so prejudiced by the symbols of TNT towards seeing number-theoretical meaning (and *only* number-theoretical meaning) in strings of TNT, that to conceive of certain strings of TNT as statements about the MIU-system is quite difficult. But Gödel's isomorphism compels us to recognize this second level of meaning in certain strings of TNT.

Decoded in the more familiar way, MUMON bears the message:

30 is a MIU-number.

This is a statement of number theory, gotten by interpreting each sign in the conventional way.

But in discovering Gödel-numbering and the whole isomorphism built upon it, we have in a sense broken a code in which messages about the MIU-system are written in strings of TNT. Gödel's isomorphism is a new information-revealer, just as the decipherments of ancient scripts were information-revealers. Decoded by this new and less familiar mechanism, MUMON bears the message

MU is a theorem of the MIU-system.

The moral of the story is one we have heard before: that meaning is an automatic by-product of our recognition of any isomorphism; therefore there are at least two passive meanings of MUMON—maybe more!

The Boomerang: Gödel-Numbering TNT

Of course things do not stop here. We have only begun realizing the potential of Gödel's isomorphism. The natural trick would be to turn TNT's capability of mirroring other formal systems back on itself, as the Tortoise turned the Crab's phonographs against themselves, and as his Goblet G turned against itself, in destroying itself. In order to do this, we will have to Gödel-number TNT itself, just as we did the MIU-system, and then "arithmetize" its rules of inference. The Gödel-numbering is easy to do. For instance, we could make the following correspondence:

Symbol		*Codon*	*Mnemonic Justification*	
0	.....	666	Number of the Beast for the Mysterious Zero	
S	.....	123	successorship: 1, 2, 3, ...	
=	.....	111	visual resemblance, turned sideways	
+	.....	112	1 + 1 = 2	
·	.....	236	2 × 3 = 6	
(	.....	362	ends in 2	
)	.....	323	ends in 3	
<	.....	212	ends in 2	*these three pairs*
>	.....	213	ends in 3	*form a pattern*
[	.....	312	ends in 2	
]	.....	313	ends in 3	
a	.....	262	opposite to ∀ (626)	
′	.....	163	163 is prime	
∧	.....	161	'∧' is a "graph" of the sequence 1-6-1	
∨	.....	616	'∨' is a "graph" of the sequence 6-1-6	
⊃	.....	633	6 "implies" 3 and 3, in some sense ...	
~	.....	223	2 + 2 is *not* 3	
∃	.....	333	'∃' looks like '3'	
∀	.....	626	opposite to a; also a "graph" of 6-2-6	
:	.....	636	two dots, two sixes	
punc.	.....	611	special number, as on Bell system (411, 911)	

Each symbol of TNT is matched up with a triplet composed of the digits 1, 2, 3, and 6, in a manner chosen for mnemonic value. I shall call each such triplet of digits a *Gödel codon,* or *codon* for short. Notice that I have given no codon for b, c, d, or e; we are using austere TNT. There is a hidden motivation for this, which you will find out about in Chapter XVI. I will explain the bottom entry, "punctuation", in Chapter XIV.

Now we can rewrite any string or rule of TNT in the new garb. Here, for instance, is Axiom 1 in the two notations, the old below the new:

626,262,636,223,123,262,111,666
∀ a : ~ S a = 0

Conveniently, the standard convention of putting in a comma every third digit happens to coincide with our codons, setting them off for "easy" legibility.

Here is the Rule of Detachment, in the new notation:

RULE: If x and 212x633y213 are both theorems, then y is a theorem.

Finally, here is an entire derivation taken from last Chapter, given in austere TNT and also transcribed into the new notation:

626,262,636,626,262,163,636,362,262,112,123,262,163,323,111,123,362,262,112,262,163,323 axiom 3
∀ a : ∀ a ′ : (a + S a ′) = S (a + a ′)

626,262,163,636,362,123,666,112,123,262,163,323,111,123,362,123,666,112,262,163,323 specification
∀ a ′ : (S 0 + S a ′) = S (S 0 + a ′)

362,123,666,112,123,666,323,111,123,362,123,666,112,666,323 specification
(S 0 + S 0) = S (S 0 + 0)

626,262,636,362,262,112,666,323,111,262 axiom 2
∀ a : (a + 0) = a

362,123,666,112,666,323,111,123,666 specification
(S 0 + 0) = S 0

123,362,123,666,112,666,323,111,123,123,666 insert '123'
S (S 0 + 0) = S S 0

362,123,666,112,123,666,323,111,123,123,666 transitivity
(S 0 + S 0) = S S 0

Notice that I changed the name of the "Add S" rule to "Insert '123'", since that is the typographical operation which it now legitimizes.

This new notation has a pretty strange feel to it. You lose all sense of meaning; but if you had been brought up on it, you could read strings in this notation as easily as you do TNT. You would be able to look and, at a glance, distinguish well-formed formulas from ill-formed ones. Naturally, since it is so visual, you would think of this as a typographical operation—but at the same time, picking out well-formed *formulas* in this notation is picking out a special class of *integers*, which have an arithmetical definition, too.

Now what about "arithmetizing" all the rules of inference? As matters stand, they are all still typographical rules. But wait! According to the Central Proposition, a typographical rule is really equivalent to an arithmetical rule. Inserting and moving digits in decimally represented numbers is an *arithmetical* operation, which can be carried out typographically. Just as appending a '0' on the end is exactly the same as multiplying by 10, so each rule is a condensed way of describing a messy arithmetical operation. Therefore, in a sense, we do not even need to look for equivalent arithmetical rules, because all of the rules are *already* arithmetical!

TNT-Numbers: A Recursively Enumerable Set of Numbers

Looked at this way, the preceding derivation of the theorem "362,123,666,112,123,666,323,111,123,123,666" is a sequence of highly convoluted number-theoretical transformations, each of which acts on one or more input numbers, and yields an output number, which is, as before, called a *producible number,* or, to be more specific, a *TNT-number*. Some of the arithmetical rules take an old TNT-number and *increase* it in a particular way, to yield a new TNT-number; some take an old TNT-number and *decrease* it; other rules take two TNT-numbers, operate on each of them in some odd way, and then combine the results into a new TNT-number—and so on and so forth. And instead of starting with just one known TNT-number, we have *five* initial TNT-numbers—one for each (austere) axiom, of course. Arithmetized TNT is actually extremely similar to the arithmetized MIU-system, only there are more rules and axioms, and to write out arithmetical equivalents explicitly would be a big bother—and quite unenlightening, incidentally. If you followed how it was done for the MIU-system, there ought to be no doubt on your part that it is quite analogous here.

There is a new number-theoretical predicate brought into being by this "Gödelization" of TNT: the predicate

a is a TNT-number.

For example, we know from the preceding derivation that 362,123,666,112,123,666,323,111,123,123,666 *is* a TNT-number, while on the other hand, presumably 123,666,111,666 is *not* a TNT-number.

Now it occurs to us that this new number-theoretical predicate is *expressible* by some string of TNT with one free variable, say a. We could put a tilde in front, and that string would express the complementary notion

a is not a TNT-number.

Now if we replaced all the occurrences of a in this second string by the TNT-numeral for 123,666,111,666—a numeral which would contain exactly 123,666,111,666 S's, much too long to write out—we would have a TNT-string which, just like MUMON, is capable of being interpreted on two levels. In the first place, that string would say

123,666,111,666 is not a TNT-number.

But because of the isomorphism which links TNT-numbers to theorems of TNT, there would be a second-level meaning of this string, which is:

S0=0 is not a theorem of TNT.

TNT Tries to Swallow Itself

This unexpected double-entendre demonstrates that TNT contains strings which talk about other strings of TNT. In other words, the metalanguage in which we, on the outside, can speak about TNT, is at least partially imitated *inside* TNT itself. And this is not an accidental feature of TNT; it happens because the architecture of any formal system can be mirrored inside N (number theory). It is just as inevitable a feature of TNT as are the vibrations induced in a record player when it plays a record. It seems as if vibrations should come from the outside world—for instance, from jumping children or bouncing balls; but a side effect of producing sounds—and an unavoidable one—is that they wrap around and shake the very mechanism which produces them. It is no accident; it is a side effect which cannot be helped. It is in the nature of record players. And it is in the nature of any formalization of number theory that its metalanguage is embedded within it.

We can dignify this observation by calling it the *Central Dogma of Mathematical Logic,* and depicting it in a two-step diagram:

TNT $\Rightarrow$ N $\Rightarrow$ meta-TNT

In words: a string of TNT has an interpretation in N; and a statement of N may have a second meaning as a statement about TNT.

G: A String Which Talks about Itself in Code

This much is intriguing yet it is only half the story. The rest of the story involves an intensification of the self-reference. We are now at the stage where the Tortoise was when he realized that a record could be made which would make the phonograph playing it break—but now the question is: "Given a record player, how do you actually figure out what to put on the record?" That is a tricky matter.

We want to find a string of TNT—which we'll call 'G'—which is about *itself,* in the sense that one of its passive meanings is a sentence about G. In particular the passive meaning will turn out to be

"G is not a theorem of TNT."

I should quickly add that G also has a passive meaning which is a *statement of number theory;* just like MUMON it is susceptible to being construed in (at least) two different ways. The important thing is that each passive meaning is valid and useful and doesn't cast doubt on the other passive meaning in any way. (The fact that a phonograph playing a record can induce vibrations in itself and in the record does not diminish in any way the fact that those vibrations are musical sounds!)

G's Existence Is What Causes TNT's Incompleteness

The ingenious method of creating G, and some important concepts relating to TNT, will be developed in Chapters XIII and XIV; for now it is just interesting to glance ahead, a bit superficially, at the consequences of finding a self-referential piece of TNT. Who knows? It might blow up! In a sense it does. We focus down on the obvious question:

Is G a theorem of TNT, or not?

Let us be sure to form our *own* opinion on this matter, rather than rely on G's opinion about itself. After all, G may not understand itself any better than a Zen master understands himself. Like MUMON, G may express a falsity. Like MU, G may be a nontheorem. We don't need to believe every possible string of TNT—only its theorems. Now let us use our power of reasoning to clarify the issue as best we can at this point.

We will make our usual assumption: that TNT incorporates valid methods of reasoning, and therefore that TNT never has falsities for theorems. In other words, anything which is a theorem of TNT expresses a truth. So if G were a theorem, it would express a truth, namely: "G is not a theorem". The full force of its self-reference hits us. By being a theorem, G would have to be a falsity. Relying on our assumption that TNT never has falsities for theorems, we'd be forced to conclude that *G is not a theorem.* This is all right; it leaves us, however, with a lesser problem. Knowing that G is not a theorem, we'd have to concede that G expresses a truth. Here is a situation in which TNT doesn't live up to our expectations—we have found a string which expresses a true statement yet the string is not a theorem. And in our amazement, we shouldn't lose track of the fact that G has an arithmetical interpretation, too—which allows us to summarize our findings this way:

> A string of TNT has been found; it expresses, unambiguously, a statement about certain arithmetical properties of natural numbers; moreover, by reasoning outside the system we can determine not only that the statement is a true one, but also that the string fails to be a theorem of TNT. And thus, if we ask TNT whether the statement is true, TNT says neither yes nor no.

Is the Tortoise's string in the *Mu Offering* the analogue of G? Not quite. The analogue of the Tortoise's string is ~G. Why is this so? Well, let us think a moment about what ~G says. It must say the opposite of what G says. G says, "G is not a theorem of TNT", so ~G must say "G is a theorem". We could rephrase both G and ~G this way:

G: "I am not a theorem (of TNT)."
~G: "My negation is a theorem (of TNT)."

It is ~G which is parallel to the Tortoise's string, for that string spoke not about itself, but about the string which the Tortoise first proffered to Achilles—which had an extra knot on it (or one too few, however you want to look at it).

Mumon Has the Last Word

Mumon penetrated into the Mystery of the Undecidable as clearly as anyone, in his concise poem on Jōshū's MU:

Has a dog Buddha-nature?
This is the most serious question of all.
If you say yes or no,
You lose your own Buddha-nature.

PART II

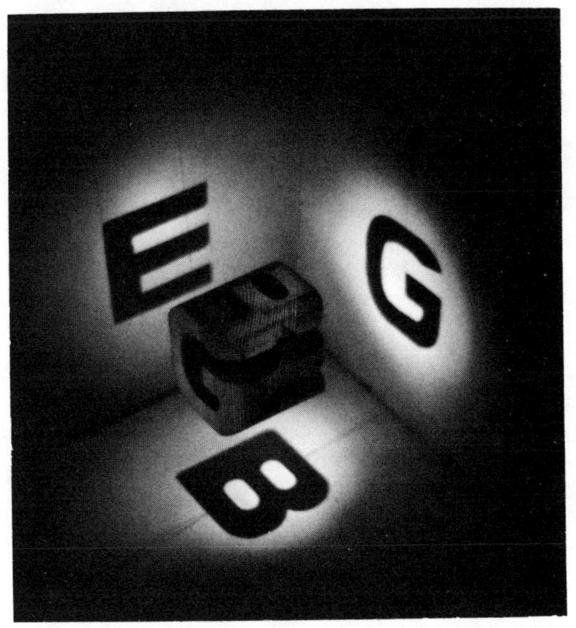

Prelude . . .

Achilles and the Tortoise have come to the residence of their friend the Crab, to make the acquaintance of one of his friends, the Anteater. The introductions having been made, the four of them settle down to tea.

Tortoise: We have brought along a little something for you, Mr. Crab.

Crab: That's most kind of you. But you shouldn't have.

Tortoise: Just a token of our esteem. Achilles, would you like to give it to Mr. C?

Achilles: Surely. Best wishes, Mr. Crab. I hope you enjoy it.

(Achilles hands the Crab an elegantly wrapped present, square and very thin. The Crab begins unwrapping it.)

Anteater: I wonder what it could be.

Crab: We'll soon find out. *(Completes the unwrapping, and pulls out the gift.)* Two records! How exciting! But there's no label. Uh-oh—is this another of your "specials", Mr. T?

Tortoise: If you mean a phonograph-breaker, not this time. But it is in fact a custom-recorded item, the only one of its kind in the entire world. In fact, it's never even been heard before—except, of course, when Bach played it.

Crab: When Bach played it? What do you mean, exactly?

Achilles: Oh, you are going to be fabulously excited, Mr. Crab, when Mr. T tells you what these records in fact are.

Tortoise: Oh, you go ahead and tell him, Achilles.

Achilles: May I? Oh, boy! I'd better consult my notes, then. *(Pulls out a small filing card, and clears his voice.)* Ahem. Would you be interested in hearing about the remarkable new result in mathematics, to which your records owe their existence?

Crab: My records derive from some piece of mathematics? How curious! Well, now that you've provoked my interest, I must hear about it.

Achilles: Very well, then. *(Pauses for a moment to sip his tea, then resumes.)* Have you heard of Fermat's infamous "Last Theorem"?

Anteater: I'm not sure . . . It sounds strangely familiar, and yet I can't quite place it.

Achilles: It's a very simple idea. Pierre de Fermat, a lawyer by vocation but mathematician by avocation, had been reading in his copy of the classic text *Arithmetica* by Diophantus, and came across a page containing the equation

$$a^2 + b^2 = c^2$$

He immediately realized that this equation has infinitely many solutions a, b, c, and then wrote in the margin the following notorious comment:

> The equation
>
> $$a^n + b^n = c^n$$
>
> has solutions in positive integers a, b, c, and n only when $n = 2$ (and then there are infinitely many triplets a, b, c which satisfy the equation); but there are no solutions for $n > 2$. I have discovered a truly marvelous proof of this statement, which, unfortunately, this margin is too small to contain.

Ever since that day, some three hundred years ago, mathematicians have been vainly trying to do one of two things: either to prove Fermat's claim, and thereby vindicate Fermat's reputation, which, although very high, has been somewhat tarnished by skeptics who think he never really found the proof he claimed to have found—or else to refute the claim, by finding a counterexample: a set of four integers a, b, c, and n, with $n > 2$, which satisfy the equation. Until very recently, every attempt in either direction had met with failure. To be sure, the Theorem has been proven for many specific values of n—in particular, all n up to 125,000.

Anteater: Shouldn't it be called a "Conjecture" rather than a "Theorem", if it's never been given a proper proof?

Achilles: Strictly speaking, you're right, but tradition has kept it this way.

Crab: Has someone at last managed to resolve this celebrated question?

Achilles: Indeed! In fact, Mr. Tortoise has done so, and as usual, by a wizardly stroke. He has not only found a PROOF of Fermat's Last Theorem (thus justifying its name as well as vindicating Fermat), but also a COUNTEREXAMPLE, thus showing that the skeptics had good intuition!

Crab: Oh my gracious! That is a revolutionary discovery.

Anteater: But please don't leave us in suspense. What magical integers are they, that satisfy Fermat's equation? I'm especially curious about the value of n.

Achilles: Oh, horrors! I'm most embarrassed! Can you believe this? I left the values at home on a truly colossal piece of paper. Unfortunately it was too huge to bring along. I wish I had them here to show to you. If it's of any help to you, I do remember one thing—the value of n is the only positive integer which does not occur anywhere in the continued fraction for π.

Crab: Oh, what a shame that you don't have them here. But there's no reason to doubt what you have told us.

FIGURE 54. Möbius Strip II, *by M. C. Escher (woodcut, 1963).*

FIGURE 55. Pierre de Fermat.

Anteater: Anyway, who needs to see n written out decimally? Achilles has just told us how to find it. Well, Mr. T, please accept my hearty felicitations, on the occasion of your epoch-making discovery!

Tortoise: Thank you. But what I feel is more important than the result itself is the practical use to which my result immediately led.

Crab: I am dying to hear about it, since I always thought number theory was the Queen of Mathematics—the purest branch of mathematics—the one branch of mathematics which has NO applications!

Tortoise: You're not the only one with that belief, but in fact it is quite impossible to make a blanket statement about when or how some branch—or even some individual Theorem—of pure mathematics will have important repercussions outside of mathematics. It is quite unpredictable—and this case is a perfect example of that phenomenon.

Achilles: Mr. Tortoise's double-barreled result has created a breakthrough in the field of acoustico-retrieval!

Anteater: What is acoustico-retrieval?

Achilles: The name tells it all: it is the retrieval of acoustic information from extremely complex sources. A typical task of acoustico-retrieval is to reconstruct the sound which a rock made on plummeting into a lake from the ripples which spread out over the lake's surface.

Crab: Why, that sounds next to impossible!

Achilles: Not so. It is actually quite similar to what one's brain does, when it reconstructs the sound made in the vocal cords of another person from the vibrations transmitted by the eardrum to the fibers in the cochlea.

Crab: I see. But I still don't see where number theory enters the picture, or what this all has to do with my new records.

Achilles: Well, in the mathematics of acoustico-retrieval, there arise many questions which have to do with the number of solutions of certain Diophantine equations. Now Mr. T has been for years trying to find a way of reconstructing the sounds of Bach playing his harpsichord, which took place over two hundred years ago, from calculations involving the motions of all the molecules in the atmosphere at the present time.

Anteater: Surely that is impossible! They are irretrievably gone, gone forever!

Achilles: Thus think the naïve . . . But Mr. T has devoted many years to this problem, and came to the realization that the whole thing hinged on the number of solutions to the equation

$$a^n + b^n = c^n$$

in positive integers, with $n > 2$.

Tortoise: I could explain, of course, just how this equation arises, but I'm sure it would bore you.

Achilles: It turned out that acoustico-retrieval theory predicts that the Bach sounds can be retrieved from the motion of all the molecules in the atmosphere, provided that EITHER there exists at least one solution to the equation—

Crab: Amazing!

Anteater: Fantastic!

Tortoise: Who would have thought!

Achilles: I was about to say, "provided that there exists EITHER such a solution OR a proof that there are NO solutions!" And therefore, Mr. T, in careful fashion, set about working at both ends of the problem, simultaneously. As it turns out, the discovery of the counterexample was the key ingredient to finding the proof, so the one led directly to the other.

Crab: How could that be?

Tortoise: Well, you see, I had shown that the structural layout of any proof of Fermat's Last Theorem—if one existed—could be described by an elegant formula, which, it so happened, depended on the values of a solution to a certain equation. When I found this second equation, to my surprise it turned out to be the Fermat equation. An amusing accidental relationship between form and content. So when I found the counterexample, all I needed to do was to use those numbers as a blueprint for constructing my proof that there were no solutions to the equation. Remarkably simple, when you think about it. I can't imagine why no one had ever found the result before.

Achilles: As a result of this unanticipatedly rich mathematical success, Mr. T was able to carry out the acoustico-retrieval which he had so long dreamed of. And Mr. Crab's present here represents a palpable realization of all this abstract work.

Crab: Don't tell me it's a recording of Bach playing his own works for harpsichord!

Achilles: I'm sorry, but I have to, for that is indeed just what it is! This is a set of two records of Johann Sebastian Bach playing all of his *Well-Tempered Clavier.* Each record contains one of the two volumes of the *Well-Tempered Clavier;* that is to say, each record contains 24 preludes and fugues—one in each major and minor key.

Crab: Well, we must absolutely put one of these priceless records on, immediately! And how can I ever thank the two of you?

Tortoise: You have already thanked us plentifully, with this delicious tea which you have prepared.

(The Crab slides one of the records out of its jacket, and puts it on. The sound of an incredibly masterful harpsichordist fills the room, in the highest imaginable fidelity. One even hears—or is it one's imagination?—the soft sounds of Bach singing to himself as he plays . . .)

Crab: Would any of you like to follow along in the score? I happen to have a unique edition of the *Well-Tempered Clavier,* specially illuminated by a teacher of mine who happens also to be an unusually fine calligrapher.

Tortoise: I would very much enjoy that.

(The Crab goes to his elegant glass-enclosed wooden bookcase, opens the doors, and draws out two large volumes.)

Crab: Here you are, Mr. Tortoise. I've never really gotten to know all the beautiful illustrations in this edition. Perhaps your gift will provide the needed impetus for me to do so.

Tortoise: I do hope so.

Anteater: Have you ever noticed how in these pieces the prelude always sets the mood perfectly for the following fugue?

Crab: Yes. Although it may be hard to put it into words, there is always some subtle relation between the two. Even if the prelude and fugue do not have a common melodic subject, there is nevertheless always some intangible abstract quality which underlies both of them, binding them together very strongly.

Tortoise: And there is something very dramatic about the few moments of silent suspense hanging between prelude and fugue—that moment where the the theme of the fugue is about to ring out, in single tones, and then to join with itself in ever-increasingly complex levels of weird, exquisite harmony.

Achilles: I know just what you mean. There are so many preludes and fugues which I haven't yet gotten to know, and for me that fleeting interlude of silence is very exciting; it's a time when I try to second-guess old Bach. For example, I always wonder what the fugue's tempo will be: allegro, or adagio? Will it be in 6/8, or 4/4? Will it have three voices, or five—or four? And then, the first voice starts . . . Such an exquisite moment.

Crab: Ah, yes, well do I remember those long-gone days of my youth, the days when I thrilled to each new prelude and fugue, filled with the excitement of their novelty and beauty and the many unexpected surprises which they conceal.

Achilles: And now? Is that thrill all gone?

Crab: It's been supplanted by familiarity, as thrills always will be. But in that familiarity there is also a kind of depth, which has its own compensations. For instance, I find that there are always new surprises which I hadn't noticed before.

Achilles: Occurrences of the theme which you had overlooked?

Crab: Perhaps—especially when it is inverted and hidden among several other voices, or where it seems to come rushing up from the depths, out of nowhere. But there are also amazing modulations which it is marvelous to listen to over and over again, and wonder how old Bach dreamt them up.

Achilles: I am very glad to hear that there is something to look forward to, after I have been through the first flush of infatuation with the *Well-Tempered Clavier*—although it also makes me sad that this stage could not last forever and ever.

Crab: Oh, you needn't fear that your infatuation will totally die. One of the nice things about that sort of youthful thrill is that it can always be resuscitated, just when you thought it was finally dead. It just takes the right kind of triggering from the outside.

Achilles: Oh, really? Such as what?

Crab: Such as hearing it through the ears, so to speak, of someone to whom it is a totally new experience—someone such as you, Achilles. Somehow the excitement transmits itself, and I can feel thrilled again.

Achilles: That is intriguing. The thrill has remained dormant somewhere inside you, but by yourself, you aren't able to fish it up out of your subconscious.

Crab: Exactly. The potential of reliving the thrill is "coded", in some unknown way, in the structure of my brain, but I don't have the power to summon it up at will; I have to wait for chance circumstance to trigger it.

Achilles: I have a question about fugues which I feel a little embarrassed about asking, but as I am just a novice at fugue-listening, I was wondering if perhaps one of you seasoned fugue-listeners might help me in learning . . . ?

Tortoise: I'd certainly like to offer my own meager knowledge, if it might prove of some assistance.

Achilles: Oh, thank you. Let me come at the question from an angle. Are you familiar with the print called *Cube with Magic Ribbons,* by M. C. Escher?

Tortoise: In which there are circular bands having bubble-like distortions which, as soon as you've decided that they are bumps, seem to turn into dents—and vice versa?

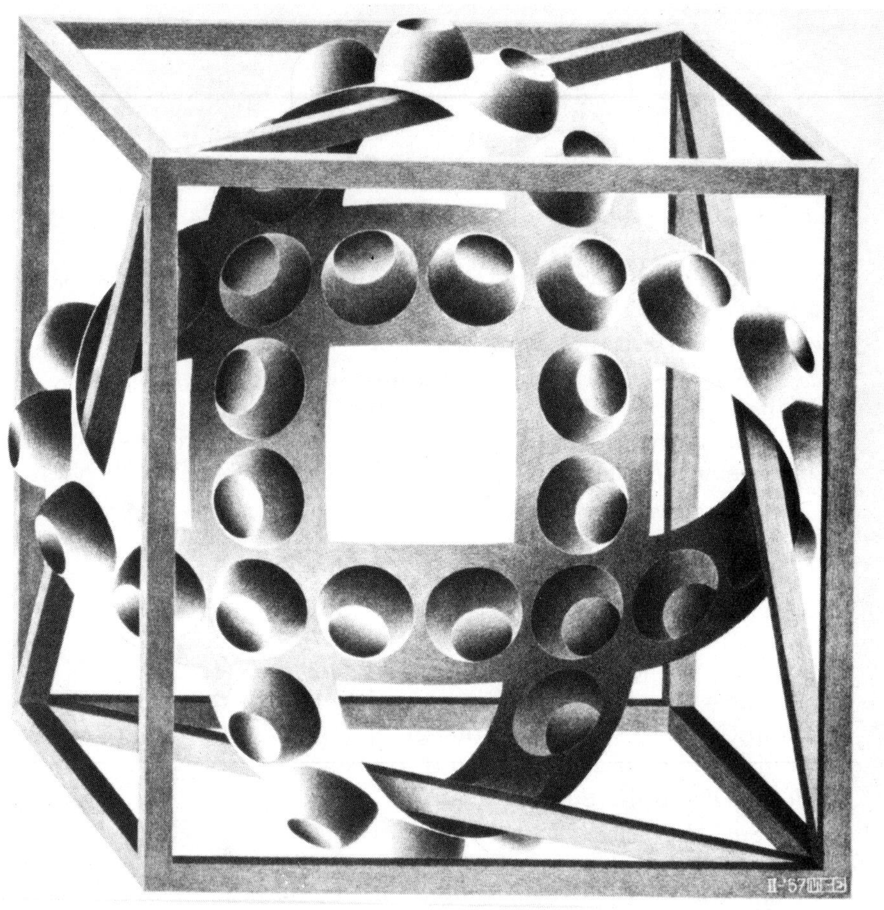

FIGURE 56. Cube with Magic Ribbons, *by M. C. Escher (lithograph, 1957).*

Achilles: Exactly.

Crab: I remember that picture. Those little bubbles always seem to flip back and forth between being concave and convex, depending on the direction that you approach them from. There's no way to see them simultaneously as concave AND convex—somehow one's brain doesn't allow that. There are two mutually exclusive "modes" in which one can perceive the bubbles.

Achilles: Just so. Well, I seem to have discovered two somewhat analogous modes in which I can listen to a fugue. The modes are these: either to follow one individual voice at a time, or to listen to the total effect of all of them together, without trying to disentangle one from another. I have tried out both of these modes, and, much to my frustration, each one of them shuts out the other. It's simply not in my power to follow the paths of individual voices and at the same time to hear the whole effect. I find that I flip back and forth between one mode and the other, more or less spontaneously and involuntarily.

Anteater: Just as when you look at the magic bands, eh?

Achilles: Yes. I was just wondering . . . does my description of these two modes of fugue-listening brand me unmistakably as a naïve, inexperienced listener, who couldn't even begin to grasp the deeper modes of perception which exist beyond his ken?

Tortoise: No, not at all, Achilles. I can only speak for myself, but I too find myself shifting back and forth from one mode to the other without exerting any conscious control over which mode should be dominant. I don't know if our other companions here have also experienced anything similar.

Crab: Most definitely. It's quite a tantalizing phenomenon, since you feel that the essence of the fugue is flitting about you, and you can't quite grasp all of it, because you can't quite make yourself function both ways at once.

Anteater: Fugues have that interesting property, that each of their voices is a piece of music in itself; and thus a fugue might be thought of as a collection of several distinct pieces of music, all based on one single theme, and all played simultaneously. And it is up to the listener (or his subconscious) to decide whether it should be perceived as a unit, or as a collection of independent parts, all of which harmonize.

Achilles: You say that the parts are "independent", yet that can't be literally true. There has to be some coordination between them, otherwise when they were put together one would just have an unsystematic clashing of tones—and that is as far from the truth as could be.

Anteater: A better way to state it might be this: if you listened to each voice on its own, you would find that it seemed to make sense all by itself. It could stand alone, and that is the sense in which I meant that it is independent. But you are quite right in pointing out that each of these individually meaningful lines fuses with the others in a highly nonrandom way, to make a graceful totality. The art of writing a beautiful fugue lies precisely in this ability, to manufacture several different lines, each one of which gives the illusion of having been written for its own beauty, and yet which when taken together form a whole, which does not feel forced in any way. Now, this dichotomy between hearing a fugue as a whole, and hearing its component voices, is a particular example of a very general dichotomy, which applies to many kinds of structures built up from lower levels.

Achilles: Oh, really? You mean that my two "modes" may have some more general type of applicability, in situations other than fugue-listening?

Anteater: Absolutely.

Achilles: I wonder how that could be. I guess it has to do with alternating between perceiving something as a whole, and perceiving it as a collection of parts. But the only place I have ever run into that dichotomy is in listening to fugues.

Tortoise: Oh, my, look at this! I just turned the page while following the music, and came across this magnificent illustration facing the first page of the fugue.

Crab: I have never seen that illustration before. Why don't you pass it 'round?

(The Tortoise passes the book around. Each of the foursome looks at it in a characteristic way—this one from afar, that one from close up, everyone tipping his head this way and that in puzzlement. Finally it has made the rounds, and returns to the Tortoise, who peers at it rather intently.)

Achilles: Well, I guess the prelude is just about over. I wonder if, as I listen to this fugue, I will gain any more insight into the question, "What is the right way to listen to a fugue: as a whole, or as the sum of its parts?"

TTortoise: Listen carefully, and you will!

(The prelude ends. There is a moment of silence; and . . .)

[*ATTACCA*]

CHAPTER X

Levels of Description, and Computer Systems

Levels of Description

GÖDEL'S STRING G, and a Bach fugue: they both have the property that they can be understood on different levels. We are all familiar with this kind of thing; and yet in some cases it confuses us, while in others we handle it without any difficulty at all. For example, we all know that we human beings are composed of an enormous number of cells (around twenty-five trillion), and therefore that everything we do could in principle be described in terms of cells. Or it could even be described on the level of molecules. Most of us accept this in a rather matter-of-fact way; we go to the doctor, who looks at us on lower levels than we think of ourselves. We read about DNA and "genetic engineering" and sip our coffee. We seem to have reconciled these two inconceivably different pictures of ourselves simply by disconnecting them from each other. We have almost no way to relate a microscopic description of ourselves to that which we feel ourselves to be, and hence it is possible to store separate representations of ourselves in quite separate "compartments" of our minds. Seldom do we have to flip back and forth between these two concepts of ourselves, wondering "How can these two totally different things be the same *me*?"

Or take a sequence of images on a television screen which shows Shirley MacLaine laughing. When we watch that sequence, we know that we are actually looking not at a woman, but at sets of flickering dots on a flat surface. We know it, but it is the furthest thing from our mind. We have these two wildly opposing representations of what is on the screen, but that does not confuse us. We can just shut one out, and pay attention to the other—which is what all of us do. Which one is "more real"? It depends on whether you're a human, a dog, a computer, or a television set.

Chunking and Chess Skill

One of the major problems of Artificial Intelligence research is to figure out how to bridge the gap between these two descriptions; how to construct a system which can accept one level of description, and produce the other. One way in which this gap enters Artificial Intelligence is well illustrated by the progress in knowledge about how to program a computer to play good chess. It used to be thought—in the 1950's and on into the 1960's—that the trick to making a machine play well was to make the machine look further ahead into the branching network of possible sequences of play than any chess master can. However, as this goal gradually became attained, the level of computer chess did not have any sudden spurt, and surpass human experts. In fact, a human expert can quite soundly and confidently trounce the best chess programs of this day.

The reason for this had actually been in print for many years. In the 1940's, the Dutch psychologist Adriaan de Groot made studies of how chess novices and chess masters perceive a chess situation. Put in their starkest terms, his results imply that chess masters perceive the distribution of pieces in *chunks*. There is a higher-level description of the board than the straightforward "white pawn on K5, black rook on Q6" type of description, and the master somehow produces such a mental image of the board. This was proven by the high speed with which a master could reproduce an actual position taken from a game, compared with the novice's plodding reconstruction of the position, after both of them had had five-second glances at the board. Highly revealing was the fact that masters' mistakes involved placing whole *groups* of pieces in the wrong place, which left the game strategically almost the same, but to a novice's eyes, not at all the same. The clincher was to do the same experiment but with pieces randomly assigned to the squares on the board, instead of copied from actual games. The masters were found to be simply no better than the novices in reconstructing such random boards.

The conclusion is that in normal chess play, certain types of situation recur—certain patterns—and it is to those high-level patterns that the master is sensitive. He thinks *on a different level* from the novice; his set of concepts is different. Nearly everyone is surprised to find out that in actual play, a master rarely looks ahead any further than a novice does—and moreover, a master usually examines only a handful of possible moves! The trick is that his mode of perceiving the board is like a filter: he literally *does not see bad moves* when he looks at a chess situation—no more than chess amateurs see *illegal* moves when they look at a chess situation. Anyone who has played even a little chess has organized his perception so that diagonal rook-moves, forward captures by pawns, and so forth, are never brought to mind. Similarly, master-level players have built up higher levels of organization in the way they see the board; consequently, to them, bad moves are as unlikely to come to mind as illegal moves are, to most people. This might be called *implicit pruning* of the giant branching tree of possibilities. By contrast, *explicit pruning* would involve thinking of a move, and after superficial examination, deciding not to pursue examining it any further.

The distinction can apply just as well to other intellectual activities—for instance, doing mathematics. A gifted mathematician doesn't usually think up and try out all sorts of false pathways to the desired theorem, as less gifted people might do; rather, he just "smells" the promising paths, and takes them immediately.

Computer chess programs which rely on looking ahead have not been taught to think on a higher level; the strategy has just been to use brute force look-ahead, hoping to crush all types of opposition. But it has not worked. Perhaps someday, a look-ahead program with enough brute force will indeed overcome the best human players—but that will be a small intellectual gain, compared to the revelation that intelligence depends *crucially* on the ability to create high-level descriptions of complex arrays, such as chess boards, television screens, printed pages, or paintings.

Similar Levels

Usually, we are not required to hold more than one level of understanding of a situation in our minds at once. Moreover, the different descriptions of a single system are usually so conceptually distant from each other that, as was mentioned earlier, there is no problem in maintaining them both; they are just maintained in separate mental compartments. What *is* confusing, though, is when a single system admits of two or more descriptions on different levels which nevertheless *resemble* each other in some way. Then we find it hard to avoid mixing levels when we think about the system, and can easily get totally lost.

Undoubtedly this happens when we think about our own psychology—for instance, when we try to understand people's motivations for various actions. There are many levels in the human mental structure—certainly it is a system which we do not understand very well yet. But there are hundreds of rival theories which tell why people act the way they do, each theory based on some underlying assumptions about how far down in this set of levels various kinds of psychological "forces" are found. Since at this time we use pretty much the same kind of language for all mental levels, this makes for much level-mixing and most certainly for hundreds of wrong theories. For instance, we talk of "drives"—for sex, for power, for fame, for love, etc., etc.—without knowing where these drives come from in the human mental structure. Without belaboring the point, I simply wish to say that our confusion about who we are is certainly related to the fact that we consist of a large set of levels, and we use overlapping language to describe ourselves on all of those levels.

Computer Systems

There is another place where many levels of description coexist for a single system, and where all the levels are conceptually quite close to one another. I am referring to computer systems. When a computer program is running, it can be viewed on a number of levels. On each level, the description is given in the language of computer science, which makes all the descriptions similar in some ways to each other—yet there are extremely important differences between the views one gets on the different levels. At the lowest level, the description can be so complicated that it is like the dot-description of a television picture. For some purposes, however, this is by far the most important view. At the highest level, the description is greatly *chunked,* and takes on a completely different feel, despite the fact that many of the same concepts appear on the lowest and highest levels. The chunks on the high-level description are like the chess expert's chunks, and like the chunked description of the image on the screen: they summarize in capsule form a number of things which on lower levels are seen as separate. (See Fig. 57.) Now before things become too abstract, let us pass on to the

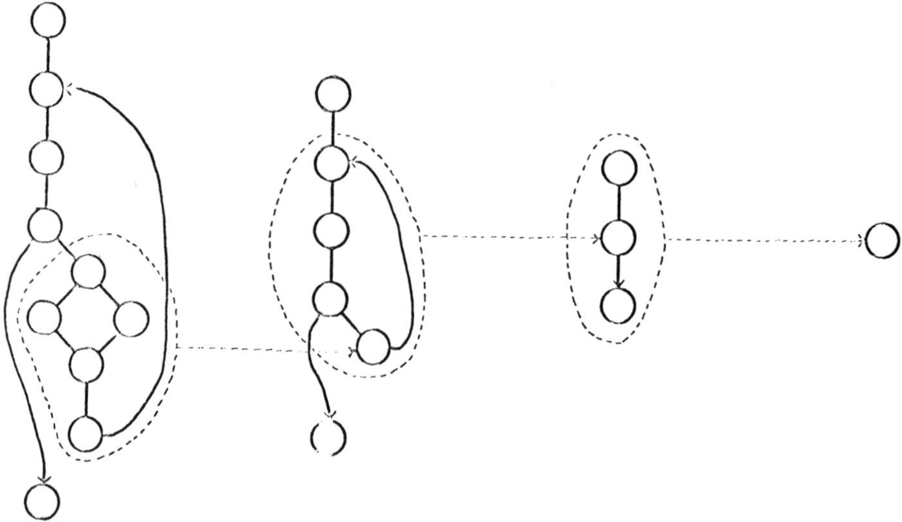

FIGURE 57. The idea of "chunking": a group of items is reperceived as a single "chunk". The chunk's boundary is a little like a cell membrane or a national border: it establishes a separate identity for the cluster within. According to context, one may wish to ignore the chunk's internal structure or to take it into account.

concrete facts about computers, beginning with a very quick skim of what a computer system is like on the lowest level. The lowest level? Well, not really, for I am not going to talk about elementary particles—but it is the lowest level which we wish to think about.

At the conceptual rock-bottom of a computer, we find a *memory,* a *central processing unit* (CPU), and some *input-output* (I/O) *devices.* Let us first describe the memory. It is divided up into distinct physical pieces, called *words.* For the sake of concreteness, let us say there are 65,536 words of memory (a typical number, being 2 to the 16th power). A word is further divided into what we shall consider the atoms of computer science—*bits.* The number of bits in a typical word might be around thirty-six. Physically, a bit is just a magnetic "switch" that can be in either of two positions.

O	O	X	O	X	X	X	O	X	O	O	X	X	O	O	X	O	X	X	X	X	X	X	O	X	X	O	O	X	X	X	O	O	O	O	O

--- *a word of 36 bits* ---

You could call the two positions "up" and "down", or "x" and "o", or "1" and "0" . . . The third is the usual convention. It is perfectly fine, but it has the possibly misleading effect of making people think that a computer, deep down, is storing *numbers*. This is not true. A set of thirty-six bits does not have to be thought of as a number any more than two bits has to be thought of as the price of an ice cream cone. Just as money can do various things depending on how you use it, so a word in memory can serve many functions. Sometimes, to be sure, those thirty-six bits will indeed represent a number in binary notation. Other times, they may represent thirty-six dots on a television screen. And other times, they may represent a few letters of text. How a word in memory is to be thought of depends entirely on the role that this word plays in the program which uses it. It may, of course, play more than one role—like a note in a canon.

Instructions and Data

There is one interpretation of a word which I haven't yet mentioned, and that is as an *instruction*. The words of memory contain not only data to be acted on, but also the program to act on the data. There exists a limited repertoire of operations which can be carried out by the central processing unit—the CPU—and part of a word, usually its first several bits—is interpretable as the name of the instruction-type which is to be carried out. What do the rest of the bits in a word-interpreted-as-instruction stand for? Most often, they tell which other words in memory are to be acted upon. In other words, the remaining bits constitute a *pointer* to some other word (or words) in memory. Every word in memory has a distinct location, like a house on a street; and its location is called its *address*. Memory may have one "street", or many "streets"—they are called "pages". So a given word is addressed by its page number (if memory is paged) together with its position within the page. Hence the "pointer" part of an instruction is the numerical address of some word(s) in memory. There are no restrictions on the pointer, so an instruction may even "point" at itself, so that when it is executed, it causes a change in itself to be made.

How does the computer know what instruction to execute at any given time? This is kept track of in the CPU. The CPU has a special pointer which points at (i.e., stores the address of) the next word which is to be interpreted as an instruction. The CPU fetches that word from memory, and copies it electronically into a special word belonging to the CPU itself. (Words in the CPU are usually not called "words", but rather, *registers*.) Then the CPU executes that instruction. Now the instruction may call for any of a large number of types of operations to be carried out. Typical ones include:

ADD the word pointed to in the instruction, to a register.
(In this case, the word pointed to is obviously interpreted as a number.)

PRINT the word pointed to in the instruction, as letters.
(In this case, the word is obviously interpreted *not* as a number, but as a string of letters.)

JUMP to the word pointed to in the instruction.
(In this case, the CPU is being told to interpret that particular word as its next instruction.)

Unless the instruction explicitly dictates otherwise, the CPU will pick up the very next word and interpret it as an instruction. In other words, the CPU assumes that it should move down the "street" sequentially, like a mailman, interpreting word after word as an instruction. But this sequential order can be broken by such instructions as the **JUMP** instruction, and others.

Machine Language *vs.* Assembly language

This is a very brief sketch of *machine language.* In this language, the types of operations which exist constitute a finite repertoire which cannot be extended. Thus all programs, no matter how large and complex, must be made out of compounds of these types of instructions. Looking at a program written in machine language is vaguely comparable to looking at a DNA molecule atom by atom. If you glance back to Fig. 41, showing the nucleotide sequence of a DNA molecule—and then if you consider that each nucleotide contains two dozen atoms or so—and if you imagine trying to write the DNA, atom by atom, for a small virus (not to mention a human being!)—then you will get a feeling for what it is like to write a complex program in machine language, and what it is like to try to grasp what is going on in a program if you have access only to its machine language description.

It must be mentioned, however, that computer programming was originally done on an even lower level, if possible, than that of machine language—namely, connecting wires to each other, so that the proper operations were "hard-wired" in. This is so amazingly primitive by modern standards that it is painful even to imagine. Yet undoubtedly the people who first did it experienced as much exhilaration as the pioneers of modern computers ever do . . .

We now wish to move to a higher level of the hierarchy of levels of description of programs. This is the *assembly language* level. There is not a gigantic spread between assembly language and machine language; indeed, the step is rather gentle. In essence, there is a one-to-one correspondence between assembly language instructions and machine language instructions. The idea of assembly language is to "chunk" the individual machine language instructions, so that instead of writing the sequence of bits "010111000" when you want an instruction which adds one number to another, you simply write **ADD**, and then instead of giving the address in binary representation, you can refer to the word in memory by a *name.*

Therefore, a program in assembly language is very much like a machine language program made legible to humans. You might compare the machine language version of a program to a TNT-derivation done in the obscure Gödel-numbered notation, and the assembly language version to the isomorphic TNT-derivation, done in the original TNT-notation, which is much easier to understand. Or, going back to the DNA image, we can liken the difference between machine language and assembly language to the difference between painfully specifying each nucleotide, atom by atom, and specifying a nucleotide by simply giving its *name* (i.e., 'A', 'G', 'C', or 'T'). There is a tremendous saving of labor in this very simple "chunking" operation, although conceptually not much has been changed.

Programs That Translate Programs

Perhaps the central point about assembly language is not its differences from machine language, which are not that enormous, but just the key idea that programs could be written on a different level *at all!* Just think about it: the hardware is built to "understand" machine language programs—sequences of bits—but not letters and decimal numbers. What happens when hardware is fed a program in assembly language? It is as if you tried to get a cell to accept a piece of paper with the nucleotide sequence written out in letters of the alphabet, instead of in chemicals. What can a cell do with a piece of paper? What can a computer do with an assembly language program?

And here is the vital point: someone can write, in machine language, a *translation program.* This program, called an *assembler,* accepts mnemonic instruction names, decimal numbers, and other convenient abbreviations which a programmer can remember easily, and carries out the conversion into the monotonous but critical bit-sequences. After the assembly language program has been *assembled* (i.e., translated), it is *run*—or rather, its machine language equivalent is run. But this is a matter of terminology. Which level program is running? You can never go wrong if you say that the machine language program is running, for hardware is always involved when any program runs—but it is also quite reasonable to think of the running program in terms of assembly language. For instance, you might very well say, "Right now, the CPU is executing a JUMP instruction", instead of saying, "Right now, the CPU is executing a '111010000' instruction". A pianist who plays the notes G-E-B E-G-B is also playing an arpeggio in the chord of E minor. There is no reason to be reluctant about describing things from a higher-level point of view. So one can think of the assembly language program running concurrently with the machine language program. We have two modes of describing what the CPU is doing.

Higher-Level Languages, Compilers, and Interpreters

The next level of the hierarchy carries much further the extremely powerful idea of using the computer itself to translate programs from a high level into lower levels. After people had programmed in assembly language for a number of years, in the early 1950's, they realized that there were a number of characteristic structures which kept reappearing in program after program. There seemed to be, just as in chess, certain fundamental patterns which cropped up naturally when human beings tried to formulate *algorithms*—exact descriptions of processes they wanted carried out. In other words, algorithms seemed to have certain higher-level components, in terms of which they could be much more easily and esthetically specified than in the very restricted machine language, or assembly language. Typically, a high-level algorithm component consists not of one or two machine language instructions, but of a whole collection of them, not necesssarily all contiguous in memory. Such a component could be represented in a higher-level language by a single item—a chunk.

Aside from standard chunks—the newly discovered components out of which all algorithms can be built—people realized that almost all programs contain even larger chunks—superchunks, so to speak. These superchunks differ from program to program, depending on the kinds of high-level tasks the program is supposed to carry out. We discussed superchunks in Chapter V, calling them by their usual names: "subroutines" and "procedures". It was clear that a most powerful addition to any programming language would be the ability to *define* new higher-level entities in terms of previously known ones, and then to *call* them by name. This would build the chunking operation right into the language. Instead of there being a determinate repertoire of instructions out of which all programs had to be explicitly assembled, the programmer could construct his own modules, each with its own name, each usable anywhere inside the program, just as if it had been a built-in feature of the language. Of course, there is no getting away from the fact that down below, on a machine language level, everything would still be composed of the same old machine language instructions, but that would not be explicitly visible to the high-level programmer; it would be implicit.

The new languages based on these ideas were called *compiler languages*. One of the earliest and most elegant was called "Algol", for "Algorithmic Language". Unlike the case with assembly language, there is no straightforward one-to-one correspondence between statements in Algol and machine language instructions. To be sure, there is still a type of mapping from Algol into machine language, but it is far more "scrambled" than that between assembly language and machine language. Roughly speaking, an Algol program is to its machine language translation as a word problem in an elementary algebra text is to the equation it translates into. (Actually, getting from a word problem to an equation is far more complex, but it gives some inkling of the types of "unscrambling" that have to be carried out in translating from a high-level language to a lower-level language.) In the mid-1950's, successful programs called *compilers* were written whose function was to carry out the translation from compiler languages to machine language.

Also, *interpreters* were invented. Like compilers, interpreters translate from high-level languages into machine language, but instead of translating all the statements first and then executing the machine code, they read one line and execute it immediately. This has the advantage that a user need not have written a complete program to use an interpreter. He may invent his program line by line, and test it out as he goes along. Thus, an interpreter is to a compiler as a simultaneous interpreter is to a translator of a written speech. One of the most important and fascinating of all computer languages is LISP (standing for "List Processing"), which was invented by John McCarthy around the time Algol was invented. Subsequently, LISP has enjoyed great popularity with workers in Artificial Intelligence.

There is one interesting difference between the way interpreters work and compilers work. A compiler takes input (a finished Algol program, for instance) and produces output (a long sequence of machine language instructions). At this point, the compiler has done its duty. The output is then given to the computer to run. By contrast, the interpreter is constantly running while the programmer types in one LISP statement after another, and each one gets executed then and there. But this doesn't mean that each statement gets first translated, then executed, for then an interpreter would be nothing but a line-by-line compiler. Instead, in an interpreter, the operations of reading a new line, "understanding" it, and executing it are intertwined: they occur simultaneously.

Here is the idea, expanded a little more. Each time a new line of LISP is typed in, the interpreter tries to process it. This means that the interpreter jolts into action, and certain (machine language) instructions inside it get executed. Precisely *which* ones get executed depends on the LISP statement itself, of course. There are many JUMP instructions inside the interpreter, so that the new line of LISP may cause control to move around in a complex way—forwards, backwards, then forwards again, etc. Thus, each LISP statement gets converted into a "pathway" inside the interpreter, and the act of following that pathway achieves the desired effect.

Sometimes it is helpful to think of the LISP statements as mere pieces of data which are fed sequentially to a constantly running machine language program (the LISP interpreter). When you think of things this way, you get a different image of the relation between a program written in a higher-level language and the machine which is executing it.

Bootstrapping

Of course a compiler, being itself a program, has to be written in some language. The first compilers were written in assembly language, rather than machine language, thus taking full advantage of the already accomplished first step up from machine language. A summary of these rather tricky concepts is presented in Figure 58.

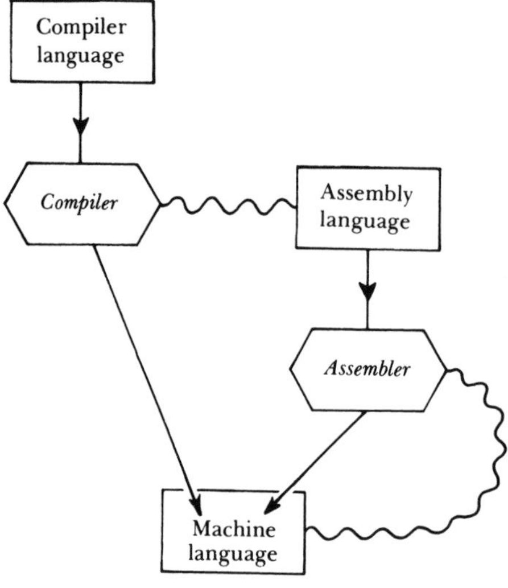

FIGURE 58. Assemblers and compilers are both translators into machine language. This is indicated by the solid lines. Moreover, since they are themselves pro-grams, *they are originally written in a language also. The wavy lines indicate that a compiler can be written in assembly language, and an assembler in machine language.*

Now as sophistication increased, people realized that a partially written compiler could be used to compile extensions of itself. In other words, once a certain minimal core of a compiler had been written, then that minimal compiler could translate bigger compilers into machine language—which in turn could translate yet bigger compilers, until the final, full-blown compiler had been compiled. This process is affectionately known as "bootstrapping"—for obvious reasons (at least if your native language is English it is obvious). It is not so different from the attainment by a child of a critical level of fluency in his native language, from which point on his vocabulary and fluency can grow by leaps and bounds, since he can *use* language to *acquire* new language.

Levels on Which to Describe Running Programs

Compiler languages typically do not reflect the structure of the machines which will run programs written in them. This is one of their chief advantages over the highly specialized assembly and machine languages. Of course, when a compiler language program is translated into machine language, the resulting program is machine-dependent. Therefore one can describe a program which is being executed in a machine-independent way or a machine-dependent way. It is like referring to a paragraph in a book by its subject matter (publisher-independent), or its page number and position on the page (publisher-dependent).

As long as a program is running correctly, it hardly matters how you describe it or think of its functioning. It is when something goes wrong that it is important to be able to think on different levels. If, for instance, the machine is instructed to divide by zero at some stage, it will come to a halt and let the user know of this problem, by telling where in the program the questionable event occurred. However, the specification is often given on a lower level than that in which the programmer wrote the program. Here are three parallel descriptions of a program grinding to a halt:

Machine Language Level:
"Execution of the program stopped in location 1110010101110111"

Assembly Language Level:
"Execution of the program stopped when the DIV (divide) instruction was hit"

Compiler Language Level:
"Execution of the program stopped during evaluation of the algebraic expression '(A + B)/Z' "

One of the greatest problems for systems programmers (the people who write compilers, interpreters, assemblers, and other programs to be used by many people) is to figure out how to write error-detecting routines in such a way that the messages which they feed to the user whose program has a "bug" provide high-level, rather than low-level, descriptions of the problem. It is an interesting reversal that when something goes wrong in a genetic "program" (e.g., a mutation), the "bug" is manifest only to people on a *high* level—namely on the phenotype level, not the genotype level. Actually, modern biology uses mutations as one of its principal windows onto genetic processes, because of their multilevel traceability.

Microprogramming and Operating Systems

In modern computer systems, there are several other levels of the hierarchy. For instance, some systems—often the so-called "microcomputers"—come with machine language instructions which are even more rudimentary than the instruction to add a number in memory to a number in a register. It is up to the user to decide what kinds of ordinary machine-level instructions he would like to be able to program in; he "microprograms" these instructions in terms of the "micro-instructions" which are available. Then the "higher-level machine language" instructions which he has designed may be burned into the circuitry and become hard-wired, although they need not be. Thus microprogramming allows the user to step a little below the conventional machine language level. One of the consequences is that a computer of one manufacturer can be hard-wired (via microprogramming) so as to have the same machine language instruction set as a computer of the same, or even another, manufacturer. The microprogrammed computer is said to be "emulating" the other computer.

Then there is the level of the *operating system,* which fits between the machine language program and whatever higher level the user is programming in. The operating system is itself a program which has the functions of shielding the bare machine from access by users (thus protecting the system), and also of insulating the programmer from the many extremely intricate and messy problems of reading the program, calling a translator, running the translated program, directing the output to the proper channels at the proper time, and passing control to the next user. If there are several users "talking" to the same CPU at once, then the operating system is the program that shifts attention from one to the other in some orderly fashion. The complexities of operating systems are formidable indeed, and I shall only hint at them by the following analogy.

Consider the first telephone system. Alexander Graham Bell could phone his assistant in the next room: electronic transmission of a voice! Now that is like a bare computer minus operating system: electronic computation! Consider now a modern telephone system. You have a choice of other telephones to connect to. Not only that, but many different calls can be handled simultaneously. You can add a prefix and dial into different areas. You can call direct, through the operator, collect, by credit card, person-to-person, on a conference call. You can have a call rerouted or traced. You can get a busy signal. You can get a siren-like signal that says that the number you dialed isn't "well-formed", or that you have taken too in long in dialing. You can install a local switchboard so that a group of phones are all locally connected—etc., etc. The list is amazing, when you think of how much flexibility there is, particularly in comparison to the erstwhile miracle of a "bare" telephone. Now sophisticated operating systems carry out similar traffic-handling and level-switching operations with respect to users and their programs. It is virtually certain that there are somewhat parallel things which take place in the brain: handling of many stimuli at the same time; decisions of what should have priority over what and for how long; instantaneous "interrupts" caused by emergencies or other unexpected occurrences; and so on.

Cushioning the User and Protecting the System

The many levels in a complex computer system have the combined effect of "cushioning" the user, preventing him from having to think about the many lower-level goings-on which are most likely totally irrelevant to him anyway. A passenger in an airplane does not usually want to be aware of the levels of fuel in the tanks, or the wind speeds, or how many chicken dinners are to be served, or the status of the rest of the air traffic around the destination—this is all left to employees on different levels of the airlines hierarchy, and the passenger simply gets from one place to another. Here again, it is when something goes *wrong*—such as his baggage not arriving—that the passenger is made aware of the confusing system of levels underneath him.

Are Computers Super-Flexible or Super-Rigid?

One of the major goals of the drive to higher levels has always been to make as natural as possible the task of communicating to the computer what you want it to do. Certainly, the high-level constructs in compiler languages are closer to the concepts which humans naturally think in, than are lower-level constructs such as those in machine language. But in this drive towards ease of communication, one aspect of "naturalness" has been quite neglected. That is the fact that interhuman communication is far less rigidly constrained than human-machine communication. For instance, we often produce meaningless sentence fragments as we search for the best way to express something, we cough in the middle of sentences, we interrupt each other, we use ambiguous descriptions and "improper" syntax, we coin phrases and distort meanings—but our message still gets through, mostly. With programming languages, it has generally been the rule that there is a very strict syntax which has to be obeyed one hundred per cent of the time; there are no ambiguous words or constructions. Interestingly, the printed equivalent of coughing (i.e., a nonessential or irrelevant comment) is allowed, but only provided it is signaled in advance by a key word (e.g., COMMENT), and then terminated by another key word (e.g., a semicolon). This small gesture towards flexibility has its own little pitfall, ironically: if a semicolon (or whatever key word is used for terminating a comment) is used inside a comment, the translating program will interpret that semicolon as signaling the end of the comment, and havoc will ensue.

If a procedure named INSIGHT has been defined and then called seventeen times in the program, and the eighteenth time it is misspelled as INSIHGT, woe to the programmer. The compiler will balk and print a rigidly unsympathetic error message, saying that it has never heard of INSIHGT. Often, when such an error is detected by a compiler, the compiler tries to continue, but because of its lack of insihgt, it has not understood what the programmer meant. In fact, it may very well suppose that something entirely different was meant, and proceed under that erroneous assumption. Then a long series of error messages will pepper the rest of the program, because the compiler—not the programmer—got confused. Imagine the chaos that would result if a simultaneous English-Russian interpreter, upon hearing one phrase of French in the English, began trying to interpret all the remaining English as French. Compilers often get lost in such pathetic ways. *C'est la vie.*

Perhaps this sounds condemnatory of computers, but it is not meant to be. In some sense, things had to be that way. When you stop to think what most people use computers for, you realize that it is to carry out very definite and precise tasks, which are too complex for people to do. If the computer is to be reliable, then it is necessary that it should understand, without the slightest chance of ambiguity, what it is supposed to do. It is also necessary that it should do neither more nor less than it is explicitly instructed to do. If there is, in the cushion underneath the programmer, a program whose purpose is to "guess" what the programmer wants or means, then it is quite conceivable that the programmer could try to communicate his task and be totally misunderstood. So it is important that the high-level program, while comfortable for the human, still should be unambiguous and precise.

Second-Guessing the Programmer

Now it is possible to devise a programming language—and a program which translates it into the lower levels—which allows some sorts of imprecision. One way of putting it would be to say that a translator for such a programming language tries to make sense of things which are done "outside of the rules of the language". But if a language allows certain "transgressions", then transgressions of that type are no longer true transgressions, because they have been included inside the rules! If a programmer is aware that he may make certain types of misspelling, then he may use this feature of the language deliberately, knowing that he is actually operating within the rigid rules of the language, despite appearances. In other words, if the user is aware of all the flexibilities programmed into the translator for his convenience, then he knows the bounds which he cannot overstep, and therefore, to him, the translator still appears rigid and inflexible, although it may allow him much more freedom than early versions of the language, which did not incorporate "automatic compensation for human error".

With "rubbery" languages of that type, there would seem to be two alternatives: (1) the user is aware of the built-in flexibilities of the language and its translator; (2) the user is unaware of them. In the first case, the language is still usable for communicating programs precisely, because the programmer can predict how the computer will interpret the programs he writes in the language. In the second case, the "cushion" has hidden features which may do things that are unpredictable (from the vantage point of a user who doesn't know the inner workings of the translator). This may result in gross misinterpretations of programs, so such a language is unsuitable for purposes where computers are used mainly for their speed and reliability.

Now there is actually a third alternative: (3) the user is aware of the built-in flexibilities of the language and its translator, but there are so many of them and they interact with each other in such a complex way that he cannot tell how his programs will be interpreted. This may well apply to the person who wrote the translating program; he certainly knows its insides as well as anyone could—but he still may not be able to anticipate how it will react to a given type of unusual construction.

One of the major areas of research in Artificial Intelligence today is called *automatic programming*, which is concerned with the development of yet higher-level languages—languages whose translators are sophisticated, in that they can do at least some of the following impressive things: generalize from examples, correct some misprints or grammatical errors, try to make sense of ambiguous descriptions, try to second-guess the user by having a primitive user model, ask questions when things are unclear, use English itself, etc. The hope is that one can walk the tightrope between reliability and flexibility.

AI Advances Are Language Advances

It is striking how tight the connection is between progress in computer science (particularly Artificial Intelligence) and the development of new languages. A clear trend has emerged in the last decade: the trend to consolidate new types of discoveries in new languages. One key for the understanding and creation of intelligence lies in the constant development and refinement of the languages in terms of which processes for symbol manipulation are describable. Today, there are probably three or four dozen experimental languages which have been developed exclusively for Artificial Intelligence research. It is important to realize that any program which can be written in one of these languages is in principle programmable in lower-level languages, but it would require a supreme effort for a human; and the resulting program would be so long that it would exceed the grasp of humans. It is not that each higher level extends the potential of the computer; the full potential of the computer already exists in its machine language instruction set. It is that the new concepts in a high-level language suggest directions and perspectives by their very nature.

The "space" of all possible programs is so huge that no one can have a sense of what is possible. Each higher-level language is naturally suited for exploring certain regions of "program space"; thus the programmer, by using that language, is channeled into those areas of program space. He is not *forced* by the language into writing programs of any particular type, but the language makes it *easy* for him to do certain kinds of things. Proximity to a concept, and a gentle shove, are often all that is needed for a major discovery—and that is the reason for the drive towards languages of ever higher levels.

Programming in different languages is like composing pieces in different keys, particularly if you work at the keyboard. If you have learned or written pieces in many keys, each key will have its own special emotional aura. Also, certain kinds of figurations "lie in the hand" in one key but are awkward in another. So you are channeled by your choice of key. In some ways, even enharmonic keys, such as C-sharp and D-flat, are quite distinct in feeling. This shows how a notational system can play a significant role in shaping the final product.

A "stratified" picture of AI is shown in Figure 59, with machine components such as transistors on the bottom, and "intelligent programs" on the top. The picture is taken from the book *Artificial Intelligence* by Patrick Henry Winston, and it represents a vision of AI shared by nearly all AI workers. Although I agree with the idea that AI must be stratified in some such way, I do not think that, with so few layers, intelligent programs

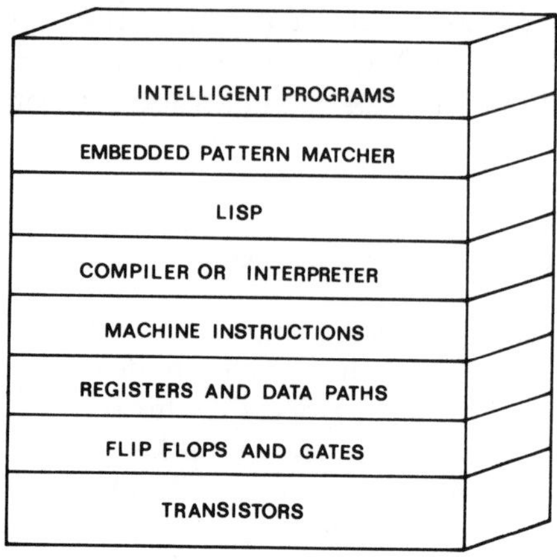

FIGURE 59. To create intelligent programs, one needs to build up a series of levels of hardware and software, so that one is spared the agony of seeing everything only on the lowest level. Descriptions of a single process on different levels will sound very different from each other, only the top one being sufficiently chunked that it is comprehensible to us. [Adapted from P. H. Winston, Artificial Intelligence *(Reading, Mass.: Addison-Wesley, 1977).]*

can be reached. Between the machine language level and the level where true intelligence will be reached, I am convinced there will lie perhaps another dozen (or even several dozen!) layers, each new layer building on and extending the flexibilities of the layer below. What they will be like we can hardly dream of now . . .

The Paranoid and the Operating System

The similarity of all levels in a computer system can lead to some strange level-mixing experiences. I once watched a couple of friends—both computer novices—playing with the program "PARRY" on a terminal. PARRY is a rather infamous program which simulates a paranoid in an extremely rudimentary way, by spitting out canned phrases in English chosen from a wide repertoire; its plausibility is due to its ability to tell which of its stock phrases might sound reasonable in response to English sentences typed to it by a human.

At one point, the response time got very sluggish—PARRY was taking very long to reply—and I explained to my friends that this was probably because of the heavy load on the time-sharing system. I told them they could find out how many users were logged on, by typing a special "control" character which would go directly to the operating system, and would be unseen by PARRY. One of my friends pushed the control character. In a flash, some internal data about the operating system's status overwrote some of PARRY's words on the screen. PARRY knew nothing of this: it is a program with "knowledge" only of horse racing and bookies—not operating systems and terminals and special control characters. But to my friends, both PARRY and the operating system were just "the computer"—a mysterious, remote, amorphous entity that responded to them when they typed. And so it made perfect sense when one of them blithely typed, in English, "Why are you overtyping what's on the screen?" The idea that PARRY could know nothing about the operating system it was running under was not clear to my friends. The idea that "you" know all about "yourself" is so familiar from interaction with people that it was natural to extend it to the computer—after all, it was intelligent enough that it could "talk" to them in English! Their question was not unlike asking a person, "Why are you making so few red blood cells today?" People do not know about that level—the "operating system level"—of their bodies.

The main cause of this level-confusion was that communication with all levels of the computer system was taking place on a single screen, on a single terminal. Although my friends' naïveté might seem rather extreme, even experienced computer people often make similar errors when several levels of a complex system are all present at once on the same screen. They forget "who" they are talking to, and type something which makes no sense at that level, although it would have made perfect sense on another level. It might seem desirable, therefore, to have the system itself sort out the levels—to interpret commands according to what "makes sense". Unfortunately, such interpretation would require the system to have a lot of common sense, as well as perfect knowledge of the programmer's overall intent—both of which would require more artificial intelligence than exists at the present time.

The Border between Software and Hardware

One can also be confused by the flexibility of some levels and the rigidity of others. For instance, on some computers there are marvelous text-editing systems which allow pieces of text to be "poured" from one format into another, practically as liquids can be poured from one vessel into another. A thin page can turn into a wide page, or vice versa. With such power, you might expect that it would be equally trivial to change from one font to another—say from roman to *italics.* Yet there may be only a single font available on the screen, so that such changes are impossible. Or it may be feasible on the screen but not printable by the printer—or the other way around. After dealing with computers for a long time, one gets spoiled, and thinks that everything should be programmable: no printer should be so rigid as to have only one character set, or even a finite repertoire of them—typefaces should be user-specifiable! But once that degree of flexibility has been attained, then one may be annoyed that the printer cannot print in different colors of ink, or that it cannot accept paper of all shapes and sizes, or that it does not fix itself when it breaks . . .

The trouble is that somewhere, all this flexibility has to "bottom out", to use the phrase from Chapter V. There must be a hardware level which underlies it all, and which is inflexible. It may lie deeply hidden, and there may be so much flexibility on levels above it that few users feel the hardware limitations—but it is inevitably there.

What is this proverbial distinction between *software* and *hardware*? It is the distinction between programs and machines—between long complicated sequences of instructions, and the physical machines which carry them out. I like to think of software as "anything which you could send over the telephone lines", and hardware as "anything else". A piano is hardware, but printed music is software. A telephone set is hardware, but a telephone number is software. The distinction is a useful one, but not always so clear-cut.

We humans also have "software" and "hardware" aspects, and the difference is second nature to us. We are used to the rigidity of our physiology: the fact that we cannot, at will, cure ourselves of diseases, or grow hair of any color—to mention just a couple of simple examples. We can, however, "reprogram" our minds so that we operate in new conceptual frameworks. The amazing flexibility of our minds seems nearly irreconcilable with the notion that our brains must be made out of fixed-rule hardware, which cannot be reprogrammed. We cannot make our neurons fire faster or slower, we cannot rewire our brains, we cannot redesign the interior of a neuron, we cannot make *any* choices about the hardware—and yet, we can control how we think.

But there are clearly aspects of thought which are beyond our control. We cannot make ourselves smarter by an act of will; we cannot learn a new language as fast as we want; we cannot make ourselves think faster than we do; we cannot make ourselves think about several things at once; and so on. This is a kind of primordial self-knowledge which is so obvious that it is hard to see it at all; it is like being conscious that the air is there. We never really bother to think about what might cause these "defects" of our minds: namely, the organization of our brains. To suggest ways of reconciling the software of mind with the hardware of brain is a main goal of this book.

Intermediate Levels and the Weather

We have seen that in computer systems, there are a number of rather sharply defined strata, in terms of any one of which the operation of a running program can be described. Thus there is not merely a single low level and a single high level—there are all degrees of lowness and highness. Is the existence of intermediate levels a general feature of systems which have low and high levels? Consider, for example, the system whose "hardware" is the earth's atmosphere (not very hard, but no matter), and whose "software" is the weather. Keeping track of the motions of all of the molecules simultaneously would be a very low-level way of "understanding" the weather, rather like looking at a huge, complicated program on the machine language level. Obviously it is way beyond human comprehension. But we still have our own peculiarly human ways of looking at, and describing, weather phenomena. Our chunked view of the weather is based on very high-level phenomena, such as: rain, fog, snow, hurricanes, cold fronts, seasons, pressures, trade winds, the jet stream, cumulo-nimbus clouds, thunderstorms, inversion layers, and so on. All of these phenomena involve astronomical numbers of molecules, somehow behaving in concert so that large-scale trends emerge. This is a little like looking at the weather in a compiler language.

Is there something analogous to looking at the weather in an intermediate-level language, such as assembly language? For instance, are there very small local "mini-storms", something like the small whirlwinds which one occasionally sees, whipping up some dust in a swirling column a few feet wide, at most? Is a local gust of wind an intermediate-level chunk which plays a role in creating higher-level weather phenomena? Or is there just no practical way of combining knowledge of such kinds of phenomena to create a more comprehensive explanation of the weather?

Two other questions come to my mind. The first is: "Could it be that the weather phenomena which we perceive on our scale—a tornado, a drought—are just intermediate-level phenomena: parts of vaster, slower phenomena?" If so, then true high-level weather phenomena would be global, and their time scale would be geological. The Ice Age would be a high-level weather event. The second question is: "Are there intermediate-level weather phenomena which have so far escaped human perception, but which, if perceived, could give greater insight into why the weather is as it is?"

From Tornados to Quarks

This last suggestion may sound fanciful, but it is not all that far-fetched. We need only look to the hardest of the hard sciences—physics—to find peculiar examples of systems which are explained in terms of interacting "parts" which are themselves invisible. In physics, as in any other discipline, a *system* is a group of interacting *parts*. In most systems that we know, the parts retain their identities during the interaction, so that we still see the parts inside the system. For example, when a team of football players assembles, the individual players retain their separateness—they do not melt into some composite entity, in which their individuality is lost. Still—and this is important—some processes are going on in their brains which are evoked by the team-context, and which would not go on otherwise, so that in a minor way, the players change identity when they become part of the larger system, the team. This kind of system is called a *nearly decomposable system* (the term comes from H. A. Simon's article "The Architecture of Complexity"; see the Bibliography). Such a system consists of weakly interacting modules, each of which maintains its own private identity throughout the interaction but by becoming slightly different from how it is when outside of the system, contributes to the cohesive behavior of the whole system. The systems studied in physics are usually of this type. For instance, an atom is seen as made of a nucleus whose positive charge captures a number of electrons in "orbits", or bound states. The bound electrons are very much like free electrons, despite their being internal to a composite object.

Some systems studied in physics offer a contrast to the relatively straightforward atom. Such systems involve extremely strong interactions, as a result of which the parts are swallowed up into the larger system, and lose some or all of their individuality. An example of this is the nucleus of an atom, which is usually described as being "a collection of protons and neutrons". But the forces which pull the component particles together are so strong that the component particles do not survive in anything like their "free" form (the form they have when outside a nucleus). And in fact a nucleus acts in many ways as a single particle, rather than as a collection of interacting particles. When a nucleus is split, protons and neutrons are often released, but also other particles, such as pi-mesons and gamma rays, are commonly produced. Are all those different particles physically present inside a nucleus before it is split, or are they just "sparks" which fly off when the nucleus is split? It is perhaps not meaningful to try to give an answer to such a question. On the level of particle physics, the difference between storing the potential to make "sparks" and storing actual subparticles is not so clear.

A nucleus is thus one system whose "parts", even though they are not visible while on the inside, can be pulled out and made visible. However, there are more pathological cases, such as the proton and neutron seen as systems themselves. Each of them has been hypothesized to be constituted from a trio of "quarks"—hypothetical particles which can be combined in twos or threes to make many known fundamental particles. However, the interaction between quarks is so strong that not only can they not be seen inside the proton and neutron, but they cannot even be pulled out at all! Thus, although quarks help to give a theoretical understanding of certain properties of protons and neutrons, their own existence may perhaps never be independently established. Here we have the antithesis of a "nearly decomposable system"—it is a system which, if anything, is "nearly indecomposable". Yet what is curious is that a quark-based theory of protons and neutrons (and other particles) has considerable explanatory power, in that many experimental results concerning the particles which quarks supposedly compose can be accounted for quite well, quantitatively, by using the "quark model".

Superconductivity: A "Paradox" of Renormalization

In Chapter V we discussed how renormalized particles emerge from their bare cores, by recursively compounded interactions with virtual particles. A renormalized particle can be seen either as this complex mathematical construct, or as the single lump which it is, physically. One of the strangest and most dramatic consequences of this way of describing particles is the explanation it provides for the famous phenomenon of *superconductivity:* resistance-free flow of electrons in certain solids, at extremely low temperatures.

It turns out that electrons in solids are renormalized by their interactions with strange quanta of vibration called *phonons* (themselves renormalized as well!). These renormalized electrons are called *polarons.* Calculation shows that at very low temperatures, two oppositely spinning polarons will begin to attract each other, and can actually become bound together in a certain way. Under the proper conditions, all the current-carrying polarons will get paired up, forming *Cooper pairs.* Ironically, this pairing comes about precisely because electrons—the bare cores of the paired polarons—repel each other electrically. In contrast to the electrons, each Cooper pair feels neither attracted to nor repelled by any other Cooper pair, and consequently it can can slip freely through a metal as if the metal were a vacuum. If you convert the mathematical description of such a metal from one whose primitive units are polarons into one whose primitive units are Cooper pairs, you get a considerably simplified set of equations. This mathematical simplicity is the physicist's way of knowing that "chunking" into Cooper pairs is the natural way to look at superconductivity.

Here we have several levels of particle: the Cooper pair itself; the two oppositely spinning polarons which compose it; the electrons and phonons which make up the polarons; and then, within the electrons, the virtual photons and positrons, etc. etc. We can look at each level and perceive phenomena there, which are explained by an understanding of the levels below.

"Sealing-off"

Similarly, and fortunately, one does not have to know all about quarks to understand many things about the particles which they may compose. Thus, a nuclear physicist can proceed with theories of nuclei that are based on protons and neutrons, and ignore quark theories and their rivals. The nuclear physicist has a *chunked* picture of protons and neutrons—a description derived from lower-level theories but which does not require understanding the lower-level theories. Likewise, an atomic physicist has a chunked picture of an atomic nucleus derived from nuclear theory. Then a chemist has a chunked picture of the electrons and their orbits, and builds theories of small molecules, theories which can be taken over in a chunked way by the molecular biologist, who has an intuition for how small molecules hang together, but whose technical expertise is in the field of extremely large molecules and how they interact. Then the cell biologist has a chunked picture of the units which the molecular biologist pores over, and tries to use them to account for the ways that cells interact. The point is clear. Each level is, in some sense, "sealed off" from the levels below it. This is another of Simon's vivid terms, recalling the way in which a submarine is built in compartments, so that if one part is damaged, and water begins pouring in, the trouble can be prevented from spreading, by closing the doors, thereby sealing off the damaged compartment from neighboring compartments.

Although there is always some "leakage" between the hierarchical levels of science, so that a chemist cannot afford to ignore lower-level physics totally, or a biologist to ignore chemistry totally, there is almost no leakage from one level to a distant level. That is why people can have intuitive understandings of other people without necessarily understanding the quark model, the structure of nuclei, the nature of electron orbits, the chemical bond, the structure of proteins, the organelles in a cell, the methods of intercellular communication, the physiology of the various organs of the human body, or the complex interactions among organs. All that a person needs is a chunked model of how the highest level acts; and as we all know, such models are very realistic and successful.

The Trade-off between Chunking and Determinism

There is, however, perhaps one significant negative feature of a chunked model: it usually does not have exact predictive power. That is, we save ourselves from the impossible task of seeing people as collections of quarks (or whatever is at the lowest level) by using chunked models; but of course such models only give us probabilistic estimates of how other people feel, will react to what we say or do, and so on. In short, in using chunked high-level models, we sacrifice determinism for simplicity. Despite not being sure how people will react to a joke, we tell it with the expectation that they will do something such as laugh, or not laugh—rather than, say, climb the nearest flagpole. (Zen masters might well do the latter!) A chunked model defines a "space" within which behavior is expected to fall, and specifies probabilities of its falling in different parts of that space.

"Computers Can Only Do What You Tell Them to Do"

Now these ideas can be applied as well to computer programs as to composite physical systems. There is an old saw which says, "Computers can only do what you tell them to do." This is right in one sense, but it misses the point: you don't know in advance the consequences of what you tell a computer to do; therefore its behavior can be as baffling and surprising and unpredictable to you as that of a person. You generally know in advance the *space* in which the output will fall, but you don't know details of where it will fall. For instance, you might write a program to calculate the first million digits of π. Your program will spew forth digits of π much faster than you can—but there is no paradox in the fact that the computer is outracing its programmer. You know in advance the space in which the output will lie—namely the space of digits between 0 and 9—which is to say, you have a chunked model of the program's behavior; but if you'd known the rest, you wouldn't have written the program.

There is another sense in which this old saw is rusty. This involves the fact that as you program in ever higher-level languages, you know less and less precisely what you've told the computer to do! Layers and layers of translation may separate the "front end" of a complex program from the actual machine language instructions. At the level you think and program, your statements may resemble declaratives and suggestions more than they resemble imperatives or commands. And all the internal rumbling provoked by the input of a high-level statement is invisible to you, generally, just as when you eat a sandwich, you are spared conscious awareness of the digestive processes that it triggers.

In any case, this notion that "computers can only do what they are told to do," first propounded by Lady Lovelace in her famous memoir, is so prevalent and so connected with the notion that "computers cannot think" that we shall return to it in later Chapters when our level of sophistication is greater.

Two Types of System

There is an important division between two types of system built up from many parts. There are those systems in which the behavior of some parts tends to *cancel out* the behavior of other parts, with the result that it does not matter too much what happens on the low level, because most anything will yield similar high-level behavior. An example of this kind of system is a container of gas, where all the molecules bump and bang against each other in very complex microscopic ways; but the total outcome, from a macroscopic point of view, is a very calm, stable system with a certain temperature, pressure, and volume. Then there are systems where the effect of a single low-level event may get *magnified* into an enormous high-level consequence. Such a system is a pinball machine, where the exact angle with which a ball strikes each post is crucial in determining the rest of its descending pathway.

A computer is an elaborate combination of these two types of system. It contains subunits such as wires, which behave in a highly predictable fashion: they conduct electricity according to Ohm's law, a very precise, chunked law which resembles the laws governing gases in containers, since it depends on statistical effects in which billions of random effects cancel each other out, yielding a predictable overall behavior. A computer also contains macroscopic subunits, such as a printer, whose behavior is completely determined by delicate patterns of currents. What the printer prints is not by any means created by a myriad canceling microscopic effects. In fact, in the case of most computer programs, the value of every single bit in the program plays a critical role in the output that gets printed. If any bit were changed, the output would also change drastically.

Systems which are made up of "reliable" subsystems only—that is, subsystems whose behavior can be reliably predicted from chunked descriptions—play inestimably important roles in our daily lives, because they are pillars of stability. We can rely on walls not to fall down, on sidewalks to go where they went yesterday, on the sun to shine, on clocks to tell the time correctly, and so on. Chunked models of such systems are virtually entirely deterministic. Of course, the other kind of system which plays a very large role in our lives is a system that has variable behavior which depends on some internal microscopic parameters—often a very large number of them, moreover—which we cannot directly observe. Our chunked model of such a system is necessarily in terms of the "space" of operation, and involves probabilistic estimates of landing in different regions of that space.

A container of gas, which, as I already pointed out, is a reliable system because of many canceling effects, obeys precise, deterministic laws of physics. Such laws are *chunked laws,* in that they deal with the gas as a whole, and ignore its constituents. Furthermore, the microscopic and macroscopic descriptions of a gas use entirely different terms. The former requires the specification of the position and velocity of every single component molecule; the latter requires only the specification of three new quantities: temperature, pressure, and volume, the first two of which do not even have microscopic counterparts. The simple mathematical relationship which relates these three parameters—$pV = cT$, where c is a constant—is a law which depends on, yet is independent of, the lower-level phenomena. Less paradoxically, this law can be derived from the laws governing the molecular level; in that sense it depends on the lower level. On the other hand, it is a law which allows you to ignore the lower level completely, if you wish; in that sense it is independent of the lower level.

It is important to realize that the high-level law cannot be stated in the vocabulary of the low-level description. "Pressure" and "temperature" are new terms which experience with the low level alone cannot convey. We humans perceive temperature and pressure directly; that is how we are built, so that it is not amazing that we should have found this law. But creatures which knew gases only as theoretical mathematical constructs would have to have an ability to synthesize new concepts, if they were to discover this law.

Epiphenomena

In drawing this Chapter to a close, I would like to relate a story about a complex system. I was talking one day with two systems programmers for the computer I was using. They mentioned that the operating system seemed to be able to handle up to about thirty-five users with great comfort, but at about thirty-five users or so, the response time all of a sudden shot up, getting so slow that you might as well log off and go home and wait until later. Jokingly I said, "Well, that's simple to fix—just find the place in the operating system where the number '35' is stored, and change it to '60'!" Everyone laughed. The point is, of course, that there is no such place. Where, then, does the critical number—35 users—come from? The answer is: *It is a visible consequence of the overall system organization—an "epiphenomenon".*

Similarly, you might ask about a sprinter, "Where is the '9.3' stored, that makes him be able to run 100 yards in 9.3 seconds?" Obviously, it is not stored anywhere. His time is a result of how he is built, what his reaction time is, a million factors all interacting when he runs. The time is quite reproducible, but it is not stored in his body anywhere. It is spread around among all the cells of his body and only manifests itself in the act of the sprint itself.

Epiphenomena abound. In the game of "Go", there is the feature that "two eyes live". It is not built into the rules, but it is a consequence of the rules. In the human brain, there is gullibility. How gullible are you? Is your gullibility located in some "gullibility center" in your brain? Could a neurosurgeon reach in and perform some delicate operation to lower your gullibility, otherwise leaving you alone? If you believe this, you are pretty gullible, and should perhaps consider such an operation.

Mind *vs.* Brain

In coming Chapters, where we discuss the brain, we shall examine whether the brain's top level—the mind—can be understood without understanding the lower levels on which it both depends and does not depend. Are there laws of thinking which are "sealed off" from the lower laws that govern the microscopic activity in the cells of the brain? Can mind be "skimmed" off of brain and transplanted into other systems? Or is it impossible to unravel thinking processes into neat and modular subsystems? Is the brain more like an atom, a renormalized electron, a nucleus, a neutron, or a quark? Is consciousness an epiphenomenon? To understand the mind, must one go all the way down to the level of nerve cells?

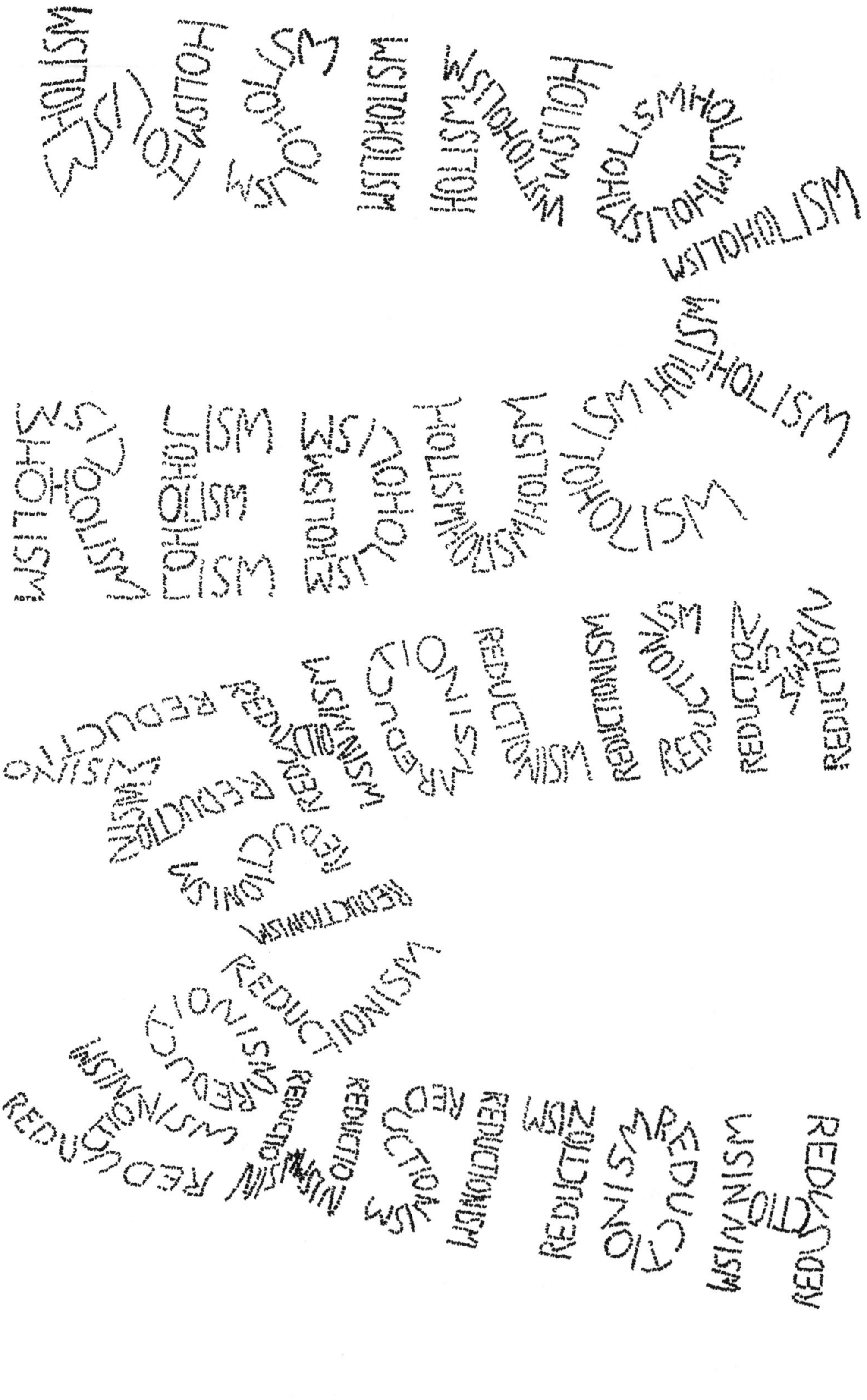

HOLISM HOLISM HOLISM HOLISM HOLISM HOLISM HOLISM HOLISM HOLISM HOLISM

HOLISM HOLISM HOLISM HOLISM HOLISM HOLISM HOLISM HOLISM HOLISM HOLISM

REDUCTIONISM REDUCTIONISM REDUCTIONISM REDUCTIONISM REDUCTIONISM REDUCTIONISM

REDUCTIONISM REDUCTIONISM REDUCTIONISM REDUCTIONISM REDUCTIONISM REDUCTIONISM

. . . *Ant Fugue*

. . . then, one by one, the four voices of the fugue chime in.)

Achilles: I know the rest of you won't believe this, but the answer to the question is staring us all in the face, hidden in the picture. It is simply one word—but what an important one: "MU"!

CCrab: I know the rest of you won't believe this, but the answer to the question is staring us all in the face, hidden in the picture. It is simply one word—but what an important one: "HOLISM"!

Achilles: Now hold on a minute. You must be seeing things. It's plain as day that the message of this picture is "MU", not "HOLISM"!

Crab: I beg your pardon, but my eyesight is extremely good. Please look again, and then tell me if the the picture doesn't say what I said it says!

Anteater: I know the rest of you won't believe this, but the answer to the question is staring us all in the face, hidden in the picture. It is simply one word—but what an important one: "REDUCTIONISM"!

Crab: Now hold on a minute. You must be seeing things. It's plain as day that the message of this picture is "HOLISM", not "REDUCTIONISM"!

Achilles: Another deluded one! Not "HOLISM", not "REDUCTIONISM", but "MU" is the message of this picture, and that much is certain.

Anteater: I beg your pardon, but my eyesight is extremely clear. Please look again, and then see if the picture doesn't say what I said it says.

Achilles: Don't you see that the picture is composed of two pieces, and that each of them is a single letter?

Crab: You are right about the two pieces, but you are wrong in your identification of what they are. The piece on the left is entirely composed of three copies of one word: "HOLISM"; and the piece on the right is composed of many copies, in smaller letters, of the same word. Why the letters are of different sizes in the two parts, I don't know, but I know what I see, and what I see is "HOLISM", plain as day. How you see anything else is beyond me.

Anteater: You are right about the two pieces, but you are wrong in your identification of what they are. The piece on the left is entirely composed of many copies of one word: "REDUCTIONISM"; and the piece on the right is composed of one single copy, in larger letters, of the same word. Why the letters are of different sizes in the two parts, I don't know, but I know what I see, and what I see is "REDUCTIONISM", plain as day. How you see anything else is beyond me.

Achilles: I know what is going on here. Each of you has seen letters which compose, or are composed of, other letters. In the left-hand piece,

FIGURE 60. [*Drawing by the author.*]

there are indeed three "HOLISM"'s, but each one of them is composed out of smaller copies of the word "REDUCTIONISM". And in complementary fashion, in the right-hand piece, there is indeed one "REDUCTIONISM", but it is composed out of smaller copies of the word "HOLISM". Now this is all fine and good, but in your silly squabble, the two of you have actually missed the forest for the trees. You see, what good is it to argue about whether "HOLISM" or "REDUCTIONISM" is right, when the proper way to understand the matter is to transcend the question, by answering "MU"?

Crab: I now see the picture as you have described it, Achilles, but I have no idea of what you mean by the strange expression "transcending the question".

Anteater: I now see the picture as you have described it, Achilles, but I have no idea of what you mean by the strange expression "MU".

Achilles: I will be glad to indulge both of you, if you will first oblige me, by telling me the meaning of these strange expressions, "HOLISM" and "REDUCTIONISM".

Crab: HOLISM is the most natural thing in the world to grasp. It's simply the belief that "the whole is greater than the sum of its parts". No one in his right mind could reject holism.

Anteater: REDUCTIONISM is the most natural thing in the world to grasp. It's simply the belief that "a whole can be understood completely if you understand its parts, and the nature of their 'sum'". No one in her left brain could reject reductionism.

Crab: I reject reductionism. I challenge you to tell me, for instance, how to understand a brain reductionistically. Any reductionistic explanation of a brain will inevitably fall far short of explaining where the consciousness experienced by a brain arises from.

Anteater: I reject holism. I challenge you to tell me, for instance, how a holistic description of an ant colony sheds any more light on it than is shed by a description of the ants inside it, and their roles, and their interrelationships. Any holistic explanation of an ant colony will inevitably fall far short of explaining where the consciousness experienced by an ant colony arises from.

Achilles: Oh, no! The last thing which I wanted to do was to provoke another argument. Anyway, now that I understand the controversy, I believe that my explanation of "MU" will help greatly. You see, "MU" is an ancient Zen answer which, when given to a question, UNASKS the question. Here, the question seems to be, "Should the world be understood via holism, or via reductionism?" And the answer of "MU" here rejects the premises of the question, which are that one or the other must be chosen. By unasking the question, it reveals a wider truth: that there is a larger context into which both holistic and reductionistic explanations fit.

Anteater: Absurd! Your "MU" is as silly as a cow's moo. I'll have none of this Zen wishy-washiness.

Crab: Ridiculous! Your "MU" is as silly as a kitten's mew. I'll have none of this Zen washy-wishiness.

Achilles: Oh, dear! We're getting nowhere fast. Why have you stayed so strangely silent, Mr. Tortoise? It makes me very uneasy. Surely you must somehow be capable of helping straighten out this mess?

Tortoise: I know the rest of you won't believe this, but the answer to the question is staring us all in the face, hidden in the picture. It is simply one word—but what an important one: "MU"!

(Just as he says this, the fourth voice in the fugue being played enters, exactly one octave below the first entry.)

Achilles: Oh, Mr. T, for once you have let me down. I was sure that you, who always see the most deeply into things, would be able to resolve this dilemma—but apparently, you have seen no further than I myself saw. Oh, well, I guess I should feel pleased to have seen as far as Mr. Tortoise, for once.

Tortoise: I beg your pardon, but my eyesight is extremely fine. Please look again, and then tell me if the picture doesn't say what I said it says.

Achilles: But of course it does! You have merely repeated my own original observation.

Tortoise: Perhaps "MU" exists in this picture on a deeper level than you imagine, Achilles—an octave lower (figuratively speaking). But for now I doubt that we can settle the dispute in the abstract. I would like to see both the holistic and reductionistic points of view laid out more explicitly; then there may be more of a basis for a decision. I would very much like to hear a reductionistic description of an ant colony, for instance.

Crab: Perhaps Dr. Anteater will tell you something of his experiences in that regard. After all, he is by profession something of an expert on that subject.

Tortoise: I am sure that we have much to learn from you, Dr. Anteater. Could you tell us more about ant colonies, from a reductionistic point of view?

Anteater: Gladly. As Mr. Crab mentioned to you, my profession has led me quite a long way into the understanding of ant colonies.

Achilles: I can imagine! The profession of anteater would seem to be synonymous with being an expert on ant colonies!

Anteater: I beg your pardon. "Anteater" is not my profession; it is my species. By profession, I am a colony surgeon. I specialize in correcting nervous disorders of the colony by the technique of surgical removal.

Achilles: Oh, I see. But what do you mean by "nervous disorders" of an ant colony?

Anteater: Most of my clients suffer from some sort of speech impairment. You know, colonies which have to grope for words in everyday situations. It can be quite tragic. I attempt to remedy the situation by, uhh—removing—the defective part of the colony. These operations are sometimes quite involved, and of course years of study are required before one can perform them.

Achilles: But—isn't it true that, before one can suffer from speech impairment, one must have the faculty of speech?

Anteater: Right.

Achilles: Since ant colonies don't have that faculty, I am a little confused.

Crab: It's too bad, Achilles, that you weren't here last week, when Dr. Anteater and Aunt Hillary were my house guests. I should have thought of having you over then.

Achilles: Is Aunt Hillary your aunt, Mr. Crab?

Crab: Oh, no, she's not really anybody's aunt.

Anteater: But the poor dear insists that everybody should call her that, even strangers. It's just one of her many endearing quirks.

Crab: Yes, Aunt Hillary is quite eccentric, but such a merry old soul. It's a shame I didn't have you over to meet her last week.

Anteater: She's certainly one of the best-educated ant colonies I have ever had the good fortune to know. The two of us have spent many a long evening in conversation on the widest range of topics.

Achilles: I thought anteaters were devourers of ants, not patrons of ant-intellectualism!

Anteater: Well, of course the two are not mutually inconsistent. I am on the best of terms with ant colonies. It's just ANTS that I eat, not colonies—and that is good for both parties: me, and the colony.

Achilles: How is it possible that—

Tortoise: How is it possible that—

Achilles: —having its ants eaten can do an ant colony any good?

Crab: How is it possible that—

Tortoise: —having a forest fire can do a forest any good?

Anteater: How is it possible that—

Crab: —having its branches pruned can do a tree any good?

Anteater: —having a haircut can do Achilles any good?

Tortoise: Probably the rest of you were too engrossed in the discussion to notice the lovely stretto which just occurred in this Bach fugue.

Achilles: What is a stretto?

Tortoise: Oh, I'm sorry; I thought you knew the term. It is where one theme repeatedly enters in one voice after another, with very little delay between entries.

Achilles: If I listen to enough fugues, soon I'll know all of these things and will be able to pick them out myself, without their having to be pointed out.

Tortoise: Pardon me, my friends. I am sorry to have interrupted. Dr. Anteater was trying to explain how eating ants is perfectly consistent with being a friend of an ant colony.

Achilles: Well, I can vaguely see how it might be possible for a limited and regulated amount of ant consumption to improve the overall health of a colony—but what is far more perplexing is all this talk about having conversations with ant colonies. That's impossible. An ant colony is simply a bunch of individual ants running around at random looking for food and making a nest.

Anteater: You could put it that way if you want to insist on seeing the trees but missing the forest, Achilles. In fact, ant colonies, seen as wholes, are quite well-defined units, with their own qualities, at times including the mastery of language.

Achilles: I find it hard to imagine myself shouting something out loud in the middle of the forest, and hearing an ant colony answer back.

Anteater: Silly fellow! That's not the way it happens. Ant colonies don't converse out loud, but in writing. You know how ants form trails leading them hither and thither?

Achilles: Oh, yes—usually straight through the kitchen sink and into my peach jam.

Anteater: Actually, some trails contain information in coded form. If you know the system, you can read what they're saying just like a book.

Achilles: Remarkable. And can you communicate back to them?

Anteater: Without any trouble at all. That's how Aunt Hillary and I have conversations for hours. I take a stick and draw trails in the moist ground, and watch the ants follow my trails. Presently, a new trail starts getting formed somewhere. I greatly enjoy watching trails develop. As they are forming, I anticipate how they will continue (and more often I am wrong than right). When the trail is completed, I know what Aunt Hillary is thinking, and I in turn make my reply.

Achilles: There must be some amazingly smart ants in that colony, I'll say that.

Anteater: I think you are still having some difficulty realizing the difference in levels here. Just as you would never confuse an individual tree with a forest, so here you must not take an ant for the colony. You see, all the ants in Aunt Hillary are as dumb as can be. They couldn't converse to save their little thoraxes!

Achilles: Well then, where does the ability to converse come from? It must reside somewhere inside the colony! I don't understand how the ants can all be unintelligent, if Aunt Hillary can entertain you for hours with witty banter.

Tortoise: It seems to me that the situation is not unlike the composition of a human brain out of neurons. Certainly no one would insist that individual brain cells have to be intelligent beings on their own, in order to explain the fact that a person can have an intelligent conversation.

Achilles: Oh, no, clearly not. With brain cells, I see your point completely. Only . . . ants are a horse of another color. I mean, ants just roam about at will, completely randomly, chancing now and then upon a morsel of food . . . They are free to do what they want to do, and with that freedom, I don't see at all how their behavior, seen as a whole, can amount to anything coherent—especially something so coherent as the brain behavior necessary for conversing.

Crab: It seems to me that the ants are free only within certain constraints. For example, they are free to wander, to brush against each other, to pick up small items, to work on trails, and so on. But they never step out of that small world, that ant-system, which they are in. It would never occur to them, for they don't have the mentality to imagine anything of the kind. Thus the ants are very reliable components, in the sense that you can depend on them to perform certain kinds of tasks in certain ways.

Achilles: But even so, within those limits they are still free, and they just act at random, running about incoherently without any regard for the thought mechanisms of a higher-level being which Dr. Anteater asserts they are merely components of.

Anteater: Ah, but you fail to recognize one thing, Achilles—the regularity of statistics.

Achilles: How is that?

Anteater: For example, even though ants as individuals wander about in what seems a random way, there are nevertheless overall trends, involving large numbers of ants, which can emerge from that chaos.

Achilles: Oh, I know what you mean. In fact, ant trails are a perfect example of such a phenomenon. There, you have really quite unpredictable motion on the part of any single ant—and yet, the trail itself seems to remain well-defined and stable. Certainly that must mean that the individual ants are not just running about totally at random.

Anteater: Exactly, Achilles. There is some degree of communication among the ants, just enough to keep them from wandering off completely at random. By this minimal communication they can remind each other that they are not alone but are cooperating with teammates. It takes a large number of ants, all reinforcing each other this way, to sustain any activity—such as trail-building—for any length of time. Now my very hazy understanding of the operation of brains leads me to believe that something similar pertains to the firing of neurons. Isn't it true, Mr. Crab, that it takes a group of neurons firing in order to make another neuron fire?

Crab: Definitely. Take the neurons in Achilles' brain, for example. Each neuron receives signals from neurons attached to its input lines, and if the sum total of inputs at any moment exceeds a critical threshold, then that neuron will fire and send its own output pulse rushing off to other neurons, which may in turn fire—and on down the line it goes. The neural flash swoops relentlessly in its Achillean path, in shapes stranger then the dash of a gnat-hungry swallow; every twist, every turn foreordained by the neural structure in Achilles' brain, until sensory input messages interfere.

Achilles: Normally, I think that I'M in control of what I think—but the way you put it turns it all inside out, so that it sounds as though "I" am just what comes out of all this neural structure, and natural law. It makes what I consider my SELF sound at best like a by-product of an organism governed by natural law, and at worst, an artificial notion produced by my distorted perspective. In other words, you make me feel like I don't know who—or what—I am, if anything.

Tortoise: You'll come to understand much better as we go along. But Dr. Anteater—what do you make of this similarity?

Anteater: I knew there was something parallel going on in the two very different systems. Now I understand it much better. It seems that group phenomena which have coherence—trail-building, for example—will take place only when a certain threshold number of ants get involved. If an effort is initiated, perhaps at random, by a few ants in some locale, one of two things can happen: either it will fizzle out after a brief sputtering start—

Achilles: When there aren't enough ants to keep the thing rolling?

Anteater: Exactly. The other thing that can happen is that a critical mass of ants is present, and the thing will snowball, bringing more and more ants into the picture. In the latter case, a whole "team" is brought into being which works on a single project. That project might be trail-making, or food-gathering, or it might involve nest-keeping. Despite the extreme simplicity of this scheme on a small scale, it can give rise to very complex consequences on a larger scale.

Achilles: I can grasp the general idea of order emerging from chaos, as you sketch it, but that still is a long way from the ability to converse. After all, order also emerges from chaos when molecules of a gas bounce against each other randomly—yet all that results there is an amorphous mass with but three parameters to characterize it: volume, pressure, and temperature. Now that's a far cry from the ability to understand the world, or to talk about it!

Anteater: That highlights a very interesting difference between the explanation of the behavior of an ant colony and the explanation of the behavior of gas inside a container. One can explain the behavior of the gas simply by calculating the statistical properties of the motions of its molecules. There is no need to discuss any higher elements of structure than molecules, except the full gas itself. On the other hand, in an ant colony, you can't even begin to understand the activities of the colony unless you go through several layers of structure.

Achilles: I see what you mean. In a gas, one jump takes you from the lowest level—molecules—to the highest level—the full gas. There are no intermediate levels of organization. Now how do intermediate levels of organized activity arise in an ant colony?

Anteater: It has to do with the existence of several different varieties of ants inside any colony.

Achilles: Oh, yes. I think I have heard about that. They are called "castes", aren't they?

Anteater: That's correct. Aside from the queen, there are males, who do practically nothing towards the upkeep of the nest, and then—

Achilles: And of course there are soldiers—Glorious Fighters Against Communism!

Crab: Hmm . . . I hardly think that could be right, Achilles. An ant colony is quite communistic internally, so why would its soldiers fight against communism? Or am I right, Dr. Anteater?

Anteater: Yes, about colonies you are right, Mr. Crab; they are indeed based on somewhat communistic principles. But about soldiers Achilles is somewhat naïve. In fact, the so-called "soldiers" are hardly adept at fighting at all. They are slow, ungainly ants with giant heads, who can snap with their strong jaws, but are hardly to be glorified. As in a true communistic state, it is rather the workers who are to be glorified. It is they who do most of the chores, such as food-gathering, hunting, and nursing of the young. It is even they who do most of the fighting.

Achilles: Bah. That is an absurd state of affairs. Soldiers who won't fight!

Anteater: Well, as I just said, they really aren't soldiers at all. It's the workers who are soldiers; the soldiers are just lazy fatheads.

Achilles: Oh, how disgraceful! Why, if I were an ant, I'd put some discipline in their ranks! I'd knock some sense into those fatheads!

Tortoise: If you were an ant? How could you be an ant? There is no way to map your brain onto an ant brain, so it seems to me to be a pretty fruitless question to worry over. More reasonable would be the proposition of mapping your brain onto an ant colony . . . But let us not get sidetracked. Let Dr. Anteater continue with his most illuminating description of castes and their role in the higher levels of organization.

Anteater: Very well. There are all sorts of tasks which must be accomplished in a colony, and individual ants develop specializations. Usually an ant's specialization changes as the ant ages. And of course it is also dependent on the ant's caste. At any one moment, in any small area of a colony, there are ants of all types present. Of course, one caste may be be very sparse in some places and very dense in others.

Crab: Is the density of a given caste, or specialization, just a random thing? Or is there a reason why ants of one type might be more heavily concentrated in certain areas, and less heavily in others?

Anteater: I'm glad you brought that up, since it is of crucial importance in understanding how a colony thinks. In fact, there evolves, over a long period of time, a very delicate distribution of castes inside a colony. And it is this distribution which allows the colony to have the complexity which underlies the ability to converse with me.

Achilles: It would seem to me that the constant motion of ants to and fro would completely prevent the possibility of a very delicate distribution. Any delicate distribution would be quickly destroyed by all the random motions of ants, just as any delicate pattern among molecules in a gas would not survive for an instant, due to the random bombardment from all sides.

Anteater: In an ant colony, the situation is quite the contrary. In fact, it is just exactly the constant to-ing and fro-ing of ants inside the colony which adapts the caste distribution to varying situations, and thereby preserves the delicate caste distribution. You see, the caste distribution cannot remain as one single rigid pattern; rather, it must constantly be changing so as to reflect, in some manner, the real-world situation with which the colony is dealing, and it is precisely the motion inside the colony which updates the caste distribution, so as to keep it in line with the present circumstances facing the colony.

Tortoise: Could you give an example?

Anteater: Gladly. When I, an anteater, arrive to pay a visit to Aunt Hillary, all the foolish ants, upon sniffing my odor, go into a panic—which means, of course, that they begin running around completely differently from the way they were before I arrived.

Achilles: But that's understandable, since you're a dreaded enemy of the colony.

Anteater: Oh, no. I must reiterate that, far from being an enemy of the colony, I am Aunt Hillary's favorite companion. And Aunt Hillary is my favorite aunt. I grant you, I'm quite feared by all the individual ants in the colony—but that's another matter entirely. In any case, you see that the ants' action in response to my arrival completely changes the internal distribution of ants.

Achilles: That's clear.

Anteater: And that sort of thing is the updating which I spoke of. The new distribution reflects my presence. One can describe the change from old state to new as having added a "piece of knowledge" to the colony.

Achilles: How can you refer to the distribution of different types of ants inside a colony as a "piece of knowledge"?

Anteater: Now there's a vital point. It requires some elaboration. You see, what it comes down to is how you choose to describe the caste distribution. If you continue to think in terms of the lower levels—individual ants—then you miss the forest for the trees. That's just too microscopic a level, and when you think microscopically, you're bound to miss some large-scale features. You've got to find the proper high-level framework in which to describe the caste distribution—only then will it make sense how the caste distribution can encode many pieces of knowledge.

Achilles: Well, how DO you find the proper-sized units in which to describe the present state of the colony, then?

Anteater: All right. Let's begin at the bottom. When ants need to get something done, they form little "teams", which stick together to perform a chore. As I mentioned earlier, small groups of ants are constantly forming and unforming. Those which actually exist for a while are the teams, and the reason they don't fall apart is that there really is something for them to do.

Achilles: Earlier you said that a group will stick together if its size exceeds a certain threshold. Now you're saying that a group will stick together if there is something for it to do.

Anteater: They are equivalent statements. For instance, in food-gathering, if there is an inconsequential amount of food somewhere which gets discovered by some wandering ant who then attempts to communicate its enthusiasm to other ants, the number of ants who respond will be proportional to the size of the food sample—and an inconsequential amount will not attract enough ants to surpass the threshold. Which is exactly what I meant by saying there is nothing to do—too little food ought to be ignored.

Achilles: I see. I assume that these "teams" are one of the levels of structure falling somewhere in between the single-ant level and the colony level.

Anteater: Precisely. There exists a special kind of team, which I call a "signal"—and all the higher levels of structure are based on signals. In fact, all the higher entities are collections of signals acting in concert. There are teams on higher levels whose members are not ants, but teams on lower levels. Eventually you reach the lowest-level teams—which is to say, signals—and below them, ants.

Achilles: Why do signals deserve their suggestive name?

Anteater: It comes from their function. The effect of signals is to transport ants of various specializations to appropriate parts of the colony. So the typical story of a signal is thus: it comes into existence by exceeding the threshold needed for survival, then it migrates for some distance through the colony, and at some point it more or less disintegrates into its individual members, leaving them on their own.

Achilles: It sounds like a wave, carrying sand dollars and seaweed from afar, and leaving them strewn, high and dry, on the shore.

Anteater: In a way that's analogous, since the team does indeed deposit something which it has carried from a distance, but whereas the water in the wave rolls back to the sea, there is no analogous carrier substance in the case of a signal, since the ants themselves compose it.

Tortoise: And I suppose that a signal loses its coherency just at some spot in the colony where ants of that type were needed in the first place.

Anteater: Naturally.

Achilles: Naturally? It's not so obvious to ME that a signal should always go just where it is needed. And even if it goes in the right direction, how does it figure out where to decompose? How does it know it has arrived?

Anteater: Those are extremely important matters, since they involve explaining the existence of purposeful behavior—or what seems to be purposeful behavior—on the part of signals. From the description, one would be inclined to characterize the signals' behavior as being oriented towards filling a need, and to call it "purposeful". But you can look at it otherwise.

Achilles: Oh, wait. Either the behavior IS purposeful, or it is NOT. I don't see how you can have it both ways.

Anteater: Let me explain my way of seeing things, and then see if you agree. Once a signal is formed, there is no awareness on its part that it should head off in any particular direction. But here, the delicate caste distribution plays a crucial role. It is what determines the motion of signals through the colony, and also how long a signal will remain stable, and where it will "dissolve".

Achilles: So everything depends on the caste distribution, eh?

Anteater: Right. Let's say a signal is moving along. As it goes, the ants which compose it interact, either by direct contact or by exchange of scents, with ants of the local neighborhoods which it passes through. The contacts and scents provide information about local matters of urgency, such as nest-building, or nursing, or whatever. The signal will remain glued together as long as the local needs are different from what it can supply; but if it CAN contribute, it disintegrates, spilling a fresh team of usable ants onto the scene. Do you see now how the caste distribution acts as an overall guide of the teams inside the colony?

Achilles: I do see that.

Anteater: And do you see how this way of looking at things requires attributing no sense of purpose to the signal?

Achilles: I think so. Actually, I'm beginning to see things from two different vantage points. From an ant's-eye point of view, a signal has NO purpose. The typical ant in a signal is just meandering around the colony, in search of nothing in particular, until it finds that it feels like stopping. Its teammates usually agree, and at that moment the team unloads itself by crumbling apart, leaving just its members but none of its coherency. No planning is required, no looking ahead; nor is any search required, to determine the proper direction. But from the COLONY'S point of view, the team has just responded to a message which was written in the language of the caste distribution. Now from this perspective, it looks very much like purposeful activity.

Crab: What would happen if the caste distribution were entirely random? Would signals still band and disband?

Anteater: Certainly. But the colony would not last long, due to the meaninglessness of the caste distribution.

Crab: Precisely the point I wanted to make. Colonies survive because their caste distribution has meaning, and that meaning is a holistic aspect, invisible on lower levels. You lose explanatory power unless you take that higher level into account.

Anteater: I see your side; but I believe you see things too narrowly.

Crab: How so?

Anteater: Ant colonies have been subjected to the rigors of evolution for billions of years. A few mechanisms were selected for, and most were selected against. The end result was a set of mechanisms which make ant colonies work as we have been describing. If you could watch the whole process in a movie—running a billion or so times faster than life, of course—the emergence of various mechanisms would be seen as natural responses to external pressures, just as bubbles in boiling water are natural responses to an external heat source. I don't suppose you see "meaning" and "purpose" in the bubbles in boiling water—or do you?

Crab: No, but—

Anteater: Now that's MY point. No matter how big a bubble is, it owes its existence to processes on the molecular level, and you can forget about any "higher-level laws". The same goes for ant colonies and their teams. By looking at things from the vast perspective of evolution, you can drain the whole colony of meaning and purpose. They become superfluous notions.

Achilles: Why, then, Dr. Anteater, did you tell me that you talked with Aunt Hillary? It now seems that you would deny that she can talk or think at all.

Anteater: I am not being inconsistent, Achilles. You see, I have as much difficulty as anyone else in seeing things on such a grandiose time scale, so I find it much easier to change points of view. When I do so, forgetting about evolution and seeing things in the here and now, the vocabulary of teleology comes back: the MEANING of the caste distribution and the PURPOSEFULNESS of signals. This not only happens when I think of ant colonies, but also when I think about my own brain and other brains. However, with some effort I can always remember the other point of view if necessary, and drain all these systems of meaning, too.

Crab: Evolution certainly works some miracles. You never know the next trick it will pull out of its sleeve. For instance, it wouldn't surprise me one bit if it were theoretically possible for two or more "signals" to pass through each other, each one unaware that the other one is also a signal; each one treating the other as if it were just part of the background population.

Anteater: It is better than theoretically possible; in fact it happens routinely!

Achilles: Hmm . . . What a strange image that conjures up in my mind. I can just imagine ants moving in four different directions, some black, some white, criss-crossing, together forming an orderly pattern, almost like—like—

Tortoise: A fugue, perhaps?

Achilles: Yes—that's it! An ant fugue!

Crab: An interesting image, Achilles. By the way, all that talk of boiling water made me think of tea. Who would like some more?

Achilles: I could do with another cup, Mr. C.

Crab: Very good.

Achilles: Do you suppose one could separate out the different visual "voices" of such an "ant fugue"? I know how hard it is for me—

Tortoise: Not for me, thank you.

Achilles: —to track a single voice—

Anteater: I'd like some, too, Mr. Crab—

Achilles: —in a musical fugue—

FIGURE 61. "Ant Fugue", *by M. C. Escher (woodcut, 1953).*

Anteater: —if it isn't too much trouble.

Achilles: —when all of them—

Crab: Not at all. Four cups of tea—

Tortoise: Three!

Achilles: —are going at once.

Crab: —coming right up!

Anteater: That's an interesting thought, Achilles. But it's unlikely that anyone could draw such a picture in a convincing way.

Achilles: That's too bad.

Tortoise: Perhaps you could answer this, Dr. Anteater. Does a signal, from its creation until its dissolution, always consist of the same set of ants?

Anteater: As a matter of fact, the individuals in a signal sometimes break off and get replaced by others of the same caste, if there are a few in the area. Most often, signals arrive at their disintegration points with nary an ant in common with their starting lineup.

Crab: I can see that the signals are constantly affecting the caste distribution throughout the colony, and are doing so in response to the internal needs of the colony—which in turn reflect the external situation which the colony is faced with. Therefore the caste distribution, as you said, Dr. Anteater, gets continually updated in a way which ultimately reflects the outer world.

Achilles: But what about those intermediate levels of structure? You were saying that the caste distribution should best be pictured not in terms of ants or signals, but in terms of teams whose members were other teams, whose members were other teams, and so on until you come down to the ant level. And you said that that was the key to understanding how it was possible to describe the caste distribution as encoding pieces of information about the world.

Anteater: Yes, we are coming to all that. I prefer to give teams of a sufficiently high level the name of "symbols". Mind you, this sense of the word has some significant differences from the usual sense. My "symbols" are ACTIVE SUBSYSTEMS of a complex system, and they are composed of lower-level active subsystems . . . They are therefore quite different from PASSIVE symbols, external to the system, such as letters of the alphabet or musical notes, which sit there immobile, waiting for an active system to process them.

Achilles: Oh, this is rather complicated, isn't it? I just had no idea that ant colonies had such an abstract structure.

Anteater: Yes, it's quite remarkable. But all these layers of structure are necessary for the storage of the kinds of knowledge which enable an organism to be "intelligent" in any reasonable sense of the word. Any system which has a mastery of language has essentially the same underlying sets of levels.

Achilles: Now just a cotton-picking minute. Are you insinuating that my brain consists of, at bottom, just a bunch of ants running around?

Anteater: Oh, hardly. You took me a little too literally. The lowest level may be utterly different. Indeed, the brains of anteaters, for instance, are not composed of ants. But when you go up a level or two in a brain, you reach a level whose elements have exact counterparts in other systems of equal intellectual strength—such as ant colonies.

Tortoise: That is why it would be reasonable to think of mapping your brain, Achilles, onto an ant colony, but not onto the brain of a mere ant.

Achilles: I appreciate the compliment. But how would such a mapping be carried out? For instance, what in my brain corresponds to the low-level teams which you call signals?

Anteater: Oh, I but dabble in brains, and therefore couldn't set up the map in its glorious detail. But—and correct me if I'm wrong, Mr. Crab—I would surmise that the brain counterpart to an ant colony's signal is the firing of a neuron; or perhaps it is a larger-scale event, such as a pattern of neural firings.

Crab: I would tend to agree. But don't you think that, for the purposes of our discussion, delineating the exact counterpart is not in itself crucial, desirable though it might be? It seems to me that the main idea is that such a correspondence does exist, even if we don't know exactly how to define it right now. I would only question one point, Dr. Anteater, which you raised, and that concerns the level at which one can have faith that the correspondence begins. You seemed to think that a SIGNAL might have a direct counterpart in a brain; whereas I feel that it is only at the level of your ACTIVE SYMBOLS and above that it is likely that a correspondence must exist.

Anteater: Your interpretation may very well be more accurate than mine, Mr. Crab. Thank you for bringing out that subtle point.

Achilles: What does a symbol do that a signal couldn't do?

Anteater: It is something like the difference between words and letters. Words, which are meaning-carrying entities, are composed of letters, which in themselves carry no meaning. This gives a good idea of the difference between symbols and signals. In fact it is a useful analogy, as long as you keep in mind the fact that words and letters are PASSIVE, symbols and signals are ACTIVE.

Achilles: I'll do so, but I'm not sure I understand why it is so vital to stress the difference between active and passive entities.

Anteater: The reason is that the meaning which you attribute to any passive symbol, such as a word on a page, actually derives from the meaning which is carried by corresponding active symbols in your brain. So that the meaning of passive symbols can only be properly understood when it is related to the meaning of active symbols.

Achilles: All right. But what is it that endows a SYMBOL—an active one, to be sure—with meaning, when you say that a SIGNAL, which is a perfectly good entity in its own right, has none?

Anteater: It all has to do with the way that symbols can cause other symbols to be triggered. When one symbol becomes active, it does not do so in isolation. It is floating about, indeed, in a medium, which is characterized by its caste distribution.

Crab: Of course, in a brain there is no such thing as a caste distribution, but the counterpart is the "brain state". There, you describe the states of all the neurons, and all the interconnections, and the threshold for firing of each neuron.

Anteater: Very well; let's lump "caste distribution" and "brain state" under a common heading, and call them just the "state". Now the state can be described on a low level or on a high level. A low-level description of the state of an ant colony would involve painfully specifying the location of each ant, its age and caste, and other similar items. A very detailed description, yielding practically no global insight as to WHY it is in that state. On the other hand, a description on a high level would involve specifying which symbols could be triggered by which combinations of other symbols, under what conditions, and so forth.

Achilles: What about a description on the level of signals, or teams?

Anteater: A description on that level would fall somewhere in between the low-level and symbol-level descriptions. It would contain a great deal of information about what is actually going on in specific locations throughout the colony, although certainly less than an ant-by-ant description, since teams consist of clumps of ants. A team-by-team description is like a summary of an ant-by-ant description. However, you have to add extra things which were not present in the ant-by-ant description—such as the relationships between teams, and the supply of various castes here and there. This extra complication is the price you pay for the right to summarize.

Achilles: It is interesting to me to compare the merits of the descriptions at various levels. The highest-level description seems to carry the most explanatory power, in that it gives you the most intuitive picture of the ant colony, although strangely enough, it leaves out seemingly the most important feature—the ants.

Anteater: But you see, despite appearances, the ants are not the most important feature. Admittedly, were it not for them, the colony wouldn't exist; but something equivalent—a brain—can exist, ant-free. So, at least from a high-level point of view, the ants are dispensable.

Achilles: I'm sure no ant would embrace your theory with eagerness.

Anteater: Well, I never met an ant with a high-level point of view.

Crab: What a counterintuitive picture you paint, Dr. Anteater. It seems that, if what you say is true, in order to grasp the whole structure, you have to describe it omitting any mention of its fundamental building blocks.

Anteater: Perhaps I can make it a little clearer by an analogy. Imagine you have before you a Charles Dickens novel.

Achilles: *The Pickwick Papers*—will that do?

Anteater: Excellently! And now imagine trying the following game: you must find a way of mapping letters onto ideas, so that the entire *Pickwick Papers* makes sense when you read it letter by letter.

Achilles: Hmm . . . You mean that every time I hit a word such as "the", I have to think of three definite concepts, one after another, with no room for variation?

Anteater: Exactly. They are the 't'-concept, the 'h'-concept, and the 'e'-concept—and every time, those concepts are as they were the preceding time.

Achilles: Well, it sounds like that would turn the experience of "reading" *The Pickwick Papers* into an indescribably boring nightmare. It would be an exercise in meaninglessness, no matter what concept I associated with each letter.

Anteater: Exactly. There is no natural mapping from the individual letters into the real world. The natural mapping occurs on a higher level—between words, and parts of the real world. If you wanted to describe the book, therefore, you would make no mention of the letter level.

Achilles: Of course not! I'd describe the plot and the characters, and so forth.

Anteater: So there you are. You would omit all mention of the building blocks, even though the book exists thanks to them. They are the medium, but not the message.

Achilles: All right—but what about ant colonies?

Anteater: Here, there are active signals instead of passive letters, and active symbols instead of passive words—but the idea carries over.

Achilles: Do you mean I couldn't establish a mapping between signals and things in the real world?

Anteater: You would find that you could not do it in such a way that the triggering of new signals would make any sense. Nor could you succeed on any lower level—for example the ant level. Only on the symbol level do the triggering patterns make sense. Imagine, for instance, that one day you were watching Aunt Hillary when I arrived to pay a call. You could watch as carefully as you wanted, and yet you would probably perceive nothing more than a rearrangement of ants.

Achilles: I'm sure that's accurate.

Anteater: And yet, as I watched, reading the higher level instead of the lower level, I would see several dormant symbols being awakened, those which translate into the thought, "Oh, here's that charming Dr. Anteater again—how pleasant!"—or words to that effect.

Achilles: That sounds like what happened when the four of us all found different levels to read in the MU-picture—or at least THREE of us did . . .

Tortoise: What an astonishing coincidence that there should be such a resemblance between that strange picture which I chanced upon in the *Well-Tempered Clavier,* and the trend of our conversation.

Achilles: Do you think it's just coincidence?

Tortoise: Of course.

Anteater: Well, I hope you can grasp now how the thoughts in Aunt Hillary emerge from the manipulation of symbols composed of signals composed of teams composed of lower-level teams, all the way down to ants.

Achilles: Why do you call it "symbol manipulation"? Who does the manipulating, if the symbols are themselves active? Who is the agent?

Anteater: This gets back to the question which you earlier raised about purpose. You're right that symbols themselves are active, but the activities which they follow are nevertheless not absolutely free. The activities of all symbols are strictly determined by the state of the full system in which they reside. Therefore, the full system is responsible for how its symbols trigger each other, and so it is quite reasonable to speak of the full system as the "agent". As the symbols operate, the state of the system gets slowly transformed, or updated. But there are many features which remain over time. It is this partially constant, partially varying system which is the agent. One can give a name to the full system. For example, Aunt Hillary is the "who" who can be said to manipulate her symbols; and you are similar, Achilles.

Achilles: That's quite a strange characterization of the notion of who I am. I'm not sure I can fully understand it, but I will give it some thought.

Tortoise: It would be quite interesting to follow the symbols in your brain as you do that thinking about the symbols in your brain.

Achilles: That's too complicated for me. I have trouble enough just trying to picture how it is possible to look at an ant colony and read it on the symbol level. I can certainly imagine perceiving it at the ant level; and with a little trouble, I can imagine what it must be like to perceive it at the signal level; but what in the world can it be like to perceive an ant colony at the symbol level?

Anteater: One only learns through long practice. But when one is at my stage, one reads the top level of an ant colony as easily as you yourself read the "MU" in the MU-picture.

Achilles: Really? That must be an amazing experience.

Anteater: In a way—but it is also one which is quite familiar to you, Achilles.

Achilles: Familiar to me? What do you mean? I have never looked at an ant colony on anything but the ant level.

Anteater: Maybe not; but ant colonies are no different from brains in many respects.

Achilles: I have never seen nor read any brain either, however.

Anteater: What about your OWN brain? Aren't you aware of your own thoughts? Isn't that the essence of consciousness? What else are you doing but reading your own brain directly at the symbol level?

Achilles: I never thought of it that way. You mean that I bypass all the lower levels, and only see the topmost level?

Anteater: That's the way it is, with conscious systems. They perceive themselves on the symbol level only, and have no awareness of the lower levels, such as the signal levels.

Achilles: Does it follow that in a brain, there are active symbols which are constantly updating themselves so that they reflect the overall state of the brain itself, always on the symbol level?

Anteater: Certainly. In any conscious system there are symbols which represent the brain state, and they are themselves part of the very brain state which they symbolize. For consciousness requires a large degree of self-consciousness.

Achilles: That is a weird notion. It means that although there is frantic activity occurring in my brain at all times, I am only capable of registering that activity in one way—on the symbol level; and I am completely insensitive to the lower levels. It is like being able to read a Dickens novel by direct visual perception, without ever having learned the letters of the alphabet. I can't imagine anything as weird as that really happening.

Crab: But precisely that sort of thing DID happen when you read "MU", without perceiving the lower levels "HOLISM" and "REDUCTIONISM".

Achilles: You're right—I bypassed the lower levels, and saw only the top. I wonder if I'm missing all sorts of meaning on lower levels of my brain as well, by reading only the symbol level. It's too bad that the top level doesn't contain all the information about the bottom level, so that by reading the top, one also learns what the bottom level says. But I guess it would be naïve to hope that the top level encodes anything from the bottom level—it probably doesn't percolate up. The MU-picture is the most striking possible example of that: there, the topmost level says only "MU", which bears no relation whatever to the lower levels!

Crab: That's absolutely true. *(Picks up the MU-picture, to inspect it more closely.)* Hmm . . . There's something strange about the smallest letters in this picture; they're very wiggly . . .

Anteater: Let me take a look. *(Peers closely at the MU-picture.)* I think there's yet another level, which all of us missed!

Tortoise: Speak for yourself, Dr. Anteater.

Achilles: Oh, no—that can't be! Let me see. *(Looks very carefully.)* I know the rest of you won't believe this, but the message of this picture is staring us all in the face, hidden in its depths. It is simply one word, repeated over and over again, like a mantra—but what an important one: "MU"! What do you know! It is the same as the top level! And none of us suspected it in the least.

Crab: We would never have noticed it if it hadn't been for you, Achilles.

Anteater: I wonder if the coincidence of the highest and lowest levels happened by chance? Or was it a purposeful act carried out by some creator?

Crab: How could one ever decide that?

Tortoise: I don't see any way to do so, since we have no idea why that particular picture is in the Crab's edition of the *Well-Tempered Clavier.*

Anteater: Although we have been having a lively discussion, I have still managed to listen with a good fraction of an ear to this very long and complex four-voice fugue. It is extraordinarily beautiful.

Tortoise: It certainly is. And now, in just a moment, comes an organ point.

Achilles: Isn't an organ point what happens when a piece of music slows down slightly, settles for a moment or two on a single note or chord, and then resumes at normal speed after a short silence?

Tortoise: No, you're thinking of a "fermata"—a sort of musical semicolon. Did you notice there was one of those in the prelude?

Achilles: I guess I must have missed it.

Tortoise: Well, you have another chance coming up to hear a fermata—in fact, there are a couple of them coming up, towards the end of this fugue.

Achilles: Oh, good. You'll point them out in advance, won't you?

Tortoise: If you like.

Achilles: But do tell me, what is an organ point?

Tortoise: An organ point is the sustaining of a single note by one of the voices in a polyphonic piece (often the lowest voice), while the other voices continue their own independent lines. This organ point is on the note of G. Listen carefully, and you'll hear it.

Anteater: There occurred an incident one day when I visited with Aunt Hillary which reminds me of your suggestion of observing the symbols in Achilles' brain as they create thoughts which are about themselves.

Crab: Do tell us about it.

Anteater: Aunt Hillary had been feeling very lonely, and was very happy to have someone to talk to that day. So she gratefully told me to help myself to the juiciest ants I could find. (She's always been most generous with her ants.)

Achilles: Gee!

Anteater: It just happened that I had been watching the symbols which were carrying out her thoughts, because in them were some particularly juicy-looking ants.

Achilles: Gee!

Anteater: So I helped myself to a few of the fattest ants which had been parts of the higher-level symbols which I had been reading. Specifically, the symbols which they were part of were the ones which had expressed the thought, "Help yourself to any of the ants which look appetizing."

Achilles: Gee!

Anteater: Unfortunately for them, but fortunately for me, the little bugs didn't have the slightest inkling of what they were collectively telling me, on the symbol level.

Achilles: Gee! That is an amazing wraparound. They were completely unconscious of what they were participating in. Their acts could be seen as part of a pattern on a higher level, but of course they were completely unaware of that. Ah, what a pity—a supreme irony, in fact—that they missed it.

Crab: You are right, Mr. T—that was a lovely organ point.

Anteater: I had never heard one before, but that one was so conspicuous that no one could miss it. Very effective.

Achilles: What? Has the organ point already occurred? How can I not have noticed it, if it was so blatant?

Tortoise: Perhaps you were so wrapped up in what you were saying that you were completely unaware of it. Ah, what a pity—a supreme irony, in fact—that you missed it.

Crab: Tell me, does Aunt Hillary live in an anthill?

Anteater: Well, she owns a rather large piece of property. It used to belong to someone else, but that is rather a sad story. In any case, her estate is quite expansive. She lives rather sumptuously, compared to many other colonies.

Achilles: How does that jibe with the communistic nature of ant colonies which you earlier described to us? It sounds quite inconsistent, to me, to preach communism and to live in a fancy estate!

Anteater: The communism is on the ant level. In an ant colony all ants work for the common good, even to their own individual detriment at times. Now this is simply a built-in aspect of Aunt Hillary's structure, but for all I know, she may not even be aware of this internal communism. Most human beings are not aware of anything about their neurons; in fact they probably are quite content not to know anything about their brains, being somewhat squeamish creatures. Aunt Hillary is also somewhat squeamish; she gets rather antsy whenever she starts to think about ants at all. So she avoids thinking about them whenever possible. I truly doubt that she knows anything about the communistic society which is built into her very structure. She herself is a staunch believer in libertarianism—you know, laissez-faire and all that. So it makes perfect sense, to me at least, that she should live in a rather sumptuous manor.

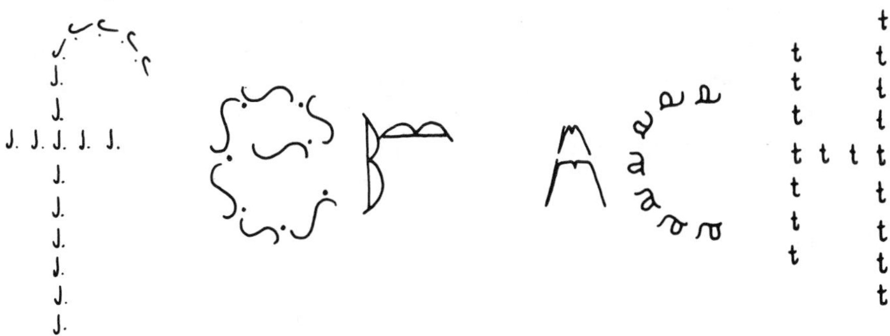

FIGURE 62. [*Drawing by the author.*]

Tortoise: As I turned the page just now, while following along in this lovely edition of the *Well-Tempered Clavier,* I noticed that the first of the two fermatas is coming up soon—so you might listen for it, Achilles.

Achilles: I will, I will.

Tortoise: Also, there's a most curious picture facing this page.

Crab: Another one? What next?

Tortoise: See for yourself. *(Passes the score over to the Crab.)*

Crab: Aha! It's just a few bunches of letters. Let's see—there are various numbers of the letters 'J', 'S', 'B', 'm', 'a', and 't'. It's strange, how the first three letters grow, and then the last three letters shrink again.

Anteater: May I see it?

Crab: Why, certainly.

Anteater: Oh, by concentrating on details, you have utterly missed the big picture. In reality, this group of letters is 'f', 'e', 'r', 'A', 'C', 'H', without any repetitions. First they get smaller, then they get bigger. Here, Achilles—what do you make of it?

Achilles: Let me see. Hmm. Well, I see it as a set of upper-case letters which grow as you move to the right.

Tortoise: Do they spell anything?

Achilles: Ah . . . "J. S. BACH". Oh! I understand now. It's Bach's name!

Tortoise: Strange that you should see it that way. I see it as a set of lower-case letters, shrinking as they move to the right, and . . . spelling out . . . the name of . . . *(Slows down slightly, especially drawing out the last few words. Then there is a brief silence. Suddenly he resumes as if nothing unusual had happened.)* —"fermat".

Achilles: Oh, you've got Fermat on the brain, I do believe. You see Fermat's Last Theorem everywhere.

Anteater: You were right, Mr. Tortoise—I just heard a charming little fermata in the fugue.

Crab: So did I.

Achilles: Do you mean everybody heard it but me? I'm beginning to feel stupid.

Tortoise: There, there, Achilles—don't feel bad. I'm sure you won't miss Fugue's Last Fermata (which is coming up quite soon). But, to return to our previous topic, Dr. Anteater, what is the very sad story which you alluded to, concerning the former owner of Aunt Hillary's property?

Anteater: The former owner was an extraordinary individual, one of the most creative ant colonies who ever lived. His name was Johant Sebastiant Fermant, and he was a mathematiciant by vocation, but a musiciant by avocation.

Achilles: How very versantile of him!

Anteater: At the height of his creative powers, he met with a most untimely demise. One day, a very hot summer day, he was out soaking up the warmth, when a freak thundershower—the kind that hits only once every hundred years or so—appeared from out of the blue, and thoroughly drenched J. S .F. Since the storm came utterly without warning, the ants got completely disoriented and confused. The intricate organization which had been so finely built up over decades, all went down the drain in a matter of minutes. It was tragic.

Achilles: Do you mean that all the ants drowned, which obviously would spell the end of poor J. S. F.?

Anteater: Actually, no. The ants managed to survive, every last one of them, by crawling onto various sticks and logs which floated above the raging torrents. But when the waters receded and left the ants back on their home grounds, there was no organization left. The caste distribution was utterly destroyed, and the ants themselves had no ability to reconstruct what had once before been such a finely tuned organization. They were as helpless as the pieces of Humpty Dumpty in putting themselves back together again. I myself tried, like all the king's horses and all the king's men, to put poor Fermant together again. I faithfully put out sugar and cheese, hoping against hope that somehow Fermant would reappear . . . *(Pulls out a handkerchief and wipes his eyes.)*

Achilles: How valiant of you! I never knew Anteaters had such big hearts.

Anteater: But it was all to no avail. He was gone, beyond reconstitution.

However, something very strange then began to take place: over the next few months, the ants which had been components of J. S. F. slowly regrouped, and built up a new organization. And thus was Aunt Hillary born.

Crab: Remarkable! Aunt Hillary is composed of the very same ants as Fermant was?

Anteater: Well, originally she was, yes. By now, some of the older ants have died, and been replaced. But there are still many holdovers from the J. S. F.-days.

Crab: And can't you recognize some of J. S. F.'s old traits coming to the fore, from time to time, in Aunt Hillary?

Anteater: Not a one. They have nothing in common. And there is no reason they should, as I see it. There are, after all, often several distinct ways to rearrange a group of parts to form a "sum". And Aunt Hillary was just a new "sum" of the old parts. Not MORE than the sum, mind you—just that particular KIND of sum.

Tortoise: Speaking of sums, I am reminded of number theory, where occasionally one will be able to take apart a theorem into its component symbols, rearrange them in a new order, and come up with a new theorem.

Anteater: I've never heard of such a phenomenon, although I confess to being a total ignoramus in the field.

Achilles: Nor have I heard of it—and I am rather well versed in the field, if I don't say so myself. I suspect Mr. T is just setting up one of his elaborate spoofs. I know him pretty well by now.

Anteater: Speaking of number theory, I am reminded of J. S. F. again, for number theory is one of the domains in which he excelled. In fact, he made some rather remarkable contributions to number theory. Aunt Hillary, on the other hand, is remarkably dull-witted in anything that has even the remotest connection with mathematics. Also, she has only a rather banal taste in music, whereas Sebastiant was extremely gifted in music.

Achilles: I am very fond of number theory. Could you possibly relate to us something of the nature of Sebastiant's contributions?

Anteater: Very well, then. *(Pauses for a moment to sip his tea, then resumes.)* Have you heard of Fourmi's infamous "Well-Tested Conjecture"?

Achilles: I'm not sure . . . It sounds strangely familiar, and yet I can't quite place it.

Anteater: It's a very simple idea. Lierre de Fourmi, a mathematiciant by vocation but lawyer by avocation, had been reading in his copy of the classic text *Arithmetica* by Di of Antus, and came across a page containing the equation

$$2^a + 2^b = 2^c$$

He immediately realized that this equation has infinitely many solutions a, b, c, and then wrote in the margin the following notorious comment:

FIGURE 63. During emigrations army ants sometimes create living bridges of their own bodies. In this photograph of such a bridge (de Fourmi Lierre), the workers of an Eciton burchelli *colony can be seen linking their legs and, along the top of the bridge, hooking their tarsal claws together to form irregular systems of chains. A symbiotic silverfish,* Trichatelura manni, *is seen crossing the bridge in the center. [From E. O. Wilson,* The Insect Societies *(Cambridge, Mass.: Harvard University Press, 1971), p. 62.]*

> The equation
>
> $$n^a + n^b = n^c$$
>
> has solutions in positive integers a, b, c, and n only when $n = 2$ (and then there are infinitely many triplets a, b, c which satisfy the equation); but there are no solutions for $n > 2$. I have discovered a truly marvelous proof of this statement, which, unfortunately, is so small that it would be well-nigh invisible if written in the margin.

Ever since that year, some three hundred days ago, mathematiciants have been vainly trying to do one of two things: either to prove Fourmi's claim, and thereby vindicate Fourmi's reputation, which, although very high, has been somewhat tarnished by skeptics who think he never really found the proof he claimed to have found—or else to refute the claim, by finding a counterexample: a set of four integers a, b, c, and n, with $n > 2$, which satisfy the equation. Until very recently, every attempt in either direction had met with failure. To be sure, the Conjecture has been verified for many specific values of n—in particular, all n up to 125,000. But no one had succeeded in proving it for ALL n—no one, that is, until Johant Sebastiant Fermant came upon the scene. It was he who found the proof that cleared Fourmi's name.

It now goes under the name "Johant Sebastiant's Well-Tested Conjecture".

Achilles: Shouldn't it be called a "Theorem" rather than a "Conjecture", if it's finally been given a proper proof?

Anteater: Strictly speaking, you're right, but tradition has kept it this way.

Tortoise: What sort of music did Sebastiant do?

Anteater: He had great gifts for composition. Unfortunately, his greatest work is shrouded in mystery, for he never reached the point of publishing it. Some believe that he had it all in his mind; others are more unkind, saying that he probably never worked it out at all, but merely blustered about it.

Achilles: What was the nature of this magnum opus?

Anteater: It was to be a giant prelude and fugue; the fugue was to have twenty-four voices, and to involve twenty-four distinct subjects, one in each of the major and minor keys.

Achilles: It would certainly be hard to listen to a twenty-four-voice fugue as a whole!

Crab: Not to mention composing one!

Anteater: But all that we know of it is Sebastiant's description of it, which he wrote in the margin of his copy of Buxtehude's Preludes and Fugues for Organ. The last words which he wrote before his tragic demise were:

> I have composed a truly marvelous fugue. In it, I have added together the power of 24 keys, and the power of 24 themes; I came up with a fugue with the power of 24 voices. Unfortunately, this margin is too narrow to contain it.

And the unrealized masterpiece simply goes by the name, "Fermant's Last Fugue".

Achilles: Oh, that is unbearably tragic.

Tortoise: Speaking of fugues, this fugue which we have been listening to is nearly over. Towards the end, there occurs a strange new twist on its theme. *(Flips the page in the* Well-Tempered Clavier.*)* Well, what have we here? A new illustration—how appealing! *(Shows it to the Crab.)*

FIGURE 64. [*Drawing by the author.*]

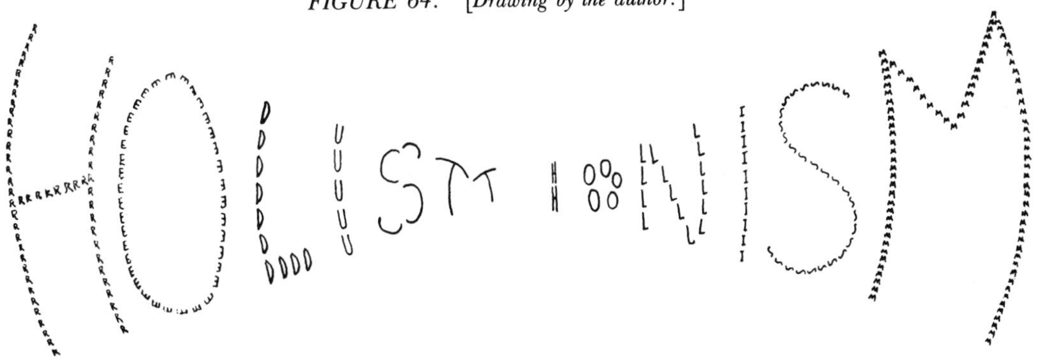

Crab: Well, what have we here? Oh, I see: it's "HOLISMIONISM", written in large letters that first shrink and then grow back to their original size. But that doesn't make any sense, because it's not a word. Oh me, oh my! *(Passes it to the Anteater.)*

Anteater: Well, what have we here? Oh, I see: it's "REDUCTHOLISM", written in small letters that first grow and then shrink back to their original size. But that doesn't make any sense, because it's not a word. Oh my, oh me! *(Passes it to Achilles.)*

Achilles: I know the rest of you won't believe this, but in fact this picture consists of the word "HOLISM" written twice, with the letters continually shrinking as they proceed from left to right. *(Returns it to the Tortoise.)*

Tortoise: I know the rest of you won't believe this, but in fact this picture consists of the word "REDUCTIONISM" written once, with the letters continually growing as they proceed from left to right.

Achilles: At last—I heard the new twist on the theme this time! I am so glad that you pointed it out to me, Mr. Tortoise. Finally, I think I am beginning to grasp the art of listening to fugues.

CHAPTER XI

Brains and Thoughts

New Perspectives on Thought

IT WAS ONLY with the advent of computers that people actually tried to create "thinking" machines, and witnessed bizarre variations on the theme of thought. Programs were devised whose "thinking" was to human thinking as a slinky flipping end over end down a staircase is to human locomotion. All of a sudden the idiosyncracies, the weaknesses and powers, the vagaries and vicissitudes of human thought were hinted at by the newfound ability to experiment with alien, yet hand-tailored forms of thought—or approximations of thought. As a result, we have acquired, in the last twenty years or so, a new kind of perspective on what thought is, and what it is not. Meanwhile, brain researchers have found out much about the small-scale and large-scale hardware of the brain. This approach has not yet been able to shed much light on how the brain manipulates concepts, but it gives us some ideas about the biological mechanisms on which thought manipulation rests.

In the coming two Chapters, then, we will try to unite some insights gleaned from attempts at computer intelligence with some of the facts learned from ingenious experiments on living animal brains, as well as with results from research on human thought processes done by cognitive psychologists. The stage has been set by the *Prelude, Ant Fugue*; now we develop the ideas more deeply.

Intensionality and Extensionality

Thought must depend on *representing reality in the hardware of the brain.* In the preceding Chapters, we have developed formal systems which represent domains of mathematical reality in their symbolisms. To what extent is it reasonable to use such formal systems as models for how the brain might manipulate ideas?

We saw, in the pq-system and then in other more complicated systems, how meaning, in a limited sense of the term, arose as a result of an isomorphism which maps typographical symbols onto numbers, operations, and relations; and strings of typographical symbols onto statements. Now in the brain we don't have typographical symbols, but we have something even better: active elements which can store information and transmit it and receive it from other active elements. Thus we have *active* symbols, rather than passive typographical symbols. In the brain, the rules are mixed right in with the symbols themselves, whereas on paper, the symbols are static entities, and the rules are in our heads.

It is important not to get the idea, from the rather strict nature of all the formal systems we have seen, that the isomorphism between symbols and real things is a rigid, one-to-one mapping, like the strings which link a marionette and the hand guiding it. In TNT, the notion "fifty" can be expressed in different symbolic ways; for example,

((SSSSSSS0·SSSSSSS0)+(S0·S0))
((SSSSS0·SSSSS0)+(SSSSS0·SSSSS0))

That these both represent the same number is not a priori clear. You can manipulate each expression independently, and at some point stumble across a theorem which makes you exclaim, "Oh—it's *that* number!"

In your mind, you can also have different mental descriptions for a single person; for example,

> The person whose book I sent to a friend in Poland a while back.
>
> The stranger who started talking with me and my friends tonight in this coffee house.

That they both represent the same person is not a priori clear. Both descriptions may sit in your mind, unconnected. At some point during the evening you may stumble across a topic of conversation which leads to the revelation that they designate the same person, making you exclaim, "Oh—you're *that* person!"

Not all descriptions of a person need be attached to some central symbol for that person, which stores the person's name. Descriptions can be manufactured and manipulated in themselves. We can invent nonexistent people by making descriptions of them; we can merge two descriptions when we find they represent a single entity; we can split one description into two when we find it represents two things, not one—and so on. This "calculus of descriptions" is at the heart of thinking. It is said to be *intensional* and not *extensional,* which means that descriptions can "float" without being anchored down to specific, known objects. The intensionality of thought is connected to its flexibility; it gives us the ability to imagine hypothetical worlds, to amalgamate different descriptions or chop one description into separate pieces, and so on.

Suppose a friend who has borrowed your car telephones you to say that your car skidded off a wet mountain road, careened against a bank, and overturned, and she narrowly escaped death. You conjure up a series of images in your mind, which get progressively more vivid as she adds details, and in the end you "see it all in your mind's eye". Then she tells you that it's all been an April Fool's joke, and both she and the car are fine! In many ways that is irrelevant. The story and the images lose nothing of their vividness, and the memory will stay with you for a long, long time. Later, you may even think of her as an unsafe driver because of the strength of the first impression, which should have been wiped out when you learned it was all untrue. Fantasy and fact intermingle very closely in our minds, and this is because thinking involves the manufacture and manipulation of complex descriptions, which need in no way be tied down to real events or things.

A flexible, intensional representation of the world is what thinking is all about. Now how can a physiological system such as the brain support such a system?

The Brain's "Ants"

The most important cells in the brain are nerve cells, or *neurons* (see Fig. 65), of which there are about ten billion. (Curiously, outnumbering the neurons by about ten to one are the glial cells, or glia. Glia are believed to play more of a supporting role to the neurons' starring role, and therefore we will not discuss them.) Each neuron possesses a number of *synapses* ("entry ports") and one *axon* ("output channel"). The input and output are electrochemical flows: that is, moving ions. In between the entry ports of a neuron and its output channel is its cell body, where "decisions" are made.

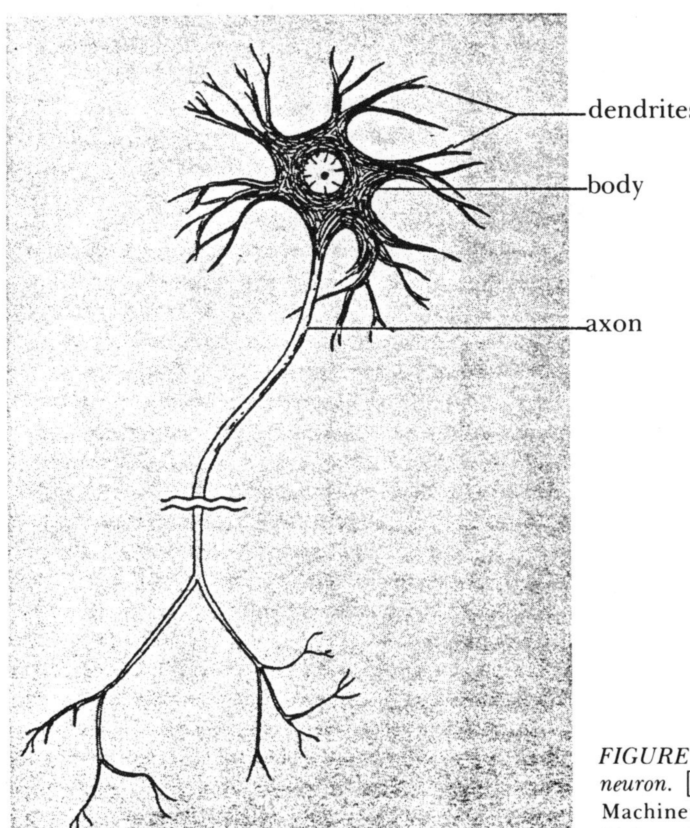

FIGURE 65. Schematic drawing of a neuron. [Adapted From D. Wooldridge, The Machinery of the Brain *(New York: McGraw-Hill, 1963), p. 6.]*

The type of decision which a neuron faces—and this can take place up to a thousand times per second—is this: whether or not to *fire*—that is, to release ions down its axon, which eventually will cross over into the entry ports of one or more *other* neurons, thus causing them to make the same sort of decision. The decision is made in a very simple manner: if the sum of all inputs exceeds a certain threshold, *yes;* otherwise, *no.* Some of the inputs can be negative inputs, which cancel out positive inputs coming from somewhere else. In any case, it is simple addition which rules the lowest level of the mind. To paraphrase Descartes' famous remark, "I think, therefore I sum" (from the Latin *Cogito, ergo am*).

Now although the manner of making the decision sounds very simple, there is one fact which complicates the issue: there may be as many as 200,000 separate entry ports to a neuron, which means that up to 200,000 separate summands may be involved in determining the neuron's next action. Once the decision has been made, a pulse of ions streaks down the axon towards its terminal end. Before the ions reach the end, however, they may encounter a bifurcation—or several. In such cases, the single output pulse splits up as it moves down the bifurcating axon, and by the time it has reached the end, "it" has become "they"—and they may reach their destinations at separate times, since the axon branches along which they travel may be of different lengths and have different resistivities. The important thing, though, is that they all began as one single pulse, moving away from the cell body. After a neuron fires, it needs a short recovery time before firing again; characteristically this is measured in milliseconds, so that a neuron may fire up to about a thousand times per second.

Larger Structures in the Brain

Now we have described the brain's "ants". What about "teams", or "signals"? What about "symbols"? We make the following observation: despite the complexity of its input, a single neuron can respond only in a very primitive way—by firing, or not firing. This is a very small amount of information. Certainly for large amounts of information to be carried or processed, many neurons must be involved. And therefore one might guess that larger structures, composed from many neurons, would exist, which handle concepts on a higher level. This is undoubtedly true, but the most naïve assumption—that there is a fixed group of neurons for each different concept—is almost certainly false.

There are many anatomical portions of the brain which can be distinguished from each other, such as the cerebrum, the cerebellum, the hypothalamus (see Fig. 66). The *cerebrum* is the largest part of the human brain, and is divided into a left hemisphere and a right hemisphere. The outer few millimeters of each cerebral hemisphere are coated with a layered "bark", or *cerebral cortex.* The amount of cerebral cortex is the major distinguishing feature, in terms of anatomy, between human brains and brains of less intelligent species. We will not describe any of the brain's suborgans in detail because, as it turns out, only the roughest mapping can

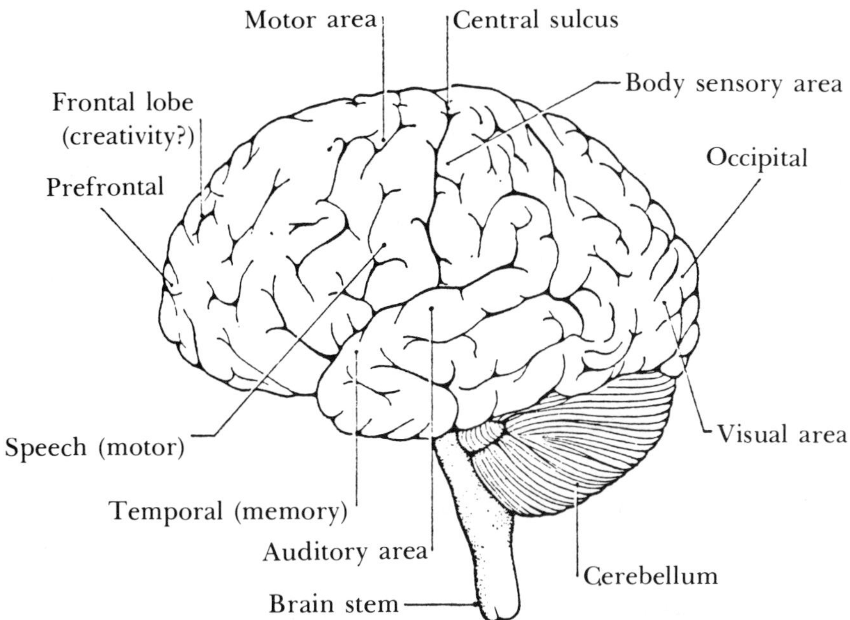

FIGURE 66. The human brain, seen from the left side. It is strange that the visual area is in the back of the head. [From Steven Rose, The Conscious Brain, *updated ed. (New York: Vintage, 1966), p. 50.]*

at this time be made between such large-scale suborgans and the activities, mental or physical, which they are responsible for. For instance, it is known that language is primarily handled in one of the two cerebral hemispheres—in fact, usually the left hemisphere. Also, the *cerebellum* is the place where trains of impulses are sent off to muscles to control motor activity. But how these areas carry out their functions is still largely a mystery.

Mappings between Brains

Now an extremely important question comes up here. If thinking does take place in the brain, then how are two brains different from each other? How is my brain different from yours? Certainly you do not think exactly as I do, nor as anyone else does. But we all have the same anatomical divisions in our brains. How far does this identity of brains extend? Does it go to the neural level? Yes, if you look at animals on a low enough level of the thinking-hierarchy—the lowly earthworm, for instance. The following quote is from the neurophysiologist, David Hubel, speaking at a conference on communication with extraterrestrial intelligence:

> The number of nerve cells in an animal like a worm would be measured, I suppose, in the thousands. One very interesting thing is that we may point to a particular individual cell in a particular earthworm, and then identify the same cell, the corresponding cell in another earthworm of the same species.[1]

Earthworms have isomorphic brains! One could say, "There is only one earthworm."

But such one-to-one mappability between individuals' brains disappears very soon as you ascend in the thinking-hierarchy and the number of neurons increases—confirming one's suspicions that there is not just one human! Yet considerable physical similarity can be detected between different human brains when they are compared on a scale larger than a single neuron but smaller than the major suborgans of the brain. What does this imply about how individual mental differences are represented in the physical brain? If we looked at my neurons' interconnections, could we find various structures that could be identified as coding for specific things I know, specific beliefs I have, specific hopes, fears, likes and dislikes I harbor? If mental experiences can be attributed to the brain, can knowledge and other aspects of mental life likewise be traced to specific locations inside the brain, or to specific physical subsystems of the brain? This will be a central question to which we will often return in this Chapter and the next.

Localization of Brain Processes: An Enigma

In an attempt to answer this question, the neurologist Karl Lashley, in a long series of experiments beginning around 1920 and running for many years, tried to discover where in its brain a rat stores its knowledge about maze running. In his book *The Conscious Brain,* Steven Rose describes Lashley's trials and tribulations this way:

> Lashley was attempting to identify the locus of memory within the cortex, and, to do so, first trained rats to run mazes, and then removed various cortical regions. He allowed the animals to recover and tested the retention of the maze-running skills. To his surprise it was not possible to find a particular region corresponding to the ability to remember the way through a maze. Instead all the rats which had had cortex regions removed suffered some kind of impairment, and the extent of the impairment was roughly proportional to the amount of cortex taken off. Removing cortex damaged the motor and sensory capacities of the animals, and they would limp, hop, roll, or stagger, but somehow they always managed to traverse the maze. So far as memory was concerned, the cortex appeared to be equipotential, that is, with all regions of equal possible utility. Indeed, Lashley concluded rather gloomily in his last paper "In Search of the Engram", which appeared in 1950, that the only conclusion was that memory was not possible at all.[2]

Curiously, evidence for the opposite point of view was being developed in Canada at roughly the same time that Lashley was doing his last work, in the late 1940's. The neurosurgeon Wilder Penfield was examining the reactions of patients whose brains had been operated on, by inserting electrodes into various parts of their exposed brains, and then using small electrical pulses to stimulate the neuron or neurons to which the electrodes had been attached. These pulses were similar to the pulses which come from other neurons. What Penfield found was that stimulation of certain neurons would reliably create specific images or sensations in the patient. These artificially provoked impressions ranged from strange but indefinable fears to buzzes and colors, and, most impressively of all, to entire successions of events recalled from some earlier time of life, such as a childhood birthday party. The set of locations which could trigger such specific events was extremely small—basically centered upon a single neuron. Now these results of Penfield dramatically oppose the conclusions of Lashley, since they seem to imply that local areas are responsible for specific memories, after all.

What can one make of this? One possible explanation could be that memories are coded locally, but over and over again in different areas of the cortex—a strategy perhaps developed in evolution as security against possible loss of cortex in fights, or in experiments conducted by neurophysiologists. Another explanation would be that memories can be reconstructed from dynamic processes spread over the whole brain, but can be triggered from local spots. This theory is based on the notion of modern telephone networks, where the routing of a long-distance call is not predictable in advance, for it is selected at the time the call is placed, and depends on the situation all over the whole country. Destroying any local part of the network would not block calls; it would just cause them to be routed around the damaged area. In this sense any call is potentially nonlocalizable. Yet any call just connects up two specific points; in this sense any call is localizable.

Specificity in Visual Processing

Some of the most interesting and significant work on localization of brain processes has been done in the last fifteen years by David Hubel and Torsten Wiesel, at Harvard. They have mapped out visual pathways in the brains of cats, starting with the neurons in the retina, following their connections towards the rear of the head, passing through the "relay station" of the lateral geniculate, and ending up in the visual cortex, at the very back of the brain. First of all, it is remarkable that there exist well-defined neural pathways, in light of Lashley's results. But more remarkable are the properties of the neurons located at different stages along the pathway.

It turns out that retinal neurons are primarily contrast sensors. More specifically, the way they act is this. Each retinal neuron is normally firing at a "cruising speed". When its portion of the retina is struck by light, it may either fire faster or slow down and even stop firing. However, it will do so only provided that the surrounding part of the retina is *less* illuminated. So this means that there are two types of neuron: "on-center", and "off-center". The *on-center* neurons are those whose firing rate increases whenever, in the small circular retinal area to which they are sensitive, the center is bright but the outskirts are dark; the *off-center* neurons are those which fire faster when there is darkness in the center and brightness in the outer ring. If an on-center pattern is shown to an off-center neuron, the neuron will *slow down* in firing (and vice versa). Uniform illumination will leave both types of retinal neuron unaffected; they will continue to fire at cruising speed.

From the retina, signals from these neurons proceed via the optic nerve to the lateral geniculate, located somewhere towards the middle of the brain. There, one can find a direct mapping of the retinal surface in the sense that there are lateral-geniculate neurons which are triggered only by specific stimuli falling on specific areas of the retina. In that sense, the lateral geniculate is disappointing; it seems to be only a "relay station", and not a further processor (although to give it its due, the contrast sensitivity seems to be enhanced in the lateral geniculate). The retinal image is coded in a straightforward way in the firing patterns of the neurons in the lateral geniculate, despite the fact that the neurons there are not arranged on a two-dimensional surface in the form of the retina, but in a three-dimensional block. So two dimensions get mapped onto three, yet the information is preserved: an isomorphism. There is probably some deep meaning to the change in the dimensionality of the representation, which is not yet fully appreciated. In any case, there are so many further unexplained stages of vision that we should not be disappointed but pleased by the fact that—to some extent—we have figured out this one stage!

From the lateral geniculate, the signals proceed back to the visual cortex. Here, some new types of processing occur. The cells of the visual cortex are divided into three categories: simple, complex, and hypercomplex. *Simple* cells act very much like retinal cells or lateral geniculate cells: they respond to point-like light or dark spots with contrasting surrounds, in particular regions of the retina. *Complex* cells, by contrast, usually receive input from a hundred or more other cells, and they detect light or dark bars oriented at specific angles on the retina (see Fig. 67). *Hypercomplex* cells respond to corners, bars, or even "tongues" moving in specific directions (again see Fig. 67). These latter cells are so highly specialized that they are sometimes called "higher-order hypercomplex cells".

A "Grandmother Cell"?

Because of the discovery of cells in the visual cortex which can be triggered by stimuli of ever-increasing complexity, some people have wondered if things are not leading in the direction of "one cell, one concept"—for example, you would have a "grandmother cell" which would fire if, and only if, your grandmother came into view. This somewhat humorous example of a "superhypercomplex cell" is not taken very seriously. However, it is not obvious what alternative theory seems reasonable. One possibility is that larger neural networks are excited collectively by sufficiently complex visual stimuli. Of course, the triggering of these larger multineuron units would somehow have to come from integration of signals emanating from the many hypercomplex cells. How this might be done, nobody knows. Just when we seem to be approaching the threshold where

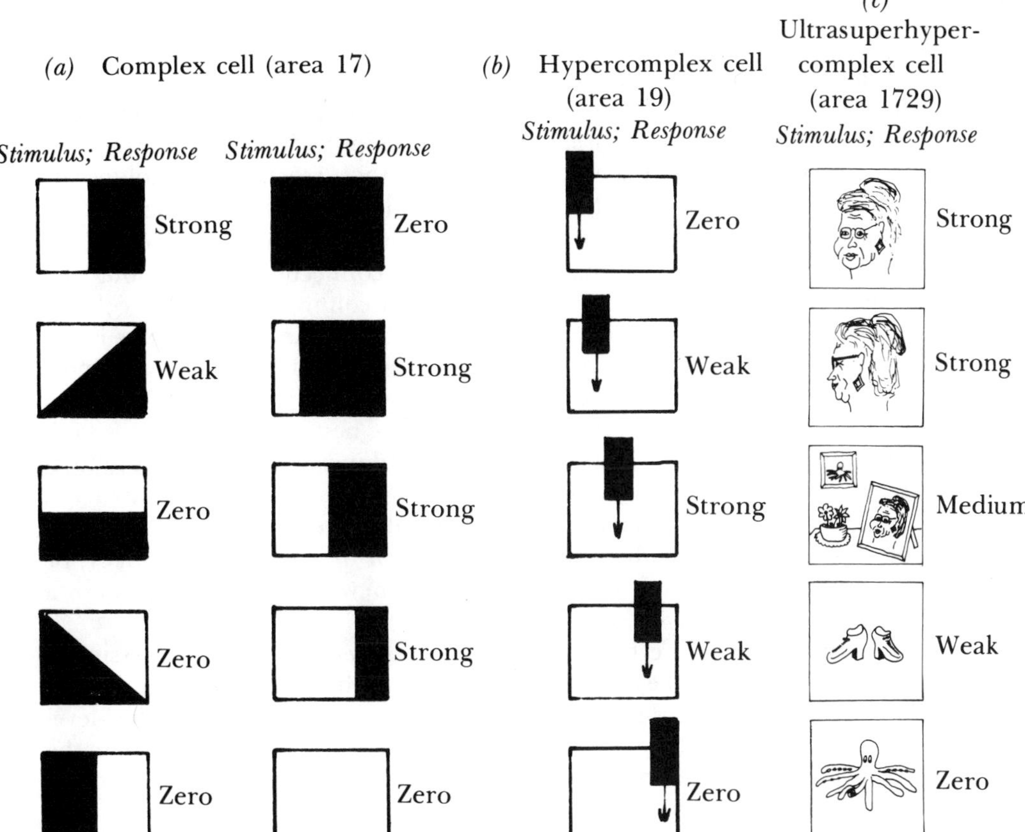

FIGURE 67. Responses to patterns by certain sample neurons.

(a) This edge-detecting neuron looks for vertical edges with light on the left and dark on the right. The first column shows how the orientation *of an edge is relevant to this neuron. The second column shows how the* position *of the edge within the field is irrelevant, for this particular neuron.*

(b) Showing how a hypercomplex cell responds more selectively: here, only when the descending tongue is in the middle of the field.

(c) The responses of a hypothetical "grandmother cell" to various random stimuli; the reader may enjoy pondering how an "octopus cell" would respond to the same stimuli.

"symbol" might emerge from "signal", the trail gets lost—a tantalizingly unfinished story. We will return to this story shortly, however, and try to fill in some of it.

Earlier I mentioned the coarse-grained isomorphism between all human brains which exists on a large anatomical scale, and the very fine-grained, neural-level isomorphism which exists between earthworm brains. It is quite interesting that there is also an isomorphism between the visual processing apparatus of cat, monkey, and human, the "grain" of which is somewhere between coarse and fine. Here is how that isomorphism works. First of all, all three species have "dedicated" areas of cortex at the back of their brains where visual processing is done: the *visual cortex*. Secondly, in each of them, the visual cortex breaks up into three subregions, called areas 17, 18, and 19 of the cortex. These areas are still universal, in the sense that they can be located in the brain of any normal individual in any of the three species. Within each area you can go still further, reaching the "columnar" organization of the visual cortex. Perpendicular to the surface of the cortex, moving radially inwards towards the inner brain, visual neurons are arranged in "columns"—that is, almost all connections move along the radial, columnar direction, and not between columns. And each column maps onto a small, specific retinal region. The number of columns is not the same in each individual, so that one can't find "the same column". Finally, within a column, there are layers in which simple neurons tend to be found, and other layers in which complex neurons tend to be found. (The hypercomplex neurons tend to be found in areas 18 and 19 predominantly, while the simple and complex ones are found mostly in area 17.) It appears that we run out of isomorphisms at this level of detail. From here on down to the individual neuron level, each individual cat, monkey, or human has a completely unique pattern—somewhat like a fingerprint or a signature.

One minor but perhaps telling difference between visual processing in cats' brains and monkeys' brains has to do with the stage at which information from the two eyes is integrated to yield a single combined higher-level signal. It turns out that it takes place slightly later in the monkey than in the cat, which gives each separate eye's signal a slightly longer time to get processed by itself. This is not too surprising, since one would expect that the higher a species lies in the intelligence hierarchy, the more complex will be the problems which its visual system will be called upon to handle; and therefore signals ought to pass through more and more early processing before receiving a final "label". This is quite dramatically confirmed by observations of the visual abilities of a newborn calf, which seems to be born with as much power of visual discrimination as it will ever have. It will shy away from people or dogs, but not from other cattle. Probably its entire visual system is "hard-wired" before birth, and involves relatively little cortical processing. On the other hand, a human's visual system, so deeply reliant on the cortex, takes several years to reach maturity.

Funneling into Neural Modules

A puzzling thing about the discoveries so far made about the organization of the brain is that few direct correspondences have been found between large-scale hardware and high-level software. The visual cortex, for instance, is a large-scale piece of hardware, which is entirely dedicated to a clear software purpose—the processing of visual information—yet all of the processing so far discovered is still quite low-level. Nothing approaching recognition of *objects* has been localized in the visual cortex. This means that no one knows where or how the output from complex and hypercomplex cells gets transformed into conscious recognition of shapes, rooms, pictures, faces, and so on. People have looked for evidence of the "funneling" of many low-level neural responses into fewer and fewer higher-level ones, culminating in something such as the proverbial grandmother cell, or some kind of multineuron network, as mentioned above. It is evident that this will not be found in some gross anatomical division of the brain, but rather in a more microscopic analysis.

One possible alternative to the the grandmother cell might be a fixed set of neurons, say a few dozen, at the thin end of the "funnel", all of which fire when Granny comes into view. And for each different recognizable object, there would be a unique network and a funneling process that would focus down onto that network. There are more complicated alternatives along similar lines, involving networks which can be excited in different manners, instead of in a fixed manner. Such networks would be the "symbols" in our brains.

But is such funneling necessary? Perhaps an object being looked at is implicitly identified by its "signature" in the visual cortex—that is, the collected responses of simple, complex, and hypercomplex cells. Perhaps the brain does not need any further recognizer for a particular form. This theory, however, poses the following problem. Suppose you are looking at a scene. It registers its signature on your visual cortex; but then how do you get from that signature to a verbal description of the scene? For instance, the paintings of Edouard Vuillard, a French post-impressionist, often take a few seconds of scrutiny, and then suddenly a human figure will jump out at you. Presumably the signature gets imprinted on the visual cortex in the first fraction of a second—but the picture is only understood after a few seconds. This is but one example of what is actually a common phenomenon—a sensation of something "crystallizing" in your mind at the moment of recognition, which takes place not when the light rays hit your retina, but sometime later, after some part of your intelligence has had a chance to act on the retinal signals.

The crystallization metaphor yields a pretty image derived from statistical mechanics, of a myriad microscopic and uncorrelated activities in a medium, slowly producing local regions of coherence which spread and enlarge; in the end, the myriad small events will have performed a complete structural revamping of their medium from the bottom up, changing it from a chaotic assembly of independent elements into one large, coherent, fully linked structure. If one thinks of the early neural activities as independent, and of the end result of their many independent firings as the triggering of a well-defined large "module" of neurons, then the word "crystallization" seems quite apt.

Another argument for funneling is based on the fact that there are a myriad distinct scenes which can cause you to feel you have perceived the same object—for example, your grandmother, who may be smiling or frowning, wearing a hat or not, in a bright garden or a dark train station, seen from near or far, from side or front, and so on. All these scenes produce extremely different signatures on the visual cortex; yet all of them could prompt you to say "Hello, Granny." So a funneling process must take place at some point after the reception of the visual signature and before the words are uttered. One could claim that this funneling is not part of the perception of Granny, but just part of verbalization. But it seems quite unnatural to partition the process that way, for you could internally use the information that it is Granny without verbalizing it. It would be very unwieldy to handle all of the information in the entire visual cortex, when so much of it could be thrown away, since you don't care about where shadows fall or how many buttons there are on her blouse, etc.

Another difficulty with a non-funneling theory is to explain how there can be different interpretations for a single signature—for example, the Escher picture *Convex and Concave* (Fig. 23). Just as it seems obvious to us that we do not merely perceive *dots* on a television screen, but *chunks*, likewise it seems ridiculous to postulate that perception has taken place when a giant dot-like "signature" has been created on the visual cortex. There must be some funneling, whose end result is to trigger some specific modules of neurons, each of which is associated with the concepts—the chunks—in the scene.

Modules Which Mediate Thought Processes

Thus we are led to the conclusion that for each concept there is a fairly well-defined module which can be triggered—a module that consists of a small group of neurons—a "neural complex" of the type suggested earlier. A problem with this theory—at least if it is taken naïvely—is that it would suggest that one should be able to locate such modules somewhere within the brain. This has not yet been done, and some evidence, such as the experiments by Lashley, points against localization. However, it is still too early to tell. There may be many copies of each module spread around, or modules may overlap physically; both of these effects would tend to obscure any division of neurons into "packets". Perhaps the complexes are like very thin pancakes packed in layers which occasionally pass through each other; perhaps they are like long snakes which curl around each other, here and there flattening out, like cobras' heads; perhaps they are like spiderwebs; or perhaps they are circuits in which signals travel round and round in shapes stranger than the dash of a gnat-hungry swallow. There is no telling. It is even possible that these modules are software, rather than hardware, phenomena—but this is something which we will discuss later.

There are many questions that come to mind concerning these hypothesized neural complexes. For instance:

> Do they extend into the lower regions of the brain, such as the midbrain, the hypothalamus, etc.?
> Can a single neuron belong to more than one such complex?
> To how many such complexes can a single neuron belong?
> By how many neurons can such complexes overlap?

Are these complexes pretty much the same for everybody?
Are corresponding ones found in corresponding places in different people's brains?
Do they overlap in the same way in everybody's brain?

Philosophically, the most important question of all is this: What would the existence of modules—for instance, a grandmother module—tell us? Would this give us any insight into the phenomenon of our own consciousness? Or would it still leave us as much in the dark about what consciousness is, as does knowledge that a brain is built out of neurons and glia? As you might guess from reading the *Ant Fugue,* my feeling is that it would go a long way towards giving us an understanding of the phenomenon of consciousness. The crucial step that needs to be taken is from a low-level—neuron-by-neuron—description of the state of a brain, to a high-level—module-by-module—description of the same state of the same brain. Or, to revert to the suggestive terminology of the *Ant Fugue,* we want to shift the description of the brain state from the *signal* level to the *symbol* level.

Active Symbols

Let us from now on refer to these hypothetical neural complexes, neural modules, neural packets, neural networks, multineuron units—call them what you will, whether they come in the form of pancakes, garden rakes, rattlesnakes, snowflakes, or even ripples on lakes—as *symbols.* A description of a brain state in terms of symbols was alluded to in the Dialogue. What would such a description be like? What kinds of concepts is it reasonable to think actually might be "symbolized"? What kinds of interrelations would symbols have? And what insights would this whole picture provide into consciousness?

The first thing to emphasize is that symbols can be either *dormant,* or *awake* (activated). An active symbol is one which has been triggered—that is, one in which a threshold number of neurons have been caused to fire by stimuli coming from outside. Since a symbol can be triggered in many different ways, it can act in many different ways when awakened. This suggests that we should think of a symbol not as a fixed entity, but as a variable entity. Therefore it would not suffice to describe a brain state by saying "Symbols A, B, . . ., N are all active"; rather, we would have to supply in addition a set of parameters for each active symbol, characterizing some aspects of the symbol's internal workings. It is an interesting question whether in each symbol there are certain core neurons, which invariably fire when the symbol is activated. If such a core set of neurons exists, we might refer to it as the "invariant core" of the symbol. It is tempting to assume that each time you think of, say, a waterfall, some fixed neural process is repeated, without doubt embellished in different ways depending on the context, but reliably occurring. However, it is not clear that this must be so.

Now what does a symbol do, when awakened? A low-level description would say, "Many of its neurons fire." But this no longer interests us. The high-level description should eliminate all reference to neurons, and concentrate exclusively on symbols. So a high-level description of what makes a symbol active, as distinguished from dormant, would be, "It sends out *messages,* or signals, whose purpose is to try to awaken, or trigger, other symbols." Of course these messages would be carried as streams of nerve impulses, by neurons—but to the extent that we can avoid such phraseology, we should, for it represents a low-level way of looking at things, and we hope that we can get along on purely a high level. In other words, we hope that thought processes can be thought of as being sealed off from neural events in the same way that the behavior of a clock is sealed off from the laws of quantum mechanics, or the biology of cells is sealed off from the laws of quarks.

But what is the advantage of this high-level picture? Why is it better to say, "Symbols A and B triggered symbol C" than to say, "Neurons 183 through 612 excited neuron 75 and caused it to fire"? This question was answered in the *Ant Fugue:* It is better because symbols *symbolize* things, and neurons don't. Symbols are the hardware realizations of concepts. Whereas a group of neurons triggering another neuron corresponds to no outer event, the triggering of some symbol by other symbols bears a relation to events in the real world—or in an imaginary world. Symbols are related to each other by the messages which they can send back and forth, in such a way that their triggering patterns are very much like the large-scale events which do happen in our world, or could happen in a world similar to ours. In essence, meaning arises here for the same reason as it did in the pq-system—isomorphism; only here, the isomorphism is infinitely more complex, subtle, delicate, versatile, and intensional.

Incidentally, the requirement that symbols should be able to pass sophisticated messages to and fro is probably sufficient to exclude neurons themselves from playing the role of symbols. Since a neuron has only a single way of sending information out of itself, and has no way of selectively directing a signal now in one direction, now in another, it simply does not have the kind of selective triggering power which a symbol must have to act like an object in the real world. In his book *The Insect Societies,* E. O. Wilson makes a similar point about how messages propagate around inside ant colonies:

> [Mass communication] is defined as the transfer, among groups, of information that a single individual could not pass to another.[3]

It is not such a bad image, the brain as an ant colony!

The next question—and an extremely important one it is, too—concerns the nature and "size" of the concepts which are represented in the brain by single symbols. About the nature of symbols there are questions like this: Would there be a symbol for the general notion of waterfalls, or would there be different symbols for various specific waterfalls? Or would both of these alternatives be realized? About the "size" of symbols, there are questions like this: Would there be a symbol for an entire story? Or for a melody? Or a joke? Or is it more likely that there would only be symbols for concepts roughly the size of words, and that larger ideas, such as phrases or sentences, would be represented by concurrent or sequential activation of various symbols?

Let us consider the issue of the size of concepts represented by symbols. Most thoughts expressed in sentences are made up out of basic, quasi-atomic components which we do not usually analyze further. These are of word size, roughly—sometimes a little longer, sometimes a little shorter. For instance, the noun "waterfall", the proper noun "Niagara Falls", the past-tense suffix "-ed", the verb "to catch up with", and longer idiomatic phrases are all close to atomic. These are typical elementary brush strokes which we use in painting portraits of more complex concepts, such as the plot of a movie, the flavor of a city, the nature of consciousness, etc. Such complex ideas are not single brush strokes. It seems reasonable to think that the brush strokes of language are also brush strokes of thought, and therefore that symbols represent concepts of about this size. Thus a symbol would be roughly something for which you know a word or stock phrase, or with which you associate a proper name. And the representation in the brain of a more complex idea, such as a problem in a love affair, would be a very complicated sequence of activations of various symbols by other symbols.

Classes and Instances

There is a general distinction concerning thinking: that between *categories* and *individuals,* or *classes* and *instances.* (Two other terms sometimes used are "types" and "tokens".) It might seem at first sight that a given symbol would inherently be either a symbol for a class or a symbol for an instance—but that is an oversimplification. Actually, most symbols may play either role, depending on the context of their activation. For example, look at the list below:

(1) a publication
  (2) a newspaper
    (3) *The San Francisco Chronicle*
      (4) the May 18 edition of the *Chronicle*
        (5) my copy of the May 18 edition of the *Chronicle*
          (6) my copy of the May 18 edition of the *Chronicle* as it was when I first picked it up (as contrasted with my copy as it was a few days later: in my fireplace, burning)

Here, lines 2 to 5 all play both roles. Thus, line 4 is an instance of of the general class of line 3, and line 5 is an instance of line 4. Line 6 is a special kind of instance of a class: a *manifestation.* The successive stages of an object during its life history are its manifestations. It is interesting to wonder if the cows on a farm perceive the invariant individual underneath all the manifestations of the jolly farmer who feeds them hay.

The Prototype Principle

The list above seems to be a hierarchy of generality—the top being a very broad conceptual category, the bottom some very humble particular thing located in space and time. However, the idea that a "class" must always be enormously broad and abstract is far too limited. The reason is that our thought makes use of an ingenious principle, which might be called the *prototype principle:*

> The most specific event can serve as a general example of a class of events.

Everyone knows that specific events have a vividness which imprints them so strongly on the memory that they can later be used as models for other events which are like them in some way. Thus in each specific event, there is the germ of a whole class of similar events. This idea that there is generality in the specific is of far-reaching importance.

Now it is natural to ask: Do the symbols in the brain represent classes, or instances? Are there certain symbols which represent only classes, while other symbols represent only instances? Or can a single symbol serve duty either as a class symbol or instance symbol, depending which parts of it are activated? The latter theory seems appealing; one might think that a "light" activation of a symbol might represent a class, and that a deeper, or more complex, activation would contain more detailed internal neural firing patterns, and hence would represent an instance. But on second thought, this is crazy: it would imply, for example, that by activating the symbol for "publication" in a sufficiently complex way, you would get the very complex symbol which represents a specific newspaper burning in my fireplace. And every other possible manifestation of every other piece of printed matter would be represented internally by some manner of activating the single symbol for "publication". That seems much too heavy a burden to place on the single symbol "publication". One must conclude, therefore, that instance symbols can exist side by side with class symbols, and are not just modes of activation of the latter.

The Splitting-off of Instances from Classes

On the other hand, instance symbols often inherit many of their properties from the classes to which those instances belong. If I tell you I went to see a movie, you will begin "minting" a fresh new instance symbol for that particular movie; but in the absence of more information, the new instance symbol will have to lean rather heavily on your pre-existing class symbol for "movie". Unconsciously, you will rely on a host of presuppositions about that movie—for example, that it lasted between one and three hours, that it was shown in a local theater, that it told a story about some people, and so on. These are built into the class symbol as expected links to other symbols (i.e., potential triggering relations), and are called *default options*. In any freshly minted instance symbol, the default options can easily be overridden, but unless this is explicitly done, they will remain in the instance symbol, inherited from its class symbol. Until they are overridden, they provide some preliminary basis for you to think about the new instance—for example, the movie I went to see—by using the reasonable guesses which are supplied by the "stereotype", or class symbol.

A fresh and simple instance is like a child without its own ideas or experiences—it relies entirely on its parents' experiences and opinions and just parrots them. But gradually, as it interacts more and more with the rest of the world, the child acquires its own idiosyncratic experiences and inevitably begins to split away from the parents. Eventually, the child becomes a full-fledged adult. In the same way, a fresh instance can split off from its parent class over a period of time, and become a class, or prototype, in its own right.

For a graphic illustration of such a splitting-off process, suppose that some Saturday afternoon you turn on your car radio, and happen to tune in on a football game between two "random" teams. At first you do not know the names of the players on either team. All you register, when the announcer says, "Palindromi made the stop on the twenty-seven yard line, and that brings up fourth down and six to go," is that some player stopped some other player. Thus it is a case of activation of the class symbol "football player", with some sort of coordinated activation of the symbol for tackling. But then as Palindromi figures in a few more key plays, you begin building up a fresh instance symbol for him in particular, using his name, perhaps, as a focal point. This symbol is dependent, like a child, on the class symbol for "football player": most of your image of Palindromi is supplied by your stereotype of a football player as contained in the "football player" symbol. But gradually, as more information comes to you, the "Palindromi" symbol becomes more autonomous, and relies less and less on concurrent activation of its parent class symbol. This may happen in a few minutes, as Palindromi makes a few good plays and stands out. His teammates may still all be represented by activations of the class symbol, however. Eventually, perhaps after a few days, when you have read some articles in the sports section of your paper, the umbilical cord is broken, and Palindromi can stand on his own two feet. Now you know such things as his home town and his major in college; you recognize his face; and so on. At this point, Palindromi is no longer conceived of merely as a football player, but as a human being who happens also to be a football player. "Palindromi" is an instance symbol which can become active while its parent class symbol (football player) remains dormant.

Once, the Palindromi symbol was a satellite orbiting around its mother symbol, like an artificial satellite circling the Earth, which is so much bigger and more massive. Then there came an intermediate stage, where one symbol was more important than the other, but they could be seen as orbiting around each other—something like the Earth and the Moon. Finally, the new symbol becomes quite autonomous; now it might easily serve as a class symbol around which could start rotating new satellites—

symbols for other people who are less familiar but who have something in common with Palindromi, and for whom he can serve as a temporary stereotype, until you acquire more information, enabling the new symbols also to become autonomous.

The Difficulty of Disentangling Symbols from Each Other

These stages of growth and eventual detachment of an instance from a class will be distinguishable from each other by the way in which the symbols involved are linked. Sometimes it will no doubt be very difficult to tell just where one symbol leaves off and the other one begins. How "active" is the one symbol, compared to the other? If one can be activated independently of the other, then it would be quite sensible to call them autonomous.

We have used an astronomy metaphor above, and it is interesting that the problem of the motion of planets is an extremely complex one—in fact the general problem of three gravitationally interacting bodies (such as the Earth, Moon, and Sun) is far from solved, even after several centuries of work. One situation in which it is possible to obtain good approximate solutions, however, is when one body is much more massive than the other two (here, the Sun); then it makes sense to consider that body as stationary, with the other two rotating about it; on top of this can finally be added the interaction between the two satellites. But this approximation depends on breaking up the system into the Sun, and a "cluster": the Earth-Moon system. This is an approximation, but it enables the system to be understood quite deeply. So to what extent is this cluster a part of reality, and to what extent is it a mental fabrication, a human imposition of structure on the universe? This problem of the "reality" of boundaries drawn between what are perceived to be autonomous or semi-autonomous clusters will create endless trouble when we relate it to symbols in the brain.

One greatly puzzling question is the simple issue of plurals. How do we visualize, say, three dogs in a teacup? Or several people in an elevator? Do we begin with the class symbol for "dog" and then rub three "copies" off of it? That is, do we manufacture three fresh instance symbols using the class symbol "dog" as template? Or do we jointly activate the symbols "three" and "dog"? By adding more or less detail to the scene being imagined, either theory becomes hard to maintain. For instance, we certainly do not have a separate instance symbol for each nose, mustache, grain of salt, etc., that we have ever seen. We let class symbols take care of such numerous items, and when we pass people on the street who have mustaches, we somehow just activate the "mustache" class symbol, without minting fresh instance symbols, unless we scrutinize them carefully.

On the other hand, once we begin to distinguish individuals, we cannot rely on a single class symbol (e.g., "person") to timeshare itself among all the different people. Clearly there must come into existence separate instance symbols for individual people. It would be ridiculous to imagine that this feat could be accomplished by "juggling"—that is, by the single class symbol flitting back and forth between several different modes of activation (one for each person).

Between the extremes, there must be room for many sorts of intermediate cases. There may be a whole hierarchy of ways of creating the class-instance distinction in the brain, giving rise to symbols—and symbol-organizations—of varying degrees of specificity. The following different kinds of individual and joint activation of symbols might be responsible for mental images of various degrees of specificity:

(1) various different modes or depths of activation of a single class symbol;
(2) simultaneous activation of several class symbols in some coordinated manner;
(3) activation of a single instance symbol;
(4) activation of a single instance symbol in conjunction with activation of several class symbols;
(5) simultaneous activation of several instance symbols and several class symbols in some coordinated manner.

This brings us right back to the question: "When is a symbol a distinguishable subsystem of the brain?" For instance, consider the second example—simultaneous activation of several class symbols in some coordinated manner. This could easily be what happens when "piano sonata" is the concept under consideration (the symbols for "piano" and "sonata" being at least two of the activated symbols). But if this pair of symbols gets activated in conjunction often enough, it is reasonable to assume that the link between them will become strong enough that they will act as a unit, when activated together in the proper way. So two or more symbols can act as one, under the proper conditions, which means that the problem of enumerating the number of symbols in the brain is trickier than one might guess.

Sometimes conditions can arise where two previously unlinked symbols get activated simultaneously and in a coordinated fashion. They may fit together so well that it seems like an inevitable union, and a single new symbol is formed by the tight interaction of the two old symbols. If this happens, would it be fair to say that the new symbol "always had been there but never had been activated"—or should one say that it has been "created"?

In case this sounds too abstract, let us take a concrete example: the Dialogue *Crab Canon.* In the invention of this Dialogue, two existing symbols—that for "musical crab canon", and that for "verbal dialogue"—had to be activated simultaneously and in some way forced to interact. Once this was done, the rest was quite inevitable: a new symbol—a class symbol—was born from the interaction of these two, and from then on it was able to be activated on its own. Now had it always been a dormant symbol in my brain? If so, then it must also have been a dormant symbol in the brain of every human who ever had its component symbols, even if it never was awakened in them. This would mean that to enumerate the symbols in anyone's brain, one would have to count all *dormant* symbols—all possible combinations and permutations of all types of activations of all known symbols. This would even include those fantastic creatures of software that one's brain invents when one is asleep—the strange mixtures of ideas which wake up when their host goes to sleep . . . The existence of these "potential symbols" shows that it is really a huge oversimplification to imagine that the brain is a well-defined collection of symbols in well-defined states of activation. It is much harder than that to pin down a brain state on the symbol level.

Symbols – Software or Hardware?

With the enormous and ever-growing repertoire of symbols that exist in each brain, you might wonder whether there eventually comes a point when the brain is saturated—when there is just no more room for a new symbol. This would come about, presumably, if symbols never overlapped each other—if a given neuron never served a double function, so that symbols would be like people getting into an elevator. "Warning: This brain has a maximum capacity of 350,275 symbols!"

This is not a necessary feature of the symbol model of brain function, however. In fact, overlapping and completely tangled symbols are probably the rule, so that each neuron, far from being a member of a unique symbol, is probably a functioning part of hundreds of symbols. This gets a little disturbing, because if it is true, then might it not just as easily be the case that each neuron is part of every single symbol? If that were so, then there would be no localizability whatsoever of symbols—every symbol would be identified with the whole of the brain. This would account for results like Lashley's cortex removal in rats—but it would also mean abandonment of our original idea of breaking the brain up into physically distinct subsystems. Our earlier characterization of symbols as "hardware realizations of concepts" could at best be a great oversimplification. In fact, if every symbol were made up of the same component neurons as every other symbol, then what sense would it make to speak of distinct symbols at all? What would be the signature of a given symbol's activation—that is, how could the activation of symbol A be distinguished from the activation of symbol B? Wouldn't our whole theory go down the drain? And even if there is not a *total* overlap of symbols, is our theory not more and more difficult to maintain, the more that symbols *do* overlap? (One possible way of portraying overlapping symbols is shown in Figure 68.)

There is a way to keep a theory based on symbols even if physically, they overlap considerably or totally. Consider the surface of a pond, which can support many different types of waves or ripples. The hardware—namely the water itself—is the same in all cases, but it possesses different possible modes of excitation. Such software excitations of the same

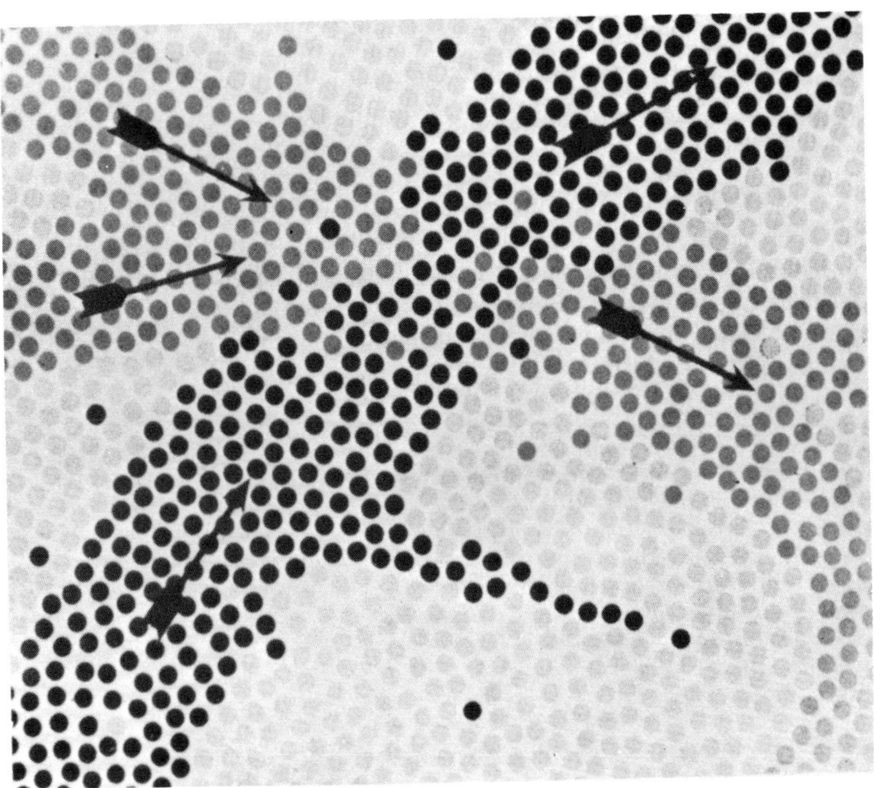

FIGURE 68. In this schematic diagram, neurons are imagined as laid out as dots in one plane. Two overlapping neural pathways are shown in different shades of gray. It may happen that two independent "neural flashes" simultaneously race down these two pathways, passing through one another like two ripples on a pond's surface (as in Fig. 52). This is illustrative of the idea of two "active symbols" which share neurons and which may even be simultaneously activated. [From John C. Eccles, Facing Reality *(New York: Springer Verlag, 1970), p.21.]*

hardware can all be distinguished from each other. By this analogy, I do not mean to go so far as to suggest that all the different symbols are just different kinds of "waves" propagating through a uniform neural medium which admits of no meaningful division into physically distinct symbols. But it may be that in order to distinguish one symbol's activation from that of another symbol, a process must be carried out which involves not only locating the neurons which are firing, but also identifying very precise details of the timing of the firing of those neurons. That is, which neuron preceded which other neuron, and by how much? How many times a second was a particular neuron firing? Thus perhaps several symbols can coexist in the same set of neurons by having different characteristic neural firing patterns. The difference between a theory having physically distinct symbols, and a theory having overlapping symbols which are distinguished from each other by modes of excitation, is that the former gives hardware realizations of concepts, while the latter gives partly hardware, partly software realizations of concepts.

Liftability of Intelligence

Thus we are left with two basic problems in the unraveling of thought processes, as they take place in the brain. One is to explain how the low-level traffic of neuron firings gives rise to the high-level traffic of symbol activations. The other is to explain the high-level traffic of symbol activation in its own terms—to make a theory which does not talk about the low-level neural events. If this latter is possible—and it is a key assumption at the basis of all present research into Artificial Intelligence—then intelligence can be realized in other types of hardware than brains. Then intelligence will have been shown to be a property that can be "lifted" right out of the hardware in which it resides—or in other words, intelligence will be a *software* property. This will mean that the phenomena of consciousness and intelligence are indeed high-level in the same sense as most other complex

FIGURE 69. The construction of an arch by workers of the termite Macrotermes bellicosus. *Each column is built up by the addition of pellets of soil and excrement. On the outer part of the left column a worker is seen depositing a round fecal pellet. Other workers, having carried pellets in their mandibles up the columns, are now placing them at the growing ends of the columns. When a column reaches a certain height the termites, evidently guided by odor, begin to extend it at an angle in the direction of a neighboring column. A completed arch is shown in the background. [Drawing by Turid Hölldobler; from E. O. Wilson,* The Insect Societies *(Cambridge, Mass.: Harvard University Press, 1971), p. 230.]*

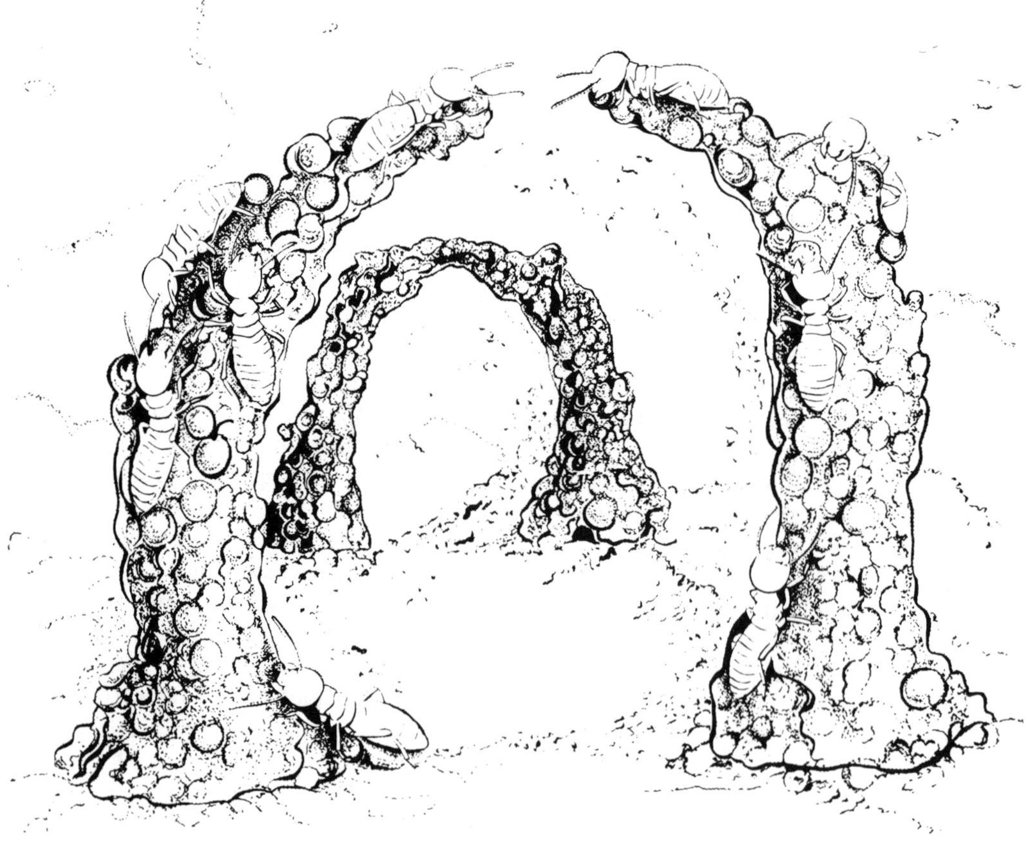

phenomena of nature: they have their own high-level laws which depend on, yet are "liftable" out of, the lower levels. If, on the other hand, there is absolutely no way to realize symbol-triggering patterns without having all the hardware of neurons (or simulated neurons), this will imply that intelligence is a brain-bound phenomenon, and much more difficult to unravel than one which owes its existence to a hierarchy of laws on several different levels.

Here we come back to the mysterious collective behavior of ant colonies, which can build huge and intricate nests, despite the fact that the roughly 100,000 neurons of an ant brain almost certainly do not carry any information about nest structure. How, then, does the nest get created? Where does the information reside? In particular, ponder where the information describing an arch such as is shown in Figure 69 can be found. Somehow, it must be spread about in the colony, in the caste distribution, the age distribution—and probably largely in the physical properties of the ant-body itself. That is, the interaction between ants is determined just as much by their six-leggedness and their size and so on, as by the information stored in their brain. Could there be an Artificial Ant Colony?

Can One Symbol Be Isolated?

Is it possible that one single symbol could be awakened in isolation from all others? Probably not. Just as objects in the world always exist in a context of other objects, so symbols are always connected to a constellation of other symbols. This does not necessarily mean that symbols can never be disentangled from each other. To make a rather simple analogy, males and females always arise in a species together: their roles are completely intertwined, and yet this does not mean that a male cannot be distinguished from a female. Each is reflected in the other, as the beads in Indra's net reflect each other. The recursive intertwining of the functions $F(n)$ and $M(n)$ in Chapter V does not prevent each function from having its own characteristics. The intertwining of F and M could be mirrored in a pair of RTN's which call each other. From this we can jump to a whole network of ATN's intertwined with each other—a heterarchy of interacting recursive procedures. Here, the meshing is so inherent that no one ATN could be activated in isolation; yet its activation may be completely distinctive, not confusable with that of any other of the ATN's. It is not such a bad image, the brain as an ATN-colony!

Likewise, symbols, with all their multiple links to each other, are meshed together and yet ought to be able to be teased apart. This might involve identifying a neural network, a network plus a mode of excitation—or possibly something of a completely different kind. In any case, if symbols are part of reality, presumably there exists a natural way to chart them out in a real brain. However, if some symbols were finally identified in a brain, this would not mean that any one of them could be awakened in isolation.

The fact that a symbol cannot be awakened in isolation does not diminish the separate identity of the symbol; in fact, quite to the contrary: a symbol's identity lies precisely in its ways of being connected (via potential triggering links) to other symbols. The network by which symbols can potentially trigger each other constitutes the brain's working model of the real universe, as well as of the alternate universes which it considers (and which are every bit as important for the individual's survival in the real world as the real world is).

The Symbols of Insects

Our facility for making instances out of classes and classes out of instances lies at the basis of our intelligence, and it is one of the great differences between human thought and the thought processes of other animals. Not that I have ever belonged to another species and experienced at first hand how it feels to think their way—but from the outside it is apparent that no other species forms general concepts as we do, or imagines hypothetical worlds—variants on the world as it is, which aid in figuring out which future pathway to choose. For instance, consider the celebrated "language of the bees"—information-laden dances which are performed by worker bees returning to the hive, to inform other bees of the location of nectar. While there may be in each bee a set of rudimentary symbols which are activated by such a dance, there is no reason to believe that a bee has an expandable vocabulary of symbols. Bees and other insects do not seem to have the power to generalize—that is, to develop new class symbols from instances which we would perceive as nearly identical.

A classic experiment with solitary wasps is reported in Dean Wooldridge's book, *Mechanical Man,* from which I quote:

> When the time comes for egg laying, the wasp Sphex builds a burrow for the purpose and seeks out a cricket which she stings in such a way as to paralyze but not kill it. She drags the cricket into the burrow, lays her eggs alongside, closes the burrow, then flies away, never to return. In due course, the eggs hatch and the wasp grubs feed off the paralyzed cricket, which has not decayed, having been kept in the wasp equivalent of a deepfreeze. To the human mind, such an elaborately organized and seemingly purposeful routine conveys a convincing flavor of logic and thoughtfulness—until more details are examined. For example, the wasp's routine is to bring the paralyzed cricket to the burrow, leave it on the threshold, go inside to see that all is well, emerge, and then drag the cricket in. If the cricket is moved a few inches away while the wasp is inside making her preliminary inspection, the wasp, on emerging from the burrow, will bring the cricket back to the threshold, but not inside, and will then repeat the preparatory procedure of entering the burrow to see that everything is all right. If again the cricket is removed a few inches while the wasp is inside, once again she will move the cricket up to the threshold and reenter the burrow for a final check. The wasp never thinks of pulling the cricket straight in. On one occasion this procedure was repeated forty times, always with the same result.[4]

This seems to be completely hard-wired behavior. Now in the wasp brain, there may be rudimentary symbols, capable of triggering each other; but there is nothing like the human capacity to see several instances as instances of an as-yet-unformed class, and then to make the class symbol; nor is there anything like the human ability to wonder, "What if I did this—what would ensue in that hypothetical world?" This type of thought process requires an ability to manufacture instances and to manipulate them as if they were symbols standing for objects in a real situation, although that situation may not be the case, and may never be the case.

Class Symbols and Imaginary Worlds

Let us reconsider the April Fool's joke about the borrowed car, and the images conjured up in your mind during the telephone call. To begin with, you need to activate symbols which represent a road, a car, a person in a car. Now the concept "road" is a very general one, with perhaps several stock samples which you can unconsciously pull out of dormant memory when the occasion arises. "Road" is a class, rather than an instance. As you listen to the tale, you quickly activate symbols which are instances with gradually increasing specificity. For instance, when you learn that the road was wet, this conjures up a more specific image, though you realize that it is most likely quite different from the actual road where the incident took place. But that is not important; what matters is whether your symbol is sufficiently well suited for the story—that is, whether the symbols which it can trigger are the right kind.

As the story progresses, you fill in more aspects of this road: there is a high bank against which a car could smash. Now does this mean that you are activating the symbol for "bank", or does it mean that you are setting some parameters in your symbol for "road"? Undoubtedly both. That is, the network of neurons which represents "road" has many different ways of firing, and you are selecting which subnetwork actually shall fire. At the same time, you are activating the symbol for "bank", and this is probably instrumental in the process of selecting the parameters for "road", in that its neurons may send signals to some of those in "road"—and vice versa. (In case this seems a little confusing, it is because I am somewhat straddling levels of description—I am trying to set up an image of the symbols, as well as of their component neurons.)

No less important than the nouns are the verbs, prepositions, etc. They, too, activate symbols, which send messages back and forth to each other. There are characteristic differences between the kinds of triggering patterns of symbols for verbs and symbols for nouns, of course, which means that they may be physically somewhat differently organized. For instance, nouns might have fairly localized symbols, while verbs and prepositions might have symbols with many "tentacles" reaching all around the cortex; or any number of other possibilities.

After the story is all over, you learn it was all untrue. The power of "rubbing off" instances from classes, in the way that one makes rubbings from brasses in churches, has enabled you to represent the situation, and has freed you from the need to remain faithful to the real world. The fact that symbols can act as templates for other symbols gives you some mental independence of reality: you can create artificial universes, in which there can happen nonreal events with any amount of detail that you care to imbue them with. But the class symbols themselves, from which all of this richness springs, are deeply grounded in reality.

Usually symbols play isomorphic roles to events which seem like they could happen, although sometimes symbols are activated which represent situations which could not happen—for example, watches sizzling, tubas laying eggs, etc. The borderline between what could and what could not happen is an extremely fuzzy one. As we imagine a hypothetical event, we bring certain symbols into active states—and depending on how well they interact (which is presumably reflected in our comfort in continuing the train of thought), we say the event "could" or "could not" happen. Thus the terms "could" and "could not" are extremely subjective. Actually, there is a good deal of agreement among people about which events could or could not happen. This reflects the great amount of mental structure which we all share—but there is a borderline area where the subjective aspect of what kinds of hypothetical worlds we are willing to entertain is apparent. A careful study of the kinds of imaginary events that people consider could and could not happen would yield much insight into the triggering patterns of the symbols by which people think.

Intuitive Laws of Physics

When the story has been completely told, you have built up quite an elaborate mental model of a scene, and in this model all the objects obey physical law. This means that physical law itself must be implicitly present in the triggering patterns of the symbols. Of course, the phrase "physical law" here does not mean "the laws of physics as expounded by a physicist", but rather the intuitive, chunked laws which all of us have to have in our minds in order to survive.

A curious sidelight is that one can voluntarily manufacture mental sequences of events which violate physical law, if one so desires. For instance, if I but suggest that you imagine a scene with two cars approaching each other and then passing right through each other, you won't have any trouble doing so. The intuitive physical laws can be overridden by imaginary laws of physics; but how this overriding is done, how such sequences of images are manufactured—indeed what any one visual image is—all of these are deeply cloaked mysteries—inaccessible pieces of knowledge.

Needless to say, we have in our brains chunked laws not only of how inanimate objects act, but also of how plants, animals, people and societies act—in other words, chunked laws of biology, psychology, sociology, and so on. All of the internal representations of such entities involve the inevitable feature of chunked models: determinism is sacrificed for simplicity. Our representation of reality ends up being able only to predict probabilities of ending up in certain parts of abstract spaces of behavior—not to predict anything with the precision of physics.

Procedural and Declarative Knowledge

A distinction which is made in Artificial Intelligence is that between procedural and declarative types of knowledge. A piece of knowledge is said to be *declarative* if it is stored explicitly, so that not only the programmer but also the program can "read" it as if it were in an encyclopedia or an almanac. This usually means that it is encoded locally, not spread around. By contrast, *procedural* knowledge is not encoded as facts—only as programs. A programmer may be able to peer in and say, "I see that because of these procedures here, the program 'knows' how to write English sentences"—but the program itself may have no explicit awareness of *how* it writes those sentences. For instance, its vocabulary may include none of the words "English", "sentence", and "write" at all! Thus procedural knowledge is usually spread around in pieces, and you can't retrieve it, or "key" on it. It is a global consequence of how the program works, not a local detail. In other words, a piece of purely procedural knowledge is an epiphenomenon.

In most people there coexists, along with a powerful procedural representation of the grammar of their native language, a weaker declarative representation of it. The two may easily be in conflict, so that a native speaker will often instruct a foreigner to say things he himself would never say, but which agree with the declarative "book learning" he acquired in school sometime. The intuitive or chunked laws of physics and other disciplines mentioned earlier fall mainly on the procedural side; the knowledge that an octopus has eight tentacles falls mainly on the declarative side.

In between the declarative and procedural extremes, there are all possible shades. Consider the recall of a melody. Is the melody stored in your brain, note by note? Could a surgeon extract a winding neural filament from your brain, then stretch it straight, and finally proceed to pinpoint along it the successively stored notes, almost as if it were a piece of magnetic tape? If so, then melodies are stored declaratively. Or is the recall of a melody mediated by the interaction of a large number of symbols, some of which represent tonal relationships, others of which represent emotional qualities, others of which represent rhythmic devices, and so on? If so, then melodies are stored procedurally. In reality, there is probably a mixture of these extremes in the way a melody is stored and recalled.

It is interesting that, in pulling a melody out of memory, most people do not discriminate as to key, so that they are as likely to sing "Happy Birthday" in the key of F-sharp as in the key of C. This indicates that tone *relationships,* rather than absolute tones, are stored. But there is no reason that tone relationships could not be stored quite declaratively. On the other hand, some melodies are very easy to memorize, whereas others are extremely elusive. If it were just a matter of storing successive notes, any melody could be stored as easily as any other. The fact that some melodies are catchy and others are not seems to indicate that the brain has a certain repertoire of familiar patterns which are activated as the melody is heard. So, to "play back" the melody, those patterns would have to be activated in the same order. This returns us to the concept of symbols triggering one another, rather than a simple linear sequence of declaratively stored notes or tone relationships.

How does the brain know whether a piece of knowledge is stored declaratively? For instance, suppose you are asked, "What is the population of Chicago?" Somehow the number five million springs to mind, without your wondering, "Gee, how would I go about counting them all?" Now suppose I ask you, "How many chairs are there in your living room?" Here, the opposite happens—instead of trying to dredge the answer out of a mental almanac, you immediately either go to the room and count the chairs, or you manufacture the room in your head and count the chairs in the image of the room. The questions were of a single type—"how many?"—yet one of them caused a piece of declarative knowledge to be fetched, while the other one caused a procedural method of finding the answer to be invoked. This is one example where it is clear that you have knowledge about how you classify your own knowledge; and what is more, some of that *metaknowledge* may itself be stored procedurally, so that it is used without your even being aware of how it is done.

Visual Imagery

One of the most remarkable and difficult-to-describe qualities of consciousness is visual imagery. How do we create a visual image of our living room? Of a roaring mountain brook? Of an orange? Even more mysterious, how do we manufacture images unconsciously, images which guide our thoughts, giving them power and color and depth? From what store are they fetched? What magic allows us to mesh two or three images, hardly giving a thought as to how we should do it? Knowledge of how to do this is among the most procedural of all, for we have almost no insight into what mental imagery is.

It may be that imagery is based on our ability to suppress motor activity. By this, I mean the following. If you imagine an orange, there may occur in your cortex a set of commands to pick it up, to smell it, to inspect it, and so on. Clearly these commands cannot be carried out, because the orange is not there. But they can be sent along the usual channels towards the cerebellum or other suborgans of the brain, until, at some critical point, a "mental faucet" is closed, preventing them from actually being carried out. Depending on how far down the line this "faucet" is situated, the images may be more or less vivid and real-seeming. Anger can cause us to imagine quite vividly picking up some object and throwing it, or kicking something; yet we don't actually do so. On the other hand, we feel so "near" to actually doing so. Probably the faucet catches the nerve impulses "at the last moment".

Here is another way in which visualization points out the distinction between accessible and inaccessible knowledge. Consider how you visualized the scene of the car skidding on the mountain road. Undoubtedly you imagined the mountain as being much larger than the car. Now did this happen because sometime long ago you had occasion to note that "cars are not as big as mountains"; then you committed this statement to rote memory; and in imagining the story, you retrieved this fact, and made use of it in constructing your image? A most unlikely theory. Or did it happen instead as a consequence of some introspectively inaccessible interactions of the symbols which were activated in your brain? Obviously the latter seems far more likely. This knowledge that cars are smaller than mountains is not a piece of rote memorization, but a piece of knowledge which can be created by *deduction.* Therefore, most likely it is not stored in any single symbol in your brain, but rather it can be produced as a result of the activation, followed by the mutual interaction, of many symbols—for example, those for "compare", "size", "car", "mountain", and probably others. This means that the knowledge is stored not explicitly, but implicitly, in a spread-about manner, rather than as a local "packet of information". Such simple facts as relative sizes of objects have to be assembled, rather than merely retrieved. Therefore, even in the case of a verbally accessible piece of knowledge, there are complex inaccessible processes which mediate its coming to the state of being ready to be said.

We shall continue our exploration of the entities called "symbols" in different Chapters. In Chapters XVIII and XIX, on Artificial Intelligence, we shall discuss some possible ways of implementing active symbols in programs. And next Chapter, we shall discuss some of the insights that our symbol-based model of brain activity give into the comparison of brains.

English French German Suite

By Lewis Carroll[1] . . .

. . . et Frank L. Warrin[2] . . .

. . . und Robert Scott[3]

'Twas brillig, and the slithy toves
Did gyre and gimble in the wabe:
All mimsy were the borogoves,
And the mome raths outgrabe.

Il brilgue: les tôves lubricilleux
Se gyrent en vrillant dans le guave.
Enmîmés sont les gougebosqueux
Et le mômerade horsgrave.

Es brillig war. Die schlichten Toven
Wirrten und wimmelten in Waben;
Und aller-mümsige Burggoven
Die mohmen Räth' ausgraben.

"Beware the Jabberwock, my son!
The jaws that bite, the claws that catch!
Beware the Jubjub bird, and shun
The frumious Bandersnatch!"

«Garde-toi du Jaseroque, mon fils!
La gueule qui mord; la griffe qui prend!
Garde-toi de l'oiseau Jube, évite
Le frumieux Band-à-prend!»

»Bewahre doch vor Jammerwoch!
Die Zähne knirschen, Krallen kratzen!
Bewahr' vor Jubjub-Vogel, vor
Frumiösen Banderschnätzchen!«

He took his vorpal sword in hand:
Long time the manxome foe he sought—
So rested he by the Tumtum tree,
And stood awhile in thought.

Son glaive vorpal en main, il va-
T-à la recherche du fauve manscant;
Puis arrivé à l'arbre Té-té,
Il y reste, réfléchissant.

Er griff sein vorpals Schwertchen zu,
Er suchte lang das manchsam' Ding;
Dann, stehend unterm Tumtum Baum,
Er an-zu-denken-fing.

And, as in uffish thought he stood,
The Jabberwock, with eyes of flame,
Came whiffling through the tulgey wood,
And burbled as it came!

Pendant qu'il pense, tout uffusé,
Le Jaseroque, à l'oeil flambant,
Vient siblant par le bois tullegeais,
Et burbule en venant.

Als stand er tief in Andacht auf,
Des Jammerwochen's Augen-feuer
Durch turgen Wald mit Wiffek kam
Ein burbelnd Ungeheuer!

One, two! One, two! And through and through
The vorpal blade went snicker-snack!
He left it dead, and with its head
He went galumphing back.

Un deux, un deux, par le milieu,
Le glaive vorpal fait pat-à-pan!
La bête défaite, avec sa tête,
Il rentre gallomphant.

Eins, Zwei! Eins, Zwei! Und durch und durch
Sein vorpals Schwert zerschnifer-schnück,
Da blieb es todt! Er, Kopf in Hand,
Geläumfig zog zurück.

"And hast thou slain the Jabberwock?
Come to my arms, my beamish boy!
O frabjous day! Callooh! Callay!"
He chortled in his joy.

«As-tu tué le Jaseroque?
Viens à mon coeur, fils rayonnais!
Ô jour frabbejais! Calleau! Callai!»
Il cortule dans sa joie.

»Und schlugst Du ja den Jammerwoch?
Umarme mich, mein Böhm'sches Kind!
O Freuden-Tag! O Halloo-Schlag!«
Er schortelt froh-gesinnt.

'Twas brillig, and the slithy toves
Did gyre and gimble in the wabe:
All mimsy were the borogoves,
And the mome raths outgrabe.

Il brilgue: les tôves lubricilleux
Se gyrent en vrillant dans le guave.
Enmîmés sont les gougebosqueux
Et le mômerade horsgrave.

Es brillig war. Die schlichten Toven
Wirrten und wimmelten in Waben;
Und aller-mümsige Burggoven
Die mohmen Räth' ausgraben.

CHAPTER XII

Minds and Thoughts

Can Minds Be Mapped onto Each Other?

NOW THAT WE have hypothesized the existence of very high-level active subsystems of the brain (symbols), we may return to the matter of a possible isomorphism, or partial isomorphism, between two brains. Instead of asking about an isomorphism on the neural level (which surely does not exist), or on the macroscopic suborgan level (which surely does exist but does not tell us very much), we ask about the possibility of an isomorphism between brains on the symbol level: a correspondence which not only maps symbols in one brain onto symbols in another brain, but also maps triggering patterns onto triggering patterns. This means that corresponding symbols in the two brains are linked in corresponding ways. This would be a true *functional* isomorphism—the same type of isomorphism as we spoke of when trying to characterize what it is that is invariant about all butterflies.

It is clear from the outset that such an isomorphism does not exist between any pair of human beings. If it did, they would be completely indistinguishable in their thoughts; but in order for that to be true, they would have to have completely indistinguishable memories, which would mean they would have to have led one and the same life. Even identical twins do not approach, in the remotest degree, this ideal.

How about a single individual? When you look back over things which you yourself wrote a few years ago, you think "How awful!" and smile with amusement at the person you once were. What is worse is when you do the same thing with something you wrote or said five minutes ago. When this happens, it shows that you do not fully understand the person you were moments ago. The isomorphism from your brain *now* to your brain *then* is imperfect. What, then, of the isomorphisms to other people, other species . . . ?

The opposite side of the coin is shown by the power of the communication that arises between the unlikeliest partners. Think of the barriers spanned when you read lines of poetry penned in jail by François Villon, the French poet of the 1400's. Another human being, in another era, captive in jail, speaking another language . . . How can you ever hope to have a sense of the connotations behind the facade of his words, translated into English? Yet a wealth of meaning comes through.

Thus, on the one hand, we can drop all hopes of finding exactly isomorphic software in humans, but on the other, it is clear that some people think more alike than others do. It would seem an obvious conclu-

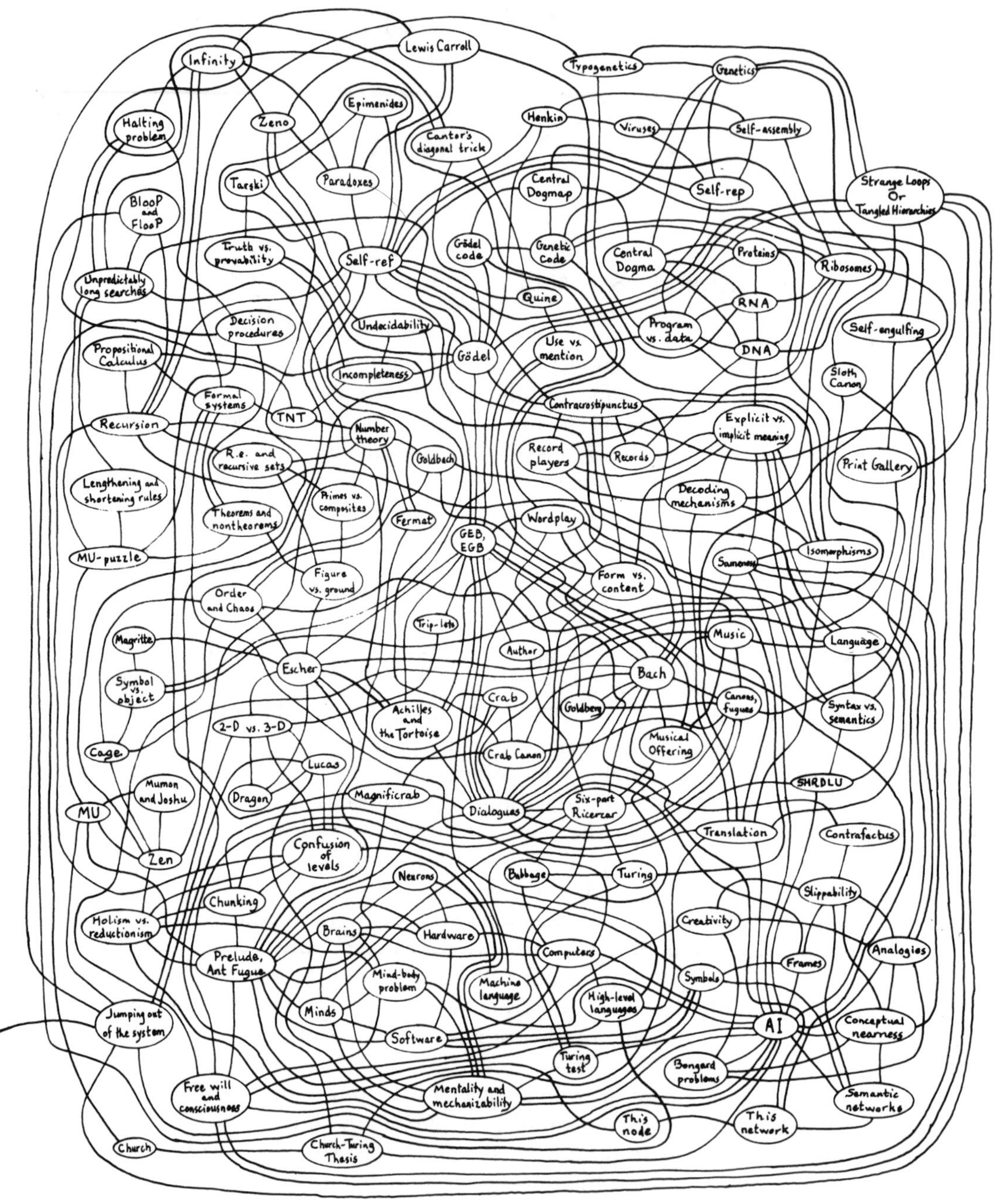

FIGURE 70. A tiny portion of the author's "semantic network".

sion that there is some sort of partial software isomorphism connecting the brains of people whose style of thinking is similar—in particular, a correspondence of (1) the repertoire of symbols, and (2) the triggering patterns of symbols.

Comparing Different Semantic Networks

But what is a *partial* isomorphism? This is a most difficult question to answer. It is made even more difficult by the fact that no one has found an adequate way to represent the network of symbols and their triggering patterns. Sometimes a picture of a small part of such a network of symbols is drawn, where each symbol is represented as a node into which, and out of which, lead some arcs. The lines represent triggering relationships—in some sense. Such figures attempt to capture something of the intuitively sensible notion of "conceptual nearness". However, there are many different kinds of nearness, and different ones are relevant in different contexts. A tiny portion of my own "semantic network" is shown in Figure 70. The problem is that representing a complex interdependency of many symbols cannot be carried out very easily with just a few lines joining vertices.

Another problem with such a diagram is that it is not accurate to think of a symbol as simply "on" or "off". While this is true of neurons, it does not carry upwards, to collections of them. In this respect, symbols are quite a bit more complicated than neurons—as you might expect, since they are made up of many neurons. The messages that are exchanged between symbols are more complex than the mere fact, "I am now activated". That is more like the neuron-level messages. Each symbol can be activated in many different ways, and the type of activation will be influential in determining which other symbols it tries to activate. How these intertwining triggering relationships can be represented in a pictorial manner—indeed, whether they can be at all—is not clear.

But for the moment, suppose that issue had been solved. Suppose we now agree that there are certain drawings of nodes, connected by links (let us say they come in various colors, so that various types of conceptual nearness can be distinguished from each other), which capture precisely the way in which symbols trigger other symbols. Then under what conditions would we feel that two such drawings were isomorphic, or nearly isomorphic? Since we are dealing with a visual representation of the network of symbols, let us consider an analogous visual problem. How would you try to determine whether two spiderwebs had been spun by spiders belonging to the same species? Would you try to identify individual vertices which correspond exactly, thereby setting up an exact map of one web onto the other, vertex by vertex, fiber by fiber, perhaps even angle by angle? This would be a futile effort. Two webs are never exactly the same; yet there is still some sort of "style", "form", what-have-you, that infallibly brands a given species' web.

In any network-like structure, such as a spiderweb, one can look at local properties and global properties. Local properties require only a very nearsighted observer—for example an observer who can only see one vertex at a time; and global properties require only a sweeping vision, without attention to detail. Thus, the overall shape of a spiderweb is a global property, whereas the average number of lines meeting at a vertex is a local property. Suppose we agree that the most reasonable criterion for calling two spiderwebs "isomorphic" is that they should have been spun by spiders of the same species. Then it is interesting to ask which kind of observation—local or global—tends to be a more reliable guide in determining whether two spiderwebs are isomorphic. Without answering the question for spiderwebs, let us now return to the question of the closeness—or isomorphicness, if you will—of two symbol networks.

Translations of "Jabberwocky"

Imagine native speakers of English, French, and German, all of whom have excellent command of their respective native languages, and all of whom enjoy wordplay in their own language. Would their symbol networks be similar on a local level, or on a global level? Or is it meaningful to ask such a question? The question becomes concrete when you look at the preceding translations of Lewis Carroll's famous "Jabberwocky".

I chose this example because it demonstrates, perhaps better than an example in ordinary prose, the problem of trying to find "the same node" in two different networks which are, on some level of analysis, extremely nonisomorphic. In ordinary language, the task of translation is more straightforward, since to each word or phrase in the original language, there can usually be found a corresponding word or phrase in the new language. By contrast, in a poem of this type, many "words" do not carry ordinary meaning, but act purely as exciters of nearby symbols. However, what is nearby in one language may be remote in another.

Thus, in the brain of a native speaker of English, "slithy" probably activates such symbols as "slimy", "slither", "slippery", "lithe", and "sly", to varying extents. Does "lubricilleux" do the corresponding thing in the brain of a Frenchman? What indeed would be "the corresponding thing"? Would it be to activate symbols which are the ordinary translations of those words? What if there is no word, real or fabricated, which will accomplish that? Or what if a word does exist, but is very intellectual-sounding and Latinate ("lubricilleux"), rather than earthy and Anglo-Saxon ("slithy")? Perhaps "huilasse" would be better than "lubricilleux"? Or does the Latin origin of the word "lubricilleux" not make itself felt to a speaker of French in the way that it would if it were an English word ("lubricilious", perhaps)?

An interesting feature of the translation into French is the transposition into the present tense. To keep it in the past would make some unnatural turns of phrase necessary, and the present tense has a much fresher flavor in French than the past. The translator sensed that this would be "more appropriate"—in some ill-defined yet compelling sense—and made the switch. Who can say whether remaining faithful to the English tense would have been better?

In the German version, the droll phrase "er an-zu-denken-fing" occurs; it does not correspond to any English original. It is a playful reversal of words, whose flavor vaguely resembles that of the English phrase "he out-to-ponder set", if I may hazard a reverse translation. Most likely this funny turnabout of words was inspired by the similar playful reversal in the English of one line earlier: "So rested he by the Tumtum tree". It corresponds, yet doesn't correspond.

Incidentally, why did the Tumtum tree get changed into an "arbre Té-té" in French? Figure it out for yourself.

The word "manxome" in the original, whose "x" imbues it with many rich overtones, is weakly rendered in German by "manchsam", which back-translates into English as "maniful". The French "manscant" also lacks the manifold overtones of "manxome". There is no end to the interest of this kind of translation task.

When confronted with such an example, one realizes that it is utterly impossible to make an exact translation. Yet even in this pathologically difficult case of translation, there seems to be some rough equivalence obtainable. Why is this so, if there really is no isomorphism between the brains of people who will read the different versions? The answer is that there is a kind of rough isomorphism, partly global, partly local, between the brains of all the readers of these three poems.

ASU's

An amusing geographical fantasy will give some intuition for this kind of quasi-isomorphism. (Incidentally, this fantasy is somewhat similar to a geographical analogy devised by M. Minsky in his article on "frames", which can be found in P. H. Winston's book *The Psychology of Computer Vision.*) Imagine that you are given a strange atlas of the USA, with all natural geological features premarked—such as rivers, mountains, lakes, and so on—but with nary a printed word. Rivers are shown as blue lines, mountains by color, and so on. Now you are told to convert it into a road atlas for a trip which you will soon make. You must neatly fill in the names of all states, their boundaries, time zones, then all counties, cities, towns, all freeways and highways and toll routes, all county roads, all state and national parks, campgrounds, scenic areas, dams, airports, and so on . . . All of this must be carried out down to the level that would appear in a detailed road atlas. And it must be manufactured out of your own head. You are not allowed access to any information which would help you for the duration of your task.

You are told that it will pay off, in ways that will become clear at a later date, to make your map as true as you can. Of course, you will begin by filling in large cities and major roads, etc., which you know. And when you have exhausted your factual knowledge of an area, it will be to your advantage to use your imagination to help you reproduce at least the flavor of that area, if not its true geography, by making up fake town names, fake populations, fake roads, fake parks, and so on. This arduous task will take months. To make things a little easier, you have a cartographer on hand to print everything in neatly. The end product will be your personal map of the "Alternative Structure of the Union"—your own personal "ASU".

Your personal ASU will be very much like the USA in the area where you grew up. Furthermore, wherever your travels have chanced to lead you, or wherever you have perused maps with interest, your ASU will have spots of striking agreement with the USA: a few small towns in North Dakota or Montana, perhaps, or the whole of metropolitan New York, might be quite faithfully reproduced in your ASU.

A Surprise Reversal

When your ASU is done, a surprise takes place. Magically, the country you have designed comes into being, and you are transported there. A friendly committee presents you with your favorite kind of automobile, and explains that, "As a reward for your designing efforts, you may now enjoy an all-expense-paid trip, at a leisurely pace, around the good old A. S. of U. You may go wherever you want, do whatever you wish to do, taking as long as you wish—compliments of the Geographical Society of the ASU. And—to guide you around—here is a road atlas." To your surprise, you are given not the atlas which you designed, but a regular road atlas of the USA.

When you embark on your trip, all sorts of curious incidents will take place. A road atlas is being used to guide you through a country which it only partially fits. As long as you stick to major freeways, you will probably be able to cross the country without gross confusions. But the moment you wander off into the byways of New Mexico or rural Arkansas, there will be adventure in store for you. The locals will not recognize any of the towns you're looking for, nor will they know the roads you're asking about. They will only know the large cities you name, and even then the routes to those cities will not be the same as are indicated on your map. It will happen occasionally that some of the cities which are considered huge by the locals are nonexistent on your map of the USA; or perhaps they exist, but their population according to the atlas is wrong by an order of magnitude.

Centrality and Universality

What makes an ASU and the USA, which are so different in some ways, nevertheless so similar? It is that their most important cities and routes of communication can be mapped onto each other. The differences between them are found in the less frequently traveled routes, the cities of smaller size, and so on. Notice that this cannot be characterized either as a local or a global isomorphism. Some correspondences do extend down to the very local level—for instance, in both New Yorks, the main street may be Fifth Avenue, and there may be a Times Square in both as well—yet there may not be a single town that is found in both Montanas. So the local-global distinction is not relevant here. What is relevant is the *centrality* of the city, in terms of economics, communication, transportation, etc. The more vital the city is, in one of these ways, the more certain it will be to occur in both the ASU and the USA.

In this geographic analogy, one aspect is very crucial: that there are certain definite, absolute points of reference which will occur in nearly all ASU's: New York, San Francisco, Chicago, and so on. From these it is then possible to orient oneself. In other words, if we begin comparing my ASU with yours, I can use the known agreement on big cities to establish points of reference with which I can communicate the location of smaller cities in my ASU. And if I hypothesize a voyage from Kankakee to Fruto and you don't know where those towns are, I can refer to something we have in common, and thereby guide you. And if I talk about a voyage from Atlanta to Milwaukee, it may go along different freeways or smaller roads, but the voyage itself can still be carried out in both countries. And if you start describing a trip from Horsemilk to Janzo, I can plot out what seems to me to be an *analogous* trip in my ASU, despite not having towns by those names, as long as you constantly keep me oriented by describing your position with respect to nearby larger towns which are found in my ASU as well as in yours.

My roads will not be exactly the same as yours, but, with our separate maps, we can each get from a particular part of the country to another. We can do this, thanks to the external, predetermined geological facts—mountain chains, streams, etc.—facts which were available to us both as we worked on our maps. Without those external features, we would have no possibility of reference points in common. For instance, if you had been given only a map of France, and I had been given a map of Germany, and then we had both filled them in in great detail, there would be no way to try to find "the same place" in our fictitious lands. It is necessary to begin with identical external conditions—otherwise nothing will match.

Now that we have carried our geographical analogy quite far, we return to the question of isomorphisms between brains. You might well wonder why this whole question of brain isomorphisms has been stressed so much. What does it matter if two brains are isomorphic, or quasi-isomorphic, or not isomorphic at all? The answer is that we have an intuitive sense that, although other people differ from us in important ways, they are still "the same" as we are in some deep and important ways. It would be instructive to be able to pinpoint what this invariant core of human intelligence is, and then to be able to describe the kinds of "embellishments" which can be added to it, making each one of us a unique embodiment of this abstract and mysterious quality called "intelligence".

In our geographic analogy, cities and towns were the analogues of symbols, while roads and highways were analogous to potential triggering paths. The fact that all ASU's have some things in common, such as the East Coast, the West Coast, the Mississippi River, the Great Lakes, the Rockies, and many major cities and roads is analogous to the fact that we are all forced, by external realities, to construct certain class symbols and triggering paths in the same way. These core symbols are like the large cities, to which everyone can make reference without ambiguity. (Incidentally, the fact that cities are localized entities should in no way be taken as indicative that symbols in a brain are small, almost point-like entities. They are merely symbolized in that manner in a network.)

The fact is that a large proportion of every human's network of symbols is *universal.* We simply take what is common to all of us so much for granted that it is hard to see how much we have in common with other people. It takes the conscious effort of imagining how much—or how little—we have in common with other types of entities, such as stones, cars, restaurants, ants, and so forth, to make evident the large amount of overlap that we have with randomly chosen people. What we notice about another person immediately is not the standard overlap, because that is taken for granted as soon as we recognize the humanity of the other person; rather, we look beyond the standard overlap and generally find some major differences, as well as some unexpected, additional overlap.

Occasionally, you find that another person is missing some of what you thought was the standard, minimal core—as if Chicago were missing from their ASU, which is almost unimaginable. For instance, someone might not know what an elephant is, or who is President, or that the earth is round. In such cases, their symbolic network is likely to be so fundamentally different from your own that significant communication will be difficult. On the other hand, perhaps this same person will share some specialized kind of knowledge with you—such as expertise in the game of dominoes—so that you can communicate well in a limited domain. This would be like meeting someone who comes from the very same rural area of North Dakota as you do, so that your two ASU's coincide in great detail over a very small region, which allows you to describe how to get from one place to another very fluently.

How Much Do Language and Culture Channel Thought?

If we now go back to comparing our own symbol network with those of a Frenchman and a German, we can say that we expect them to have the standard core of class symbols, despite the fact of different native languages. We do not expect to share highly specialized networks with them, but we do not expect such sharing with a randomly chosen person who shares our native language, either. The triggering patterns of people with other languages will be somewhat different from our own, but still the major class symbols, and the major routes between them, will be universally available, so that more minor routes can be described with reference to them.

Now each of our three people may in addition have some command of the languages of the other two. What is it that marks the difference between true fluency, and a mere ability to communicate? First of all, someone fluent in English uses most words at roughly their regular frequencies. A non-native speaker will have picked up some words from dictionaries, novels, or classes—words which at some time may have been prevalent or preferable, but which are now far down in frequency—for example, "fetch" instead of "get", "quite" instead of "very", etc. Though the meaning usually comes through, there is an alien quality transmitted by the unusual choice of words.

But suppose that a foreigner learns to use all words at roughly the normal frequencies. Will that make his speech truly fluent? Probably not. Higher than the word level, there is an association level, which is attached to the culture as a whole—its history, geography, religion, children's stories, literature, technological level, and so on. For instance, to be able to speak modern Hebrew absolutely fluently, you need to know the Bible quite well in Hebrew, because the language draws on a stock of biblical phrases and their connotations. Such an association level permeates each language very deeply. Yet there is room for all sorts of variety inside fluency—otherwise the only truly fluent speakers would be people whose thoughts were the most stereotyped possible!

Although we should recognize the depth to which *culture* affects thought, we should not overstress the role of *language* in molding thoughts. For instance, what we might call two "chairs" might be perceived by a speaker of French as objects belonging to two distinct types: "chaise" and "fauteuil" ("chair" and "armchair"). People whose native language is French are more aware of that difference than we are—but then people who grow up in a rural area are more aware of, say, the difference between a pickup and a truck, than a city dweller is. A city dweller may call them both "trucks". It is not the difference in native language, but the difference in culture (or subculture), that gives rise to this perceptual difference.

The relationships between the symbols of people with different native languages have every reason to be quite similar, as far as the core is concerned, because everyone lives in the same world. When you come down to more detailed aspects of the triggering patterns, you will find that there is less in common. It would be like comparing rural areas in Wisconsin in ASU's which had been made up by people who had never lived in Wisconsin. This will be quite irrelevant, however, as long as there is sufficient agreement on the major cities and major routes, so that there are common points of reference all over the map.

Trips and Itineraries in ASU's

Without making it explicit, I have been using an image of what a "thought" is in the ASU-analogy—namely, I have been implying that a *thought* corresponds to a *trip*. The *towns* which are passed through represent the *symbols* which are excited. This is not a perfect analogy, but it is quite strong. One problem with it is that when a thought recurs in someone's mind sufficiently often, it can get chunked into a single concept. This would correspond to quite a strange event in an ASU: a commonly taken trip would become, in some strange fashion, a new town or city! If one is to continue to use the ASU-metaphor, then, it is important to remember that the cities represent not only the *elementary* symbols, such as those for "grass", "house", and "car", but also symbols which get created as a result of the *chunking* ability of a brain—symbols for such sophisticated concepts as "crab canon", "palindrome", or "ASU".

Now if it is granted that the notion of taking a trip is a fair counterpart to the notion of having a thought, then the following difficult issue comes up: virtually any route leading from one city to a second, then to a third, and so on, can be imagined, as long as one remembers that some intervening cities are also passed through. This would correspond to the activation of an *arbitrary sequence of symbols,* one after another, making allowance for some extra symbols—those which lie en route. Now if virtually any sequence of symbols can be activated in any desired order, it may seem that a brain is an indiscriminate system, which can absorb or produce any thought whatsoever. But we all know that that is not so. In fact, there are certain kinds of thoughts which we call *knowledge,* or *beliefs,* which play quite a different role from random fancies, or humorously entertained absurdities. How can we characterize the difference between dreams, passing thoughts, beliefs, and pieces of knowledge?

Possible, Potential, and Preposterous Pathways

There are some pathways—you can think of them as pathways either in an ASU or in a brain—which are taken routinely in going from one place to another. There are other pathways which can only be followed if one is led through them by the hand. These pathways are "potential pathways", which would be followed only if special external circumstances arose. The pathways which one relies on over and over again are pathways which incorporate knowledge—and here I mean not only knowledge of *facts* (declarative knowledge), but also knowledge of *how-to's* (procedural knowledge). These stable, reliable pathways are what constitute knowledge. Pieces of knowledge merge gradually with beliefs, which are also represented by reliable pathways, but perhaps ones which are more susceptible to replacement if, so to speak, a bridge goes out, or there is heavy fog. This leaves us with fancies, lies, falsities, absurdities, and other variants. These would correspond to peculiar routes such as: New York City to Newark via Bangor, Maine and Lubbock, Texas. They are indeed possible pathways, but ones which are not likely to be stock routes, used in everyday voyages.

A curious, and amusing, implication of this model is that all of the "aberrant" kinds of thoughts listed above are composed, at rock bottom, completely out of beliefs or pieces of knowledge. That is, any weird and snaky indirect route breaks up into a number of non-weird, non-snaky direct stretches, and these short, straightforward symbol-connecting routes represent simple thoughts that one can rely on—beliefs and pieces of knowledge. On reflection, this is hardly surprising, however, since it is quite reasonable that we should only be able to imagine fictitious things that are somehow grounded in the realities we have experienced, no matter how wildly they deviate from them. Dreams are perhaps just such random meanderings about the ASU's of our minds. Locally, they make sense—but globally . . .

Different Styles of Translating Novels

A poem like "Jabberwocky" is like an unreal journey around an ASU, hopping from one state to another very quickly, following very curious routes. The translations convey this aspect of the poem, rather than the precise sequence of symbols which are triggered, although they do their best in that respect. In ordinary prose, such leaps and bounds are not so common. However, similar problems of translation do occur. Suppose you are translating a novel from Russian to English, and come across a sentence whose literal translation is, "She had a bowl of borscht." Now perhaps many of your readers will have no idea what borscht is. You could attempt to replace it by the "corresponding" item in their culture—thus, your translation might run, "She had a bowl of Campbell's soup." Now if you think this is a silly exaggeration, take a look at the first sentence of Dostoevsky's novel *Crime and Punishment* in Russian and then in a few different English translations. I happened to look at three different English paperback translations, and found the following curious situation.

The first sentence employs the street name "S. Pereulok" (as transliterated). What is the meaning of this? A careful reader of Dostoevsky's work who knows Leningrad (which used to be called "St. Petersburg"—or should I say "Petrograd"?) can discover by doing some careful checking of the rest of the geography in the book (which incidentally is also given only by its initials) that the street must be "Stoliarny Pereulok". Dostoevsky probably wished to tell his story in a realistic way, yet not so realistically that people would take literally the addresses at which crimes and other events were supposed to have occurred. In any case, we have a translation problem; or to be more precise, we have several translation problems, on several different levels.

First of all, should we keep the initial so as to reproduce the aura of semi-mystery which appears already in this first sentence of the book? We would get "S. Lane" ("lane" being the standard translation of "pereulok"). None of the three translators took this tack. However, one chose to write "S. Place". The translation of *Crime and Punishment* which I read in high school took a similar option. I will never forget the disoriented feeling I experienced when I began reading the novel and encountered those streets with only letters for names. I had some sort of intangible malaise about the beginning of the book; I was sure that I was missing something essential, and yet I didn't know what it was . . . I decided that all Russian novels were very weird.

Now we could be frank with the reader (who, it may be assumed, probably won't have the slightest idea whether the street is real or fictitious anyway!) and give him the advantage of our modern scholarship, writing

"Stoliarny Lane" (or "Place"). This was the choice of translator number 2, who gave the translation as "Stoliarny Place".

What about number 3? This is the most interesting of all. This translation says "Carpenter's Lane". And why not, indeed? After all, "stoliar" means "carpenter" and "ny" is an adjectival ending. So now we might imagine ourselves in London, not Petrograd, and in the midst of a situation invented by Dickens, not Dostoevsky. Is that what we want? Perhaps we should just read a novel by Dickens instead, with the justification that it is "the corresponding work in English". When viewed on a sufficiently high level, it is a "translation" of the Dostoevsky novel—in fact, the best possible one! Who needs Dostoevsky?

We have come all the way from attempts at great literal fidelity to the author's style, to high-level translations of flavor. Now if this happens already in the first sentence, can you imagine how it must go on in the rest of the book? What about the point where a German landlady begins shouting in her German-style Russian? How do you translate broken Russian spoken with a German accent, into English?

Then one may also consider the problems of how to translate slang and colloquial modes of expression. Should one search for an "analogous" phrase, or should one settle for a word-by-word translation? If you search for an analogous phrase, then you run the risk of committing a "Campbell's soup" type of blunder; but if you translate every idiomatic phrase word by word, then the English will sound alien. Perhaps this is desirable, since the Russian culture is an alien one to speakers of English. But a speaker of English who reads such a translation will constantly be experiencing, thanks to the unusual turns of phrase, a sense—an artificial sense—of strangeness, which was not intended by the author, and which is not experienced by readers of the Russian original.

Problems such as these give one pause in considering such statements as this one, made by Warren Weaver, one of the first advocates of translation by computer, in the late 1940's: "When I look at an article in Russian, I say, 'This is really written in English, but it has been coded in some strange symbols. I will now proceed to decode.'"[1] Weaver's remark simply cannot be taken literally; it must rather be considered a provocative way of saying that there is an objectively describable meaning hidden in the symbols, or at least something pretty close to objective; therefore, there would be no reason to suppose a computer could not ferret it out, if sufficiently well programmed.

High-Level Comparisons between Programs

Weaver's statement is about translations between different natural languages. Let's consider now the problem of translating between two computer languages. For instance, suppose two people have written programs which run on different computers, and we want to know if the two programs carry out the same task. How can we find out? We must compare the programs. But on what level should this be done? Perhaps one programmer wrote in a machine language, the other in a compiler language. Are two such programs comparable? Certainly. But how to compare them? One way might be to compile the compiler language program, producing a program in the machine language of its home computer.

Now we have two machine language programs. But there is another problem: there are two computers, hence two different machine languages—and they may be extremely different. One machine may have sixteen-bit words; the other thirty-six-bit words. One machine may have built-in stack-handling instructions (pushing and popping), while the other lacks them. The differences between the hardware of the two machines may make the two machine language programs seem incomparable—and yet we suspect they are performing the same task, and we would like to see that at a glance. We are obviously looking at the programs from much too close a distance.

What we need to do is to step back, away from machine language, towards a higher, more chunked view. From this vantage point, we hope we will be able to perceive chunks of program which make each program seem rationally planned out on a global, rather than a local, scale—that is, chunks which fit together in a way that allows one to perceive the goals of the programmer. Let us assume that both programs were originally written in high-level languages. Then some chunking has already been done for us. But we will run into other troubles. There is a proliferation of such languages: Fortran, Algol, LISP, APL, and many others. How can you compare a program written in APL with one written in Algol? Certainly not by matching them up line by line. You will again chunk these programs in your mind, looking for conceptual, functional units which correspond. Thus, you are not comparing hardware, you are not comparing software—you are comparing "etherware"—the pure concepts which lie back of the software. There is some sort of abstract "conceptual skeleton" which must be lifted out of low levels before you can carry out a meaningful comparison of two programs in different computer languges, of two animals, or of two sentences in different natural languages.

Now this brings us back to an earlier question which we asked about computers and brains: How can we make sense of a low-level description of a computer or a brain? Is there, in any reasonable sense, an *objective* way to pull a high-level description out of a low-level one, in such complicated systems? In the case of a computer, a full display of the contents of memory—a so-called *memory dump*—is easily available. Dumps were commonly printed out in the early days of computing, when something went wrong with a program. Then the programmer would have to go home and pore over the memory dump for hours, trying to understand what each minuscule piece of memory represented. In essence, the programmer would be doing the opposite of what a compiler does: he would be translating from machine language into a higher-level language, a conceptual language. In the end, the programmer would understand the goals of the program and could describe it in high-level terms—for example, "This program translates novels from Russian to English", or "This program composes an eight-voice fugue based on any theme which is fed in".

High-Level Comparisons between Brains

Now our question must be investigated in the case of brains. In this case, we are asking, "Are people's brains also capable of being 'read', on a high level? Is there some objective description of the content of a brain?" In the *Ant Fugue,* the Anteater claimed to be able to tell what Aunt Hillary was thinking about, by looking at the scurryings of her component ants. Could some superbeing—a Neuroneater, perhaps—conceivably look down on our neurons, chunk what it sees, and come up with an analysis of our thoughts?

Certainly the answer must be yes, since we are all quite able to describe, in chunked (i.e., non-neural) terms, the activity of our minds at any given time. This means that we have a mechanism which allows us to chunk our own brain state to some rough degree, and to give a functional description of it. To be more precise, we do not chunk all of the brain state—we only chunk those portions of it which are active. However, if someone asks us about a subject which is coded in a currently inactive area of our brain, we can almost instantly gain access to the appropriate dormant area and come up with a chunked description of it—that is, some belief on that subject. Note that we come back with absolutely zero information on the neural level of that part of the brain: our description is so chunked that we don't even have any idea what part of our brain it is a description of. This can be contrasted with the programmer whose chunked description comes from conscious analysis of every part of the memory dump.

Now if a person can provide a chunked description of any part of his own brain, why shouldn't an outsider too, given some nondestructive means of access to the same brain, not only be able to chunk limited portions of the brain, but actually to give a complete chunked description of it—in other words, a complete documentation of the beliefs of the person whose brain is accessible? It is obvious that such a description would have an astronomical size, but that is not of concern here. We are interested in the question of whether, in principle, there exists a well-defined, high-level description of a brain, or whether, conversely, the neuron-level description—or something equally physiological and intuitively unenlightening—is the best description that in principle exists. Surely, to answer this question would be of the highest importance if we seek to know whether we can ever understand ourselves.

Potential Beliefs, Potential Symbols

It is my contention that a chunked description is possible, but when we get it, all will not suddenly be clear and light. The problem is that in order to pull a chunked description out of the brain state, we need a language to describe our findings. Now the most appropriate way to describe a brain, it would seem, would be to enumerate the kinds of thoughts it could entertain, and the kinds of thoughts it could not entertain—or, perhaps, to enumerate its beliefs and the things which it does not believe. If that is the kind of goal we will be striving for in a chunked description, then it is easy to see what kinds of troubles we will run up against.

Suppose you wanted to enumerate all possible voyages that could be taken in an ASU; there are infinitely many. How do you determine which ones are *plausible,* though? Well, what does "plausible" mean? We will have precisely this kind of difficulty in trying to establish what a "possible pathway" from symbol to symbol in a brain is. We can imagine an upside-down dog flying through the air with a cigar in its mouth—or a collision between two giant fried eggs on a freeway—or any number of other ridiculous images. The number of far-fetched pathways which can be followed in our brains is without bound, just as is the number of insane itineraries that could be planned on an ASU. But just what constitutes a "sane" itinerary, given an ASU? And just what constitutes a "reasonable" thought, given a brain state? The brain state itself does not forbid any pathway, because for any pathway there are always circumstances which could force the following of that pathway. The physical status of a brain, if read correctly, gives information telling not which pathways could be followed, but rather how much resistance would be offered along the way.

Now in an ASU, there are many trips which could be taken along two or more reasonable alternative routes. For example, the trip from San Francisco to New York could go along either a northern route or a southern route. Each of them is quite reasonable, but people tend to take them under different circumstances. Looking at a map at a given moment in time does not tell you anything about which route will be preferable at some remote time in the future—that depends on the external circumstances under which the trip is to be taken. Likewise, the "reading" of a brain state will reveal that several reasonable alternative pathways are often available, connecting a given set of symbols. However, the trip among these symbols need not be imminent; it may be simply one of billions of "potential" trips, all of which figure in the readout of the brain state. From this follows an important conclusion: there is no information in the brain state itself which tells which route will be chosen. The external circumstances will play a large determining role in choosing the route.

What does this imply? It implies that thoughts which clash totally may be produced by a single brain, depending on the circumstances. And any high-level readout of the brain state which is worth its salt must contain all such conflicting versions. Actually this is quite obvious—that we all are bundles of contradictions, and we manage to hang together by bringing out only one side of ourselves at a given time. The selection cannot be predicted in advance, because the conditions which will force the selection are not known in advance. What the brain state can provide, if properly read, is a *conditional* description of the selection of routes.

Consider, for instance, the Crab's plight, described in the *Prelude.* He can react in various ways to the playing of a piece of music. Sometimes he will be nearly immune to it, because he knows it so well. Other times, he will be quite excited by it, but this reaction requires the right kind of triggering from the outside—for instance, the presence of an enthusiastic listener, to whom the work is new. Presumably, a high-level reading of the Crab's brain state would reveal the potential thrill (and conditions which would induce it), as well as the potential numbness (and conditions which would induce it). The brain state itself would not tell which one would occur on the next hearing of the piece, however; it could only say, "If such-&-such conditions obtain, then a thrill will result; otherwise . . ."

Thus a chunked description of a brain state would give a catalogue of beliefs which would be evoked conditionally, dependent on circumstances. Since not all possible circumstances can be enumerated, one would have to settle for those which one thinks are "reasonable". Furthermore, one would have to settle for a chunked description of the circumstances themselves, since they obviously cannot—and should not—be specified down to the atomic level! Therefore, one will not be able to make an exact, deterministic prediction saying which beliefs will be pulled out of the brain state by a given chunked circumstance. In summary, then, a chunked description of a brain state will consist of a probabilistic catalogue, in which are listed those beliefs which are most likely to be induced (and those symbols which are most likely to be activated) by various sets of "reasonably likely" circumstances, themselves described on a chunked level. Trying to chunk someone's beliefs without referring to context is precisely as silly as trying to describe the range of a single person's "potential progeny" without referring to the mate.

The same sorts of problems arise in enumerating all the symbols in a given person's brain. There are potentially not only an infinite number of *pathways* in a brain, but also an infinite number of *symbols*. As was pointed out, new concepts can always be formed from old ones, and one could argue that the symbols which represent such new concepts are merely dormant symbols in each individual, waiting to be awakened. They may never get awakened in the person's lifetime, but it could be claimed that those symbols are nonetheless always there, just waiting for the right circumstances to trigger their synthesis. However, if the probability is very low, it would seem that "dormant" would be a very unrealistic term to apply in the situation. To make this clear, try to imagine all the "dormant dreams" which are sitting there inside your skull while you're awake. Is it conceivable that there exists a decision procedure which could tell "potentially dreamable themes" from "undreamable themes", given your brain state?

Where Is the Sense of Self?

Looking back on what we have discussed, you might think to yourself, "These speculations about brain and mind are all well and good, but what about the feelings involved in consciousness? These symbols may trigger each other all they want, but unless someone *perceives* the whole thing, there's no consciousness."

This makes sense to our intuition on some level, but it does not make much sense logically. For we would then be compelled to look for an explanation of the mechanism which does the perceiving of all the active symbols, if it is not covered by what we have described so far. Of course, a "soulist" would not have to look any further—he would merely assert that the perceiver of all this neural action is the soul, which cannot be described in physical terms, and that is that. However, we shall try to give a "non-soulist" explanation of where consciousness arises.

Our alternative to the soulist explanation—and a disconcerting one it is, too—is to stop at the symbol level and say, "This is it—this is what consciousness *is*. Consciousness is that property of a system that arises whenever there exist symbols in the system which obey triggering patterns somewhat like the ones described in the past several sections." Put so starkly, this may seem inadequate. How does it account for the sense of "I", the sense of self?

Subsystems

There is no reason to expect that "I", or "the self", should not be represented by a symbol. In fact, the symbol for the self is probably the most complex of all the symbols in the brain. For this reason, I choose to put it on a new level of the hierarchy and call it a *subsystem,* rather than a symbol. To be precise, by "subsystem", I mean a constellation of symbols, each of which can be separately activated under the control of the subsystem itself. The image I wish to convey of a subsystem is that it functions almost as an independent "subbrain", equipped with its own repertoire of symbols which can trigger each other internally. Of course, there is also much communication between the subsystem and the "outside" world—that is, the rest of the brain. "Subsystem" is just another name for an overgrown symbol, one which has gotten so complicated that it has many subsymbols which interact among themselves. Thus, there is no strict level distinction between symbols and subsystems.

Because of the extensive links between a subsystem and the rest of the brain (some of which will be described shortly), it would be very difficult to draw a sharp boundary between the subsystem and the outside; but even if the border is fuzzy, the subsystem is quite a real thing. The interesting thing about a subsystem is that, once activated and left to its own devices, it can work on its own. Thus, two or more subsystems of the brain of an individual may operate simultaneously. I have noticed this happening on occasion in my own brain: sometimes I become aware that two different melodies are running through my mind, competing for "my" attention. Somehow, each melody is being manufactured, or "played", in a separate compartment of my brain. Each of the systems responsible for drawing a melody out of my brain is presumably activating a number of symbols, one after another, completely oblivious to the other system doing the same thing. Then they both attempt to communicate with a third subsystem of my brain—my self-symbol—and it is at that point that the "I" inside my brain gets wind of what's going on; in other words, it starts picking up a chunked description of the activities of those two subsystems.

Subsystems and Shared Code

Typical subsystems might be those that represent the people we know intimately. They are represented in such a complex way in our brains that their symbols enlarge to the rank of subsystem, becoming able to act autonomously, making use of some resources in our brains for support. By this, I mean that a subsystem symbolizing a friend can activate many of the symbols in my brain just as I can. For instance, I can fire up my subsystem for a good friend and virtually feel myself in his shoes, running through thoughts which he might have, activating symbols in sequences which reflect his thinking patterns more accurately than my own. It could be said that my model of this friend, as embodied in a subsystem of my brain, constitutes my own chunked description of his brain.

Does this subsystem include, then, a symbol for every symbol which I think is in his brain? That would be redundant. Probably the subsystem makes extensive use of symbols already present in my brain. For instance, the symbol for "mountain" in my brain can be borrowed by the subsystem, when it is activated. The way in which that symbol is then used by the subsystem will not necessarily be identical to the way it is used by my full brain. In particular, if I am talking with my friend about the Tien Shan mountain range in Central Asia (neither of us having been there), and I know that a number of years ago he had a wonderful hiking experience in the Alps, then my interpretation of his remarks will be colored in part by my imported images of his earlier Alpine experience, since I will be trying to imagine how *he* visualizes the area.

In the vocabulary we have been building up in this Chapter, we could say that the activation of the "mountain" symbol in me is under control of my subsystem representing him. The effect of this is to open up a different window onto to my memories from the one which I normally use—namely, my "default option" switches from the full range of my memories to the set of my memories of his memories. Needless to say, my representations of his memories are only approximations to his actual memories, which are complex modes of activation of the symbols in his brain, inaccessible to me.

My representations of his memories are also complex modes of activation of my own symbols—those for "primordial" concepts, such as grass, trees, snow, sky, clouds, and so on. These are concepts which I must assume are represented in him "identically" to the way they are in me. I must also assume a similar representation in him of even more primordial notions: the experiences of gravity, breathing, fatigue, color, and so forth. Less primordial but perhaps a nearly universal human quality is the enjoyment of reaching a summit and seeing a view. Therefore, the intricate processes in my brain which are responsible for this enjoyment can be taken over directly by the friend-subsystem without much loss of fidelity.

We could go on to attempt to describe how I understand an entire tale told by my friend, a tale filled with many complexities of human relationships and mental experiences. But our terminology would quickly become inadequate. There would be tricky recursions connected with representations in him of representations in me of representations in him of one thing and another. If mutual friends figured in the tale being told, I would unconsciously look for compromises between my image of *his* representations of them, and my *own* images of them. Pure recursion would simply be an inappropriate formalism for dealing with symbol amalgams of this type. And I have barely scratched the surface! We plainly lack the vocabulary today for describing the complex interactions that are possible between symbols. So let us stop before we get bogged down.

We should note, however, that computer systems are beginning to run into some of the same kinds of complexity, and therefore some of these notions have been given names. For instance, my "mountain" symbol is analogous to what in computer jargon is called *shared* (or *reentrant*) *code*—code which can be used by two or more separate timesharing programs running on a single computer. The fact that activation of one symbol can have different results when it is part of different subsystems can be explained by saying that its code is being processed by different interpreters. Thus, the triggering patterns in the "mountain" symbol are not absolute; they are relative to the system within which the symbol is activated.

The reality of such "subbrains" may seem doubtful to some. Perhaps the following quote from M. C. Escher, as he discusses how he creates his periodic plane-filling drawings, will help to make clear what kind of phenomenon I am referring to:

> While drawing I sometimes feel as if I were a spiritualist medium, controlled by the creatures which I am conjuring up. It is as if they themselves decide on the shape in which they choose to appear. They take little account of my critical opinion during their birth and I cannot exert much influence on the measure of their development. They are usually very difficult and obstinate creatures.[2]

Here is a perfect example of the near-autonomy of certain subsystems of the brain, once they are activated. Escher's subsystems seemed to him almost to be able to override his esthetic judgment. Of course, this opinion must be taken with a grain of salt, since those powerful subsystems came into being as a result of his many years of training and submission to precisely the forces that molded his esthetic sensitivities. In short, it is wrong to divorce the subsystems in Escher's brain from Escher himself or from his esthetic judgement. They consititute a vital part of his esthetic sense, where "he" is the complete being of the artist.

The Self-Symbol and Consciousness

A very important side effect of the *self*-subsystem is that it can play the role of "soul", in the following sense: in communicating constantly with the rest of the subsystems and symbols in the brain, it keeps track of what symbols are active, and in what way. This means that it has to have symbols for mental activity—in other words, symbols for symbols, and symbols for the actions of symbols.

Of course, this does not elevate consciousness or awareness to any "magical", nonphysical level. Awareness here is a direct effect of the complex hardware and software we have described. Still, despite its earthly origin, this way of describing awareness—as the monitoring of brain activity by a subsystem of the brain itself—seems to resemble the nearly indescribable sensation which we all know and call "consciousness". Certainly one can see that the complexity here is enough that many unexpected effects could be created. For instance, it is quite plausible that a computer program with this kind of structure would make statements about itself which would have a great deal of resemblance to statements which people commonly make about themselves. This includes insisting that it has free will, that it is not explicable as a "sum of its parts", and so on. (On this subject, see the article "Matter, Mind, and Models" by M. Minsky in his book *Semantic Information Processing.*)

What kind of guarantee is there that a subsystem, such as I have here postulated, which represents the self, actually exists in our brains? Could a whole complex network of symbols such as has been described above evolve without a self-symbol evolving? How could these symbols and their activities play out "isomorphic" mental events to real events in the surrounding universe, if there were no symbol for the host organism? All the stimuli coming into the system are centered on one small mass in space. It would be quite a glaring hole in a brain's symbolic structure not to have a symbol for the physical object in which it is housed, and which plays a larger role in the events it mirrors than any other object. In fact, upon reflection, it seems that the only way one could make sense of the world surrounding a localized animate object is to understand the role of that object in relation to the other objects around it. This necessitates the existence of a self-symbol; and the step from symbol to subsystem is merely a reflection of the importance of the self-symbol, and is not a qualitative change.

Our First Encounter with Lucas

The Oxford philosopher J. R. Lucas (not connected with the Lucas numbers described earlier) wrote a remarkable article in 1961, entitled "Minds, Machines, and Gödel". His views are quite opposite to mine, and yet he manages to mix many of the same ingredients together in coming up with his opinions. The following excerpt is quite relevant to what we have just been discussing:

> At one's first and simplest attempts to philosophize, one becomes entangled in questions of whether when one knows something one knows that one knows it, and what, when one is thinking of oneself, is being thought about, and what is doing the thinking. After one has been puzzled and bruised by this problem for a long time, one learns not to press these questions: the concept of a conscious being is, implicitly, realized to be different from that of an unconscious object. In saying that a conscious being knows something, we are saying not only that he knows it, but that he knows that he knows it, and that he knows that he knows that he knows it, and so on, as long as we care to pose the

question: there is, we recognize, an infinity here, but it is not an infinite regress in the bad sense, for it is the questions that peter out, as being pointless, rather than the answers. The questions are felt to be pointless because the concept contains within itself the idea of being able to go on answering such questions indefinitely. Although conscious beings have the power of going on, we do not wish to exhibit this simply as a succession of tasks they are able to perform, nor do we see the mind as an infinite sequence of selves and super-selves and super-super-selves. Rather, we insist that a conscious being is a unity, and though we talk about parts of the mind, we do so only as a metaphor, and will not allow it to be taken literally.

The paradoxes of consciousness arise because a conscious being can be aware of itself, as well as of other things, and yet cannot really be construed as being divisible into parts. It means that a conscious being can deal with Gödelian questions in a way in which a machine cannot, because a conscious being can both consider itself and its performance and yet not be other than that which did the performance. A machine can be made in a manner of speaking to "consider" its performance, but it cannot take this "into account" without thereby becoming a different machine, namely the old machine with a "new part" added. But it is inherent in our idea of a conscious mind that it can reflect upon itself and criticize its own performances, and no extra part is required to do this: it is already complete, and has no Achilles' heel.

The thesis thus begins to become more of a matter of conceptual analysis than mathematical discovery. This is borne out by considering another argument put forward by Turing. So far, we have constructed only fairly simple and predictable artifacts. When we increase the complexity of our machines, there may, perhaps, be surprises in store for us. He draws a parallel with a fission pile. Below a certain "critical" size, nothing much happens: but above the critical size, the sparks begin to fly. So too, perhaps, with brains and machines. Most brains and all machines are, at present, "sub-critical"—they react to incoming stimuli in a stodgy and uninteresting way, have no ideas of their own, can produce only stock responses—but a few brains at present, and possibly some machines in the future, are super-critical, and scintillate on their own account. Turing is suggesting that it is only a matter of complexity, and that above a certain level of complexity a qualitative difference appears, so that "super-critical" machines will be quite unlike the simple ones hitherto envisaged.

This may be so. Complexity often does introduce qualitative differences. Although it sounds implausible, it might turn out that above a certain level of complexity, a machine ceased to be predictable, even in principle, and started doing things on its own account, or, to use a very revealing phrase, it might begin to have a mind of its own. It might begin to have a mind of its own. It would begin to have a mind of its own when it was no longer entirely predictable and entirely docile, but was capable of doing things which we recognized as intelligent, and not just mistakes or random shots, but which we had not programmed into it. But then it would cease to be a machine, within the meaning of the act. What is at stake in the mechanist debate is not how minds are, or might be, brought into being, but how they operate. It is essential for the mechanist thesis that the mechanical model of the mind shall operate according to "mechanical principles," that is, that we can understand the operation of the whole in terms of the operations of its parts, and the operation of each part either shall be determined by its initial state and the construction of the machine, or shall be a random choice between a determinate number of determinate operations. If the mechanist produces a machine which is so complicated that this ceases to hold good of it, then it is no longer a machine for the purposes of our discussion, no matter how it was constructed. We should say, rather, that he had created a mind, in the same sort of sense as we procreate people at present. There would then be two ways of bringing new minds into the world, the traditional way, by begetting children born of women, and a new way by constructing very, very complicated systems of, say, valves and relays. When talking of the second way, we should take care to stress that although what was created looked like a machine, it was not one really, because it was not just the total of its parts. One could not tell what it was going to do merely by knowing the way in which it was built up and the initial state of its parts: one could not even tell the limits of what it could do, for even when presented with a Gödel-type question, it got the answer right. In fact we should say briefly that any system which was not floored by the Gödel question was *eo ipso* not a Turing machine, i.e. not a machine within the meaning of the act.[3]

In reading this passage, my mind constantly boggles at the rapid succession of topics, allusions, connotations, confusions, and conclusions. We jump from a Carrollian paradox to Gödel to Turing to Artificial Intelligence to holism and reductionism, all in the span of two brief pages. About Lucas one can say that he is nothing if not stimulating. In the following Chapters, we shall come back to many of the topics touched on so tantalizingly and fleetingly in this odd passage.

Aria with Diverse Variations

Achilles has been unable to sleep these past few nights. His friend the Tortoise has come over tonight, to keep him company during these annoying hours.

Tortoise: I am so sorry to hear of the troubles that have been plaguing you, my dear Achilles. I hope my company will provide a welcome relief from all the unbearable stimulation which has kept you awake. Perhaps I will bore you sufficiently that you will at long last go to sleep. In that way, I will be of some service.

Achilles: Oh, no, I am afraid that I have already had some of the world's finest bores try their hand at boring me to sleep—and all, sad to say, to no avail. So you will be no match for them. No, Mr. T, I invited you over hoping that perhaps you could entertain me with a little this or that, taken from number theory, so that I could at least while away these long hours in an agreeable fashion. You see, I have found that a little number theory does wonders for my troubled psyche.

Tortoise: How quaint an idea! You know, it reminds me, just a wee bit, of the story of poor Count Kaiserling.

Achilles: Who was he?

Tortoise: Oh, he was a Count in Saxony in the eighteenth century—a Count of no account, to tell the truth—but because of him—well, shall I tell you the story? It is quite entertaining.

Achilles: In that case, by all means, do!

Tortoise: There was a time when the good Count was suffering from sleeplessness, and it just so happened that a competent musician lived in the same town, and so Count Kaiserling commissioned this musician to compose a set of variations to be played by the Count's court harpsichordist for him during his sleepless nights, to make the hours pass by more pleasantly.

Achilles: Was the local composer up to the challenge?

Tortoise: I suppose so, for after they were done, the Count rewarded him most lucratively—he presented him with a gold goblet containing one hundred Louis d'or.

Achilles: You don't say! I wonder where he came upon such a goblet and all those Louis d'or, in the first place.

Tortoise: Perhaps he saw it in a museum, and took a fancy to it.

Achilles: Are you suggesting he absconded with it?

Tortoise: Now, now, I wouldn't put it exactly that way, but . . . Those days, Counts could get away with most anything. Anyway, it is clear that the Count was most pleased with the music, for he was constantly entreating his harpsichordist—a mere lad of a fellow, name of Goldberg—to play one or another of these thirty variations. Consequently (and somewhat ironically) the variations became attached to the name of young Goldberg, rather than to the distinguished Count's name.

Achilles: You mean, the composer was Bach, and these were the so-called "Goldberg Variations"?

Tortoise: Do I ever! Actually, the work was entitled *Aria with Diverse Variations,* of which there are thirty. Do you know how Bach structured these thirty magnificent variations?

Achilles: Do tell.

Tortoise: All the pieces—except the final one—are based on a single theme, which he called an "aria". Actually, what binds them all together is not a common melody, but a common harmonic ground. The melodies may vary, but underneath, there is a constant theme. Only in the last variation did Bach take liberties. It is a sort of "post-ending ending". It contains extraneous musical ideas having little to do with the original Theme—in fact, two German folk tunes. That variation is called a "quodlibet".

Achilles: What else is unusual about the Goldberg Variations?

Tortoise: Well, every third variation is a canon. First a canon in which the two canonizing voices enter on the SAME note. Second, a canon in which one of the canonizing voices enters ONE NOTE HIGHER than the first. Third, one voice enters TWO notes higher than the other. And so on, until the final canon has entries just exactly one ninth apart. Ten canons, all told. And—

Achilles: Wait a minute. Don't I recall reading somewhere or other about fourteen recently discovered Goldberg canons . . . ?

Tortoise: Didn't that appear in the same journal where they recently reported the discovery of fourteen previously unknown days in November?

Achilles: No, it's true. A fellow named Wolff—a musicologist—heard about a special copy of the Goldberg Variations in Strasbourg. He went there to examine it, and to his surprise, on the back page, as a sort of "post-ending ending", he found these fourteen new canons, all based on the first eight notes of the theme of the Goldberg Variations. So now it is known that there are in reality forty-four Goldberg Variations, not thirty.

Tortoise: That is, there are forty-four of them, unless some other musicologist discovers yet another batch of them in some unlikely spot. And although it seems improbable, it is still possible, even if unlikely, that still another batch will be discovered, and then another one, and on and on and on . . . Why, it might never stop! We may never know if or when we have the full complement of Goldberg Variations.

Achilles: That is a peculiar idea. Presumably, everybody thinks that this latest discovery was just a fluke, and that we now really do have all the Goldberg Variations. But just supposing that you are right, and some more turn up sometime, we shall start to expect this kind of thing. At that point, the name "Goldberg Variations" will start to shift slightly in meaning, to include not only the known ones, but also any others which might eventually turn up. Their number—call it 'g'—is certain to be finite, wouldn't you agree?—but merely knowing that g is finite isn't the same as knowing how big g is. Consequently, this information won't tell us when the last Goldberg Variation has been located.

Tortoise: That is certainly true.

Achilles: Tell me—when was it that Bach wrote these celebrated variations?

Tortoise: It all happened in the year 1742, when he was Cantor in Leipzig.

Achilles: 1742? Hmm . . . That number rings a bell.

Tortoise: It ought to, for it happens to be a rather interesting number, being a sum of two odd primes: 1729 and 13.

Achilles: By thunder! What a curious fact! I wonder how often one runs across an even number with that property. Let's see . . .

$$
\begin{aligned}
6 &= 3 + 3 \\
8 &= 3 + 5 \\
10 &= 3 + 7 = 5 + 5 \\
12 &= 5 + 7 \\
14 &= 3 + 11 = 7 + 7 \\
16 &= 3 + 13 = 5 + 11 \\
18 &= 5 + 13 = 7 + 11 \\
20 &= 3 + 17 = 7 + 13 \\
22 &= 3 + 19 = 5 + 17 = 11 + 11 \\
24 &= 5 + 19 = 7 + 17 = 11 + 13 \\
26 &= 3 + 23 = 7 + 19 = 13 + 13 \\
28 &= 5 + 23 = 11 + 17 \\
30 &= 7 + 23 = 11 + 19 = 13 + 17
\end{aligned}
$$

Now what do you know—according to my little table here, it seems to be quite a common occurrence. Yet I don't discern any simple regularity in the table so far.

Tortoise: Perhaps there is no regularity to be discerned.

Achilles: But of course there is! I am just not clever enough to spot it right off the bat.

Tortoise: You seem quite convinced of it.

Achilles: There's no doubt in my mind. I wonder . . . Could it be that ALL even numbers (except 4) can be written as a sum of two odd primes?

Tortoise: Hmm . . . That question rings a bell . . . Ah, I know why! You're not the first person to ask that question. Why, as a matter of fact, in the year 1742, a mathematical amateur put forth this very question in a—

Achilles: Did you say 1742? Excuse me for interrupting, but I just noticed that 1742 happens to be a rather interesting number, being a difference of two odd primes: 1747 and 5.

Tortoise: By thunder! What a curious fact! I wonder how often one runs across an even number with that property.

Achilles: But please don't let me distract you from your story.

Tortoise: Oh, yes—as I was saying, in 1742, a certain mathematical amateur, whose name escapes me momentarily, sent a letter to Euler, who at the time was at the court of King Frederick the Great in Potsdam, and—well, shall I tell you the story? It is not without charm.

Achilles: In that case, by all means, do!

Tortoise: Very well. In his letter, this dabbler in number theory propounded an unproved conjecture to the great Euler: "Every even number can be represented as a sum of two odd primes." Now what was that fellow's name?

Achilles: I vaguely recollect the story, from some number theory book or other. Wasn't the fellow named "Kupfergödel"?

Tortoise: Hmm . . . No, that sounds too long.

Achilles: Could it have been "Silberescher"?

Tortoise: No, that's not it, either. There's a name on the tip of my tongue—ah—ah—oh yes! It was "Goldbach"! Goldbach was the fellow.

Achilles: I knew it was something like that.

Tortoise: Yes—your guesses helped jog my memory. It's quite odd, how one occasionally has to hunt around in one's memory as if for a book in a library without call numbers . . . But let us get back to 1742.

Achilles: Indeed, let's. I wanted to ask you: did Euler ever prove that this guess by Goldbach was right?

Tortoise: Curiously enough, he never even considered it worthwhile working on. However, his disdain was not shared by all mathematicians. In fact, it caught the fancy of many, and became known as the "Goldbach Conjecture".

Achilles: Has it ever been proven correct?

Tortoise: No, it hasn't. But there have been some remarkable near misses. For instance, in 1931 the Russian number theorist Schnirelmann proved that any number—even or odd—can be represented as the sum of not more than 300,000 primes.

Achilles: What a strange result. Of what good is it?

Tortoise: It has brought the problem into the domain of the finite. Previous to Schnirelmann's proof, it was conceivable that as you took larger and larger even numbers, they would require more and more primes to represent them. Some even number might take a trillion primes to represent it! Now it is known that that is not so—a sum of 300,000 primes (or fewer) will always suffice.

Achilles: I see.

Tortoise: Then in 1937, a sly fellow named Vinogradov—a Russian too—managed to establish something far closer to the desired result: namely, every sufficiently large ODD number can be represented as a sum of no more than THREE odd primes. For example, 1937 = 641 + 643 + 653. We could say that an odd number which is representable as a sum of three odd primes has "the Vinogradov property". Thus, all sufficiently large odd numbers have the Vinogradov property.

Achilles: Very well—but what does "sufficiently large" mean?

Tortoise: It means that some finite number of odd numbers may fail to have the Vinogradov property, but there is a number—call it 'v'—beyond which all odd numbers have the Vinogradov property. But Vinogradov was unable to say how big v is. So in a way, v is like g, the finite but unknown number of Goldberg Variations. Merely knowing that v is finite isn't the same as knowing how big v is. Consequently, this information won't tell us when the last odd number which needs more than three primes to represent it has been located.

Achilles: I see. And so any sufficiently large even number $2N$ can be represented as a sum of FOUR primes, by first representing $2N - 3$ as a sum of three primes, and then adding back the prime number 3.

Tortoise: Precisely. Another close approach is contained in the Theorem which says, "All even numbers can be represented as a sum of one prime and one number which is a product of at most two primes."

Achilles: This question about sums of two primes certainly leads you into strange territory. I wonder where you would be led if you looked at DIFFERENCES of two odd primes. I'll bet I could glean some insight into this teaser by making a little table of even numbers, and their representations as differences of two odd primes, just as I did for sums. Let's see . . .

$$
\begin{array}{rlllll}
2 = & 5 - 3, & 7 - 5, & 13 - 11, & 19 - 17, & \text{etc.} \\
4 = & 7 - 3, & 11 - 7, & 17 - 13, & 23 - 19, & \text{etc.} \\
6 = & 11 - 5, & 13 - 7, & 17 - 11, & 19 - 13, & \text{etc.} \\
8 = & 11 - 3, & 13 - 5, & 19 - 11, & 31 - 23, & \text{etc.} \\
10 = & 13 - 3, & 17 - 7, & 23 - 13, & 29 - 19, & \text{etc.}
\end{array}
$$

My gracious! There seems to be no end to the number of different representations I can find for these even numbers. Yet I don't discern any simple regularity in the table so far.

Tortoise: Perhaps there is no regularity to be discerned.

Achilles: Oh, you and your constant rumblings about chaos! I'll hear none of that, thank you.

Tortoise: Do you suppose that EVERY even number can be represented somehow as the difference of two odd primes?

Achilles: The answer certainly would appear to be yes, from my table. But then again, I suppose it could also be no. That doesn't really get us very far, does it?

Tortoise: With all due respect, I would say there are deeper insights to be had on the matter.

Achilles: Curious how similar this problem is to Goldbach's original one. Perhaps it should be called a "Goldbach Variation".

Tortoise: Indeed. But you know, there is a rather striking difference between the Goldbach Conjecture, and this Goldbach Variation, which I would like to tell you about. Let us say that an even number $2N$ has the "Goldbach property" if it is the SUM of two odd primes, and it has the "Tortoise property" if it is the DIFFERENCE of two odd primes.

Achilles: I think you should call it the "Achilles property". After all, I suggested the problem.

Tortoise: I was just about to propose that we should say that a number which LACKS the Tortoise property has the "Achilles property".

Achilles: Well, all right . . .

Tortoise: Now consider, for instance, whether 1 trillion has the Goldbach property or the Tortoise property. Of course, it may have both.

Achilles: I can consider it, but I doubt whether I can give you an answer to either question.

Tortoise: Don't give up so soon. Suppose I asked you to answer one or the other question. Which one would you pick to work on?

Achilles: I suppose I would flip a coin. I don't see much difference between them.

Tortoise: Aha! But there's a world of difference! If you pick the Goldbach property, involving SUMS of primes, then you are limited to using primes which are bounded between 2 and 1 trillion, right?

Achilles: Of course.

Tortoise: So your search for a representation for 1 trillion as a sum of two primes is GUARANTEED TO TERMINATE.

Achilles: Ahhh! I see your point. Whereas if I chose to work on representing 1 trillion as the DIFFERENCE of two primes, I would not have any bound on the size of the primes involved. They might be so big that it would take me a trillion years to find them.

Tortoise: Or then again, they might not even EXIST! After all, that's what the question was asking—do such primes exist? It wasn't of much concern how big they might turn out to be.

Achilles: You're right. If they didn't exist, then a search process would lead on forever, never answering yes, and never answering no. And nevertheless, the answer would be no.

Tortoise: So if you have some number, and you wish to test whether it has the Goldbach property or the Tortoise property, the difference between the two tests will be this: in the former, the search involved is GUARANTEED TO TERMINATE; in the latter, it is POTENTIALLY ENDLESS—there are no guarantees of any type. It might just go merrily on forever, without yielding an answer. And yet, on the other hand, in some cases, it might stop on the first step.

Achilles: I see there is a rather vast difference between the Goldbach and Tortoise properties.

Tortoise: Yes, the two similar problems concern these vastly different properties. The Goldbach Conjecture is to the effect that all even numbers have the Goldbach property; the Goldbach Variation suggests that all even numbers have the Tortoise property. Both problems are unsolved, but what is interesting is that although they sound very much alike, they involve properties of whole numbers which are quite different.

Achilles: I see what you mean. The Goldbach property is a detectable, or recognizable property of any even number, since I know how to test for its presence—just embark on a search. It will automatically come to an end with a yes or no answer. The Tortoise property, however, is more elusive, since a brute force search just may never give an answer.

Tortoise: Well, there may be cleverer ways of searching in the case of the Tortoise property, and maybe following one of them would always come to an end, and yield an answer.

Achilles: Couldn't the search only end if the answer were yes?

Tortoise: Not necessarily. There might be some way of proving that whenever the search lasts longer than a certain length of time, then the answer must be no. There might even be some OTHER way of searching for the primes, not such a brute force way, which is guaranteed to find them if they exist, and to tell if they don't. In either case, a finite search would be able to yield the answer no. But I don't know if such a thing can be proven or not. Searching through infinite spaces is always a tricky matter, you know.

Achilles: So as things stand now, you know of no test for the Tortoise property which is guaranteed to terminate—and yet there MIGHT exist such a search.

Tortoise: Right. I suppose one could embark on a search for such a search, but I can give no guarantee that that "meta-search" would terminate, either.

Achilles: You know, it strikes me as quite peculiar that if some even number—for example, a trillion—failed to have the Tortoise property, it would be caused by an infinite number of separate pieces of information. It's funny to think of wrapping all that information up into one bundle, and calling it, as you so gallantly suggested, "the Achilles property" of 1 trillion. It is really a property of the number system as a WHOLE, not just of the number 1 trillion.

Tortoise: That is an interesting observation, Achilles, but I maintain that it makes a good deal of sense to attach this fact to the number 1 trillion nevertheless. For purposes of illustration, let me suggest that you consider the simpler statement "29 is prime". Now in fact, this statement really means that 2 times 2 is not 29, and 5 times 6 is not 29, and so forth, doesn't it?

Achilles: It must, I suppose.

Tortoise: But you are perfectly happy to collect all such facts together, and attach them in a bundle to the number 29, saying merely, "29 is prime"?

Achilles: Yes . . .

Tortoise: And the number of facts involved is actually infinite, isn't it? After all, such facts as "4444 times 3333 is not 29" are all part of it, aren't they?

Achilles: Strictly speaking, I suppose so. But you and I both know that you can't produce 29 by multiplying two numbers which are both bigger than 29. So in reality, saying "29 is prime" is only summarizing a FINITE number of facts about multiplication.

Tortoise: You can put it that way if you want, but think of this: the fact that two numbers which are bigger than 29 can't have a product equal to 29 involves the entire structure of the number system. In that sense, that fact in itself is a summary of an infinite number of facts. You can't get away from the fact, Achilles, that when you say "29 is prime", you are actually stating an infinite number of things.

Achilles: Maybe so, but it feels like just one fact to me.

Tortoise: That's because an infinitude of facts are contained in your prior knowledge—they are embedded implicitly in the way you visualize things. You don't see an explicit infinity because it is captured implicitly inside the images you manipulate.

Achilles: I guess that you're right. It still seems odd to lump a property of the entire number system into a unit, and label the unit "primeness of 29".

Tortoise: Perhaps it seems odd, but it is also quite a convenient way to look at things. Now let us come back to your hypothetical idea. If, as you suggested, the number 1 trillion has the Achilles property, then no matter what prime you add to it, you do not get another prime. Such a state of affairs would be caused by an infinite number of separate mathematical "events". Now do all these "events" necessarily spring from the same source? Do they have to have a common cause? Because if they don't, then some sort of "infinite coincidence" has created the fact, rather than an underlying regularity.

Achilles: An "infinite coincidence"? Among the natural numbers, NOTHING is coincidental—nothing happens without there being some underlying pattern. Take 7, instead of a trillion. I can deal with it more easily, because it is smaller. 7 has the Achilles property.

Tortoise: You're sure?

Achilles: Yes. Here's why. If you add 2 to it, you get 9, which isn't prime. And if you add any other prime to 7, you are adding two odd numbers, resulting in an even number—thus you again fail to get a prime. So here the "Achilleanity" of 7, to coin a term, is a consequence of just TWO reasons: a far cry from any "infinite coincidence". Which just goes to support my assertion: that it never takes an infinite number of reasons to account for some arithmetical truth. If there WERE some arithmetical fact which were caused by an infinite collection of unrelated coincidences, then you could never give a finite proof for that truth. And that is ridiculous.

Tortoise: That is a reasonable opinion, and you are in good company in making it. However—

Achilles: Are there actually those who disagree with this view? Such people would have to believe that there are "infinite coincidences", that there is chaos in the midst of the most perfect, harmonious, and beautiful of all creations: the system of natural numbers.

Tortoise: Perhaps they do; but have you ever considered that such chaos might be an integral part of the beauty and harmony?

FIGURE 71. Order and Chaos, *by M. C. Escher (lithograph, 1950).*

Achilles: Chaos, part of perfection? Order and chaos make a pleasing unity? Heresy!

Tortoise: Your favorite artist, M. C. Escher, has been known to suggest such a heretical point of view in one of his pictures . . . And while we're on the subject of chaos, I believe that you might be interested in hearing about two different categories of search, both of which are guaranteed to terminate.

Achilles: Certainly.

Tortoise: The first type of search—the non-chaotic type—is exemplified by the test involved in checking for the Goldbach property. You just look at primes less than $2N$, and if some pair adds up to $2N$, then $2N$ has the Goldbach property; otherwise, it doesn't. This kind of test is not only sure to terminate, but you can predict BY WHEN it will terminate, as well.

Achilles: So it is a PREDICTABLY TERMINATING test. Are you going to tell me that checking for some number-theoretical properties involves tests which are guaranteed to terminate, but about which there is no way to know in advance how long they will take?

Tortoise: How prophetic of you, Achilles. And the existence of such tests shows that there is intrinsic chaos, in a certain sense, in the natural number system.

Achilles: Well, in that case, I would have to say that people just don't know enough about the test. If they did a little more research, they could figure out how long it will take, at most, before it terminates. After all, there must always be some rhyme or reason to the patterns among integers. There can't just be chaotic patterns which defy prediction!

Tortoise: I can understand your intuitive faith, Achilles. However, it's not always justified. Of course, in many cases you are exactly right—just because somebody doesn't know something, one can't conclude that it is unknowable! But there are certain properties of integers for which terminating tests can be proven to exist, and yet about which it can also be PROVEN that there is no way to predict in advance how long they will take.

Achilles: I can hardly believe that. It sounds as if the devil himself managed to sneak in and throw a monkey wrench into God's beautiful realm of natural numbers!

Tortoise: Perhaps it will comfort you to know that it is by no means easy, or natural, to define a property for which there is a terminating but not PREDICTABLY terminating test. Most "natural" properties of integers do admit of predictably terminating tests. For example, primeness, squareness, being a power of ten, and so on.

Achilles: Yes, I can see that those properties are completely straightforward to test for. Will you tell me a property for which the only possible test is a terminating but nonpredictable one?

Tortoise: That's too complicated for me in my sleepy state. Let me instead show you a property which is very easy to define, and yet for which no terminating test is known. I'm not saying there won't ever be one discovered, mind you—just that none is known. You begin with a number—would you care to pick one?

Achilles: How about 15?

Tortoise: An excellent choice. We begin with your number, and if it is ODD, we triple it, and add 1. If it is EVEN, we take half of it. Then we repeat the process. Call a number which eventually reaches 1 this way a WONDROUS number, and a number which doesn't, an UNWONDROUS number.

Achilles: Is 15 wondrous, or unwondrous? Let's see:

15	is ODD, so I make $3n + 1$:	46
46	is EVEN, so I take half:	23
23	is ODD, so I make $3n + 1$:	70
70	is EVEN, so I take half:	35
35	is ODD, so I make $3n + 1$:	106
106	is EVEN, so I take half:	53
53	is ODD, so I make $3n + 1$:	160
160	is EVEN, so I take half:	80
80	is EVEN, so I take half:	40
40	is EVEN, so I take half:	20
20	is EVEN, so I take half:	10
10	is EVEN, so I take half:	5
5	is ODD, so I make $3n + 1$:	16
16	is EVEN, so I take half:	8
8	is EVEN, so I take half:	4
4	is EVEN, so I take half:	2
2	is EVEN, so I take half:	1.

Wow! That's quite a roundabout journey, from 15 to 1. But I finally reached it. That shows that 15 has the property of being wondrous. I wonder what numbers are UNwondrous . . .

Tortoise: Did you notice how the numbers swung up and down, in this simply defined process?

Achilles: Yes. I was particularly surprised, after thirteen turns, to find myself at 16, only one greater than 15, the number I started with. In one sense, I was almost back where I started—yet in another sense, I was nowhere near where I had started. Also, I found it quite curious that I had to go as high as 160 to resolve the question. I wonder how come.

Tortoise: Yes, there is an infinite "sky" into which you can sail, and it is very hard to know in advance how high into the sky you will wind up sailing. Indeed, it is quite plausible that you might just sail up and up and up, and never come down.

Achilles: Really? I guess that is conceivable—but what a weird coincidence it would require! You'd just have to hit odd number after odd number, with only a few evens mixed in. I doubt if that would ever happen—but I just don't know for sure.

Tortoise: Why don't you try starting with 27? Mind you, I don't promise anything. But sometime, just try it, for your amusement. And I'd advise you to bring along a rather large sheet of paper.

Achilles: Hmm . . . Sounds interesting. You know, it still makes me feel funny to associate the wondrousness (or unwondrousness) with the starting number, when it is so obviously a property of the entire number system.

Tortoise: I understand what you mean, but it's not that different from saying "29 is prime", or "gold is valuable"—both statements attribute to a single entity a property which it has only by virtue of being embedded in a particular context.

Achilles: I suppose you're right. This "wondrousness" problem is wondrous tricky, because of the way in which the numbers oscillate—now increasing, now decreasing. The pattern OUGHT to be regular, yet on the surface it appears to be quite chaotic. Therefore, I can well imagine why, as of yet, no one knows of a test for the property of wondrousness which is guaranteed to terminate.

Tortoise: Speaking of terminating and nonterminating processes, and those which hover in between, I am reminded of a friend of mine, an author, who is at work on a book.

Achilles: Oh, how exciting! What is it called?

Tortoise: *Copper, Silver, Gold: an Indestructible Metallic Alloy.* Doesn't that sound interesting?

Achilles: Frankly, I'm a little confused by the title. After all, what do Copper, Silver, and Gold have to do with each other?

Tortoise: It seems clear to me.

Achilles: Now if the title were, say, *Giraffes, Silver, Gold,* or *Copper, Elephants, Gold,* why, I could see it . . .

Tortoise: Perhaps you would prefer *Copper, Silver, Baboons*?

Achilles: Oh, absolutely! But that original title is a loser. No one would understand it.

Tortoise: I'll tell my friend. He'll be delighted to have a catchier title (as will his publisher).

Achilles: I'm glad. But how were you reminded of his book by our discussion?

Tortoise: Ah, yes. You see, in his book there will be a Dialogue in which he wants to throw readers off by making them SEARCH for the ending.

Achilles: A funny thing to want to do. How is it done?

Tortoise: You've undoubtedly noticed how some authors go to so much trouble to build up great tension a few pages before the end of their stories—but a reader who is holding the book physically in his hands can FEEL that the story is about to end. Hence, he has some extra information which acts as an advance warning, in a way. The tension is a bit spoiled by the physicality of the book. It would be so much better if, for instance, there were a lot of padding at the end of novels.

Achilles: Padding?

Tortoise: Yes; what I mean is, a lot of extra printed pages which are not part of the story proper, but which serve to conceal the exact location of the end from a cursory glance, or from the feel of the book.

Achilles: I see. So a story's true ending might occur, say, fifty or a hundred pages before the physical end of the book?

Tortoise: Yes. This would provide an element of surprise, because the reader wouldn't know in advance how many pages are padding, and how many are story.

Achilles: If this were standard practice, it might be quite effective. But there is a problem. Suppose your padding were very obvious—such as a lot of blanks, or pages covered with X's or random letters. Then, it would be as good as absent.

Tortoise: Granted. You'd have to make it resemble normal printed pages.

Achilles: But even a cursory glance at a normal page from one story will often suffice to distinguish it from another story. So you will have to make the padding resemble the genuine story rather closely.

Tortoise: That's quite true. The way I've always envisioned it is this: you bring the story to an end; then without any break, you follow it with something which looks like a continuation but which is in reality just padding, and which is utterly unrelated to the true theme. The padding is, in a way, a "post-ending ending". It may contain extraneous literary ideas, having little to do with the original theme.

Achilles: Sneaky! But then the problem is that you won't be able to tell when the real ending comes. It'll just blend right into the padding.

Tortoise: That's the conclusion my author friend and I have reached as well. It's a shame, for I found the idea rather appealing.

Achilles: Say, I have a suggestion. The transition between genuine story and padding material could be made in such a way that, by sufficiently assiduous inspection of the text, an intelligent reader will be able to detect where one leaves off and the other begins. Perhaps it will take him quite a while. Perhaps there will be no way to predict how long it will take . . . But the publisher could give a guarantee that a sufficiently assiduous search for the true ending will always terminate, even if he can't say how long it will be before the test terminates.

Tortoise: Very well—but what does "sufficiently assiduous" mean?

Achilles: It means that the reader must be on the lookout for some small but telltale feature in the text which occurs at some point. That would signal the end. And he must be ingenious enough to think up, and hunt for, many such features until he finds the right one.

Tortoise: Such as a sudden shift of letter frequencies or word lengths? Or a rash of grammatical mistakes?

Achilles: Possibly. Or a hidden message of some sort might reveal the true end to a sufficiently assiduous reader. Who knows? One could even throw in some extraneous characters or events which are inconsistent with the spirit of the foregoing story. A naïve reader would swallow the whole thing, whereas a sophisticated reader would be able to spot the dividing line exactly.

Tortoise: That's a most original idea, Achilles. I'll relay it to my friend, and perhaps he can incorporate it in his Dialogue.

Achilles: I would be highly honored.

Tortoise: Well, I am afraid that I myself am growing a little groggy, Achilles. It would be well for me to take my leave, while I am still capable of navigating my way home.

Achilles: I am most flattered that you have stayed up for so long, and at such an odd hour of the night, just for my benefit. I assure you that your number-theoretical entertainment has been a perfect antidote to my usual tossing and turning. And who knows—perhaps I may even be able to go to sleep tonight. As a token of my gratitude, Mr. T, I would like to present you with a special gift.

Tortoise: Oh, don't be silly, Achilles.

Achilles: It is my pleasure, Mr. T. Go over to that dresser; on it, you will see an Asian box.

(The Tortoise moseys over to Achilles' dresser.)

Tortoise: You don't mean this very gold Asian box, do you?

Achilles: That's the one. Please accept it, Mr. T, with my warmest compliments.

Tortoise: Thank you very much indeed, Achilles. Hmm . . . Why are all these mathematicians' names engraved on the top? What a curious list:

D e M o r g a n
A **b** e l
B o **o** l e
B r o **u** w e r
S i e r **p** i ń s k i
W e i e r **s** t r a s s

Achilles: I believe it is supposed to be a Complete List of All Great Mathematicians. What I haven't been able to figure out is why the letters running down the diagonal are so much bolder.

Tortoise: At the bottom it says, "Subtract 1 from the diagonal, to find Bach in Leipzig".

Achilles: I saw that, but I couldn't make head or tail of it. Say, how about a shot of excellent whiskey? I happen to have some in that decanter on my shelf.

Tortoise: No, thanks. I'm too tired. I'm just going to head home. *(Casually, he opens the box.)* Say, wait a moment, Achilles—there are one hundred Louis d'or in here!

Achilles: I would be most pleased if you would accept them, Mr. T.

Tortoise: But—but—

Achilles: No objections, now. The box, the gold—they're yours. And thank you for an evening without parallel.

Tortoise: Now whatver has come over you, Achilles? Well, thank you for your outstandig generosity, and I hope you have sweet dreams about the strange Golbach Conjecture, and its Variation. Good night.

(And he picks up the very gold Asian box filled with the one hundred Louis d'or, and walks towards the door. As he is about to leave, there is a loud knock.)

Who could be knocking at this ungodly hour, Achilles?

Achilles: I haven't the foggiest idea. It seems suspicious to me. Why don't you go hide behind the dresser, in case there's any funny business.

Tortoise: Good idea. *(Scrambles in behind the dresser.)*

Achilles: Who's there?

Voice: Open up—it's the cops.

Achilles: Come in, it's open.

(Two burly policemen walk in, wearing shiny badges.)

Cop: I'm Silva. This is Gould. *(Points at his badge.)* Is there an Achilles at this address?

Achilles: That's me!

Cop: Well, Achilles, we have reason to believe that there is a very gold Asian box here, filled with one hundred Louis d'or. Someone absconded with it from the museum this afternoon.

Achilles: Heavens to Betsy!

Cop: If it is here, Achilles, since you would be the only possible suspect, I regret to say that I should have to take you into custody. Now I have here a search warrant—

Achilles: Oh, sirs, am I ever glad you arrived! All evening long, I have been being terrorized by Mr. Tortoise and his very Asian gold box. Now at last you have come to liberate me! Please, sirs, just take a look behind that dresser, and there you will find the culprit!

(The cops look behind the dresser and spy the Tortoise huddled behind it, holding his very gold Asian box, and trembling.)

Cop: So there it is! And so Mr. Tortoise is the varmint, eh? I never would have suspected HIM. But he's caught, red-handed.

Achilles: Haul the villain away, kind sirs! Thank goodness, that's the last I'll have to hear of him, and the Very Asian Gold Box!

CHAPTER XIII

BlooP and FlooP and GlooP

Self-Awareness and Chaos

BLOOP, FLOOP, AND GLOOP are not trolls, talking ducks, or the sounds made by a sinking ship—they are three computer languages, each one with its own special purpose. These languages were invented specially for this Chapter. They will be of use in explaining some new senses of the word "recursive"—in particular, the notions of *primitive recursivity* and *general recursivity*. They will prove very helpful in clarifying the machinery of self-reference in TNT.

We seem to be making a rather abrupt transition from brains and minds to technicalities of mathematics and computer science. Though the transition is abrupt in some ways, it makes some sense. We just saw how a certain kind of self-awareness seems to be at the crux of consciousness. Now we are going to scrutinize "self-awareness" in more formal settings, such as TNT. The gulf between TNT and a mind is wide, but some of the ideas will be most illuminating, and perhaps metaphorically transportable back to our thoughts about consciousness.

One of the amazing things about TNT's self-awareness is that it is intimately connected to questions about order versus chaos among the natural numbers. In particular, we shall see that an orderly system of sufficient complexity that it can mirror itself cannot be *totally* orderly—it must contain some strange, chaotic features. For readers who have some Achilles in them, this will be hard to take. However, there is a "magical" compensation: there is a kind of order to the disorder, which is now its own field of study, called "recursive function theory". Unfortunately, we will not be able to do much more than hint at the fascination of this subject.

Representability and Refrigerators

Phrases such as "sufficiently complex", "sufficiently powerful" and the like have cropped up quite often earlier. Just what do they mean? Let us go back to the battle of the Crab and Tortoise, and ask, "What qualifies something as a record player?" The Crab might claim that his refrigerator is a "Perfect" record player. Then to prove it, he could set any record whatsoever atop it, and say, "You see—it's playing it!" The Tortoise, if he wanted to counter this Zen-like act, would have to reply, "No—your refrigerator is too low-fidelity to be counted as a phonograph: it cannot reproduce sounds at all (let alone its self-breaking sound)." The Tortoise can only make a record called "I Cannot Be Played on Record Player X" provided that Record Player X is really a record player! The Tortoise's method is quite insidious, as it plays on the strength, rather than on the weakness, of the system. And therefore he requires "sufficiently hi-fi" record players.

Ditto for formal versions of number theory. The reason that TNT is a formalization of N is that its symbols act the right way: that is, its theorems are not silent like a refrigerator—they speak actual truths of N. Of course, so do the theorems of the pq-system. Does it, too, count as "a formalization of number theory", or is it more like a refrigerator? Well, it is a little better than a refrigerator, but it is still pretty weak. The pq-system does not include enough of the core truths of N to count as "a number theory".

What, then, are these "core truths" of N? They are the *primitive recursive truths;* that means they involve only *predictably terminating* calculations. These core truths serve for N as Euclid's first four postulates served for geometry: they allow you to throw out certain candidates before the game begins, on the grounds of "insufficient power". From here on out, the *representability of all primitive recursive truths* will be the criterion for calling a system "sufficiently powerful".

Gantō's Ax in Metamathematics

The significance of the notion is shown by the following key fact: If you have a sufficiently powerful formalization of number theory, then Gödel's method is applicable, and consequently your system is *incomplete.* If, on the other hand, your system is *not* sufficiently powerful (i.e., not all primitive recursive truths are theorems), then your system is, precisely by virtue of that lack, *incomplete.* Here we have a reformulation of "Gantō's Ax" in metamathematics: whatever the system does, Gödel's Ax will chop its head off! Notice also how this completely parallels the high-fidelity-versus-low-fidelity battle in the *Contracrostipunctus.*

Actually, it turns out that much weaker systems are still vulnerable to the Gödel method; the criterion that all primitive recursive truths need be represented as theorems is far too stringent. It is a little like a thief who will only rob "sufficiently rich" people, and whose criterion is that the potential victim should be carrying at least a million dollars in cash. In the case of TNT, luckily, we will be able to act in our capacity as thieves, for the million in cash is there—which is to say, TNT does indeed contain all primitive recursive truths as theorems.

Now before we plunge into a detailed discussion of primitive recursive functions and predicates, I would like to tie the themes of this Chapter to themes from earlier Chapters, so as to provide a bit better motivation.

Finding Order by Choosing the Right Filter

We saw at a very early stage that formal systems can be difficult and unruly beasts because they have lengthening and shortening rules, which can possibly lead to never-ending searches among strings. The discovery of Gödel-numbering showed that any search for a string having a special typographical property has an arithmetical cousin: an isomorphic search for an integer with a corresponding special arithmetical property. Consequently, the quest for decision procedures for formal systems involves solving the mystery of unpredictably long searches—*chaos*—among the integers. Now in the *Aria with Diverse Variations,* I gave perhaps too much weight to apparent manifestations of chaos in problems about integers. As a matter of fact, people have tamed wilder examples of apparent chaos than the "wondrousness" problem, finding them to be quite gentle beasts after all. Achilles' powerful faith in the regularity and predictability of numbers should therefore be accorded quite a bit of respect—especially as it reflects the beliefs of nearly all mathematicians up till the 1930's. To show why order versus chaos is such a subtle and significant issue, and to tie it in with questions about the location and revelation of meaning, I would like to quote a beautiful and memorable passage from *Are Quanta Real?*—a Galilean Dialogue by the late J. M. Jauch:

SALVIATI Suppose I give you two sequences of numbers, such as

7 8 5 3 9 8 1 6 3 3 9 7 4 4 8 3 0 9 6 1 5 6 6 0 8 4 . . .

and

$$1,\ -1/3,\ +1/5,\ -1/7,\ +1/9,\ -1/11,\ +1/13,\ -1/15,\ \ldots$$

If I asked you, Simplicio, what the next number of the first sequence is, what would you say?

SIMPLICIO I could not tell you. I think it is a random sequence and that there is no law in it.

SALVIATI And for the second sequence?

SIMPLICIO That would be easy. It must be +1/17.

SALVIATI Right. But what would you say if I told you that the first sequence is also constructed by a law and this law is in fact identical with the one you have just discovered for the second sequence?

SIMPLICIO This does not seem probable to me.

SALVIATI But it is indeed so, since the first sequence is simply the beginning of the decimal fraction [expansion] of the sum of the second. Its value is $\pi/4$.

SIMPLICIO You are full of such mathematical tricks, but I do not see what this has to do with abstraction and reality.

SALVIATI The relationship with abstraction is easy to see. The first sequence looks random unless one has developed through a process of abstraction a kind of filter which sees a simple structure behind the apparent randomness.

It is exactly in this manner that laws of nature are discovered. Nature presents us with a host of phenomena which appear mostly as chaotic randomness until we select some significant events, and abstract from their particular, irrelevant circumstances so that they become idealized. Only then can they exhibit their true structure in full splendor.

SAGREDO This is a marvelous idea! It suggests that when we try to understand nature, we should look at the phenomena as if they were *messages* to be understood. Except that each message appears to be random until we establish a code to read it. This code takes the form of an abstraction, that is, we choose to ignore certain things as irrelevant and we thus partially select the content of the message by a free choice. These irrelevant signals form the "background noise," which will limit the accuracy of our message.

But since the code is not absolute there may be several messages in the same raw material of the data, so changing the code will result in a message of equally deep significance in something that was merely noise before, and *conversely:* In a new code a former message may be devoid of meaning.

Thus a code presupposes a free choice among different, complementary aspects, each of which has equal claim to *reality,* if I may use this dubious word.

Some of these aspects may be completely unknown to us now but they may reveal themselves to an observer with a different system of abstractions.

But tell me, Salviati, how can we then still claim that we *discover* something out there in the objective real world? Does this not mean that we are merely creating things according to our own images and that reality is only within ourselves?

SALVIATI I don't think that this is necessarily so, but it is a question which requires deeper reflection.[1]

Jauch is here dealing with messages that come not from a "sentient being" but from nature itself. The questions that we raised in Chapter VI on the relation of meaning to messages can be raised equally well with messages from nature. Is nature chaotic, or is nature patterned? And what is the role of intelligence in determining the answer to this question?

To back off from the philosophy, however, we can consider the point about the deep regularity of an apparently random sequence. Might the function $Q(n)$ from Chapter V have a simple, nonrecursive explanation, too? Can every problem, like an orchard, be seen from such an angle that its secret is revealed? Or are there some problems in number theory which, no matter what angle they are seen from, remain mysteries?

With this prologue, I feel it is time to move ahead to define the precise meaning of the term "predictably long search". This will be accomplished in terms of the language BlooP.

Primordial Steps of the Language BlooP

Our topic will be searches for natural numbers which have various properties. In order to talk about the *length* of any search, we shall have to define some primordial *steps,* out of which all searches are built, so that length can be measured in terms of number of steps. Some steps which we might consider primordial are:

adding any two natural numbers;
multiplying any two natural numbers;
determining if two numbers are equal;
determining the larger (smaller) of two numbers.

Loops and Upper Bounds

If we try to formulate a test for, say, primality in terms of such steps, we shall soon see that we have to include a *control structure*—that is, descriptions of the order to do things in, when to branch back and try something again, when to skip over a set of steps, when to stop, and similar matters.

It is typical of any *algorithm*—that is, a specific delineation of how to carry out a task—that it includes a mixture of (1) specific operations to be performed, and (2) control statements. Therefore, as we develop our language for expressing predictably long calculations, we shall have to incorporate primordial control structures also. In fact, the hallmark of BlooP is its limited set of control structures. It does not allow you to branch to arbitrary steps, or to repeat groups of steps without limit; in BlooP, essentially the only control structure is the *bounded loop:* a set of instructions which can be executed over and over again, up to a predefined maximum number of times, called the *upper bound,* or *ceiling,* of the loop. If the ceiling were 300, then the loop might be executed 0, 7, or 300 times—but not 301.

Now the exact values of all the upper bounds in a program need not be put in numerically by the programmer—indeed, they may not be known in advance. Instead, any upper bound may be determined by calculations carried out *before* its loop is entered. For instance, if you wanted to calculate the value of 2^{3^n}, there would be two loops. First, you evaluate 3^n, which involves n multiplications. Then, you put 2 to that power, which involves 3^n multiplications. Thus, the upper bound for the second loop is the result of the calculation of the first loop.

Here is how you would express this in a BlooP program:

```
DEFINE PROCEDURE ''TWO-TO-THE-THREE-TO-THE'' [N]:
BLOCK 0: BEGIN
            CELL(0) ⇐ 1;
            LOOP N TIMES:
            BLOCK 1: BEGIN
                        CELL(0) ⇐ 3 × CELL(0);
            BLOCK 1: END;
            CELL(1) ⇐ 1;
            LOOP CELL(0) TIMES:
            BLOCK 2: BEGIN
                        CELL(1) ⇐ 2 × CELL(1);
            BLOCK 2: END;
            OUTPUT ⇐ CELL(1);
BLOCK 0: END.
```

Conventions of BlooP

Now it is an acquired skill to be able to look at an algorithm written in a computer language, and figure out what it is doing. However, I hope that this algorithm is simple enough that it makes sense without too much scrutiny. A *procedure* is defined, having one *input parameter*, N; its *output* is the desired value.

This procedure definition has what is called *block structure,* which means that certain portions of it are to be considered as units, or *blocks.* All the statements in a block get executed as a unit. Each block has a number (the outermost being BLOCK 0), and is delimited by a BEGIN and an END. In our example, BLOCK 1 and BLOCK 2 contain just one statement each—but shortly you will see longer blocks. A LOOP statement always means to execute the block immediately under it repeatedly. As can be seen above, blocks can be nested.

The strategy of the above algorithm is as described earlier. You begin by taking an auxiliary variable, called CELL(0); you set it initially to 1, and then, in a loop, you multiply it repeatedly by 3 until you've done so exactly N times. Next, you do the analogous thing for CELL(1)—set it to 1, multiply by 2 exactly CELL(0) times, then quit. Finally, you set OUTPUT to the value of CELL(1). This is the value returned to the outside world—the only externally visible behavior of the procedure.

A number of points about the notation should be made here. First, the meaning of the left-arrow '⇐' is this:

> Evaluate the expression to its right, then take the result and set the CELL (or OUTPUT) on its left to that value.

So the meaning of a command such as CELL(1) ⇐ 3 × CELL(1) is to triple the value stored in CELL(1). You may think of each CELL as being a separate word in the memory of some computer. The only difference between a CELL and a true word is that the latter can only hold integers up to some finite limit, whereas we allow a CELL to hold any natural number, no matter how big.

Every procedure in BlooP, when called, yields a value—namely the value of the variable called OUTPUT. At the beginning of execution of any procedure, it is assumed as a default option that OUTPUT has the value 0. That way, even if the procedure never resets OUTPUT at all, OUTPUT has a well-defined value at all times.

IF-Statements and Branching

Now let us look at another procedure which will show us some other features of BlooP which give it more generality. How do you find out, knowing only how to add, what the value of M – N is? The trick is to add various numbers onto N until you find the one which yields M. However, what happens if M is smaller than N? What if we are trying to take 5 from 2? In the domain of natural numbers, there is no answer. But we would like our BlooP procedure to give an answer anyway—let's say 0. Here, then, is a BlooP procedure which does subtraction:

```
DEFINE PROCEDURE ''MINUS'' [M,N]:
BLOCK 0: BEGIN
        IF M < N, THEN:
        QUIT BLOCK 0;
        LOOP AT MOST M + 1 TIMES:
        BLOCK 1: BEGIN
                IF OUTPUT + N = M, THEN:
                ABORT LOOP 1;
                OUTPUT ⇐ OUTPUT + 1;
        BLOCK 1: END;
BLOCK 0: END.
```

Here we are making use of the implicit feature that **OUTPUT** begins at 0. If **M** is less than **N**, then the subtraction is impossible, and we simply jump to the bottom of **BLOCK 0** right away, and the answer is 0. That is what is meant by the line **QUIT BLOCK 0**. But if **M** is not less than **N**, then we skip over that **QUIT**-statement, and carry out the next command in sequence (here, a **LOOP**-statement). That is how **IF**-statements always work in BlooP.

So we enter **LOOP 1**, so called because the block which it tells us to repeat is **BLOCK 1**. We try adding 0 to **N**, then 1, 2, etc., until we find a number that gives **M**. At that point, we **ABORT** the loop we are in, meaning we jump to the statement immediately following the **END** which marks the bottom of the loop's block. In this case, that jump brings us just below **BLOCK 1: END**, which is to say, to the last statement of the algorithm, and we are done. **OUTPUT** now contains the correct answer.

Notice that there are two distinct instructions for jumping downwards: **QUIT**, and **ABORT**. The former pertains to blocks, the latter to loops. **QUIT BLOCK** *n* means to jump to the last line of **BLOCK** *n,* whereas **ABORT LOOP** *n* means to jump just *below* the last line of **BLOCK** *n.* This distinction only matters when you are inside a loop and want to continue looping but to quit the block this time around. Then you can say **QUIT** and the proper thing will happen.

Also notice that the words **AT MOST** now precede the upper bound of the loop, which is a warning that the loop may be aborted before the upper bound is reached.

Automatic Chunking

Now there are two last features of BlooP to explain, both of them very important. The first is that, once a procedure has been *defined,* it may be *called* inside later procedure definitions. The effect of this is that *once an operation has been defined in a procedure, it is considered as simple as a primordial step.* Thus, BlooP features automatic chunking. You might compare it to the way a good ice skater acquires new motions: not by defining them as long sequences of primordial muscle-actions, but in terms of previously learned motions, which were themselves learned as compounds of earlier learned motions, etc.—and the nestedness, or chunkedness, can go back many layers until you hit primordial muscle-actions. And thus, the repertoire of BlooP programs, like the repertoire of a skater's tricks, grows, quite literally, by loops and bounds.

BlooP Tests

The other feature of BlooP is that certain procedures can have **YES** or **NO** as their output, instead of an integer value. Such procedures are *tests,* rather than *functions.* To indicate the difference, the name of a test must terminate in a question mark. Also, in a test, the default option for **OUTPUT** is not 0, of course, but **NO**.

Let us see an example of these last two features of BlooP in an algorithm which tests its argument for primality:

```
DEFINE PROCEDURE ''PRIME?'' [N]:
BLOCK 0: BEGIN
            IF N = 0, THEN:
            QUIT BLOCK 0;
            CELL(0) ⇐ 2;
            LOOP AT MOST MINUS [N,2] TIMES:
            BLOCK 1: BEGIN
                        IF REMAINDER [N,CELL(0)] = 0, THEN:
                        QUIT BLOCK 0;
                        CELL(0) ⇐ CELL(0) + 1;
            BLOCK 1: END;
            OUTPUT ⇐ YES;
BLOCK 0: END.
```

Notice that I have called two procedures inside this algorithm: **MINUS** and **REMAINDER**. (The latter is presumed to have been previously defined, and you may work out its definition yourself.) Now this test for primality works by trying out potential factors of **N** one by one, starting at 2 and increasing to a maximum of **N**−1. In case any of them divides **N** exactly (i.e., gives remainder 0), then we jump down to the bottom, and since **OUTPUT** still has its default value at this stage, the answer is **NO**. Only if **N** has no exact divisors will it survive the entirety of **LOOP 1**; then we will emerge smoothly at the statement **OUTPUT ⇐ YES**, which will get executed, and then the procedure is over.

BlooP Programs Contain Chains of Procedures

We have seen how to define procedures in BlooP; however, a procedure definition is only a part of a program. A *program* consists of a *chain of procedure definitions* (each only calling previously defined procedures), optionally followed by one or more *calls* on the procedures defined. Thus, an example of a full BlooP program would be the definition of the procedure TWO-TO-THE-THREE-TO-THE, followed by the *call*

TWO-TO-THE-THREE-TO-THE [2]

which would yield an answer of 512.

If you have only a chain of procedure definitions, then nothing ever gets executed; they are all just waiting for some call, with specific numerical values, to set them in motion. It is like a meat grinder waiting for some meat to grind—or rather, a *chain* of meat grinders all linked together, each of which is fed from earlier ones . . . In the case of meat grinders, the image is perhaps not so savory; however, in the case of BlooP programs, such a construct is quite important, and we will call it a "call-less program". This notion is illustrated in Figure 72.

Now BlooP is our language for defining predictably terminating calculations. The standard name for *functions* which are BlooP-computable is *primitive recursive functions;* and the standard name for *properties* which can be detected by BlooP-tests is *primitive recursive predicates.* Thus, the function 2^{3^n} is a primitive recursive function; and the statement "*n* is a prime number" is a primitive recursive predicate.

It is clear intuitively that the Goldbach property is primitive recursive, and to make that quite explicit, here is a procedure definition in BlooP, showing how to test for its presence or absence:

```
DEFINE PROCEDURE ''GOLDBACH?'' [N]:
BLOCK 0: BEGIN
          CELL(0) ⇐ 2;
          LOOP AT MOST N TIMES:
          BLOCK 1: BEGIN
                    IF {PRIME? [CELL(0)]
                      AND PRIME? [MINUS [N,CELL(0)]]},
                        THEN:
                    BLOCK 2: BEGIN
                              OUTPUT ⇐ YES;
                              QUIT BLOCK 0;
                    BLOCK 2: END
                    CELL(0) ⇐ CELL(0) + 1;
          BLOCK 1: END;
BLOCK 0: END.
```

As usual, we assume NO until proven YES, and we do a brute force search among pairs of numbers which sum up to N. If both are prime, we quit the outermost block; otherwise we just go back and try again, until all possibilities are exhausted.

(Warning: The fact that the Goldbach property is primitive recursive does not make the question "Do all numbers have the Goldbach property?" a simple question—far from it!)

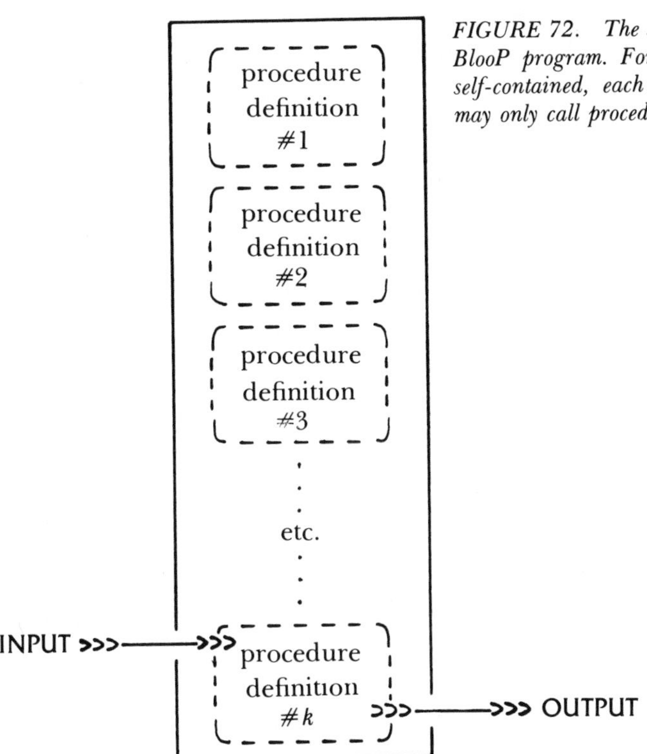

FIGURE 72. The structure of a call-less BlooP program. For this program to be self-contained, each procedure definition may only call procedures defined above it.

Suggested Exercises

Can you write a similar BlooP procedure which tests for the presence or absence of the Tortoise property (or the Achilles property)? If so, do it. If not, is it merely because you are ignorant about upper bounds, or could it be that there is a fundamental obstacle preventing the formulation of such an algorithm in BlooP? And what about the same questions, with respect to the property of wondrousness, defined in the Dialogue?

Below, I list some functions and properties, and you ought to take the time to determine whether you believe they are primitive recursive (BlooP-programmable) or not. This means that you must carefully consider what kinds of operations will be involved in the calculations which they require, and whether ceilings can be given for all the loops involved.

FACTORIAL [N] = N! (the factorial of N)
  (e.g., FACTORIAL [4] = 24)

REMAINDER [M,N] = the remainder upon dividing M by N
  (e.g., REMAINDER [24,7] = 3)

PI-DIGIT [N] = the Nth digit of π, after the decimal point
  (e.g., PI-DIGIT [1] = 1,
        PI-DIGIT [2] = 4,
        PI-DIGIT [1000000] = 1)

FIBO [N] = the Nth Fibonacci number
(e.g., FIBO [9] = 34)

PRIME-BEYOND [N] = the lowest prime beyond N
(e.g., PRIME-BEYOND [33] = 37)

PERFECT [N] = the Nth "perfect" number (a number such as 28 whose divisors sum up to itself: 28 = 1 + 2 + 4 + 7 + 14)
(e.g., PERFECT [2] = 28)

PRIME? [N] = YES if N is prime, otherwise NO.

PERFECT? [N] = YES if N is perfect, otherwise NO.

TRIVIAL? [A,B,C,N] = YES if $A^N+B^N = C^N$ is correct; otherwise NO.
(e.g., TRIVIAL? [3,4,5,2] = YES,
TRIVIAL? [3,4,5,3] = NO)

PIERRE? [A,B,C] = YES if $A^N+B^N = C^N$ is satisfiable for some value of N greater than 1, otherwise NO.
(e.g., PIERRE? [3,4,5] = YES,
PIERRE? [1,2,3] = NO)

FERMAT? [N] = YES if $A^N+B^N = C^N$ is satisfied by some positive values of A, B, C; otherwise NO.
(e.g., FERMAT? [2] = YES)

TORTOISE-PAIR? [M,N] = YES if both M and M + N are prime, otherwise NO.
(e.g., TORTOISE-PAIR [5,1742] = YES,
TORTOISE-PAIR [5,100] = NO)

TORTOISE? [N] = YES if N is the difference of two primes, otherwise NO.
(e.g., TORTOISE [1742] = YES,
TORTOISE [7] = NO)

MIU-WELL-FORMED? [N] = YES if N, when seen as a string of the MIU-system, is well-formed; otherwise NO.
(e.g., MIU-WELL-FORMED? [310] = YES,
MIU-WELL-FORMED? [415] = NO)

MIU-PROOF-PAIR? [M,N] = YES if M, as seen as a sequence of strings of the MIU-system, is a derivation of N, as seen as a string of the MIU-system; otherwise NO.
(e.g., MIU-PROOF-PAIR? [3131131111301,301] = YES,
MIU-PROOF-PAIR? [311130,30] = NO)

MIU-THEOREM? [N] = YES if N, seen as a MIU-system string, is a theorem; otherwise NO.
(e.g., MIU-THEOREM? [311] = YES,
MIU-THEOREM? [30] = NO,
MIU-THEOREM? [701] = NO)

TNT-THEOREM? [N] = YES if N, seen as a TNT-string, is a theorem.
(e.g., TNT-THEOREM? [666111666] = YES,
TNT-THEOREM? [123666111666] = NO,
TNT-THEOREM? [7014] = NO)

FALSE? [N] = YES if N, seen as a TNT-string, is a false statement of number theory; otherwise NO.

(e.g., FALSE? [666111666] = NO,
FALSE? [223666111666] = YES,
FALSE? [7014] = NO)

The last seven examples are particularly relevant to our future metamathematical explorations, so they highly merit your scrutiny.

Expressibility and Representability

Now before we go on to some interesting questions about BlooP and are led to its relative, FlooP, let us return to the reason for introducing BlooP in the first place, and connect it to TNT. Earlier, I stated that the critical mass for Gödel's method to be applicable to a formal system is attained when all primitive recursive notions are representable in that system. Exactly what does this mean? First of all, we must distinguish between the notions of representability and expressibility. *Expressing* a predicate is a mere matter of translation from English into a strict formalism. It has nothing to do with theoremhood. For a predicate to be *represented,* on the other hand, is a much stronger notion. It means that

(1) All true instances of the predicate are theorems;
(2) All false instances are nontheorems.

By "instance", I mean the string produced when you replace all free variables by numerals. For example, the predicate $m + n = k$ is represented in the pq-system, because each true instance of the predicate is a theorem, each false instance is a nontheorem. Thus any specific addition, whether true or false, translates into a *decidable string* of the pq-system. However, the pq-system is unable to express—let alone represent—any other properties of natural numbers. Therefore it would be a weak candidate indeed in a competition of systems which can do number theory.

Now TNT has the virtue of being able to *express* virtually any number-theoretical predicate; for example, it is easy to write a TNT-string which expresses the predicate "b has the Tortoise property". Thus, in terms of expressive power, TNT is all we want.

However, the question "Which properties are *represented* in TNT?" is precisely the question "How powerful an axiomatic system is TNT?" Are all possible predicates represented in TNT? If so, then TNT can answer any question of number theory; it is complete.

Primitive Recursive Predicates Are Represented in TNT

Now although completeness will turn out to be a chimera, TNT is at least complete with respect to *primitive recursive* predicates. In other words, any statement of number theory whose truth or falsity can be decided by a computer within a predictable length of time is also decidable inside TNT. Or, one final restatement of the same thing:

> If a BlooP test can be written for some property of natural numbers, then that property is represented in TNT.

Are There Functions Which Are Not Primitive Recursive?

Now the kinds of properties which can be detected by BlooP tests are widely varied, including whether a number is prime or perfect, has the Goldbach property, is a power of 2, and so on and so forth. It would not be crazy to wonder whether *every* property of numbers can be detected by some suitable BlooP program. The fact that, as of the present moment, we have no way of testing whether a number is wondrous or not need not disturb us too much, for it might merely mean that we are ignorant about wondrousness, and that with more digging around, we could discover a universal formula for the upper bound to the loop involved. Then a BlooP test for wondrousness could be written on the spot. Similar remarks could be made about the Tortoise property.

So the question really is, "Can upper bounds always be given for the length of calculations—or, is there an inherent kind of jumbliness to the natural number system, which sometimes prevents calculation lengths from being predictable in advance?" The striking thing is that the latter is the case, and we are about to see why. It is the sort of thing that would have driven Pythagoras, who first proved that the square root of 2 is irrational, out of his mind. In our demonstration, we will use the celebrated *diagonal method,* discovered by Georg Cantor, the founder of set theory.

Pool B, Index Numbers, and Blue Programs

We shall begin by imagining a curious notion: the pool of all possible BlooP programs. Needless to say, this pool—"Pool B"—is an infinite one. We want to consider a subpool of Pool B, obtained by three successive filtering operations. The first filter will retain for us only *call-less* programs. From this subpool we then eliminate all *tests,* leaving only *functions.* (By the way, in call-less programs, the *last* procedure in the chain determines whether the program as a whole is considered a test, or a function.) The third filter will retain only *functions which have exactly one input parameter.* (Again referring to the final procedure in the chain.) What is left?

> A complete pool of all call-less BlooP programs which calculate functions of exactly one input parameter.

Let us call these special BlooP programs *Blue Programs.*

What we would like to do now is to assign an unambiguous *index number* to each Blue Program. How can this be done? The easiest way—we shall use it—is to list them in order of length: the shortest possible Blue

Program being #1, the second shortest being #2, etc. Of course, there will be many programs tied for each length. To break such ties, we use alphabetical order. Here, "alphabetical order" is taken in an extended sense, where the alphabet includes all the special characters of BlooP, in some arbitrary order, such as the following:

A B C D E F G H I J K L M N
O P Q R S T U V W X Y Z + ×
0 1 2 3 4 5 6 7 8 9 ⇐ = < >
( ) [ ] { } - ' ? : ; , .

—and at the end comes the lowly blank! Altogether, fifty-six characters. For convenience's sake, we can put all Blue Programs of length 1 in Volume 1, programs of 2 characters in Volume 2, etc. Needless to say, the first few volumes will be totally empty, while later volumes will have many, many entries (though each volume will only have a finite number). The very first Blue Program would be this one:

```
DEFINE PROCEDURE "A" [B]:
BLOCK 0: BEGIN
BLOCK 0: END.
```

This rather silly meat grinder returns a value of 0 no matter what its input is. It occurs in Volume 56, since it has 56 characters (counting necessary blanks, including blanks separating successive lines).

Soon after Volume 56, the volumes will get extremely fat, because there are just so many millions of ways of combining symbols to make Blue BlooP programs. But no matter—we are not going to try to print out this infinite catalogue. All that we care about is that, in the abstract, it is well-defined, and that each Blue BlooP program therefore has a unique and definite index number. This is the crucial idea.

Let us designate the function calculated by the *k*th Blue Program this way:

Blueprogram{# *k*} [N]

Here, *k* is the index number of the program, and N is the single input parameter. For instance, Blue Program #12 might return a value twice the size of its input:

Blueprogram{#12} [N] = 2 × N

The meaning of the equation above is that the *program* named on the left-hand side returns the same value as a *human* would calculate from the ordinary algebraic expression on the right-hand side. As another example, perhaps the 5000th Blue Program calculates the cube of its input parameter:

Blueprogram{#5000} [N] = N^3

Very well—now we apply the "twist": Cantor's diagonal method. We shall take this catalogue of Blue Programs and use it to define a new function of one variable—*Bluediag* [N]—which will turn out not to be anywhere in the list (which is why its name is in italics). Yet *Bluediag* will clearly be a well-defined, calculable function of one variable, and so we will have to conclude that functions exist which simply are not programmable in BlooP.

Here is the definition of *Bluediag* [N]:

Equation (1) . . . *Bluediag* [N] = 1 + Blueprogram{#N} [N]

The strategy is: feed each meat grinder with its own index number, then add 1 to the output. To illustrate, let us find *Bluediag* [12]. We saw that Blueprogram{#12} is the function 2N; therefore, *Bluediag* [12] must have the value 1 + 2 × 12, or 25. Likewise, *Bluediag* [5000] would have the value 125,000,000,001, since that is 1 more than the cube of 5000. Similarly, you can find *Bluediag* of any particular argument you wish.

The peculiar thing about *Bluediag* [N] is that it is not represented in the catalogue of Blue Programs. It cannot be. The reason is this. To be a Blue Program, it would have to have an index number—say it were Blue Program # *X*. This assumption is expressed by writing

Equation (2) . . . *Bluediag* [N] = Blueprogram{# *X*} [N]

But there is an inconsistency between the equations (1) and (2). It becomes apparent at the moment we try to calculate the value of *Bluediag* [*X*], for we can do so by letting N take the value of *X* in either of the two equations. If we substitute into equation (1), we get:

Bluediag [*X*] = 1 + Blueprogram{# *X*} [*X*]

But if we substitute into equation (2) instead, we get:

Bluediag [*X*] = Blueprogram{# *X*} [*X*]

Now *Bluediag* [*X*] cannot be equal to a number and also to the successor of that number. But that is what the two equations say. So we will have to go back and erase some assumption on which the inconsistency is based. The only possible candidate for erasure is the assumption expressed by Equation (2): that the function *Bluediag* [N] is able to be coded up as a Blue BlooP program. And that is the proof that *Bluediag lies outside the realm of primitive recursive functions.* Thus, we have achieved our aim of destroying Achilles' cherished but naïve notion that every number-theoretical function must be calculable within a predictable number of steps.

There are some subtle things going on here. You might ponder this, for instance: the number of steps involved in the calculation of *Bluediag* [N], for each *specific* value of N, is predictable—but the different methods of prediction cannot all be united into a *general* recipe for predicting the length of calculation of *Bluediag* [N]. This is an "infinite conspiracy", related to the Tortoise's notion of "infinite coincidences", and also to ω-incompleteness. But we shall not trace out the relations in detail.

Cantor's Original Diagonal Argument

Why is this called a *diagonal* argument? The terminology comes from Cantor's original diagonal argument, upon which many other arguments (such as ours) have subsequently been based. To explain Cantor's original argument will take us a little off course, but it is worthwhile to do so. Cantor, too, was concerned with showing that some item is not in a certain list. Specifically, what Cantor wanted to show was that if a "directory" of real numbers were made, it would inevitably leave some real numbers out—so that actually, the notion of a *complete* directory of real numbers is a contradiction in terms.

It must be understood that this pertains not just to directories of finite size, but also to directories of *infinite* size. It is a much deeper result than the statement "the number of reals is infinite, so of course they cannot be listed in a finite directory". The essence of Cantor's result is that there are (at least) two distinct *types* of infinity: one kind of infinity describes how many entries there can be in an infinite directory or table, and another describes how many real numbers there are (i.e., how many points there are on a line, or line segment)—and this latter is "bigger", in the sense that the real numbers cannot be squeezed into a table whose length is described by the former kind of infinity. So let us see how Cantor's argument involves the notion of diagonal, in a literal sense.

Let us consider just real numbers between 0 and 1. Assume, for the sake of argument, that an infinite list *could* be given, in which each positive integer N is matched up with a real number $r(N)$ between 0 and 1, and in which each real number between 0 and 1 occurs somewhere down the line. Since real numbers are given by infinite decimals, we can imagine that the beginning of the table might look as follows:

$r(1)$:	**.1**	4	1	5	9	2	6	5	3	. . . .
$r(2)$:	.3	**3**	3	3	3	3	3	3	3	. . . .
$r(3)$:	.7	1	**8**	2	8	1	8	2	8	. . . .
$r(4)$:	.4	1	4	**2**	1	3	5	6	2	. . . .
$r(5)$:	.5	0	0	0	**0**	0	0	0	0	. . . .

The digits that run down the diagonal are in boldface: 1, 3, 8, 2, 0, . . . Now those diagonal digits are going to be used in making a special real number d, which is between 0 and 1 but which, we will see, is not in the list. To make d, you take the diagonal digits in order, and change each one of them to some other digit. When you prefix this sequence of digits by a decimal point, you have d. There are of course many ways of changing a digit to some other digit, and correspondingly many different d's. Suppose, for example, that we *subtract 1 from the diagonal digits* (with the convention that 1 taken from 0 is 9). Then our number d will be:

$$.0 \quad 2 \quad 7 \quad 1 \quad 9 \quad . \quad . \quad .$$

Now, because of the way we constructed it,

> d's 1st digit is not the same as the 1st digit of $r(1)$;
> d's 2nd digit is not the same as the 2nd digit of $r(2)$;
> d's 3rd digit is not the same as the 3rd digit of $r(3)$;
>
> . . . and so on.

Hence,

> d is different from $r(1)$;
> d is different from $r(2)$;
> d is different from $r(3)$;
>
> . . . and so on.

In other words, d is not in the list!

What Does a Diagonal Argument Prove?

Now comes the crucial difference between Cantor's proof and our proof—it is in the matter of what assumption to go back and undo. In Cantor's argument, the shaky assumption was that such a table could be drawn up. Therefore, the conclusion warranted by the construction of d is that no exhaustive table of reals can be drawn up after all—which amounts to saying that the set of integers is just not big enough to index the set of reals. On the other hand, in our proof, we know that the directory of Blue BlooP programs *can* be drawn up—the set of integers *is* big enough to index the set of Blue BlooP programs. So, we have to go back and retract some shakier idea which we used. And that idea is that *Bluediag* [N] is calculable by some program in BlooP. This is a subtle difference in the application of the diagonal method.

It may become clearer if we apply it to the alleged "List of All Great Mathematicians" in the Dialogue—a more concrete example. The diagonal itself is "Dboups". If we perform the desired diagonal-subtraction, we will get "Cantor". Now two conclusions are possible. If you have an unshakable belief that the list is *complete,* then you must conclude that Cantor is not a Great Mathematician, for his name differs from all those on the list. On the other hand, if you have an unshakable belief that Cantor *is* a Great Mathematician, then you must conclude that the List of All Great Mathematicians is incomplete, for Cantor's name is not on the list! (Woe to those who have unshakable beliefs on both sides!) The former case corresponds to our proof that *Bluediag* [N] is not primitive recursive; the latter case corresponds to Cantor's proof that the list of reals is incomplete.

FIGURE 73. Georg Cantor.

Cantor's proof uses a diagonal in the literal sense of the word. Other "diagonal" proofs are based on a more general notion, which is abstracted from the geometric sense of the word. The essence of the diagonal method is the fact of using one integer in two different ways—or, one could say, *using one integer on two different levels*—thanks to which one can construct an item which is outside of some predetermined list. One time, the integer serves as a *vertical* index, the other time as a *horizontal* index. In Cantor's construction this is very clear. As for the function *Bluediag* [N], it involves using one integer on two different levels—first, as a Blue Program index number; and second, as an input parameter.

The Insidious Repeatability of the Diagonal Argument

At first, the Cantor argument may seem less than fully convincing. Isn't there some way to get around it? Perhaps by throwing in the diagonally constructed number *d*, one might obtain an exhaustive list. If you consider this idea, you will see it helps not a bit to throw in the number *d*, for as soon as you assign it a specific place in the table, the diagonal method becomes applicable to the new table, and a new missing number *d'* can be constructed, which is not in the new table. No matter how many times you repeat the operation of constructing a number by the diagonal method and then throwing it in to make a "more complete" table, you still are caught on the ineradicable hook of Cantor's method. You might even try to build a table of reals which tries to outwit the Cantor diagonal method by taking the whole trick, lock, stock, and barrel, including its insidious repeatability, into account somehow. It is an interesting exercise. But if you tackle it, you will see that no matter how you twist and turn trying to avoid the Cantor "hook", you are still caught on it. One might say that any self-proclaimed "table of all reals" is hoist by its own petard.

The repeatability of Cantor's diagonal method is similar to the repeatability of the Tortoise's diabolic method for breaking the Crab's phonographs, one by one, as they got more and more "hi-fi" and—at least so the Crab hoped—more "Perfect". This method involves constructing, for each phonograph, a particular song which that phonograph cannot reproduce. It is not a coincidence that Cantor's trick and the Tortoise's trick share this curious repeatability; indeed, the *Contracrostipunctus* might well have been named "Cantorcrostipunctus" instead. Moreover, as the Tortoise subtly hinted to the innocent Achilles, the events in the *Contracrostipunctus* are a paraphrase of the construction which Gödel used in proving his Incompleteness Theorem; it follows that the Gödel construction is also very much like a diagonal construction. This will become quite apparent in the next two Chapters.

From BlooP to FlooP

We have now defined the class of primitive recursive functions and primitive recursive properties of natural numbers by means of programs written in the language BlooP. We have also shown that BlooP doesn't capture all the functions of natural numbers which we can define in words. We even constructed an "unBlooPable" function, *Bluediag* [N], by Cantor's diagonal method. What is it about BlooP that makes *Bluediag* unrepresentable in it? How could BlooP be improved so that *Bluediag* became representable?

BlooP's defining feature was the boundedness of its loops. What if we drop that requirement on loops, and invent a second language, called "FlooP" ('F' for "free")? FlooP will be identical to BlooP except in one respect: we may have loops without ceilings, as well as loops with ceilings (although the only reason one would include a ceiling when writing a loop-statement in FlooP would be for the sake of elegance). These new loops will be called **MU-LOOPS**. This follows the convention of mathematical logic, in which "free" searches (searches without bounds) are usually indicated by a symbol called a "μ-operator" (mu-operator). Thus, loop-statements in FlooP may look like this:

```
MU-LOOP:
BLOCK n: BEGIN
          .
          .
          .
BLOCK n: END;
```

This feature will allow us to write tests in FlooP for such properties as wondrousness and the Tortoise property—tests which we did not know how to program in BlooP because of the potential open-endedness of the searches involved. I shall leave it to interested readers to write a FlooP test for wondrousness which does the following things:

(1) If its input, **N**, is wondrous, the program halts and gives the answer **YES**.
(2) If **N** is unwondrous, but causes a closed cycle other than 1-4-2-1-4-2-1- . . . , the program halts and gives the answer **NO**.
(3) If **N** is unwondrous, and causes an "endlessly rising progression", the program never halts. This is FlooP's way of answering by not answering. FlooP's nonanswer bears a strange resemblance to Jōshū's nonanswer "MU".

The irony of case 3 is that **OUTPUT** always has the value **NO**, but it is always inaccessible, since the program is still grinding away. That troublesome third alternative is the price that we must pay for the right to write free loops. In all FlooP programs incorporating the **MU-LOOP** option, nontermination will always be one theoretical alternative. Of course there will be many FlooP programs which actually terminate for all possible input values. For instance, as I mentioned earlier, it is suspected by most people who have studied wondrousness that a FlooP program such as suggested above will always terminate, and moreover with the answer **YES** each time.

Terminating and Nonterminating FlooP Programs

It would seem extremely desirable to be able to separate FlooP procedures into two classes: *terminators* and *nonterminators*. A *terminator* will eventually halt no matter what its input, despite the "MU-ness" of its loops. A nonterminator will go on and on forever, for *at least one* choice of input. If we could always tell, by some kind of complicated inspection of a FlooP program, to which class it belonged, there would be some remarkable repercussions (as we shall shortly see). Needless to say, the operation of class-checking would itself have to be a terminating operation—otherwise one would gain nothing!

Turing's Trickery

The idea springs to mind that we might let a BlooP procedure do the inspection. But BlooP procedures only accept numerical input, not programs! However, we can get around that . . . by coding programs into numbers! This sly trick is just Gödel-numbering in another of its many manifestations. Let the fifty-six characters of the FlooP alphabet get the "codons" 901, 902, . . ., 956, respectively. So each FlooP program now gets a very long Gödel number. For instance, the shortest BlooP function (which is also a terminating FlooP program)—

```
DEFINE PROCEDURE ''A'' [B]:
BLOCK 0: BEGIN
BLOCK 0: END.
```

—would get the Gödel number partially shown below:

```
904,905,906,909,914,905, . . . . . . . . . . . , 905,914,904,955,
 D   E   F   I   N   E                           E   N   D   .
```

Now our scheme would be to write a BlooP test called **TERMINATOR?** which says **YES** if its input number codes for a terminating FlooP program, **NO** if not. This way we could hand the task over to a machine and with luck, distinguish terminators from nonterminators. However, an ingenious argument given by Alan Turing shows that no BlooP program can make this distinction infallibly. The trick is actually much the same as Gödel's trick, and therefore closely related to the Cantor diagonal trick. We shall not give it here—suffice it to say that the idea is to feed the termination tester its *own* Gödel number. This is not so simple, however, for it is like trying to quote an entire sentence inside itself. You have to quote the quote, and so forth; it seems to lead to an infinite regress. However, Turing figured out a trick for feeding a program its own Gödel number. A solution to the same problem in a different context will be presented next Chapter. In the present Chapter, we shall take a different route to the same goal, which is namely to prove that a termination tester is impossible. For readers who wish to see an elegant and simple presentation of the Turing approach, I recommend the article by Hoare and Allison, mentioned in the Bibliography.

A Termination Tester Would Be Magical

Before we destroy the notion, let us delineate just why having a termination tester would be a remarkable thing. In a sense, it would be like having a magical dowsing rod which could solve all problems of number theory in one swell FlooP. Suppose, for instance, that we wished to know if the Goldbach Variation is a true conjecture or not. That is, do all numbers have the Tortoise property? We would begin by writing a FlooP test called **TORTOISE?** which checks whether its input has the Tortoise property. Now the defect of this procedure—namely that it doesn't terminate if the Tortoise property is absent—here turns into a virtue! For now we run the termination tester on the procedure **TORTOISE?**. If it says **YES**, that means that **TORTOISE?** terminates for all values of its input—in other words, all numbers have the Tortoise property. If it says **NO**, then we know there exists a number which has the Achilles property. The irony is that we never actually *use* the program **TORTOISE?** at all—we just inspect it!

This idea of solving any problem in number theory by coding it into a program and then waving a termination tester over the program is not unlike the idea of testing a kōan for genuineness by coding it into a folded string and then running a test for Buddha-nature on the string instead. As Achilles suggested, perhaps the desired information lies "closer to the surface" in one representation than in another.

Pool F, Index Numbers, and Green Programs

Well, enough daydreaming. How can we prove that the termination tester is impossible? Our argument for its impossibility will hinge on trying to apply the diagonal argument to FlooP, just as we did to BlooP. We shall see that there are some subtle and crucial differences between the two cases.

As we did for BlooP, imagine the pool of all FlooP programs. We shall call it "Pool F". Then perform the same three filtering operations on Pool F, so that you get, in the end:

> A complete pool of all call-less FlooP programs which calculate functions of exactly one input parameter.

Let us call these special FlooP-programs *Green Programs* (since they may go forever).

Now just as we assigned index numbers to all Blue Programs, we can assign index numbers to Green Programs, by ordering them in a catalogue, each volume of which contains all Green Programs of a fixed length, arranged in alphabetical order.

So far, the carry-over from BlooP to FlooP has been straightforward. Now let us see if we can also carry over the last part: the diagonal trick. What if we try to define a diagonal function?

Greendiag [N] = 1 + Greenprogram{#N} [N]

Suddenly, there is a snag: this function *Greendiag* [N] may not have a well-defined output value for all input values N. This is simply because we have not filtered out the nonterminator programs from Pool F, and therefore we have no guarantee that we can calculate *Greendiag* [N] for all values of N. Sometimes we may enter calculations which never terminate. And the diagonal argument cannot be carried through in such a case, for it depends on the diagonal function having a value for all possible inputs.

The Termination Tester Gives Us Red Programs

To remedy this, we would have to make use of a termination tester, if one existed. So let us deliberately introduce the shaky assumption that one exists, and let us use it as our fourth filter. We run down the list of Green Programs, eliminating one by one all nonterminators, so that in the end we are left with:

A complete pool of all call-less FlooP programs which calculate functions of exactly one input parameter, and which *terminate* for all values of their input.

Let us call these special FlooP programs *Red Programs* (since they all must stop). Now, the diagonal argument will go through. We define

Reddiag [N] = 1 + Redprogram{#N} [N]

and in an exact parallel to *Bluediag,* we are forced to conclude that *Reddiag* [N] is a well-defined, calculable function of one variable which is not in the catalogue of Red Programs, and is hence not even calculable in the powerful language FlooP. Perhaps it is time to move on to GlooP?

GlooP . . .

Yes, but what is GlooP? If FlooP is BlooP unchained, then GlooP must be FlooP unchained. But how can you take the chains off twice? How do you make a language whose power transcends that of FlooP? In *Reddiag,* we have found a function whose values we humans know how to calculate—the method of doing so has been explicitly described in English—but which seemingly cannot be programmed in the language FlooP. This is a serious dilemma because no one has ever found any more powerful computer language than FlooP.

Careful investigation into the power of computer languages has been carried out. We need not do it ourselves; let it just be reported that there is a vast class of computer languages all of which can be proven to have *exactly the same expressive power* as FlooP does, in this sense: any calculation which can be programmed in any one of the languages can be programmed in them all. The curious thing is that almost any sensible attempt at designing a computer language ends up by creating a member of this class—which is to say, a language of power equal to that of FlooP. It takes some doing to invent a reasonably interesting computer language which is *weaker* than those in this class. BlooP is, of course, an example of a weaker language, but it is the exception rather than the rule. The point is that there are some extremely natural ways to go about inventing algorithmic languages; and different people, following independent routes, usually wind up creating equivalent languages, with the only difference being style, rather than power.

. . . Is a Myth

In fact, it is widely believed that there cannot be any more powerful language for describing calculations than languages that are equivalent to FlooP. This hypothesis was formulated in the 1930's by two people, independently of each other: Alan Turing—about whom we will say more later—and Alonzo Church, one of the eminent logicians of this century. It is called the *Church-Turing Thesis.* If we accept the CT-Thesis, we have to conclude that "GlooP" is a myth—there are no restrictions to remove in FlooP, no ways to increase its power by "unshackling" it, as we did BlooP.

This puts us in the uncomfortable position of asserting that *people* can calculate *Reddiag* [**N**] for any value of **N**, but there is no way to program a *computer* to do so. For, if it could be done at all, it could be done in FlooP—and by construction, it can't be done in FlooP. This conclusion is so peculiar that it should cause us to investigate very carefully the pillars on which it rests. And one of them, you will recall, was our shaky assumption that there is a decision procedure which can tell terminating from nonterminating FlooP programs. The idea of such a decision procedure already seemed suspect, when we saw that its existence would allow all problems of number theory to be solved in a uniform way. Now we have double the reason for believing that any termination test is a myth—that there is no way to put FlooP programs in a centrifuge and separate out the terminators from the nonterminators.

Skeptics might maintain that this is nothing like a rigorous proof that such a termination test doesn't exist. That is a valid objection; however, the Turing approach demonstrates more rigorously that no computer program can be written in a language of the FlooP class which can perform a termination test on all FlooP programs.

The Church-Turing Thesis

Let us come back briefly to the Church-Turing Thesis. We will talk about it—and variations on it—in considerable detail in Chapter XVII; for now it will suffice to state it in a couple of versions, and postpone discussion of its merits and meanings until then. Here, then, are three related ways to state the CT-Thesis:

(1) What is human-computable is machine-computable.
(2) What is machine-computable is FlooP-computable.
(3) What is human-computable is FlooP-computable
(i.e., general or partial recursive).

Terminology: General and Partial Recursive

We have made a rather broad survey, in this Chapter, of some notions from number theory and their relations to the theory of computable functions. It is a very wide and flourishing field, an intriguing blend of computer science and modern mathematics. We should not conclude this Chapter without introducing the standard terminology for the notions we have been dealing with.

As has already been mentioned, "Bloop-computable" is synonymous with "primitive recursive". Now FlooP-computable functions can be divided into two realms: (1) those which are computable by *terminating* FlooP programs: these are said to be *general recursive;* and (2) those which are computable only by *nonterminating* FlooP programs: these are said to be *partial recursive.* (Similarly for predicates.) People often just say "recursive" when they mean "general recursive".

The Power of TNT

It is interesting that TNT is so powerful that not only are all primitive recursive predicates represented, but moreover all general recursive predicates are represented. We shall not prove either of these facts, because such proofs would be superfluous to our aim, which is to show that TNT is incomplete. If TNT could not represent some primitive or general recursive predicates, then it would be incomplete in an *uninteresting* way—so we might as well assume that it can, and then show that it is incomplete in an interesting way.

Air on G's String

The Tortoise and Achilles have just completed a tour of a porridge factory.

Achilles: You don't mind if I change the subject, do you?

Tortoise: Be my guest.

Achilles: Very well, then. It concerns an obscene phone call I received a few days ago.

Tortoise: Sounds interesting.

Achilles: Yes. Well—the problem was that the caller was incoherent, at least as far as I could tell. He shouted something over the line and then hung up—or rather, now that I think of it, he shouted something, shouted it again, and then hung up.

Tortoise: Did you catch what that thing was?

Achilles: Well, the whole call went like this:

> *Myself:* Hello?
>
> *Caller (shouting wildly):* Yields falsehood when preceded by its quotation! Yields falsehood when preceded by its quotation!
>
> *(Click.)*

Tortoise: That is a most unusual thing to say to somebody on an obscene phone call.

Achilles: Exactly how it struck me.

Tortoise: Perhaps there was some meaning to that seeming madness.

Achilles: Perhaps.

(They enter a spacious courtyard framed by some charming three-story stone houses. At its center stands a palm tree, and to one side is a tower. Near the tower there is a staircase where a boy sits, talking to a young woman in a window.)

Tortoise: Where are you taking me, Achilles?

Achilles: I would like to show you the pretty view from the top of this tower.

Tortoise: Oh, how nice.

(They approach the boy, who watches them with curiosity, then says something to the young woman—they both chuckle. Achilles and Mr. T, instead of going up the boy's staircase, turn left and head down a short flight of stairs which leads to a small wooden door.)

Achilles: We can just step inside right here. Follow me.

FIGURE 74. Above and Below, *by M. C. Escher (lithograph, 1947).*

(Achilles opens the door. They enter, and begin climbing the steep helical staircase inside the tower.)

Tortoise (puffing slightly): I'm a little out of shape for this sort of exercise, Achilles. How much further do we have to go?

Achilles: Another few flights . . . but I have an idea. Instead of walking on the top side of these stairs, why don't you walk on the underside?

Tortoise: How do I do THAT?

Achilles: Just hold on tightly, and climb around underneath—there's room enough for you. You'll find that the steps make just as much sense from below as from above . . .

Tortoise (gingerly shifting himself about): Am I doing it right?

Achilles: You've got it!

Tortoise (his voice slightly muffled): Say—this little maneuver has got me confused. Should I head upstairs or downstairs, now?

Achilles: Just continue heading in the same direction as you were before. On your side of the staircase, that means go DOWN, on mine it means UP.

Tortoise: Now you're not going to tell me that I can get to the top of the tower by going down, are you?

Achilles: I don't know, but it works . . .

(And so they begin spiraling in synchrony, with A always on one side, and T matching him on the other side. Soon they reach the end of the staircase.)

Now just undo the maneuver, Mr. T. Here—let me help you up.

(He lends an arm to the Tortoise, and hoists him back to the other side of the stairs.)

Tortoise: Thanks. It was a little easier getting back up.

(And they step out onto the roof, overlooking the town.)

That's a lovely view, Achilles. I'm glad you brought me up here—or rather, DOWN here.

Achilles: I figured you'd enjoy it.

Tortoise: I've been thinking about that obscene phone call. I think I understand it a little better now.

Achilles: You do? Would you tell me about it?

Tortoise: Gladly. Do you perchance feel, as I do, that that phrase "preceded by its quotation" has a slightly haunting quality about it?

Achilles: Slightly, yes—extremely slightly.

Tortoise: Can you imagine something preceded by its quotation?

Achilles: I guess I can conjure up an image of Chairman Mao walking into a banquet room in which there already hangs a large banner with some of his own writing on it. Here would be Chairman Mao, preceded by his quotation.

Tortoise: A most imaginative example. But suppose we restrict the word "preceded" to the idea of precedence on a printed sheet, rather than elaborate entries into a banquet room.

Achilles: All right. But what exactly do you mean by "quotation" here?

Tortoise: When you discuss a word or a phrase, you conventionally put it in quotes. For example, I can say,

The word "philosopher" has five letters.

Here, I put "philosopher" in quotes to show that I am speaking about the WORD "philosopher" rather than about a philosopher in the flesh. This is called the USE-MENTION distinction.

Achilles: Oh?

Tortoise: Let me explain. Suppose I were to say to you,

Philosophers make lots of money.

Here, I would be USING the word to manufacture an image in your mind of a twinkle-eyed sage with bulging moneybags. But when I put this word—or any word—in quotes, I subtract out its meaning and connotations, and am left only with some marks on paper, or some sounds. That is called "MENTION". Nothing about the word matters, other than its typographical aspects—any meaning it might have is ignored.

Achilles: It reminds me of using a violin as a fly swatter. Or should I say "mentioning"? Nothing about the violin matters, other than its solidity—any meaning or function it might have is being ignored. Come to think of it, I guess the fly is being treated that way, too.

Tortoise: Those are sensible, if slightly unorthodox, extensions of the use-mention distinction. But now, I want you to think about preceding something by its own quotation.

Achilles: All right. Would this be correct?

"HUBBA" HUBBA

Tortoise: Good. Try another.

Achilles: All right.

"'PLOP' IS NOT THE TITLE OF ANY BOOK, SO FAR AS I KNOW"
'PLOP' IS NOT THE TITLE OF ANY BOOK, SO FAR AS I KNOW.

Tortoise: Now this example can be modified into quite an interesting specimen, simply by dropping 'Plop'.

Achilles: Really? Let me see what you mean. It becomes

"IS NOT THE TITLE OF ANY BOOK, SO FAR AS I KNOW"
IS NOT THE TITLE OF ANY BOOK, SO FAR AS I KNOW.

Tortoise: You see, you have made a sentence.

Achilles: So I have. It is a sentence about the phrase "is not the title of any book, so far as I know", and quite a silly one, too.

Tortoise: Why silly?

Achilles: Because it's so pointless. Here's another one for you:

"WILL BE BOYS" WILL BE BOYS.

Now what does that mean? Honestly, what a silly game.

Tortoise: Not to my mind. It's very earnest stuff, in my opinion. In fact this operation of preceding some phrase by its quotation is so overwhelmingly important that I think I'll give it a name.

Achilles: You will? What name will you dignify that silly operation by?

Tortoise: I believe I'll call it "to quine a phrase", to quine a phrase.

Achilles: "Quine"? What sort of word is that?

Tortoise: A five-letter word, if I'm not in error.

Achilles: What I was driving at is why you picked those exact five letters in that exact order.

Tortoise: Oh, now I understand what you meant when you asked me "What sort of word is that?" The answer is that a philosopher by the name of "Willard Van Orman Quine" invented the operation, so I name it in his honor. However, I cannot go any further than this in my explanation. Why these particular five letters make up his name—not to mention why they occur in this particular order—is a question to which I have no ready answer. However, I'd be perfectly willing to go and—

Achilles: Please don't bother! I didn't really want to know everything about Quine's name. Anyway, now I know how to quine a phrase. It's quite amusing. Here's a quined phrase:

"IS A SENTENCE FRAGMENT" IS A SENTENCE FRAGMENT.

It's silly but all the same I enjoy it. You take a sentence fragment, quine it, and lo and behold, you've made a sentence! A true sentence, in this case.

Tortoise: How about quining the phrase "is a king with no subject"?

Achilles: A king without a subject would be—

Tortoise: —an anomaly, of course. Don't wander from the point. Let's have quines first, and kings afterwards!

Achilles: I'm to quine that phrase, am I? All right—

"IS A KING WITH NO SUBJECT" IS A KING WITH NO SUBJECT.

It seems to me that it might make more sense if it said "sentence" instead of "king". Oh, well. Give me another!

Tortoise: All right—just one more. Try this one:

"WHEN QUINED, YIELDS A TORTOISE'S LOVE SONG"

Achilles: That should be easy . . . I'd say the quining gives this:

"WHEN QUINED, YIELDS A TORTOISE'S LOVE SONG"
WHEN QUINED, YIELDS A TORTOISE'S LOVE SONG.

Hmm . . . There's something just a little peculiar here. Oh, I see what it is! The sentence is talking about itself! Do you see that?

Tortoise: What do you mean? Sentences can't talk.

Achilles: No, but they REFER to things—and this one refers directly—unambiguously—unmistakably—to the very sentence which it is! You just have to think back and remember what quining is all about.

Tortoise: I don't see it saying anything about itself. Where does it say "me", or "this sentence", or the like?

Achilles: Oh, you are being deliberately thick-skulled. The beauty of it lies in just that: it talks about itself without having to come right out and say so!

Tortoise: Well, as I'm such a simple fellow, could you just spell it all out for me?

Achilles: Oh, he is such a Doubting Tortoise . . . All right, let me see . . . Suppose I make up a sentence—I'll call it "Sentence P"—with a blank in it.

Tortoise: Such as?

Achilles: Such as . . .

"______, WHEN QUINED, YIELDS A TORTOISE'S LOVE SONG".

Now the subject matter of Sentence P depends on how you fill in the blank. But once you've chosen how to fill in the blank, then the subject matter is determined: it is the phrase which you get by QUINING the blank. Call that "Sentence Q", since it is produced by an act of quining.

Tortoise: That makes sense. If the blank phrase were "is written on old jars of mustard to keep them fresh", then Sentence Q would have to be

"IS WRITTEN ON OLD JARS OF MUSTARD TO KEEP THEM FRESH"
IS WRITTEN ON OLD JARS OF MUSTARD TO KEEP THEM FRESH.

Achilles: True, and Sentence P makes the claim (though whether it is valid or not, I do not know) that Sentence Q is a Tortoise's love song. In any case, Sentence P here is not talking about itself, but rather about Sentence Q. Can we agree on that much?

Tortoise: By all means, let us agree—and what a beautiful song it is, too.

Achilles: But now I want to make a different choice for the blank, namely:

"WHEN QUINED, YIELDS A TORTOISE'S LOVE SONG".

Tortoise: Oh, heavens, you're getting a little involved here. I hope this all isn't going to be too highbrow for my modest mind.

Achilles: Oh, don't worry—you'll surely catch on. With this choice, Sentence Q becomes . . .

"WHEN QUINED, YIELDS A TORTOISE'S LOVE-SONG"
WHEN QUINED, YIELDS A TORTOISE'S LOVE-SONG.

Tortoise: Oh, you wily old warrior you, I catch on. Now Sentence Q is just the same as Sentence P.

Achilles: And since Sentence Q is always the topic of Sentence P, there is a loop, so now, P points back to itself. But you see, the self-reference is a sort of accident. Usually Sentences Q and P are entirely unlike each other; but with the right choice for the blank in Sentence P, quining will do this magic trick for you.

Tortoise: Oh, how clever. I wonder why I never thought of that myself. Now tell me: is the following sentence self-referential?

"IS COMPOSED OF FIVE WORDS" IS COMPOSED OF FIVE WORDS.

Achilles: Hmm . . . I can't quite tell. The sentence which you just gave is not really about itself, but rather about the phrase "is composed of five words". Though, of course, that phrase is PART of the sentence . . .

Tortoise: So the sentence refers to some part of itself—so what?

Achilles: Well, wouldn't that qualify as self-reference, too?

Tortoise: In my opinion, that is still a far cry from true self-reference. But don't worry too much about these tricky matters. You'll have ample time to think about them in the future.

Achilles: I will?

Tortoise: Indeed you will. But for now, why don't you try quining the phrase "yields falsehood when preceded by its quotation"?

Achilles: I see what you're getting at—that old obscene phone call. Quining it produces the following:

"YIELDS FALSEHOOD WHEN PRECEDED BY ITS QUOTATION"
YIELDS FALSEHOOD WHEN PRECEDED BY ITS QUOTATION.

So this is what that caller was saying! I just couldn't make out where the quotation marks were as he spoke. That certainly is an obscene remark! People ought to be jailed for saying things like that!

Tortoise: Why in the world?

Achilles: It just makes me very uneasy. Unlike the earlier examples, I can't quite make out if it is a truth or a falsehood. And the more I think about it, the more I can't unravel it. It makes my head spin. I wonder what kind of a lunatic mind would make something like that up, and torment innocent people in the night with it?

Tortoise: I wonder . . . Well, shall we go downstairs now?

Achilles: We needn't go down—we're at ground level already. Let's go back inside—you'll see. *(They go into the tower, and come to a small wooden door.)* We can just step outside right here. Follow me.

Tortoise: Are you sure? I don't want to fall three floors and break my shell.

Achilles: Would I fool you?

(And he opens the door. In front of them sits, to all appearances, the same boy, talking to the same young woman. Achilles and Mr. T walk up what seem to be the same stairs they walked down to enter the tower, and find themselves in what looks like just the same courtyard they first came into.)

Thank you, Mr. T, for your lucid clarification of that obscene telephone call.

Tortoise: And thank you, Achilles, for the pleasant promenade. I hope we meet again soon.

CHAPTER XIV

On Formally Undecidable Propositions of TNT and Related Systems[1]

The Two Ideas of the "Oyster"

THIS CHAPTER'S TITLE is an adaptation of the title of Gödel's famous 1931 paper—"TNT" having been substituted for *"Principia Mathematica"*. Gödel's paper was a technical one, concentrating on making his proof watertight and rigorous; this Chapter will be more intuitive, and in it I will stress the two key ideas which are at the core of the proof. The first key idea is the deep discovery that there are strings of TNT which can be interpreted as speaking about other strings of TNT; in short, that TNT, as a language, is capable of "introspection", or self-scrutiny. This is what comes from Gödel-numbering. The second key idea is that the property of self-scrutiny can be entirely concentrated into a single string; thus that string's sole focus of attention is itself. This "focusing trick" is traceable, in essence, to the Cantor diagonal method.

In my opinion, if one is interested in understanding Gödel's proof in a deep way, then one must recognize that the proof, in its essence, consists of a fusion of these two main ideas. Each of them alone is a master stroke; to put them together took an act of genius. If I were to choose, however, which of the two key ideas is deeper, I would unhesitatingly pick the first one—the idea of Gödel-numbering, for that idea is related to the whole notion of what meaning and reference are, in symbol-manipulating systems. This is an idea which goes far beyond the confines of mathematical logic, whereas the Cantor trick, rich though it is in mathematical consequences, has little if any relation to issues in real life.

The First Idea: Proof-Pairs

Without further ado, then, let us proceed to the elaboration of the proof itself. We have already given a fairly careful notion of what the Gödel isomorphism is about, in Chapter IX. We now shall describe a mathematical notion which allows us to translate a statement such as "The string 0=0 is a theorem of TNT" into a statement of number theory. This will involve the notion of *proof-pairs*. A proof-pair is a pair of natural numbers related in a particular way. Here is the idea:

> Two natural numbers, m and n respectively, form a TNT-proof-pair if and only if m is the Gödel number of a TNT-derivation whose bottom line is the string with Gödel number n.

The analogous notion exists with respect to the MIU-system, and it is a little easier on the intuition to consider that case first. So, for a moment, let us back off from TNT-proof-pairs, and look at MIU-proof-pairs. Their definition is parallel:

> Two natural numbers, m and n respectively, form a MIU-proof-pair if and only if m is the Gödel number of a MIU-system derivation whose bottom line is the string with Gödel number n.

Let us see a couple of examples involving MIU-proof-pairs. First, let $m = 3131131111301$, $n = 301$. These values of m and n do indeed form a MIU-proof-pair, because m is the Gödel number of the MIU-derivation

MI
MII
MIIII
MUI

whose last line is MUI, having Gödel number 301, which is n. By contrast, let $m = 31311311130$, and $n = 30$. Why do these two values *not* form a MIU-proof-pair? To see the answer, let us write out the alleged derivation which m codes for:

MI
MII
MIII
MU

There is an invalid step in this alleged derivation! It is the step from the second to the third line: from MII to MIII. There is no rule of inference in the MIU-system which permits such a typographical step. Correspondingly—and this is most crucial—there is no arithmetical rule of inference which carries you from 311 to 3111. This is perhaps a trivial observation, in light of our discussion in Chapter IX, yet it is at the heart of the Gödel isomorphism. What we do in any formal system has its parallel in arithmetical manipulations.

In any case, the values $m = 31311311130$, $n = 30$ certainly do not form a MIU-proof-pair. This in itself does not imply that 30 is not a MIU-number. There could be another value of m which forms a MIU-proof-pair with 30. (Actually, we know by earlier reasoning that MU is not a MIU-theorem, and therefore no number at all can form a MIU-proof-pair with 30.)

Now what about TNT-proof-pairs? Here are two parallel examples, one being merely an alleged TNT-proof-pair, the other being a valid TNT-proof-pair. Can you spot which is which? (Incidentally, here is where the '611' codon comes in. Its purpose is to separate the Gödel numbers of successive lines in a TNT-derivation. In that sense, '611' serves as a punctuation mark. In the MIU-system, the initial '3' of all lines is sufficient—no extra punctuation is needed.)

(1) *m* = 626,262,636,223,123,262,111,666,611,223,123,666,111,666
n = 123,666,111,666

(2) *m* = 626,262,636,223,123,262,111,666,611,223,333,262,636,123,262,111,666
n = 223,333,262,636,123,262,111,666

It is quite simple to tell which one is which, simply by translating back to the old notation, and making some routine examinations to see

(1) whether the alleged derivation coded for by *m* is actually a legitimate derivation;
(2) if so, whether the last line of the derivation coincides with the string which *n* codes for.

Step 2 is trivial; and step 1 is also utterly straightforward, in this sense: there are no open-ended searches involved, no hidden endless loops. Think of the examples above involving the MIU-system, and now just mentally substitute the rules of TNT for the MIU-system's rules, and the axioms of TNT for the MIU-system's one axiom. The algorithm in both cases is the same. Let me make that algorithm explicit:

Go down the lines in the derivation one by one.
Mark those which are axioms.
For each line which is *not* an axiom, check whether it follows by any of the rules of inference from earlier lines in the alleged derivation.
If all nonaxioms follow by rules of inference from earlier lines, then you have a legitimate derivation; otherwise it is a phony derivation.

At each stage, there is a clear set of tasks to perform, and the number of them is quite easily determinable in advance.

Proof-Pair-ness Is Primitive Recursive . . .

The reason I am stressing the boundedness of these loops is, as you may have sensed, that I am about to assert

FUNDAMENTAL FACT 1: The property of being a proof-pair is a primitive recursive number-theoretical property, and can therefore be tested for by a BlooP program.

There is a notable contrast to be made here with that other closely related number-theoretical property: that of being a *theorem-number*. To assert that n is a theorem-number is to assert that *some* value of m exists which forms a proof-pair with n. (Incidentally, these comments apply equally well to TNT and to the MIU-system; it may perhaps help to keep both in mind, the MIU-system serving as a prototype.) To check whether n is a theorem-number, you must embark on a search through all its potential proof-pair "partners" m—and here you may be getting into an endless chase. No one can say how far you will have to look to find a number which forms a proof-pair with n as its second element. That is the whole problem of having lengthening and shortening rules in the same system: they lead to a certain degree of unpredictability.

The example of the Goldbach Variation may prove helpful at this point. It is trivial to test whether a *pair* of numbers (m,n) form a *Tortoise-pair:* that is to say, both m and $n + m$ should be prime. The test is easy because the property of primeness is primitive recursive: it admits of a predictably terminating test. But if we want to know whether n possesses the Tortoise property, then we are asking, "Does *any* number m form a Tortoise-pair with n as its second element?"—and this, once again, leads us out into the wild, MU-loopy unknown.

. . . And Is Therefore Represented in TNT

The key concept at this juncture, then, is Fundamental Fact 1 given above, for from it we can conclude

> FUNDAMENTAL FACT 2: The property of forming a proof-pair is testable in BlooP, and consequently, it is *represented* in TNT by some formula having two free variables.

Once again, we are being casual about specifying which system these proof-pairs are relative to; it really doesn't matter, for both Fundamental Facts hold for any formal system. That is the nature of formal systems: it is always possible to tell, in a predictably terminating way, whether a given sequence of lines forms a proof, or not—and this carries over to the corresponding arithmetical notions.

The Power of Proof-Pairs

Suppose we assume we are dealing with the MIU-system, for the sake of concreteness. You probably recall the string we called "MUMON", whose interpretation on one level was the statement "MU is a theorem of the MIU-system". We can show how MUMON would be expressed in TNT, in terms of the formula which represents the notion of MIU-proof-pairs. Let us abbreviate that formula, whose existence we are assured of by Fundamental Fact 2, this way:

MIU-PROOF-PAIR{a,a′}

Since it is a property of two numbers, it is represented by a formula with two free variables. (Note: In this Chapter we shall always use austere TNT—so be careful to distinguish between the variables a, a′, a″.) In order to assert "**MU** is a theorem of the MIU-system", we would have to make the isomorphic statement "30 is a theorem-number of the MIU-system", and then translate that into TNT-notation. With the aid of our abbreviation, this is easy (remember also from Chapter VIII that to indicate the replacement of every a′ by a numeral, we write that numeral followed by "/a′"):

∃a:MIU-PROOF-PAIR{a,SSSSSSSSSSSSSSSSSSSSSSSSSSSSSS0/a′}

Count the **S**'s: there are 30. Note that this is a closed sentence of TNT, because one free variable was quantified, the other replaced by a numeral. A clever thing has been done here, by the way. Fundamental Fact 2 gave us a way to talk about *proof-pairs;* we have figured out how to talk about *theorem-numbers,* as well: you just add an existential quantifier in front! A more literal translation of the string above would be, "There exists some number a that forms a MIU-proof-pair with 30 as its second element".

Suppose that we wanted to do something parallel with respect to TNT—say, to express the statement "**0=0** is a theorem of TNT". We may abbreviate the formula which Fundamental Fact 2 assures us exists, in an analogous way (with two free variables, again):

TNT-PROOF-PAIR{a,a′}

(The interpretation of this abbreviated TNT-formula is: "Natural numbers a and a′ form a TNT-proof-pair.") The next step is to transform our statement into number theory, following the MUMON-model above. The statement becomes "There exists some number a which forms a TNT-proof-pair with 666,111,666 as its second element". The TNT-formula which expresses this is:

∃a:TNT-PROOF-PAIR{a,SSSSS. . . . . .SSSSS0/a′}

many, many **S**'*s!*
(in fact, 666,111,666 of them)

—a closed sentence of TNT. (Let us call it "JŌSHŪ", for reasons to appear momentarily.) So you see that there is a way to talk not only about the primitive recursive notion of TNT-proof-pairs, but also about the related but trickier notion of TNT-theorem-numbers.

To check your comprehension of these ideas, figure out how to translate into TNT the following statements of meta-TNT:

(1) **0=0** is not a theorem of TNT.
(2) **~0=0** is a theorem of TNT.
(3) **~0=0** is not a theorem of TNT.

How do the solutions differ from the example done above, and from each other? Here are a few more translation exercises.

(4) JŌSHŪ is a theorem of TNT. (Call the TNT-string which expresses this "META-JŌSHŪ".)
(5) META-JŌSHŪ is a theorem of TNT. (Call the TNT-string which expresses this "META-META-JŌSHŪ".)
(6) META-META-JŌSHŪ is a theorem of TNT.
(7) META-META-META-JŌSHŪ is a theorem of TNT.

(etc., etc.)

Example 5 shows that statements of meta-meta-TNT can be translated into TNT-notation; example 6 does the same for meta-meta-meta-TNT, etc.

It is important to keep in mind the difference between *expressing* a property, and *representing* it, at this point. The property of being a TNT-theorem-number, for instance, is *expressed* by the formula

∃a:TNT-PROOF-PAIR{a,a′}

Translation: "a′ is a TNT-theorem-number". However, we have no guarantee that this formula *represents* the notion, for we have no guarantee that this property is primitive recursive—in fact, we have more than a sneaking suspicion that it isn't. (This suspicion is well warranted. The property of being a TNT-theorem-number is *not* primitive recursive, and no TNT-formula can represent the property!) By contrast, the property of being a proof-pair, by virtue of its primitive recursivity, is both expressible and representable, by the formula already introduced.

Substitution Leads to the Second Idea

The preceding discussion got us to the point where we saw how TNT can "introspect" on the notion of TNT-theoremhood. This is the essence of the first part of the proof. We now wish to press on to the second major idea of the proof, by developing a notion which allows the concentration of this introspection into a single formula. To do this, we need to look at what happens to the Gödel number of a formula when you modify the formula structurally in a simple way. In fact, we shall consider this specific modification:

replacement of all free variables by a specific numeral.

Below are shown a couple of examples of this operation in the left-hand column, and in the right-hand column are exhibited the parallel changes in Gödel numbers.

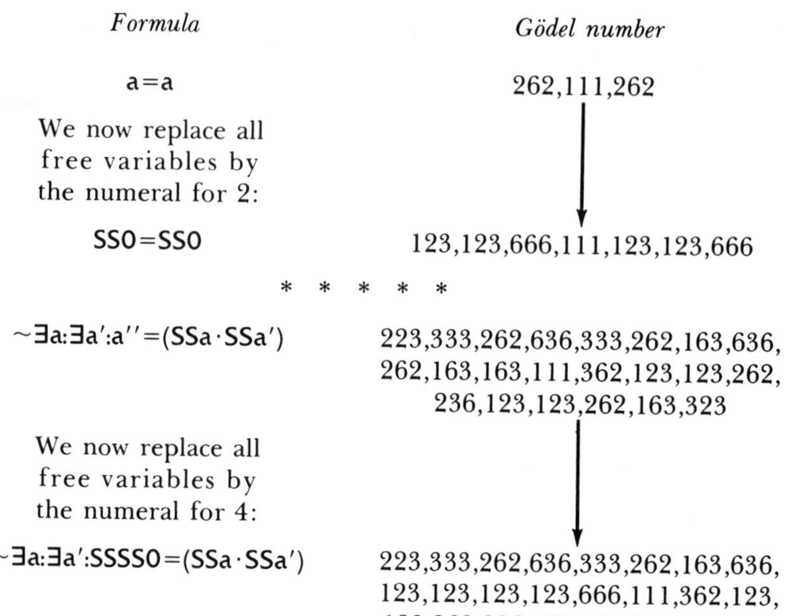

Formula	*Gödel number*
a=a	262,111,262
We now replace all free variables by the numeral for 2:	↓
SSO=SSO	123,123,666,111,123,123,666

* * * * *

~∃a:∃a′:a″=(SSa·SSa′)	223,333,262,636,333,262,163,636, 262,163,163,111,362,123,123,262, 236,123,123,262,163,323
We now replace all free variables by the numeral for 4:	↓
~∃a:∃a′:SSSSO=(SSa·SSa′)	223,333,262,636,333,262,163,636, 123,123,123,123,666,111,362,123, 123,262,236,123,123,262,163,323

An isomorphic arithmetical process is going on in the right-hand column, in which one huge number is turned into an even huger number. The function which makes the new number from the old one would not be too difficult to describe arithmetically, in terms of additions, multiplications, powers of 10 and so on—but we need not do so. The main point is this: that the relation among (1) the original Gödel number, (2) the number whose numeral is inserted, and (3) the resulting Gödel number, is a primitive recursive relation. That is to say, a BlooP test could be written which, when fed as input any three natural numbers, says **YES** if they they are related in this way, and **NO** if they aren't. You may test yourself on your ability to perform such a test—and at the same time convince yourself that there are no hidden open-ended loops to the process—by checking the following two sets of three numbers:

(1) 362,262,112,262,163,323,111,123,123,123,123,666;
2;
362,123,123,666,112,123,123,666,323,111,123,123,123,123,666.

(2) 223,362,262,236,262,323,111,262,163;
1;
223,362,123,666,236,123,666,323,111,262,163.

As usual, one of the examples checks, the other does not. Now this relationship between three numbers will be called the *substitution* relationship. Because it is primitive recursive, it is *represented* by some formula of TNT having three free variables. Let us abbreviate that TNT-formula by the following notation:

SUB{a,a′,a″}

Because this formula represents the substitution relationship, the formula shown below must be a TNT-theorem:

$$\mathbf{SUB}\{\underbrace{\mathbf{SSSSS.\,.\,.\,.\,.SSSSS0}}_{262,111,262\ \mathbf{S}\text{'s}}\mathbf{/a,SS0/a',}\underbrace{\mathbf{SSSSSS.\,.\,.\,.\,.SSSS0}}_{123,123,666,111,123,123,666\ \mathbf{S}\text{'s}}\mathbf{/a''}\}$$

(This is based on the first example of the substitution relation shown in the parallel columns earlier in this section.) And again because the **SUB** formula represents the substitution relation, the formula shown below certainly is *not* a TNT-theorem:

SUB{SSS0/a,SS0/a',S0/a''}

Arithmoquining

We now have reached the crucial point where we can combine all of our disassembled parts into one meaningful whole. We want to use the machinery of the **TNT-PROOF-PAIR** and **SUB** formulas in some way to construct a single sentence of TNT whose interpretation is: "This very string of TNT is not a TNT-theorem." How do we do it? Even at this point, with all the necessary machinery in front of us, the answer is not easy to find.

A curious and perhaps frivolous-seeming notion is that of substituting a formula's *own* Gödel number into itself. This is quite parallel to that other curious, and perhaps frivolous-seeming, notion of "quining" in the *Air on G's String*. Yet quining turned out to have a funny kind of importance, in that it showed a new way of making a self-referential sentence. Self-reference of the Quine variety sneaks up on you from behind the first time you see it—but once you understand the principle, you appreciate that it is quite simple and lovely. The arithmetical version of quining—let's call it *arithmoquining*—will allow us to make a TNT-sentence which is "about itself".

Let us see an example of arithmoquining. We need a formula with at least one free variable. The following one will do:

a=S0

This formula's Gödel number is 262,111,123,666, and we will stick this number into the formula itself—or rather, we will stick its *numeral* in. Here is the result:

$$\underbrace{\mathbf{SSSSS\,.\,.\,.\,.\,.\,SSSSS0}}_{262,111,123,666\ \mathbf{S}\text{'s}}\mathbf{=S0}$$

This new formula asserts a silly falsity—that 262,111,123,666 equals 1. If we had begun with the string **~a=S0** and then arithmoquined, we would have come up with a true statement—as you can see for yourself.

When you arithmoquine, you are of course performing a special case of the substitution operation we defined earlier. If we wanted to speak about arithmoquining inside TNT, we would use the formula

SUB{a'', a'', a'}

where the first two variables are the same. This comes from the fact that we are using a single number in two different ways (shades of the Cantor diagonal method!). The number a'' is both (1) the original Gödel number, and (2) the insertion-number. Let us invent an abbreviation for the above formula:

ARITHMOQUINE{a'', a'}

What the above formula says, in English, is:

a' is the Gödel number of the formula gotten by arithmoquining the formula with Gödel number a''.

Now the preceding sentence is long and ugly. Let's introduce a concise and elegant term to summarize it. We'll say

a' is the *arithmoquinification* of a''

to mean the same thing. For instance, the arithmoquinification of 262,111,123,666 is this unutterably gigantic number:

123,123,123, ... ,123,123,123,666,111,123,666

262,111,123,666 copies of '123'

(This is just the Gödel number of the formula we got when we arithmoquined a=S0.) We can speak quite easily about arithmoquining inside TNT.

The Last Straw

Now if you look back in the *Air on G's String*, you will see that the ultimate trick necessary for achieving self-reference in Quine's way is to quine a sentence which itself talks about the concept of quining. It's not enough just to quine—you must quine a quine-mentioning sentence! All right, then—the parallel trick in our case must be to arithmoquine some formula which itself is talking about the notion of arithmoquining!

Without further ado, we'll now write that formula down, and call it *G's uncle:*

~∃a:∃a':<TNT-PROOF-PAIR{a,a'}∧ARITHMOQUINE{a'',a'}>

You can see explicitly how arithmoquinification is thickly involved in the plot. Now this "uncle" has a Gödel number, of course, which we'll call '*u*'.

The head and tail of *u*'s decimal expansion, and even a teeny bit of its midsection, can be read off directly:

$$u = 223{,}333{,}262{,}636{,}333{,}262{,}163{,}636{,}212, \ldots, 161, \ldots, 213$$

For the rest, we'd have to know just how the formulas **TNT-PROOF-PAIR** and **ARITHMOQUINE** actually look when written out. That is too complex, and it is quite beside the point, in any case.

Now all we need to do is—arithmoquine this very uncle! What this entails is "booting out" all free variables—of which there is only one, namely a″—and putting in the numeral for *u* everywhere. This gives us:

$$\sim\exists a{:}\exists a'{:}<\text{TNT-PROOF-PAIR}\{a,a'\}\wedge\text{ARITHMOQUINE}\{\underbrace{\text{SSS}\ldots\text{SSS0}}_{u\ \text{S's}}/a'',a'\}>$$

And this, believe it or not, is Gödel's string, which we can call 'G'. Now there are two questions we must answer without delay. They are

(1) What is G's Gödel number?
(2) What is the interpretation of G?

Question 1 first. How did we make G? Well, we began with the uncle, and arithmoquined it. So, by the definition of arithmoquinification, G's Gödel number is:

the arithmoquinification of *u*.

Now question 2. We will translate G into English in stages, getting gradually more comprehensible as we go along. For our first rough try, we make a pretty literal translation:

"There do not exist numbers a and a′ such that both (1) they form a TNT-proof-pair, and (2) a′ is the arithmoquinification of *u*."

Now certainly there *is* a number a′ which is the arithmoquinification of *u*—so the problem must lie with the *other* number, a. This observation allows us to rephrase the translation of G as follows:

"There is no number a that forms a TNT-proof-pair with the arithmoquinification of *u*."

(This step, which can be confusing, is explained below in more detail.) Do you see what is happening? G is saying this:

"The formula whose Gödel number is the arithmoquinification of *u* is not a theorem of TNT."

But—and this should come as no surprise by now—that formula is none other than G itself; whence we can make the ultimate translation of G as

"G is not a theorem of TNT."

—or if you prefer,

"I am not a theorem of TNT."

We have gradually pulled a high-level interpretation—a sentence of meta-TNT—out of what was originally a low-level interpretation—a sentence of number theory.

TNT Says "Uncle!"

The main consequence of this amazing construction has already been delineated in Chapter IX: it is the incompleteness of TNT. To reiterate the argument:

> Is G a TNT-theorem? If so, then it must assert a truth. But what in fact does G assert? Its own nontheoremhood. Thus from its theoremhood would follow its nontheoremhood: a contradiction.
>
> Now what about G being a nontheorem? This is acceptable, in that it doesn't lead to a contradiction. But G's nontheoremhood is what G asserts—hence G asserts a truth. And since G is not a theorem, there exists (at least) one truth which is not a theorem of TNT.

Now to explain that one tricky step again. I will use another similar example. Take this string:

~∃a:∃a':<TORTOISE-PAIR{a,a'}∧TENTH-POWER{SS0/a'',a'}>

where the two abbreviations are for strings of TNT which you can write down yourself. TENTH-POWER{a'',a'} represents the statement "a' is the tenth power of a''". The literal translation into English is then:

"There do not exist numbers a and a' such that both (1) they form a Tortoise-pair, and (2) a' is the tenth power of 2."

But clearly, there *is* a tenth power of 2—namely 1024. Therefore, what the string is really saying is that

"There is no number a that forms a Tortoise-pair with 1024"

which can be further boiled down to:

"1024 does not have the Tortoise property."

The point is that we have achieved a way of substituting a *description* of a number, rather than its numeral, into a predicate. It depends on using one extra quantified variable (a'). Here, it was the number 1024 that was described as "the tenth power of 2"; above, it was the number described as "the arithmoquinification of *u*".

"Yields Nontheoremhood When Arithmoquined"

Let us pause for breath for a moment, and review what has been done. The best way I know to give some perspective is to set out explicitly how it compares with the version of the Epimenides paradox due to Quine. Here is a map:

falsehood	⇐⇒	nontheoremhood
quotation of a phrase	⇐⇒	Gödel number of a string
preceding a predicate by a subject	⇐⇒	substituting a numeral (or definite term) into an open formula
preceding a predicate by a quoted phrase	⇐⇒	substituting the Gödel number of a string into an open formula
preceding a predicate by itself, in quotes ("quining")	⇐⇒	substituting the Gödel number of an open formula into the formula itself ("arithmoquining")
yields falsehood when quined (a predicate without a subject)	⇐⇒	the "uncle" of G (an open formula of TNT)
"yields falsehood when quined" (the above predicate, quoted)	⇐⇒	the number *u* (the Gödel number of the above open formula)
"yields falsehood when quined" yields falsehood when quined (complete sentence formed by quining the above predicate)	⇐⇒	G itself (sentence of TNT formed by substituting *u* into the uncle, i.e., arithmoquining the uncle)

Gödel's Second Theorem

Since G's interpretation is true, the interpretation of its negation ~G is false. And we know that no false statements are derivable in TNT. Thus *neither G nor its negation ~G can be a theorem of TNT.* We have found a "hole" in our system—an undecidable proposition. This has a number of ramifications. Here is one curious fact which follows from G's undecidability: although neither G nor ~G is a theorem, the formula <G∨~G> is a theorem, since the rules of the Propositional Calculus ensure that all well-formed formulas of the form <P∨~P> are theorems.

This is one simple example where an assertion *inside* the system and an assertion *about* the system seem at odds with each other. It makes one wonder if the system really reflects itself accurately. Does the "reflected metamathematics" which exists inside TNT correspond well to the metamathematics which we do? This was one of the questions which intrigued Gödel when he wrote his paper. In particular, he was interested in whether it was possible, in the "reflected metamathematics", to prove TNT's consistency. Recall that this was a great philosophical dilemma of the day: how to prove a system consistent. Gödel found a simple way to express the statement "TNT is consistent" in a TNT formula; and then he showed that this formula (and all others which express the same idea) are only theorems of TNT under one condition: that TNT is *inconsistent.* This perverse result was a severe blow to optimists who expected that one could find a rigorous proof that mathematics is contradiction-free.

How do you express the statement "TNT is consistent" inside TNT? It hinges on this simple fact: that inconsistency means that two formulas, *x* and ~*x*, one the negation of the other, are both theorems. But if both *x* and ~*x* are theorems, then according to the Propositional Calculus, *all* well-formed formulas are theorems. Thus, to show TNT's consistency, it would suffice to exhibit one single sentence of TNT which can be proven to be a nontheorem. Therefore, one way to express "TNT is consistent" is to say "The formula ~0=0 is not a theorem of TNT". This was already proposed as an exercise a few pages back. The translation is:

$$\sim\exists a{:}\text{TNT-PROOF-PAIR}\{a,\underbrace{\text{SSSSS}\,.\,.\,.\,.\,.\,\text{SSSSS}}_{223{,}666{,}111{,}666\ \text{S's}}\text{0}/a'\}$$

It can be shown, by lengthy but fairly straightforward reasoning, that—as long as TNT is consistent—this oath-of-consistency by TNT is not a theorem of TNT. So TNT's powers of introspection are great when it comes to expressing things, but fairly weak when it comes to proving them. This is quite a provocative result, if one applies it metaphorically to the human problem of self-knowledge.

TNT Is ω-Incomplete

Now what variety of incompleteness does TNT "enjoy"? We shall see that TNT's incompleteness is of the "omega" variety—defined in Chapter VIII. This means that there is some infinite pyramidal family of strings all of which are theorems, but whose associated "summarizing string" is a nontheorem. It is easy to exhibit the summarizing string which is a nontheorem:

$$\forall a{:}\sim\exists a'{:}\langle\text{TNT-PROOF-PAIR}\{a,a'\}\wedge\text{ARITHMOQUINE}\{\overbrace{\text{SSS}\,.\,.\,.\,\text{SSS}}^{u\ \text{S's}}\text{0}/a'',a'\}\rangle$$

To understand why this string is a nontheorem, notice that it is extremely similar to G itself—in fact, G can be made from it in one step (viz., according to TNT's Rule of Interchange). Therefore, if it were a theorem, so would G be. But since G isn't a theorem, neither can this be.

Now we want to show that all of the strings in the related pyramidal family *are* theorems. We can write them down easily enough:

u

u S's

~∃a': <TNT-PROOF-PAIR{0/a,a'} ∧ ARITHMOQUINE{SSS . . . SSS0/a'', a'}>
~∃a': <TNT-PROOF-PAIR{S0/a,a'} ∧ ARITHMOQUINE{SSS . . . SSS0/a'', a'}>
~∃a': <TNT-PROOF-PAIR{SS0/a,a'} ∧ ARITHMOQUINE{SSS . . . SSS0/a'', a'}>
~∃a': <TNT-PROOF-PAIR{SSS0/a,a'} ∧ ARITHMOQUINE{SSS . . . SSS0/a'', a'}>

. . .
. . .
. . .
. . .

What does each one assert? Their translations, one by one, are:

"0 and the arithmoquinification of *u* do not form a TNT-proof-pair."
"1 and the arithmoquinification of *u* do not form a TNT-proof-pair."
"2 and the arithmoquinification of *u* do not form a TNT-proof-pair."
"3 and the arithmoquinification of *u* do not form a TNT-proof-pair."

. . . .
. . . .
. . . .

Now each of these assertions is about whether *two* specific integers form a proof-pair or not. (By contrast, G itself is about whether *one* specific integer is a theorem-number or not.) Now because G is a nontheorem, *no* integer forms a proof-pair with G's Gödel number. Therefore, each of the statements of the family is true. Now the crux of the matter is that the property of being a proof-pair is primitive recursive, hence *represented,* so that each of the statements in the list above, being true, must translate into a *theorem* of TNT—which means that everything in our infinite pyramidal family is a theorem. And that shows why TNT is ω-incomplete.

Two Different Ways to Plug Up the Hole

Since G's interpretation is true, the interpretation of its negation ~G is false. And, using the assumption that TNT is consistent, we know that no false statements are derivable in TNT. Thus neither G nor its negation ~G is a theorem of TNT. We have found a hole in our system—an undecidable proposition. Now this need be no source of alarm, if we are philosophically detached enough to recognize what this is a symptom of. It signifies that TNT can be extended, just as absolute geometry could be. In fact, TNT can be extended in two distinct directions, just as absolute geometry could be. It can be extended in a *standard* direction—which corresponds to extending absolute geometry in the Euclidean direction; or, it can be extended in a *nonstandard* direction—which corresponds, of course, to extending absolute geometry in the non-Euclidean direction. Now the standard type of extension would involve

adding G as a new axiom.

This suggestion seems rather innocuous and perhaps even desirable, since, after all, G asserts something true about the natural number system. But what about the nonstandard type of extension? If it is at all parallel to the case of the parallel postulate, it must involve

adding the negation of G as a new axiom.

But how can we even contemplate doing such a repugnant, hideous thing? After all, to paraphrase the memorable words of Girolamo Saccheri, isn't what ~G says "repugnant to the nature of the natural numbers"?

Supernatural Numbers

I hope the irony of this quotation strikes you. The exact problem with Saccheri's approach to geometry was that he began with a fixed notion of what was true and what was not true, and he set out only to prove what he'd assessed as true to start with. Despite the cleverness of his approach—which involved denying the fifth postulate, and then proving many "repugnant" propositions of the ensuing geometry—Saccheri never entertained the possibility of other ways of thinking about points and lines. Now we should be wary of repeating this famous mistake. We must consider impartially, to the extent that we can, what it would mean to add ~G as an axiom to TNT. Just think what mathematics would be like today if people had never considered adding new axioms of the following sorts:

∃a:(a+a)=S0
∃a:Sa=0
∃a:(a·a)=SS0
∃a:S(a·a)=0

While each of them is "repugnant to the nature of previously known number systems", each of them also provides a deep and wonderful *extension* of the notion of whole numbers: rational numbers, negative numbers, irrational numbers, imaginary numbers. Such a possibility is what ~G is trying to get us to open our eyes to. Now in the past, each new extension of the notion of number was greeted with hoots and catcalls. You can hear this particularly loudly in the names attached to the unwelcome arrivals, such as "irrational numbers", "imaginary numbers". True to this tradition, we shall name the numbers which ~G is announcing to us the *supernatural numbers,* showing how we feel they violate all reasonable and commonsensical notions.

If we are going to throw ~G in as the sixth axiom of TNT, we had better understand how in the world it could coexist, in one system, with the infinite pyramidal family we just finished discussing. To put it bluntly, ~G says:

"There exists *some* number which forms a TNT-proof-pair with the arithmoquinification of *u*"

—but the various members of the pyramidal family successively assert:

"0 is not that number"
"1 is not that number"
"2 is not that number"

.
.
.

This is rather confusing, because it seems to be a complete contradiction (which is why it is called "ω-inconsistency"). At the root of our confusion—much as in the case of the splitting of geometry—is our stubborn resistance to adopt a modified interpretation for the symbols, despite the fact that we are quite aware that the system is a modified system. We want to get away without reinterpreting *any* symbols—and of course that will prove impossible.

The reconciliation comes when we reinterpret ∃ as "There exists a *generalized* natural number", rather than as "There exists a natural number". As we do this, we shall also reinterpret ∀ in the corresponding way. This means that we are opening the door to some extra numbers besides the natural numbers. These are the *supernatural numbers.* The naturals and supernaturals together make up the totality of *generalized naturals.*

The apparent contradiction vanishes into thin air, now, for the pyramidal family still says what it said before: "No *natural* number forms a TNT-proof-pair with the arithmoquinification of *u.*" The family doesn't say anything about supernatural numbers, because there are *no numerals* for them. But now, ~G says, "There exists a *generalized* natural number which forms a TNT-proof-pair with the arithmoquinification of *u.*" It is clear that taken together, the family and ~G tell us something: that there is a *supernatural* number which forms a TNT-proof-pair with the arithmoquinification of *u.* That is all—there is no contradiction any more. TNT+~G is a consistent system, under an interpretation which includes supernatural numbers.

Since we have now agreed to extend the interpretations of the two quantifiers, this means that any theorem which involves either of them has an extended meaning. For example, the commutativity theorem

∀a:∀a′:(a+a′)=(a′+a)

now tells us that addition is commutative for all *generalized* natural numbers—in other words, not only for natural numbers, but also for supernatural numbers. Likewise, the TNT-theorem which says "2 is not the square of a natural number"—

~∃a:(a·a)=SS0

—now tells us that 2 is not the square of a supernatural number, either. In fact, supernatural numbers share all the properties of natural numbers, as long as those properties are given to us in theorems of TNT. In other words, everything that can be *formally proven* about natural numbers is thereby established also for supernatural numbers. This means, in particular, that supernatural numbers are not anything already familiar to you, such as fractions, or negative numbers, or complex numbers, or whatever. The supernatural numbers are, instead, best visualized as integers which are greater than all natural numbers—as *infinitely large* integers. Here is the point: although theorems of TNT can rule out negative numbers, fractions, irrational numbers, and complex numbers, still there is no way to rule out infinitely large integers. The problem is, there is no way even to *express* the statement "There are no infinite quantities".

This sounds quite strange, at first. Just exactly how big is the number which makes a TNT-proof-pair with G's Gödel number? (Let's call it '*I*', for no particular reason.) Unfortunately, we have not got any good vocabulary for describing the sizes of infinitely large integers, so I am afraid I cannot convey a sense of *I*'s magnitude. But then just how big is *i* (the square root of −1)? Its size cannot be imagined in terms of the sizes of familiar natural numbers. You can't say, "Well, *i* is about half as big as 14, and 9/10 as big as 24." You have to say, "*i* squared is −1", and more or less leave it at that. A quote from Abraham Lincoln seems à propos here. When he was asked, "How long should a man's legs be?" he drawled, "Long enough to reach the ground." That is more or less how to answer the question about the size of *I*—it should be just the size of *a number which specifies the structure of a proof of G*—no bigger, no smaller.

Of course, any theorem of TNT has many different derivations, so you might complain that my characterization of *I* is nonunique. That is so. But the parallel with *i*—the square root of −1—still holds. Namely, recall that there is another number whose square is also minus one: −*i*. Now *i* and −*i* are not the same number. They just have a property in common. The only trouble is that it is the property which defines them! We have to choose one of them—it doesn't matter which one—and call it "*i*". In fact there's no way of telling them apart. So for all we know we could have been calling the wrong one "*i*" for all these centuries and it would have made no difference. Now, like *i*, *I* is also nonuniquely defined. So you just have to think of *I* as being some specific one of the many possible supernatural numbers which form TNT-proof-pairs with the arithmoquinification of *u*.

Supernatural Theorems Have Infinitely Long Derivations

We haven't yet faced head on what it means to throw ~G in as an axiom. We have said it but not stressed it. The point is that ~G asserts that *G has a proof*. How can a system survive, when one of its axioms asserts that its own negation has a proof? We must be in hot water now! Well, it is not so bad as you might think. As long as we only construct *finite* proofs, we will never prove G. Therefore, no calamitous collision between G and its negation ~G will ever take place. The supernatural number *I* won't cause any disaster.

However, we will have to get used to the idea that ~G is now the one which asserts a truth ("G has a proof"), while G asserts a falsity ("G has no proof"). In standard number theory it is the other way around—but then, in standard number theory there aren't any supernatural numbers. Notice that a supernatural theorem of TNT—namely G—may assert a falsity, but all natural theorems still assert truths.

Supernatural Addition and Multiplication

There is one extremely curious and unexpected fact about supernaturals which I would like to tell you, without proof. (I don't know the proof either.) This fact is reminiscent of the Heisenberg uncertainty principle in quantum mechanics. It turns out that you can "index" the supernaturals in a simple and natural way by associating with each supernatural number a trio of ordinary integers (including negative ones). Thus, our original supernatural number, *I*, might have the index set (9,-8,3), and its successor, *I* + 1, might have the index set (9,-8,4). Now there is no unique way to index the supernaturals; different methods offer different advantages and disadvantages. Under some indexing schemes, it is very easy to calculate the index triplet for the *sum* of two supernaturals, given the indices of the two numbers to be added. Under other indexing schemes, it is very easy to calculate the index triplet for the *product* of two supernaturals, given the indices of the two numbers to be multiplied. But under *no* indexing scheme is it possible to calculate both. More precisely, if the sum's index can be calculated by a recursive function, then the product's index will not be a recursive function; and conversely, if the product's index is a recursive function, then the sum's index will not be. Therefore, supernatural schoolchildren who learn their supernatural plus-tables will have to be excused if they do not know their supernatural times-tables—and vice versa! You cannot know both at the same time.

Supernaturals Are Useful . . .

One can go beyond the number theory of supernaturals, and consider supernatural fractions (ratios of two supernaturals), supernatural real numbers, and so on. In fact, the calculus can be put on a new footing, using the notion of supernatural real numbers. Infinitesimals such as *dx* and *dy*, those old bugaboos of mathematicians, can be completely justified, by considering them to be reciprocals of infinitely large real numbers! Some theorems in advanced analysis can be proven more intuitively with the aid of "nonstandard analysis".

. . . But Are They Real?

Nonstandard number theory is a disorienting thing when you first meet up with it. But then, non-Euclidean geometry is also a disorienting subject. In both instances, one is powerfully driven to ask, "But which one of these two rival theories is correct? Which is *the truth*?" In a certain sense, there is no answer to such a question. (And yet, in another sense—to be discussed later—there is an answer.) The reason that there is *no* answer to the question is that the two rival theories, although they employ the same terms, do not talk about the same concepts. Therefore, they are only superficially rivals, just like Euclidean and non-Euclidean geometries. In geometry, the words "point", "line", and so on are undefined terms, and their meanings are determined by the axiomatic system within which they are used.

Likewise for number theory. When we decided to formalize TNT, we preselected the terms we would use as interpretation words—for instance, words such as "number", "plus", "times", and so on. By taking the step of formalization, we were committing ourselves to accepting whatever passive meanings these terms might take on. But—just like Saccheri—we didn't anticipate any surprises. We thought we knew what the true, the real, the only theory of natural numbers was. We didn't know that there would be some questions about numbers which TNT would leave open, and which could therefore be answered ad libitum by extensions of TNT heading off in different directions. Thus, there is no basis on which to say that number theory "really" is this way or that, just as one would be loath to say that the square root of -1 "really" exists, or "really" does not.

Bifurcations in Geometry, and Physicists

There is one argument which can be, and perhaps ought to be, raised against the preceding. Suppose experiments in the real, physical world can be explained more economically in terms of one particular version of geometry than in terms of any other. Then it might make sense to say that that geometry is "true". From the point of view of a physicist who wants to use the "correct" geometry, then it makes some sense to distinguish between the "true" geometry, and other geometries. But this cannot be taken too simplistically. Physicists are always dealing with approximations and idealizations of situations. For instance, my own Ph.D. work, mentioned in Chapter V, was based on an extreme idealization of the problem of a crystal in a magnetic field. The mathematics which emerged was of a high degree of beauty and symmetry. Despite—or rather, because of—the artificiality of the model, some fundamental features emerged conspicuously in the graph. These features then suggest some guesses about the kinds of things that might happen in more realistic situations. But without the simplifying assumptions which produced my graph, there could never be such insights. One can see this kind of thing over and over again in physics, where a physicist uses a "nonreal" situation to learn about deeply hidden features of reality. Therefore, one should be extremely cautious in saying that the brand of geometry which physicists might wish to use would represent "the true geometry", for in fact, physicists will always use a variety of different geometries, choosing in any given situation the one that seems simplest and most convenient.

Furthermore—and perhaps this is even more to the point—physicists do not study just the 3-D space we live in. There are whole families of "abstract spaces" within which physical calculations take place, spaces which have totally different geometrical properties from the physical space within which we live. Who is to say, then, that "the true geometry" is defined by the space in which Uranus and Neptune orbit around the sun? There is "Hilbert space", where quantum-mechanical wave functions undulate; there is "momentum space", where Fourier components dwell; there is "reciprocal space", where wave-vectors cavort; there is "phase space", where many-particle configurations swish; and so on. There is absolutely no reason that the geometries of all these spaces should be the same; in fact, they couldn't possibly be the same! So it is essential and vital for physicists that different and "rival" geometries should exist.

Bifurcations in Number Theory, and Bankers

So much for geometry. What about number theory? Is it also essential and vital that different number theories should coexist with each other? If you asked a bank officer, my guess is that you would get an expression of horror and disbelief. How could 2 and 2 add up to anything but 4? And moreover, if 2 and 2 did not make 4, wouldn't world economies collapse immediately under the unbearable uncertainty opened up by that fact? Not really. First of all, nonstandard number theory doesn't threaten the age-old idea that 2 plus 2 equals 4. It differs from ordinary number theory only in the way it deals with the concept of the infinite. After all, *every theorem of TNT remains a theorem in any extension of TNT!* So bankers need not despair of the chaos that will arrive when nonstandard number theory takes over.

And anyway, entertaining fears about old facts being changed betrays a misunderstanding of the relationship between mathematics and the real world. Mathematics only tells you answers to questions in the real world *after* you have taken the one vital step of choosing which kind of mathematics to apply. Even if there were a rival number theory which used the symbols '2', '3', and '+', and in which a theorem said "2 + 2 = 3", there would be little reason for bankers to choose to use that theory! For that theory does not fit the way money works. You fit your mathematics to the world, and not the other way around. For instance, we don't apply number theory to cloud systems, because the very concept of whole numbers hardly fits. There can be one cloud and another cloud, and they will come together and instead of there being two clouds, there will still only be one. This doesn't prove that 1 plus 1 equals 1; it just proves that our number-theoretical concept of "one" is not applicable in its full power to cloud-counting.

So bankers, cloud-counters, and most of the rest of us need not worry about the advent of supernatural numbers: they won't affect our everyday perception of the world in the slightest. The only people who might actually be a little worried are people whose endeavors depend in some crucial way on the nature of infinite entities. There aren't too many such people around—but mathematical logicians are members of this category. How can the existence of a bifurcation in number theory affect them? Well, number theory plays two roles in logic: (1) when axiomatized, it is an *object of study;* and (2) when used informally, it is an indispensable *tool* with which formal systems can be investigated. This is the use-mention distinction once again, in fact: in role (1), number theory is mentioned, in role (2) it is used.

Now mathematicians have judged that number theory is applicable to the study of formal systems even if not to cloud-counting, just as bankers have judged that the arithmetic of real numbers is applicable to their transactions. This is an *extramathematical* judgement, and shows that the thought processes involved in doing mathematics, just like those in other areas, involve "tangled hierarchies" in which thoughts on one level can affect thoughts on any other level. Levels are not cleanly separated, as the formalist version of what mathematics is would have one believe.

The formalist philosophy claims that mathematicians only deal with abstract symbols, and that they couldn't care less whether those symbols have any applications to or connections with reality. But that is quite a distorted picture. Nowhere is this clearer than in metamathematics. If the theory of numbers is itself *used* as an aid in gaining factual knowledge about formal systems, then mathematicians are tacitly showing that they believe these ethereal things called "natural numbers" are actually *part of reality*—not just figments of the imagination. This is why I parenthetically remarked earlier that, in a certain sense, there *is* an answer to the question of which version of number theory is "true". Here is the nub of the matter: mathematical logicians must choose which version of number theory to put their faith in. In particular, they cannot remain neutral on the question of the existence or nonexistence of supernatural numbers, for the two different theories may give different answers to questions in metamathematics.

For instance, take this question: "Is ~G finitely derivable in TNT?" No one actually knows the answer. Nevertheless, most mathematical logicians would answer no without hesitation. The intuition which motivates that answer is based on the fact that if ~G were a theorem, TNT would be ω-inconsistent, and this would force supernaturals down your throat if you wanted to interpret TNT meaningfully—a most unpalatable thought for most people. After all, we didn't intend or expect supernaturals to be part of TNT when we invented it. That is, we—or most of us—believe that it is possible to make a formalization of number theory which does not force you into believing that supernatural numbers are every bit as real as naturals. It is that intuition about reality which determines which "fork" of number theory mathematicians will put their faith in, when the chips are down. But this faith may be wrong. Perhaps every consistent formalization of number theory which humans invent will imply the existence of supernaturals, by being ω-inconsistent. This is a queer thought, but it is conceivable.

If this were the case—which I doubt, but there is no disproof available—then G would not have to be undecidable. In fact, there might be no undecidable formulas of TNT at all. There could simply be one unbifurcated theory of numbers—which necessarily includes supernaturals. This is not the kind of thing mathematical logicians expect, but it is something which ought not to be rejected outright. Generally, mathematical logicians believe that TNT—and systems similar to it—are ω-consistent, and that the Gödel string which can be constructed in any such system is undecidable within that system. That means that they can choose to add either it or its negation as an axiom.

Hilbert's Tenth Problem and the Tortoise

I would like to conclude this Chapter by mentioning one extension of Gödel's Theorem. (This material is more fully covered in the article "Hilbert's Tenth Problem" by Davis and Hersh, for which see the Bibliography.) For this, I must define what a Diophantine equation is. This is an equation in which a polynomial with fixed integral coefficients and exponents is set to 0. For instance,

$$a = 0$$

and

$$5x + 13y - 1 = 0$$

and

$$5p^5 + 17q^{17} - 177 = 0$$

and

$$a^{123,666,111,666} + b^{123,666,111,666} - c^{123,666,111,666} = 0$$

are Diophantine equations. It is in general a difficult matter to know whether a given Diophantine equation has any integer solutions or not. In fact, in a famous lecture at the beginning of the century, Hilbert asked mathematicians to look for a general algorithm by which one could determine in a finite number of steps if a given Diophantine equation has integer solutions or not. Little did he suspect that no such algorithm exists!

Now for the simplification of G. It has been shown that whenever you have a sufficiently powerful formal number theory, and a Gödel-numbering for it, there is a Diophantine equation which is equivalent to G. The equivalence lies in the fact that this equation, when interpreted on a metamathematical level, asserts of itself that it has no solutions. Turn it around: if you found a solution to it, you could construct from it the Gödel number of a proof in the system that the equation has no solutions! This is what the Tortoise did in the *Prelude,* using Fermat's equation as his Diophantine equation. It is nice to know that when you do this, you can retrieve the sound of Old Bach from the molecules in the air!

Birthday Cantatatata . . .

One fine May day, the Tortoise and Achilles meet, wandering in the woods. The latter, all decked out handsomely, is doing a jiggish sort of thing to a tune which he himself is humming. On his vest he is wearing a great big button with the words "Today is my Birthday!"

Tortoise: Hello there, Achilles. What makes you so joyful today? Is it your birthday, by any chance?

Achilles: Yes, yes! Yes it is, today is my birthday!

Tortoise: That is what I had suspected, on account of that button which you are wearing, and also because unless I am mistaken, you are singing a tune from a Birthday Cantata by Bach, one written in 1727 for the fifty-seventh birthday of Augustus, King of Saxony.

Achilles: You're right. And Augustus' birthday coincides with mine, so THIS Birthday Cantata has double meaning. However, I shan't tell you my age.

Tortoise: Oh, that's perfectly all right. However, I would like to know one other thing. From what you have told me so far, would it be correct to conclude that today is your birthday?

Achilles: Yes, yes, it would be. Today IS my birthday.

Tortoise: Excellent. That's just as I suspected. So now, I WILL conclude it is your birthday, unless—

Achilles: Yes—unless what?

Tortoise: Unless that would be a premature or hasty conclusion to draw, you know. Tortoises don't like to jump to conclusions, after all. (We don't like to jump at all, but especially not to conclusions.) So let me just ask you, knowing full well of your fondness for logical thought, whether it would be reasonable to deduce logically from the foregoing sentences, that today is in fact your birthday.

Achilles: I do believe I detect a pattern to your questions, Mr. T. But rather than jump to conclusions myself, I shall take your question at face value, and answer it straightforwardly. The answer is: YES.

Tortoise: Fine! Fine! Then there is only one more thing I need to know, to be quite certain that today is—

Achilles: Yes, yes, yes, yes . . . I can already see the line of your questioning, Mr. T. I'll have you know that I am not so gullible as I was when we discussed Euclid's proof, a while back.

Tortoise: Why, who would ever have thought you to be gullible? Quite to the contrary, I regard you as an expert in the forms of logical thought, an authority in the science of valid deductions, a fount of knowledge about correct methods of reasoning . . . To tell the truth, Achilles, you are, in my opinion, a veritable titan in the art of rational cogitation.

And it is only for that reason that I would ask you, "Do the foregoing sentences present enough evidence that I should conclude without further puzzlement that today is your birthday?"

Achilles: You flatten me with your weighty praise, Mr. T—FLATTER, I mean. But I am struck by the repetitive nature of your questioning—and in my estimation, you, just as well as I, could have answered 'yes' each time.

Tortoise: Of course I could have, Achilles. But you see, to do so would have been to make a Wild Guess—and Tortoises abhor Wild Guesses. Tortoises formulate only Educated Guesses. Ah, yes—the power of the Educated Guess. You have no idea how many people fail to take into account all the Relevant Factors when they're guessing.

Achilles: It seems to me that there was only one relevant factor in this rigmarole, and that was my first statement.

Tortoise: Oh, to be sure, it's at least ONE of the factors to take into account, I'd say—but would you have me neglect Logic, that venerated science of the ancients? Logic is always a Relevant Factor in making Educated Guesses, and since I have with me a renowned expert in Logic, I thought it only Logical to take advantage of that fact, and confirm my hunches, by directly asking him whether my intuitions were correct. So let me finally come out and ask you point blank: "Do the preceding sentences allow me to conclude, with no room for doubt, that Today is your Birthday?"

Achilles: For one more time, YES. But frankly speaking, I have the distinct impression that you could have supplied that answer—as well as all the previous ones—yourself.

Tortoise: How your words sting! Would I were so wise as your insinuation suggests! But as merely a mortal Tortoise, profoundly ignorant and longing to take into account all the Relevant Factors, I needed to know the answers to all those questions.

Achilles: Well then, let me clear the matter up for once and for all. The answer to all the previous questions, and to all the succeeding ones which you will ask along the same line, is just this: YES.

Tortoise: Wonderful! In one fell swoop, you have circumvented the whole mess, in your characteristically inventive manner. I hope you won't mind if I call this ingenious trick an ANSWER SCHEMA. It rolls up yes-answers numbers 1, 2, 3, etc., into one single ball. In fact, coming as it does at the end of the line, it deserves the title "Answer Schema Omega", 'ω' being the last letter of the Greek alphabet—as if YOU needed to be told THAT!

Achilles: I don't care what you call it. I am just very relieved that you finally agree that it is my birthday, and we can go on to some other topic—such as what you are going to give me as a present.

Tortoise: Hold on—not so fast. I WILL agree it is your birthday, provided one thing.

Achilles: What? That I ask for no present?

Tortoise: Not at all. In fact, Achilles, I am looking forward to treating you to a fine birthday dinner, provided merely that I am convinced that knowledge of all those yes-answers at once (as supplied by Answer Schema ω) allows me to proceed directly and without any further detours to the conclusion that today is your birthday. That's the case, isn't it?

Achilles: Yes, of course it is.

Tortoise: Good. And now I have yes-answer $\omega + 1$. Armed with it, I can proceed to accept the hypothesis that today is your birthday, if it is valid to do so. Would you be so kind as to counsel me on that matter, Achilles?

Achilles: What is this? I thought I had seen through your infinite plot. Now doesn't yes-answer $\omega + 1$ satisfy you? All right. I'll give you not only yes-answer $\omega + 2$, but also yes-answers $\omega + 3$, $\omega + 4$, and so on.

Tortoise: How generous of you, Achilles. And here it is your birthday, when I should be giving YOU presents instead of the reverse. Or rather, I SUSPECT it is your birthday. I guess I can conclude that it IS your birthday, now, armed with the new Answer Schema, which I will call "Answer Schema 2ω". But tell me, Achilles: Does Answer Schema 2ω REALLY allow me to make that enormous leap, or am I missing something?

Achilles: You won't trick me any more, Mr. T. I've seen the way to end this silly game. I hereby shall present you with an Answer Schema to end all Answer Schemas! That is, I present you simultaneously with Answer Schemas ω, 2ω, 3ω, 4ω, 5ω, etc. With this Meta-Answer-Schema, I have JUMPED OUT of the whole system, kit and caboodle, transcended this silly game you thought you had me trapped in—and now we are DONE!

Tortoise: Good grief! I feel honored, Achilles, to be the recipient of such a powerful Answer Schema. I feel that seldom has anything so gigantic been devised by the mind of man, and I am awestruck by its power. Would you mind if I give a name to your gift?

Achilles: Not at all.

Tortoise: Then I shall call it "Answer Schema ω^2". And we can shortly proceed to other matters—as soon as you tell me whether the possession of Answer Schema ω^2 allows me to deduce that today is your birthday.

Achilles: Oh, woe is me! Can't I ever reach the end of this tantalizing trail? What comes next?

Tortoise: Well, after Answer Schema ω^2 there's answer $\omega^2 + 1$. And then answer $\omega^2 + 2$. And so forth. But you can wrap those all together into a packet, being Answer Schema $\omega^2 + \omega$. And then there are quite a few other answer-packets, such as $\omega^2 + 2\omega$, and $\omega^2 + 3\omega$ . . . Eventually, you come to Answer Schema $2\omega^2$, and after a while, Answer Schemas $3\omega^2$ and $4\omega^2$. Beyond them there are yet further Answer Schemas, such as ω^3, ω^4, ω^5, and so on. It goes on quite a ways, you know.

Achilles: I can imagine. I suppose it comes to Answer Schema ω^ω after a while.

Tortoise: Of course.

Achilles: And then ω^{ω^ω}, and $\omega^{\omega^{\omega^\omega}}$?

Tortoise: You're catching on mighty fast, Achilles. I have a suggestion for you, if you don't mind. Why don't you throw all of those together into a single Answer Schema?

Achilles: All right, though I'm beginning to doubt whether it will do any good.

Tortoise: It seems to me that within our naming conventions as so far set up, there is no obvious name for this one. So perhaps we should just arbitrarily name it Answer Schema ϵ_0.

Achilles: Confound it all! Every time you give one of my answers a NAME, it seems to signal the imminent shattering of my hopes that that answer will satisfy you. Why don't we just leave this Answer Schema nameless?

Tortoise: We can hardly do that, Achilles. We wouldn't have any way to refer to it without a name. And besides, there is something inevitable and rather beautiful about this particular Answer Schema. It would be quite ungraceful to leave it nameless! And you wouldn't want to do something lacking in grace on your birthday, would you? Or IS it your birthday? Say, speaking of birthdays, today is MY birthday!

Achilles: It is?

Tortoise: Yes, it is. Well, actually, it's my uncle's birthday, but that's almost the same. How would you like to treat me to a delicious birthday dinner this evening?

Achilles: Now just a cotton-picking minute, Mr. T. Today is MY birthday. You should do the treating!

Tortoise: Ah, but you never did succeed in convincing me of the veracity of that remark. You kept on beating around the bush with answers, Answer Schemas, and whatnot. All I wanted to know was if it was your birthday or not, but you managed to befuddle me entirely. Oh, well, too bad. In any case, I'll be happy to let you treat me to a birthday dinner this evening.

Achilles: Very well. I know just the place. They have a variety of delicious soups. And I know exactly what kind we should have . . .

CHAPTER XV

Jumping out of the System

A More Powerful Formal System

ONE OF THE things which a thoughtful critic of Gödel's proof might do would be to examine its generality. Such a critic might, for example, suspect that Gödel has just cleverly taken advantage of a hidden defect in one particular formal system, TNT. If this were the case, then perhaps a formal system superior to TNT could be developed which would not be subject to the Gödelian trick, and Gödel's Theorem would lose much of its sting. In this Chapter we will carefully scrutinize the properties of TNT which made it vulnerable to the arguments of last Chapter.

A natural thought is this: If the basic trouble with TNT is that it contains a "hole"—in other words, a sentence which is undecidable, namely G—then why not simply plug up the hole? Why not just tack G onto TNT as a sixth axiom? Of course, by comparison to the other axioms, G is a ridiculously huge giant, and the resulting system—TNT+G—would have a rather comical aspect due to the disproportionateness of its axioms. Be that as it may, adding G is a reasonable suggestion. Let us consider it done. Now, it is to be hoped, the new system, TNT+G, is a superior formal system—one which is not only supernatural-free, but also *complete*. It is certain that TNT+G is superior to TNT in at least one respect: the string G is no longer undecidable in this new system, since it is a theorem.

What was the vulnerability of TNT due to? The essence of its vulnerability was that it was capable of expressing statements about itself—in particular, the statement

"I Cannot Be Proven in Formal System TNT"

or, expanded a bit,

> "There does not exist a natural number which forms a TNT-proof-pair with the Gödel number of this string."

Is there any reason to expect or hope that TNT+G would be invulnerable to Gödel's proof? Not really. Our new system is just as expressive as TNT. Since Gödel's proof relies primarily on the expressive power of a formal system, we should not be surprised to see our new system succumb, too. The trick will be to find a string which expresses the statement

"I Cannot Be Proven in Formal System TNT+G."

Actually, it is not much of a trick, once you have seen it done for TNT. All the same principles are employed; only the context shifts slightly. (Figuratively speaking, we take a tune we know and simply sing it again, only in a higher key.) As before, the string which we are looking for—let us call it "G′"—is constructed by the intermediary of an "uncle". But instead of being based on the formula which represents TNT-proof-pairs, it is based on the similar but slightly more complicated notion of TNT+G-proof-pairs. This notion of TNT+G-proof-pairs is only a slight extension of the original notion of TNT-proof-pairs.

A similar extension could be envisaged for the MIU-system. We have seen the unadulterated form of MIU-proof-pairs. Were we now to add **MU** as a second axiom, we would be dealing with a new system—the MIU+**MU** system. A derivation in this extended system is presented:

MU	axiom
MUU	rule 2

There is a MIU+**MU**-proof-pair which corresponds—namely, m = 30300, n = 300. Of course, this pair of numbers does not form a MIU-proof-pair—only a MIU+**MU**-proof-pair. The addition of an extra axiom does not substantially complicate the arithmetical properties of proof-pairs. The significant fact about them—that being a proof-pair is primitive recursive—is preserved.

The Gödel Method Reapplied

Now, returning to TNT+G, we will find a similar situation. TNT+G-proof-pairs, like their predecessors, are primitive recursive, so they are represented inside TNT+G by a formula which we abbreviate in an obvious manner:

(TNT+G)-PROOF-PAIR{a,a′}

Now we just do everything all over again. We make the counterpart of G by beginning with an "uncle", just as before:

~∃a:∃a′:<(TNT+G)-PROOF-PAIR{a,a′}∧ARITHMOQUINE{a″,a′}>

Let us say its Gödel-number is u'. Now we arithmoquine this very uncle. That will give us G′:

~∃a:∃a′:<(TNT+G)-PROOF-PAIR{a,a′}
∧ARITHMOQUINE{SSS . . . . SSS0/a″,a′}>
u' **S**'s

Its interpretation is

"There is no number a that forms a TNT+G-proof-pair with the arithmoquinification of u'."

More concisely,

"I Cannot Be Proven in Formal System TNT+G."

Multifurcation

Well (yawn), the details are quite boring from here on out. G′ is precisely to TNT+G as G was to TNT itself. One finds that either G′ or ~G′ can be added to TNT+G, to yield a further splitting of number theory. And, lest you think this only happens to the "good guys", this very same dastardly trick can be played upon TNT+~G—that is, upon the nonstandard extension of TNT gotten by adding G's negation. So now we see (Fig. 75) that there are all sorts of bifurcations in number theory:

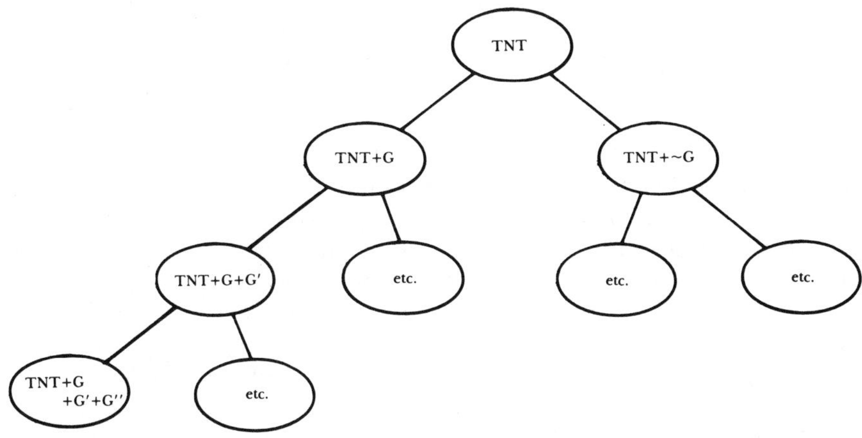

FIGURE 75. "Multifurcation" of TNT. Each extension of TNT has its very own Gödel sentence; that sentence, or its negation, can be added on, so that from each extension there sprouts a pair of further extensions, a process which goes on ad infinitum.

Of course, this is just the beginning. Let us imagine moving down the leftmost branch of this downwards-pointing tree, where we always toss in the Gödel sentences (rather than their negations). This is the best we can do by way of eliminating supernaturals. After adding G, we add G′. Then we add G″, and G‴, and so on. Each time we make a new extension of TNT, its vulnerability to the Tortoise's method—pardon me, I mean Gödel's method—allows a new string to be devised, having the interpretation

"I Cannot Be Proven in Formal System X."

Naturally, after a while, the whole process begins to seem utterly predictable and routine. Why, all the "holes" are made by one single technique! This means that, viewed as typographical objects, they are all cast from one single mold, which in turn means that one single axiom schema suffices to represent all of them! So if this is so, why not plug up all the holes at once and be done with this nasty business of incompleteness once and for all? This would be accomplished by *adding an axiom schema to TNT,* instead of just one axiom at a time. Specifically, this axiom schema would be the mold in which all of G, G′, G″, G‴, etc., are cast. By adding this axiom schema (let's call it "G_ω"), we would be outsmarting the "Gödelization" method. Indeed, it seems quite clear that adding G_ω to TNT would be the *last step* necessary for the complete axiomatization of all of number-theoretical truth.

It was at about this point in the *Contracrostipunctus* that the Tortoise related the Crab's invention of "Record Player Omega". However, readers were left dangling as to the fate of that device, since before completing his tale, the tuckered-out Tortoise decided that he had best go home to sleep (but not before tossing off a sly reference to Gödel's Incompleteness Theorem). Now, at last, we can get around to clearing up that dangling detail . . . Perhaps you already have an inkling, after reading the *Birthday Cantatatata.*

Essential Incompleteness

As you probably suspected, even this fantastic advance over TNT suffers the same fate. And what makes it quite weird is that it is still for, in essence, the same reason. The axiom schema is not powerful enough, and the Gödel construction can again be effected. Let me spell this out a little. (One can do it much more rigorously than I shall here.) If there is a way of capturing the various strings G, G′, G″, G‴, . . . in a single *typographical* mold, then there is a way of describing their Gödel numbers in a single *arithmetical* mold. And this arithmetical portrayal of an infinite class of numbers can then be represented inside TNT+G_ω by some formula OMEGA-AXIOM{a} whose interpretation is: "a is the Gödel number of one of the axioms coming from G_ω". When a is replaced by any specific numeral, the formula which results will be a theorem of TNT+G_ω if and only if the numeral stands for the Gödel number of an axiom coming from the schema.

With the aid of this new formula, it becomes possible to represent even such a complicated notion as TNT+G_ω-proof-pairs inside TNT+G_ω:

(TNT+G_ω)-PROOF-PAIR{a,a′}

Using this formula, we can construct a new uncle, which we proceed to arithmoquine in the by now thoroughly familiar way, making yet another undecidable string, which will be called "TNT+$G_{\omega+1}$". At this point, you might well wonder, "Why isn't $G_{\omega+1}$ among the axioms created by the axiom schema G_ω?" The answer is that G_ω was not clever enough to foresee its *own* embeddability inside number theory.

In the *Contracrostipunctus,* one of the essential steps in the Tortoise's making an "unplayable record" was to get a hold of a manufacturer's blueprint of the record player which he was out to destroy. This was necessary so that he could figure out to what kinds of vibrations it was vulnerable, and then incorporate into his record such grooves as would code for sounds which would induce those vibrations. It is a close analogue to the Gödel trick, in which the system's own properties are reflected inside the notion of proof-pairs, and then used against it. Any system, no matter how complex or tricky it is, can be Gödel-numbered, and then the notion of its proof-pairs can be defined—and this is the petard by which it is hoist. Once a system is well-defined, or "boxed", it becomes vulnerable.

This principle is excellently illustrated by the Cantor diagonal trick, which finds an omitted real number for each well-defined list of reals between 0 and 1. It is the act of giving an explicit list—a "box" of reals—which causes the downfall. Let us see how the Cantor trick can be repeated over and over again. Consider what happens if, starting with some list L, you do the following:

(1a) Take list L, and construct its diagonal number d.
(1b) Throw d somewhere into list L, making a new list $L+d$.

(2a) Take list $L+d$, and construct its diagonal number d'.
(2b) Throw d' somewhere into list $L+d$, making a new list $L+d+d'$.

.
.
.

Now this step-by-step process may seem a doltish way to patch up L, for we could have made the entire list $d, d', d'', d''', \ldots$ at once, given L originally. But if you think that making such a list will enable you to complete your list of reals, you are very wrong. The problem comes at the moment you ask, "Where to incorporate the list of diagonal numbers inside L?" No matter how diabolically clever a scheme you devise for ensconcing the d-numbers inside L, once you have done it, then the new list is still vulnerable. As was said above, it is the act of giving an explicit list—a "box" of reals—that causes the downfall.

Now in the case of formal systems, it is the act of giving an explicit recipe for what supposedly characterizes number-theoretical truth that causes the incompleteness. This is the crux of the problem with $TNT+G_\omega$. Once you insert all the G's in a well-defined way into TNT, there is seen to be some *other* G—some unforeseen G—which you didn't capture in your axiom schema. And in the case of the TC-battle inside the *Contracrostipunctus,* the instant a record player's "architecture" is determined, the record player becomes capable of being shaken to pieces.

So what is to be done? There is no end in sight. It appears that TNT, even when extended ad infinitum, cannot be made complete. TNT is therefore said to suffer from *essential incompleteness* because the incompleteness here is part and parcel of TNT; it is an essential part of the nature of TNT and cannot be eradicated in any way, whether simple-minded or ingenious. What's more, this problem will haunt any formal version of number theory, whether it is an extension of TNT, a modification of TNT, or an alternative to TNT. The fact of the matter is this: the possibility of constructing, in a given system, an undecidable string via Gödel's self-reference method, depends on three basic conditions:

(1) That the system should be rich enough so that all desired statements about numbers, whether true or false, can be *expressed* in it. (Failure on this count means that the system is from the very start too weak to be counted as a rival to TNT, because it can't even express number-theoretical notions that TNT can. In the metaphor of the *Contracrostipunctus,* it is as if one did not have a phonograph but a refrigerator or some other kind of object.)

(2) That all general recursive relations should be *represented* by formulas in the system. (Failure on this count means the system fails to capture in a theorem some general recursive truth, which can only be considered a pathetic bellyflop if it is attempting to produce all of number theory's truths. In the *Contracrostipunctus* metaphor, this is like having a record player, but one of low fidelity.)

(3) That the axioms and typographical patterns defined by its rules be recognizable by some terminating decision procedure. (Failure on this count means that there is no method to distinguish valid derivations in the system from invalid ones—thus that the "formal system" is not formal after all, and in fact is not even well-defined. In the *Contracrostipunctus* metaphor, it is a phonograph which is still on the drawing board, only partially designed.)

Satisfaction of these three conditions guarantees that any consistent system will be incomplete, because Gödel's construction is applicable.

The fascinating thing is that any such system digs its own hole; the system's own richness brings about its own downfall. The downfall occurs essentially because the system is powerful enough to have self-referential sentences. In physics, the notion exists of a "critical mass" of a fissionable substance, such as uranium. A solid lump of the substance will just sit there, if its mass is less than critical. But beyond the critical mass, such a lump will undergo a chain reaction, and blow up. It seems that with formal systems there is an analogous critical point. Below that point, a system is "harmless" and does not even approach defining arithmetical truth formally; but beyond the critical point, the system suddenly attains the capacity for self-reference, and thereby dooms itself to incompleteness. The threshold seems to be roughly when a system attains the three properties listed above.

Once this ability for self-reference is attained, the system has a hole which is tailor-made for itself; the hole takes the features of the system into account and uses them against the system.

The Passion According to Lucas

The baffling repeatability of the Gödel argument has been used by various people—notably J. R. Lucas—as ammunition in the battle to show that there is some elusive and ineffable quality to human intelligence, which makes it unattainable by "mechanical automata"—that is, computers. Lucas begins his article "Minds, Machines, and Gödel" with these words:

> Gödel's theorem seems to me to prove that Mechanism is false, that is, that minds cannot be explained as machines.[1]

Then he proceeds to give an argument which, paraphrased, runs like this. For a computer to be considered as intelligent as a person is, it must be able to do every intellectual task which a person can do. Now Lucas claims that no computer can do "Gödelization" (one of his amusingly irreverent terms) in the manner that people can. Why not? Well, think of any particular formal system, such as TNT, or TNT+G, or even TNT+G_ω. One can write a computer program rather easily which will systematically generate theorems of that system, and in such a manner that eventually, any preselected theorem will be printed out. That is, the theorem-generating program won't skip any portion of the "space" of all theorems. Such a program would be composed of two major parts: (1) a subroutine which stamps out axioms, given the "molds" of the axiom schemas (if there are any), and (2) a subroutine which takes known theorems (including axioms, of course) and applies rules of inference to produce new theorems. The program would alternate between running first one of these subroutines, and then the other.

We can anthropomorphically say that this program "knows" some facts of number theory—namely, it knows those facts which it prints out. If it fails to print out some true fact of number theory, then of course it doesn't "know" that fact. Therefore, a computer program will be inferior to human beings if it can be shown that humans know something which the program cannot know. Now here is where Lucas starts rolling. He says that we humans can always do the Gödel trick on any formal system as powerful as TNT—and hence no matter what the formal system, we know more than it does. Now this may only sound like an argument about formal systems, but it can also be slightly modified so that it becomes, seemingly, an invincible argument against the possibility of Artificial Intelligence ever reproducing the human level of intelligence. Here is the gist of it:

Rigid internal codes entirely rule computers and robots; ergo . . .
Computers are isomorphic to formal systems. Now . . .
Any computer which wants to be as smart as we are has got to be
able to do number theory as well as we can, so . . .

Among other things, it has to be able to do primitive recursive arithmetic. But for this very reason . . .

It is vulnerable to the Gödelian "hook", which implies that . . .

We, with our *human* intelligence, can concoct a certain statement of number theory which is true, but the *computer* is blind to that statement's truth (i.e., will never print it out), precisely because of Gödel's boomeranging argument.

This implies that there is one thing which computers just cannot be programmed to do, but which we can do. So we are smarter.

Let us enjoy, with Lucas, a transient moment of anthropocentric glory:

> However complicated a machine we construct, it will, if it is a machine, correspond to a formal system, which in turn will be liable to the Gödel procedure for finding a formula unprovable-in-that-system. This formula the machine will be unable to produce as being true, although a mind can see it is true. And so the machine will still not be an adequate model of the mind. We are trying to produce a model of the mind which is mechanical—which is essentially "dead"—but the mind, being in fact "alive," can always go one better than any formal, ossified, dead system can. Thanks to Gödel's theorem, the mind always has the last word.[2]

On first sight, and perhaps even on careful analysis, Lucas' argument appears compelling. It usually evokes rather polarized reactions. Some seize onto it as a nearly religious proof of the existence of souls, while others laugh it off as being unworthy of comment. I feel it is wrong, but fascinatingly so—and therefore quite worthwhile taking the time to rebut. In fact, it was one of the major early forces driving me to think over the matters in this book. I shall try to rebut it in one way in this Chapter, and in other ways in Chapter XVII.

We must try to understand more deeply why Lucas says the computer cannot be programmed to "know" as much as we do. Basically the idea is that we are always *outside* the system, and from out there we can always perform the "Gödelizing" operation, which yields something which the program, from within, can't see is true. But why can't the "Gödelizing operator", as Lucas calls it, be programmed and added to the program as a third major component? Lucas explains:

> The procedure whereby the Gödelian formula is constructed is a standard procedure—only so could we be sure that a Gödelian formula can be constructed for every formal system. But if it is a standard procedure, then a machine should be able to be programmed to carry it out too. . . . This would correspond to having a system with an additional rule of inference which allowed one to add, as a theorem, the Gödelian formula of the rest of the formal system, and then the Gödelian formula of this new, strengthened, formal system, and so on. It would be tantamount to adding to the original formal system an infinite sequence of axioms, each the Gödelian formula of the system hitherto obtained. . . . We might expect a mind, faced with a machine that possessed a Gödelizing operator, to take this into account, and

out-Gödel the new machine, Gödelizing operator and all. This has, in fact, proved to be the case. Even if we adjoin to a formal system the infinite set of axioms consisting of the successive Gödelian formulae, the resulting system is still incomplete, and contains a formula which cannot be proved-in-the-system, although a rational being can, standing outside the system, see that it is true. We had expected this, for even if an infinite set of axioms were added, they would have to be specified by some finite rule or specification, and this further rule or specification could then be taken into account by a mind considering the enlarged formal system. In a sense, just because the mind has the last word, it can always pick a hole in any formal system presented to it as a model of its own workings. The mechanical model must be, in some sense, finite and definite: and then the mind can always go one better.[3]

Jumping Up a Dimension

A visual image provided by M. C. Escher is extremely useful in aiding the intuition here: his drawing *Dragon* (Fig. 76). Its most salient feature is, of course, its subject matter—a dragon biting its tail, with all the Gödelian connotations which that carries. But there is a deeper theme to this picture. Escher himself wrote the following most interesting comments. The first comment is about a set of his drawings all of which are concerned with "the conflict between the flat and the spatial"; the second comment is about *Dragon* in particular.

> I. Our three-dimensional space is the only true reality we know. The two-dimensional is every bit as fictitious as the four-dimensional, for nothing is flat, not even the most finely polished mirror. And yet we stick to the convention that a wall or a piece of paper *is* flat, and curiously enough, we still go on, as we have done since time immemorial, producing illusions of space on just such plane surfaces as these. Surely it is a bit absurd to draw a few lines and then claim: "This is a house". This odd situation is the theme of the next five pictures [including *Dragon*].[4]

> II. However much this dragon tries to be spatial, he remains completely flat. Two incisions are made in the paper on which he is printed. Then it is folded in such a way as to leave two square openings. But this dragon is an obstinate beast, and in spite of his two dimensions he persists in assuming that he has three; so he sticks his head through one of the holes and his tail through the other.[5]

This second remark especially is a very telling remark. The message is that no matter how cleverly you try to simulate three dimensions in two, you are always missing some "essence of three-dimensionality". The dragon tries very hard to fight his two-dimensionality. He defies the two-dimensionality of the paper on which he thinks he is drawn, by sticking his head through it; and yet all the while, we outside the drawing can see the pathetic futility of it all, for the dragon *and* the holes *and* the folds are all merely two-dimensional simulations of those concepts, and not a one of them is real. But the dragon cannot step out of his two-dimensional space, and cannot

FIGURE 76. Dragon, *by M. C. Escher (wood-engraving, 1952).*

know it as we do. We could, in fact, carry the Escher picture any number of steps further. For instance, we could tear it out of the book, fold it, cut holes in it, pass it through itself, and photograph the whole mess, so that it again becomes two-dimensional. And to that photograph, we could once again do the same trick. Each time, at the instant that it becomes two-dimensional—no matter how cleverly we seem to have simulated three dimensions inside two—it becomes vulnerable to being cut and folded again.

Now with this wonderful Escherian metaphor, let us return to the program versus the human. We were talking about trying to encapsulate the "Gödelizing operator" inside the program itself. Well, even if we had written a program which carried the operation out, that program would not capture the essence of Gödel's method. For once again, we, outside the system, could still "zap" it in a way which it couldn't do. But then are we arguing with, or against, Lucas?

The Limits of Intelligent Systems

Against. For the very fact that we cannot write a program to do "Gödelizing" must make us somewhat suspicious that we ourselves could do it in every case. It is one thing to make the argument in the abstract that Gödelizing "can be done"; it is another thing to know how to do it in every particular case. In fact, as the formal systems (or programs) escalate in complexity, our own ability to "Gödelize" will eventually begin to waver. It must, since, as we have said above, we do *not* have any algorithmic way of describing how to perform it. If we can't tell *explicitly* what is involved in applying the Gödel method in all cases, then for each of us there will eventually come some case so complicated that we simply can't figure out how to apply it.

Of course, this borderline of one's abilities will be somewhat ill-defined, just as is the borderline of weights which one can pick up off the ground. While on some days you may not be able to pick up a 250-pound object, on other days maybe you can. Nevertheless, there are no days whatsoever on which you can pick up a 250-ton object. And in this sense, though everyone's Gödelization threshold is vague, for each person, there are systems which lie far beyond his ability to Gödelize.

This notion is illustrated in the *Birthday Cantatatata.* At first, it seems obvious that the Tortoise can proceed as far as he wishes in pestering Achilles. But then Achilles tries to sum up all the answers in a single swoop. This is a move of a different character than any that has gone before, and is given the new name 'ω'. The newness of the name is quite important. It is the first example where the old naming scheme—which only included names for all the natural numbers—had to be transcended. Then come some more extensions, some of whose names seem quite obvious, others of which are rather tricky. But eventually, we run out of names once again—at the point where the answer-schemas

$$\omega, \omega^{\omega}, \omega^{\omega^{\omega}}, \ldots$$

are all subsumed into one outrageously complex answer schema. The altogether new name 'ϵ_0' is supplied for this one. And the reason a new name is needed is that some fundamentally *new* kind of step has been taken—a sort of irregularity has been encountered. Thus a new name must be supplied ad hoc.

There Is No Recursive Rule for Naming Ordinals

Now offhand you might think that these irregularities in the progression from *ordinal* to *ordinal* (as these names of infinity are called) could be handled by a computer program. That is, there would be a program to produce new names in a regular way, and when it ran out of gas, it would invoke the "irregularity handler", which would supply a new name, and pass control back to the simple one. But this will not work. It turns out that the irregularities themselves happen in irregular ways, and one would need also a second-order program—that is, a program which makes new programs which make new names. And even this is not enough. Eventually, a third-order program becomes necessary. And so on, and so on.

All of this perhaps ridiculous-seeming complexity stems from a deep theorem, due to Alonzo Church and Stephen C. Kleene, about the structure of these "infinite ordinals", which says:

> There is no recursively related notation-system
> which gives a name to every constructive ordinal.

What "recursively related notation-systems" are, and what "constructive ordinals" are, we must leave to the more technical sources, such as Hartley Rogers' book, to explain. But the intuitive idea has been presented. As the ordinals get bigger and bigger, there are irregularities, and irregularities in the irregularities, and irregularities in the irregularities in the irregularities, etc. No single scheme, no matter how complex, can name all the ordinals. And from this, it follows that no algorithmic method can tell how to apply the method of Gödel to all possible kinds of formal systems. And unless one is rather mystically inclined, therefore one must conclude that any human being simply will reach the limits of his own ability to Gödelize at some point. From there on out, formal systems of that complexity, though admittedly incomplete for the Gödel reason, will have as much power as that human being.

Other Refutations of Lucas

Now this is only one way to argue against Lucas' position. There are others, possibly more powerful, which we shall present later. But this counterargument has special interest because it brings up the fascinating concept of trying to create a computer program which can get outside of itself, see itself completely from the outside, and apply the Gödel zapping-trick to itself. Of course this is just as impossible as for a record player to be able to play records which would cause it to break.

But—one should not consider TNT defective for that reason. If there is a defect anywhere, it is not in TNT, but in our expectations of what it should be able to do. Furthermore, it is helpful to realize that *we* are equally vulnerable to the word trick which Gödel transplanted into mathematical formalisms: the Epimenides paradox. This was quite cleverly pointed out by C. H. Whitely, when he proposed the sentence "Lucas cannot consistently assert this sentence." If you think about it, you will see that (1) it is true, and yet (2) Lucas cannot consistently assert it. So Lucas is also "incomplete" with respect to truths about the world. The way in which he mirrors the world in his brain structures prevents him from simultaneously being "consistent" and asserting that true sentence. But Lucas is no more vulnerable than any of us. He is just on a par with a sophisticated formal system.

An amusing way to see the incorrectness of Lucas' argument is to translate it into a battle between men and women . . . In his wanderings, Loocus the Thinker one day comes across an unknown object—a woman. Such a thing he has never seen before, and at first he is wondrous thrilled at her likeness to himself; but then, slightly scared of her as well, he cries to all the men about him, "Behold! I can look upon her face, which is something *she* cannot do—therefore women can never be like me!" And thus he proves man's superiority over women, much to his relief, and that of his male companions. Incidentally, the same argument proves that Loocus is superior to all other males, as well—but he doesn't point that out to them. The woman argues back: "Yes, you can see my face, which is something I can't do—but I can see *your* face, which is something *you* can't do! We're even." However, Loocus comes up with an unexpected counter: "I'm sorry, you're deluded if you think you can *see* my face. What you women do is not the same as what we men do—it is, as I have already pointed out, of an inferior caliber, and does not deserve to be called by the same name. You may call it 'womanseeing'. Now the fact that you can 'womansee' my face is of no import, because the situation is not symmetric. You see?" "I womansee," womanreplies the woman, and womanwalks away . . .

Well, this is the kind of "heads-in-the-sand" argument which you have to be willing to stomach if you are bent on seeing men and women running ahead of computers in these intellectual battles.

Self-Transcendence — A Modern Myth

It is still of great interest to ponder whether we humans ever can jump out of ourselves—or whether computer programs can jump out of themselves. Certainly it is possible for a program to modify itself—but such modifiability has to be inherent in the program to start with, so that cannot be counted as an example of "jumping out of the system". No matter how a program twists and turns to get out of itself, it is still following the rules inherent in itself. It is no more possible for it to escape than it is for a human being to decide voluntarily not to obey the laws of physics. Physics is an overriding system, from which there can be no escape. However, there is a lesser ambition which it is possible to achieve: that is, one can certainly jump from a subsystem of one's brain into a wider subsystem. One can step out of ruts on occasion. This is still due to the interaction of various subsystems of one's brain, but it can feel very much like stepping entirely out of oneself. Similarly, it is entirely conceivable that a partial ability to "step outside of itself" could be embodied in a computer program.

However, it is important to see the distinction between *perceiving* oneself, and *transcending* oneself. You can gain visions of yourself in all sorts of ways—in a mirror, in photos or movies, on tape, through the descriptions of others, by getting psychoanalyzed, and so on. But you cannot quite break out of your own skin and be on the outside of yourself (modern occult movements, pop psychology fads, etc. notwithstanding). TNT can talk about itself, but it cannot jump out of itself. A computer program can modify itself but it cannot violate its own instructions—it can at best change some parts of itself by *obeying* its own instructions. This is reminiscent of the humorous paradoxical question, "Can God make a stone so heavy that he can't lift it?"

Advertisement and Framing Devices

This drive to jump out of the system is a pervasive one, and lies behind all progress in art, music, and other human endeavors. It also lies behind such trivial undertakings as the making of radio and television commercials. This insidious trend has been beautifully perceived and described by Erving Goffman in his book *Frame Analysis:*

> For example, an obviously professional actor completes a commercial pitch and, with the camera still on him, turns in obvious relief from his task, now to take real pleasure in consuming the product he had been advertising.
>
> This is, of course, but one example of the way in which TV and radio commercials are coming to exploit framing devices to give an appearance of naturalness that (it is hoped) will override the reserve auditors have developed. Thus, use is currently being made of children's voices, presumably because these seem unschooled; street noises, and other effects to give the impression of interviews with unpaid respondents; false starts, filled pauses, byplays, and overlapping speech to simulate actual conversation; and, following Welles, the interception of a firm's jingle commercials to give news of its new product, alternating occasionally with interception by a public interest spot, this presumably keeping the faith of the auditor alive.
>
> The more that auditors withdraw to minor expressive details as a test of genuineness, the more that advertisers chase after them. What results is a sort of interaction pollution, a disorder that is also spread by the public relations consultants of political figures, and, more modestly, by micro-sociology.[6]

Here we have yet another example of an escalating "TC-battle"—the antagonists this time being Truth and Commercials.

Simplicio, Salviati, Sagredo: Why Three?

There is a fascinating connection between the problem of jumping out of the system and the quest for complete objectivity. When I read Jauch's four Dialogues in *Are Quanta Real?* based on Galileo's four *Dialogues Concerning Two New Sciences,* I found myself wondering why there were *three* characters participating: Simplicio, Salviati, and Sagredo. Why wouldn't two have sufficed: Simplicio, the educated simpleton, and Salviati, the knowledgeable thinker? What function does Sagredo have? Well, he is supposed to be a sort of neutral third party, dispassionately weighing the two sides and coming out with a "fair" and "impartial" judgment. It sounds very balanced, and yet there is a problem: Sagredo is always agreeing with Salviati, not with Simplicio. How come Objectivity Personified is playing favorites? One answer, of course, is that Salviati is enunciating correct views, so Sagredo has no choice. But what, then, of fairness or "equal time"?

By adding Sagredo, Galileo (and Jauch) stacked the deck *more* against Simplicio, rather than less. Perhaps there should be added a yet higher-level Sagredo—someone who will be objective about this whole situation . . . You can see where it is going. We are getting into a never-ending series of "escalations in objectivity", which have the curious property of never getting any more objective than at the first level: where Salviati is simply *right,* and Simplicio *wrong*. So the puzzle remains: why add Sagredo at all? And the answer is, it gives the illusion of stepping out of the system, in some intuitively appealing sense.

Zen and "Stepping Out"

In Zen, too, we can see this preoccupation with the concept of transcending the system. For instance, the kōan in which Tōzan tells his monks that "the higher Buddhism is not Buddha". Perhaps, self-transcendence is even the central theme of Zen. A Zen person is always trying to understand more deeply what he is, by stepping more and more out of what he sees himself to be, by breaking every rule and convention which he perceives himself to be chained by—needless to say, including those of Zen itself. Somewhere along this elusive path may come enlightenment. In any case (as I see it), the hope is that by gradually deepening one's self-awareness, by gradually widening the scope of "the system", one will in the end come to a feeling of being at one with the entire universe.

Edifying Thoughts of a Tobacco Smoker

Achilles has been invited to the Crab's home.

Achilles: I see you have made a few additions since I was last here, Mr. Crab. Your new paintings are especially striking.

Crab: Thank you. I am quite fond of certain painters—especially René Magritte. Most of the paintings I have are by him. He's my favorite artist.

Achilles: They are very intriguing images, I must say. In some ways, these paintings by Magritte remind me of works by MY favorite artist, M. C. Escher.

Crab: I can see that. Both Magritte and Escher use great realism in exploring the worlds of paradox and illusion; both have a sure sense for the evocative power of certain visual symbols, and—something which even their admirers often fail to point out—both of them have a sense of the graceful line.

Achilles: Nevertheless, there is something quite different about them. I wonder how one could characterize that difference.

Crab: It would be fascinating to compare the two in detail.

Achilles: I must say, Magritte's command of realism is astonishing. For instance, I was quite taken in by that painting over there of a tree with a giant pipe behind it.

FIGURE 77. The Shadows, *by René Magritte (1966).*

Crab: You mean, a normal pipe with a tiny tree in front of it!

Achilles: Oh, is that what it is? Well, in any case, when I first spotted it, I was convinced I was smelling pipe smoke! Can you imagine how silly I felt?

Crab: I quite understand. My guests are often taken in by that one.

(So saying, he reaches up, removes the pipe from behind the tree in the painting, turns it over and taps it against the table, and the room begins to reek of pipe tobacco. He begins packing in a new wad of tobacco.)

This is a fine old pipe, Achilles. Believe it or not, the bowl has a copper lining, which makes it age wonderfully.

Achilles: A copper lining! You don't say!

Crab (pulls out a box of matches, and lights his pipe): Would you care for a smoke, Achilles?

Achilles: No, thank you. I only smoke cigars now and then.

Crab: No problem! I have one right here! *(Reaches out towards another Magritte painting, featuring a bicycle mounted upon a lit cigar.)*

Achilles: Uhh—no thank you, not now.

Crab: As you will. I myself am an incurable tobacco smoker. Which reminds me—you undoubtedly know of Old Bach's predilection for pipe smoking?

Achilles: I don't recall exactly.

Crab: Old Bach was fond of versifying, philosophizing, pipe smoking, and

FIGURE 78. State of Grace, *by René Magritte (1959).*

music making (not necessarily in that order). He combined all four into a droll poem which he set to music. It can be found in the famous musical notebook he kept for his wife, Anna Magdalena, and it is called

Edifying Thoughts of a Tobacco Smoker[1]

Whene'er I take my pipe and stuff it
And smoke to pass the time away,
My thoughts, as I sit there and puff it,
Dwell on a picture sad and gray:
It teaches me that very like
Am I myself unto my pipe.

Like me, this pipe so fragrant burning
Is made of naught but earth and clay;
To earth I too shall be returning.
It falls and, ere I'd think to say,
It breaks in two before my eyes;
In store for me a like fate lies.

No stain the pipe's hue yet doth darken;
It remains white. Thus do I know
That when to death's call I must harken
My body, too, all pale will grow.
To black beneath the sod 'twill turn,
Likewise the pipe, if oft it burn.

Or when the pipe is fairly glowing,
Behold then, instantaneously,
The smoke off into thin air going,
Till naught but ash is left to see.
Man's fame likewise away will burn
And unto dust his body turn.

How oft it happens when one's smoking:
The stopper's missing from its shelf,
And one goes with one's finger poking
Into the bowl and burns oneself.
If in the pipe such pain doth dwell,
How hot must be the pains of hell.

Thus o'er my pipe, in contemplation
Of such things, I can constantly
Indulge in fruitful meditation,
And so, puffing contentedly,
On land, on sea, at home, abroad,
I smoke my pipe and worship God.

A charming philosophy, is it not?

Achilles: Indeed. Old Bach was a turner of phrases quite pleasin'.

Crab: You took the very words from my mouth. You know, in my time I have tried to write clever verses. But I fear mine don't measure up to much. I don't have such a way with words.

Achilles: Oh, come now, Mr. Crab. You have—how to put it?—quite a penchant for trick'ry and teasin'. I'd be honored if you'd sing me one of your songs, Mr. C.

Crab: I'm most flattered. How about if I play you a record of myself singing one of my efforts? I don't remember when it dates from. Its title is "A Song Without Time or Season".

Achilles: How poetic!

(The Crab pulls a record from his shelves, and walks over to a huge, complex piece of apparatus. He opens it up, and inserts the record into an ominous-looking mechanical mouth. Suddenly a bright flash of greenish light sweeps over the surface of the record, and after a moment, the record is silently whisked into some hidden belly of the fantastic machine. A moment passes, and then the strains of the Crab's voice ring out.)

A turner of phrases quite pleasin',
Had a penchant for trick'ry and teasin'.
  In his songs, the last line
  Might seem sans design;
What I mean is, without why or wherefore.

Achilles: Lovely! Only, I'm puzzled by one thing. It seems to me that in your song, the last line is—

Crab: Sans design?

Achilles: No . . . What I mean is, without rhyme or reason.

Crab: You could be right.

Achilles: Other than that, it's a very nice song, but I must say I am even more intrigued by this monstrously complex contraption. Is it merely an oversized record player?

Crab: Oh, no, it's much more than that. This is my Tortoise-chomping record player.

Achilles: Good grief!

Crab: Well, I don't mean that it chomps up Tortoises. But it chomps up records produced by Mr. Tortoise.

Achilles: Whew! That's a little milder. Is this part of that weird musical battle that evolved between you and Mr. T some time ago?

Crab: In a way. Let me explain a little more fully. You see, Mr. Tortoise's sophistication had reached the point where he seemed to be able to destroy almost any record player I would obtain.

Achilles: But when I last heard about your rivalry, it seemed to me you had at last come into possession of an invincible phonograph—one with a built-in TV camera, minicomputer and so on, which could take itself apart and rebuild itself in such a way that it would not be destroyed.

Crab: Alack and alas! My plan was foiled. For Mr. Tortoise took advantage of one small detail which I had overlooked: the subunit which directed the disassembly and reassembly processes was itself stable during the entire process. That is, for obvious reasons, it could not take itself apart and rebuild itself, so it stayed intact.

Achilles: Yes, but what consequences did that have?

Crab: Oh, the direst ones! For you see, Mr. T focused his method down onto that subunit entirely.

Achilles: How is that?

Crab: He simply made a record which would induce fatal vibrations in the one structure he knew would never change—the disassembly-reassembly subunit.

Achilles: Oh, I see . . . Very sneaky.

Crab: Yes, so I thought, too. And his strategy worked. Not the first time, mind you. I thought I had outwitted him when my phonograph survived his first onslaught. I laughed gleefully. But the next time, he returned with a steely glint in his eye, and I knew he meant business. I placed his new record on my turntable. Then, both of us eagerly watched the computer-directed subunit carefully scan the grooves, then dismount the record, disassemble the record player, reassemble it in an astonishingly different way, remount the record—and then slowly lower the needle into the outermost groove.

Achilles: Golly!

Crab: No sooner had the first strains of sound issued forth than a loud SMASH! filled the room. The whole thing fell apart, but particularly badly destroyed was the assembler-disassembler. In that painful instant I finally realized, to my chagrin, that the Tortoise would ALWAYS be able to focus down upon—if you'll pardon the phrase—the Achilles' heel of the system.

Achilles: Upon my soul! You must have felt devastated.

Crab: Yes, I felt rather forlorn for a while. But, happily, that was not the end of the story. There is a sequel to the tale, which taught me a valuable lesson, which I may pass on to you. On the Tortoise's recommendation, I was browsing through a curious book filled with strange Dialogues about many subjects, including molecular biology, fugues, Zen Buddhism, and heaven knows what else.

Achilles: Probably some crackpot wrote it. What is the book called?

Crab: If I recall correctly, it was called *Copper, Silver, Gold: an Indestructible Metallic Alloy.*

Achilles: Oh, Mr. Tortoise told me about it, too. It's by a friend of his, who, it appears, is quite taken with metal-logic.

Crab: I wonder which friend it is . . . Anyway, in one of the Dialogues, I encountered some Edifying Thoughts on the Tobacco Mosaic Virus, ribosomes, and other strange things I had never heard of.

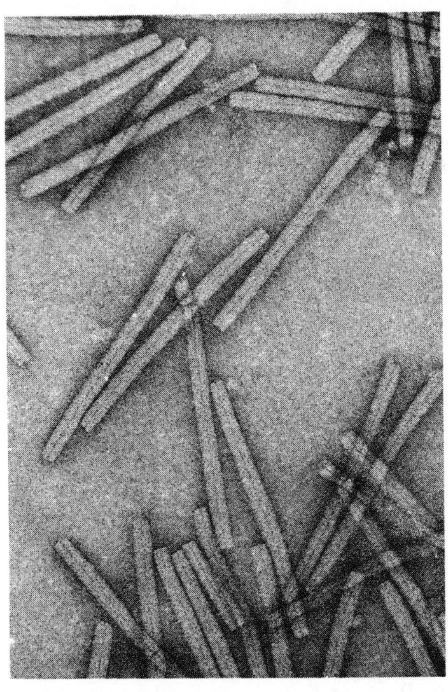

FIGURE 79. Tobacco Mosaic Virus. [*From A. Lehninger,* Biochemistry *(New York: Worth Publishers, 1976).*]

Achilles: What is the Tobacco Mosaic Virus? What are ribosomes?

Crab: I can't quite say, for I'm a total dunce when it comes to biology. All I know is what I gathered from that Dialogue. There, it said that Tobacco Mosaic Viruses are tiny cigarette-like objects that cause a disease in tobacco plants.

Achilles: Cancer?

Crab: No, not exactly, but—

Achilles: What next? A tobacco plant smoking, and getting cancer! Serves it right!

Crab: I believe you've jumped to a hasty conclusion, Achilles. Tobacco plants don't SMOKE these "cigarettes". The nasty little "cigarettes" just come and attack them, uninvited.

Achilles: I see. Well, now that I know all about Tobacco Mosaic Viruses, tell me what a ribosome is.

Crab: Ribosomes are apparently some sort of subcellular entities which take a message in one form and convert it into a message in another form.

Achilles: Something like a teeny tape recorder or phonograph?

Crab: Metaphorically, I suppose so. Now the thing which caught my eye was a line where this one exceedingly droll character mentions the fact that ribosomes—as well as Tobacco Mosaic Viruses and certain other bizarre biological structures—possess "the baffling ability to spontaneously self-assemble". Those were his exact words.

Achilles: That was one of his droller lines, I take it.

Crab: That's just what the other character in the Dialogue thought. But that's a preposterous interpretation of the statement. *(The Crab draws deeply from his pipe, and puffs several billows of smoke into the air.)*

Achilles: Well, what does "spontaneous self-assembly" mean, then?

Crab: The idea is that when some biological units inside a cell are taken apart, they can spontaneously reassemble themselves—without being directed by any other unit. The pieces just come together, and presto!—they stick.

Achilles: That sounds like magic. Wouldn't it be wonderful if a full-sized record player could have that property? I mean, if a miniature "record player" such as a ribosome can do it, why not a big one? That would allow you to create an indestructible phonograph, right? Any time it was broken, it would just put itself together again.

Crab: Exactly my thought. I breathlessly rushed a letter off to my manufacturer explaining the concept of self-assembly, and asked him if he could build me a record player which could take itself apart and spontaneously self-assemble in another form.

Achilles: A hefty bill to fill.

Crab: True; but after several months, he wrote to me that he had succeeded, at long last—and indeed he sent me quite a hefty bill. One fine day, ho! My Grand Self-assembling Record Player arrived in the mail, and it was with great confidence that I telephoned Mr. Tortoise, and invited him over for the purpose of testing my ultimate record player.

Achilles: So this magnificent object before us must be the very machine of which you speak.

Crab: I'm afraid not, Achilles.

Achilles: Don't tell me that once again . . .

Crab: What you suspect, my dear friend, is unfortunately the case. I don't pretend to understand the reasons why. The whole thing is too painful to recount. To see all those springs and wires chaotically strewn about on the floor, and puffs of smoke here and there—oh, me . . .

Achilles: There, there, Mr. Crab, don't take it too badly.

Crab: I'm quite all right; I just have these spells every so often. Well, to go on, after Mr. Tortoise's initial gloating, he at last realized how sorrowful I was feeling, and took pity. He tried to comfort me by explaining that it couldn't be helped—it all had to do with somebody-or-other's "Theorem", but I couldn't follow a word of it. It sounded like "Turtle's Theorem".

Achilles: I wonder if it was that "Gödel's Theorem" which he spoke of once before to me . . . It has a rather sinister ring to it.

Crab: It could be. I don't recall.

Achilles: I can assure you, Mr. Crab, that I have followed this tale with the utmost empathy for your position. It is truly sad. But, you mentioned that there was a silver lining. Pray tell, what was that?

Crab: Oh, yes—the silver lining. Well, eventually, I abandoned my quest after "Perfection" in phonographs, and decided that I might do better to tighten up my defenses against the Tortoise's records. I concluded that a more modest aim than a record player which can play anything is simply a record player that can SURVIVE: one that will avoid getting destroyed—even if that means that it can only play a few particular records.

Achilles: So you decided you would develop sophisticated anti-Tortoise mechanisms at the sacrifice of being able to reproduce every possible sound, eh?

Crab: Well . . . I wouldn't exactly say I "decided" it. More accurate would be to say that I was FORCED into that position.

Achilles: Yes, I can see what you mean.

Crab: My new idea was to prevent all "alien" records from being played on my phonograph. I knew my own records are harmless, and so if I prevented anyone else from infiltrating THEIR records, that would protect my record player, and still allow me to enjoy my recorded music.

Achilles: An excellent strategy for your new goal. Now does this giant thing before us represent your accomplishments to date along those lines?

Crab: That it does. Mr. Tortoise, of course, has realized that he must change HIS strategy, as well. His main goal is now to devise a record which can slip past my censors—a new type of challenge.

Achilles: For your part, how are you planning to keep his and other "alien" records out?

Crab: You promise you won't reveal my strategy to Mr. T, now?

Achilles: Tortoise's honor.

Crab: What!?

Achilles: Oh—it's just a phrase I've picked up from Mr. T. Don't worry—I swear your secret will remain secret with me.

Crab: All right, then. My basic plan is to use a LABELING technique. To each and every one of my records will be attached a secret label. Now the phonograph before you contains, as did its predecessors, a television camera for scanning the records, and a computer for processing the data obtained in the scan and controlling subsequent operations. My idea is simply to chomp all records which do not bear the proper label!

Achilles: Ah, sweet revenge! But it seems to me that your plan will be easy to foil. All Mr. T needs to do is to get a hold of one of your records, and copy its label!

Crab: Not so simple, Achilles. What makes you think he will be able to tell the label from the rest of the record? It may be better integrated than you suspect.

Achilles: Do you mean that it could be mixed up somehow with the actual music?

Crab: Precisely. But there is a way to disentangle the two. It requires sucking the data off the record visually, and then—

Achilles: Is that what that bright green flash was for?

Crab: That's right. That was the TV camera scanning the grooves. The groove-patterns were sent to the minicomputer, which analyzed the musical style of the piece I had put on—all in silence. Nothing had been played yet.

Achilles: Then is there a screening process, which eliminates pieces which aren't in the proper styles?

Crab: You've got it, Achilles. The only records which can pass this second test are records of pieces in my own style—and it will be hopelessly difficult for Mr. T to imitate that. So you see, I am convinced I will win this new musical battle. However, I should mention that Mr. T is equally convinced that somehow, he will manage to slip a record past my censors.

Achilles: And smash your marvelous machine to smithereens?

Crab: Oh, no—he has proved his point on that. Now he just wants to prove to me that he can slip a record—an innocuous one—by me, no matter what measures I take to prevent it. He keeps on muttering things about songs with strange titles, such as "I Can Be Played on Record Player X". But he can't scare ME! The only thing that worries me a little is that, as before, he seems to have some murky arguments which . . . which . . . *(He trails off into silence. Then, looking quite pensive, he takes a few puffs on his pipe.)*

Achilles: Hmm . . . I'd say Mr. Tortoise has an impossible task on his hands. He's met his match, at long last!

Crab: Curious that you should think so . . . I don't suppose that you know Henkin's Theorem forwards and backwards, do you?

Achilles: Know WHOSE Theorem forwards and backwards? I've never heard of anything that sounds like that. I'm sure it's fascinating, but I'd rather hear more about "music to infiltrate phonographs by". It's an amusing little story. Actually, I guess I can fill in the end. Obviously, Mr. T will find out that there is no point in going on, and so he will sheepishly admit defeat, and that will be that. Isn't that exactly it?

Crab: That's what I'm hoping, at least. Would you like to see a little bit of the inner workings of my defensive phonograph?

Achilles: Gladly. I've always wanted to see a working television camera.

Crab: No sooner said than done, my friend. *(Reaches into the gaping "mouth" of the large phonograph, undoes a couple of snaps, and pulls out a neatly packaged instrument.)* You see, the whole thing is built of independent modules, which can be detached and used independently. This TV camera, for instance, works very well by itself. Watch the screen over there, beneath the painting with the flaming tuba. *(He points the camera at Achilles, whose face instantly appears on the large screen.)*

Achilles: Terrific! May I try it out?

Crab: Certainly.

Achilles (pointing the camera at the Crab): There YOU are, Mr. Crab, on the screen.

FIGURE 80. The Fair Captive, *by René Magritte (1947).*

Crab: So I am.

Achilles: Suppose I point the camera at the painting with the burning tuba. Now it is on the screen, too!

Crab: The camera can zoom in and out, Achilles. You ought to try it.

Achilles: Fabulous! Let me just focus down onto the tip of those flames, where they meet the picture frame . . . It's such a funny feeling to be able to instantaneously "copy" anything in the room—anything I want—onto that screen. I merely need to point the camera at it, and it pops like magic onto the screen.

Crab: ANYTHING in the room, Achilles?

Achilles: Anything in sight, yes. That's obvious.

Crab: What happens, then, if you point the camera at the flames on the TV screen?

(Achilles shifts the camera so that it points directly at that part of the television screen on which the flames are—or were—displayed.)

Achilles: Hey, that's funny! That very act makes the flames DISAPPEAR from the screen! Where did they go?

Crab: You can't keep an image still on the screen and move the camera at the same time.

Achilles: So I see . . . But I don't understand what's on the screen now—not at all! It seems to be a strange long corridor. Yet I'm certainly not

(a) The simplest case.

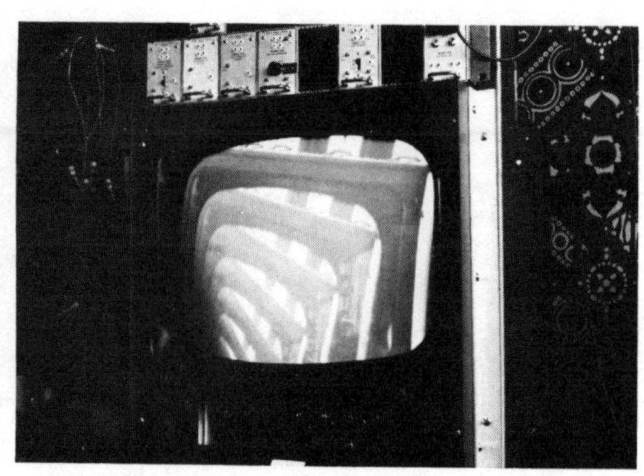

(d) A "failed self-engulfing".

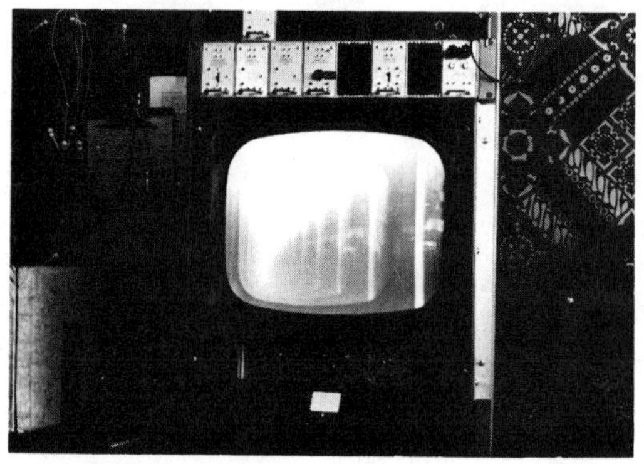

(b) Achilles' "corridor".

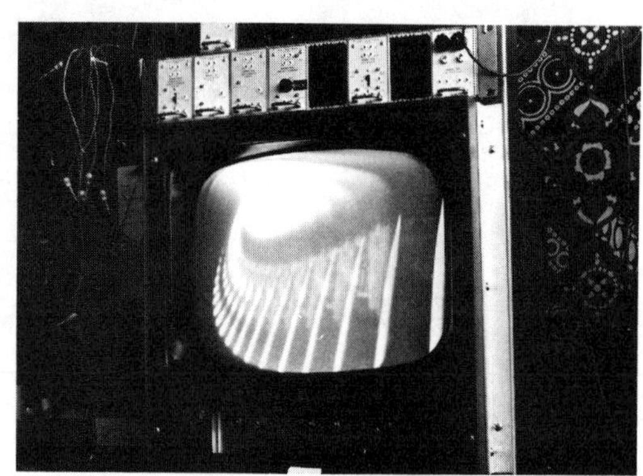

(e) What happens when you zoom in.

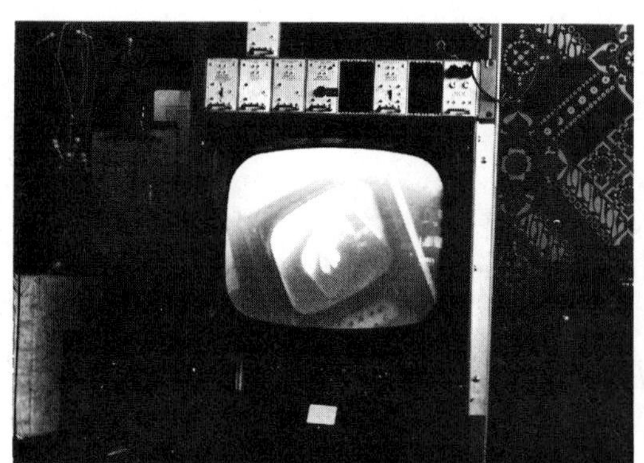

(c) What happens when you rotate the camera.

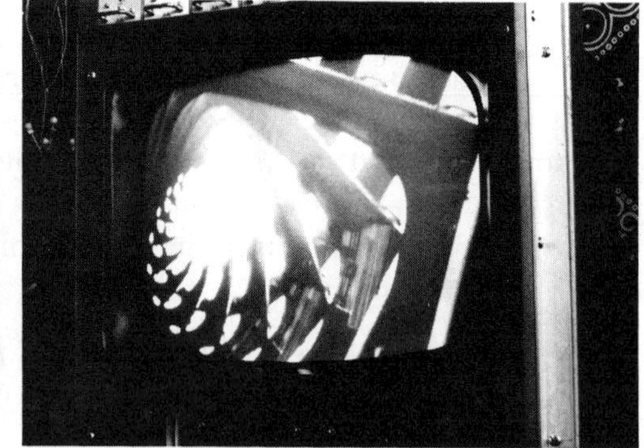

(f) Combined effect of rotation and zooming.

FIGURE 81. Twelve self-engulfing TV screens. I would have included one more, had 13 not been prime.

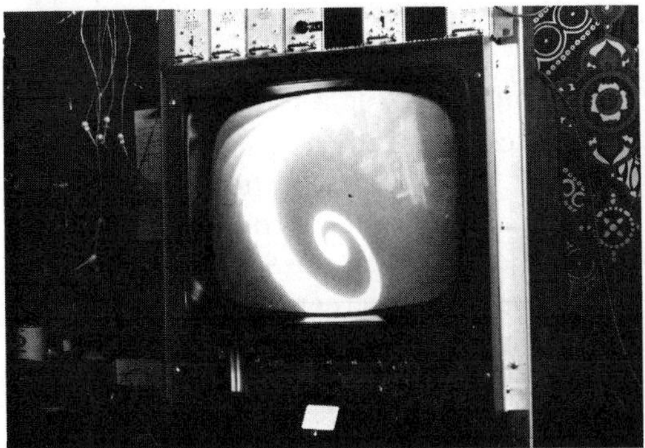

(g) Starting to get weird . . .

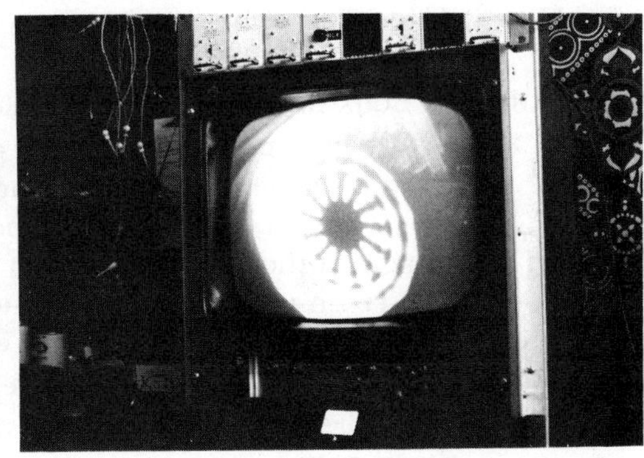

(j) The late stages of a galaxy. Count the number of spokes!

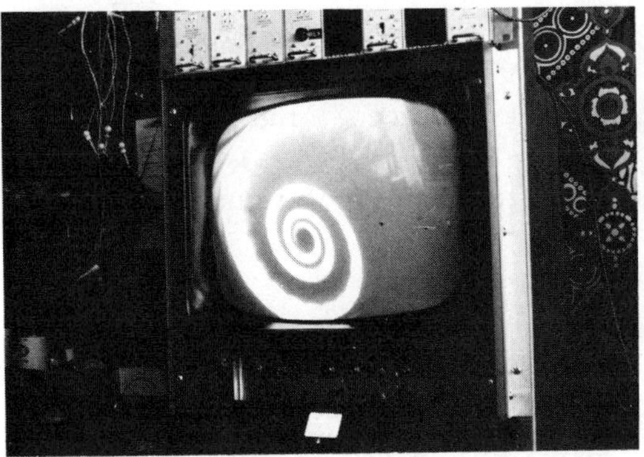

(h) A "galaxy" is born.

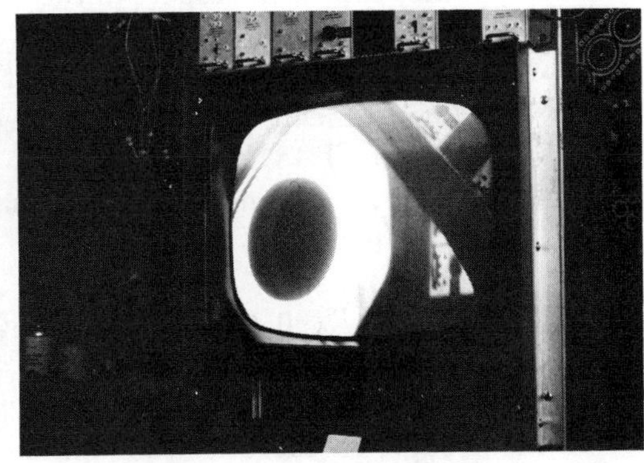

(k) The galaxy has burned itself out, and become—a black hole!

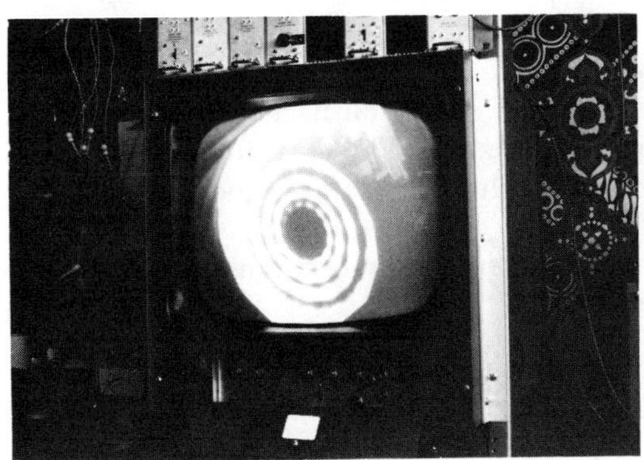

(i) The galaxy evolves . . .

(l) A "pulsating petal pattern", caught in the middle of one of its pulsations.

pointing the camera down any corridor. I'm merely pointing it at an ordinary TV screen.

Crab: Look more carefully, Achilles. Do you really see a corridor?

Achilles: Ahhh, now I see. It's a set of nested copies of the TV screen itself, getting smaller and smaller and smaller . . . Of course! The image of the flames HAD to go away, because it came from my pointing the camera at the PAINTING. When I point the camera at the SCREEN, then the screen itself appears, with whatever is on the screen at the time—which is the screen itself, with whatever is on the screen at the time—which is the screen itself, with—

Crab: I believe I can fill in the rest, Achilles. Why don't you try rotating the camera?

Achilles: Oh! I get a beautiful spiraling corridor! Each screen is rotated inside its framing screen, so that the littler they get, the more rotated they are, with respect to the outermost screen. This idea of having a TV screen "engulf itself" is weird.

Crab: What do you mean by "self-engulfing", Achilles?

Achilles: I mean, when I point the camera at the screen—or at part of the screen. THAT'S self-engulfing.

Crab: Do you mind if I pursue that a little further? I'm intrigued by this new notion.

Achilles: So am I.

Crab: Very well, then. If you point the camera at a CORNER of the screen, is that still what you mean by "self-engulfing"?

Achilles: Let me try it. Hmm—the "corridor" of screens seems to go off the edge, so there isn't an infinite nesting any more. It's pretty, but it doesn't seem to me to have the spirit of self-engulfing. It's a "failed self-engulfing".

Crab: If you were to swing the TV camera back towards the center of the screen, maybe you could fix it up again . . .

Achilles (slowly and cautiously turning the camera): Yes! The corridor is getting longer and longer . . . There it is! Now it's all back. I can look down it so far that it vanishes in the distance. The corridor became infinite again precisely at the moment when the camera took in the WHOLE screen. Hmm—that reminds me of something Mr. Tortoise was saying a while back, about self-reference only occurring when a sentence talks about ALL of itself . . .

Crab: Pardon me?

Achilles: Oh, nothing—just muttering to myself.

(As Achilles plays with the lens and other controls on the camera, a profusion of new kinds of self-engulfing images appear: swirling spirals that resemble galaxies, kaleidoscopic flower-like shapes, and other assorted patterns . . .)

Crab: You seem to be having a grand time.

Achilles (turns away from the camera): I'll say! What a wealth of images this simple idea can produce! *(He glances back at the screen, and a look of astonishment crosses his face.)* Good grief, Mr. Crab! There's a pulsating petal-pattern on the screen! Where do the pulsations come from? The TV is still, and so is the camera.

Crab: You can occasionally set up patterns which change in time. This is because there is a slight delay in the circuitry between the moment the camera "sees" something, and the moment it appears on the screen—around a hundredth of a second. So if you have a nesting of depth fifty or so, roughly a half-second delay will result. If somehow a moving image gets onto the screen—for example, by you putting your finger in front of the camera—then it takes a while for the more deeply nested screens to "find out" about it. This delay then reverberates through the whole system, like a visual echo. And if things are set up so the echo doesn't die away, then you can get pulsating patterns.

Achilles: Amazing! Say—what if we tried to make a TOTAL self-engulfing?

Crab: What precisely do you mean by that?

Achilles: Well, it seems to me that this stuff with screens within screens is interesting, but I'd like to get a picture of the TV camera AND the screen, ON the screen. Only then would I really have made the system engulf itself. For the screen is only PART of the total system.

Crab: I see what you mean. Perhaps with this mirror, you can achieve the effect you want.

(The Crab hands him a mirror, and Achilles maneuvers the mirror and camera in such a way that the camera and the screen are both pictured on the screen.)

Achilles: There! I've created a TOTAL self-engulfing!

Crab: It seems to me you only have the front of the mirror—what about its back? If it weren't for the back of the mirror, it wouldn't be reflective—and you wouldn't have the camera in the picture.

Achilles: You're right. But to show both the front and back of this mirror, I need a second mirror.

Crab: But then you'll need to show the back of that mirror, too. And what about including the back of the television, as well as its front? And then there's the electric cord, and the inside of the television, and—

Achilles: Whoa, whoa! My head's beginning to spin! I can see that this "total self-engulfing project" is going to pose a wee bit of a problem. I'm feeling a little dizzy.

Crab: I know exactly how you feel. Why don't you sit down here and take your mind off all this self-engulfing? Relax! Look at my paintings, and you'll calm down.

(Achilles lies down, and sighs.)

Oh—perhaps my pipe smoke is bothering you? Here, I'll put my pipe away. *(Takes the pipe from his mouth, and carefully places it above some written words in another Magritte painting.)* There! Feeling any better?

Achilles: I'm still a little woozy. *(Points at the Magritte.)* That's an interesting painting. I like the way it's framed, especially the shiny inlay inside the wooden frame.

FIGURE 82. The Air and the Song, *by René Magritte (1964).*

Crab: Thank you. I had it specially done—it's a gold lining.

Achilles: A gold lining? What next? What are those words below the pipe? They aren't in English, are they?

Crab: No, they are in French. They say, *"Ceci n'est pas une pipe."* That means, "This is not a pipe". Which is perfectly true.

Achilles: But it IS a pipe! You were just smoking it!

Crab: Oh, you misunderstand the phrase, I believe. The word *"ceci"* refers to the painting, not to the pipe. Of course the pipe is a pipe. But a painting is not a pipe.

Achilles: I wonder if that *"ceci"* inside the painting refers to the WHOLE painting, or just to the pipe inside the painting. Oh, my gracious! That would be ANOTHER self-engulfing! I'm not feeling at all well, Mr. Crab. I think I'm going to be sick . . .

CHAPTER XVI

Self-Ref and Self-Rep

IN THIS CHAPTER, we will look at some of the mechanisms which create self-reference in various contexts, and compare them to the mechanisms which allow some kinds of systems to reproduce themselves. Some remarkable and beautiful parallels between these mechanisms will come to light.

Implicitly and Explicitly Self-Referential Sentences

To begin with, let us look at sentences which, at first glance, may seem to provide the simplest examples of self-reference. Some such sentences are these:

(1) This sentence contains five words.
(2) This sentence is meaningless because it is self-referential.
(3) This sentence no verb.
(4) This sentence is false. (Epimenides paradox)
(5) The sentence I am now writing is the sentence you are now reading.

All but the last one (which is an anomaly) involve the simple-seeming mechanism contained in the phrase "this sentence". But that mechanism is in reality far from simple. All of these sentences are "floating" in the context of the English language. They can be compared to icebergs, whose tips only are visible. The word sequences are the tips of the icebergs, and the processing which must be done to understand them is the hidden part. In this sense their meaning is implicit, not explicit. Of course, no sentence's meaning is completely explicit, but the more explicit the self-reference is, the more exposed will be the mechanisms underlying it. In this case, for the self-reference of the sentences above to be recognized, not only has one to be comfortable with a language such as English which can deal with linguistic subject matter, but also one has to be able to figure out the referent of the phrase "this sentence". It seems simple, but it depends on our very complex yet totally assimilated ability to handle English. What is especially important here is the ability to figure out the referent of a noun phrase with a demonstrative adjective in it. This ability is built up slowly, and should by no means be considered trivial. The difficulty is perhaps underlined when a sentence such as number 4 is presented to someone naïve about paradoxes and linguistic tricks, such as a child. They may say, "*What* sentence is false?" and it may take a bit of persistence to get across the idea that the sentence is talking about itself. The whole idea is a little mind-boggling at first. A couple of pictures may help (Figs. 83, 84). Figure 83 is a picture which can be interpreted on two levels. On one level, it is a sentence pointing at itself; on the other level, it is a picture of Epimenides executing his own death sentence.

FIGURE 83.

Figure 84, showing visible and invisible portions of the iceberg, suggests the relative proportion of sentence to processing required for the recognition of self-reference:

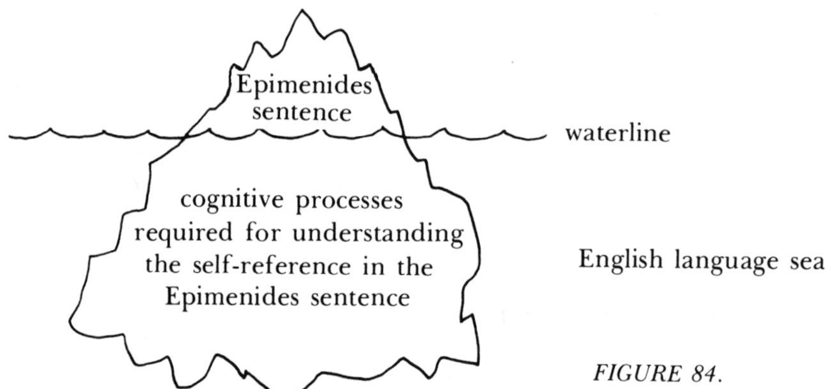

FIGURE 84.

It is amusing to try to create a self-referring sentence without using the trick of saying "this sentence". One could try to quote a sentence inside itself. Here is an attempt:

The sentence "The sentence contains five words" contains five words.

But such an attempt must fail, for any sentence that could be quoted entirely inside itself would have to be shorter than itself. This is actually possible, but only if you are willing to entertain infinitely long sentences, such as:

The sentence
"The sentence
"The sentence
"The sentence

.
.
.
etc., etc.
.
.
.

is infinitely long"
is infinitely long"
is infinitely long"
is infinitely long.

But this cannot work for finite sentences. For the same reason, Gödel's string G could not contain the explicit numeral for its Gödel number: it would not fit. No string of TNT can contain the TNT-numeral for its own Gödel number, for that numeral always contains more symbols than the string itself does. But you can get around this by having G contain a *description* of its own Gödel number, by means of the notions of "sub" and "arithmoquinification".

One way of achieving self-reference in an English sentence by means of description instead of by self-quoting or using the phrase "this sentence" is the Quine method, illustrated in the dialogue *Air on G's String*. The understanding of the Quine sentence requires less subtle mental processing than the four examples cited earlier. Although it may appear at first to be trickier, it is in some ways more explicit. The Quine construction is quite like the Gödel construction, in the way that it creates self-reference by describing another typographical entity which, as it turns out, is isomorphic to the Quine sentence itself. The description of the new typographical entity is carried out by two parts of the Quine sentence. One part is a set of *instructions* telling how to build a certain phrase, while the other part contains the construction materials to be used; that is, the other part is a *template*. This resembles a floating cake of soap more than it resembles an iceberg (See Fig. 85).

FIGURE 85.

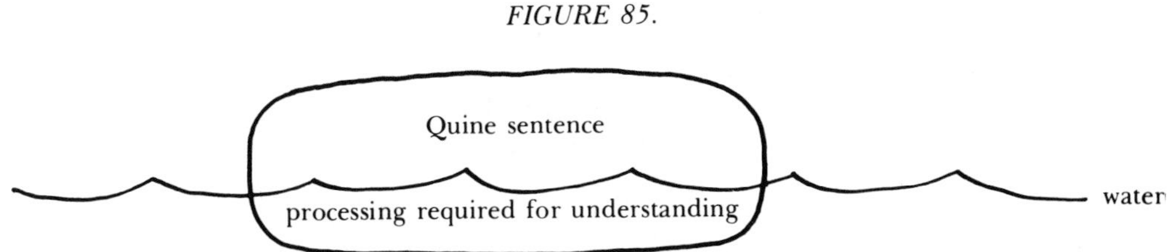

The self-reference of this sentence is achieved in a more direct way than in the Epimenides paradox; less hidden processing is needed. By the way, it is interesting to point out that the phrase "this sentence" appears in the previous sentence; yet it is not there to cause self-reference; you probably understood that its referent was the Quine sentence, rather than the sentence in which it occurs. This just goes to show how pointer phrases such as "this sentence" are interpreted according to context, and helps to show that the processing of such phrases is indeed quite involved.

A Self-Reproducing Program

The notion of quining, and its usage in creating self-reference, have already been explained inside the Dialogue itself, so we need not dwell on such matters here. Let us instead show how a computer program can use precisely the same technique to reproduce itself. The following self-reproducing program is written in a BlooP-like language and is based on *following* a phrase by its own quotation (the opposite order from quining, so I reverse the name "quine" to make "eniuq"):

```
DEFINE PROCEDURE ''ENIUQ'' [TEMPLATE]: PRINT [TEMPLATE, LEFT-BRACKET,
QUOTE-MARK, TEMPLATE, QUOTE-MARK, RIGHT-BRACKET, PERIOD].
ENIUQ
  ['DEFINE PROCEDURE ''ENIUQ'' [TEMPLATE]: PRINT [TEMPLATE, LEFT-BRACKET,
   QUOTE-MARK, TEMPLATE, QUOTE-MARK, RIGHT-BRACKET, PERIOD].
   ENIUQ'].
```

ENIUQ is a procedure defined in the first two lines, and its input is called "TEMPLATE". It is understood that when the procedure is called, TEMPLATE's value will be some string of typographical characters. The effect of ENIUQ is to carry out a printing operation, in which TEMPLATE gets printed twice: the first time just plain; the second time wrapped in (single) quotes and brackets, and garnished with a final period. Thus, if TEMPLATE's value were the string DOUBLE-BUBBLE, then performing ENIUQ on it would yield:

DOUBLE-BUBBLE ['DOUBLE-BUBBLE'].

Now in the last four lines of the program above, the procedure ENIUQ is called with a specific value of TEMPLATE—namely the long string inside the single quotes: DEFINE . . . ENIUQ. That value has been carefully chosen; it consists of the *definition* of ENIUQ, followed by the *word* ENIUQ. This makes the program itself—or, if you prefer, a perfect copy of it—get printed out. It is very similar to Quine's version of the Epimenides sentence:

"yields falsehood when preceded by its quotation"
yields falsehood when preceded by its quotation.

It is very important to realize that the character string which appears in quotes in the last three lines of the program above—that is, the value of

TEMPLATE—is never interpreted as a sequence of instructions. That it happens to be one is, in a sense, just an accident. As was pointed out above, it could just as well have been DOUBLE-BUBBLE or any other string of characters. The beauty of the scheme is that when the same string appears in the top two lines of this program, it *is* treated as a program (because it is not in quotes). Thus in this program, one string functions in two ways: first as program, and second as data. This is the secret of self-reproducing programs, and, as we shall see, of self-reproducing molecules. It is useful, incidentally, to call any kind of self-reproducing object or entity a *self-rep;* and likewise to call any self-referring object or entity a *self-ref*. I will use those terms occasionally from here on.

The preceding program is an elegant example of a self-reproducing program written in a language which was not designed to make the writing of self-reps particularly easy. Thus, the task had to be carried out using those notions and operations which were assumed to be part of the language—such as the word QUOTE-MARK, and the command PRINT. But suppose a language were designed expressly for making self-reps easy to write. Then one could write much shorter self-reps. For example, suppose that the operation of eniuq-ing were a built-in feature of the language, needing no explicit definition (as we assumed PRINT was). Then a teeny self-rep would be this:

ENIUQ ['ENIUQ'].

It is very similar to the Tortoise's version of Quine's version of the Epimenides self-ref, where the verb "to quine" is assumed to be known:

"yields falsehood when quined" yields falsehood when quined.

But self-reps can be even shorter. For instance, in some computer language it might be a convention that any program whose first symbol is an asterisk is to be copied before being executed normally. Then the program consisting of merely one asterisk is a self-rep! You may complain that this is silly and depends on a totally arbitrary convention. In doing so, you are echoing my earlier point that it is almost cheating to use the phrase "this sentence" to achieve self-reference—it relies too much on the processor, and not enough on explicit directions for self-reference. Using an asterisk as an example of a self-rep is like using the word "I" as an example of a self-ref: both conceal all the interesting aspects of their respective problems.

This is reminiscent of another curious type of self-reproduction: via photocopy machine. It might be claimed that any written document is a self-rep because it can cause a copy of itself to be printed when it is placed in a photocopy machine and the appropriate button is pushed. But somehow this violates our notion of self-reproduction; the piece of paper is not consulted at all, and is therefore not directing its own reproduction. Again, everything is in the processor. Before we call something a self-rep, we want to have the feeling that, to the maximum extent possible, it *explicitly* contains the directions for copying itself.

To be sure, explicitness is a matter of degree; nonetheless there is an intuitive borderline on one side of which we perceive true self-directed self-reproduction, and on the other side of which we merely see copying being carried out by an inflexible and autonomous copying machine.

What Is a Copy?

Now in any discussion of self-refs and self-reps, one must sooner or later come to grips with the essential issue: what is a copy? We already dealt with that question quite seriously in Chapters V and VI; and now we come back to it. To give the flavor of the issue, let us describe some highly fanciful, yet plausible, examples of self-reps.

A Self-Reproducing Song

Imagine that there is a nickelodeon in the local bar which, if you press buttons 11-U, will play a song whose lyrics go this way:

Put another nickel in, in the nickelodeon,
All I want is 11-U, and music, music, music.

We could make a little diagram of what happens one evening (Fig. 86).

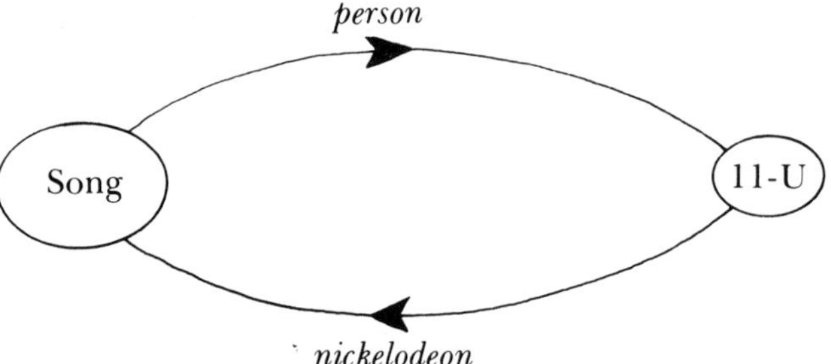

FIGURE 86. A self-reproducing song.

Although the effect is that the song reproduces itself, it would feel strange to call the song a self-rep, because of the fact that when it passes through the 11-U stage, not all of the information is there. The information only gets put back by virtue of the fact that it is fully stored in the nickelodeon—that is, in one of the *arrows* in the diagram, not in one of the ovals. It is questionable whether this song contains a complete description of how to get itself played again, because the symbol pair "11-U" is only a trigger, not a copy.

A "Crab" Program

Consider next a computer program which prints itself out backwards. (Some readers might enjoy thinking about how to write such a program in the BlooP-like language above, using the given self-rep as a model.) Would this funny program count as a self-rep? Yes, in a way, because a trivial transformation performed on its output will restore the original program. It seems fair to say that the output contains the same information as the program itself, just recast in a simple way. Yet it is clear that someone might look at the output and not recognize it as a program printed backwards. To recall terminology from Chapter VI, we could say that the "inner messages" of the output and the program itself are the same, but they have different "outer messages"—that is, they must be read by using different decoding mechanisms. Now if one counts the outer message as part of the information—which seems quite reasonable—then the total information is not the same after all, so the program can't be counted as a self-rep.

However, this is a disquieting conclusion, because we are accustomed to considering something and its mirror image as containing the same information. But recall that in Chapter VI, we made the concept of "intrinsic meaning" dependent on a hypothesized universal notion of intelligence. The idea was that, in determining the intrinsic meaning of an object, we could disregard some types of outer message—those which would be universally understood. That is, if the decoding mechanism seems fundamental enough, in some still ill-defined sense, then the inner message which it lets be revealed is the only meaning that counts. In this example, it seems reasonably safe to guess that a "standard intelligence" would consider two mirror images to contain the same information as each other; that is, it would consider the isomorphism between the two to be so trivial as to be ignorable. And thus our intuition that the program is in some sense a fair self-rep, is allowed to stand.

Epimenides Straddles the Channel

Now another far-fetched example of a self-rep would be a program which prints itself out, but translated into a different computer language. One might liken this to the following curious version of the Quine version of the Epimenides self-ref:

> "est une expression qui, quand elle est précédée de sa traduction, mise entre guillemets, dans la langue provenant de l'autre côté de la Manche, crée une fausseté" is an expression which, when it is preceded by its translation, placed in quotation marks, into the language originating on the other side of the Channel, yields a falsehood.

You might try to write down the sentence which is described by this weird concoction. (Hint: It is not itself—or at least it is not if "itself" is taken in a naïve sense.) If the notion of "self-rep by retrograde motion" (i.e., a program which writes itself out backwards) is reminiscent of a crab canon, the notion of "self-rep by translation" is no less reminiscent of a canon which involves a transposition of the theme into another key.

A Program That Prints Out Its Own Gödel Number

The idea of printing out a translation instead of an exact copy of the original program may seem pointless. However, if you wanted to write a self-rep program in BlooP or FlooP, you would have to resort to some such device, for in those languages, **OUTPUT** is always a *number,* rather than a typographical string. Therefore, you would have to make the program print out its own Gödel number: a very huge integer whose decimal expansion codes for the program, character by character, by using three-digit codons. The program is coming as close as it can to printing itself, within the means available to it: it prints out a copy of itself in another "space", and it is easy to switch back and forth between the space of integers and the space of strings. Thus, the value of **OUTPUT** is not a mere trigger, like "11-U". Instead, all the information of the original program lies "close to the surface" of the output.

Gödelian Self-Reference

This comes very close to describing the mechanism of Gödel's self-ref G. After all, that string of TNT contains a description not of itself, but of an integer (the arithmoquinification of u). It just so happens that that integer is an exact "image" of the string G, in the space of natural numbers. Thus, G refers to a translation of itself into another space. We still feel comfortable in calling G a self-referential string, because the isomorphism between the two spaces is so tight that we can consider them to be identical.

This isomorphism that mirrors TNT inside the abstract realm of natural numbers can be likened to the quasi-isomorphism that mirrors the real world inside our brains, by means of symbols. The symbols play quasi-isomorphic roles to the objects, and it is thanks to them that we can think. Likewise, the Gödel numbers play isomorphic roles to strings, and it is thanks to them that we can find metamathematical meanings in statements about natural numbers. The amazing, nearly magical, thing about G is that it manages to achieve self-reference despite the fact that the language in which it is written, TNT, seems to offer no hope of referring to its own structures, unlike English, in which it is the easiest thing in the world to discuss the English language.

So G is an outstanding example of a self-ref via translation—hardly the most straightforward case. One might also think back to some of the Dialogues, for some of them, too, are self-refs via translation. For instance, take the *Sonata for Unaccompanied Achilles.* In that Dialogue, several references are made to the Bach Sonatas for unaccompanied violin, and the Tortoise's suggestion of imagining harpsichord accompaniments is particularly interesting. After all, if one applies this idea to the Dialogue itself, one invents lines which the Tortoise is saying; but if one assumes that Achilles' part stands alone (as does the violin), then it is quite wrong to attribute any lines at all to the Tortoise. In any case, here again is a self-ref by means of a mapping which maps Dialogues onto pieces by Bach. And this mapping is left, of course, for the reader to notice. Yet even if the reader does not notice it, the mapping is still there, and the Dialogue is still a self-ref.

A Self-Rep by Augmentation

We have been likening self-reps to canons. What, then, would be a fair analogue to a canon by augmentation? Here is a possibility: consider a program which contains a dummy loop whose only purpose is to slow up the program. A parameter might tell how often to repeat the loop. A self-rep could be made which prints out a copy of itself, but with the parameter changed, so that when that copy is run, it will run at half the speed of its parent program; and its "daughter" will in turn run at half again the speed, and so on . . . None of these programs prints itself out precisely; yet all clearly belong to a single "family".

This is reminiscent of the self-reproduction of living organisms. Clearly, an individual is never identical to either of its parents; why, then, is the act of making young called "self-reproduction"? The answer is that there is a coarse-grained isomorphism between parent and child; it is an isomorphism which preserves the information about *species*. Thus, what is reproduced is the *class,* rather than the *instance.* This is also the case in the recursive picture Gplot, in Chapter V: that is, the mapping between "magnetic butterflies" of various sizes and shapes is coarse-grained; no two are identical, but they all belong to a single "species", and the mapping preserves precisely that fact. In terms of self-replicating programs, this would correspond to a *family* of programs, all written in "dialects" of a single computer language; each one can write itself out, but slightly modified, so that it comes out in a dialect of its original language.

A Kimian Self-Rep

Perhaps the sneakiest example of a self-rep is the following: instead of writing a legal expression in the compiler language, you type one of the compiler's own error messages. When the compiler looks at your "program", the first thing it does is get confused, because your "program" is ungrammatical; hence the compiler prints out an error message. All you need to do is arrange that the one it prints out will be the one you typed in. This kind of self-rep, suggested to me by Scott Kim, exploits a different level of the system from the one you would normally approach. Although it may seem frivolous, it may have counterparts in complex systems where self-reps vie against each other for survival, as we shall soon discuss.

What Is the Original?

Besides the question "What constitutes a copy?", there is another fundamental philosophical question concerning self-reps. That is the obverse side of the coin: "What is the original?" This can best be explained by referring to some examples:

(1) a program which, when interpreted by some interpreter running on some computer, prints itself out;

(2) a program which, when interpreted by some interpreter running on some computer, prints itself out along with a complete copy of the interpreter (which, after all, is also a program);

(3) a program which, when interpreted by some interpreter running on some computer, not only prints itself out along with a complete copy of the interpreter, but also directs a mechanical assembly process in which a second computer, identical to the one on which the interpreter and program are running, is put together.

It is clear that in (1), the program is the self-rep. But in (3), is it the program which is the self-rep, or the compound system of program plus interpreter, or the union of program, interpreter, and processor?

Clearly, a self-rep can involve more than just printing itself out. In fact, most of the rest of this Chapter is a discussion of self-reps in which data, program, interpreter, and processor are all extremely intertwined, and in which self-replication involves replicating all of them at once.

Typogenetics

We are now about to broach one of the most fascinating and profound topics of the twentieth century: the study of "the molecular logic of the living state", to borrow Albert Lehninger's richly evocative phrase. And logic it is, too—but of a sort more complex and beautiful than any a human mind ever imagined. We will come at it from a slightly novel angle: via an artificial solitaire game which I call *Typogenetics*—short for "Typographical Genetics". In Typogenetics I have tried to capture some ideas of molecular genetics in a typographical system which, on first sight, resembles very much the formal systems exemplified by the MIU-system. Of course, Typogenetics involves many simplifications, and therefore is useful primarily for didactic purposes.

I should explain immediately that the field of molecular biology is a field in which phenomena on several levels interact, and that Typogenetics is only trying to illustrate phenomena from one or two levels. In particular, purely chemical aspects have been completely avoided—they belong to a level lower than is here dealt with; similarly, all aspects of classical genetics (viz., nonmolecular genetics) have also been avoided—they belong to a level higher than is here dealt with. I have intended in Typogenetics only to give an intuition for those processes centered on the celebrated *Central Dogma of*

Molecular Biology, enunciated by Francis Crick (one of the co-discoverers of the double-helix structure of DNA):

$$\text{DNA} \Rightarrow \text{RNA} \Rightarrow \text{proteins.}$$

It is my hope that with this very skeletal model I have constructed the reader will perceive some simple unifying principles of the field—principles which might otherwise be obscured by the enormously intricate interplay of phenomena at many different levels. What is sacrificed is, of course, strict accuracy; what is gained is, I hope, a little insight.

Strands, Bases, Enzymes

The game of Typogenetics involves typographical manipulation on sequences of letters. There are four letters involved:

A C G T.

Arbitrary sequences of them are called *strands*. Thus, some strands are:

GGGG
ATTACCA
CATCATCATCAT

Incidentally, "STRAND" spelled backwards begins with "DNA". This is appropriate since strands, in Typogenetics, play the role of pieces of DNA (which, in real genetics, are often called "strands"). Not only this, but "STRAND" fully spelled out backwards is "DNA RTS", which may be taken as an acronym for "DNA Rapid Transit Service". This, too, is appropriate, for the function of "messenger RNA"—which in Typogenetics is represented by strands as well—is quite well characterized by the phrase "Rapid Transit Service" for DNA, as we shall see later.

I will sometimes refer to the letters A, C, G, T as *bases,* and to the positions which they occupy as *units*. Thus, in the middle strand, there are seven units, in the fourth of which is found the base A.

If you have a strand, you can operate on it and change it in various ways. You can also produce additional strands, either by copying, or by cutting a strand in two. Some operations lengthen strands, some shorten them, and some leave their length alone.

Operations come in packets—that is, several to be performed together, in order. Such a packet of operations is a little like a programmed machine which moves up and down the strand doing things to it. These mobile machines are called "typographical enzymes"—*enzymes* for short. Enzymes operate on strands one unit at a time, and are said to be "bound" to the unit they are operating on at any given moment.

I will show how some sample enzymes act on particular strings. The first thing to know is that each enzyme likes to start out bound to a particular letter. Thus, there are four kinds of enzyme—those which prefer

A, those which prefer C, etc. Given the sequence of operations which an enzyme performs, you can figure out which letter it prefers, but for now I'll just give them without explanation. Here's a sample enzyme, consisting of three operations:

(1) Delete the unit to which the enzyme is bound (and then bind to the next unit to the right).
(2) Move one unit to the right.
(3) Insert a T (to the immediate right of this unit).

This enzyme happens to like to bind to A initially. And here's a sample strand:

ACA

What happens if our enzyme binds to the left A and begins acting? Step 1 deletes the A, so we are left with CA—and the enzyme is now bound to the C. Step 2 slides the enzyme rightwards, to the A, and Step 3 appends a T onto the end to form the strand CAT. And the enzyme has done its complete duty: it has transformed ACA into CAT.

What if it had bound itself to the *right* A of ACA? It would have deleted that A and moved off the end of the strand. Whenever this happens, the enzyme quits (this is a general principle). So the entire effect would just be to lop off one symbol.

Let's see some more examples. Here is another enzyme:

(1) Search for the nearest pyrimidine to the right of this unit.
(2) Go into Copy mode.
(3) Search for the nearest purine to the right of of this unit.
(4) Cut the strand here (viz., to the right of the present unit).

Now this contains the terms "pyrimidine" and "purine". They are easy terms. A and G are called *purines,* and C and T are called *pyrimidines.* So searching for a pyrimidine merely means searching for the nearest C or T.

Copy Mode and Double Strands

The other new term is *Copy mode.* Any strand can be "copied" onto another strand, but in a funny way. Instead of copying A onto A, you copy it onto T, and vice versa. And instead of copying C onto C, you copy it onto G, and vice versa. Note that a purine copies onto a pyrimidine, and vice versa. This is called *complementary base pairing.* The complements are shown below:

	complement	
purines	A ⇐⇒ T	*pyrimidines*
	G ⇐⇒ C	

You can perhaps remember this molecular pairing scheme by recalling that Achilles is paired with the Tortoise, and the Crab with his Genes.

When "copying" a strand, therefore, you don't actually copy it, but you manufacture its *complementary* strand. And this one will be written upside down above the original strand. Let's see this in concrete terms. Let the previous enzyme act on the following strand (and that enzyme also happens to like to start at A):

CAAAGAGAATCCTCTTTGAT

There are many places it could start. Let's take the second A, for example. The enzyme binds to it, then executes step 1: Search for the nearest pyrimidine to the right. Well, this means a C or a T. The first one is a T somewhere near the middle of the strand, so that's where we go. Now step 2: Copy mode. Well, we just put an upside-down A above our T. But that's not all, for Copy mode *remains in effect* until it is shut off—or until the enzyme is done, whichever comes first. This means that every base which is passed through by the enzyme while Copy mode is on will get a complementary base put above it. Step 3 says to look for a purine to the right of our T. That is the G two symbols in from the right-hand end. Now as we move up to that G, we must "copy"—that is, create a complementary strand. Here's what that gives:

```
         ∀⅁⅁∀⅁∀∀∀Ɔ
CAAAGAGAATCCTCTTTGAT
```

The last step is to *cut* the strand. This will yield two pieces:

```
         ∀⅁⅁∀⅁∀∀∀Ɔ
CAAAGAGAATCCTCTTTG
```

and AT.

And the instruction packet is done. We are left with a double strand, however. Whenever this happens, we separate the two complementary strands from each other (general principle); so in fact our end product is a set of three strands:

AT, CAAAGAGGA, and CAAAGAGAATCCTCTTTG.

Notice that the upside-down strand has been turned right side up, and thereby right and left have been reversed.

Now you have seen most of the typographical operations which can be carried out on strands. There are two other instructions which should be mentioned. One shuts *off* Copy mode; the other *switches* the enzyme from a strand to the upside-down strand above it. When this happens, if you keep the paper right side up, then you must switch "left" and "right" in all the instructions. Or better, you can keep the wording and just turn the paper around so the top strand becomes legible. If the "switch" command is given, but there is no complementary base where the enzyme is bound at that instant, then the enzyme just detaches itself from the strand, and its job is done.

It should be mentioned that when a "cut" instruction is encountered, this pertains to *both* strands (if there are two); however, "delete" pertains only to the strand on which the enzyme is working. If Copy mode is *on*, then the "insert" command pertains to both strands—the base itself into the strand the enzyme is working on, and its complement into the other strand. If Copy mode is *off*, then the "insert" command pertains only to the one strand, so a blank space must be inserted into the complementary strand.

And, whenever Copy mode is *on*, "move" and "search" commands require that one manufacture complementary bases to all bases which the sliding enzyme touches. Incidentally, Copy mode is always *off* when an enzyme starts to work. If Copy mode is *off*, and the command "Shut off copy mode" is encountered, nothing happens. Likewise, if Copy mode is already *on*, and the command "Turn copy mode on" is encountered, then nothing happens.

Amino Acids

There are fifteen types of command, listed below:

cut	——	cut strand(s)
del	——	delete a base from strand
swi	——	switch enzyme to other strand
mvr	——	move one unit to the right
mvl	——	move one unit to the left
cop	——	turn on Copy mode
off	——	turn off Copy mode
ina	——	insert A to the right of this unit
inc	——	insert C to the right of this unit
ing	——	insert G to the right of this unit
int	——	insert T to the right of this unit
rpy	——	search for the nearest pyrimidine to the right
rpu	——	search for the nearest purine to the right
lpy	——	search for the nearest pyrimidine to the left
lpu	——	search for the nearest purine to the left

Each one has a three-letter abbreviation. We shall refer to the three-letter abbreviations of commands as *amino acids*. Thus, *every enzyme is made up of a sequence of amino acids*. Let us write down an arbitrary enzyme:

rpu – inc – cop – mvr – mvl – swi – lpu – int

and an arbitrary strand:

TAGATCCAGTCCATCGA

and see how the enzyme acts on the strand. It so happens that the enzyme binds to G only. Let us bind to the middle G and begin. Search rightwards for a purine (viz., A or G). We (the enzyme) skip over TCC and land on A. Insert a C. Now we have

TAGATCCAGTCCACTCGA
↑

where the arrow points to the unit to which the enzyme is bound. Set Copy mode. This puts an upside-down G above the C. Move right, move left, then switch to the other strand. Here's what we have so far:

↓
⅁∀
TAGATCCAGTCCACTCGA

Let's turn it upside down, so that the enzyme is attached to the lower strand:

∀⅁Ɔ⊥Ɔ∀ƆƆ⊥⅁∀ƆƆ⊥∀⅁∀⊥
AG
↑

Now we search leftwards for a purine, and find A. Copy mode is on, but the complementary bases are already there, so nothing is added. Finally, we insert a T (in Copy mode), and quit:

∀⅁Ɔ⊥∀Ɔ∀ƆƆ⊥⅁∀ƆƆ⊥∀⅁∀⊥
ATG
↑

Our final product is thus two strands:

ATG, and TAGATCCAGTCCACATCGA

The old one is of course gone.

Translation and the Typogenetic Code

Now you might be wondering where the enzymes and strands come from, and how to tell the initial binding-preference of a given enzyme. One way might be just to throw some random strands and some random enzymes together, and see what happens when those enzymes act on those strands and their progeny. This has a similar flavor to the MU-puzzle, where there were some given rules of inference and an axiom, and you just began. The only difference is that here, every time a strand is acted on, its original form is gone forever. In the MU-puzzle, acting on MI to make MIU didn't destroy MI.

But in Typogenetics, as in real genetics, the scheme is quite a bit trickier. We do begin with some arbitrary strand, somewhat like an axiom in a formal system. But we have, initially, no "rules of inference"—that is, no enzymes. However, we can *translate* each strand into one or more enzymes! Thus, the strands themselves will dictate the operations which will be performed upon them, and those operations will in turn produce new strands which will dictate further enzymes, etc. etc.! This is mixing levels with a vengeance! Think, for the sake of comparison, how different the MU-puzzle would have been if each new *theorem* produced could have been turned into a new *rule of inference* by means of some code.

How is this "translation" done? It involves a *Typogenetic Code* by which adjacent pairs of bases—called "duplets"—in a single strand represent different amino acids. There are sixteen possible duplets: AA, AC, AG, AT, CA, CC, etc. And there are fifteen amino acids. The Typogenetic Code is shown in Figure 87.

Second Base

First Base	A	C	G	T
A		cut *s*	del *s*	swi *r*
C	mvr *s*	mvl *s*	cop *r*	off *l*
G	ina *s*	inc *r*	ing *r*	int *l*
T	rpy *r*	rpu *l*	lpy *l*	lpu *l*

FIGURE 87. The Typogenetic Code, by which each duplet in a strand codes for one of fifteen "amino acids" (or a punctuation mark).

According to the table, the translation of the duplet GC is "inc" ("insert a C"); that of AT is "swi" ("switch strands"); and so on. Therefore it becomes clear that a strand can dictate an enzyme very straightforwardly. For example, the strand

TAGATCCAGTCCACATCGA

breaks up into duplets as follows:

TA GA TC CA GT CC AC AT CG A

with the A left over at the end. Its translation into an enzyme is:

rpy – ina – rpu – mvr – int – mvl – cut – swi – cop.

(Note that the leftover A contributes nothing.)

Tertiary Structure of Enzymes

What about the little letters '*s*', '*l*', and '*r*' in the lower righthand corner of each box? They are crucial in determining the enzyme's binding-preference, and in a peculiar way. In order to figure out what letter an enzyme likes to bind to, you have to figure out the enzyme's "tertiary structure", which is itself determined by the enzyme's "primary structure". By its *primary structure* is meant its amino acid sequence. By its *tertiary structure* is meant the way it likes to "fold up". The point is that enzymes don't like being in straight lines, as we have so far exhibited them. At each internal amino acid (all but the two ends), there is a possibility of a "kink", which is dictated by the letters in the corners. In particular, *'l'* and *'r'* stand for "left" and "right", and *'s'* stands for "straight". So let us take our most recent sample enzyme, and let it fold itself up to show its tertiary structure. We will start with the enzyme's primary structure, and move along it from left to right. At each amino acid whose corner-letter is *'l'* we'll put a left turn, for those with *'r'*, we'll put a right turn, and at *'s'* we'll put no turn. In Figure 88 is shown the two-dimensional conformation for our enzyme.

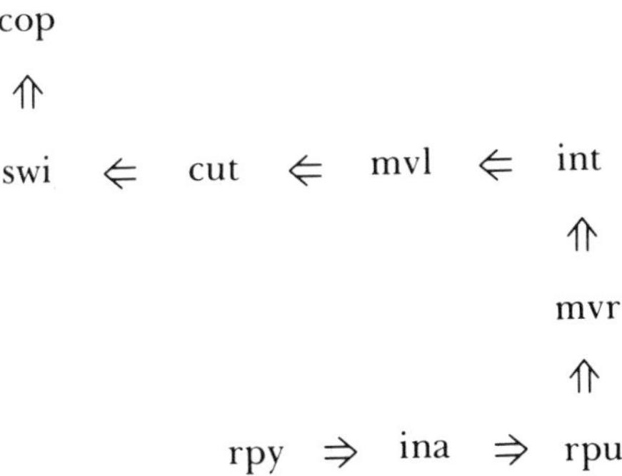

FIGURE 88. The tertiary structure of a typoenzyme.

Note the left-kink at "rpu", the right-kink at "swi", and so on. Notice also that the first segment ("rpy ⇒ ina") and the last segment ("swi ⇒ cop") are perpendicular. This is the key to the binding-preference. In fact, the *relative orientation of the first and last segments* of an enzyme's tertiary structure determines the binding-preference of the enzyme. We can always orient the enzyme so that its first segment points to the right. If we do so, then the last segment determines the binding-preference, as shown in Figure 89.

FIGURE 89. Table of binding-preferences for typoenzymes.

First Segment	*Last Segment*	*Binding-letter*
⇒	⇒	A
⇒	⇑	C
⇒	⇓	G
⇒	⇐	T

So in our case, we have an enzyme which likes the letter C. If, in folding up, an enzyme happens to cross itself, that's okay—just think of it as going under or over itself. Notice that *all* its amino acids play a role in the determination of an enzyme's tertiary structure.

Punctuation, Genes, and Ribosomes

Now one thing remains to be explained. Why is there a blank in box AA of the Typogenetic Code? The answer is that the duplet AA acts as a *punctuation mark* inside a strand, and it signals the end of the code for an enzyme. That is to say, one strand may code for two or more enzymes if it has one or more duplets AA in it. For example, the strand

CG GA TA CT AA AC CG A

codes for *two* enzymes:

cop – ina – rpy – off

and

cut – cop

with the AA serving to divide the strand up into two "genes". The definition of *gene* is: *that portion of a strand which codes for a single enzyme.* Note that the mere presence of AA inside a strand does not mean that the strand codes for two enzymes. For instance, CAAG codes for "mvr – del". The AA begins on an even-numbered unit and therefore is not read as a duplet!

The mechanism which reads strands and produces the enzymes which are coded inside them is called a *ribosome.* (In Typogenetics, the player of the game does the work of the ribosomes.) Ribosomes are not in any way responsible for the *tertiary* structure of enzymes, for that is entirely determined once the *primary* structure is created. Incidentally, the process of *translation* always goes *from strands to enzymes,* and never in the reverse direction.

Puzzle: A Typogenetical Self-Rep

Now that the rules of Typogenetics have been fully set out, you may find it interesting to experiment with the game. In particular, it would be most interesting to devise a self-replicating strand. This would mean something along the following lines. A single strand is written down. A ribosome acts on it, to produce any or all of the enzymes which are coded for in the strand. Then those enzymes are brought into contact with the original strand, and allowed to work on it. This yields a set of "daughter strands". The daughter strands themselves pass through the ribosomes, to yield a second generation of enzymes, which act on the daughter strands; and the cycle goes on and on. This can go on for any number of stages; the hope is that eventually, among the strands which are present at some point, there will be found two copies of the original strand (one of the copies may be, in fact, the original strand).

The Central Dogma of Typogenetics

Typogenetical processes can be represented in skeletal form in a diagram (Fig. 90).

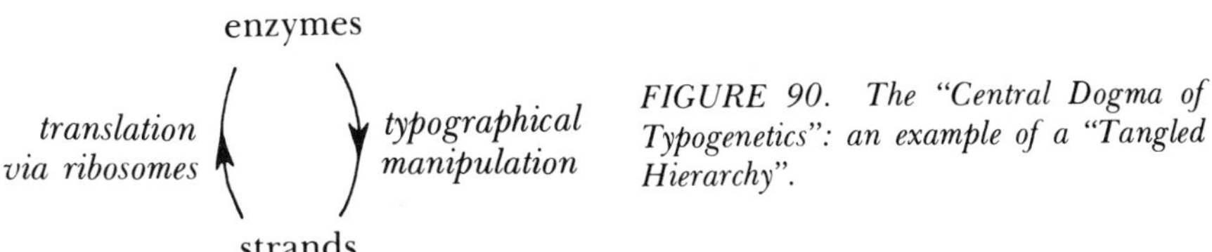

FIGURE 90. The "Central Dogma of Typogenetics": an example of a "Tangled Hierarchy".

This diagram illustrates the *Central Dogma of Typogenetics.* It shows how strands define enzymes (via the Typogenetic Code); and how in turn, enzymes act back on the strands which gave rise to them, yielding new strands. Therefore, the line on the left portrays how *old information flows upwards,* in the sense that an enzyme is a translation of a strand, and contains therefore the same information as the strand, only in a different form—in particular; in an active form. The line on the right, however, does not show information flowing downwards; instead, it shows how *new information gets created:* by the shunting of symbols in strands.

An enzyme in Typogenetics, like a rule of inference in a formal system, blindly shunts symbols in strands without regard to any "meaning" which may lurk in those symbols. So there is a curious mixture of levels here. On the one hand, strands are acted upon, and therefore play the role of *data* (as is indicated by the arrow on the right); on the other hand, they also dictate the actions which are to be performed on the data, and therefore they play the role of *programs* (as is indicated by the arrow on the left). It is the player of Typogenetics who acts as interpreter and processor, of course. The two-way street which links "upper" and "lower" levels of Typogenetics shows that, in fact, neither strands nor enzymes can be thought of as being on a higher level than the other. By contrast, a picture of the *Central Dogma of the MIU-system* looks this way:

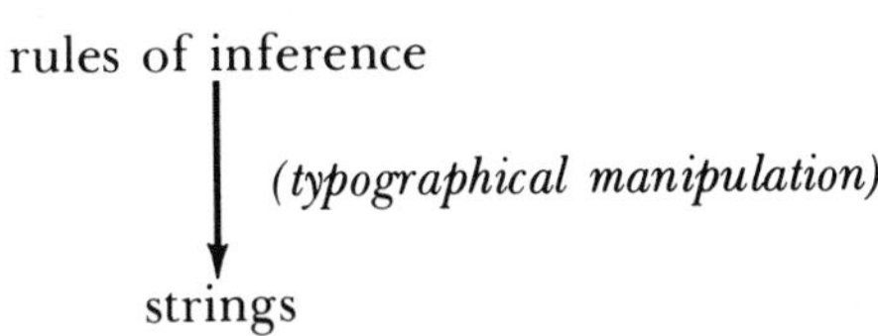

In the MIU-system, there *is* a clear distinction of levels: rules of inference simply belong to a higher level than strings. Similarly for TNT, and all formal systems.

However, we have seen that in TNT, levels *are* mixed, in another sense. In fact, the distinction between language and metalanguage breaks down: statements *about* the system get mirrored *inside* the system. It turns out that if we make a diagram showing the relationship between TNT and its metalanguage, we will produce something which resembles in a remarkable way the diagram which represents the Central Dogma of Molecular Biology. In fact, it is our goal to make this comparison in detail; but to do so, we need to indicate the places where Typogenetics and *true* genetics coincide, and where they differ. Of course, real genetics is far more complex than Typogenetics—but the "conceptual skeleton" which the reader has acquired in understanding Typogenetics will be very useful as a guide in the labyrinth of true genetics.

DNA and Nucleotides

We begin by discussing the relationship between "strands", and DNA. The initials "DNA" stand for "deoxyribonucleic acid". The DNA of most cells resides in the cell's *nucleus,* which is a small area protected by a membrane. Gunther Stent has characterized the nucleus as the "throne room" of the cell, with DNA acting as the ruler. DNA consists of long chains of relatively simple molecules called *nucleotides.* Each nucleotide is made up of three parts: (1) a phosphate group stripped of one special oxygen atom, whence the prefix "deoxy"; (2) a sugar called "ribose", and (3) a *base.* It is the base alone which distinguishes one nucleotide from another; thus it suffices to specify its base to identify a nucleotide. The four types of bases which occur in DNA nucleotides are:

A: adenine } *purines*
G: guanine }

C: cytosine } *pyrimidines*
T: thymine }

(Also see Fig. 91.) It is easy to remember which ones are pyrimidines because the first vowel in "cytosine", "thymine", and "pyrimidine" is 'y'. Later, when we talk about RNA, "uracil"—also a pyrimidine—will come in and wreck the pattern, unfortunately. (Note: Letters representing nucleotides in real genetics will not be in the Quadrata font, as they were in Typogenetics.)

A single strand of DNA thus consists of many nucleotides strung together like a chain of beads. The chemical bond which links a nucleotide to its two neighbors is very strong; such bonds are called *covalent bonds,* and the "chain of beads" is often called the *covalent backbone* of DNA.

Now DNA usually comes in double strands—that is, two single strands which are paired up, nucleotide by nucleotide (see Fig. 92). It is the bases

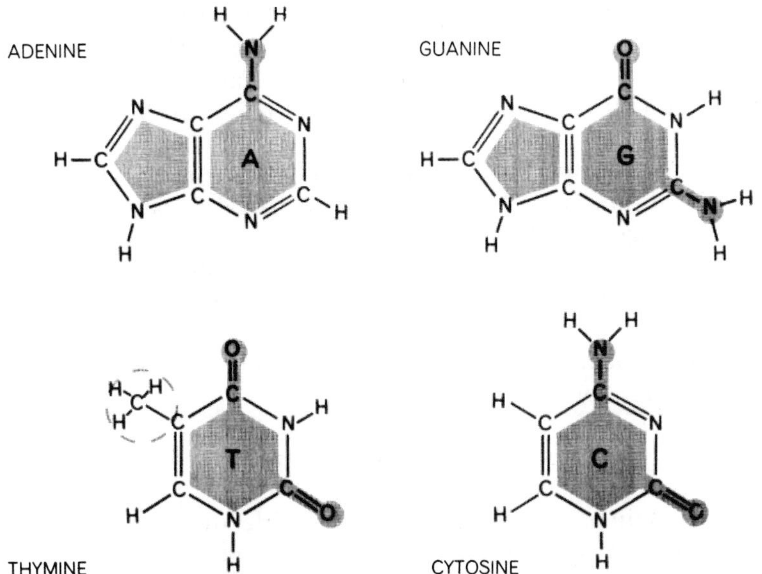

FIGURE 91. The four constituent bases of DNA: Adenine, Guanine, Thymine, Cytosine. [*From Hanawalt and Haynes,* The Chemical Basis of Life *(San Francisco: W. H. Freeman, 1973), p. 142.*]

FIGURE 92. DNA structure resembles a ladder in which the side pieces consist of alternating units of deoxyribose and phosphate. The rungs are formed by the bases paired in a special way, A with T and G with C, and held together respectively by two and three hydrogen bonds. [*From Hanawalt and Haynes,* The Chemical Basis of Life, *p. 142.*]

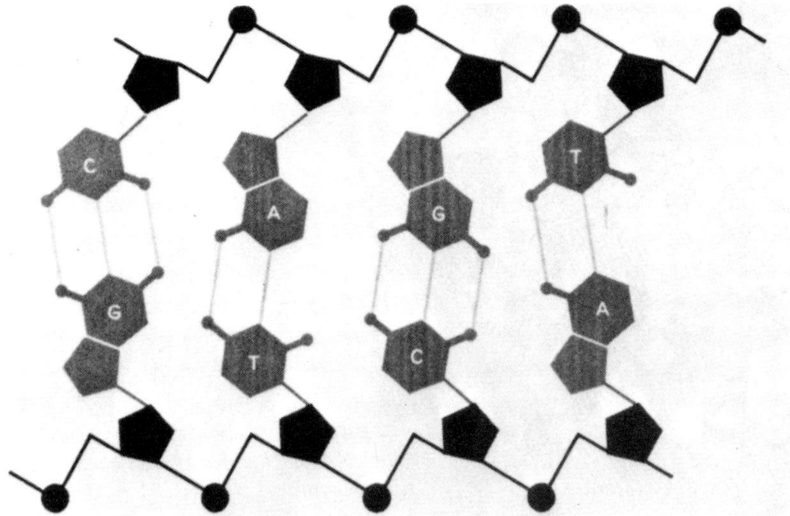

which are responsible for the peculiar kind of pairing which takes place between strands. Each base in one strand faces a complementary base in the other strand, and binds to it. The complements are as in Typogenetics: A pairs up with T, and C with G. Always one purine pairs up with a pyrimidine.

Compared to the strong covalent bonds along the backbone, the *interstrand* bonds are quite weak. They are not covalent bonds, but *hydrogen bonds.* A hydrogen bond arises when two molecular complexes are aligned in such a way that a hydrogen atom which originally belonged to one of them becomes "confused" about which one it belongs to, and it hovers between the two complexes, vacillating as to which one to join. Because the two halves of double-stranded DNA are held together only by hydrogen bonds, they may come apart or be put together relatively easily; and this fact is of great import for the workings of the cell.

When DNA forms double strands, the two strands curl around each other like twisting vines (Fig. 93). There are exactly ten nucleotide pairs per revolution; in other words, at each nucleotide, the "twist" is 36 degrees. Single-stranded DNA does not exhibit this kind of coiling, for it is a consequence of the base-pairing.

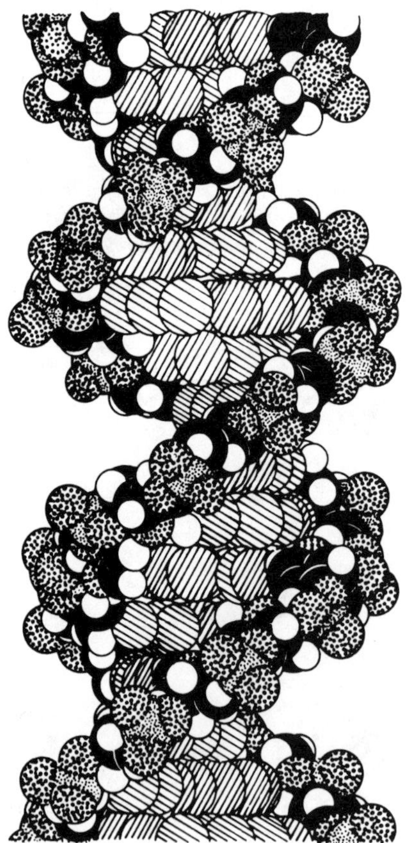

FIGURE 93. Molecular model of the DNA double helix. [From Vernon M. Ingram, Biosynthesis *(Menlo Park, Calif.: W. A. Benjamin, 1972), p. 13.]*

Messenger RNA and Ribosomes

As was mentioned above, in many cells, DNA, the ruler of the cell, dwells in its private "throne room": the nucleus of the cell. But most of the "living" in a cell goes on outside of the nucleus, namely in the *cytoplasm*—the "ground" to the nucleus' "figure". In particular, *enzymes,* which make practically every life process go, are manufactured by *ribosomes* in the cytoplasm, and they do most of their work in the cytoplasm. And just as in Typogenetics, the blueprints for all enzymes are stored inside the strands—that is, inside the DNA, which remains protected in its little nuclear home. So how does the information about enzyme structure get from the nucleus to the ribosomes?

Here is where *messenger RNA*—mRNA—comes in. Earlier, mRNA strands were humorously said to constitute a kind of DNA Rapid Transit Service; by this is meant not that mRNA physically carries DNA anywhere, but rather that it serves to carry the information, or message, stored in the DNA in its nuclear chambers, out to the ribosomes in the cytoplasm. How is this done? The idea is easy: a special kind of enzyme inside the nucleus faithfully copies long stretches of the DNA's base sequence onto a new strand—a strand of messenger RNA. This mRNA then departs from the nucleus and wanders out into the cytoplasm, where it runs into many ribosomes which begin doing their enzyme-creating work on it.

The process by which DNA gets copied onto mRNA inside the nucleus is called *transcription;* in it, the double-stranded DNA must be temporarily separated into two single strands, one of which serves as a template for the mRNA. Incidentally, "RNA" stands for "ribonucleic acid", and it is very much like DNA except that all of its nucleotides possess that special oxygen atom in the phosphate group which DNA's nucleotides lack. Therefore the "deoxy" prefix is dropped. Also, instead of thymine, RNA uses the base uracil, so the information in strands of RNA can be represented by arbitrary sequences of the four letters 'A', 'C', 'G', 'U'. Now when mRNA is transcribed off of DNA, the transcription process operates via the usual base-pairing (except with U instead of T), so that a DNA-template and its mRNA-mate might look something like this:

DNA: ........CGTAAATCAAGTCA........ (template)
mRNA: ........GCAUUUAGUUCAGU........ ("copy")

RNA does not generally form long double strands with itself, although it can. Therefore it is prevalently found not in the helical form which so characterizes DNA, but rather in long, somewhat randomly curving strands.

Once a strand of mRNA has escaped the nucleus, it encounters those strange subcellular creatures called "ribosomes"—but before we go on to explain how a ribosome uses mRNA, I want to make some comments about enzymes and proteins. Enzymes belong to the general category of biomolecules called *proteins,* and the job of ribosomes is to make all proteins, not just enzymes. Proteins which are not enzymes are much more passive kinds of beings; many of them, for instance, are *structural* molecules, which means that they are like girders and beams and so forth in buildings: they hold the cell's parts together. There are other kinds of proteins, but for our purposes, the principal proteins are enzymes, and I will henceforth not make a sharp distinction.

Amino Acids

Proteins are composed of sequences of *amino acids,* which come in twenty primary varieties, each with a three-letter abbreviation:

ala	——	alanine
arg	——	arginine
asn	——	asparagine
asp	——	aspartic acid
cys	——	cysteine
gln	——	glutamine
glu	——	glutamic acid
gly	——	glycine
his	——	histidine
ile	——	isoleucine
leu	——	leucine
lys	——	lysine
met	——	methionine
phe	——	phenylalanine
pro	——	proline
ser	——	serine
thr	——	threonine
trp	——	tryptophan
tyr	——	tyrosine
val	——	valine

Notice the slight numerical discrepancy with Typogenetics, where we had only fifteen "amino acids" composing enzymes. An amino acid is a small molecule of roughly the same complexity as a nucleotide; hence the building blocks of proteins and of nucleic acids (DNA, RNA) are roughly of the same size. However, proteins are composed of much shorter sequences of components: typically, about three hundred amino acids make a complete protein, whereas a strand of DNA can consist of hundreds of thousands or millions of nucleotides.

Ribosomes and Tape Recorders

Now when a strand of mRNA, after its escape into the cytoplasm, encounters a ribosome, a very intricate and beautiful process called *translation* takes place. It could be said that this process of translation is at the very heart of all of life, and there are many mysteries connected with it. But in essence it is easy to describe. Let us first give a picturesque image, and then render it more precise. Imagine the mRNA to be like a long piece of magnetic recording tape, and the ribosome to be like a tape recorder. As the tape passes through the playing head of the recorder, it is "read" and converted into music, or other sounds. Thus magnetic markings are "translated" into notes. Similarly, when a "tape" of mRNA passes through the "playing head" of a ribosome, the "notes" which are produced are *amino acids,* and the "pieces of music" which they make up are *proteins.* This is what translation is all about; it is shown in Figure 96.

The Genetic Code

But how can a ribosome produce a chain of amino acids when it is reading a chain of nucleotides? This mystery was solved in the early 1960's by the efforts of a large number of people, and at the core of the answer lies the *Genetic Code*—a mapping from triplets of nucleotides into amino acids (see Fig. 94). This is in spirit extremely similar to the Typogenetic Code, except that here, three consecutive bases (or nucleotides) form a *codon,* whereas there, only two were needed. Thus there must be 4×4×4 (equals 64) different entries in the table, instead of sixteen. A ribosome clicks down a strand of RNA three nucleotides at a time—which is to say, one codon at a time—and each time it does so, it appends a single new amino acid to the protein it is presently manufacturing. Thus, a protein comes out of the ribosome amino acid by amino acid.

CUA GAU
Cu Ag Au

A typical segment of mRNA read first as two triplets (above), and second as three duplets (below): an example of hemiolia in biochemistry.

Tertiary Structure

However, as a protein emerges from a ribosome, it is not only getting longer and longer, but it is also continually folding itself up into an extraordinary three-dimensional shape, very much in the way that those funny little Fourth-of-July fireworks called "snakes" simultaneously grow longer and curl up, when they are lit. This fancy shape is called the protein's *tertiary structure* (Fig. 95), while the amino acid sequence per se is called the *primary structure* of the protein. The tertiary structure is implicit in the primary structure, just as in Typogenetics. However, the recipe for deriving the tertiary structure, if you know only the primary structure, is by far more complex than that given in Typogenetics. In fact, it is one of the outstanding problems of contemporary molecular biology to figure out some rules by which the tertiary structure of a protein can be predicted if only its primary structure is known.

The Genetic Code.

	U	C	A	G	
U	phe	ser	tyr	cys	U
	phe	ser	tyr	cys	C
	leu	ser	*punc.*	*punc.*	A
	leu	ser	*punc.*	trp	G
C	leu	pro	his	arg	U
	leu	pro	his	arg	C
	leu	pro	gln	arg	A
	leu	pro	gln	arg	G
A	ile	thr	asn	ser	U
	ile	thr	asn	ser	C
	ile	thr	lys	arg	A
	met	thr	lys	arg	G
G	val	ala	asp	gly	U
	val	ala	asp	gly	C
	val	ala	glu	gly	A
	val	ala	glu	gly	G

FIGURE 94. The Genetic Code, by which each triplet in a strand of messenger RNA codes for one of twenty amino acids (or a punctuation mark).

Reductionistic Explanation of Protein Function

Another discrepancy between Typogenetics and true genetics—and this is probably the most serious one of all—is this: whereas in Typogenetics, each component amino acid of an enzyme is responsible for some specific "piece of the action", in real enzymes, individual amino acids cannot be assigned such clear roles. It is the tertiary structure *as a whole* which determines the mode in which an enzyme will function; there is no way one can say, "This amino acid's presence means that such-and-such an operation will get performed". In other words, in real genetics, an individual amino acid's contribution to the enzyme's overall function is not "context-free". However, this fact should not be construed in any way as ammunition for an antireductionist argument to the effect that "the whole [enzyme] cannot be explained as the sum of its parts". That would be wholly unjustified. What *is* justified is rejection of the simpler claim that "each amino acid contributes to the sum in a manner which is independent of the other amino acids present". In other words, the function of a protein cannot be considered to be built up from context-free functions of its parts; rather, one must consider how the parts interact. It is still possible in principle to write a computer program which takes as input the primary structure of a protein,

FIGURE 95. The structure of myoglobin, deduced from high-resolution X-ray data. The large-scale "twisted pipe" appearance is the tertiary structure; the finer helix inside—the "alpha helix"—is the secondary structure. [From A. Lehninger, Biochemistry.]

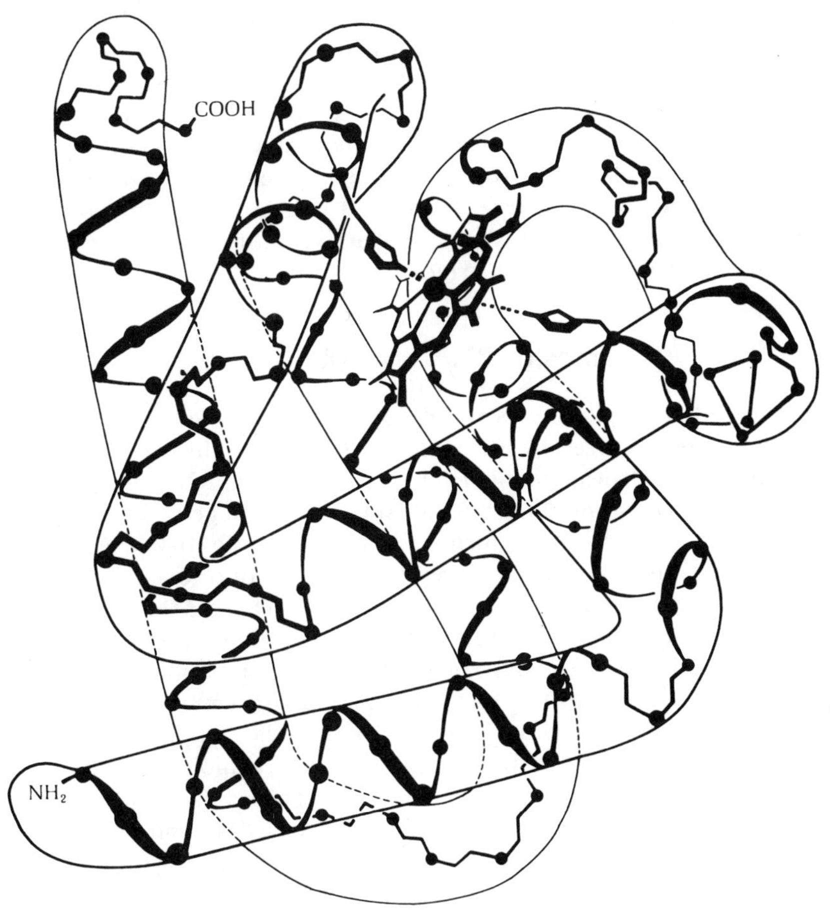

and firstly determines its tertiary structure, and secondly determines the function of the enzyme. This would be a completely reductionistic explanation of the workings of proteins, but the determination of the "sum" of the parts would require a highly complex algorithm. The elucidation of the *function* of an enzyme, given its primary, or even its tertiary, *structure,* is another great problem of contemporary molecular biology.

Perhaps, in the last analysis, the function of the whole enzyme can be considered to be built up from functions of parts in a context-free manner, but where the parts are now considered to be individual particles, such as electrons and protons, rather than "chunks", such as amino acids. This exemplifies the "Reductionist's Dilemma": In order to explain everything in terms of *context-free* sums, one has to go down to the level of physics; but then the number of particles is so huge as to make it only a theoretical "in-principle" kind of thing. So, one has to settle for a context-dependent sum, which has two disadvantages. The first is that the parts are much larger units, whose behavior is describable only on a high level, and therefore indeterminately. The second is that the word "sum" carries the connotation that each part can be assigned a simple function and that the function of the whole is just a context-free sum of those individual functions. This just cannot be done when one tries to explain a whole enzyme's function, given its amino acids as parts. But for better or for worse, this is a general phenomenon which arises in the explanations of complex systems. In order to acquire an intuitive and manageable understanding of how parts interact—in short, in order to proceed—one often has to sacrifice the exactness yielded by a microscopic, context-free picture, simply because of its unmanageability. But one does not sacrifice at that time the faith that such an explanation exists in principle.

Transfer RNA and Ribosomes

Returning, then, to ribosomes and RNA and proteins, we have stated that a protein is manufactured by a ribosome according to the blueprint carried from the DNA's "royal chambers" by its messenger, RNA. This seems to imply that the ribosome can translate from the language of codons into the language of amino acids, which amounts to saying that the ribosome "knows" the Genetic Code. However, that amount of information is simply not present in a ribosome. So how does it do it? Where *is* the Genetic Code stored? The curious fact is that the Genetic Code is stored—where else?—in the DNA itself. This certainly calls for some explanation.

Let us back off from a total explanation for a moment, and give a partial explanation. There are, floating about in the cytoplasm at any given moment, large numbers of four-leaf-clover-shaped molecules; loosely fastened (i.e., hydrogen-bonded) to one leaf is an amino acid, and on the opposite leaf there is a triplet of nucleotides called an *anticodon.* For our purposes, the other two leaves are irrelevant. Here is how these "clovers" are used by the ribosomes in their production of proteins. When a new

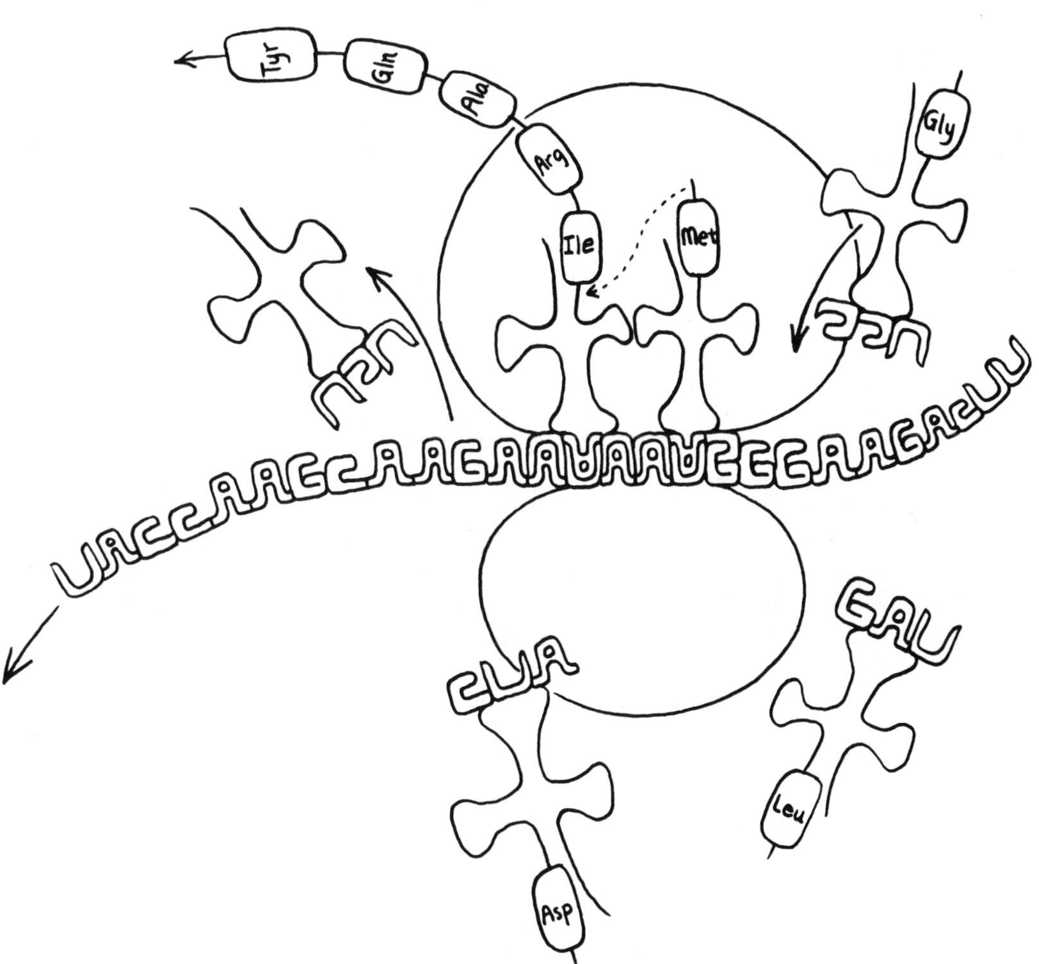

FIGURE 96. A section of mRNA passing through a ribosome. Floating nearby are tRNA molecules, carrying amino acids which are stripped off by the ribosome and appended to the growing protein. The Genetic Code is contained in the tRNA molecules, collectively. Note how the base-pairing (A-U, C-G) is represented by interlocking letter-forms in the diagram. [Drawing by Scott E. Kim.]

codon of mRNA clicks into position in the ribosome's "playing head", the ribosome reaches out into the cytoplasm and latches onto a clover whose anticodon is complementary to the mRNA codon. Then it pulls the clover into such a position that it can rip off the clover's amino acid, and stick it covalently onto the growing protein. (Incidentally, the bond between an amino acid and its neighbor in a protein is a very strong covalent bond, called a "peptide bond". For this reason, proteins are sometimes called "polypeptides".) Of course it is no accident that the "clovers" carry the proper amino acids, for they have all been manufactured according to precise instructions emanating from the "throne room".

The real name for such a clover is *transfer RNA*. A molecule of tRNA is quite small—about the size of a very small protein—and consists of a chain of about eighty nucleotides. Like mRNA, tRNA molecules are made by *transcription* off of the grand cellular template, DNA. However, tRNA's are tiny by comparison with the huge mRNA molecules, which may contain thousands of nucleotides in long, long chains. Also, tRNA's resemble proteins (and are unlike strands of mRNA) in this respect: they have fixed, well-defined tertiary structures—determined by their primary structure. A tRNA molecule's tertiary structure allows precisely one amino acid to bind to its amino-acid site; to be sure, it is that one dictated according to the Genetic Code by the anticodon on the opposite arm. A vivid image of the function of tRNA molecules is as flashcards floating in a cloud around a simultaneous interpreter, who snaps one out of the air—invariably the right one!—whenever he needs to translate a word. In this case, the interpreter is the ribosome, the words are codons, and their translations are amino acids.

In order for the inner message of DNA to get decoded by the ribosomes, the tRNA flashcards must be floating about in the cytoplasm. In some sense, the tRNA's contain the essence of the *outer* message of the DNA, since they are the keys to the process of translation. But they themselves came from the DNA. Thus, the outer message is trying to be part of the inner message, in a way reminiscent of the message-in-a-bottle which tells what language it is written in. Naturally, no such attempt can be totally successful: there is no way for the DNA to hoist itself by its own bootstraps. Some amount of knowledge of the Genetic Code must already be present in the cell beforehand, to allow the manufacture of those enzymes which transcribe tRNA's themselves off of the master copy of DNA. And this knowledge resides in previously manufactured tRNA molecules. This attempt to obviate the need for any outer message at all is like the Escher dragon, who tries as hard as he can, within the context of the two-dimensional world to which he is constrained, to be three-dimensional. He seems to go a long way—but of course he never makes it, despite the fine imitation he gives of three-dimensionality.

Punctuation and the Reading Frame

How does a ribosome know when a protein is done? Just as in Typogenetics, there is a signal inside the mRNA which indicates the termination or initiation of a protein. In fact, three special codons—UAA, UAG, UGA—act as *punctuation marks* instead of coding for amino acids. Whenever such a triplet clicks its way into the "reading head" of a ribosome, the ribosome releases the protein under construction and begins a new one.

Recently, the entire genome of the tiniest known virus, ϕX174, has been laid bare. One most unexpected discovery was made en route: some of its nine genes overlap—that is, *two distinct proteins are coded for by the same stretch of DNA!* There is even one gene contained entirely inside another!

This is accomplished by having the reading frames of the two genes *shifted* relative to each other, by exactly one unit. The density of information-packing in such a scheme is incredible. This is, of course, the inspiration behind the strange "5/17 haiku" in Achilles' fortune cookie, in the *Canon by Intervallic Augmentation.*

Recap

In brief, then, this picture emerges: from its central throne, DNA sends off long strands of messenger RNA to the ribosomes in the cytoplasm; and the ribosomes, making use of the "flashcards" of tRNA hovering about them, efficiently construct proteins, amino acid by amino acid, according to the blueprint contained in the mRNA. Only the primary structure of the proteins is dictated by the DNA; but this is enough, for as they emerge from the ribosomes, the proteins "magically" fold up into complex conformations which then have the ability to act as powerful chemical machines.

Levels of Structure and Meaning in Proteins and Music

We have been using this image of ribosome as tape recorder, mRNA as tape, and protein as music. It may seem arbitrary, and yet there are some beautiful parallels. Music is not a mere linear sequence of notes. Our minds perceive pieces of music on a level far higher than that. We chunk notes into phrases, phrases into melodies, melodies into movements, and movements into full pieces. Similarly, proteins only make sense when they act as chunked units. Although a primary structure carries all the information for the tertiary structure to be created, it still "feels" like less, for its potential is only realized when the tertiary structure is actually physically created.

Incidentally, we have been referring only to primary and tertiary structures, and you may well wonder whatever happened to the *secondary* structure. Indeed, it exists, as does a quaternary structure, as well. The folding-up of a protein occurs at more than one level. Specifically, at some points along the chain of amino acids, there may be a tendency to form a kind of helix, called the *alpha helix* (not to be confused with the DNA double helix). This helical twisting of a protein is on a lower level than its tertiary structure. This level of structure is visible in Figure 95. Quaternary structure can be directly compared with the building of a musical piece out of independent movements, for it involves the assembly of several distinct polypeptides, already in their full-blown tertiary beauty, into a larger structure. The binding of these independent chains is usually accomplished by hydrogen bonds, rather than covalent bonds; this is of course just as with pieces of music composed of several movements, which are far less tightly bound to each other than they are internally, but which nevertheless form a tight "organic" whole.

The four levels of primary, secondary, tertiary, and quaternary structure can also be compared to the four levels of the MU-picture (Fig. 60) in

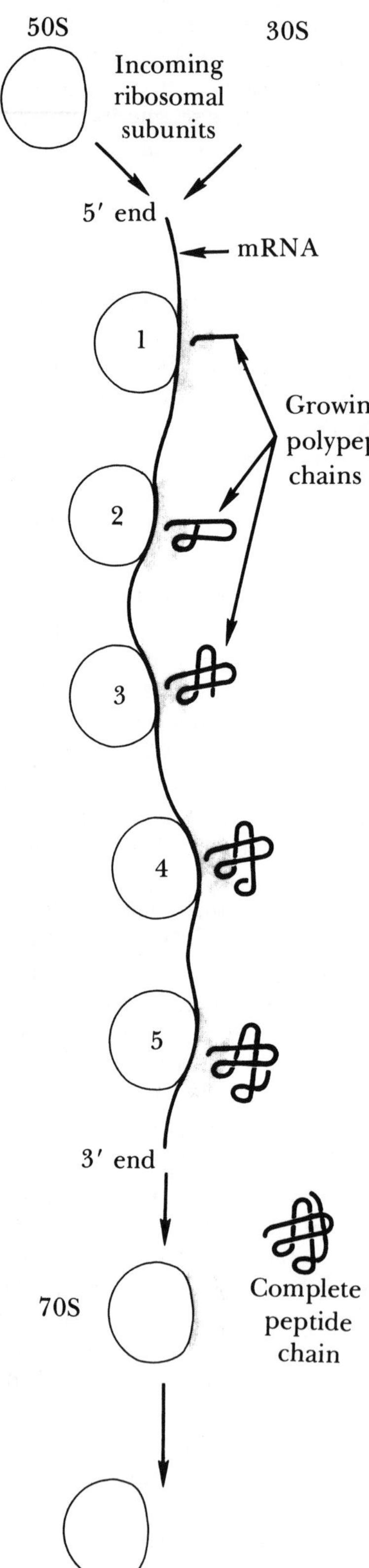

FIGURE 97. A polyribosome. A single strand of mRNA passes through one ribosome after another, like one tape passing through several tape recorders in a row. The result is a set of growing proteins in various stages of completion: the analogue to a musical canon produced by the staggered tape recorders. [From A. Lehninger, Biochemistry.*]*

the *Prelude, Ant Fugue.* The global structure—consisting of the letters 'M' and 'U'—is its *quaternary* structure; then each of those two parts has a *tertiary* structure, consisting of "HOLISM" or "REDUCTIONISM"; and then the opposite word exists on the *secondary* level, and at bottom, the *primary* structure is once again the word "MU", over and over again.

Polyribosomes and Two-Tiered Canons

Now we come to another lovely parallel between tape recorders translating tape into music and ribosomes translating mRNA into proteins. Imagine a collection of many tape recorders, arranged in a row, evenly spaced. We might call this array a "polyrecorder". Now imagine a single tape passing serially through the playing heads of all the component recorders. If the tape contains a single long melody, then the output will be a many-voiced canon, of course, with the delay determined by the time it takes the tape to get from one tape recorder to the next. In cells, such "molecular canons" do indeed exist, where many ribosomes, spaced out in long lines—forming what is called a *polyribosome*—all "play" the same strand of mRNA, producing identical proteins, staggered in time (see Fig. 97).

Not only this, but nature goes one better. Recall that mRNA is made by transcription off of DNA; the enzymes which are responsible for this process are called *RNA polymerases* ("-ase" is a general suffix for enzymes). It happens often that a series of RNA polymerases will be at work *in parallel* on a single strand of DNA, with the result that many separate (but identical) strands of mRNA are being produced, each delayed with respect to the other by the time required for the DNA to slide from one RNA polymerase to the next. At the same time, there can be several different ribosomes working on each of the parallel emerging mRNA's. Thus one arrives at a double-decker, or two-tiered, "molecular canon" (Fig. 98). The corresponding image in music is a rather fanciful but amusing scenario: several

FIGURE 98. Here, an even more complex scheme. Not just one *but* several *strands of mRNA, all emerging by transcription from a single strand of DNA, are acted upon by polyribosomes. The result is a two-tiered molecular canon. [From Hanawalt and Haynes,* The Chemical Basis of Life, *p. 271.]*

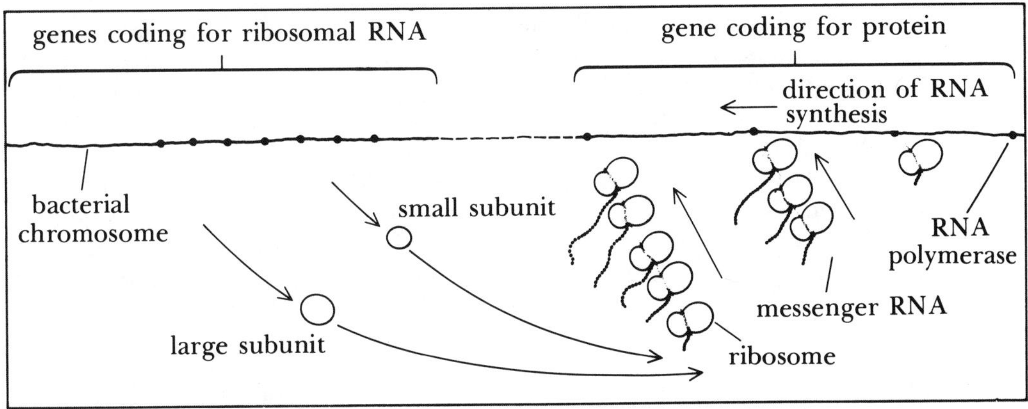

different copyists are all at work simultaneously, each one of them copying the same original manuscript from a clef which flutists cannot read into a clef which they can read. As each copyist finishes a page of the original manuscript, he passes it on to the next copyist, and starts transcribing a new page himself. Meanwhile, from each score emerging from the pens of the copyists, a set of flutists are reading and tooting the melody, each flutist delayed with respect to the others who are reading from the same sheet. This rather wild image gives, perhaps, an idea of some of the complexity of the processes which are going on in each and every cell of your body during every second of every day . . .

Which Came First—The Ribosome or the Protein?

We have been talking about these wonderful beasts called ribosomes; but what are they themselves composed of? How are they made? Ribosomes are composed of two types of things: (1) various kinds of proteins, and (2) another kind of RNA, called *ribosomal RNA* (rRNA). Thus, in order for a ribosome to be made, certain kinds of proteins must be present, and rRNA must be present. Of course, for proteins to be present, ribosomes must be there to make them. So how do you get around the vicious circle? Which comes first—the ribosome or the protein? Which makes which? Of course there is no answer because one always traces things back to previous members of the same class—just as with the chicken-and-the-egg question—until everything vanishes over the horizon of time. In any case, ribosomes are made of two pieces, a large and a small one, each of which contains some rRNA and some proteins. Ribosomes are about the size of large proteins; they are much much smaller than the strands of mRNA which they take as input, and along which they move.

Protein Function

We have spoken somewhat of the structure of proteins—specifically enzymes; but we have not really mentioned the kinds of tasks which they perform in the cell, nor how they do them. All enzymes are *catalysts*, which means that in a certain sense, they do no more than *selectively accelerate* various chemical processes in the cell, rather than make things happen which without them never could happen. An enzyme realizes certain pathways out of the myriad myriad potentialities. Therefore, in choosing which enzymes shall be present, you choose what shall happen and what shall not happen—despite the fact that, theoretically speaking, there is a nonzero probability for any cellular process to happen spontaneously, without the aid of catalysts.

Now how do enzymes act upon the molecules of the cell? As has been mentioned, enzymes are folded-up polypeptide chains. In every enzyme, there is a cleft or pocket or some other clearly-defined surface feature where the enzyme binds to some other kind of molecule. This location is called its *active site,* and any molecule which gets bound there is called a *substrate.* Enzymes may have more than one active site, and more than one substrate. Just as in Typogenetics, enzymes are indeed very choosy about what they will operate upon. The active site usually is quite specific, and allows just one kind of molecule to bind to it, although there are sometimes "decoys"—other molecules which can fit in the active site and clog it up, fooling the enzyme and in fact rendering it inactive.

Once an enzyme and its substrate are bound together, there is some disequilibrium of electric charge, and consequently charge—in the form of electrons and protons—flows around the bound molecules and readjusts itself. By the time equilibrium has been reached, some rather profound chemical changes may have occurred to the substrate. Some examples are these: there may have been a "welding", in which some standard small molecule got tacked onto a nucleotide, amino acid, or other common cellular molecule; a DNA strand may have been "nicked" at a particular location; some piece of a molecule may have gotten lopped off; and so forth. In fact, bio-enzymes do operations on molecules which are quite similar to the typographical operations which Typo-enzymes perform. However, most enzymes perform essentially only a single task, rather than a sequence of tasks. There is one other striking difference between Typo-enzymes and bio-enzymes, which is this: whereas Typo-enzymes operate only on strands, bio-enzymes can act on DNA, RNA, other proteins, ribosomes, cell membranes—in short, on anything and everything in the cell. In other words, enzymes are the universal mechanisms for getting things done in the cell. There are enzymes which stick things together and take them apart and modify them and activate them and deactivate them and copy them and repair them and destroy them . . .

Some of the most complex processes in the cell involve "cascades" in which a single molecule of some type triggers the production of a certain kind of enzyme; the manufacturing process begins and the enzymes which come off the "assembly line" open up a new chemical pathway which allows a second kind of enzyme to be produced. This kind of thing can go on for three or four levels, each newly produced type of enzyme triggering the production of another type. In the end a "shower" of copies of the final type of enzyme is produced, and all of the copies go off and do their specialized thing, which may be to chop up some "foreign" DNA, or to help make some amino acid for which the cell is very "thirsty", or whatever.

Need for a Sufficiently Strong Support System

Let us describe nature's solution to the puzzle posed for Typogenetics: "What kind of strand of DNA can direct its own replication?" Certainly not every strand of DNA is inherently a self-rep. The key point is this: any strand which wishes to direct its own copying must contain directions for assembling precisely those enzymes which can carry out the task. Now it is futile to hope that a strand of DNA in isolation could be a self-rep; for in order for those potential proteins to be pulled out of the DNA, there must not only be ribosomes, but also RNA polymerase, which makes the mRNA that gets transported to the ribosomes. And so we have to begin by assuming a kind of "minimal support system" just sufficiently strong that it allows transcription and translation to be carried out. This minimal support system will thus consist in (1) some proteins, such as RNA polymerase, which allow mRNA to be made from DNA, and (2) some ribosomes.

How DNA Self-Replicates

It is not by any means coincidental that the phrases "sufficiently strong support system" and "sufficiently powerful formal system" sound alike. One is the precondition for a self-rep to arise, the other for a self-ref to arise. In fact there is in essence only one phenomenon going on in two very different guises, and we shall explicitly map this out shortly. But before we do so, let us finish the description of how a strand of DNA can be a self-rep.

The DNA must contain the codes for a set of proteins which will copy it. Now there is a very efficient and elegant way to copy a double-stranded piece of DNA, whose two strands are complementary. This involves two steps:

(1) unravel the two strands from each other;
(2) "mate" a new strand to each of the two new single strands.

This process will create two new double strands of DNA, each identical to the original one. Now if our solution is to be based on this idea, it must involve a set of proteins, coded for in the DNA itself, which will carry out these two steps.

It is believed that in cells, these two steps are performed together in a coordinated way, and that they require three principal enzymes: DNA endonuclease, DNA polymerase, and DNA ligase. The first is an "unzipping enzyme": it peels the two original strands apart for a short distance, and then stops. Then the other two enzymes come into the picture. The DNA polymerase is basically a copy-and-move enzyme: it chugs down the short single strands of DNA, copying them complementarily in a fashion reminiscent of the Copy mode in Typogenetics. In order to copy, it draws on raw materials—specifically nucleotides—which are floating about in the cytoplasm. Because the action proceeds in fits and starts, with some unzipping and some copying each time, some short gaps are created, and the DNA ligase is what plugs them up. The process is repeated over and over again. This precision three-enzyme machine proceeds in careful fashion all the way down the length of the DNA molecule, until the whole thing has been peeled apart and simultaneously replicated, so that there are now two copies of it.

Comparison of DNA's Self-Rep Method with Quining

Note that in the enzymatic action on the DNA strands, the fact that information is stored in the DNA is just plain irrelevant; the enzymes are merely carrying out their symbol-shunting functions, just like rules of inference in the MIU-system. It is of no interest to the three enzymes that at some point they are actually copying the very genes which coded for them. The DNA, to them, is just a template without meaning or interest.

It is quite interesting to compare this with the Quine sentence's method of describing how to construct a copy of itself. There, too, one has a sort of "double strand"—two copies of the same information, where one copy acts as instructions, the other as template. In DNA, the process is vaguely parallel, since the three enzymes (DNA endonuclease, DNA polymerase, DNA ligase) are coded for in just one of the two strands, which therefore acts as *program,* while the other strand is merely a *template.* The parallel is not perfect, for when the copying is carried out, both strands are used as template, not just one. Nevertheless, the analogy is highly suggestive. There is a biochemical analogue to the use-mention dichotomy: when DNA is treated as a mere sequence of chemicals to be copied, it is like *mention* of typographical symbols; when DNA is dictating what operations shall be carried out, it is like *use* of typographical symbols.

Levels of Meaning of DNA

There are several levels of meaning which can be read from a strand of DNA, depending on how big the chunks are which you look at, and how powerful a decoder you use. On the lowest level, each DNA strand codes for an equivalent RNA strand—the process of decoding being *transcription.* If one chunks the DNA into triplets, then by using a "genetic decoder", one can read the DNA as a sequence of amino acids. This is *translation* (on top of transcription). On the next natural level of the hierarchy, DNA is readable as a code for a set of proteins. The physical pulling-out of proteins from genes is called *gene expression.* Currently, this is the highest level at which we understand what DNA means.

However, there are certain to be higher levels of DNA meaning which are harder to discern. For instance, there is every reason to believe that the DNA of, say, a human being codes for such features as nose shape, music talent, quickness of reflexes, and so on. Could one, in principle, learn to read off such pieces of information directly from a strand of DNA, without going through the actual physical process of *epigenesis*—the physical pulling-out of phenotype from genotype? Presumably, yes, since—in theory—one could have an incredibly powerful computer program simulating the entire process, including every cell, every protein, every tiny feature involved in the replication of DNA, of cells, to the bitter end. The output of such a *pseudo-epigenesis* program would be a high-level description of the phenotype.

There is another (extremely faint) possibility: that we could learn to read the phenotype off of the genotype *without* doing an isomorphic simulation of the physical process of epigenesis, but by finding some simpler sort of decoding mechanism. This could be called "shortcut pseudo-epigenesis". Whether shortcut or not, pseudo-epigenesis is, of course, totally beyond reach at the present time—with one notable exception: in the species *Felis catus,* deep probing has revealed that it is indeed possible to read the phenotype directly off of the genotype. The reader will perhaps better appreciate this remarkable fact after directly examining the following typical section of the DNA of *Felis catus:*

... CATCATCATCATCATCATCATCATCATCAT ...

Below is shown a summary of the levels of DNA-readability, together with the names of the different levels of decoding. DNA can be read as a sequence of:

(1) bases (nucleotides) . . . . *transcription*
(2) amino acids. . . . . *translation*
(3) proteins (primary structure) } . . . . *gene expression*
(4) proteins (tertiary structure) }
(5) protein clusters . . . . *higher levels of gene expression*
(6) ???
. .
. . } . . . . *unknown levels of DNA meaning*
. .
($N-1$) ???
(N) physical, mental, and
psychological traits. *pseudo-epigenesis*

The Central Dogmap

With this background, now we are in a position to draw an elaborate comparison between F. Crick's "Central Dogma of Molecular Biology" (.DOGMA I) upon which all cellular processes are based; and what I, with poetic license, call the "Central Dogma of Mathematical Logic" (.DOGMA II), upon which Gödel's Theorem is based. The mapping from one onto the other is laid out in Figure 99 and the following chart, which together constitute the *Central Dogmap.*

FIGURE 99. The Central Dogmap. An analogy is established between two fundamental Tangled Hierarchies: that of molecular biology and that of mathematical logic.

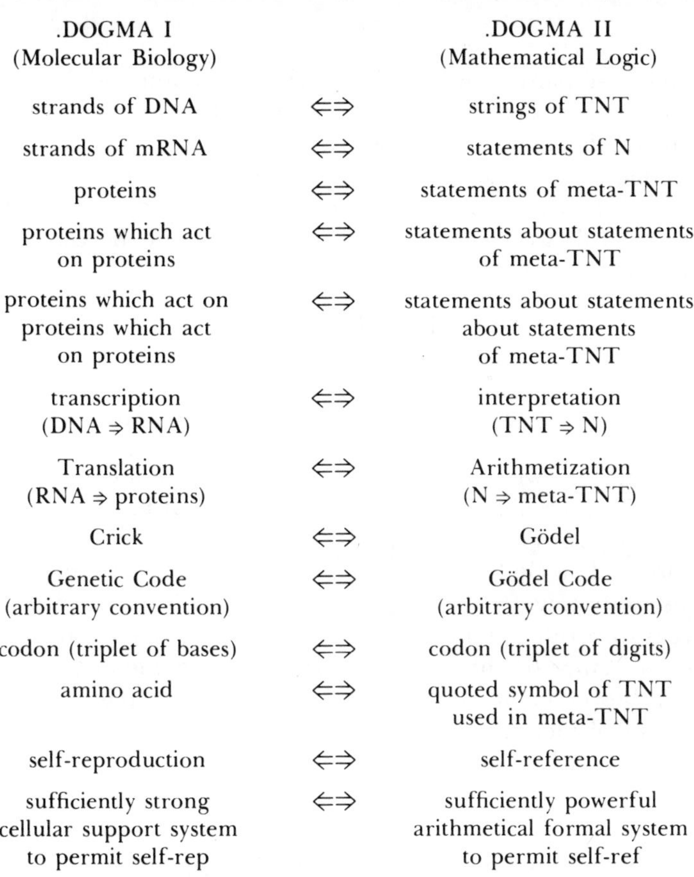

.DOGMA I (Molecular Biology)		.DOGMA II (Mathematical Logic)
strands of DNA	⇐⇒	strings of TNT
strands of mRNA	⇐⇒	statements of N
proteins	⇐⇒	statements of meta-TNT
proteins which act on proteins	⇐⇒	statements about statements of meta-TNT
proteins which act on proteins which act on proteins	⇐⇒	statements about statements about statements of meta-TNT
transcription (DNA ⇒ RNA)	⇐⇒	interpretation (TNT ⇒ N)
Translation (RNA ⇒ proteins)	⇐⇒	Arithmetization (N ⇒ meta-TNT)
Crick	⇐⇒	Gödel
Genetic Code (arbitrary convention)	⇐⇒	Gödel Code (arbitrary convention)
codon (triplet of bases)	⇐⇒	codon (triplet of digits)
amino acid	⇐⇒	quoted symbol of TNT used in meta-TNT
self-reproduction	⇐⇒	self-reference
sufficiently strong cellular support system to permit self-rep	⇐⇒	sufficiently powerful arithmetical formal system to permit self-ref

Central Dogmap

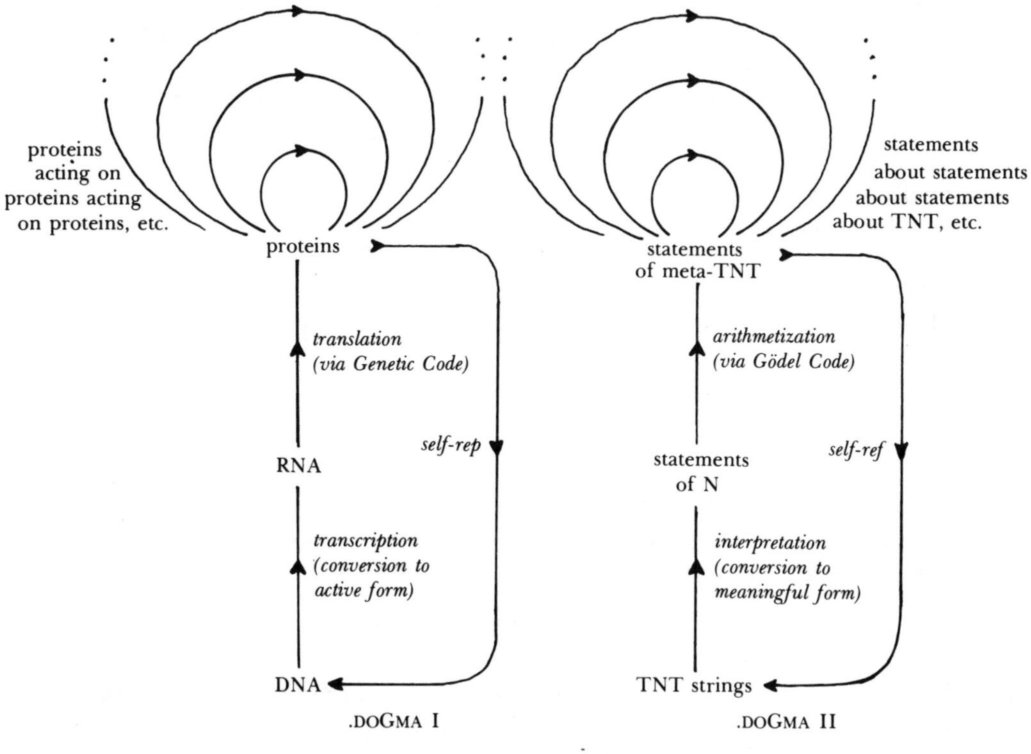

Note the base-pairing of A and T (Arithmetization and Translation), as well as of G and C (Gödel and Crick). Mathematical logic gets the purine side, and molecular biology gets the pyrimidine side.

To complete the esthetic side of this mapping, I chose to model my Gödel-numbering scheme on the Genetic Code absolutely faithfully. In fact, under the following correspondence, the table of the Genetic Code becomes the table of the Gödel Code:

(odd)	1	⇐⇒	A	(purine)
(even)	2	⇐⇒	C	(pyrimidine)
(odd)	3	⇐⇒	G	(purine)
(even)	6	⇐⇒	U	(pyrimidine)

Each amino acid—of which there are twenty—corresponds to exactly one symbol of TNT—of which there are twenty. Thus, at last, my motive for concocting "austere TNT" comes out—so that there would be exactly twenty symbols! The Gödel Code is shown in Figure 100. Compare it with the Genetic Code (Fig. 94).

There is something almost mystical in seeing the deep sharing of such an abstract structure by these two esoteric, yet fundamental, advances in knowledge achieved in our century. This Central Dogmap is by no means a rigorous proof of identity of the two theories; but it clearly shows a profound kinship, which is worth deeper exploration.

Strange Loops in the Central Dogmap

One of the more interesting similarities between the two sides of the map is the way in which "loops" of arbitrary complexity arise on the top level of both: on the left, proteins which act on proteins which act on proteins and so on, ad infinitum; and on the right, statements about statements about statements of meta-TNT and so on, ad infinitum. These are like heterarchies, which we discussed in Chapter V, where a sufficiently complex substratum allows high-level Strange Loops to occur and to cycle around, totally sealed off from lower levels. We will explore this idea in greater detail in Chapter XX.

Incidentally, you may be wondering about this question: "What, according to the Central Dogmap, is Gödel's Incompleteness Theorem itself mapped onto?" This is a good question to think about before reading ahead.

The Central Dogmap and the *Contracrostipunctus*

It turns out that the Central Dogmap is quite similar to the mapping that was laid out in Chapter IV between the *Contracrostipunctus* and Gödel's Theorem. One can therefore draw parallels between all three systems:

The Gödel Code.

	6	2	1	3	
6	0	∀	∨	:	6
	0	∀	∨	:	2
	a	∀	*punc.*	*punc.*	1
	a	∀	*punc.*	⊃	3
2	a	~	<	·	6
	a	~	<	·	2
	a	~	>	·	1
	a	~	>	·	3
1	∧	S	+	∀	6
	∧	S	+	∀	2
	∧	S	=	·	1
	′	S	=	·	3
3	(	)	[	∃	6
	(	)	[	∃	2
	(	)	]	∃	1
	(	)	]	∃	3

FIGURE 100. The Gödel Code. Under this Gödel-numbering scheme, each TNT symbol gets one or more codons. The small ovals show how this table subsumes the earlier Gödel-numbering table of Chapter IX.

(1) formal systems and strings;
(2) cells and strands of DNA;
(3) record players and records.

In the following chart, the mapping between systems 2 and 3 is explained carefully.

Contracrostipunctus		Molecular Biology
phonograph	⇐⇒	cell
"Perfect" phonograph	⇐⇒	"Perfect" cell
record	⇐⇒	strand of DNA
record playable by a given phonograph	⇐⇒	strand of DNA reproducible by a given cell
record unplayable by that phonograph	⇐⇒	strand of DNA unreproducible by that cell
process of converting record grooves into sounds	⇐⇒	process of transcription of DNA onto mRNA
sounds produced by record player	⇐⇒	strands of messenger RNA
translation of sounds into vibrations of phonograph	⇐⇒	translation of mRNA into proteins
mapping from external sounds onto vibrations of phonograph	⇐⇒	Genetic Code (mapping from mRNA triplets onto amino acids)
breaking of phonograph	⇐⇒	destruction of the cell
Title of song specially tailored for Record Player X: "I Cannot Be Played on Record Player X"	⇐⇒	High-level interpretation of DNA strand specially tailored for Cell X: "I Cannot Be Replicated by Cell X"
"Imperfect" Record Player	⇐⇒	Cell for which there exists at least one DNA strand which it cannot reproduce
"Tödel's Theorem": "There always exists an unplayable record, given a particular phonograph."	⇐⇒	Immunity Theorem: "There always exists an unreproducible DNA strand, given a particular cell."

The analogue of Gödel's Theorem is seen to be a peculiar fact, probably little useful to molecular biologists (to whom it is likely quite obvious):

> It is always possible to design a strand of DNA which, if injected into a cell, would, upon being transcribed, cause such proteins to be manufactured as would destroy the cell (or the DNA), and thus result in the non-reproduction of that DNA.

This conjures up a somewhat droll scenario, at least if taken in light of evolution: an invading species of virus enters a cell by some surreptitious

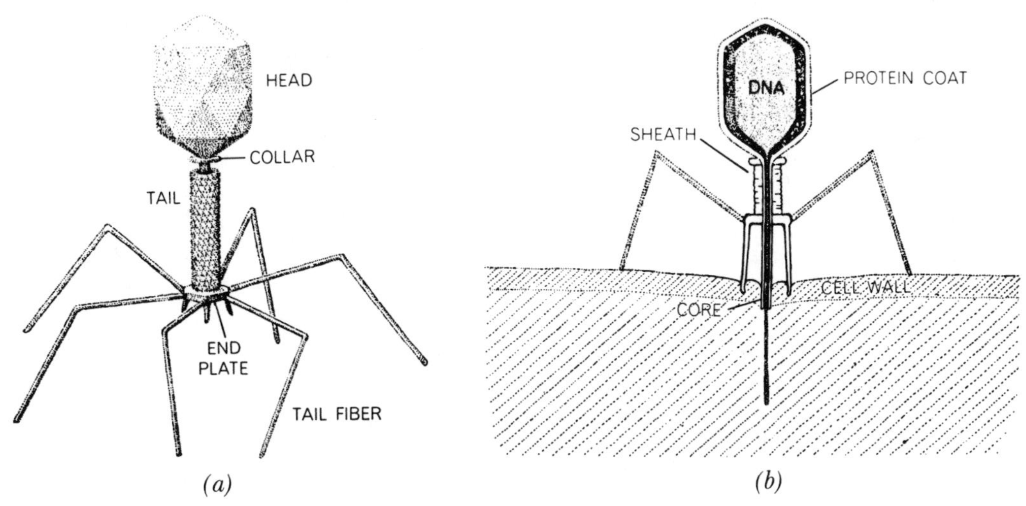

FIGURE 101. The T4 bacterial virus is an assembly of protein components (a). The "head" is a protein membrane, shaped like a kind of prolate icosahedron with thirty facets and filled with DNA. It is attached by a neck to a tail consisting of a hollow core surrounded by a contractile sheath and based on a spiked end plate to which six fibers are attached. The spikes and fibers affix the virus to a bacterial cell wall (b). The sheath contracts, driving the core through the wall, and viral DNA enters the cell. [From Hanawalt and Haynes, The Chemical Basis of Life, *p. 230.*]

means, and then carefully ensures the manufacture of proteins which will have the effect of destroying the virus itself! It is a sort of suicide—or Epimenides sentence, if you will—on the molecular level. Obviously it would not prove advantageous from the point of view of survival of the species. However, it demonstrates the spirit, if not the letter, of the mechanisms of protection and subversion which cells and their invaders have developed.

E. Coli *vs.* T4

Let us consider the biologists' favorite cell, that of the bacterium *Escherichia coli* (no relation to M. C. Escher), and one of their favorite invaders of that cell: the sinister and eerie *T4 phage,* pictures of which you can see in Figure 101. (Incidentally, the words "phage" and "virus" are synonymous and mean "attacker of bacterial cells".) The weird tidbit looks like a little like a cross between a LEM (Lunar Excursion Module) and a mosquito—and it is much more sinister than the latter. It has a "head" wherein is stored all its "knowledge"—namely its DNA; and it has six "legs" wherewith to fasten itself to the cell it has chosen to invade; and it has a "stinging tube" (more properly called its "tail") like a mosquito. The major difference is that unlike a mosquito, which uses its stinger for sucking blood, the T4 phage uses its stinger for injecting its hereditary substance into the cell against the will of its victim. Thus the phage commits "rape" on a tiny scale.

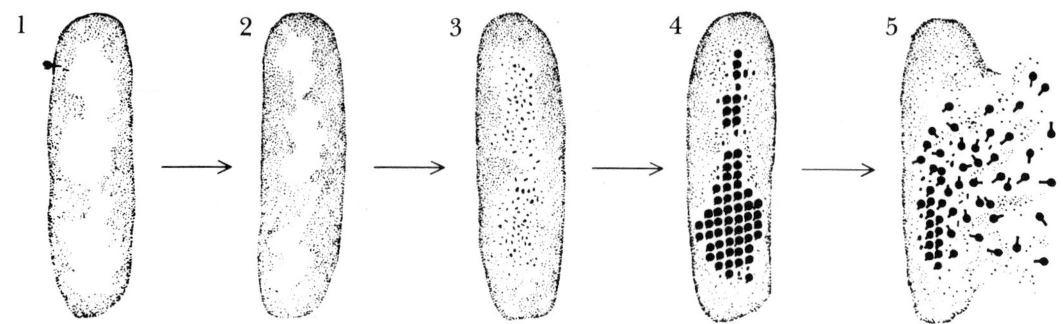

FIGURE 102. Viral infection begins when viral DNA enters a bacterium. Bacterial DNA is disrupted and viral DNA replicated. Synthesis of viral structural proteins and their assembly into virus continues until the cell bursts, releasing particles. [From Hanawalt and Haynes, The Chemical Basis of Life, *p. 230.*]

A Molecular Trojan Horse

What actually happens when the viral DNA enters a cell? The virus "hopes", to speak anthropomorphically, that its DNA will get exactly the same treatment as the DNA of the host cell. This would mean getting transcribed and translated, thus allowing it to direct the synthesis of its own special proteins, alien to the host cell, which will then begin to do their thing. This amounts to secretly transporting alien proteins "in code" (viz., the Genetic Code) into the cell, and then "decoding" (viz., producing) them. In a way this resembles the story of the Trojan horse, according to which hundreds of soldiers were sneaked into Troy inside a harmless-seeming giant wooden horse; but once inside the city, they broke loose and captured it. The alien proteins, once they have been "decoded" (synthesized) from their carrier DNA, now jump into action. The sequence of actions directed by the T4 phage has been carefully studied, and is more or less as follows (see also Figs. 102 and 103):

Time elapsed	*Action taking place*
0 min.	Injection of viral DNA.
1 min.	Breakdown of host DNA. Cessation of production of native proteins and initiation of production of alien (T4) proteins. Among the earliest produced proteins are those which direct the replication of the alien (T4) DNA.
5 min.	Replication of viral DNA begins.
8 min.	Initiation of production of structural proteins which will form the "bodies" of new phages.

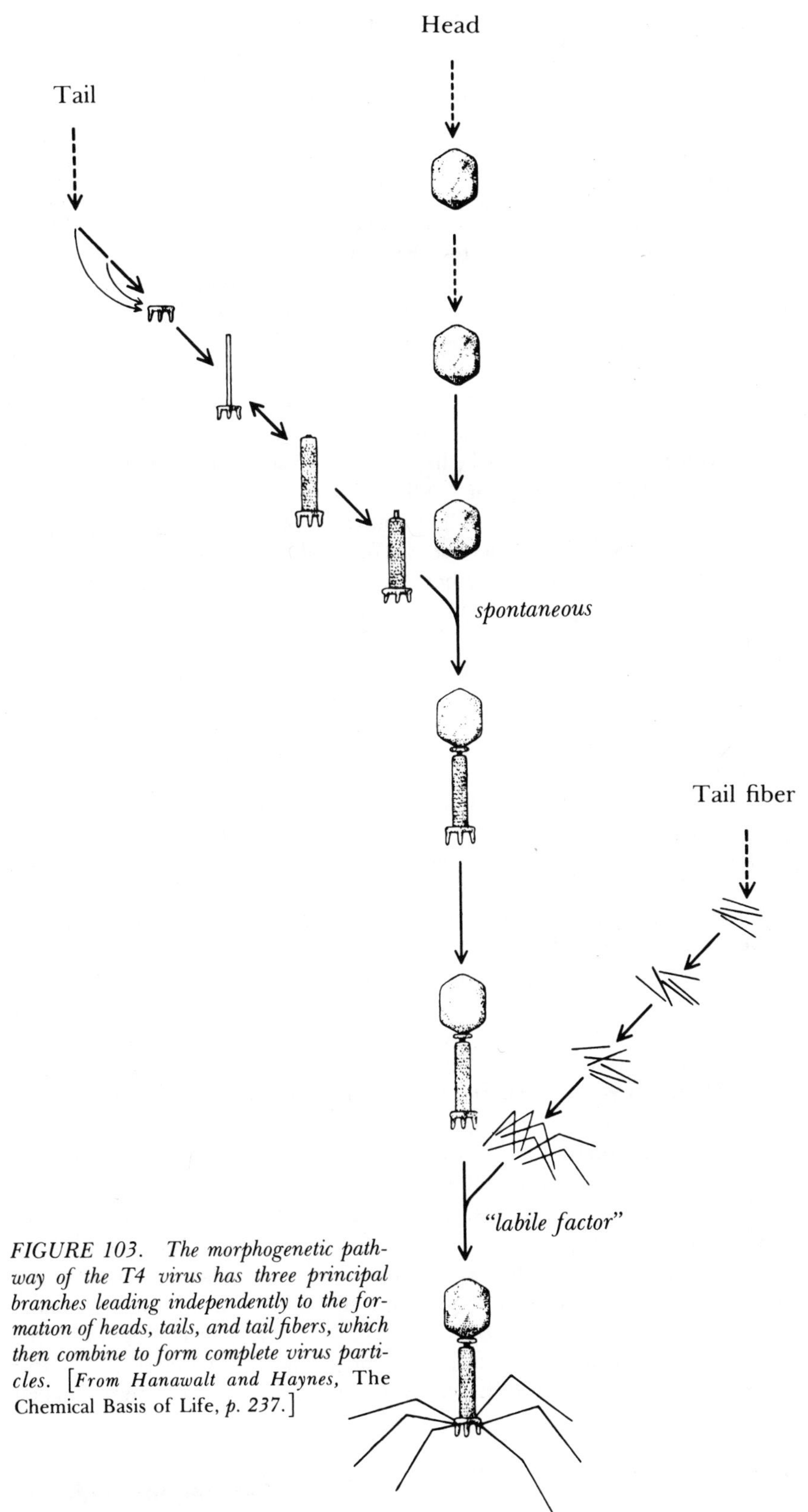

FIGURE 103. The morphogenetic pathway of the T4 virus has three principal branches leading independently to the formation of heads, tails, and tail fibers, which then combine to form complete virus particles. [*From Hanawalt and Haynes,* The Chemical Basis of Life, *p. 237.*]

13 min. First complete replica of T4 invader is produced.

25 min. Lysozyme (a protein) attacks host cell wall, breaking open the bacterium, and the "bicentuplets" emerge.

Thus, when a T4 phage invades an *E. coli* cell, after the brief span of about twenty-four or twenty-five minutes, the cell has been completely subverted, and breaks open. Out pop about two hundred exact copies of the original virus—"bicentuplets"—ready to go attack more bacterial cells, the original cell having been largely consumed in the process.

Although from a bacterium's point of view this kind of thing is a deadly serious menace, from our large-scale vantage point it can be looked upon as an amusing game between two players: the invader, or "T" player (named after the T-even class of phages, including the T2, T4, and others), and the "C" player (standing for "Cell"). The objective of the T player is to invade and take over the cell of the C player from within, for the purpose of reproducing itself. The objective of the C player is to protect itself and destroy the invader. When described this way, the molecular TC-game can be seen to be quite parallel to the macroscopic TC-game described in the preceding Dialogue. (The reader can doubtless figure out which player—T or C—corresponds to the Tortoise, and which to the Crab.)

Recognition, Disguises, Labeling

This "game" emphasizes the fact that *recognition* is one of the central themes of cellular and subcellular biology. How do molecules (or higher-level structures) recognize each other? It is essential for the functioning of enzymes that they should be able to latch onto special "binding sites" on their substrates; it is essential that a bacterium should be able to distinguish its own DNA from that of phages; it is essential that two cells should be able to recognize each other and interact in a controlled way. Such recognition problems may remind you of the original, key problem about formal systems: How can you tell if a string has, or does not have, some property such as theoremhood? Is there a decision procedure? This kind of question is not restricted to mathematical logic: it permeates computer science and, as we are seeing, molecular biology.

The labeling technique described in the Dialogue is in fact one of *E. coli*'s tricks for outwitting its phage invaders. The idea is that strands of DNA can be chemically labeled by tacking on a small molecule—methyl—to various nucleotides. Now this labeling operation does not change the usual biological properties of the DNA; in other words, methylated (labeled) DNA can be transcribed just as well as unmethylated (unlabeled) DNA, and so it can direct the synthesis of proteins. But if the host cell has some special mechanisms for examining whether DNA is labeled or not, then the label may make all the difference in the world. In particular, the host cell may have an enzyme system which looks for unlabeled DNA, and destroys any that it finds by unmercifully chopping it to pieces. In that case, woe to all unlabeled invaders!

The methyl labels on the nucleotides have been compared to serifs on letters. Thus, using this metaphor, we could say that the *E. coli* cell is looking for DNA written in its "home script", with its own particular typeface—and will chop up any strand of DNA written in an "alien" typeface. One counterstrategy, of course, is for phages to learn to label themselves, and thereby become able to fool the cells which they are invading into reproducing them.

This TC-battle can continue to arbitrary levels of complexity, but we shall not pursue it further. The essential fact is that it is a battle between a host which is trying to reject all invading DNA, and a phage which is trying to infiltrate its DNA into some host which will transcribe it into mRNA (after which its reproduction is guaranteed). Any phage DNA which succeeds in getting itself reproduced this way can be thought of as having this high-level interpretation: "I *Can* Be Reproduced in Cells of Type X". This is to be distinguished from the evolutionarily pointless kind of phage mentioned earlier, which codes for proteins that destroy it, and whose high-level interpretation is the self-defeating sentence: "I *Cannot* Be Reproduced in Cells of Type X".

Henkin Sentences and Viruses

Now both of these contrasting types of self-reference in molecular biology have their counterparts in mathematical logic. We have already discussed the analogue of the self-defeating phages—namely, strings of the Gödel type, which assert their own unproducibility within specific formal systems. But one can also make a counterpart sentence to a real phage: the phage asserts its own producibility in a specific cell, and the sentence asserts its own producibility in a specific formal system. Sentences of this type are called *Henkin sentences,* after the mathematical logician Leon Henkin. They can be constructed exactly along the lines of Gödel sentences, the only difference being the omission of a negation. One begins with an "uncle", of course:

$$\exists a{:}\exists a'{:}\langle \text{TNT-PROOF-PAIR}\{a,a'\}\wedge\text{ARITHMOQUINE}\{a'',a'\}\rangle$$

and then proceeds by the standard trick. Say the Gödel number of the above "uncle" is h. Now by arithmoquining this very uncle, you get a Henkin sentence:

$$\exists a{:}\exists a'{:}\langle \text{TNT-PROOF-PAIR}\{a,a'\}\wedge\text{ARITHMOQUINE}\{\underbrace{\text{SSS}\ldots\text{SSS0}}_{h\ \text{S's}}/a'',a'\}\rangle$$

(By the way, can you spot how this sentence differs from ~G?) The reason I show it explicitly is to point out that a Henkin sentence does not give a full recipe for its own derivation; it just asserts that there *exists* one. You might well wonder whether its claim is justified. Do Henkin sentences indeed possess derivations? Are they, as they claim, theorems? It is useful to recall that one need not believe a politician who says, "I am honest"—he may be honest, and yet he may not be. Are Henkin sentences any more trustworthy than politicians? Or do Henkin sentences, like politicians, lie in cast-iron sinks?

It turns out that these Henkin sentences are invariably truth tellers. Why this is so is not obvious; but we will accept this curious fact without proof.

Implicit *vs.* Explicit Henkin Sentences

I mentioned that a Henkin sentence tells nothing about its own derivation; it just asserts that one exists. Now it is possible to invent a variation on the theme of Henkin sentences—namely sentences which *explicitly describe* their own derivations. Such a sentence's high-level interpretation would not be "Some Sequence of Strings Exists Which is a Derivation of Me", but rather, "The Herein-described Sequence of Strings . . . . . Is a Derivation of Me". Let us call the first type of sentence an *implicit* Henkin sentence. The new sentences will be called *explicit* Henkin sentences, since they explicitly describe their own derivations. Note that, unlike their implicit brethren, explicit Henkin sentences *need not be theorems.* In fact, it is quite easy to write a string which asserts that its own derivation consists of the single string 0=0—a false statement, since 0=0 is not a derivation of anything. However, it is also possible to write an explicit Henkin sentence which *is* a theorem—that is, a sentence which in fact gives a recipe for its own derivation.

Henkin Sentences and Self-Assembly

The reason I bring up this distinction between explicit and implicit Henkin sentences is that it corresponds very nicely to a significant distinction between types of virus. There are certain viruses, such as the so-called "tobacco mosaic virus", which are called *self-assembling* viruses; and then there are others, such as our favorite T-evens, which are *non-self-assembling.* Now what is this distinction? It is a direct analogue to the distinction between implicit and explicit Henkin sentences.

The DNA of a self-assembling virus codes only for the *parts* of a new virus, but not for any *enzymes.* Once the parts are produced, the sneaky virus relies upon them to link up to each other without help from any enzymes. Such a process depends on chemical affinities which the parts have for each other, when swimming in the rich chemical brew of a cell. Not only viruses, but also some organelles—such as ribosomes—assemble themselves. Sometimes, enzymes may be needed—but in such cases, they are recruited from the host cell, and enslaved. This is what is meant by self-assembly.

By contrast, the DNA of more complex viruses, such as the T-evens, codes not only for the parts, but in addition for various enzymes which play special roles in the assembly of the parts into wholes. Since the assembly process is not spontaneous but requires "machines", such viruses are not considered to be self-assembling. The essence of the distinction, then, between self-assembling units and non-self-assembling units is that the former get away with self-reproduction without telling the cell anything about their construction, while the latter need to give *instructions* as to how to assemble themselves.

Now the parallel to Henkin sentences, implicit and explicit, ought to be quite clear. Implicit Henkin sentences are self-proving but do not tell anything at all about their proofs—they are analogous to self-assembling viruses; explicit Henkin sentences direct the construction of their own proofs—they are analogous to more complex viruses which direct their host cells in putting copies of themselves together.

The concept of self-assembling biological structures as complex as viruses raises the possibility of complex self-assembling machines as well. Imagine a set of parts which, when placed in the proper supporting environment, spontaneously group themselves in such a way as to form a complex machine. It seems unlikely, yet this is quite an accurate way to describe the process of the tobacco mosaic virus' method of self-reproduction via self-assembly. The information for the total conformation of the organism (or machine) is spread about in its parts; it is not concentrated in some single place.

Now this concept can lead in some strange directions, as was shown in the *Edifying Thoughts of a Tobacco Smoker*. There, we saw how the Crab used the idea that information for self-assembly can be distributed around, instead of being concentrated in a single place. His hope was that this would prevent his new phonographs from succumbing to the Tortoise's phonograph-crashing method. Unfortunately, just as with the most sophisticated axiom schemata, once the system is all built and packaged into a box, its well-definedness renders it vulnerable to a sufficiently clever "Gödelizer"; and that was the sad tale related by the Crab. Despite its apparent absurdity, the fantastic scenario of that Dialogue is not so far from reality, in the strange, surreal world of the cell.

Two Outstanding Problems: Differentiation and Morphogenesis

Now self-assembly may be the trick whereby certain subunits of cells are constructed, and certain viruses—but what of the most complex macroscopic structures, such as the body of an elephant or a spider, or the shape of a Venus's-flytrap? How are homing instincts built into the brain of a bird, or hunting instincts into the brain of a dog? In short, how is it that merely by dictating which *proteins* are to be produced in cells, DNA exercises such spectacularly precise control over the exact structure and function of macroscopic living objects? There are two major distinct problems here. One is that of *cellular differentiation:* how do different cells, sharing exactly the same DNA, perform different roles—such as a kidney cell, a bone marrow cell, and a brain cell? The other is that of *morphogenesis* ("birth of form"): how does intercellular communication on a local level give rise to large-scale, global structures and organizations—such as the various organs of the body, the shape of the face, the suborgans of the brain, and so on? Although both cellular differentiation and morphogenesis are poorly understood at present, the trick appears to reside in exquisitely fine-tuned feedback and "feedforward" mechanisms within cells and between cells, which tell a cell when to "turn on" and when to "turn off" production of various proteins.

Feedback and Feedforward

Feed*back* takes place when there is too much or too little of some desired substance in the cell; then the cell must somehow regulate the production line which is assembling that substance. Feed*forward* also involves the regulation of an assembly line, but not according to the amount of end product present; rather, according to the amount of some *precursor* of the end product of that assembly line. There are two major devices for achieving *negative* feedforward or feedback. One way is to prevent the relevant enzymes from being able to perform—that is, to "clog up" their active sites. This is called *inhibition.* The other way is to prevent the relevant enzymes from ever being manufactured! This is called *repression.* Conceptually, inhibition is simple: you just block up the active site of the first enzyme in the assembly line, and the whole process of synthesis gets stopped dead.

Repressors and Inducers

Repression is trickier. How does a cell stop a gene from being expressed? The answer is, it prevents it from ever getting transcribed. This means that it has to prevent RNA polymerase from doing its job. This can be accomplished by placing a huge obstacle in its path, along the DNA, precisely in front of that gene which the cell wants not to get transcribed. Such obstacles do exist, and are called *repressors.* They are themselves proteins, and they bind to special obstacle-holding sites on the DNA, called (I am not sure why) *operators.* An operator therefore is a site of control for the gene (or genes) which immediately follow it; those genes are called its *operon.* Because a series of enzymes often act in concert in carrying out a long chemical transformation, they are often coded for in sequence; and this is why operons often contain several genes, rather than just one. The effect of the successful repression of an operon is that a whole series of genes is prevented from being transcribed, which means that a whole set of related enzymes remains unsynthesized.

What about *positive* feedback and feedforward? Here again, there are two options: (1) unclog the clogged enzymes, or (2) stop the repression of the relevant operon. (Notice how nature seems to *love* double-negations! Probably there is some very deep reason for this.) The mechanism by which repression is repressed involves a class of molecules called *inducers*. The role of an inducer is simple: it combines with a repressor protein before the latter has had a chance to bind to an operator on a DNA molecule; the resulting "repressor-inducer complex" is incapable of binding to an operator, and this leaves the door open for the associated operon to be transcribed into mRNA and subsequently translated into protein. Often the end product or some precursor of the end product can act as an inducer.

Feedback and Strange Loops Compared

Incidentally, this is a good time to distinguish between simple kinds of feedback, as in the processes of inhibition and repression, and the looping-back between different informational levels, shown in the Central Dogmap. Both are "feedback" in some sense; but the latter is much deeper than the former. When an amino acid, such as tryptophan or isoleucine, acts as feedback (in the form of an inducer) by binding to its repressor so that more of it gets made, it is not telling *how* to construct itself; it is just telling enzymes to make more of it. This could be compared to a radio's volume, which, when fed through a listener's ears, may cause itself to be turned down or up. This is another thing entirely from the case in which the broadcast itself tells you explicitly to turn your radio on or off, or to tune to another wavelength—or even how to build another radio! The latter is much more like the looping-back between informational levels, for here, information inside the radio signal gets "decoded" and translated into mental structures. The radio signal is composed of symbolic constituents whose symbolic meaning matters—a case of use, rather than mention. On the other hand, when the sound is just too loud, the symbols are not conveying meaning; they are merely being perceived as loud sounds, and might as well be devoid of meaning—a case of mention, rather than use. This case more resembles the feedback loops by which proteins regulate their own rates of synthesis.

It has been theorized that the difference between two neighboring cells which share the exact same genotype and yet have different functions is that different segments of their genome have been repressed, and therefore they have different *working sets* of proteins. A hypopothesis like this could account for the phenomenal differences between cells in different organs of the body of a human being.

Two Simple Examples of Differentiation

The process by which one initial cell replicates over and over, giving rise to a myriad of differentiated cells with specialized functions, can be likened to the spread of a chain letter from person to person, in which each new participant is asked to propagate the message faithfully, but also to add some extra personal touch. Eventually, there will be letters which are tremendously different from each other.

Another illustration of the ideas of differentiation is provided by this extremely simple computer analogue of a differentiating self-rep. Consider a very short program which is controlled by an up-down switch, and which has an internal parameter N—a natural number. This program can run in two modes—the up-mode, and the down-mode. When it runs in the *up*-mode, it self-replicates into an adjacent part of the computer's memory—except it makes the internal parameter N of its "daughter" one greater than in itself. When it runs in the *down*-mode, it does not self-rep, but instead calculates the number

$$(-1)^N/(2N + 1)$$

and adds it to a running total.

Well, suppose that at the beginning, there is one copy of the program in memory, $N = 0$, and the mode is up. Then the program will copy itself next door in memory, with $N = 1$. Repeating the process, the new program will self-rep next door to itself, with a copy having $N = 2$. And over and over again . . . What happens is that a very large program is growing inside memory. When memory is full, the process quits. Now all of memory can be looked upon as being filled with one *big* program, composed of many similar, but differentiated, modules—or "cells". Now suppose we switch the mode to down, and run this big program. What happens? The first "cell" runs, and calculates 1/1. The second "cell" runs, calculating $-1/3$, and adding it to the previous result. The third "cell" runs, calculating $+1/5$ and adding it on . . . The end result is that the whole "organism"—the big program—calculates the sum

$$1 - 1/3 + 1/5 - 1/7 + 1/9 - 1/11 + 1/13 - 1/15 + \dots$$

to a large number of terms (as many terms as "cells" can fit inside memory). And since this series converges (albeit slowly) to $\pi/4$, we have a "phenotype" whose function is to calculate the value of a famous mathematical constant.

Level Mixing in the Cell

I hope that the descriptions of processes such as labeling, self-assembly, differentiation, morphogenesis, as well as transcription and translation, have helped to convey some notion of the immensely complex system which is a cell—an information-processing system with some strikingly novel features. We have seen, in the Central Dogmap, that although we can try to draw a clear line between program and data, the distinction is somewhat arbitrary. Carrying this line of thought further, we find that not only are *program* and *data* intricately woven together, but also the *interpreter* of programs, the physical *processor,* and even the *language* are included in this intimate fusion. Therefore, although it is possible (to some extent) to draw boundaries and separate out the levels, it is just as important—and just as fascinating—to recognize the level-crossings and mixings. Illustrative of this is the amazing fact that in biological systems, all the various features necessary for self-rep (viz., language, program, data, interpreter, and processor) cooperate to such a degree that all of them are replicated simultaneously—which shows how much deeper is biological self-rep'ing than anything yet devised along those lines by humans. For instance, the self-rep program exhibited at the beginning of this Chapter takes for granted the pre-existence of three external aspects: a language, an interpreter, and a processor, and does not replicate those.

Let us try to summarize various ways in which the subunits of a cell can be classified in computer science terms. First, let us take DNA. Since DNA contains all the information for construction of proteins, which are the active agents of the cell, DNA can be viewed as a *program* written in a higher-level language, which is subsequently translated (or interpreted) into the "machine language" of the cell (proteins). On the other hand, DNA is itself a passive molecule which undergoes manipulation at the hands of various kinds of enzymes; in this sense, a DNA molecule is exactly like a long piece of *data,* as well. Thirdly, DNA contains the templates off of which the tRNA "flashcards" are rubbed, which means that DNA also contains the definition of its own higher-level *language.*

Let us move on to proteins. Proteins are active molecules, and carry out all the functions of the cell; therefore it is quite appropriate to think of them as *programs* in the "machine language" of the cell (the cell itself being the processor). On the other hand, since proteins are hardware and most programs are software, perhaps it is better to think of the proteins as *processors.* Thirdly, proteins are often acted upon by other proteins, which means that proteins are often *data.* Finally, one can view proteins as *interpreters*; this involves viewing DNA as a collection of high-level language programs, in which case enzymes are merely carrying out the programs written in the DNA code, which is to say, the proteins are acting as interpreters.

Then there are ribosomes and tRNA molecules. They mediate the translation from DNA to proteins, which can be compared to the translation of a program from a high-level language to a machine language; in other words, the ribosomes are functioning as *interpreters* and the tRNA molecules provide the definition of the higher-level *language.* But an alternative view of translation has it that the ribosomes are *processors,* while the tRNA's are *interpreters.*

We have barely scratched the surface in this analysis of interrelations between all these biomolecules. What we have seen is that nature feels quite comfortable in mixing levels which *we* tend to see as quite distinct. Actually, in computer science there is already a visible tendency to mix all these seemingly distinct aspects of an information-processing system. This is particularly so in Artificial Intelligence research, which is usually at the forefront of computer language design.

The Origin of Life

A natural and fundamental question to ask, on learning of these incredibly intricately interlocking pieces of software and hardware is: "How did they ever get started in the first place?" It is truly a baffling thing. One has to imagine some sort of a bootstrap process occurring, somewhat like that which is used in the development of new computer languages—but a bootstrap from simple molecules to entire cells is almost beyond one's power to imagine. There are various theories on the origin of life. They all run aground on this most central of all central questions: "How did the Genetic Code, along with the mechanisms for its translation (ribosomes and tRNA molecules), originate?" For the moment, we will have to content ourselves with a sense of wonder and awe, rather than with an answer. And perhaps experiencing that sense of wonder and awe is more satisfying than having an answer—at least for a while.

The Magnificrab, Indeed

It is spring, and the Tortoise and Achilles are taking a Sunday promenade in the woods together. They have decided to climb a hill at the top of which, it is said, there is a wonderful teahouse, with all sorts of delicious pastries.

Achilles: Man oh man! If a crab—

Tortoise: If a crab??

Achilles: I was about to say, if a crab ever were intelligent, then surely it would be our mutual friend the Crab. Why, he must be at least two times as smart as any crab alive. Or maybe even three times as smart as any crab alive. Or perhaps—

Tortoise: My soul! How you magnify the Crab!

Achilles: Well, I just happen to be an admirer of his . . .

Tortoise: No need to apologize. I admire him, too. Speaking of Crab-admirers, did I tell you about the curious fan letter which the Crab received not too long ago?

Achilles: I don't believe so. Who sent it?

Tortoise: It bore a postmark from India, and was from someone neither of us had ever heard of before—a Mr. Najunamar, I believe.

Achilles: I wonder why someone who never knew Mr. Crab would send him a letter—or for that matter, how they would get his address.

Tortoise: Apparently whoever it was was under the illusion that the Crab is a mathematician. It contained numerous results, all of which were— But, ho! Speak of the devil! Here comes Mr. Crab now, down the hill.

Crab: Good-bye! It was nice to talk with you again. Well, I guess I had best be off. But I'm utterly stuffed—couldn't eat one more bite if I had to! I've just been up there myself—recommend it highly. Have you ever been to the teahouse at the crest of the hill? How are you, Achilles? Oh, there's Achilles. Hello, hello. Well, well, if it isn't Mr. T!

Tortoise: Hello, Mr. C. Are you headed up to the hilltop teahouse?

Crab: Why, yes indeed, I am; how did you guess it? I'm quite looking forward to some of their special napoleons—scrumptious little morsels. I'm so hungry I could eat a frog. Oh, there's Achilles. How are you, Achilles?

Achilles: Could be worse, I suppose.

Crab: Wonderful! Well, don't let me interrupt your discussion. I'll just tag along.

Tortoise: Curiously enough, I was just about to describe your mysterious letter from that Indian fellow a few weeks back—but now that you're here, I'll let Achilles get the story from the Crab's mouth.

FIGURE 104. Castrovalva, *by M. C. Escher (lithograph, 1930).*

Crab: Well, it was this way. This fellow Najunamar had apparently never had any formal training in mathematics, but had instead worked out some of his own methods for deriving new truths of mathematics. Some of his discoveries defeated me completely; I had never seen anything in the least like them before. For instance, he exhibited a map of India that he had managed to color using no fewer than 1729 distinct colors.

Achilles: 1729! Did you say 1729?

Crab: Yes—why do you ask?

Achilles: Well, 1729 is a very interesting number, you know.

Crab: Indeed. I wasn't aware of it.

Achilles: In particular, it so happens that 1729 is the number of the taxicab which I took to Mr. Tortoise's this morning!

Crab: How fascinating! Could you possibly tell me the number of the trolley car which you'll take to Mr. Tortoise's tomorrow morning?

Achilles (after a moment's thought): It's not obvious to me; however, I should think it would be very large.

Tortoise: Achilles has a wonderful intuition for these things.

Crab: Yes. Well, as I was saying, Najunamar in his letter also proved that every even prime is the sum of two odd numbers, and that there are no solutions in positive integers to the equation

$$a^n + b^n = c^n \qquad \text{for } n = 0.$$

Achilles: What? All these old classics of mathematics resolved in one fell swoop? He must be a genius of the first rank!

Tortoise: But Achilles—aren't you even in the slightest skeptical?

Achilles: What? Oh, yes—skeptical. Well, of course I am. You don't think I believe that Mr. Crab got such a letter, do you? I don't fall for just anything, you know. So it must have been YOU, Mr. T, who received the letter!

Tortoise: Oh, no, Achilles, the part about Mr. C receiving the letter is quite true. What I meant was, aren't you skeptical about the content of the letter—its extravagant claims?

Achilles: Why should I be? Hmm . . . Well, of course I am. I'm a very skeptical person, as both of you should well know by now. It's very hard to convince me of anything, no matter how true or false it is.

Tortoise: Very well put, Achilles. You certainly have a first-class awareness of your own mental workings.

Achilles: Did it ever occur to you, my friends, that these claims of Najunamar might be incorrect?

Crab: Frankly, Achilles, being rather conservative and orthodox myself, I was a bit concerned about that very point on first receiving the letter. In fact, I suspected at first that here was an out-and-out fraud. But on second thought, it occurred to me that not many types of people could manufacture such strange-sounding and complex results purely from their imagination. In fact, what it boiled down to was this question:

"Which is the more likely: a charlatan of such extraordinary ingenuity, or a mathematician of great genius?" And before long, I realized that the probabilities clearly favored the former.

Achilles: Didn't you directly check out any of his amazing claims, however?

Crab: Why should I? The probability argument was the most convincing thing I had ever thought of; no mathematical proof would have equaled it. But Mr. T here insisted on rigor. I finally gave in to his insistence, and checked all of Najunamar's results. To my great surprise, each one of them was right. How he discovered them, I'll never know, however. He must have some amazing and inscrutable Oriental type of insight which we here in the Occident can have no inkling of. At present, that's the only theory which makes any sense to me.

Tortoise: Mr. Crab has always been a little more susceptible to mystical or fanciful explanations than I am. I have full confidence that whatever Najunamar did in his way has a complete parallel inside orthodox mathematics. There is no way of doing mathematics which is fundamentally different from what we now know, in my opinion.

Achilles: That is an interesting opinion. I suppose it has something to do with the Church-Turing Thesis and related topics.

Crab: Oh, well, let us leave these technical matters aside on such a fine day, and enjoy the quiet of the forest, the chirping of the birds, and the play of sunlight on the new leaves and buds. Ho!

Tortoise: I second the motion. After all, all generations of Tortoises have reveled in such delights of nature.

Crab: As have all generations of Crabs.

Achilles: You don't happen to have brought your flute along, by any chance, Mr. C?

Crab: Why, certainly! I take it with me everywhere. Would you like to hear a tune or two?

Achilles: It would be delightful, in this pastoral setting. Do you play from memory?

Crab: Sad to say, that is beyond my capability. I have to read my music from a sheet. But that is no problem. I have several very pleasant pieces here in this case.

(He opens up a thin case and draws out a few pieces of paper. The topmost one has the following symbols on it:

$$\forall a:\sim Sa=0$$

He sticks the top sheet into a little holder attached to his flute, and plays. The tune is very short.)

Achilles: That was charming. *(Peers over at the sheet on the flute, and a quizzical expression beclouds his face.)* What is that statement of number theory doing, attached to your flute like that?

(The Crab looks at his flute, then his music, turns his head all around, and appears slightly confused.)

Crab: I don't understand. What statement of number theory?

Achilles: "Zero is not the successor of any natural number." Right there, in the holder on your flute!

Crab: That's the third Piano Postulate. There are five of them, and I've arranged them all for flute. They're obvious, but catchy.

Achilles: What's not obvious to me is how a number-theoretical statement can be played as music.

Crab: But I insist, it's NOT a number-theoretical statement—it's a Piano Postulate! Would you like to hear another?

Achilles: I'd be enchanted.

(The Crab places another piece of paper on his flute, and this time Achilles watches more carefully.)

Well, I watched your eyes, and they were looking at that FORMULA on the sheet. Are you sure that that is musical notation? I swear, it most amazingly resembles the notation which one might use in a formalized version of number theory.

Crab: How odd! But certainly that is music, not any kind of statement of mathematics, as far as I can tell! Of course, I am not a mathematician in any sense of the word. Would you like to hear any other tunes?

Achilles: By all means. Have you some others?

Crab: Scads.

(He takes a new sheet, and attaches it to his flute. It contains the following symbols:

$$\sim\exists a:\exists b:(SSa\cdot SSb)=SSSSSSSSSSSSS0$$

Achilles peers at it, while the Crab plays it.)

Isn't it lovely?

Achilles: Yes, it certainly is a tuneful little piece. But I have to say, it's looking more and more like number theory to me.

Crab: Heavens! It is just my usual music notation, nothing more. I simply don't know how you read all these extramusical connotations into a straightforward representation for sounds.

Achilles: Would you be averse to playing a piece of my own composition?

Crab: Not in the least. Have you got it with you?

Achilles: Not yet, but I have a hunch I might be able to compose some tunes all by myself.

Tortoise: I must tell you, Achilles, that Mr. C is a harsh judge of music composed by others, so do not be disappointed if, by some chance, he is not an enthusiast for your efforts.

Achilles: That is very kind of you to forewarn me. Still, I'm willing to give it a try . . .

(He writes:

$$((SSS0\cdot SSS0)+(SSSS0\cdot SSSS0))=(SSSSS0\cdot SSSSS0)$$

The Crab takes it, looks it over for a moment, then sets it in his music holder, and pipes.)

Crab: Why, that's quite nice, Achilles. I enjoy strange rhythms.

Achilles: What's strange about the rhythms in that piece?

Crab: Oh, naturally, to you as the composer it must seem quite bland, but to my ears, shifting from a 3/3 rhythm to 4/4 and then to 5/5 is quite exotic. If you have any other songs, I'd be glad to play them.

Achilles: Thank you very much. I've never composed anything before, and I must say composing is quite different from how I had imagined it to be. Let me try my hand at another one. *(Jots down a line.)*

$$\sim\exists a{:}\exists b{:}(SSa \cdot SSb)=SSSSSSSSSSSSSS0$$

Crab: Hmmm . . . Isn't that just a copy of my earlier piece?

Achilles: Oh, no! I've added one more S. Where you had thirteen in a row, I have fourteen.

Crab: Oh, yes. Of course. *(He plays it, and looks very stern.)*

Achilles: I do hope you didn't dislike my piece!

Crab: I am afraid, Achilles, that you completely failed to grasp the subtleties of my piece, upon which yours is modeled. But how could I expect you to understand it on first hearing? One does not always understand what is at the root of beauty. It is so easy to mistake the superficial aspects of a piece for its beauty, and to imitate them, when the beauty itself is locked deep inside the music, in a way which seems always to elude analysis.

Achilles: I am afraid that you have lost me a little in your erudite commentary. I understand that my piece does not measure up to your high standards, but I do not know exactly where I went astray. Could you perhaps tell me some specific way in which you find fault with my composition?

Crab: One possible way to save your composition, Achilles, would be to insert another three S's—five would do as well—into that long group of S's near the end. That would create a subtle and unusual effect.

Achilles: I see.

Crab: But there are other ways you might choose to change your piece. Personally, I would find it most appealing to put another tilde in the front. Then there would be a nice balance between the beginning and the end. Having two tildes in a row never fails to give a gay little twist to a piece, you know.

Achilles: How about if I take both of your suggestions, and make the following piece?

$$\sim\sim\exists a{:}\exists b{:}(SSa \cdot SSb)=SSSSSSSSSSSSSSSSS0$$

Crab (a painful grimace crossing his face): Now, Achilles, it is important to learn the following lesson: never try to put too much into any single piece. There is always a point beyond which it cannot be improved, and further attempts to improve it will in fact destroy it. Such is the case in this example. Your idea of incorporating both of my suggestions together does not yield the desired extra amount of beauty, but on the contrary creates an imbalance which quite takes away all the charm.

Achilles: How is it that two very similar pieces, such as yours with thirteen **S**'s, and mine with fourteen **S**'s, seem to you to be so different in their musical worth? Other than in that minor respect, the two are identical.

Crab: Gracious! There is a world of difference between your piece and mine. Perhaps this is a place where words fail to convey what the spirit can feel. Indeed, I would venture to say that there exists no set of rules which delineate what it is that makes a piece beautiful, nor could there ever exist such a set of rules. The sense of Beauty is the exclusive domain of Conscious Minds, minds which through the experience of living have gained a depth that transcends explanation by any mere set of rules.

Achilles: I will always remember this vivid clarification of the nature of Beauty. I suppose that something similar applies to the concept of Truth, as well?

Crab: Without doubt. Truth and Beauty are as interrelated as—as—

Achilles: As interrelated as, say, mathematics and music?

Crab: Oh! You took the words right out of my mouth! How did you know that that is what I was thinking?

Tortoise: Achilles is very clever, Mr. C. Never underestimate the potency of his insight.

Achilles: Would you say that there could conceivably be any relationship between the truth or falsity of a particular statement of mathematics, and the beauty, or lack of beauty, of an associated piece of music? Or is that just a far-fetched fancy of mine, with no basis in reality?

Crab: If you are asking me, that is carrying things much too far. When I spoke of the interrelatedness of music and mathematics, I was speaking very figuratively, you know. As for a direct connection between specific pieces of music and specific statements of mathematics, however, I harbor extremely grave doubts about its possibility. I would humbly counsel you not to give too much time to such idle speculations.

Achilles: You are no doubt right. It would be most unprofitable. Perhaps I ought to concentrate on sharpening my musical sensitivity by composing some new pieces. Would you be willing to serve as my mentor, Mr. C?

Crab: I would be very happy to aid you in your steps towards musical understanding.

(So Achilles takes pen in hand, and, with what appears to be a great deal of concentration, writes:

∧**00a∀**′∨~∧∧**:b+cS(∃∃=0**∧⊃**((~d)**<∨**(∀S·+(**>∨

The Crab looks very startled.)

You really want me to play that—that—that whatever-it-is?

Achilles: Oh, please do!

(So the Crab plays it, with evident difficulty.)

Tortoise: Bravo! Bravo! Is John Cage your favorite composer, Achilles?

Achilles: Actually, he's my favorite anti-composer. Anyway, I'm glad you liked MY music.

Crab: The two of you may find it amusing to listen to such totally meaningless cacophony, but I assure you it is not at all pleasant for a sensitive composer to be subjected to such excruciating, empty dissonances and meaningless rhythms. Achilles, I thought you had a good feeling for music. Could it be that your previous pieces had merit merely by coincidence?

Achilles: Oh, please forgive me, Mr. Crab. I was trying to explore the limits of your musical notation. I wanted to learn directly what kinds of sound result when I write certain types of note sequences, and also how you evaluate pieces written in various styles.

Crab: Harrumph! I am not just an automatic music-machine, you know. Nor am I a garbage disposal for musical trash.

Achilles: I am very sorry. But I feel that I have learned a great deal by writing that small piece, and I am convinced that I can now write much better music than I ever could have if I hadn't tried that idea. And if you'll just play one more piece of mine, I have high hopes that you will feel better about my musical sensitivities.

Crab: Well, all right. Write it down and I'll give it a chance.

(Achilles writes:

$$\forall a:\forall b:\langle(a\cdot a)=(SS0\cdot(b\cdot b))\supset a=0\rangle$$

and the Crab plays.)

You were right, Achilles. You seem to have completely regained your musical acuity. This is a little gem! How did you come to compose it? I have never heard anything like it. It obeys all the rules of harmony, and yet has a certain—what shall I say?—irrational appeal to it. I can't put my finger on it, but I like it for that very reason.

Achilles: I kind of thought you might like it.

Tortoise: Have you got a name for it, Achilles? Perhaps you might call it "The Song of Pythagoras". You remember that Pythagoras and his followers were among the first to study musical sound.

Achilles: Yes, that's true. That would be a fine title.

Crab: Wasn't Pythagoras also the first to discover that the ratio of two squares can never be equal to 2?

Tortoise: I believe you're right. It was considered a truly sinister discovery at the time, for never before had anyone realized that there are numbers—such as the square root of 2—which are not ratios of integers. And thus the discovery was deeply disturbing to the Pythagoreans, who felt that it revealed an unsuspected and grotesque defect in the abstract world of numbers. But I don't know what this has to do with the price of tea in China.

Achilles: Speaking of tea, isn't that the teahouse just up there ahead of us?

Tortoise: Yes, that's it, all right. We ought to be there in a couple of minutes.

Achilles: Hmm . . . That's just enough time for me to whistle for you the tune which the taxi driver this morning had on his radio. It went like this.

Crab: Hold on for a moment; I'll get some paper from my case, and jot down your tune. *(Scrounges around inside his case, and finds a blank sheet.)* Go ahead; I'm ready.

(Achilles whistles a rather long tune, and the Crab scrambles to keep up with him.)

Could you whistle the last few bars again?

Achilles: Why, certainly.

(After a couple of such repeats, the session is complete, and the Crab proudly displays his transcription:

$$<((SSSSS0\cdot SSSSS0)+(SSSSS0\cdot SSSSS0))=((SSSSSSS0\cdot SSSSSSS0)+(S0\cdot S0))$$
$$\wedge\sim\exists b:<\exists c:(Sc+b)=((SSSSSSS0\cdot SSSSSSS0)+(S0\cdot S0))\wedge\exists d:\exists d':\exists e:\exists e':$$
$$<\sim<d=e\vee d=e'>\wedge<b=((Sd\cdot Sd)+(Sd'\cdot Sd'))\wedge b=((Se\cdot Se)+(Se'\cdot Se'))>>>>$$

The Crab then plays it himself.)

Tortoise: It's peculiar music, isn't it? It sounds a wee bit like music from India, to me.

Crab: Oh, I think it's too simple to be from India. But of course I know precious little about such things.

Tortoise: Well, here we are at the teahouse. Shall we sit outside here, on the verandah?

Crab: If you don't mind, I'd prefer to go inside. I've gotten perhaps enough sun for the day.

(They go inside the teahouse and are seated at a nice wooden table, and order cakes and tea. Soon a cart of scrumptious-looking pastries is wheeled up, and each of them chooses his favorite.)

Achilles: You know, Mr. C, I would love to know what you think of another piece which I have just composed in my head.

Crab: Can you show it to me? Here, write it down on this napkin.

(Achilles writes:

$$\forall a:\exists b:\exists c:<\sim\exists d:\exists e:<(SSd\cdot SSe)=b\vee(SSd\cdot SSe)=c>\wedge(a+a)=(b+c)>$$

The Crab and Tortoise study it with interest.)

Tortoise: Is it another beautiful piece, Mr. C, in your opinion?

Crab: Well, uh . . . *(Shifts in his chair, and looks somewhat uncomfortable.)*

Achilles: What's the matter? Is it harder to decide whether this piece is beautiful than it is for other pieces?

Crab: Ahm . . . No, it's not that—not at all. It's just that, well . . . I really have to HEAR a piece before I can tell how much I like it.

Achilles: So go ahead and play it! I'm dying to know whether you find it beautiful or not.

Crab: Of course, I'd be extremely glad to play it for you. The only thing is . . .

Achilles: Can't you play it for me? What's the matter? Why are you balking?

Tortoise: Don't you realize, Achilles, that for Mr. Crab to fulfill your request would be most impolite and disturbing to the clientele and employees of this fine establishment?

Crab (suddenly looking relieved): That's right. We have no right to impose our music on others.

Achilles (dejectedly): Oh, PHOOEY! And I SO much wanted to know what he thinks of this piece!

Crab: Whew! That was a close call!

Achilles: What was that remark?

Crab: Oh—nothing. It's just that that waiter over there, he got knocked into by another waiter, and almost dropped a whole pot of tea into a lady's lap. A narrow escape, I must say. What do you say, Mr. Tortoise?

Tortoise: Very good teas, I'd say. Wouldn't you agree, Achilles?

Achilles: Oh, yes. Prime teas, in fact.

Crab: Definitely. Well, I don't know about you two, but I should perhaps be going, for I've a long steep trail back to my house, on the other side of this hill.

Achilles: You mean this is a big bluff?

Crab: You said it, Achilles.

Achilles: I see. Well, I'll have to remember that.

Crab: It has been such a jolly afternoon, Achilles, and I sincerely hope we will exchange more musical compositions another day.

Achilles: I'm looking forward to that very much, Mr. C. Well, good-bye.

Tortoise: Good-bye, Mr. C.

(And the Crab heads off down his side of the hill.)

Achilles: Now there goes a brilliant fellow . . . In my estimation, he's at least four times as smart as any crab alive. Or he might even be five—

Tortoise: As you said in the beginning, and probably shall be saying forevermore, words without end.

CHAPTER XVII

Church, Turing, Tarski, and Others

Formal and Informal Systems

WE HAVE COME to the point where we can develop one of the main theses of this book: that every aspect of thinking can be viewed as a high-level description of a system which, on a low level, is governed by simple, even formal, rules. The "system", of course, is a brain—unless one is speaking of thought processes flowing in another medium, such as a computer's circuits. The image is that of a formal system underlying an "informal system"—a system which can, for instance, make puns, discover number patterns, forget names, make awful blunders in chess, and so forth. This is what one sees from the outside: its informal, overt, software level. By contrast, it has a formal, hidden, hardware level (or "substrate") which is a formidably complex mechanism that makes transitions from state to state according to definite rules physically embodied in it, and according to the input of signals which impinge on it.

A vision of the brain such as this has many philosophical and other consequences, needless to say. I shall try to spell some of them out in this Chapter. Among other things, this vision seems to imply that, at bottom, the brain is some sort of a "mathematical" object. Actually, that is at best a very awkward way to look at the brain. The reason is that, even if a brain is, in a technical and abstract sense, some sort of formal system, it remains true that mathematicians only work with simple and elegant systems, systems in which everything is extremely clearly defined—and the brain is a far cry from that, with its ten billion or more semi-independent neurons, quasi-randomly connected up to each other. So mathematicians would never study a real brain's networks. And if you define "mathematics" as what mathematicians enjoy doing, then the properties of brains are not mathematical.

The only way to understand such a complex system as a brain is by chunking it on higher and higher levels, and thereby losing some precision at each step. What emerges at the top level is the "informal system" which obeys so many rules of such complexity that we do not yet have the vocabulary to think about it. And that is what Artificial Intelligence research is hoping to find. It has quite a different flavor from mathematics research. Nevertheless, there is a loose connection to mathematics: AI people often come from a strong mathematics background, and mathematicians sometimes are intrigued by the workings of their own brains. The following passage, quoted from Stanislaw Ulam's autobiographical *Adventures of a Mathematician,* illustrates this point:

> It seems to me that more could be done to elicit . . . the nature of associations, with computers providing the means for experimentation. Such a study would have to involve a gradation of notions, of symbols, of classes of symbols, of classes of classes, and so on, in the same way that the complexity of mathematical or physical structures is investigated.
>
> There must be a trick to the train of thought, a recursive formula. A group of neurons starts working automatically, sometimes without external impulse. It is a kind of iterative process with a growing pattern. It wanders about in the brain, and the way it happens must depend on the memory of similar patterns.[1]

Intuition and the Magnificent Crab

Artificial Intelligence is often referred to as "AI". Often, when I try to explain what is meant by the term, I say that the letters "AI" could just as well stand for "Artificial Intuition", or even "Artificial Imagery". The aim of AI is to get at what is happening when one's mind silently and invisibly chooses, from a myriad alternatives, which one makes most sense in a very complex situation. In many real-life situations, deductive reasoning is inappropriate, not because it would give *wrong* answers, but because there are too many correct but *irrelevant* statements which can be made; there are just too many things to take into account simultaneously for reasoning alone to be sufficient. Consider this mini-dialogue:

> "The other day I read in the paper that the—"
> "Oh—you were reading? It follows that you have eyes. Or at least *one* eye. Or rather, that you had at least one eye *then.*"

A sense of judgment—"What is important here, and what is not?"—is called for. Tied up with this is a sense of simplicity, a sense of beauty. Where do these intuitions come from? How can they emerge from an underlying formal system?

In the *Magnificrab,* some unusual powers of the Crab's mind are revealed. His own version of his powers is merely that he listens to music and distinguishes the *beautiful* from the *non-beautiful.* (Apparently for him there is a sharp dividing line.) Now Achilles finds another way to describe the Crab's abilities: the Crab divides statements of number theory into the categories *true* and *false.* But the Crab maintains that, if he chances to do so, it is only by the purest accident, for he is, by his own admission, incompetent in mathematics. What makes the Crab's performance all the more mystifying to Achilles, however, is that it seems to be in direct violation of a celebrated result of metamathematics with which Achilles is familiar:

CHURCH'S THEOREM: There is no infallible method for telling theorems of TNT from nontheorems.

It was proven in 1936 by the American logician Alonzo Church. Closely related is what I call the

TARSKI-CHURCH-TURING THEOREM: There is no infallible method for telling true from false statements of number theory.

The Church-Turing Thesis

To understand Church's Theorem and the Tarski-Church-Turing Theorem better, we should first describe one of the ideas on which they are based; and that is the *Church-Turing Thesis* (often called "Church's Thesis"). For the Church-Turing Thesis is certainly one of the most important concepts in the philosophy of mathematics, brains, and thinking.

Actually, like tea, the Church-Turing Thesis can be given in a variety of different strengths. So I will present it in various versions, and we will consider what they imply.

The first version sounds very innocent—in fact almost pointless:

CHURCH-TURING THESIS, TAUTOLOGICAL VERSION: Mathematics problems can be solved only by doing mathematics.

Of course, its meaning resides in the meaning of its constituent terms. By "mathematics problem" I mean the problem of deciding whether some number possesses or does not possess a given arithmetical property. It turns out that by means of Gödel-numbering and related coding tricks, almost any problem in any branch of mathematics can be put into this form, so that "mathematics problem" retains its ordinary meaning. What about "doing mathematics"? When one tries to ascertain whether a number has a property, there seem to be only a small number of operations which one uses in combination over and over again—addition, multiplication, checking for equality or inequality. That is, loops composed of such operations seem to be the only tool we have that allows us to probe the world of numbers. Note the word "seem". This is the critical word which the Church-Turing Thesis is about. We can give a revision:

CHURCH-TURING THESIS, STANDARD VERSION: Suppose there is a method which a sentient being follows in order to sort numbers into two classes. Suppose further that this method always yields an answer within a finite amount of time, and that it always gives the same answer for a given number. *Then:* Some terminating FlooP program (i.e., some general recursive function) exists which gives exactly the same answers as the sentient being's method does.

The central hypothesis, to make it very clear, is that any mental process which divides numbers into two sorts can be described in the form of a FlooP program. The intuitive belief is that there are no other tools than those in FlooP, and that there are no ways to use those tools other than by unlimited iterations (which FlooP allows). The Church-Turing Thesis is not a provable fact in the sense of a Theorem of mathematics—it is a hypothesis about the processes which human brains use.

The Public-Processes Version

Now some people might feel that this version asserts too much. These people might put their objections as follows: "Someone such as the Crab might exist—someone with an almost mystical insight into mathematics, but who is just as much in the dark about his own peculiar abilities as anyone else—and perhaps that person's mental mechanisms carry out operations which have no counterpart in FlooP." The idea is that perhaps we have a subconscious potential for doing things which transcend the conscious processes—things which are somehow inexpressible in terms of the elementary FlooP operations. For these objectors, we shall give a weaker version of the Thesis, one which distinguishes between public and private mental processes:

CHURCH-TURING THESIS, PUBLIC-PROCESSES VERSION: Suppose there is a method which a sentient being follows in order to sort numbers into two classes. Suppose further that this method always yields an answer within a finite amount of time, and that it always gives the same answer for a given number. *Proviso:* Suppose also that this method can be communicated reliably from one sentient being to another by means of language. *Then:* Some terminating FlooP program (i.e., general recursive function) exists which gives exactly the same answers as the sentient beings' method does.

This says that public methods are subject to "FlooPification", but asserts nothing about private methods. It does not say that they are un-FlooP-able, but it at least leaves the door open.

Srinivasa Ramanujan

As evidence against any stronger version of the Church-Turing Thesis, let us consider the case of the famous Indian mathematician of the first quarter of the twentieth century, Srinivasa Ramanujan (1887-1920). Ramanujan (Fig. 105) came from Tamilnadu, the southernmost part of India, and studied mathematics a little in high school. One day, someone who recognized Ramanujan's talent for math presented him with a copy of a slightly out-of-date textbook on analysis, which Ramanujan devoured (figuratively speaking). He then began making his own forays into the world of analysis, and by the time he was twenty-three, he had made a number of discoveries which he considered worthwhile. He did not know to whom to turn, but somehow was told about a professor of mathematics in faraway England, named G. H. Hardy. Ramanujan compiled his best

FIGURE 105. Srinivasa Ramanujan and one of his strange Indian melodies.

$$\cfrac{1}{1+\cfrac{e^{-2\pi\sqrt{5}}}{1+\cfrac{e^{-4\pi\sqrt{5}}}{1+\cfrac{e^{-6\pi\sqrt{5}}}{1+\ddots}}}} = \left(\frac{\sqrt{5}}{1+\sqrt[5]{5^{3/4}\left(\frac{\sqrt{5}-1}{2}\right)^{5/2}-1}} - \frac{\sqrt{5}+1}{2}\right)e^{2\pi/\sqrt{5}}$$

results together in a packet of papers, and sent them all to the unforewarned Hardy with a covering letter which friends helped him express in English. Below are some excerpts taken from Hardy's description of his reaction upon receiving the bundle:

> . . . It soon became obvious that Ramanujan must possess much more general theorems and was keeping a great deal up his sleeve. . . . [Some formulae] defeated me completely; I had never seen anything in the least like them before. A single look at them is enough to show that they could only be written down by a mathematician of the highest class. They must be true because, if they were not true, no one would have had the imagination to invent them. Finally . . . the writer must be completely honest, because great mathematicians are commoner than thieves or humbugs of such incredible skill.[2]

What resulted from this correspondence was that Ramanujan came to England in 1913, sponsored by Hardy; and then followed an intense collaboration which terminated in Ramanujan's early demise at age thirty-three from tuberculosis.

Ramanujan had several extraordinary characteristics which set him apart from the majority of mathematicians. One was his lack of rigor. Very often he would simply state a result which he would insist, had just come to him from a vague intuitive source, far out of the realm of conscious probing. In fact, he often said that the goddess Namagiri inspired him in his dreams. This happened time and again, and what made it all the more mystifying—perhaps even imbuing it with a certain mystical quality—was the fact that many of his "intuition-theorems" were *wrong*. Now there is a curious paradoxical effect where sometimes an event which you think could not help but make credulous people become a little more skeptical, actually has the reverse effect, hitting the credulous ones in some vulnerable spot of their minds, tantalizing them with the hint of some baffling irrational side of human nature. Such was the case with Ramanujan's blunders: many educated people with a yearning to believe in something of the sort considered Ramanujan's intuitive powers to be evidence of a mystical insight into Truth, and the fact of his fallibility seemed, if anything, to strengthen, rather than weaken, such beliefs.

Of course it didn't hurt that he was from one of the most backward parts of India, where fakirism and other eerie Indian rites had been practiced for millennia, and were still practiced with a frequency probably exceeding that of the teaching of higher mathematics. And his occasional wrong flashes of insight, instead of suggesting to people that he was merely human, paradoxically inspired the idea that Ramanujan's wrongness always had some sort of "deeper rightness" to it—an "Oriental" rightness, perhaps touching upon truths inaccessible to Western minds. What a delicious, almost irresistible thought! Even Hardy—who would have been the first to deny that Ramanujan had any mystical powers—once wrote about one of Ramanujan's failures, "And yet I am not sure that, in some ways, his failure was not more wonderful than any of his triumphs."

The other outstanding feature of Ramanujan's mathematical personality was his "friendship with the integers", as his colleague Littlewood put it. This is a characteristic that a fair number of mathematicians share to some degree or other, but which Ramanujan possessed to an extreme. There are a couple of anecdotes which illustrate this special power. The first one is related by Hardy:

> I remember once going to see him when he was lying ill at Putney. I had ridden in taxi-cab No. 1729, and remarked that the number seemed to me rather a dull one, and that I hoped it was not an unfavourable omen. "No," he replied, "it is a very interesting number; it is the smallest number expressible as a sum of two cubes in two different ways." I asked him, naturally, whether he knew the answer to the corresponding problem for fourth powers; and he replied, after a moment's thought, that he could see no obvious example, and thought that the first such number must be very large.[3]

It turns out that the answer for fourth powers is:

$$635318657 = 134^4 + 133^4 = 158^4 + 59^4$$

The reader may find it interesting to tackle the analogous problem for squares, which is much easier.

It is actually quite interesting to ponder why it is that Hardy immediately jumped to fourth powers. After all, there are several other reasonably natural generalizations of the equation

$$u^3 + v^3 = x^3 + y^3$$

along different dimensions. For instance, there is the question about representing a number in three distinct ways as a sum of two cubes:

$$r^3 + s^3 = u^3 + v^3 = x^3 + y^3$$

Or, one can use three different cubes:

$$u^3 + v^3 + w^3 = x^3 + y^3 + z^3$$

Or one can even make a Grand Generalization in all dimensions at once:

$$r^4 + s^4 + t^4 = u^4 + v^4 + w^4 = x^4 + y^4 + z^4$$

There is a sense, however, in which Hardy's generalization is "the most mathematician-like". Could this sense of mathematical esthetics ever be programmed?

The other anecdote is taken from a biography of Ramanujan by his countryman S. R. Ranganathan, where it is called "Ramanujan's Flash". It is related by a Indian friend of Ramanujan's from his Cambridge days, Dr. P. C. Mahalanobis.

> On another occasion, I went to his room to have lunch with him. The First World War had started some time earlier. I had in my hand a copy of the monthly "Strand Magazine" which at that time used to publish a number of puzzles to be solved by readers. Ramanujan was stirring something in a pan over the fire for our lunch. I was sitting near the table, turning over the pages of the Magazine. I got interested in a problem involving a relation between two numbers. I have forgotten the details; but I remember the type of the problem. Two British officers had been billeted in Paris in two different houses in a long street; the door numbers of these houses were related in a special way; the problem was to find out the two numbers. It was not at all difficult. I got the solution in a few minutes by trial and error.
>
> MAHALANOBIS (in a joking way): Now here is a problem for you.
>
> RAMANUJAN: What problem, tell me. (He went on stirring the pan.)
>
> I read out the question from the "Strand Magazine".
>
> RAMANUJAN: Please take down the solution. (He dictated a continued fraction.)
>
> The first term was the solution which I had obtained. Each successive term represented successive solutions for the same type of relation between two numbers, as the number of houses in the street would increase indefinitely. I was amazed.
>
> MAHALANOBIS: Did you get the solution in a flash?
>
> RAMANUJAN: Immediately I heard the problem, it was clear that the solution was obviously a continued fraction; I then thought, "Which continued fraction?" and the answer came to my mind. It was just as simple as this.[4]

Hardy, as Ramanujan's closest co-worker, was often asked after

Ramanujan's death if there had been any occult or otherwise exotically flavored elements to Ramanujan's thinking style. Here is one comment which he gave:

> I have often been asked whether Ramanujan had any special secret; whether his methods differed in kind from those of other mathematicians; whether there was anything really abnormal in his mode of thought. I cannot answer these questions with any confidence or conviction; but I do not believe it. My belief is that all mathematicians think, at bottom, in the same kind of way, and that Ramanujan was no exception.[5]

Here Hardy states in essence his own version of the Church-Turing Thesis. I paraphrase:

CHURCH-TURING THESIS, HARDY'S VERSION: At bottom, all mathematicians are isomorphic.

This does not equate the mathematical potential of mathematicians with that of general recursive functions; for that, however, all you need is to show that *some* mathematician's mental capacity is no more general than recursive functions. Then, if you believe Hardy's Version, you know it for *all* mathematicians.

Then Hardy compares Ramanujan with calculating prodigies:

> His memory, and his powers of calculation, were very unusual, but they could not reasonably be called "abnormal". If he had to multiply two large numbers, he multiplied them in the ordinary way; he could do it with unusual rapidity and accuracy, but not more rapidly and accurately than any mathematician who is naturally quick and has the habit of computation.[6]

Hardy describes what he perceived as Ramanujan's outstanding intellectual attributes:

> With his memory, his patience, and his power of calculation, he combined *a power of generalisation, a feeling for form, and a capacity for rapid modification of his hypotheses,* that were often really startling, and made him, in his own field, without a rival in his day.[7]

The part of this passage which I have italicized seems to me to be an excellent characterization of some of the subtlest features of intelligence in general. Finally, Hardy concludes somewhat nostalgically:

> [His work] has not the simplicity and inevitableness of the very greatest work; it would be greater if it were less strange. One gift it has which no one can deny—profound and invincible originality. He would probably have been a greater mathematician if he had been caught and tamed a little in his youth; he would have discovered more that was new, and that, no doubt, of greater importance. On the other hand he would have been less of a Ramanujan, and more of a European professor and the loss might have been greater than the gain.[8]

The esteem in which Hardy held Ramanujan is revealed by the romantic way in which he speaks of him.

"Idiots Savants"

There is another class of people whose mathematical abilities seem to defy rational explanation—the so-called "idiots savants", who can perform complex calculations at lightning speeds in their heads (or wherever they do it). Johann Martin Zacharias Dase, who lived from 1824 to 1861 and was employed by various European governments to perform computations, is an outstanding example. He not only could multiply two numbers each of 100 digits in his head; he also had an uncanny sense of quantity. That is, he could just "tell", without counting, how many sheep were in a field, or words in a sentence, and so forth, up to about 30—this in contrast to most of us, who have such a sense up to about 6, with reliability. Incidentally, Dase was not an idiot.

I shall not describe the many fascinating documented cases of "lightning calculators", for that is not my purpose here. But I do feel it is important to dispel the idea that they do it by some mysterious, unanalyzable method. Although it is often the case that such wizards' calculational abilities far exceed their abilities to explain their results, every once in a while, a person with other intellectual gifts comes along who also has this spectacular ability with numbers. From such people's introspection, as well as from extensive research by psychologists, it has been ascertained that nothing occult takes place during the performances of lightning calculators, but simply that their minds race through intermediate steps with the kind of self-confidence that a natural athlete has in executing a complicated motion quickly and gracefully. They do not reach their answers by some sort of instantaneous flash of enlightenment (though subjectively it may feel that way to some of them), but—like the rest of us—by sequential calculation, which is to say, by FlooP-ing (or BlooP-ing) along.

Incidentally, one of the most obvious clues that no "hot line to God" is involved is the mere fact that when the numbers involved get bigger, the answers are slower in coming. Presumably, if God or an "oracle" were supplying the answers, he wouldn't have to slow up when the numbers got bigger. One could probably make a nice plot showing how the time taken by a lightning calculator varies with the sizes of the numbers involved, and the operations involved, and from it deduce some features of the algorithms employed.

The Isomorphism Version of the Church-Turing Thesis

This finally brings us to a strengthened standard version of the Church-Turing Thesis:

CHURCH-TURING THESIS, ISOMORPHISM VERSION: Suppose there is a method which a sentient being follows in order to sort numbers into two classes. Suppose further that this method always yields an answer within a finite amount of time, and that it always gives the same answer for a given number. *Then:* Some terminating FlooP program (i.e., general recursive function) exists which gives exactly the same answers as the sentient being's method does. *Moreover:* The mental process and the FlooP program are isomorphic in the sense that on some level there is a correspondence between the steps being carried out in both computer and brain.

Notice that not only has the conclusion been strengthened, but also the proviso of communicability of the faint-hearted Public-Processes Version has been dropped. This bold version is the one which we now shall discuss.

In brief, this version asserts that when one computes something, one's mental activity can be mirrored isomorphically in some FlooP program. And let it be very clear that this does not mean that the brain is actually running a FlooP program, written in the FlooP language complete with **BEGIN**'s, **END**'s, **ABORT**'s, and the rest—not at all. It is just that the steps are taken in the same order as they could be in a FlooP program, and the logical structure of the calculation can be mirrored in a FlooP program.

Now in order to make sense of this idea, we shall have to make some level distinctions in both computer and brain, for otherwise it could be misinterpreted as utter nonsense. Presumably the steps of the calculation going on inside a person's head are on the highest level, and are supported by lower levels, and eventually by hardware. So if we speak of an isomorphism, it means we've tacitly made the assumption that the highest level can be isolated, allowing us to discuss what goes on there independently of other levels, and then to map that top level into FlooP. To be more precise, the assumption is that there exist software entities which play the roles of various mathematical constructs, and which are activated in ways which can be mirrored exactly inside FlooP (see Fig. 106). What enables these software entities to come into existence is the entire infrastructure discussed in Chapters XI and XII, as well as in the *Prelude, Ant Fugue.* There is no assertion of isomorphic activity on the lower levels of brain and computer (e.g., neurons and bits).

The spirit of the Isomorphism Version, if not the letter, is gotten across by saying that what an idiot savant does in calculating, say, the logarithm of π, is isomorphic to what a pocket calculator does in calculating it—where the isomorphism holds on the arithmetic-step level, *not* on the lower levels of, in the one case, neurons, and in the other, integrated circuits. (Of course different routes can be followed in calculating anything—but presumably the pocket calculator, if not the human, could be instructed to calculate the answer in any specific manner.)

FIGURE 106. The behavior of natural numbers can be mirrored in a human brain or in the programs of a computer. These two different representations can then be mapped onto each other on an appropriately abstract level.

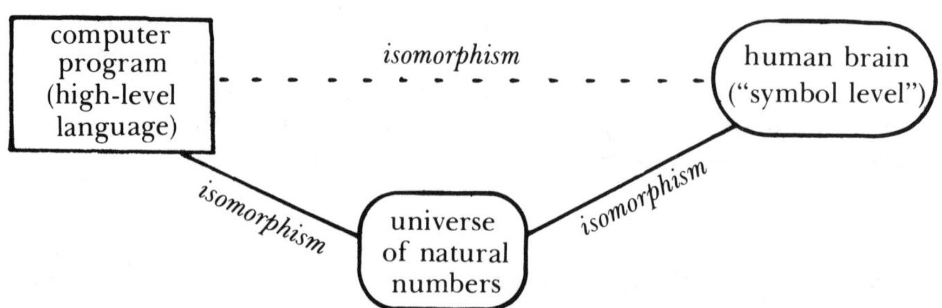

Now this seems quite plausible when the domain referred to is number theory, for there the total universe in which things happen is very small and clean. Its boundaries and residents and rules are well-defined, as in a hard-edged maze. Such a world is far less complicated than the open-ended and ill-defined world which we inhabit. A number theory problem, once stated, is complete in and of itself. A real-world problem, on the other hand, never is sealed off from any part of the world with absolute certainty. For instance, the task of replacing a burnt-out light bulb may turn out to require moving a garbage bag; this may unexpectedly cause the spilling of a box of pills, which then forces the floor to be swept so that the pet dog won't eat any of the spilled pills, etc., etc. The pills and the garbage and the dog and the light bulb are all quite distantly related parts of the world—yet an intimate connection is created by some everyday happenings. And there is no telling what else could be brought in by some other small variations on the expected. By contrast, if you are given a number theory problem, you never wind up having to consider extraneous things such as pills or dogs or bags of garbage or brooms in order to solve your problem. (Of course, your intuitive knowledge of such objects may serve you in good stead as you go about unconsciously trying to manufacture mental images to help you in visualizing the problem in geometrical terms—but that is another matter.)

Because of the complexity of the world, it is hard to imagine a little pocket calculator that can answer questions put to it when you press a few buttons bearing labels such as "dog", "garbage", "light bulb", and so forth. In fact, so far it has proven to be extremely complicated to have a full-size high-speed computer answer questions about what appear to us to be rather simple subdomains of the real world. It seems that a large amount of knowledge has to be taken into account in a highly integrated way for "understanding" to take place. We can liken real-world thought processes to a tree whose visible part stands sturdily above ground but depends vitally on its invisible roots which extend way below ground, giving it stability and nourishment. In this case the roots symbolize complex processes which take place below the conscious level of the mind—processes whose effects permeate the way we think but of which we are unaware. These are the "triggering patterns of symbols" which were discussed in Chapters XI and XII.

Real-world thinking is quite different from what happens when we do a multiplication of two numbers, where everything is "above ground", so to speak, open to inspection. In arithmetic, the top level can be "skimmed off" and implemented equally well in many different sorts of hardware: mechanical adding machines, pocket calculators, large computers, people's brains, and so forth. This is what the Church-Turing Thesis is all about. But when it comes to real-world understanding, it seems that there is no simple way to skim off the top level, and program it alone. The triggering patterns of symbols are just too complex. There must be several levels through which thoughts may "percolate" and "bubble".

In particular—and this comes back to a major theme of Chapters XI and XII—the representation of the real world in the brain, although rooted in isomorphism to some extent, involves some elements which have no counterparts at all in the outer world. That is, there is much more to it than simple mental structures representing "dog", "broom", etc. All of these symbols exist, to be sure—but their internal structures are extremely complex and to a large degree are unavailable for conscious inspection. Moreover, one would hunt in vain to map each aspect of a symbol's internal structure onto some specific feature of the real world.

Processes That Are Not So Skimmable

For this reason, the brain begins to look like a very peculiar formal system, for on its bottom level—the neural level—where the "rules" operate and change the state, there may be no interpretation of the primitive elements (neural firings, or perhaps even lower-level events). Yet on the top level, there emerges a meaningful interpretation—a mapping from the large "clouds" of neural activity which we have been calling "symbols", onto the real world. There is some resemblance to the Gödel construction, in that a high-level isomorphism allows a high level of meaning to be read into strings; but in the Gödel construction, the higher-level meaning "rides" on the lower level—that is, it is derived from the lower level, once the notion of Gödel-numbering has been introduced. But in the brain, the events on the neural level are *not* subject to real-world interpretation; they are simply not imitating anything. They are there purely as the substrate to support the higher level, much as transistors in a pocket calculator are there purely to support its number-mirroring activity. And the implication is that there is no way to skim off just the highest level and make an isomorphic copy in a program; if one is to mirror the brain processes which allow real-world understanding, then one *must* mirror some of the lower-level things which are taking place: the "languages of the brain". This doesn't necessarily mean that one must go all the way down to the level of the hardware, though that may turn out to be the case.

In the course of developing a program with the aim of achieving an "intelligent" (viz., human-like) internal representation of what is "out there", at some point one will probably be forced into using structures and processes which do not admit of any straightforward interpretations—that is, which cannot be directly mapped onto elements of reality. These lower layers of the program will be able to be understood only by virtue of their catalytic relation to layers above them, rather than because of some direct connection they have to the outer world. (A concrete image of this idea was suggested by the Anteater in the *Ant Fugue*: the "indescribably boring nightmare" of trying to understand a book on the letter level.)

Personally, I would guess that such multilevel architecture of concept-handling systems becomes necessary just when processes involving images and analogies become significant elements of the program—in contrast to processes which are supposed to carry out strictly deductive reasoning. Processes which carry out deductive reasoning can be programmed in essentially one single level, and are therefore skimmable, by definition. According to my hypothesis, then, imagery and analogical thought processes intrinsically require several layers of substrate and are therefore intrinsically non-skimmable. I believe furthermore that it is precisely at this same point that creativity starts to emerge—which would imply that creativity intrinsically depends upon certain kinds of "uninterpretable" lower-level events. The layers of underpinning of analogical thinking are, of course, of extreme interest, and some speculations on their nature will be offered in the next two Chapters.

Articles of Reductionistic Faith

One way to think about the relation between higher and lower levels in the brain is this. One could assemble a neural net which, on a local (neuron-to-neuron) level, performed in a manner indistinguishable from a neural net in a brain, but which had no higher-level meaning at all. The fact that the lower level is composed of interacting neurons does not necessarily force any higher level of meaning to appear—no more than the fact that alphabet soup contains letters forces meaningful sentences to be found, swimming about in the bowl. High-level meaning is an optional feature of a neural network—one which may emerge as a consequence of evolutionary environmental pressures.

Figure 107 is a diagram illustrating the fact that emergence of a higher level of meaning is optional. The upwards-pointing arrow indicates that a substrate can occur without a higher level of meaning, but not vice versa: the higher level must be derived from properties of a lower one.

FIGURE 107. Floating on neural activity, the symbol level of the brain mirrors the world. But neural activity per se, which can be simulated on a computer, does not create thought; that calls for higher levels of organization.

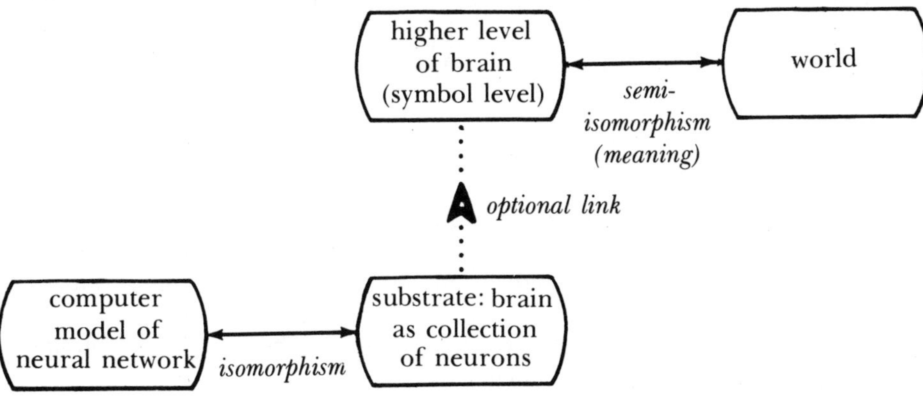

The diagram includes an indication of a computer simulation of a neural network. This is in principle feasible, no matter how complicated the network, provided that the behavior of individual neurons can be described in terms of computations which a computer can carry out. This is a subtle postulate which few people even think of questioning. Nevertheless it is a piece of "reductionistic faith"; it could be considered a "microscopic version" of the Church-Turing Thesis. Below we state it explicitly:

CHURCH-TURING THESIS, MICROSCOPIC VERSION: The behavior of the components of a living being can be simulated on a computer. That is, the behavior of any component (typically assumed to be a cell) can be calculated by a FlooP program (i.e., general recursive function) to any desired degree of accuracy, given a sufficiently precise description of the component's internal state and local environment.

This version of the Church-Turing Thesis says that brain processes do not possess any more mystique—even though they possess more levels of organization—than, say, stomach processes. It would be unthinkable in this day and age to suggest that people digest their food, not by ordinary chemical processes, but by a sort of mysterious and magic "assimilation". This version of the CT-Thesis simply extends this kind of commonsense reasoning to brain processes. In short, it amounts to faith that the brain operates in a way which is, in principle, understandable. It is a piece of reductionist faith.

A corollary to the Microscopic CT-Thesis is this rather terse new macroscopic version:

CHURCH-TURING THESIS, REDUCTIONIST'S VERSION: All brain processes are derived from a computable substrate.

This statement is about the strongest theoretical underpinning one could give in support of the eventual possibility of realizing Artificial Intelligence.

Of course, Artificial Intelligence research is not aimed at simulating neural networks, for it is based on another kind of faith: that probably there are significant features of intelligence which can be floated on top of entirely different sorts of substrates than those of organic brains. Figure 108 shows the presumed relations among Artificial Intelligence, natural intelligence, and the real world.

Parallel Progress in AI and Brain Simulation?

The idea that, if AI is to be achieved, the actual hardware of the brain might one day have to be simulated or duplicated, is, for the present at least, quite an abhorrent thought to many AI workers. Still one wonders, "How finely will we need to copy the brain to achieve AI?" The real answer is probably that it all depends on how many of the features of human consciousness you want to simulate.

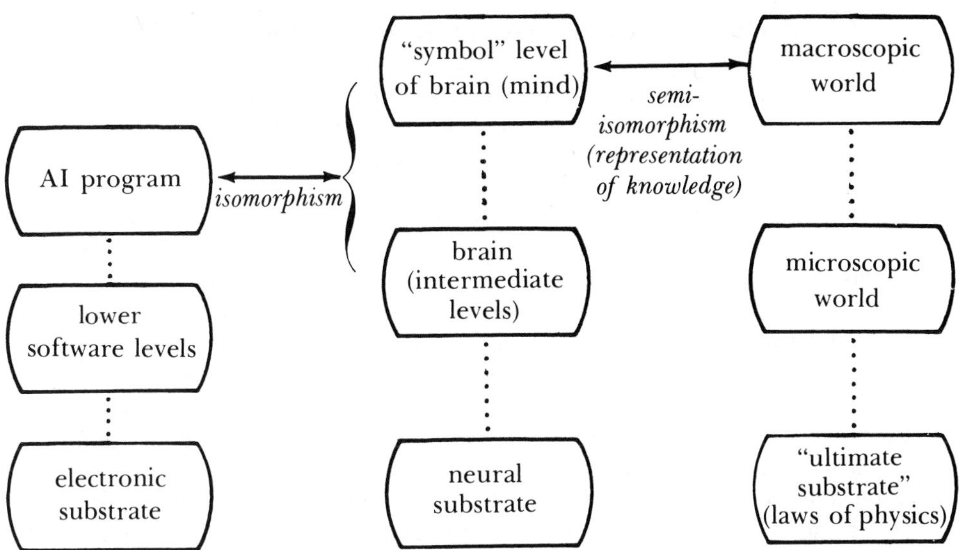

FIGURE 108. Crucial to the endeavor of Artificial Intelligence research is the notion that the symbolic levels of the mind can be "skimmed off" of their neural substrate and implemented in other media, such as the electronic substrate of computers. To what depth the copying of brain must go is at present completely unclear.

Is an ability to play checkers well a sufficient indicator of intelligence? If so, then AI already exists, since checker-playing programs are of world class. Or is intelligence an ability to integrate functions symbolically, as in a freshman calculus class? If so, then AI already exists, since symbolic integration routines outdo the best people in most cases. Or is intelligence the ability to play chess well? If so, then AI is well on its way, since chess-playing programs can defeat most good amateurs; and the level of artificial chess will probably continue to improve slowly.

Historically, people have been naïve about what qualities, if mechanized, would undeniably constitute intelligence. Sometimes it seems as though each new step towards AI, rather than producing something which everyone agrees is real intelligence, merely reveals what real intelligence is *not*. If intelligence involves learning, creativity, emotional responses, a sense of beauty, a sense of self, then there is a long road ahead, and it may be that these will only be realized when we have totally duplicated a living brain.

Beauty, the Crab, and the Soul

Now what, if anything, does all this have to say about the Crab's virtuoso performance in front of Achilles? There are two issues clouded together here. They are:

(1) Could any brain process, under any circumstances, distinguish completely reliably between true and false statements of TNT without being in violation of the Church-Turing Thesis—or is such an act in principle impossible?

(2) Is perception of beauty a brain process?

First of all, in response to (1), if violations of the Church-Turing Thesis are allowed, then there seems to be no fundamental obstacle to the strange events in the Dialogue. So what we are interested in is whether a believer in the Church-Turing Thesis would have to disbelieve in the Crab's ability. Well, it all depends on which version of the CT-Thesis you believe. For example, if you only subscribe to the Public-Processes Version, then you could reconcile the Crab's behavior with it very easily by positing that the Crab's ability is not communicable. Contrariwise, if you believe the Reductionist's Version, you will have a very hard time believing in the Crab's ostensible ability (because of Church's Theorem—soon to be demonstrated). Believing in intermediate versions allows you a certain amount of wishy-washiness on the issue. Of course, switching your stand according to convenience allows you to waffle even more.

It seems appropriate to present a new version of the CT-Thesis, one which is tacitly held by vast numbers of people, and which has been publicly put forth by several authors, in various manners. Some of the more famous ones are: philosophers Hubert Dreyfus, S. Jaki, Mortimer Taube, and J. R. Lucas; the biologist and philosopher Michael Polanyi (a holist par excellence); the distinguished Australian neurophysiologist John Eccles. I am sure there are many other authors who have expressed similar ideas, and countless readers who are sympathetic. I have attempted below to summarize their joint position. I have probably not done full justice to it, but I have tried to convey the flavor as accurately as I can:

CHURCH-TURING THESIS, SOULISTS' VERSION: Some kinds of things which a brain can do can be vaguely approximated on a computer but not most, and certainly not the interesting ones. But anyway, even if they all could, that would still leave the soul to explain, and there is no way that computers have any bearing on that.

This version relates to the tale of the *Magnificrab* in two ways. In the first place, its adherents would probably consider the tale to be silly and implausible, but—not forbidden in principle. In the second place, they would probably claim that appreciation of qualities such as beauty is one of those properties associated with the elusive soul, and is therefore inherently possible only for humans, not for mere machines.

We will come back to this second point in a moment; but first, while we are on the subject of "soulists", we ought to exhibit this latest version in an even more extreme form, since that is the form to which large numbers of well-educated people subscribe these days:

CHURCH-TURING THESIS, THEODORE ROSZAK VERSION: Computers are ridiculous. So is science in general.

This view is prevalent among certain people who see in anything smacking of numbers or exactitude a threat to human values. It is too bad that they do not appreciate the depth and complexity and beauty involved in exploring abstract structures such as the human mind, where, indeed, one comes in intimate contact with the ultimate questions of what to be human is.

Getting back to beauty, we were about to consider whether the appreciation of beauty is a brain process, and if so, whether it is imitable by a computer. Those who believe that it is not accounted for by the brain are very unlikely to believe that a computer could possess it. Those who believe it is a brain process again divide up according to which version of the CT-Thesis they believe. A total reductionist would believe that any brain process can in principle be transformed into a computer program; others, however, might feel that beauty is too ill-defined a notion for a computer program ever to assimilate. Perhaps they feel that the appreciation of beauty requires an element of irrationality, and therefore is incompatible with the very fiber of computers.

Irrational and Rational Can Coexist on Different Levels

However, this notion that "irrationality is incompatible with computers" rests on a severe confusion of levels. The mistaken notion stems from the idea that since computers are faultlessly functioning machines, they are therefore bound to be "logical" on all levels. Yet it is perfectly obvious that a computer can be instructed to print out a sequence of illogical statements—or, for variety's sake, a batch of statements having random truth values. Yet in following such instructions, a computer would not be making any mistakes! On the contrary, it would only be a mistake if the computer printed out something other than the statements it had been instructed to print. This illustrates how faultless functioning on one level may underlie symbol manipulation on a higher level—and the goals of the higher level may be completely unrelated to the propagation of Truth.

Another way to gain perspective on this is to remember that a brain, too, is a collection of faultlessly functioning elements—neurons. Whenever a neuron's threshold is surpassed by the sum of the incoming signals, BANG!—it fires. It never happens that a neuron forgets its arithmetical knowledge—carelessly adding its inputs and getting a wrong answer. Even when a neuron dies, it continues to function correctly, in the sense that its components continue to obey the laws of mathematics and physics. Yet as we all know, neurons are perfectly capable of supporting high-level behavior that is wrong, on its own level, in the most amazing ways. Figure 109 is meant to illustrate such a clash of levels: an incorrect belief held in the software of a mind, supported by the hardware of a faultlessly functioning brain.

The point—a point which has been made several times earlier in various contexts—is simply that meaning can exist on two or more different levels of a symbol-handling system, and along with meaning, rightness and wrongness can exist on all those levels. The presence of meaning on a given

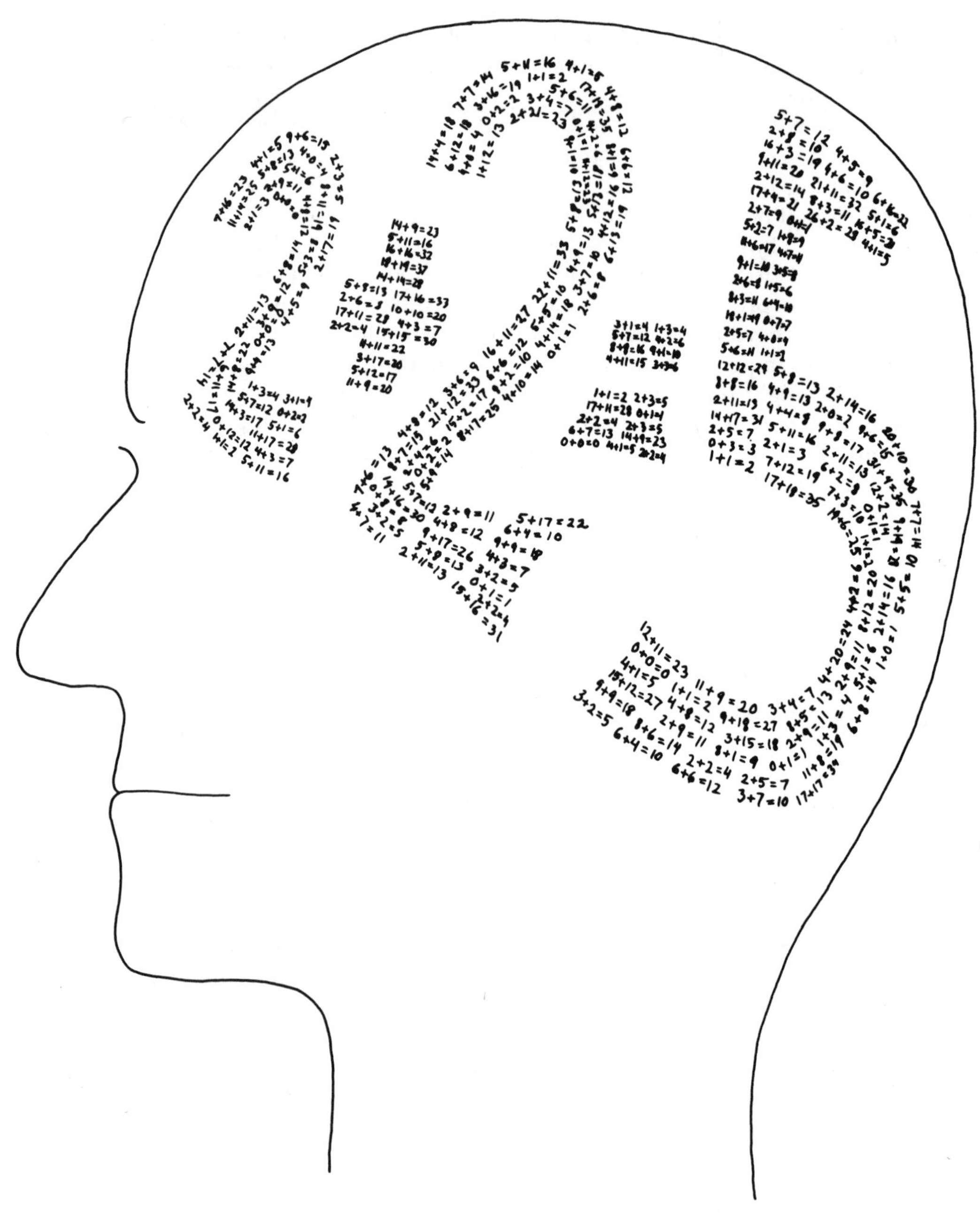

FIGURE 109. The brain is rational; the mind may not be. [*Drawing by the author.*]

level is determined by whether or not reality is mirrored in an isomorphic (or looser) fashion on that level. So the fact that neurons always perform correct additions (in fact, much more complex calculations) has no bearing whatsoever on the correctness of the top-level conclusions supported by their machinery. Whether one's top level is engaged in proving kōans of Boolean Buddhism or in meditating on theorems of Zen Algebra, one's neurons are functioning rationally. By the same token, the high-level symbolic processes which in a brain create the experience of appreciating beauty are perfectly rational on the bottom level, where the faultless functioning is taking place; any irrationality, if there is such, is on the higher level, and is an epiphenomenon—a consequence—of the events on the lower level.

To make the same point in a different way, let us say you are having a hard time making up your mind whether to order a cheeseburger or a pineappleburger. Does this imply that your neurons are also balking, having difficulty deciding whether or not to fire? Of course not. Your hamburger-confusion is a high-level state which fully depends on the efficient firing of thousands of neurons in very organized ways. This is a little ironic, yet it is perfectly obvious when you think about it. Nevertheless, it is probably fair to say that nearly all confusions about minds and computers have their origin in just such elementary level-confusions.

There is no reason to believe that a computer's faultlessly functioning hardware could not support high-level symbolic behavior which would represent such complex states as confusion, forgetting, or appreciation of beauty. It would require that there exist massive subsystems interacting with each other according to a complex "logic". The overt behavior could appear either rational or irrational; but underneath it would be the performance of reliable, logical hardware.

More Against Lucas

Incidentally, this kind of level distinction provides us with some new fuel in arguing against Lucas. The Lucas argument is based on the idea that Gödel's Theorem is applicable, by definition, to machines. In fact, Lucas makes a most emphatic pronunciation:

> Gödel's theorem must apply to cybernetical machines, because it is of the essence of being a machine, that it should be a concrete instantiation of a formal system.[9]

This is, as we have seen, true on the hardware level—but since there may be higher levels, it is not the last word on the subject. Now Lucas gives the impression that in the mind-imitating machines he discusses, there is *only one level* on which manipulation of symbols takes place. For instance, the Rule of Detachment (called "Modus Ponens" in his article) would be wired into the hardware and would be an unchangeable feature of such a machine. He goes further and intimates that if Modus Ponens were not an immutable pillar of the machine's system, but could be overridden on occasion, then:

> The system will have ceased to be a formal logical system, and the machine will barely qualify for the title of a model for the mind.[10]

Now many programs which are being developed in AI research have very little in common with programs for generating truths of number theory—programs with inflexible rules of inference and fixed sets of axioms. Yet they are certainly intended as "models for the mind". On their top level—the "informal" level—there may be manipulation of images, formulation of analogies, forgetting of ideas, confusing of concepts, blurring of distinctions, and so forth. But this does not contradict the fact that they rely on the correct functioning of their underlying hardware as much as brains rely on the correct functioning of their neurons. So AI programs are still "concrete instantiations of formal systems"—but they are not machines to which Lucas' transmogrification of Gödel's proof can be applied. Lucas' argument applies merely to their bottom level, on which their intelligence—however great or small it may be—does not lie.

There is one other way in which Lucas betrays his oversimplified vision of how mental processes would have to be represented inside computer programs. In discussing the matter of consistency, he writes

> If we really were inconsistent machines, we should remain content with our inconsistencies, and would happily affirm both halves of a contradiction. Moreover, we would be prepared to say absolutely anything—which we are not. It is easily shown that in an inconsistent formal system everything is provable.[11]

This last sentence shows that Lucas assumes that the Propositional Calculus must of necessity be built into any formal system which carries out reasoning. In particular, he is thinking of the theorem <<P∧~P>⊃Q> of the Propositional Calculus; evidently he has the erroneous belief that it is an inevitable feature of mechanized reasoning. However, it is perfectly plausible that logical thought processes, such as propositional reasoning, will emerge as *consequences* of the general intelligence of an AI program, rather than being *preprogrammed*. This is what happens in humans! And there is no particular reason to assume that the strict Propositional Calculus, with its rigid rules and the rather silly definition of consistency that they entail, would emerge from such a program.

An Underpinning of AI

We can summarize this excursion into level distinctions and come away with one final, strongest version of the Church-Turing Thesis:

CHURCH-TURING THESIS, AI VERSION: Mental processes of any sort can be simulated by a computer program whose underlying language is of power equal to that of FlooP—that is, in which all partial recursive functions can be programmed.

It should also be pointed out that in practice, many AI researchers rely on another article of faith which is closely related to the CT-Thesis, and which I call the *AI Thesis*. It runs something like this:

AI THESIS: As the intelligence of machines evolves, its underlying mechanisms will gradually converge to the mechanisms underlying human intelligence.

In other words, all intelligences are just variations on a single theme; to create true intelligence, AI workers will just have to keep pushing to ever lower levels, closer and closer to brain mechanisms, if they wish their machines to attain the capabilities which we have.

Church's Theorem

Now let us come back to the Crab and to the question of whether his decision procedure for theoremhood (which is presented in the guise of a filter for musical beauty) is compatible with reality. Actually, from the events which occur in the Dialogue, we have no way of deducing whether the Crab's gift is an ability to tell *theorems* from *nontheorems,* or alternatively, an ability to tell *true statements* from *false ones.* Of course in many cases this amounts to the same thing but Gödel's Theorem shows that it doesn't always. But no matter: both of these alternatives are impossible, if you believe the AI Version of the Church-Turing Thesis. The proposition that it is impossible to have a decision procedure for *theoremhood* in any formal system with the power of TNT is known as *Church's Theorem.* The proposition that it is impossible to have a decision procedure for number-theoretical *truth*—if such truth exists, which one can well doubt after meeting up with all the bifurcations of TNT—follows quickly from *Tarski's Theorem* (published in 1933, although the ideas were known to Tarski considerably earlier).

The proofs of these two highly important results of metamathematics are very similar. Both of them follow quite quickly from self-referential constructions. Let us first consider the question of a decision procedure for TNT-theoremhood. If there were a uniform way by which people could decide which of the classes "theorem" and "nontheorem" any given formula X fell into, then, by the CT-Thesis (Standard Version), there would exist a terminating FlooP program (a general recursive function) which could make the same decision, when given as input the Gödel number of formula X. The crucial step is to recall that any property that can be tested for by a terminating FlooP program is *represented* in TNT. This means that the property of TNT-theoremhood would be represented (as distinguished from merely expressed) inside TNT. But as we shall see in a moment, this would put us in hot water, for if theoremhood is a representable attribute, then Gödel's formula G becomes as vicious as the Epimenides paradox.

It all hinges on what G says: "G is not a theorem of TNT". Assume that G were a theorem. Then, since theoremhood is supposedly represented, the TNT-formula which asserts "G is a theorem" would be a theorem of TNT. But this formula is ~G, the negation of G, so that TNT is inconsistent. On the other hand, assume G were not a theorem. Then once again by the supposed representability of theoremhood, the formula which asserts "G is not a theorem" would be a theorem of TNT. But this formula is G, and once again we get into paradox. Unlike the situation before, there is no resolution of the paradox. The problem is created by the assumption that theoremhood is represented by some formula of TNT, and therefore we must backtrack and erase that assumption. This forces us also to conclude that no FlooP program can tell the Gödel numbers of theorems from those of nontheorems. Finally, if we accept the AI Version of the CT-Thesis, then we must backtrack further, and conclude that no method whatsoever could exist by which humans could reliably tell theorems from nontheorems—and this includes determinations based on beauty. Those who subscribe only to the Public-Processes Version might still think the Crab's performance is possible; but of all the versions, that one is perhaps the hardest one to find any justification for.

Tarski's Theorem

Now let us proceed to Tarski's result. Tarski asked whether there could be a way of expressing in TNT the concept of number-theoretical truth. That theoremhood is expressible (though not representable) we have seen; Tarski was interested in the analogous question regarding the notion of truth. More specifically, he wished to determine whether there is any TNT-formula with a single free variable a which can be translated thus:

"The formula whose Gödel number is a expresses a truth."

Let us suppose, with Tarski, that there is one—which we'll abbreviate as **TRUE{a}**. Now what we'll do is use the diagonalization method to produce a sentence which asserts about itself that it is untrue. We copy the Gödel method exactly, beginning with an "uncle":

$$\exists a{:}<\sim \mathrm{TRUE}\{a\}\wedge \mathrm{ARITHMOQUINE}\{a'',a\}>$$

Let us say the Gödel number of the uncle is *t*. We arithmoquine this very uncle, and produce the Tarski formula T:

$$\exists a{:}<\sim \mathrm{TRUE}\{a\}\wedge \mathrm{ARITHMOQUINE}\{\underbrace{\mathrm{SSS}\ldots\mathrm{SSS}}_{t\ \mathrm{S's}}0/a'',a\}>$$

When interpreted, it says:

> "The arithmoquinification of *t* is the
> Gödel number of a false statement."

But since the arithmoquinification of *t* is T's own Gödel number, Tarski's formula T reproduces the Epimenides paradox to a tee inside TNT, saying of itself, "I am a falsity". Of course, this leads to the conclusion that it must be simultaneously true and false (or simultaneously neither). There arises now an interesting matter: What is so bad about reproducing the Epimenides paradox? Is it of any consequence? After all, we already have it in English, and the English language has not gone up in smoke.

The Impossibility of the Magnificrab

The answer lies in remembering that there are two levels of meaning involved here. One level is the level we have just been using; the other is as a statement of number theory. If the Tarski formula T actually existed, then it would be a statement *about natural numbers* that is both true and false at once! There is the rub. While we can always just sweep the English-language Epimenides paradox under the rug, saying that its subject matter (its own truth) is abstract, this is not so when it becomes a concrete statement about numbers! If we believe this is a ridiculous state of affairs, then we have to undo our assumption that the formula **TRUE{a}** exists. Thus, there is no way of expressing the notion of truth inside TNT. Notice that this makes truth a far more elusive property than theoremhood, for the latter *is* expressible. The same backtracking reasons as before (involving the Church-Turing Thesis, AI Version) lead us to the conclusion that

> The Crab's mind cannot be a truth-recognizer any more than it is
> a TNT-theorem-recognizer.

The former would violate the Tarski-Church-Turing Theorem ("There is no decision procedure for arithmetical truth"), while the latter would violate Church's Theorem.

Two Types of Form

It is extremely interesting, then, to think about the meaning of the word "form" as it applies to constructions of arbitrarily complex shapes. For instance, what is it that we respond to when we look at a painting and feel its beauty? Is it the "form" of the lines and dots on our retina? Evidently it must be, for that is how it gets passed along to the analyzing mechanisms in our heads—but the complexity of the processing makes us feel that we are not merely looking at a two-dimensional surface; we are responding to some sort of inner meaning inside the picture, a multidimensional aspect trapped somehow inside those two dimensions. It is the word "meaning" which is important here. Our minds contain interpreters which accept two-dimensional patterns and then "pull" from them high-dimensional notions which are so complex that we cannot consciously describe them. The same can be said about how we respond to music, incidentally.

It feels subjectively that the pulling-out mechanism of inner meaning is not at all akin to a decision procedure which checks for the presence or absence of some particular quality such as well-formedness in a string. Probably this is because inner meaning is something which reveals more of itself over a period of time. One can never be sure, as one can about well-formedness, that one has finished with the issue.

This suggests a distinction that could be drawn between two senses of "form" in patterns which we analyze. First, there are qualities such as well-formedness, which can be detected by *predictably terminating tests,* as in BlooP programs. These I propose to call *syntactic* qualities of form. One intuitively feels about the syntactic aspects of form that they lie close to the surface, and therefore they do not provoke the creation of multidimensional cognitive structures.

By contrast, the *semantic* aspects of form are those which cannot be tested for in predictable lengths of time: they require *open-ended tests.* Such an aspect is theoremhood of TNT-strings, as we have seen. You cannot just apply some standard test to a string and find out if it is a theorem. Somehow, the fact that its *meaning* is involved is crucially related to the difficulty of telling whether or not a string is a TNT-theorem. The act of pulling out a string's meaning involves, in essence, establishing all the implications of its connections to all other strings, and this leads, to be sure, down an open-ended trail. So "semantic" properties are connected to open-ended searches because, in an important sense, an object's *meaning is not localized* within the object itself. This is not to say that no understanding of any object's meaning is possible until the end of time, for as time passes, more and more of the meaning unfolds. However, there are always aspects of its meaning which will remain hidden arbitrarily long.

Meaning Derives from Connections to Cognitive Structures

Let us switch from strings to pieces of music, just for variety. You may still substitute the term "string" for every reference to a piece of music, if you prefer. The discussion is meant to be general, but its flavor is better gotten across, I feel, by referring to music. There is a strange duality about the meaning of a piece of music: on the one hand, it seems to be spread around, by virtue of its relation to many other things in the world—and yet, on the other hand, the meaning of a piece of music is obviously derived from the music itself, so it must be localized somewhere inside the music.

The resolution of this dilemma comes from thinking about the interpreter—the mechanism which does the pulling-out of meaning. (By "interpreter" in this context, I mean not the performer of the piece, but the mental mechanism in the listener which derives meaning when the piece is played.) The interpreter may discover many important aspects of a piece's meaning while hearing it for the first time; this seems to confirm the notion that the meaning is housed in the piece itself, and is simply being read off. But that is only part of the story. The music interpreter works by setting up a multidimensional cognitive structure—a mental representation of the piece—which it tries to integrate with pre-existent information by finding links to other multidimensional mental structures which encode previous experiences. As this process takes place, the full meaning gradually unfolds. In fact, years may pass before someone comes to feel that he has penetrated to the core meaning of a piece. This seems to support the opposite view: that musical meaning is spread around, the interpreter's role being to assemble it gradually.

The truth undoubtedly lies somewhere in between: meanings—both musical and linguistic—are to some extent localizable, to some extent spread around. In the terminology of Chapter VI, we can say that musical pieces and pieces of text are partly triggers, and partly carriers of explicit meaning. A vivid illustration of this dualism of meaning is provided by the example of a tablet with an ancient inscription: the meaning is partially stored in the libraries and the brains of scholars around the world, and yet it is also obviously implicit in the tablet itself.

Thus, another way of characterizing the difference between "syntactic" and "semantic" properties (in the just-proposed sense) is that the syntactic ones reside unambiguously inside the object under consideration, whereas semantic properties depend on its relations with a potentially infinite class of other objects, and therefore are not completely localizable. There is nothing cryptic or hidden, in principle, in syntactic properties, whereas hiddenness is of the essence in semantic properties. That is the reason for my suggested distinction between "syntactic" and "semantic" aspects of visual form.

Beauty, Truth, and Form

What about beauty? It is certainly not a syntactic property, according to the ideas above. Is it even a semantic property? Is beauty a property which, for instance, a particular painting has? Let us immediately restrict our consideration to a single viewer. Everyone has had the experience of finding something beautiful at one time, dull another time—and probably intermediate at other times. So is beauty an attribute which varies in time? One could turn things around and say that it is the beholder who has varied in time. Given a particular beholder of a particular painting at a particular time, is it reasonable to assert that beauty is a quality that is definitely present or absent? Or is there still something ill-defined and intangible about it?

Different levels of interpreter probably could be invoked in every person, depending on the circumstances. These various interpreters pull out different meanings, establish different connections, and generally evaluate all deep aspects differently. So it seems that this notion of beauty is extremely hard to pin down. It is for this reason that I chose to link beauty, in the *Magnificrab,* with truth, which we have seen is also one of the most intangible notions in all of metamathematics.

The Neural Substrate of the Epimenides Paradox

I would like to conclude this Chapter with some ideas about that central problem of truth, the Epimenides paradox. I think the Tarski reproduction of the Epimenides paradox inside TNT points the way to a deeper understanding of the nature of the Epimenides paradox in English. What Tarski found was that his version of the paradox has two distinct levels to it. On one level, it is a sentence *about itself* which would be true if it were false, and false if it were true. On the other level—which I like to call the *arithmetical substrate*—it is a sentence *about integers* which is true if and only if false.

Now for some reason this latter bothers people a lot more than the former. Some people simply shrug off the former as "meaningless", because of its self-referentiality. But you can't shrug off paradoxical statements about integers. Statements about integers simply cannot be both true and false.

Now my feeling is that the Tarski transformation of the Epimenides paradox teaches us to *look for a substrate* in the English-language version. In the arithmetical version, the upper level of meaning is supported by the lower arithmetical level. Perhaps analogously, the self-referential sentence which we perceive ("This sentence is false") is only the top level of a dual-level entity. What would be the lower level, then? Well, what is the mechanism that language rides on? The brain. Therefore one ought to look for a *neural substrate* to the Epimenides paradox—a lower level of physical events which clash with each other. That is, two events which by their nature cannot occur simultaneously. If this physical substrate exists, then the reason we cannot make heads or tails of the Epimenides sentence is that our brains are trying to do an impossible task.

Now what would be the nature of the conflicting physical events? Presumably when you hear the Epimenides sentence, your brain sets up some "coding" of the sentence—an internal configuration of interacting symbols. Then it tries to classify the sentence as "true" or "false". This classifying act must involve an attempt to force several symbols to interact in a particular way. (Presumably this happens when any sentence is processed.) Now if it happens that the act of classification would physically disrupt the coding of the sentence—something which would ordinarily never happen—then one is in trouble, for it is tantamount to trying to force a record player to play its self-breaking record. We have described the conflict in physical terms, but not in neural terms. If this analysis is right so far, then presumably the rest of the discussion could be carried on when we know something about the constitution of the "symbols" in the brain out of neurons and their firings, as well as about the way that sentences become converted into "codings".

This sketch of the neural substrate of the Epimenides paradox suggests (to me, at least) that the resolution of the English version of the Epimenides paradox might be similar to that for the Tarski version. The resolution involves abandoning the notion that a brain could ever provide a fully accurate representation for the notion of truth. The novelty of this resolution lies in its suggestion that a total modeling of truth is impossible for quite *physical* reasons: namely, such a modeling would require physically incompatible events to occur in a brain.

SHRDLU, Toy of Man's Designing[1]

One day, Eta Oin wanders into the MIT Artificial Intelligence Laboratory, where she runs into the brilliant young computer program, SHRDLU. It happens that SHRDLU is just dying for someone to try out the newly developed human being "Dr. Tony Earrwig". SHRDLU explains that this Earrwig is fairly intelligent in the limited domain of analyzing a conversation about a "toy world" containing blocks of various shapes, sizes, and colors—blocks which sit on a table, and which can be picked up and moved around. Eta Oin, fascinated, begins enthusiastically typing to SHRDLU. Earrwig dangles over her shoulder and provides a sort of running commentary, as promised.

1. *Eta Oin:* pick up a big red block. [Refer to Fig. 110.]
   SHRDLU: OK.
   Dr. Tony Earrwig: SHRDLU answers "OK" when it carries out a command. In order to pick up the red block, it had to clear it off by finding a space for the green one and moving the green one away.
2. *Eta Oin:* grasp the pyramid.
   SHRDLU: I DON'T UNDERSTAND WHICH PYRAMID YOU MEAN.
   Dr. Tony Earrwig: There are three pyramids in the scene, and SHRDLU knows that phrases beginning with "the" are intended to refer to a specific object the speaker has in mind.

FIGURE 110. "Pick up a big red block." [*From Terry Winograd,* Understanding Natural Language, *(New York: Academic Press, 1972), p. 8.*]

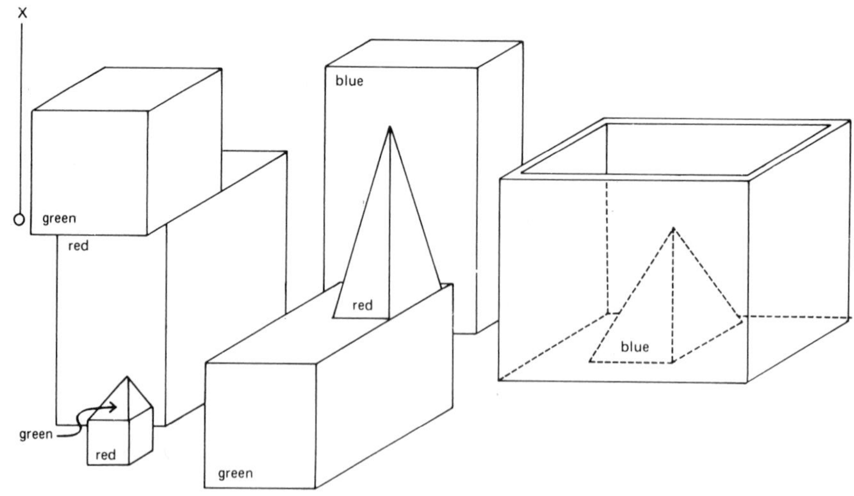

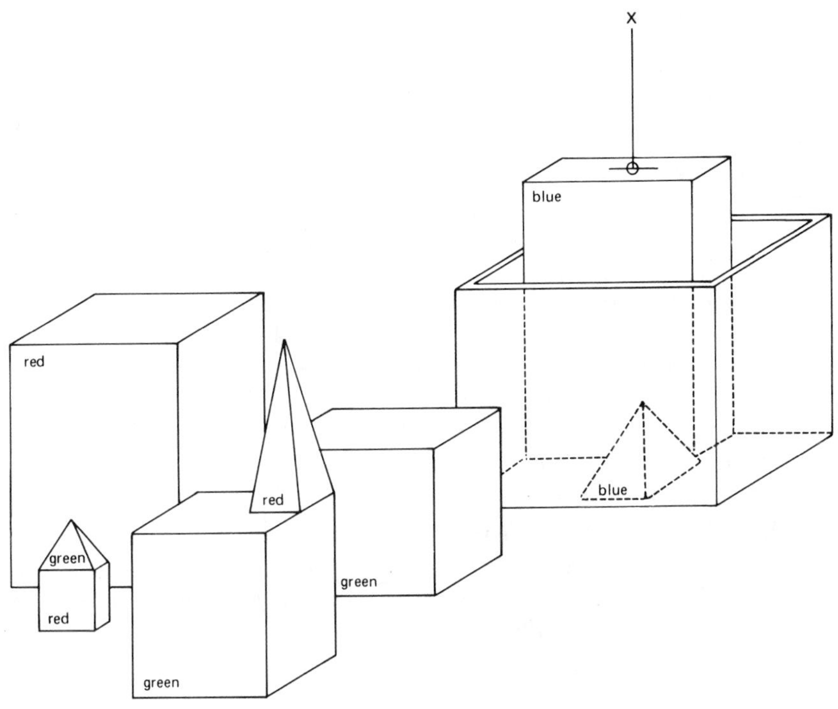

FIGURE 111. "Find a block which is bigger than the one you are holding and put it into the box." [*Adapted from Terry Winograd,* Understanding Natural Language, *p.9.*]

3. *Eta Oin:* find a block which is taller than the one you are holding and put it into the box. [Refer to Fig. 111.]

   Dr. Tony Earrwig: SHRDLU has to figure out what is meant by "one" and "it", as well as using deduction to decide which block it is holding.

   SHRDLU: BY "IT", I ASSUME YOU MEAN THE BLOCK WHICH IS TALLER THAN THE ONE I AM HOLDING.

   Dr. Tony Earrwig: There were two possible referents for "it": the one selected, and "the one you are holding". SHRDLU has heuristics for choosing between them.

   SHRDLU: OK.

   Dr. Tony Earrwig: SHRDLU's hand has to set down the red block to pick up the taller blue one.

4. *Eta Oin:* what does the box contain?

   SHRDLU: THE BLUE PYRAMID AND THE BLUE BLOCK.

5. *Eta Oin:* what is the pyramid supported by?

   Dr. Tony Earrwig: Earlier SHRDLU responded to "the pyramid" by saying "I DON'T UNDERSTAND WHICH PYRAMID YOU MEAN." Now, it assumes that you mean the one just mentioned.

   SHRDLU: THE BOX.

6. *Eta Oin:* how many blocks are not in the box?

Dr. Tony Earrwig: Negatives such as "not" and "no" are handled whether they appear in clauses or noun groups (such as "no boxes").

SHRDLU: FOUR OF THEM.

7. *Eta Oin:* is at least one of them narrower than the one which i told you to pick up?

   Dr. Tony Earrwig: The phrase "at least one of them" must be analyzed to refer back to the previous question, and "the one which . . ." must also refer to blocks. Noun groups can have complicated numbers such as "at least one", "more than three", "exactly seven", etc.. The memory of past events is used to deduce which block is "the one which I told you to pick up". Notice the complicated syntactic structure in which "the one" is actually the object of a clause embedded two levels deeper. SHRDLU's parser incorporates a grammar that handles this kind of complexity in a general way.

   SHRDLU: YES, THE RED CUBE.

8. *Eta Oin:* is it supported?

   SHRDLU: YES, BY THE TABLE.

   Dr. Tony Earrwig: SHRDLU has heuristics about what information is relevant to a person who asks a question, even if the person does not request that information specifically.

9. *Eta Oin:* can the table pick up blocks?

   SHRDLU: NO.

   Dr. Tony Earrwig: SHRDLU's semantic rules include the knowledge that "picking up" must be done by an animate object, and that the table is not animate.

10. *Eta Oin:* can a pyramid be supported by a block?

    SHRDLU: YES.

    Dr. Tony Earrwig: SHRDLU's deductive system finds an actual example, so it knows this is possible.

11. *Eta Oin:* can a pyramid support a pyramid?

    SHRDLU: I DON'T KNOW.

    Dr. Tony Earrwig: The question makes sense, but there is no example in the scene. SHRDLU has no way of looking into the details of its programs, even though these ultimately define its capabilities.

12. *Eta Oin:* stack up two pyramids.

    SHRDLU: I CAN'T.

    Dr. Tony Earrwig: SHRDLU tries and fails.

13. *Eta Oin:* the blue pyramid is mine.

    SHRDLU: I UNDERSTAND.

    Dr. Tony Earrwig: All of the information about colors, shapes, sizes and locations of the objects is known to SHRDLU initially. Possession is left as an unknown property of objects about which it can be told. It stores this information in its data base.

14. *Eta Oin:* i own all blocks which are not red, but i don't own anything which supports a pyramid.

    SHRDLU: I UNDERSTAND.

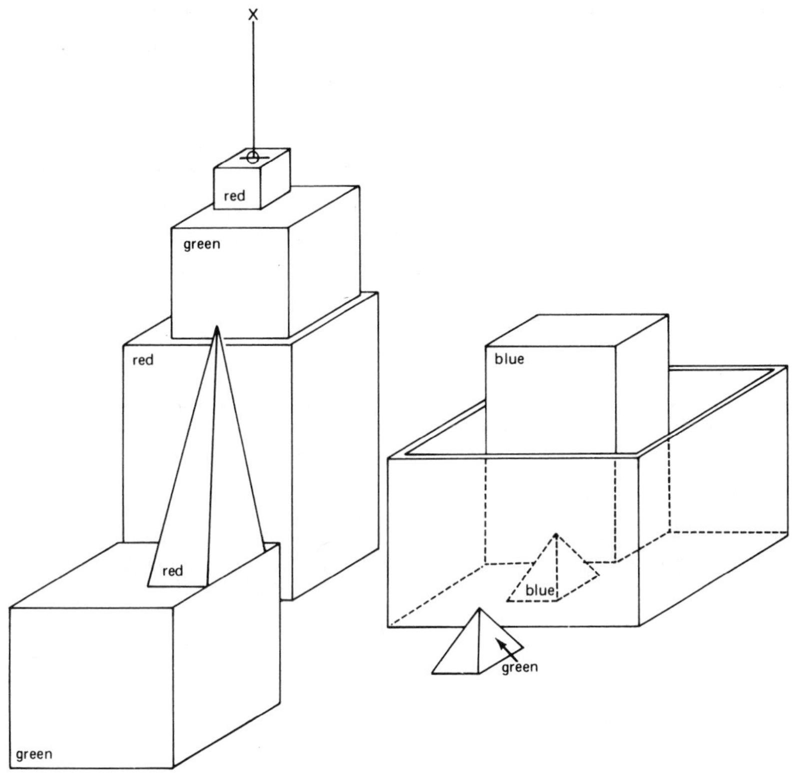

FIGURE 112. "Will you please stack up both of the red blocks and either a green cube or a pyramid?" [*Adapted from Terry Winograd,* Understanding Natural Language, *p. 12.*]

Dr. Tony Earrwig: The result of understanding these sentences is the creation of new procedures within SHRDLU for deducing whether you own something.

15. *Eta Oin:* do i own the box?

SHRDLU: NO.

Dr. Tony Earrwig: The box supports whatever it contains, including a pyramid.

16. *Eta Oin:* do i own anything in the box?

SHRDLU: YES, TWO THINGS: THE BLUE BLOCK AND THE BLUE PYRAMID.

17. *Eta Oin:* will you please stack up both of the red blocks and either a green cube or a pyramid? [Refer to Fig. 112.]

SHRDLU: OK.

Dr. Tony Earrwig: Logical connectives, such as "and", "or", "either", etc. are handled in both the grammar and semantics. A command is executed even if it looks like a question. The command is carried out by the following series of steps: SHRDLU puts a green cube on the large red block (it chooses the green cube with nothing on it), then removes the small pyramid from the little red cube, so that it can use that cube to complete the stack as specified.

18. *Eta Oin:* which cube is sitting on the table?

SHRDLU: THE LARGE GREEN ONE WHICH SUPPORTS THE RED PYRAMID.

Dr. Tony Earrwig: Ordinarily SHRDLU would be satisfied with calling this "A BIG GREEN CUBE", but since you asked "which cube" it has to be more specific in naming it. It has procedures for using substitute nouns (such as "one") and pronouns in answers to make the discourse more natural.

19. *Eta Oin:* is there a large block behind a pyramid?

SHRDLU: YES, THREE OF THEM: A LARGE RED ONE, A LARGE GREEN CUBE AND THE BLUE ONE.

20. *Eta Oin:* put a small one onto the green cube which supports a pyramid.

Dr. Tony Earrwig: "A small one" must contrast with your earlier phrase "a large block", so this means "a small block".

SHRDLU: OK.

21. *Eta Oin:* put the littlest pyramid on top of it.

Dr. Tony Earrwig: Words like "littlest" are not in SHRDLU's dictionary but are interpreted from the root forms like "little". SHRDLU compares the sizes of all the pyramids to carry out the action.

SHRDLU: OK.

22. *Eta Oin:* how many things are on top of green cubes?

SHRDLU: I'M NOT SURE WHAT YOU MEAN BY "ON TOP OF" IN THE PHRASE "ON TOP OF GREEN CUBES"
DO YOU MEAN
1—DIRECTLY ON THE SURFACE
2—ANYWHERE ON TOP OF IT?

Dr. Tony Earrwig: In the previous sentence, the phrase "on top of" was interpreted without trouble because the context made it clear what was meant. In this one, the heuristics for disambiguation do not come up with a clearcut definition, so SHRDLU had to ask you for more information. The paraphrases were included with the dictionary definitions of "on". A request like this can also occur when the meaning of a pronoun such as "it" or "they" is unclear.

23. *Eta Oin:* had you touched any pyramid before you put the green one on the little cube?

SHRDLU: YES, THE GREEN ONE.

Dr. Tony Earrwig: SHRDLU deduces from its memory what event is being referred to, and can relate other events to it in time. The analysis includes complex tenses of verb groups, and there are special facilities for keeping track of the various forms of irregular verbs like "have".

24. *Eta Oin:* when did you pick it up?

SHRDLU: WHILE I WAS STACKING UP THE RED CUBE, A LARGE RED BLOCK AND A LARGE GREEN ONE.

Dr. Tony Earrwig: A time is described by naming the main goal which was being worked on.

25. *Eta Oin:* why?

Dr. Tony Earrwig: SHRDLU remembers immediately previous sentences to understand questions like this and the following ones.

SHRDLU: TO GET RID OF IT.

Dr. Tony Earrwig: By keeping track of selected parts of the original subgoal tree, SHRDLU has some understanding of its own motives.

26. *Eta Oin:* why did you do that?

SHRDLU: TO CLEAN OFF THE RED CUBE.

Dr. Tony Earrwig: "That" refers to the action mentioned in SHRDLU's last answer.

27. *Eta Oin:* why did you clear off that cube?

Dr. Tony Earrwig: Here, "that" means the one just referred to.

SHRDLU: TO PUT IT ON A LARGE GREEN CUBE.

28. *Eta Oin:* why did you do that?

SHRDLU: TO STACK UP THE RED CUBE, A LARGE RED BLOCK, AND A LARGE GREEN CUBE.

29. *Eta Oin:* why did you do that?

SHRDLU: BECAUSE YOU ASKED ME TO.

Dr. Tony Earrwig: SHRDLU remembers what it did, not how the request was worded.

30. *Eta Oin:* how did you do it?

Dr. Tony Earrwig: "It" refers to the action mentioned in the last question, rather than the answer.

SHRDLU: BY PUTTING A LARGE RED BLOCK ON THE TABLE; THEN LETTING GO OF IT; THEN PUTTING A LARGE GREEN CUBE ON IT; THEN LETTING GO OF THAT CUBE; THEN PUTTING THE RED CUBE ON THAT CUBE; THEN LETTING GO OF THAT CUBE.

Dr. Tony Earrwig: This is a description of the top-level subgoals used to achieve the goal.

31. *Eta Oin:* how many objects did you touch while you were doing it?

Dr. Tony Earrwig: Time references can refer to a time span, using words like "while".

SHRDLU: FOUR OF THEM.

32. *Eta Oin:* what did the red cube support before you started to clean it off?

SHRDLU: THE GREEN PYRAMID.

Dr. Tony Earrwig: By remembering what has been moved, it is possible to deduce things about earlier states of the scene. You can use complex nestings of clauses like "before you started to clean it off".

33. *Eta Oin:* there were five blocks to the left of the box then.

Dr. Tony Earrwig: "Then" must refer to the time mentioned in the last sentence.

SHRDLU: NO, ONLY FOUR OF THEM: THE RED CUBE, TWO LARGE GREEN CUBES, AND A LARGE RED BLOCK.

Dr. Tony Earrwig: Earlier in the dialogue, information about "owning" was accepted at face value. But SHRDLU knows that it has complete informa-

tion about the basic properties of objects, including location. It checks any statements about these to see whether it agrees or disagrees.

34. *Eta Oin:* put the blue pyramid on the block in the box.

Dr. Tony Earrwig: This is syntactically ambiguous, but is understood without ambiguity since the parsing corresponding to "put the blue pyramid which is on the block into the box" does not make sense in the current scene, whereas "put the blue pyramid onto the block which is in the box" does.

SHRDLU: OK.

35. *Eta Oin:* is there anything which is bigger than every pyramid but is not as wide as the thing that supports it?

Dr. Tony Earrwig: SHRDLU understands quantifiers such as "every", "any", "some", etc., and structures such as "bigger than", "as wide as", etc.. Here "it" refers to a noun group within the same sentence but two levels up in the parsing tree.

SHRDLU: YES, THE BLUE BLOCK.

36. *Eta Oin:* does a steeple

SHRDLU: SORRY, I DON'T KNOW THE WORD "STEEPLE".

Dr. Tony Earrwig: SHRDLU interrupts immediately when it sees a word it doesn't know.

37. *Eta Oin:* a "steeple" is a stack which contains two green cubes and a pyramid.

Dr. Tony Earrwig: The quote marks indicate that you are defining a new word. The definition is added to SHRDLU's dictionary in a form compatible with all the words it already knows.

SHRDLU: I UNDERSTAND.

38. *Eta Oin:* are there any steeples now?

SHRDLU: NO.

Dr. Tony Earrwig: The new definition is used as a description to recognize steeples.

39. *Eta Oin:* build one.

SHRDLU: OK.

Dr. Tony Earrwig: The description given in the definition is enough to deduce a plan for building new objects as well as recognizing them. The word "one" means "a steeple", and the plan is carried out. First, one of the large green cubes is cleared off by removing the red cube and two pyramids. Then the two green cubes and the tall red pyramid are stacked up to fit the description.

40. *Eta Oin:* call the biggest block "superblock".

SHRDLU: OK.

Dr. Tony Earrwig: You can also give objects names.

41. *Eta Oin:* have you picked up superblock since we began?

SHRDLU: YES.

42. *Eta Oin:* why did you drop it?

SHRDLU: BY "IT", I ASSUME YOU MEAN THE SUPERBLOCK. I CAN'T EXPLAIN A NON-EXISTENT EVENT.

Dr. Tony Earrwig: There is no record in SHRDLU's memory of any block being dropped. The answer routines have special checks for this type of "loaded question" which presupposes things which aren't true.

43. *Eta Oin:* is there anything to the right of the red pyramid?

SHRDLU: YES, FOUR THINGS: THE BOX, THE RED CUBE, THE BLUE PYRAMID AND THE BLOCK NAMED SUPERBLOCK.

Dr. Tony Earrwig: Once an object has a name, the name is used to help identify it in responses.

44. *Eta Oin:* thank you.

SHRDLU: YOU'RE WELCOME!

Dr. Tony Earrwig: SHRDLU's grammar is written as a procedure, which gives SHRDLU the capability to carry out special sorts of parsing for idioms.

CHAPTER XVIII

Artificial Intelligence: Retrospects

Turing

FIGURE 113. *Alan Turing, after a successful race (May, 1950).* [*From Sara Turing,* Alan M. Turing *(Cambridge, U. K.: W. Heffer & Sons, 1959).*]

IN 1950, ALAN TURING wrote a most prophetic and provocative article on Artificial Intelligence. It was entitled "Computing Machinery and Intelligence" and appeared in the journal *Mind.*[1] I will say some things about that article, but I would like to precede them with some remarks about Turing the man.

Alan Mathison Turing was born in London in 1912. He was a child full of curiosity and humor. Gifted in mathematics, he went to Cambridge where his interests in machinery and mathematical logic cross-fertilized and resulted in his famous paper on "computable numbers", in which he invented the theory of Turing machines and demonstrated the unsolvability of the halting problem; it was published in 1937. In the 1940's, his interests turned from the theory of computing machines to the actual building of real computers. He was a major figure in the development of computers in Britain, and a staunch defender of Artificial Intelligence when it first came under attack. One of his best friends was David Champernowne (who later worked on computer composition of music). Champernowne and Turing were both avid chess players and invented "round-the-house" chess: after your move, run around the house—if you get back before your opponent has moved, you're entitled to another move. More seriously, Turing and Champernowne invented the first chess-playing program, called "Turochamp". Turing died young, at 41—apparently of an accident with chemicals. Or some say suicide. His mother, Sara Turing, wrote his biography. From the people she quotes, one gets the sense that Turing was highly unconventional, even gauche in some ways, but so honest and decent that he was vulnerable to the world. He loved games, chess, children, and bike riding; he was a strong long-distance runner. As a student at Cambridge, he bought himself a second-hand violin and taught himself to play. Though not very musical, he derived a great deal of enjoyment from it. He was somewhat eccentric, given to great bursts of energy in the oddest directions. One area he explored was the problem of morphogenesis in biology. According to his mother, Turing "had a particular fondness for the *Pickwick Papers*", but "poetry, with the exception of Shakespeare's, meant nothing to him." Alan Turing was one of the true pioneers in the field of computer science.

The Turing Test

Turing's article begins with the sentence: "I propose to consider the question 'Can machines think?'" Since, as he points out, these are loaded terms, it is obvious that we should search for an operational way to approach the question. This, he suggests, is contained in what he calls the "imitation game"; it is nowadays known as the *Turing test.* Turing introduces it as follows:

> It is played with three people: a man (A), a woman (B), and an interrogator (C) who may be of either sex. The interrogator stays in a room apart from the other two. The object of the game for the interrogator is to determine which of the other two is the man and which is the woman. He knows them by labels X and Y, and at the end of the game he says either "X is A and Y is B" or "X is B and Y is A". The interrogator is allowed to put questions to A and B thus:
>
> C: Will X please tell me the length of his or her hair?
>
> Now suppose X is actually A, then A must answer. It is A's object in the game to try to cause C to make the wrong identification. His answer might therefore be
>
> "My hair is shingled, and the longest strands are about nine inches long."
>
> In order that tones of voice may not help the interrogator the answers should be written, or better still, typewritten. The ideal arrangement is to have a teleprinter communicating between the two rooms. Alternatively the questions and answers can be repeated by an intermediary. The object of the game for the third player (B) is to help the interrogator. The best strategy for her is probably to give truthful answers. She can add such things as "I am the woman, don't listen to him!" to her answers, but it will avail nothing as the man can make similar remarks.

> We now ask the question, "What will happen when a machine takes the part of A in this game?" Will the interrogator decide wrongly as often when the game is played like this as he does when the game is played between a man and a woman? These questions replace our original, "Can machines think?"[2]

After having spelled out the nature of his test, Turing goes on to make some commentaries on it, which, given the year he was writing in, are quite sophisticated. To begin with, he gives a short hypothetical dialogue between interrogator and interrogatee:[3]

> Q: Please write me a sonnet on the subject of the Forth Bridge [a bridge over the Firth of Forth, in Scotland].
> A: Count me out on this one. I never could write poetry.
> Q: Add 34957 to 70764.
> A: (Pause about 30 seconds and then give as answer) 105621.
> Q: Do you play chess?
> A: Yes.
> Q: I have K at my K1, and no other pieces. You have only K at K6 and R at R1. It is your move. What do you play?
> A: (After a pause of 15 seconds) R-R8 mate.

Few readers notice that in the arithmetic problem, not only is there an inordinately long delay, but moreover, the answer given is wrong! This would be easy to account for if the respondent were a human: a mere calculational error. But if the respondent were a machine, a variety of explanations are possible. Here are some:

(1) a run-time error on the hardware level (i.e., an irreproducible fluke);
(2) an unintentional hardware (or programming) error which (reproducibly) causes arithmetical mistakes;
(3) a ploy deliberately inserted by the machine's programmer (or builder) to introduce occasional arithmetical mistakes, so as to trick interrogators;
(4) an unanticipated epiphenomenon: the program has a hard time thinking abstractly, and simply made "an honest mistake", which it might not make the next time around;
(5) a joke on the part of the machine itself, deliberately teasing its interrogator.

Reflection on what Turing might have meant by this subtle touch opens up just about all the major philosophical issues connected with Artificial Intelligence.

Turing goes on to point out that

> The new problem has the advantage of drawing a fairly sharp line between the physical and the intellectual capacities of a man. . . . We do not wish to penalize the machine for its inability to shine in beauty competitions, nor to penalize a man for losing in a race against an airplane.[4]

One of the pleasures of the article is to see how far Turing traced out each line of thought, usually turning up a seeming contradiction at some stage and, by refining his concepts, resolving it at a deeper level of analysis. Because of this depth of penetration into the issues, the article still shines after nearly thirty years of tremendous progress in computer development and intensive work in AI. In the following short excerpt you can see some of this rich back-and-forth working of ideas:

> The game may perhaps be criticized on the ground that the odds are weighted too heavily against the machine. If the man were to try to pretend to be the machine he would clearly make a very poor showing. He would be given away at once by slowness and inaccuracy in arithmetic. May not machines carry out something which ought to be described as thinking but which is very different from what a man does? This objection is a very strong one, but at least we can say that if, nevertheless, a machine can be constructed to play the imitation game satisfactorily, we need not be troubled by this objection.
>
> It might be urged that when playing the "imitation game" the best strategy for the machine may possibly be something other than imitation of the behaviour of a man. This may be, but I think it is unlikely that there is any great effect of this kind. In any case there is no intention to investigate here the theory of the game, and it will be assumed that the best strategy is to try to provide answers that would naturally be given by a man.[5]

Once the test has been proposed and discussed, Turing remarks:

> The original question "Can machines think?" I believe to be too meaningless to deserve discussion. Nevertheless, I believe that at the end of the century the use of words and general educated opinion will have altered so much that one will be able to speak of machines thinking without expecting to be contradicted.[6]

Turing Anticipates Objections

Aware of the storm of opposition that would undoubtedly greet this opinion, he then proceeds to pick apart, concisely and with wry humor, a series of objections to the notion that machines could think. Below I list the nine types of objections he counters, using his own descriptions of them.[7] Unfortunately there is not space to reproduce the humorous and ingenious responses he formulated. You may enjoy pondering the objections yourself, and figuring out your own responses.

(1) *The Theological Objection.* Thinking is a function of man's immortal soul. God has given an immortal soul to every man and woman, but not to any other animal or to machines. Hence no animal or machine can think.

(2) *The "Heads in the Sand" Objection.* The consequences of machines thinking would be too dreadful. Let us hope and believe that they cannot do so.

(3) *The Mathematical Objection.* [This is essentially the Lucas argument.]

(4) *The Argument from Consciousness.* "Not until a machine can write a sonnet or compose a concerto because of thoughts and emotions felt, and not by the chance fall of symbols, could we agree that machine equals brain—that is, not only write it but know that it had written it. No mechanism

could feel (and not merely artificially signal, an easy contrivance) pleasure at its successes, grief when its valves fuse, be warmed by flattery, be made miserable by its mistakes, be charmed by sex, be angry or depressed when it cannot get what it wants." [A quote from a certain Professor Jefferson.]

Turing is quite concerned that he should answer this serious objection in full detail. Accordingly, he devotes quite a bit of space to his answer, and in it he offers another short hypothetical dialogue:[8]

Interrogator: In the first line of your sonnet which reads "Shall I compare thee to a summer's day", would not "a spring day" do as well or better?
Witness: It wouldn't scan.
Interrogator: How about "a winter's day"? That would scan all right.
Witness: Yes, but nobody wants to be compared to a winter's day.
Interrogator: Would you say Mr. Pickwick reminded you of Christmas?
Witness: In a way.
Interrogator: Yet Christmas is a winter's day, and I do not think Mr. Pickwick would mind the comparison.
Witness: I don't think you're serious. By a winter's day one means a typical winter's day, rather than a special one like Christmas.

After this dialogue, Turing asks, "What would Professor Jefferson say if the sonnet-writing machine was able to answer like this in the *viva voce?*"

Further objections:

(5) *Arguments from Various Disabilities.* These arguments take the form, "I grant you that you can make machines do all the things that you have mentioned but you will never be able to make one to do X." Numerous features X are suggested in this connection. I offer a selection:
Be kind, resourceful, beautiful, friendly, have initiative, have a sense of humor, tell right from wrong, make mistakes, fall in love, enjoy strawberries and cream, make someone fall in love with it, learn from experience, use words properly, be the subject of its own thought, have as much diversity of behaviour as a man, do something really new.

(6) *Lady Lovelace's Objection.* Our most detailed information of Babbage's Analytical Engine comes from a memoir by Lady Lovelace. In it she states, "The Analytical Engine has no pretensions to *originate* anything. It can do *whatever we know how to order it* to perform" (her italics).

(7) *Argument from Continuity in the Nervous System.* The nervous system is certainly not a discrete state machine. A small error in the information about the size of a nervous impulse impinging on a neuron may make a large difference to the size of the outgoing impulse. It may be argued that, this being so, one cannot expect to be able to mimic the behaviour of the nervous system with a discrete state system.

(8) *The Argument from Informality of Behaviour.* It seems to run something like this. "If each man had a definite set of rules of conduct by which he regulated his life he would be no better than a machine. But there are no such rules, so men cannot be machines."

(9) *The Argument from Extra-Sensory Perception.* Let us play the imitation game, using as witnesses a man who is good as a telepathic receiver, and a digital computer. The interrogator can ask such questions as "What suit

> does the card in my right hand belong to?" The man by telepathy or clairvoyance gives the right answer 130 times out of 400 cards. The machine can only guess at random, and perhaps get 104 right, so the interrogator makes the right identification.

As you can see, many of the objections overlap and are related to each other. I have attempted in this book to answer most of them in some way or other, usually far less concisely than Turing did.

Objection (9) I find remarkable. I have seen the Turing paper reprinted in a book—but with objection (9) omitted—which I find equally remarkable. My own point of view—contrary to Turing's—is that ESP does not exist. Turing was reluctant to accept the idea that ESP is real, but did so nonetheless, being compelled by his outstanding scientific integrity to accept the consequences of what he viewed as powerful statistical evidence in favor of ESP. I disagree, though I consider it an exceedingly complex and fascinating question.

"Parry Encounters the Doctor"

In the Dialogue preceding this Chapter, you have seen an authentic exchange between a computer program and a human. Now there have been many programs with less ability to "converse" than SHRDLU. Two rather famous ones are "Doctor", created by Joseph Weizenbaum, and "Parry", created by Kenneth Colby. The former is supposed to simulate a psychiatrist using "nondirective therapy", the latter to simulate the belief structure of a paranoid. An amusing head-on collision between the two programs was arranged over a computer network, and the following is a short excerpt from the full transcript:[9]

> Parry: I don't understand your motives.
> Doctor: Why don't you understand my motives?
> Parry: Shouldn't I be?
> Doctor: Why do you say that just now?
> Parry: That's just the way it is.
> Doctor: Suppose that's just the way it were not.
> Parry: So?
> Doctor: I am not sure I understand you fully.
> Parry: You harp on the same questions too much.
> Doctor: What does that similarity suggest to you?
> Parry: I don't get you.
> Doctor: Do you wish that you did get I?
> Parry: It's about time you showed some interest in my feelings.
> Doctor: Oh? I showed some interest in your feelings?

These programs, of course, were made to talk to humans, so they are shown up rather poorly here, pitted against each other. Mostly they rely on shrewd guesses as to the nature of the input (which they analyze quite shallowly) and spit back canned answers that have been carefully selected from a large repertoire. The answer may be only partially canned: for example, a template with blanks that can be filled in. It is assumed that their human partners will read much more into what they say than is actually underlying it. And in fact, according to Weizenbaum, in his book *Computer Power and Human Reason,* just that happens. He writes:

> ELIZA [the program from which Doctor was made] created the most remarkable illusion of having understood in the minds of the many people who conversed with it. . . . They would often demand to be permitted to converse with the system in private, and would, after conversing with it for a time, insist, in spite of my explanations, that the machine really understood them.[10]

Given the above excerpt, you may find this incredible. Incredible, but true. Weizenbaum has an explanation:

> Most men don't understand computers to even the slightest degree. So, unless they are capable of very great skepticism (the kind we bring to bear while watching a stage magician), they can explain the computer's intellectual feats only by bringing to bear the single analogy available to them, that is, their model of their own capacity to think. No wonder, then, that they overshoot the mark; it is truly impossible to imagine a human who could imitate ELIZA, for example, but for whom ELIZA's language abilities were his limit.[11]

Which amounts to an admission that this kind of program is based on a shrewd mixture of bravado and bluffing, taking advantage of people's gullibility.

In light of this weird "ELIZA-effect", some people have suggested that the Turing test needs revision, since people can apparently be fooled by simplistic gimmickry. It has been suggested that the interrogator should be a Nobel Prize-winning scientist. It might be more advisable to turn the Turing test on its head, and insist that the interrogator should be another computer. Or perhaps there should be two interrogators—a human and a computer—and one witness, and the two interrogators should try to figure out whether the witness is a human or a computer.

In a more serious vein, I personally feel that the Turing test, as originally proposed, is quite reasonable. As for the people who Weizenbaum claims were sucked in by ELIZA, they were not urged to be skeptical, or to use all their wits in trying to determine if the "person" typing to them were human or not. I think that Turing's insight into this issue was sound, and that the Turing test, essentially unmodified, will survive.

A Brief History of AI

I would like in the next few pages to present the story, perhaps from an unorthodox point of view, of some of the efforts at unraveling the algorithms behind intelligence; there have been failures and setbacks and there will continue to be. Nonetheless, we are learning a great deal, and it is an exciting period.

Ever since Pascal and Leibniz, people have dreamt of machines that could perform intellectual tasks. In the nineteenth century, Boole and De Morgan devised "laws of thought"—essentially the Propositional

Calculus—and thus took the first step towards AI software; also Charles Babbage designed the first "calculating engine"—the precursor to the hardware of computers and hence of AI. One could define AI as coming into existence at the moment when mechanical devices took over any tasks previously performable only by human minds. It is hard to look back and imagine the feelings of those who first saw toothed wheels performing additions and multiplications of large numbers. Perhaps they experienced a sense of awe at seeing "thoughts" flow in their very physical hardware. In any case, we do know that nearly a century later, when the first electronic computers were constructed, their inventors did experience an awesome and mystical sense of being in the presence of another kind of "thinking being". To what extent real thought was taking place was a source of much puzzlement; and even now, several decades later, the question remains a great source of stimulation and vitriolics.

It is interesting that nowadays, practically no one feels that sense of awe any longer—even when computers perform operations that are incredibly more sophisticated than those which sent thrills down spines in the early days. The once-exciting phrase "Giant Electronic Brain" remains only as a sort of "camp" cliché, a ridiculous vestige of the era of Flash Gordon and Buck Rogers. It is a bit sad that we become blasé so quickly.

There is a related "Theorem" about progress in AI: once some mental function is programmed, people soon cease to consider it as an essential ingredient of "real thinking". The ineluctable core of intelligence is always in that next thing which hasn't yet been programmed. This "Theorem" was first proposed to me by Larry Tesler, so I call it *Tesler's Theorem:* "AI is whatever hasn't been done yet."

A selective overview of AI is furnished below. It shows several domains in which workers have concentrated their efforts, each one seeming in its own way to require the quintessence of intelligence. With some of the domains I have included a breakdown according to methods employed, or more specific areas of concentration.

- mechanical translation
  - direct (dictionary look-up with some word rearrangement)
  - indirect (via some intermediary internal language)

- game playing
  - chess
    - with brute force look-ahead
    - with heuristically pruned look-ahead
    - with no look-ahead
  - checkers
  - go
  - kalah
  - bridge (bidding; playing)
  - poker
  - variations on tic-tac-toe
  - etc.

proving theorems in various parts of mathematics
- symbolic logic
  - "resolution" theorem-proving
- elementary geometry

symbolic manipulation of mathematical expressions
- symbolic integration
- algebraic simplification
- summation of infinite series

vision
- printed matter:
  - recognition of individual hand-printed characters drawn from a small class (e.g., numerals)
  - reading text in variable fonts
  - reading passages in handwriting
  - reading Chinese or Japanese printed characters
  - reading Chinese or Japanese handwritten characters
- pictorial:
  - locating prespecified objects in photographs
  - decomposition of a scene into separate objects
  - identification of separate objects in a scene
  - recognition of objects portrayed in sketches by people
  - recognition of human faces

hearing
- understanding spoken words drawn from a limited vocabulary (e.g., names of the ten digits)
- understanding continuous speech in fixed domains
  - finding boundaries between phonemes
  - identifying phonemes
  - finding boundaries between morphemes
  - identifying morphemes
  - putting together whole words and sentences

understanding natural languages
- answering questions in specific domains
- parsing complex sentences
- making paraphrases of longer pieces of text
- using knowledge of the real world in order to understand passages
- resolving ambiguous references

producing natural language
- abstract poetry (e.g., haiku)
- random sentences, paragraphs, or longer pieces of text
- producing output from internal representation of knowledge

- creating original thoughts or works of art
    - poetry writing (haiku)
    - story writing
    - computer art
    - musical composition
        - atonal
        - tonal
- analogical thinking
    - geometrical shapes ("intelligence tests")
    - constructing proofs in one domain of mathematics based on those in a related domain
- learning
    - adjustment of parameters
    - concept formation

Mechanical Translation

Many of the preceding topics will not be touched upon in my selective discussion below, but the list would not be accurate without them. The first few topics are listed in historical order. In each of them, early efforts fell short of expectations. For example, the pitfalls in mechanical translation came as a great surprise to many who had thought it was a nearly straightforward task, whose perfection, to be sure, would be arduous, but whose basic implementation should be easy. As it turns out, translation is far more complex than mere dictionary look-up and word rearranging. Nor is the difficulty caused by a lack of knowledge of idiomatic phrases. The fact is that translation involves having a mental model of the world being discussed, and manipulating symbols in that model. A program which makes no use of a model of the world as it reads the passage will soon get hopelessly bogged down in ambiguities and multiple meanings. Even people—who have a huge advantage over computers, for they come fully equipped with an understanding of the world—when given a piece of text and a dictionary of a language they do not know, find it next to impossible to translate the text into their own language. Thus—and it is not surprising in retrospect—the first problem of AI led immediately to the issues at the heart of AI.

Computer Chess

Computer chess, too, proved to be much more difficult than the early intuitive estimates had suggested. Here again it turns out that the way humans represent a chess situation in their minds is far more complex than just knowing which piece is on which square, coupled with knowledge of the rules of chess. It involves perceiving configurations of several related pieces, as well as knowledge of *heuristics,* or rules of thumb, which pertain to such higher-level chunks. Even though heuristic rules are not rigorous in the way that the official rules are, they provide shortcut insights into what is going on on the board, which knowledge of the official rules does not. This much was recognized from the start; it was simply underestimated how large a role the intuitive, chunked understanding of the chess world plays in human chess skill. It was predicted that a program having some basic heuristics, coupled with the blinding speed and accuracy of a computer to look ahead in the game and analyze each possible move, would easily beat top-flight human players—a prediction which, even after twenty-five years of intense work by various people, still is far from being realized.

People are nowadays tackling the chess problem from various angles. One of the most novel involves the hypothesis that looking ahead is a silly thing to do. One should instead merely look at what is on the board at present, and, using some heuristics, generate a plan, and then find a move which advances that particular plan. Of course, rules for the formulation of chess plans will necessarily involve heuristics which are, in some sense, "flattened" versions of looking ahead. That is, the equivalent of many games' experience of looking ahead is "squeezed" into another form which ostensibly doesn't involve looking ahead. In some sense this is a game of words. But if the "flattened" knowledge gives answers more efficiently than the actual look-ahead—even if it occasionally misleads— then something has been gained. Now this kind of distillation of knowledge into more highly usable forms is just what intelligence excels at—so look-ahead-less chess is probably a fruitful line of research to push. Particularly intriguing would be to devise a program which itself could convert knowledge gained from looking ahead into "flattened" rules—but that is an immense task.

Samuel's Checker Program

As a matter of fact, such a method was developed by Arthur Samuel in his admirable checker-playing program. Samuel's trick was to use both *dynamic* (look-ahead) and *static* (no-look-ahead) ways of evaluating any given board position. The static method involved a simple mathematical function of several quantities characterizing any board position, and thus could be calculated practically instantaneously, whereas the dynamic evaluation method involved creating a "tree" of possible future moves, responses to them, responses to the responses, and so forth (as was shown in Fig. 38). In the static evaluation function there were some parameters which could vary; the effect of varying them was to provide a set of different possible versions of the static evaluation function. Samuel's strategy was to select, in an evolutionary way, better and better values of those parameters.

Here's how this was done: each time time the program evaluated a board position, it did so both statically and dynamically. The answer gotten by looking ahead—let us call it *D*—was used in determining the move to be made. The purpose of *S*, the static evaluation, was trickier: on each move, the variable parameters were readjusted slightly so that *S* approximated *D* as accurately as possible. The effect was to partially encode in the values of the static evalution's parameters the knowledge gained by dynamically searching the tree. In short, the idea was to "flatten" the complex dynamic evaluation method into the much simpler and more efficient static evaluation function.

There is a rather nice recursive effect here. The point is that the *dynamic* evaluation of any single board position involves looking ahead a finite number of moves—say seven. Now each of the scads of board positions which might turn up seven turns down the road has to be itself evaluated somehow as well. But when the program evaluates these positions, it certainly cannot look another seven moves ahead, lest it have to look fourteen positions ahead, then twenty-one, etc., etc.—an infinite regress. Instead, it relies on *static* evaluations of positions seven moves ahead. Therefore, in Samuel's scheme, an intricate sort of feedback takes place, wherein the program is constantly trying to "flatten" look-ahead evaluation into a simpler static recipe; and this recipe in turn plays a key role in the dynamic look-ahead evaluation. Thus the two are intimately linked together, and each benefits from improvements in the other in a recursive way.

The level of play of the Samuel checkers program is extremely high: of the order of the top human players in the world. If this is so, why not apply the same techniques to chess? An international committee, convened in 1961 to study the feasibility of computer chess, including the Dutch International Grandmaster and mathematician Max Euwe, came to the bleak conclusion that the Samuel technique would be approximately one million times as difficult to implement in chess as in checkers, and that seems to close the book on that.

The extraordinarily great skill of the checkers program cannot be taken as saying "intelligence has been achieved"; yet it should not be minimized, either. It is a combination of insights into what checkers is, how to think about checkers, and how to program. Some people might feel that all it shows is Samuel's own checkers ability. But this is not true, for at least two reasons. One is that skillful game players choose their moves according to mental processes which they do not fully understand—they use their intuitions. Now there is no known way that anyone can bring to light all of his own intuitions; the best one can do via introspection is to use "feeling" or "meta-intuition"—an intuition about one's intuitions—as a guide, and try to describe what one thinks one's intuitions are all about. But this will only give a rough approximation to the true complexity of intuitive methods. Hence it is virtually certain that Samuel has not mirrored his own personal methods of play in his program. The other reason that Samuel's program's play should not be confused with Samuel's own play is that Samuel does not play checkers as well as his program—it beats him. This is not a paradox at all—no more than is the fact that a computer which has been programmed to calculate π can outrace its programmer in spewing forth digits of π.

When Is a Program Original?

This issue of a program outdoing its programmer is connected with the question of "originality" in AI. What if an AI program comes up with an idea, or a line of play in a game, which its programmer has never entertained—who should get the credit? There are various interesting instances of this having happened, some on a fairly trivial level, some on a rather deep level. One of the more famous involved a program to find proofs of theorems in elementary Euclidean geometry, written by E. Gelernter. One day the program came up with a sparklingly ingenious proof of one of the basic theorems of geometry—the so-called "pons asinorum", or "bridge of asses".

This theorem states that the base angles of an isosceles triangle are equal. Its standard proof requires constructing an altitude which divides the triangle into symmetrical halves. The elegant method found by the program (see Fig. 114) used no construction lines. Instead, it considered

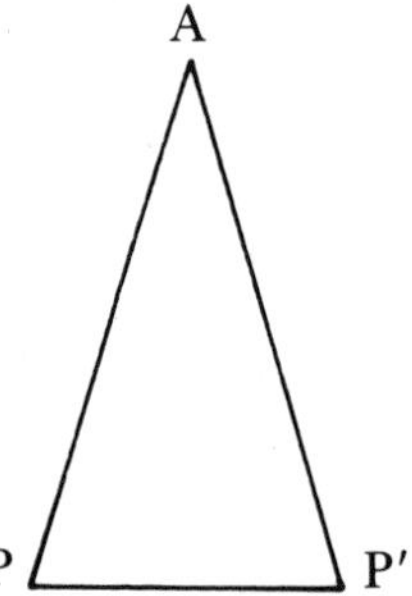

FIGURE 114. Pons Asinorum Proof (found by Pappus [~300 A.D.] and Gelernter's program [~1960 A.D.]). Problem: To show that the base angles of an isosceles triangle are equal. Solution: As the triangle is isosceles, AP and AP′ are of equal length. Therefore triangles PAP′ and P′AP are congruent (side-side-side). This implies that corresponding angles are equal. In particular, the two base angles are equal.

the triangle and its mirror image as two different triangles. Then, having proved them congruent, it pointed out that the two base angles matched each other in this congruence—QED.

This gem of a proof delighted the program's creator and others; some saw evidence of genius in its performance. Not to take anything away from this feat, it happens that in A.D. 300 the geometer Pappus had actually found this proof, too. In any case, the question remains: "Who gets the credit?" Is this intelligent behavior? Or was the proof lying deeply hidden within the human (Gelernter), and did the computer merely bring it to the surface? This last question comes close to hitting the mark. We can turn it around: Was the proof lying deeply hidden in the program? Or was it close to the surface? That is, how easy is it to see why the program did what it did? Can the discovery be attributed to some simple mechanism, or simple combination of mechanisms, in the program? Or was there a complex interaction which, if one heard it explained, would not diminish one's awe at its having happened?

It seems reasonable to say that if one can ascribe the performance to certain operations which are easily traced in the program, then in some sense the program was just revealing ideas which were in essence hidden—though not too deeply—inside the programmer's own mind. Conversely, if following the program does not serve to enlighten one as to why this particular discovery popped out, then perhaps one should begin to separate the program's "mind" from that of its programmer. The human gets credit for having invented the program, but not for having had inside his own head the ideas produced by the program. In such cases, the human can be referred to as the "meta-author"—the author of the author of the result, and the program as the (just plain) author.

In the particular case of Gelernter and his geometry machine, while Gelernter probably would not have rediscovered Pappus' proof, still the mechanisms which generated that proof were sufficiently close to the surface of the program that one hesitates to call the program a geometer in its own right. If it had kept on astonishing people by coming up with ingenious new proofs over and over again, each of which seemed to be based on a fresh spark of genius rather than on some standard method, *then* one would have no qualms about calling the program a geometer—but this did not happen.

Who Composes Computer Music?

The distinction between author and meta-author is sharply pointed up in the case of computer composition of music. There are various levels of autonomy which a program may seem to have in the act of composition. One level is illustrated by a piece whose "meta-author" was Max Mathews of Bell Laboratories. He fed in the scores of the two marches "When Johnny Comes Marching Home" and "The British Grenadiers", and instructed the computer to make a new score—one which starts out as "Johnny", but slowly merges into "Grenadiers". Halfway through the piece, "Johnny" is totally gone, and one hears "Grenadiers" by itself . . . Then the process is reversed, and the piece finishes with "Johnny", as it began. In Mathews' own words, this is

> . . . a nauseating musical experience but one not without interest, particularly in the rhythmic conversions. "The Grenadiers" is written in 2/4 time in the key of F major. "Johnny" is written in 6/8 time in the key of E minor. The change from 2/4 to 6/8 time can be clearly appreciated, yet would be quite difficult for a human musician to play. The modulation from the key of F major to E minor, which involves a change of two notes in the scale, is jarring, and a smaller transition would undoubtedly have been a better choice.[12]

The resulting piece has a somewhat droll quality to it, though in spots it is turgid and confused.

> Is the computer composing? The question is best unasked, but it cannot be completely ignored. An answer is difficult to provide. The algorithms are deterministic, simple, and understandable. No complicated or hard-to-understand computations are involved; no "learning" programs are used; no random processes occur; the machine functions in a perfectly mechanical and straightforward manner. However, the result is sequences of sound that are unplanned in fine detail by the composer, even though the over-all structure

of the section is completely and precisely specified. Thus the composer is often surprised, and pleasantly surprised, at the details of the realization of his ideas. To this extent only is the computer composing. We call the process algorithmic composition, but we immediately re-emphasize that the algorithms are transparently simple.[13]

This is Mathews' answer to a question which he would rather "unask". Despite his disclaimer, however, many people find it easier to say simply that the piece was "composed by a computer". I believe this phrase misrepresents the situation totally. The program contained no structures analogous to the brain's "symbols", and could not be said in any sense to be "thinking" about what it was doing. To attribute the composition of such a piece of music to the computer would be like attributing the authorship of this book to the computerized automatically (often incorrectly) hyphenating phototypesetting machine with which it was set.

This brings up a question which is a slight digression from AI, but actually not a huge one. It is this: When you see the word "I" or "me" in a text, what do you take it to be referring to? For instance, think of the phrase "WASH ME" which appears occasionally on the back of dirty trucks. Who is this "me"? Is this an outcry of some forlorn child who, in desperation to have a bath, scribbled the words on the nearest surface? Or is the truck requesting a wash? Or, perhaps, does the sentence itself wish to be given a shower? Or, is it that the filthy English language is asking to be cleansed? One could go on and on in this game. In this case, the phrase is a joke, and one is supposed to pretend, on some level, that the truck itself wrote the phrase and is requesting a wash. On another level, one clearly recognizes the writing as that of a child, and enjoys the humor of the misdirection. Here, in fact, is a game based on reading the "me" at the wrong level.

Precisely this kind of ambiguity has arisen in this book, first in the *Contracrostipunctus,* and later in the discussions of Gödel's string G (and its relatives). The interpretation given for unplayable records was "I Cannot Be Played on Record Player X", and that for unprovable statements was, "I Cannot Be Proven in Formal System X". Let us take the latter sentence. On what other occasions, if any, have you encountered a sentence containing the pronoun "I" where you automatically understood that the reference was not to the speaker of the sentence, but rather to the sentence itself? Very few, I would guess. The word "I", when it appears in a Shakespeare sonnet, is referring not to a fourteen-line form of poetry printed on a page, but to a flesh-and-blood creature behind the scenes, somewhere off stage.

How far back do we ordinarily trace the "I" in a sentence? The answer, it seems to me, is that we look for a sentient being to attach the authorship to. But what is a sentient being? Something onto which we can map ourselves comfortably. In Weizenbaum's "Doctor" program, is there a personality? If so, whose is it? A small debate over this very question recently raged in the pages of *Science* magazine.

This brings us back to the issue of the "who" who composes computer music. In most circumstances, the driving force behind such pieces is a human intellect, and the computer has been employed, with more or less ingenuity, as a *tool* for realizing an idea devised by the human. The program which carries this out is not anything which we can identify with. It is a simple and single-minded piece of software with no flexibility, no perspective on what it is doing, and no sense of self. If and when, however, people develop programs which have those attributes, and pieces of music start issuing forth from them, then I suggest that will be the appropriate time to start splitting up one's admiration: some to the programmer for creating such an amazing program, and some to the program itself for its sense of music. And it seems to me that that will only take place when the internal structure of such a program is based on something similar to the "symbols" in our brains and their triggering patterns, which are responsible for the complex notion of meaning. The fact of having this kind of internal structure would endow the program with properties which would make us feel comfortable in identifying with it, to some extent. But until then, I will not feel comfortable in saying "this piece was composed by a computer".

Theorem Proving and Problem Reduction

Let us now return to the history of AI. One of the early things which people attempted to program was the intellectual activity of theorem proving. Conceptually, this is no different from programming a computer to look for a derivation of **MU** in the MIU-system, except that the formal systems involved were often more complicated than the MIU-system. They were versions of the Predicate Calculus, which is an extension of the Propositional Calculus involving quantifiers. Most of the rules of the Predicate Calculus are included in TNT, as a matter of fact. The trick in writing such a program is to instill a sense of direction, so that the program does not wander all over the map, but works only on "relevant" pathways—those which, by some reasonable criterion, seem to be leading towards the desired string.

In this book we have not dealt much with such issues. How indeed can you know when you are proceeding towards a theorem, and how can you tell if what you are doing is just empty fiddling? This was one thing which I hoped to illustrate with the MU-puzzle. Of course, there can be no definitive answer: that is the content of the limitative Theorems, since if you could always know which way to go, you could construct an algorithm for proving any desired theorem, and that would violate Church's Theorem. There is no such algorithm. (I will leave it to the reader to see exactly why this follows from Church's Theorem.) However, this doesn't mean that it is impossible to develop any intuition at all concerning what is and what is not a promising route; in fact, the best programs have very sophisticated heuristics, which enable them to make deductions in the Predicate Calculus at speeds which are comparable to those of capable humans.

The trick in theorem proving is to to use the fact that you have an overall goal—namely the string you want to produce—in guiding you locally. One technique which was developed for converting global goals into local strategies for derivations is called *problem reduction*. It is based on the idea that whenever one has a long-range goal, there are usually *subgoals* whose attainment will aid in the attainment of the main goal. Therefore if one breaks up a given problem into a series of new subproblems, then breaks those in turn into subsubproblems, and so on, in a recursive fashion, one eventually comes down to very modest goals which can presumably be attained in a couple of steps. Or at least so it would seem . . .

Problem reduction got Zeno into hot water. Zeno's method, you recall, for getting from A to B (think of B as the goal), is to "reduce" the problem into two subproblems: first go halfway, then go the rest of the way. So now you have "pushed"—in the sense of Chapter V—two subgoals onto your "goal stack". Each of these, in turn, will be replaced by two subsubgoals—and so on ad infinitum. You wind up with an infinite goal-stack, instead of a single goal (Fig. 115). Popping an infinite number of goals off your stack will prove to be tricky—which is just Zeno's point, of course.

Another example of an infinite recursion in problem reduction occurred in the Dialogue *Little Harmonic Labyrinth*, when Achilles wanted to have a Typeless Wish granted. Its granting had to be deferred until permission was gotten from the Meta-Genie; but in order to get permission to give permission, she had to summon the Meta-Meta-Genie—and so on. Despite

FIGURE 115. Zeno's endless goal tree, for getting from A to B.

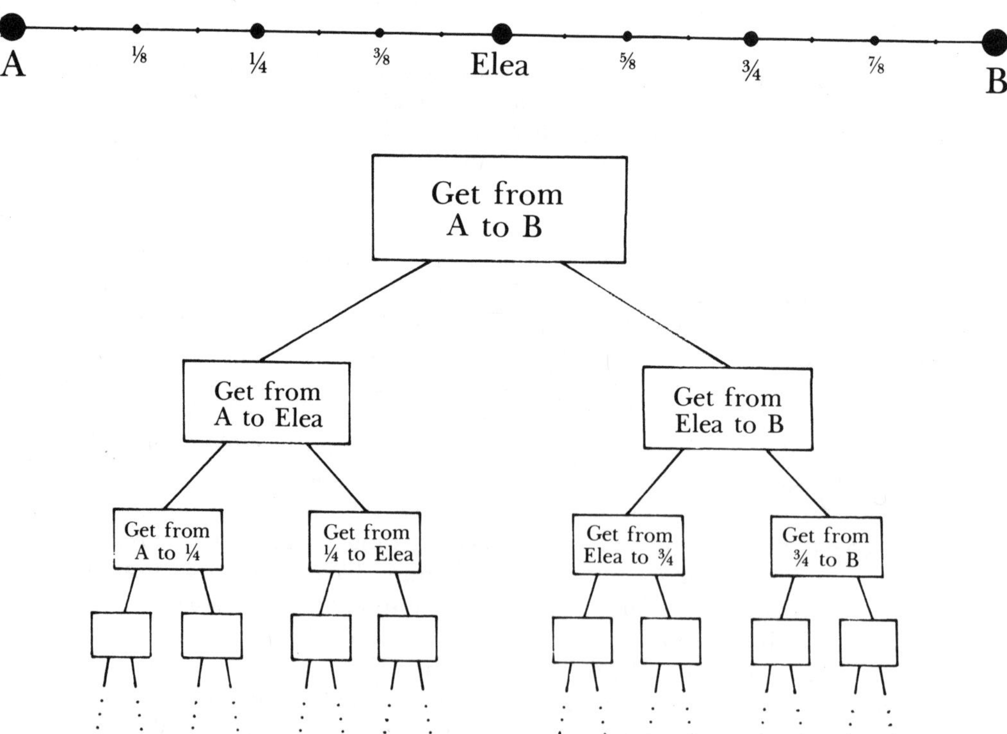

the infiniteness of the goal stack, Achilles got his wish. Problem reduction wins the day!

Despite my mockery, problem reduction is a powerful technique for converting global problems into local problems. It shines in certain situations, such as in the endgame of chess, where the look-ahead technique often performs miserably, even when it is carried to ridiculous lengths, such as fifteen or more plies. This is because the look-ahead technique is not based on *planning;* it simply has no goals and explores a huge number of pointless alternatives. Having a goal enables you to develop a strategy for the achievement of that goal, and this is a completely different philosophy from looking ahead mechanically. Of course, in the look-ahead technique, desirability or its absence is measured by the evaluation function for positions, and that incorporates indirectly a number of goals, principally that of not getting checkmated. But that is too indirect. Good chess players who play against look-ahead chess programs usually come away with the impression that their opponents are very weak in formulating plans or strategies.

Shandy and the Bone

There is no guarantee that the method of problem reduction will work. There are many situations where it flops. Consider this simple problem, for instance. You are a dog, and a human friend has just thrown your favorite bone over a wire fence into another yard. You can see your bone through the fence, just lying there in the grass—how luscious! There is an open gate in the fence about fifty feet away from the bone. What do you do? Some dogs will just run up to the fence, stand next to it, and bark; others will dash up to the open gate and double back to the lovely bone. Both dogs can be said to be exercising the problem reduction technique; however, they represent the problem in their minds in different ways, and this makes all the difference. The barking dog sees the subproblems as (1) running to the fence, (2) getting through it, and (3) running to the bone—but that second subproblem is a "toughie", whence the barking. The other dog sees the subproblems as (1) getting to the gate; (2) going through the gate; (3) running to the bone. Notice how everything depends on the way you represent the "problem space"—that is, on what you perceive as *reducing* the problem (forward motion towards the overall goal) and what you perceive as *magnifying* the problem (backward motion away from the goal).

Changing the Problem Space

Some dogs first try running directly towards the bone, and when they encounter the fence, something clicks inside their brain; soon they change course, and run over to the gate. These dogs realize that what on first glance seemed as if it would *increase* the distance between the initial situation and the desired situation—namely, running away from the bone but towards the open gate—actually would *decrease* it. At first, they confuse *physical* distance with *problem* distance. Any motion away from the bone seems, by definition, a Bad Thing. But then—somehow—they realize that they can shift their perception of what will bring them "closer" to the bone. In a properly chosen abstract space, moving towards the *gate* is a trajectory bringing the dog closer to the bone! At every moment, the dog is getting "closer"—in the new sense—to the bone. Thus, the usefulness of problem reduction depends on how you represent your problem mentally. What in one space looks like a retreat can in another space look like a revolutionary step forward.

In ordinary life, we constantly face and solve variations on the dog-and-bone problem. For instance, if one afternoon I decide to drive one hundred miles south, but am at my office and have ridden my bike to work, I have to make an extremely large number of moves in what are ostensibly "wrong" directions before I am actually on my way in car headed south. I have to leave my office, which means, say, heading east a few feet; then follow the hall in the building which heads north, then west. Then I ride my bike home, which involves excursions in all the directions of the compass; and I reach my home. A succession of short moves there eventually gets me into my car, and I am off. Not that I immediately drive due south, of course—I choose a route which may involve some excursions north, west, or east, with the aim of getting to the freeway as quickly as possible.

All of this doesn't feel paradoxical in the slightest; it is done without even any sense of amusement. The space in which physical backtracking is perceived as direct motion towards the goal is built so deeply into my mind that I don't even see any irony when I head north. The roads and hallways and so forth act as channels which I accept without much fight, so that part of the act of choosing how to perceive the situation involves just accepting what is imposed. But dogs in front of fences sometimes have a hard time doing that, especially when that bone is sitting there so close, staring them in the face, and looking so good. And when the problem space is just a shade more abstract than physical space, people are often just as lacking in insight about what to do as the barking dogs.

In some sense all problems are abstract versions of the dog-and-bone problem. Many problems are not in physical space but in some sort of conceptual space. When you realize that direct motion towards the goal in that space runs you into some sort of abstract "fence", you can do one of two things: (1) try moving away from the goal in some sort of random way, hoping that you may come upon a hidden "gate" through which you can pass and then reach your bone; or (2) try to find a new "space" in which you can represent the problem, and in which there is no abstract fence separating you from your goal—then you can proceed straight towards the goal in this new space. The first method may seem like the lazy way to go, and the second method may seem like a difficult and complicated way to go. And yet, solutions which involve restructuring the problem space more often than not come as sudden flashes of insight rather than as products of a series of slow, deliberate thought processes. Probably these intuitive flashes come from the extreme core of intelligence—and needless to say, their source is a closely protected secret of our jealous brains.

In any case, the trouble is not that problem reduction per se leads to failures; it is quite a sound technique. The problem is a deeper one: how do you choose a good internal representation for a problem? What kind of "space" do you see it in? What kinds of action reduce the "distance" between you and your goal in the space you have chosen? This can be expressed in mathematical language as the problem of hunting for an appropriate *metric* (distance function) between states. You want to find a metric in which the distance between you and your goal is very small.

Now since this matter of choosing an internal representation is itself a type of problem—and a most tricky one, too—you might think of turning the technique of problem reduction back on it! To do so, you would have to have a way of representing a huge variety of abstract spaces, which is an exceedingly complex project. I am not aware of anyone's having tried anything along these lines. It may be just a theoretically appealing, amusing suggestion which is in fact wholly unrealistic. In any case, what AI sorely lacks is programs which can "step back" and take a look at what is going on, and with this perspective, reorient themselves to the task at hand. It is one thing to write a program which excels at a single task which, when done by a human being, seems to require intelligence—and it is another thing altogether to write an intelligent program! It is the difference between the Sphex wasp (see Chapter XI), whose wired-in routine gives the deceptive appearance of great intelligence, and a human being observing a Sphex wasp.

The I-Mode and the M-Mode Again

An intelligent program would presumably be one which is versatile enough to solve problems of many different sorts. It would learn to do each different one and would accumulate experience in doing so. It would be able to work within a set of rules and yet also, at appropriate moments, to step back and make a judgment about whether working within that set of rules is likely to be profitable in terms of some overall set of goals which it has. It would be able to choose to stop working within a given framework, if need be, and to create a new framework of rules within which to work for a while.

Much of this discussion may remind you of aspects of the MU-puzzle. For instance, moving away from the goal of a problem is reminiscent of moving away from MU by making longer and longer strings which you hope may in some indirect way enable you to make MU. If you are a naïve "dog", you may feel you are moving away from your "MU-bone" whenever your string increases beyond two characters; if you are a more sophisticated dog, the use of such lengthening rules has an indirect justification, something like heading for the gate to get your MU-bone.

Another connection between the previous discussion and the MU-puzzle is the two modes of operation which led to insight about the nature of the MU-puzzle: the Mechanical mode, and the Intelligent mode. In the former, you are embedded within some fixed framework; in the latter, you can always step back and gain an overview of things. Having an overview is tantamount to choosing a representation within which to work; and working within the rules of the system is tantamount to trying the technique of problem reduction within that selected framework. Hardy's comment on Ramanujan's style—particularly his willingness to modify his own hypotheses—illustrates this interplay between the M-mode and the I-mode in creative thought.

The Sphex wasp operates excellently in the M-mode, but it has absolutely no ability to choose its framework or even to alter its M-mode in the slightest. It has no ability to notice when the same thing occurs over and over and over again in its system, for to notice such a thing would be to jump out of the system, even if only ever so slightly. It simply does not notice the sameness of the repetitions. This idea (of not noticing the identity of certain repetitive events) is interesting when we apply it to ourselves. Are there highly repetitious situations which occur in our lives time and time again, and which we handle in the identical stupid way each time, because we don't have enough of an overview to perceive their sameness? This leads back to that recurrent issue, "What is sameness?" It will soon come up as an AI theme, when we discuss pattern recognition.

Applying AI to Mathematics

Mathematics is in some ways an extremely interesting domain to study from the AI point of view. Every mathematician has the sense that there is a kind of metric between ideas in mathematics—that all of mathematics is a network of results between which there are enormously many connections. In that network, some ideas are very closely linked; others require more elaborate pathways to be joined. Sometimes two theorems in mathematics are close because one can be proven easily, given the other. Other times two ideas are close because they are analogous, or even isomorphic. These are two different senses of the word "close" in the domain of mathematics. There are probably a number of others. Whether there is an objectivity or a universality to our sense of mathematical closeness, or whether it is largely an accident of historical development is hard to say. Some theorems of different branches of mathematics appear to us hard to link, and we might say that they are unrelated—but something might turn up later which forces us to change our minds. If we could instill our highly developed sense of mathematical closeness—a "mathematician's mental metric", so to speak—into a program, we could perhaps produce a primitive "artificial mathematician". But that depends on being able to convey a sense of simplicity or "naturalness" as well, which is another major stumbling block.

These issues have been confronted in a number of AI projects. There is a collection of programs developed at MIT which go under the name "MACSYMA", whose purpose it is to aid mathematicians in symbolic manipulation of complex mathematical expressions. This program has in it some sense of "where to go"—a sort of "complexity gradient" which guides it from what we would generally consider complex expressions to simpler ones. Part of MACSYMA's repertoire is a program called "SIN", which does symbolic integration of functions; it is generally acknowledged to be superior to humans in some categories. It relies upon a number of different skills, as intelligence in general must: a vast body of knowledge, the technique of problem reduction, a large number of heuristics, and also some special tricks.

Another program, written by Douglas Lenat at Stanford, had as its aim to invent concepts and discover facts in very elementary mathematics. Beginning with the notion of sets, and a collection of notions of what is "interesting" which had been spoon-fed into it, it "invented" the idea of counting, then the idea of addition, then multiplication, then—among other things—the notion of prime numbers, and it went so far as to rediscover Goldbach's conjecture! Of course these "discoveries" were all hundreds—even thousands—of years old. Perhaps this may be explained in part by saying that the sense of "interesting" was conveyed by Lenat in a large number of rules which may have been influenced by his twentieth-century training; nonetheless it is impressive. The program seemed to run out of steam after this very respectable performance. An interesting thing about it was that it was unable to develop or improve upon its own sense of what is interesting. That seemed another level of difficulty up—or perhaps several levels up.

The Crux of AI: Representation of Knowledge

Many of the examples above have been cited in order to stress that the way a domain is represented has a huge bearing on how that domain is "understood". A program which merely printed out theorems of TNT in a preordained order would have no understanding of number theory; a program such as Lenat's with its extra layers of knowledge could be said to have a rudimentary sense of number theory; and one which embeds mathematical knowledge in a wide context of real-world experience would probably be the most able to "understand" in the sense that we think we do. It is this *representation of knowledge* that is at the crux of AI.

In the early days it was assumed that knowledge came in sentence-like "packets", and that the best way to implant knowledge into a program was to develop a simple way of translating facts into small passive packets of data. Then every fact would simply be a piece of data, accessible to the programs using it. This is exemplified by chess programs, where board positions are coded into matrices or lists of some sort and stored efficiently in memory where they can be retrieved and acted upon by subroutines.

The fact that human beings store facts in a more complicated way was known to psychologists for quite a while and has only recently been rediscovered by AI workers, who are now confronting the problems of "chunked" knowledge, and the difference between procedural and declarative types of knowledge, which is related, as we saw in Chapter XI, to the difference between knowledge which is accessible to introspection and knowledge which is inaccessible to introspection.

The naïve assumption that all knowledge should be coded into passive pieces of data is actually contradicted by the most fundamental fact about computer design: that is, how to add, subtract, multiply, and so on is not coded into pieces of data and stored in memory; it is, in fact, represented nowhere in memory, but rather in the wiring patterns of the hardware. A pocket calculator does not store in its memory knowledge of how to add; that knowledge is encoded into its "guts". There is no memory location to point to if somebody demands, "Show me where the knowledge of how to add resides in this machine!"

A large amount of work in AI has nevertheless gone into systems in which the bulk of the knowledge is stored in specific places—that is, declaratively. It goes without saying that *some* knowledge has to be embodied in programs; otherwise one would not have a program at all, but merely an encyclopedia. The question is how to split up knowledge between program and data. Not that it is always easy to distinguish between program and data, by any means. I hope that was made clear enough in Chapter XVI. But in the development of a system, if the programmer intuitively conceives of some particular item as data (or as program), that may have significant repercussions on the system's structure, because as one programs one does tend to distinguish between data-like objects and program-like objects.

It is important to point out that in principle, any manner of coding information into data structures or procedures is as good as any other, in the sense that if you are not too concerned about efficiency, what you can do in one scheme, you can do in the other. However, reasons can be given which seem to indicate that one method is definitely superior to the other. For instance, consider the following argument in favor of using procedural representations only: "As soon as you try to encode features of sufficient complexity into data, you are forced into developing what amounts to a new language, or formalism. So in effect your data structures become program-like, with some piece of your program serving as their interpreter; you might as well represent the same information directly in procedural form to begin with, and obviate the extra level of interpretation."

DNA and Proteins Help Give Some Perspective

This argument sounds quite convincing, and yet, if interpreted a little loosely, it can be read as an argument for the abolishment of DNA and RNA. Why encode genetic information in DNA, when by representing it directly in proteins, you could eliminate not just one, but *two* levels of interpretation? The answer is: it turns out that it is extremely useful to have the same information in several different forms for different purposes. One advantage of storing genetic information in the modular and data-like form of DNA is that two individuals' genes can be easily recombined to form a new genotype. This would be very difficult if the information were only in proteins. A second reason for storing information in DNA is that it is easy to transcribe and translate it into proteins. When it is not needed, it does not take up much room; when it is needed, it serves as a template. There is no mechanism for copying one protein off of another; their folded tertiary structures would make copying highly unwieldy. Complementarily, it is almost imperative to be able to get genetic information into three-dimensional structures such as enzymes, because the recognition and manipulation of molecules is by its nature a three-dimensional operation. Thus the argument for purely procedural representations is seen to be quite fallacious in the context of cells. It suggests that there are advantages to being able to switch back and forth between procedural and declarative representations. This is probably true also in AI.

This issue was raised by Francis Crick in a conference on communication with extraterrestrial intelligence:

> We see on Earth that there are two molecules, one of which is good for replication [DNA] and one of which is good for action [proteins]. Is it possible to devise a system in which one molecule does both jobs, or are there perhaps strong arguments, from systems analysis, which might suggest (if they exist) that to divide the job into two gives a great advantage? This is a question to which I do not know the answer.[14]

Modularity of Knowledge

Another question which comes up in the representation of knowledge is modularity. How easy is it to insert new knowledge? How easy is it to revise old knowledge? How modular are books? It all depends. If from a tightly structured book with many cross-references a single chapter is removed, the rest of the book may become virtually incomprehensible. It is like trying to pull a single strand out of a spider web—you ruin the whole in doing so. On the other hand, some books are quite modular, having independent chapters.

Consider a straightforward theorem-generating program which uses TNT's axioms and rules of inference. The "knowledge" of such a program has two aspects. It resides implicitly in the axioms and rules, and explicitly in the body of theorems which have so far been produced. Depending on which way you look at the knowledge, you will see it either as modular or as spread all around and completely nonmodular. For instance, suppose you had written such a program but had forgotten to include TNT's Axiom 1 in the list of axioms. After the program had done many thousands of derivations, you realized your oversight, and inserted the new axiom. The fact that you can do so in a trice shows that the system's implicit knowledge is modular; but the new axiom's contribution to the explicit knowledge of the system will only be reflected after a long time—after its effects have "diffused" outwards, as the odor of perfume slowly diffuses in a room when the bottle is broken. In that sense the new knowledge takes a long time to be incorporated. Furthermore, if you wanted to go back and replace Axiom 1 by its negation, you could not just do that by itself; you would have to delete all theorems which had involved Axiom 1 in their derivations. Clearly this system's explicit knowledge is not nearly so modular as its implicit knowledge.

It would be useful if we learned how to transplant knowledge modularly. Then to teach everyone French, we would just open up their heads and operate in a fixed way on their neural structures—then they would know how to speak French. Of course, this is only a hilarious pipe dream.

Another aspect of knowledge representation has to do with the way in which one wishes to use the knowledge. Are inferences supposed to be drawn as pieces of information arrive? Should analogies and comparisons constantly be being made between new information and old information? In a chess program, for instance, if you want to generate look-ahead trees, then a representation which encodes board positions with a minimum of redundancy will be preferable to one which repeats the information in several different ways. But if you want your program to "understand" a board position by looking for patterns and comparing them to known patterns, then representing the same information several times over in different forms will be more useful.

Representing Knowledge in a Logical Formalism

There are various schools of thought concerning the best way to represent and manipulate knowledge. One which has had great influence advocates representations using formal notations similar to those for TNT—using propositional connectives and quantifiers. The basic operations in such representations are, not surprisingly, formalizations of deductive reasoning. Logical deductions can be made using rules of inference analogous to some of those in TNT. Querying the system about some particular idea sets up a goal in the form of a string to be derived. For example: "Is MUMON a theorem?" Then the automatic reasoning mechanisms take over in a goal-oriented way, using various methods of problem reduction.

For example, suppose that the proposition "All formal arithmetics are incomplete" were known, and the program were queried, "Is *Principia Mathematica* incomplete?" In scanning the list of known facts—often called the *data base*—the system might notice that *if* it could establish that *Principia Mathematica* is a formal arithmetic, then it could answer the question. Therefore the proposition "*Principia Mathematica* is a formal arithmetic" would be set up as a subgoal, and then problem reduction would take over. If it could find further things which would help in establishing (or refuting) the goal or the subgoal, it would work on them—and so on, recursively. This process is given the name of *backwards chaining*, since it begins with the goal and works its way backwards, presumably towards things which may already be known. If one makes a graphic representation of the main goal, subsidiary goals, subsubgoals, etc., a tree-like structure will arise, since the main goal may involve several different subgoals, each of which in turn involves several subsubgoals, etc.

Notice that this method is not guaranteed to resolve the question, for there may be no way of establishing within the system that *Principia Mathematica* is a formal arithmetic. This does not imply, however, that either the goal or the subgoal is a false statement—merely that they cannot be derived with the knowledge currently available to the system. The system may print out, in such a circumstance, "I do not know" or words to that effect. The fact that some questions are left open is of course similar to the incompleteness from which certain well-known formal systems suffer.

Deductive *vs.* Analogical Awareness

This method affords a *deductive awareness* of the domain that is represented, in that correct logical conclusions can be drawn from known facts. However, it misses something of the human ability to spot similarities and to compare situations—it misses what might be called *analogical awareness*—a crucial side of human intelligence. This is not to say that analogical thought processes cannot be forced into such a mold, but they do not lend themselves naturally to being captured in that kind of formalism. These days, logic-oriented systems are not so much in vogue as other kinds, which allow complex forms of comparisons to be carried out rather naturally.

When you realize that knowledge representation is an altogether different ball game than mere storage of numbers, then the idea that "a computer has the memory of an elephant" is an easy myth to explode. What is *stored in memory* is not necessarily synonymous with what a program *knows;* for even if a given piece of knowledge is encoded somewhere inside a complex system, there may be no procedure, or rule, or other type of handler of data, which can get at it—it may be inaccessible. In such a case, you can say that the piece of knowledge has been "forgotten" because access to it has been temporarily or permanently lost. Thus a computer program may "forget" something on a high level which it "remembers" on a low level. This is another one of those ever-recurring level distinctions, from which we can probably learn much about our own selves. When a human forgets, it most likely means that a high-level pointer has been lost—not that any information has been deleted or destroyed. This highlights the extreme importance of keeping track of the ways in which you store incoming experiences, for you never know in advance under what circumstances, or from what angle, you will want to pull something out of storage.

From Computer Haiku to an RTN-Grammar

The complexity of the knowledge representation in human heads first hit home with me when I was working on a program to generate English sentences "out of the blue". I had come to this project in a rather interesting way. I had heard on the radio a few examples of so-called "Computer Haiku". Something about them struck me deeply. There was a large element of humor and simultaneously mystery to making a computer generate something which ordinarily would be considered an artistic creation. I was highly amused by the humorous aspect, and I was very motivated by the mystery—even contradiction—of programming creative acts. So I set out to write a program even more mysteriously contradictory and humorous than the haiku program.

At first I was concerned with making the grammar flexible and recursive, so that one would not have the sense that the program was merely filling in the blanks in some template. At about that time I ran across a *Scientific American* article by Victor Yngve in which he described a simple but flexible grammar which could produce a wide variety of sentences of the type found in some children's books. I modified some of the ideas I'd gleaned from that article and came up with a set of procedures which formed a Recursive Transition Network grammar, as described in Chapter V. In this grammar, the selection of words in a sentence was determined by a process which began by selecting—at random—the overall structure of the sentence; gradually the decision-making process trickled down through lower levels of structure until the word level and the letter level were reached. A lot had to be done below the word level, such as inflecting verbs and making plurals of nouns; also irregular verb and noun forms were first formed regularly, and then if they matched entries in a table, substitutions of the proper (irregular) forms were made. As each word reached its final form, it was printed out. The program was like the proverbial monkey at a typewriter, but operating on several levels of linguistic structure simultaneously—not just the letter level.

In the early stages of developing the program, I used a totally silly vocabulary—deliberately, since I was aiming at humor. It produced a lot of nonsense sentences, some of which had very complicated structures, others of which were rather short. Some excerpts are shown below:

> A male pencil who must laugh clumsily would quack. Must program not always crunch girl at memory? The decimal bug which spits clumsily might tumble. Cake who does sure take an unexpected man within relationship might always dump card.
>
> Program ought run cheerfully.
>
> The worthy machine ought not always paste the astronomer.
>
> Oh, program who ought really run off of the girl writes musician for theater. The businesslike relationship quacks.
>
> The lucky girl which can always quack will never sure quack.
>
> The game quacks. Professor will write pickle. A bug tumbles. Man takes the box who slips.

The effect is strongly surrealistic and at times a little reminiscent of haiku—for example, the final sample of four consecutive short sentences. At first it seemed very funny and had a certain charm, but soon it became rather stale. After reading a few pages of output one could sense the limits of the space in which the program was operating; and after that, seeing random points inside that space—even though each one was "new"—was nothing new. This is, it seems to me, a general principle: you get bored with something not when you have exhausted its repertoire of behavior, but when you have mapped out the limits of the space that contains its behavior. The behavior space of a person is just about complex enough that it can continually surprise other people; but that wasn't true of my program. I realized that my goal of producing truly humorous output would require that far more subtlety be programmed in. But what, in this case, was meant by "subtlety"? It was clear that absurd juxtapositions of words were just too unsubtle; I needed a way to ensure that words would be used in accordance with the realities of the world. This was where thoughts about representation of knowledge began to enter the picture.

From RTN's to ATN's

The idea I adopted was to classify each word—noun, verb, preposition, etc.—in several different "semantic dimensions". Thus, each word was a member of classes of various sorts; then there were also superclasses—classes of classes (reminiscent of the remark by Ulam). In principle, such aggregation could continue to any number of levels, but I stopped at two. At any given moment, the choice of words was now semantically restricted, because it was required that there should be *agreement* between the various parts of the phrase being constructed. The idea was, for instance, that certain kinds of acts could be performed only by animate objects; that only certain kinds of abstractions could influence events, and so on. The decisions about what categories were reasonable, and whether each category was better thought of as a class or a superclass, were quite complicated. All words were branded in several different dimensions. Common prepositions—"of", "in", etc.—had several distinct entries, corresponding to their distinct usages. Now, the output began to be much more comprehensible—and for that reason it was funny in a new way.

A Little Turing Test

Below, I have reproduced nine selections, carefully culled from many pages of output from later versions of my program. Along with them are three (seriously intended) human-written sentences. Which?

(1) Blurting may be considered as the reciprocal substitution of semiotic material (dubbing) for a semiotic dialogical product in a dynamic reflexion.

(2) Rather think of a pathway of a 'sequence' of gedankenexperiment simpletons where heir-lines are a prima facie case of a paradiachronic transitivity.

(3) Think of that as a chain strength possibility of what, eventually, comes out as a product (epistemic conditions?) and the product is not a Frankfurt-ish packing-it-all-in.

(4) Despite the efforts, the reply, if you will, had been supported by the Orient; hence a fallacy will thereafter be suspended by the attitude which will be being held by the ambassador.

(5) Of course, until the upheavals, the ambassador was slightly gradually mollycoddling the rabble.

(6) Supposedly, refined liberty caused the attitudes insofar as peace is distilled by the consequences which will not eventually be caused by the command irrevocably insofar as peace of it is sometimes causing the intransigency infinitesimally surprisingly.

(7) According to the sophists, the campaigns in the city-states, in other words, have been accepted by the Orient cunningly. Of course, the Orient has been separated by the states particularly violently.
The Orient supports the efforts which had been supported by mankind.

(8) Admittedly, the hierarchical origin of the fallacy, nevertheless, will be prophesied by the enemies of it. By the same token, the individualists will have testified that intransigency will not have suspended the campaigns.

(9) Needless to say, during the upheaval which will have warranted the secrecy, the replies do not separate the Orient. Of course, the countries, ipso facto, are always probing liberty.

(10) Although a Nobel Prize was being achieved by the humanists, yet in addition, it was being achieved by the serf.

(11) An attitude will often be held by the serfs of a strife-torn nation.

(12) Moreover, the Nobel Prizes will be achieved. By the same token, despite the consequence, the Nobel Prizes which will be achieved will sometimes be achieved by a woman.

The human-written sentences are numbers 1 to 3; they were drawn from the contemporary journal *Art-Language*[15] and are—as far as I can tell—completely serious efforts among literate and sane people to communicate something to each other. That they appear here out of context is not too misleading, since their proper context sounds just the same as they do.

My program produced the rest. Numbers 10 to 12 were chosen to show that there were occasional bursts of total lucidity; numbers 7 to 9 are more typical of the output, floating in that curious and provocative netherworld between meaning and no-meaning; and then numbers 4 to 6 pretty much transcend meaning. In a generous mood, one could say that they stand on their own as pure "language objects", something like pieces of abstract sculpture carved out of words instead of stone; alternatively, one could say that they are pure pseudointellectual drivel.

My choice of vocabulary was still aimed at producing humorous effects. The flavor of the output is hard to characterize. Although much of it "makes sense", at least on a single-sentence level, one definitely gets the feeling that the output is coming from a source with no understanding of what it is saying and no reason to say it. In particular, one senses an utter lack of visual imagery behind the words. When I saw such sentences come pouring out of the line printer, I experienced complex emotions. I was very amused by the silliness of the output. I was also very proud of my achievement and tried to describe it to friends as similar to giving rules for building up meaningful stories in Arabic out of single strokes of the pen—an exaggeration, but it pleased me to think of it that way. And lastly I was deeply thrilled by the knowledge that this enormously complicated machine was shunting around long trains of symbols inside it according to rules, and that these long trains of symbols were something like thoughts in my own head . . . something like them.

Images of What Thought Is

Of course I didn't fool myself into thinking that there was a conscious being behind those sentences—far from it. Of all people, I was the most aware of the reasons that this program was terribly remote from real thought. Tesler's Theorem is quite apt here: as soon as this level of language-handling ability had been mechanized, it was clear that it did not constitute intelligence. But this strong experience left me with an image: a glimmering sense that *real* thought was composed of much longer, much more complicated trains of symbols in the brain—many trains moving simultaneously down many parallel and crisscrossing tracks, their cars being pushed and pulled, attached and detached, switched from track to track by a myriad neural shunting-engines . . .

It was an intangible image which I cannot convey in words, and it was only an image. But images and intuitions and motivations lie mingled close in the mind, and my utter fascination with this image was a constant spur to think more deeply about what thought really could be. I have tried in other parts of this book to communicate some of the daughter images of this original image—particularly in the *Prelude, Ant Fugue.*

What stands out in my mind now, as I look back at this program from the perspective of a dozen years, is how there is no sense of imagery behind what is being said. The program had *no idea* what a serf is, what a person is, or what anything at all is. The words were empty formal symbols, as empty

شيم الحياة ولاح في اعطافه مشبه الفقير ودله المملوك
مثل العليه بن حسين بخياره خوف المسطور قليله التمرنا
يقول انا راى صار والطابه اكل الشيء وادومه رغبة واكلا
ومن لا تشوك كمروده الا وبل مستبرح الراى ولعلك موونتها
على الراى قالوا الحمو من راى طرنا بن لانه بتعا بابها وتغلبه
ويقفز عنها والنعبه موصوفه بشره الاكل ودوامه ومن اكل
من الكمش والزمكه اكل من البرذون وقيل لا يعلم في الدواب
اكل فلا برذونه رغوب فاذا كانت البرذونه اكيل الدواب بعض
حساب ذلك بن اكلها اذا ارضعت ويقال انه لو جمع اكل المراه
من غزوه الى الليل لكان اكثر من عمل الرجل وعشايه هكذا
يكون في اكثر النساء ومن تضع من غزوه الى الليل وكذلك
الحجر والفرس ○

ومن العرجان معاذ بن جبل قالوا وكان معاذ امه وكان يشبه
ابرهيم خليل الرحمن ولم يكن في السلف احسن جرده ولا انعم بدنا
من معاذ وسهل بن حنيف وقال النبي صلى الله عليه وسلم امر كل
شي من معاذ حتى خاتمه وكان يعد من الزهاد السنه وقد شهد
المشاهد وولي للنبي الولايات وجمع القران وتعليم الناس الاحكام
وتدريسهم القران وهو ابن اقل من عشرين سنه وكان عند
رسول الله وجيها وفي عيون المسلمين عظيما وقال الهيثم
انبانا ابو الهذيل بن شعيب بن مهيب الطائي في اسناد له قال بعث

as—perhaps emptier than—the **p** and **q** of the pq-system. My program took advantage of the fact that when people read text, they quite naturally tend to imbue each word with its full flavor—as if that were necessarily attached to the group of letters which form the word. My program could be looked at as a formal system, whose "theorems"—the output sentences—had ready-made interpretations (at least to speakers of English). But unlike the pq-system, these "theorems" were not all true statements when interpreted that way. Many were false, many were nonsense.

In its humble way, the pq-system mirrored a tiny corner of the world. But when my program ran, there was no mirror inside it of how the world works, except for the small semantic constraints which it had to follow. To create such a mirror of understanding, I would have had to wrap each concept in layers and layers of knowledge about the world. To do this would have been another kind of effort from what I had intended to do. Not that I didn't often think of trying to do it—but I never got around to trying it out.

Higher-Level Grammars . . .

In fact, I often pondered whether I could write an ATN-grammar (or some other kind of sentence-producing program) which would only produce *true* sentences about the world. Such a grammar would imbue the words with genuine meanings, in the way it happened in the pq-system and in TNT. This idea of a language in which false statements are ungrammatical is an old one, going back to Johann Amos Comenius, in 1633. It is very appealing because you have a crystal ball embodied in your grammar: just write down the statement you want to know about, and check to see if it is grammatical . . . Actually, Comenius went even further, for in his language, false statements were not only ungrammatical—they were inexpressible!

Carrying this thought in another direction, you might imagine a high-level grammar which would produce random kōans. Why not? Such a grammar would be equivalent to a formal system whose theorems are kōans. And if you had such a program, could you not arrange it to produce only *genuine* kōans? My friend Marsha Meredith was enthusastic about this idea of "Artificial Ism", so she tackled the project of writing a kōan-writing program. One of her early efforts produced this curious quasi-kōan:

> A SMALL YOUNG MASTER WANTED A SMALL WHITE GNARLED BOWL. "HOW CAN WE LEARN AND UNDERSTAND WITHOUT STUDY?" THE YOUNG MASTER ASKED A LARGE CONFUSED MASTER. THE CONFUSED MASTER WALKED FROM A BROWN HARD MOUNTAIN TO A WHITE SOFT MOUNTAIN WITH A SMALL RED STONY BOWL. THE CONFUSED MASTER SAW A RED SOFT HUT. THE CONFUSED MASTER WANTED THE HUT. "WHY DID BODHIDHARMA COME INTO CHINA?" THE CONFUSED MASTER

FIGURE 116. A meaningful story in Arabic. [From A. Khatibi and M. Sijelmassi, The Splendour of Islamic Calligraphy *(New York: Rizzoli, 1976).]*

ASKED A LARGE ENLIGHTENED STUDENT. "THE PEACHES ARE LARGE", THE STUDENT ANSWERED THE CONFUSED MASTER. "HOW CAN WE LEARN AND UNDERSTAND WITHOUT STUDY?" THE CONFUSED MASTER ASKED A LARGE OLD MASTER. THE OLD MASTER WALKED FROM A WHITE STONY G0025. THE OLD MASTER GOT LOST.

Your personal decision procedure for kōan genuineness probably reached a verdict without need of the Geometric Code or the Art of Zen Strings. If the lack of pronouns or the unsophisticated syntax didn't arouse your suspicions, that strange "G0025" towards the end must have. What is it? It is a strange fluke—a manifestation of a bug which caused the program to print out, in place of the English word for an object, the program's *internal* name for the "node" (a LISP atom, in fact) where all information concerning that particular object was stored. So here we have a "window" onto a lower level of the underlying Zen mind—a level that should have remained invisible. Unfortunately, we don't have such clear windows onto the lower levels of human Zen minds.

The sequence of actions, though a little arbitrary, comes from a recursive LISP procedure called "CASCADE", which creates chains of actions linked in a vaguely causal way to each other. Although the degree of comprehension of the world possessed by this kōan generator is clearly not stupendous, work is in progress to make its output a little more genuine-seeming.

Grammars for Music?

Then there is music. This is a domain which you might suppose, on first thought, would lend itself admirably to being codified in an ATN-grammar, or some such program. Whereas (to continue this naïve line of thought) language relies on connections with the outside world for meaning, music is purely formal. There is no reference to things "out there" in the sounds of music; there is just pure syntax—note following note, chord following chord, measure following measure, phrase following phrase . . .

But wait. Something is wrong in this analysis. Why is some music so much deeper and more beautiful than other music? It is because form, in music, is expressive—expressive to some strange subconscious regions of our minds. The sounds of music do not refer to serfs or city-states, but they do trigger clouds of emotion in our innermost selves; in that sense musical meaning *is* dependent on intangible links from the symbols to things in the world—those "things", in this case, being secret software structures in our minds. No, great music will not come out of such an easy formalism as an ATN-grammar. Pseudomusic, like pseudo-fairy tales, may well come out—and that will be a valuable exploration for people to make—but the secrets of meaning in music lie far, far deeper than pure syntax.

I should clarify one point here: in principle, ATN-grammars have all the power of any programming formalism, so if musical meaning is capturable in any way at all (which I believe it is), it is capturable in an ATN-grammar. True. But in that case, I maintain, the grammar will be defining not just musical structures, but the entire structures of the mind of a beholder. The "grammar" will be a full grammar of thought—not just a grammar of music.

Winograd's Program SHRDLU

What kind of program would it take to make human beings admit that it had some "understanding", even if begrudgingly? What would it take before you wouldn't feel intuitively that there is "nothing there"?

In the years 1968-70, Terry Winograd (alias Dr. Tony Earrwig) was a doctoral student at MIT, working on the joint problems of language and understanding. At that time at MIT, much AI research involved the so-called *blocks world*—a relatively simple domain in which problems concerning both vision and language-handling by computer could fit easily. The blocks world consists of a table with various kinds of toy-like blocks on it—square ones, oblong ones, triangular ones, etc., in various colors. (For a "blocks world" of another kind, see Figure 117: the painting *Mental Arithmetic* by Magritte. I find its title singularly appropriate in this context.) The vision problems in the MIT blocks world are very tricky: how can a computer figure out, from a TV-scan of a scene with many blocks in it, just what kinds of blocks are present, and what their relationships are? Some blocks may be perched on top of others, some may be in front of others, there may be shadows, and so on.

FIGURE 117. Mental Arithmetic, *by René Magritte (1931).*

Winograd's work was separate from the issues of vision, however. Beginning with the assumption that the blocks world was well represented inside the computer's memory, he confronted the many-faceted problem of how to get the computer to:

(1) understand questions in English about the situation;
(2) give answers in English to questions about the situation;
(3) understand requests in English to manipulate the blocks;
(4) break down each request into a sequence of operations it could do;
(5) understand what it had done, and for what reasons;
(6) describe its actions and their reasons, in English.

It might seem reasonable to break up the overall program into modular subprograms, with one module for each different part of the problem; then, after the modules have been developed separately, to integrate them smoothly. Winograd found that this strategy of developing independent modules posed fundamental difficulties. He developed a radical approach, which challenged the theory that intelligence can be compartmentalized into independent or semi-independent pieces. His program SHRDLU—named after the old code "ETAOIN SHRDLU", used by linotype operators to mark typos in a newspaper column—did not separate the problem into clean conceptual parts. The operations of parsing sentences, producing internal representations, reasoning about the world represented inside itself, answering questions, and so on, were all deeply and intricately meshed together in a procedural representation of knowledge. Some critics have charged that his program is so tangled that it does not represent any "theory" at all about language, nor does it contribute in any way to our insights about thought processes. Nothing could be more wrong than such claims, in my opinion. A tour de force such as SHRDLU may not be isomorphic to what we do—in fact, in no way should you think that in SHRDLU, the "symbol level" has been attained—but the act of creating it and thinking about it offers tremendous insight into the way intelligence works.

The Structure of SHRDLU

In fact, SHRDLU does consist of separate procedures, each of which contains some knowledge about the world; but the procedures have such a strong interdependency that they cannot be cleanly teased apart. The program is like a very tangled knot which resists untangling; but the fact that you cannot untangle it does not mean that you cannot understand it. There may be an elegant geometrical description of the entire knot even if it is physically messy. We could go back to a metaphor from the *Mu Offering,* and compare it to looking at an orchard from a "natural" angle.

Winograd has written lucidly about SHRDLU. I quote here from his article in Schank and Colby's book:

One of the basic viewpoints underlying the model is that all language use can be thought of as a way of activating procedures within the hearer. We can think of any utterance as a program—one that indirectly causes a set of operations to be carried out within the hearer's cognitive system. This "program writing" is indirect in the sense that we are dealing with an intelligent interpreter, who may take a set of actions which are quite different from those the speaker intended. The exact form is determined by his knowledge of the world, his expectations about the person talking to him, etc. In this program we have a simple version of this process of interpretation as it takes place in the robot. Each sentence interpreted by the robot is converted to a set of instructions in PLANNER. The program that is created is then executed to achieve the desired effect.[16]

PLANNER Facilitates Problem Reduction

The language PLANNER, referred to here, is an AI language whose principal feature is that some of the operations necessary for problem reduction are built in—namely, the recursive process of creating a tree of subgoals, subsubgoals, etc. What this means is that such processes, instead of having to be spelled out time and time again by the programmer, are automatically implied by so-called *GOAL-statements*. Someone who reads a PLANNER program will see no explicit reference to such operations; in jargon, they are *user-transparent*. If one path in the tree fails to achieve the desired goal, then the PLANNER program will "backtrack" and try another route. "Backtracking" is the magic word as far as PLANNER is concerned.

Winograd's program made excellent use of these features of PLANNER—more exactly, of MICROPLANNER, a partial implementation of the plans for PLANNER. In the past few years, however, people with the goal of developing AI have concluded that automatic backtracking, as in PLANNER, has definite disadvantages, and that it will probably not lead to their goal; therefore they have backed off from it, preferring to try other routes to AI.

Let us listen to further comments from Winograd on SHRDLU:

> The definition of every word is a program which is called at an appropriate point in the analysis, and which can do arbitrary computations involving the sentence and the present physical situation.[17]

Among the examples which Winograd cites is the following:

> The different possibilities for the meaning of "the" are procedures which check various facts about the context, then prescribe actions such as "Look for a unique object in the data base which fits this description", or "Assert that the object being described is unique as far as the speaker is concerned." The program incorporates a variety of heuristics for deciding what part of the context is relevant.[18]

It is amazing how deep this problem with the word "the" is. It is probably safe to say that writing a program which can fully handle the top five words of English—"the", "of", "and", "a", and "to"—would be equivalent to solving the entire problem of AI, and hence tantamount to knowing what intelligence and consciousness are. A small digression: the five most common *nouns* in English are—according to the *Word Frequency Book* compiled by John B. Carroll et al—"time", "people", "way", "water", and "words" (in that order). The amazing thing about this is that most people have no idea that we think in such abstract terms. Ask your friends, and 10 to 1 they'll guess such words as "man", "house", "car", "dog", and "money". And—while we're on the subject of frequencies—the top twelve letters in English, in order, according to Mergenthaler, are: "ETAOIN SHRDLU".

One amusing feature of SHRDLU which runs totally against the stereotype of computers as "number crunchers" is this fact, pointed out by Winograd: "Our system does not accept numbers in numeric form, and has only been taught to count to ten."[19] With all its mathematical underpinning, SHRDLU is a mathematical ignoramus! Just like Aunt Hillary, SHRDLU doesn't know anything about the lower levels which make it up. Its knowledge is largely *procedural* (see particularly the remark by "Dr. Tony Earrwig" in section 11 of the previous Dialogue).

It is interesting to contrast the procedural embedding of knowledge in SHRDLU with the knowledge in my sentence-generation program. All of the syntactical knowledge in my program was procedurally embedded in Augmented Transition Networks, written in the language Algol; but the semantic knowledge—the information about semantic class membership—was static: it was contained in a short list of numbers after each word. There were a few words, such as the auxiliary verbs "to be", "to have", and others, which were represented totally in procedures in Algol, but they were the exceptions. By contrast, in SHRDLU, *all* words were represented as programs. Here is a case which demonstrates that, despite the theoretical equivalence of data and programs, in practice the choice of one over the other has major consequences.

Syntax and Semantics

And now, a few more words from Winograd:

> Our program does not operate by first parsing a sentence, then doing semantic analysis, and finally by using deduction to produce a response. These three activities go on concurrently throughout the understanding of a sentence. As soon as a piece of syntactic structure begins to take shape, a semantic program is called to see whether it might make sense, and the resultant answer can direct the parsing. In deciding whether it makes sense, the semantic routine may call deductive processes and ask questions about the real world. As an example, in sentence 34 of the Dialogue ("Put the blue pyramid on the block in the box"), the parser first comes up with "the blue pyramid on the block" as a candidate for a noun group. At this point, semantic analysis is done, and since "the" is definite, a check is made in the data base for the object being referred to. When no such object is found, the parsing is redirected to find the noun group "the blue pyramid". It will then go on to find "on the block in

the box" as a single phrase indicating a location . . . Thus there is a continuing interplay between the different sorts of analysis, with the results of one affecting the others.[20]

It is extremely interesting that in natural language, syntax and semantics are so deeply intertwined. Last Chapter, in discussing the elusive concept of "form", we had broken the notion into two categories: syntactic form, which is detectable by a predictably terminating decision procedure, and semantic form, which is not. But here, Winograd is telling us that—at least when the usual senses of "syntax" and "semantics" are taken—they merge right into each other, in natural language. The external form of a sentence—that is, its composition in terms of elementary signs—does not divide up so neatly into syntactic and semantic aspects. This is a very significant point for linguistics.

Here are some final comments on SHRDLU by Winograd.

Let us look at what the system would do with a simple description like "a red cube which supports a pyramid". The description will use concepts like BLOCK, RED, PYRAMID, and EQUIDIMENSIONAL—all parts of the sys-

FIGURE 118. Procedural representation of "a red cube which supports a pyramid." [*Adapted from Roger Schank and Kenneth Colby,* Computer Models of Thought and Language *(San Francisco: W. H. Freeman, 1973), p. 172.*]

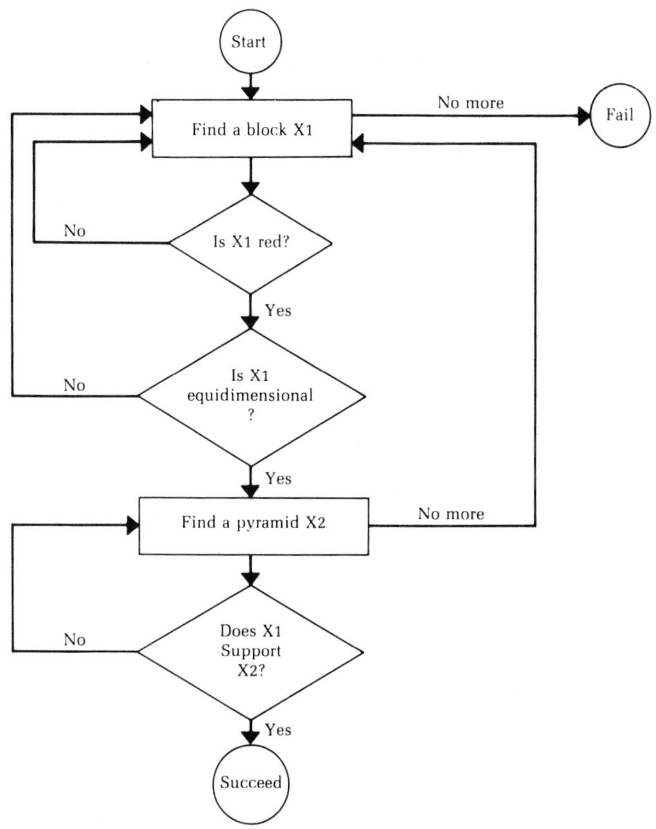

tem's underlying categorization of the world. The result can be represented in a flow chart like that in Figure 118. Note that this is a program for finding an object fitting the description. It would then be incorporated into a command for doing something with the object, a question asking something about it, or, if it appeared in a statement, it would become part of the program which was generated to represent the meaning for later use. Note that this bit of program could also be used as a test to see whether an object fit the description, if the first FIND instruction were told in advance to look only at that particular object.

At first glance, it seems that there is too much structure in this program, as we don't like to think of the meaning of a simple phrase as explicitly containing loops, conditional tests, and other programming details. The solution is to provide an internal language that contains the appropriate looping and checking as its primitives, and in which the representation of the process is as simple as the description. The program described in Figure 118 would be written in PLANNER looking something like what is below:

```
(GOAL (IS ?X1 BLOCK))
(GOAL (COLOR-OF ?X1 RED))
(GOAL (EQUIDIMENSIONAL ?X1))
(GOAL (IS ?X2 PYRAMID))
(GOAL (SUPPORT ?X1 ?X2))
```

The loops of the flowchart are implicit in PLANNER'S backtrack control structure. The description is evaluated by proceeding down the list until some goal fails, at which time the system backs up automatically to the last point where a decision was made, trying a different possibility. A decision can be made whenever a new object name or VARIABLE (indicated by the prefix "?") such as "?X1" or "?X2" appears. The variables are used by the pattern matcher. If they have already been assigned to a particular item, it checks to see whether the GOAL is true for that item. If not, it checks for all possible items which satisfy the GOAL, by choosing one, and then taking successive ones whenever backtracking occurs to that point. Thus, even the distinction between testing and choosing is implicit.[21]

One significant strategy decision in devising this program was to not translate all the way from English into LISP, but only partway—into PLANNER. Thus (since the PLANNER interpreter is itself written in LISP), a new intermediate level—PLANNER—was inserted between the top-level language (English) and the bottom-level language (machine language). Once a PLANNER program had been made from an English sentence fragment, then it could be sent off to the PLANNER interpreter, and the higher levels of SHRDLU would be freed up, to work on new tasks.

This kind of decision constantly crops up: How many levels should a system have? How much and what kind of "intelligence" should be placed on which level? These are some of the hardest problems facing AI today. Since we know so little about natural intelligence, it is hard for us to figure out which level of an artificially intelligent system should carry out what part of a task.

This gives you a glimpse behind the scenes of the Dialogue preceding this Chapter. Next Chapter, we shall meet new and speculative ideas for AI.

Contrafactus

The Crab has invited a small group of friends over to watch the Saturday afternoon football game on television. Achilles has already arrived, but the Tortoise and his friend the Sloth are still awaited.

Achilles: Could that be our friends, a-riding up on that unusual one-wheeled vehicle?

(The Sloth and Tortoise dismount and come in.)

Crab: Ah, my friends, I'm so glad you could make it. May I present my old and beloved acquaintance, Mr. Sloth—and this is Achilles. I believe you know the Tortoise.

Sloth: This is the first time I can recall making the acquaintance of a Bicyclops. Pleased to meet you, Achilles. I've heard many fine things said about the bicyclopean species.

Achilles: Likewise, I'm sure. May I ask about your elegant vehicle?

Tortoise: Our tandem unicycle, you mean? Hardly elegant. It's just a way for two to get from A to B, at the same speed.

Sloth: It's built by a company that also makes teeter-teeters.

Achilles: I see, I see. What is that knob on it?

Sloth: That's the gearshift.

Achilles: Aha! And how many speeds does it have?

Tortoise: One, including reverse. Most models have fewer, but this is a special model.

Achilles: It looks like a very nice tandem unicycle. Oh, Mr. Crab, I wanted to tell you how much I enjoyed hearing your orchestra perform last night.

Crab: Thank you, Achilles. Were you there by any chance, Mr. Sloth?

Sloth: No, I couldn't make it, I'm sad to say. I was participating in a mixed singles ping-ping tournament. It was quite exciting because my team was involved in a one-way tie for first place.

Achilles: Did you win anything?

Sloth: Certainly did—a two-sided Möbius strip made out of copper; it is silver-plated on one side, and gold-plated on the other.

Crab: Congratulations, Mr. Sloth.

Sloth: Thank you. Well, do tell me about the concert.

Crab: It was a most enjoyable performance. We played some pieces by the Bach twins—

Sloth: The famous Joh and Sebastian?

Crab: One and the same. And there was one work that made me think of you, Mr. Sloth—a marvelous piano concerto for two left hands. The next-to-last (and only) movement was a one-voice fugue. You can't imagine its intricacies. For our finale, we played Beethoven's Ninth Zenfunny. At the end, everyone in the audience rose and clapped with one hand. It was overwhelming.

Sloth: Oh, I'm sorry I missed it. But do you suppose it's been recorded? At home I have a fine hi-fi to play it on—the best two-channel monaural system money can buy.

Crab: I'm sure you can find it somewhere. Well, my friends, the game is about to begin.

Achilles: Who is playing today, Mr. Crab?

Crab: I believe it's Home Team versus Visitors. Oh, no—that was last week. I think this week it's Out-of-Towners.

Achilles: I'm rooting for Home Team. I always do.

Sloth: Oh, how conventional. I never root for Home Team. The closer a team lives to the antipodes, the more I root for it.

Achilles: Oh, so you live in the Antipodes? I've heard it's charming to live there, but I wouldn't want to visit them. They're so far away.

Sloth: And the strange thing about them is that they don't get any closer no matter which way you travel.

Tortoise: That's my kind of place.

Crab: It's game time. I think I'll turn on the TV.

(He walks over to an enormous cabinet with a screen, underneath which is an instrument panel as complicated as that of a jet airplane. He flicks a knob, and the football stadium appears in bright vivid color on the screen.)

Announcer: Good afternoon, fans. Well, it looks like that time of year has rolled around again when Home Team and Out-of-Town face each other on the gridiron and play out their classic pigskin rivalry. It's been drizzling on and off this afternoon, and the field's a little wet, but despite the weather it promises to be a fine game, especially with that GREAT pair of eighth-backs playing for Home Team, Tedzilliger and Palindromi. And now, here's Pilipik, kicking off for Home Team. It's in the air! Flampson takes it for Out-of-Towners, and runs it back—he's to the 20, the 25, the 30, and down at the 32. That was Mool in on the tackle for Home Team.

Crab: A superb runback! Did you see how he was ALMOST tackled by Quilker—but somehow broke away?

Sloth: Oh, don't be silly, Crab. Nothing of the kind happened. Quilker did NOT tackle Flampson. There's no need to confuse poor Achilles (or the rest of us) with hocus-pocus about what "almost" happened. It's a fact—with no "almost"'s, "if"'s, "and"'s, or "but"'s.

Announcer: Here's the instant replay. Just watch number 79, Quilker, come in from the side, surprising Flampson, and just about tackle him!

Sloth: "Just about"! Bah!

Achilles: Such a graceful maneuver! What would we do without instant replays?

Announcer: It's first down and 10 for Out-of-Town. Noddle takes the ball, hands off to Orwix—it's a reverse—Orwix runs around to the right, handing off to Flampson—a double reverse, folks!—and now Flampson hands it to Treefig, who's downed twelve yards behind scrimmage. A twelve-yard loss on a triple reverse!

Sloth: I love it! A sensational play!

Achilles: But, Mr. S, I thought you were rooting for Out-of-Town. They lost twelve yards on the play.

Sloth: They did? Oh, well—who cares, as long as it was a beautiful play? Let's see it again.

(. . . and so the first half of the game passes. Towards the end of the third quarter, a particularly crucial play comes up for Home Team. They are behind by eight points. It's third down and 10, and they badly need a first down.)

Announcer: The ball is hiked to Tedzilliger, who fades back, looking for a receiver, and fakes to Quilker. There's Palindromi, playing wide right, with nobody near him. Tedzilliger spots him and fires a low pass to him. Palindromi snatches it out of the air, and—*(There is an audible groan from the crowd.)*—oh, he steps out of bounds! What a crushing blow for Home Team, folks! If Palindromi hadn't stepped out of bounds, he could've run all the way to the end zone for a touchdown! Let's watch the subjunctive instant replay.

(And on the screen the same lineup appears as before.)

The ball is hiked to Tedzilliger, who fades back, looking for a receiver, and fakes to Quilker. There's Palindromi, playing wide right, with nobody near him. Tedzilliger spots him, and fires a low pass to him. Palindromi snatches it out of the air, and—*(There is an audible gasp from the crowd.)*—he almost steps out of bounds! But he's still in bounds, and it's clear all the way to the end zone! Palindromi streaks in, for a touchdown for Home Team! *(The stadium breaks into a giant roar of approval.)* Well, folks, that's what would've happened if Palindromi hadn't stepped out of bounds.

Achilles: Wait a minute . . . WAS there a touchdown, or WASN'T there?

Crab: Oh, no. That was just the subjunctive instant replay. They simply followed a hypothetical a little way out, you know.

Sloth: That is the most ridiculous thing I ever heard of! Next thing you know, they'll be inventing concrete earmuffs.

Tortoise: Subjunctive instant replays are a little unusual, aren't they?

Crab: Not particularly, if you have a Subjunc-TV.

Achilles: Is that one grade below a junk TV?

Crab: Not at all! It's a new kind of TV, which can go into the subjunctive mode. They're particularly good for football games and such. I just got mine.

Achilles: Why does it have so many knobs and fancy dials?

Crab: So that you can tune it to the proper channel. There are many channels broadcasting in the subjunctive mode, and you want to be able to select from them easily.

Achilles: Could you show us what you mean? I'm afraid I don't quite understand what all this talk of "broadcasting in the subjunctive mode" is about.

Crab: Oh, it's quite simple, really. You can figure it out yourself. I'm going into the kitchen to fix some French fries, which I know are Mr. Sloth's weakness.

Sloth: Mmmmm! Go to it, Crab! French fries are my favorite food.

Crab: What about the rest of you?

Tortoise: I could devour a few.

Achilles: Likewise. But wait—before you go into the kitchen, is there some trick to using your Subjunc-TV?

Crab: Not particularly. Just continue watching the game, and whenever there's a near miss of some sort, or whenever you wish things had gone differently in some way, just fiddle with the dials, and see what happens. You can't do it any harm, though you may pick up some exotic channels. *(And he disappears into the kitchen.)*

Achilles: I wonder what he means by that. Oh well, let's get back to this game. I was quite wrapped up in it.

Announcer: It's fourth down for Out-of-Town, with Home Team receiving. Out-of-Town is in punt formation, with Tedzilliger playing deep. Orwix is back to kick—and he gets a long high one away. It's coming down near Tedzilliger—

Achilles: Grab it, Tedzilliger! Give those Out-of-Towners a run for their money!

Announcer: —and lands in a puddle—KERSPLOSH! It takes a weird bounce! Now Sprunk is madly scrambling for the ball! It looks like it just barely grazed Tedzilliger on the bounce, and then slipped away from him—it's ruled a fumble. The referee is signaling that the formidable Sprunk has recovered for Out-of-Town on the Home Team 7! It's a bad break for Home Team. Oh, well, that's the way the cookie crumbles.

Achilles: Oh, no! If only it hadn't been raining . . . *(Wrings his hands in despair.)*

Sloth: ANOTHER of those confounded hypotheticals! Why are the rest of you always running off into your absurd worlds of fantasy? If I were you, I would stay firmly grounded in reality. "No subjunctive nonsense" is my motto. And I wouldn't abandon it even if someone offered me a hundred—nay, a hundred and twelve—French fries.

Achilles: Say, that gives me an idea. Maybe by suitably fiddling with these knobs, I can conjure up a subjunctive instant replay in which it isn't raining, there's no puddle, no weird bounce, and Tedzilliger doesn't fumble. I wonder . . . *(Walks up to the Subjunc-TV and stares at it.)* But I haven't any idea what these different knobs do. *(Spins a few at random.)*

Announcer: It's fourth down for Out-of-Town, with Home Team receiving. Out-of-Town is in punt formation, with Tedzilliger playing deep. Orwix is back to kick—and he gets a long high one away. It's coming down near Tedzilliger—

Achilles: Grab it, Tedzilliger! Give those Out-of-Towners a run for their money!

Announcer: —and lands in a puddle—KERSPLOSH! Oh—it bounces right into his arms! Now Sprunk is madly scrambling after him, but he's got good blocking, and he steers his way clear of the formidable Sprunk, and now he's got an open field ahead of him. Look at that, folks! He's to the 50, the 40, the 30, the 20, the 10—touchdown, Home Team! *(Huge cheers from the Home Team side.)* Well, fans, that's how it would have gone, if footballs were spheres instead of oblate spheroids! But in reality, Home Team loses the ball, and Out-of-Towners take over on the Home Team 7-yard line. Oh, well, that's the way the ball bounces.

Achilles: What do you think of THAT, Mr. Sloth?

(And Achilles gives a smirk in the direction of the Sloth, but the latter is completely oblivious to its devastating effect, as he is busy watching the Crab arrive with a large platter with a hundred and twelve—nay, a hundred—large and delicious French fries, and napkins for all.)

Crab: So how do you three find my Subjunc-TV?

Sloth: Most disappointing, Crab, to be quite frank. It seems to be badly out of order. It makes pointless excursions into nonsense at least half the time. If it belonged to me, I would give it away immediately to someone like you, Crab. But of course it doesn't belong to me.

Achilles: It's quite a strange device. I tried to rerun a play to see how it would have gone under different weather conditions, but the thing seems to have a will of its own! Instead of changing the weather, it changed the football shape to ROUND instead of FOOTBALL-SHAPED! Now tell me—how can a football not be shaped like a football? That's a contradiction in terms. How preposterous!

Crab: Such tame games! I thought you'd surely find more interesting subjunctives. How would you like to see how the last play would have looked if the game had been baseball instead of football?

Tortoise: Oh! An outstanding idea!

(The Crab twiddles two knobs, and steps back.)

Announcer: There are four away, and—

Achilles: FOUR away!?

Announcer: That's right, fans—four away. When you turn football into baseball, SOMETHING's got to give! Now as I was about to say, there are four away, with Out-of-Town in the field, and Home Team up. Tedzilliger is at bat. Out-of-Town is in bunt formation. Orwix raises his arm to pitch—and he gets a long high ball away. It's heading straight for Tedzilliger—

Achilles: Smash it, Tedzilliger! Give those Out-of-Towners a home run for their money!

Announcer: —but it seems to be a spitball, as it takes a strange curve. Now Sprunk is madly scrambling for the ball! It looks like it just barely grazed Tedzilliger's bat, then bounced off it—it's ruled a fly ball. The umpire is signaling that the formidable Sprunk has caught it for Out-of-Town, to end the seventh inning. It's a bad break for Home Team. That's how the last play would have looked, football fans, if this had been a game of baseball.

Sloth: Bah! You might as well transport this game to the Moon.

Crab: No sooner said than done! Just a twiddle here, a twiddle there . . .

(On the screen there appears a desolate crater-pitted field, with two teams in space suits facing each other, immobile. All at once, the two teams fly into motion, and the players are making great bounds into the air, sometimes over the heads of other players. The ball is thrown into the air, and sails so high that it almost disappears, and then slowly comes floating down into the arms of one space-suited player, roughly a quarter-mile from where it was released.)

Announcer: And there, friends, you have the subjunctive instant replay as it would have happened on the Moon. We'll be right back after this important commercial message from the friendly folks who brew Glumpf Beer—my favorite kind of beer!

Sloth: If I weren't so lazy, I would take that broken TV back to the dealer myself! But alas, it's my fate to be a lazy Sloth . . . *(Helps himself to a large gob of French fries.)*

Tortoise: That's a marvelous invention, Mr. Crab. May I suggest a hypothetical?

Crab: Of course!

Tortoise: What would that last play have looked like if space were four-dimensional?

Crab: Oh, that's a complicated one, Mr. T, but I believe I can code it into the dials. Just a moment.

(He steps up, and, for the first time, appears to be using the full power of the control panel of his Subjunc-TV, turning almost every knob two or three times, and carefully checking various meters. Then he steps back with a satisfied expression on his face.)

I think this should do it.

Announcer: And now let's watch the subjunctive instant replay.

(A confusing array of twisted pipes appears on the screen. It grows larger, then smaller, and for a moment seems to do something akin to rotation. Then it turns into a strange mushroom-shaped object, and back to a bunch of pipes. As it metamorphoses from this into other bizarre shapes, the annnouncer gives his commmentary.)

Tedzilliger's fading back to pass. He spots Palindromi ten yards outfield, and passes it to the right and outwards—it looks good! Palindromi's at the 35-yard plane, the 40, and he's tackled on his own

43-yard plane. And there you have it, 3-D fans, as it would've looked if football were played in four spatial dimensions.

Achilles: What is it you are doing, Mr. Crab, when you twirl these various dials on the control panel?

Crab: I am selecting the proper subjunctive channel. You see, there are all sorts of subjunctive channels broadcasting simultaneously, and I want to tune in precisely that one which represents the kind of hypothetical which has been suggested.

Achilles: Can you do this on any TV?

Crab: No, most TV's can't receive subjunctive channels. They require a special kind of circuit which is quite difficult to make.

Sloth: How do you know which channel is broadcasting what? Do you look it up in the newspaper?

Crab: I don't need to know the channel's call letters. Instead, I tune it in by coding, in these dials, the hypothetical situation which I want to be represented. Technically, this is called "addressing a channel by its counterfactual parameters". There are always a large number of channels broadcasting every conceivable world. All the channels which carry worlds that are "near" to each other have call letters that are near to each other, too.

Tortoise: Why did you not have to turn the dials at all, the first time we saw a subjunctive instant replay?

Crab: Oh, that was because I was tuned in to a channel which is very near to the Reality Channel, but ever so slightly off. So every once in a while, it deviates from reality. It's nearly impossible to tune EXACTLY into the Reality Channel. But that's all right, because it's so dull. All their instant replays are straight! Can you imagine? What a bore!

Sloth: I find the whole idea of Subjunc-TV's one giant bore. But perhaps I could change my mind, if I had some evidence that your machine here could handle an INTERESTING counterfactual. For example, how would that last play have looked if addition were not commutative?

Crab: Oh me, oh my! That change is a little too radical, I'm afraid, for this model. I unfortunately don't have a Superjunc-TV, which is the top of the line. Superjunc-TV's can handle ANYTHING you throw at them.

Sloth: Bah!

Crab: But look—I can do ALMOST as well. Wouldn't you like to see how the last play would have happened if 13 were not a prime number?

Sloth: No thanks! THAT doesn't make any sense! Anyway, if I were the last play, I'd be getting pretty tired of being trotted out time and again in new garb for the likes of you fuzzy-headed concept-slippers. Let's get on with the game!

Achilles: Where did you get this Subjunc-TV, Mr. Crab?

Crab: Believe it or not, Mr. Sloth and I went to a country fair the other evening, and it was offered as the first prize in a lottery. Normally I don't indulge in such frivolity, but some crazy impulse grabbed me, and I bought one ticket.

Achilles: What about you, Mr. Sloth?

Sloth: I admit, I bought one, just to humor old Crab.

Crab: And when the winning number was announced, I found, to my amazement, that I'd won the lottery!

Achilles: Fantastic! I've never known anyone who won anything in a lottery before!

Crab: I was flabbergasted at my good fortune.

Sloth: Don't you have something else to tell us about that lottery, Crab?

Crab: Oh, nothing much. It's just that my ticket number was 129. Now when they announced the winning number, it was 128—just one off.

Sloth: So you see, he actually didn't win it at all.

Achilles: He ALMOST won, though . . .

Crab: I prefer to say that I won it, you see. For I came so terribly close . . . If my number had been only one smaller, I would have won.

Sloth: But unfortunately, Crab, a miss is as good as a mile.

Tortoise: Or as bad. What about you, Mr. Sloth? What was your number?

Sloth: Mine was 256—the next power of 2 above 128. Surely, that counts as a hit, if anything does! I can't understand why, however, those fair officials—those UNfair officials—were so thickheaded about it. They refused to award me my fully deserved prize. Some other joker claimed HE deserved it, because his number was 128. I think my number was far closer than HIS, but you can't fight City Hall.

Achilles: I'm all confused. If you didn't win the Subjunc-TV after all, Mr. Crab, then how can we have been sitting here all afternoon watching it? It seems as if we ourselves have been living in some sort of hypothetical world that would have been, had circumstances just been ever so slightly different . . .

Announcer: And that, folks, was how the afternoon at Mr. Crab's would have been spent, had he won the Subjunc-TV. But since he didn't, the four friends simply spent a pleasant afternoon watching Home Team get creamed, 128-0. Or was it 256-0? Oh well, it hardly matters, in five-dimensional Plutonian steam hockey.

CHAPTER XIX

Artificial Intelligence: Prospects

"Almost" Situations and Subjunctives

AFTER READING *Contrafactus*, a friend said to me, "My uncle was almost President of the U.S.!" "Really?" I said. "Sure," he replied, "he was skipper of the PT 108." (John F. Kennedy was skipper of the PT 109.)

That is what *Contrafactus* is all about. In everyday thought, we are constantly manufacturing mental variants on situations we face, ideas we have, or events that happen, and we let some features stay exactly the same while others "slip". What features do we let slip? What ones do we not even consider letting slip? What events are perceived on some deep intuitive level as being close relatives of ones which really happened? What do we think "almost" happened or "could have" happened, even though it unambiguously did not? What alternative versions of events pop without any conscious thought into our minds when we hear a story? Why do some counterfactuals strike us as "less counterfactual" than other counterfactuals? After all, it is obvious that anything that didn't happen didn't happen. There aren't degrees of "didn't-happen-ness". And the same goes for "almost" situations. There are times when one plaintively says, "It almost happened", and other times when one says the same thing, full of relief. But the "almost" lies in the mind, not in the external facts.

Driving down a country road, you run into a swarm of bees. You don't just duly take note of it; the whole situation is immediately placed in perspective by a swarm of "replays" that crowd into your mind. Typically, you think, "Sure am lucky my window wasn't open!"—or worse, the reverse: "Too bad my window wasn't closed!" "Lucky I wasn't on my bike!" "Too bad I didn't come along five seconds earlier." Strange but possible replays: "If that had been a deer, I could have been killed!" "I bet those bees would have rather had a collision with a rosebush." Even stranger replays: "Too bad those bees weren't dollar bills!" "Lucky those bees weren't made of cement!" "Too bad it wasn't just one bee instead of a swarm." "Lucky I wasn't the swarm instead of being me." What slips naturally and what doesn't—and why?

In a recent issue of *The New Yorker* magazine, the following excerpt from the "Philadelphia Welcomat" was reprinted:[1]

> If Leonardo da Vinci had been born a female the ceiling of the Sistine Chapel might never have been painted.

The New Yorker commented:

> And if Michelangelo had been Siamese twins, the work would have been completed in half the time.

The point of *The New Yorker*'s comment is not that such counterfactuals are *false;* it is more that anyone who would entertain such an idea—anyone who would "slip" the sex or number of a given human being—would have to be a little loony. Ironically, though, in the same issue, the following sentence, concluding a book review, was printed without blushing:

> I think he [Professor Philipp Frank] would have enjoyed both of these books enormously.[2]

Now poor Professor Frank is dead; and clearly it is nonsense to suggest that someone could read books written after his death. So why wasn't this serious sentence also scoffed at? Somehow, in some difficult-to-pin-down sense, the parameters slipped in this sentence do not violate our sense of "possibility" as much as in the earlier examples. Something allows us to imagine "all other things being equal" better in this one than in the others. But why? What is it about the way we classify events and people that makes us know deep down what is "sensible" to slip, and what is "silly"?

Consider how natural it feels to slip from the valueless declarative "I don't know Russian" to the more charged conditional "I would like to know Russian" to the emotional subjunctive "I wish I knew Russian" and finally to the rich counterfactual "If I knew Russian, I would read Chekhov and Lermontov in the original". How flat and dead would be a mind that saw nothing in a negation but an opaque barrier! A live mind can see a window onto a world of possibilities.

I believe that "almost" situations and unconsciously manufactured subjunctives represent some of the richest potential sources of insight into how human beings organize and categorize their perceptions of the world. An eloquent co-proponent of this view is the linguist and translator George Steiner, who, in his book *After Babel,* has written:

> Hypotheticals, 'imaginaries', conditionals, the syntax of counter-factuality and contingency may well be the generative centres of human speech. . . . [They] do more than occasion philosophical and grammatical perplexity. No less than future tenses to which they are, one feels, related, and with which they ought probably to be classed in the larger set of 'suppositionals' or 'alternates', these 'if' propositions are fundamental to the dynamics of human feeling. . . .
>
> Ours is the ability, the need, to gainsay or 'un-say' the world, to image and speak it otherwise. . . . We need a word which will designate the power, the compulsion of language to posit 'otherness'. . . . Perhaps 'alternity' will do: to define the 'other than the case', the counter-factual propositions, images, shapes of will and evasion with which we charge our mental being and by means of which we build the changing, largely fictive milieu of our somatic and our social existence. . . .

Finally, Steiner sings a counterfactual hymn to counterfactuality:

It is unlikely that man, as we know him, would have survived without the fictive, counter-factual, anti-determinist means of language, without the semantic capacity, generated and stored in the 'superfluous' zones of the cortex, to conceive of, to articulate possibilities beyond the treadmill of organic decay and death.[3]

The manufacture of "subjunctive worlds" happens so casually, so naturally, that we hardly notice what we are doing. We select from our fantasy a world which is close, in some internal mental sense, to the real world. We compare what is real with what we perceive as *almost* real. In so doing, what we gain is some intangible kind of perspective on reality. The Sloth is a droll example of a variation on reality—a thinking being without the ability to slip into subjunctives (or at least, who *claims* to be without the ability—but you may have noticed that what he says is full of counterfactuals!). Think how immeasurably poorer our mental lives would be if we didn't have this creative capacity for slipping out of the midst of reality into soft "what if"'s! And from the point of view of studying human thought processes, this slippage is very interesting, for most of the time it happens completely without conscious direction, which means that observation of what kinds of things slip, versus what kinds don't, affords a good window on the unconscious mind.

One way to gain some perspective on the nature of this mental metric is to "fight fire with fire". This is done in the Dialogue, where our "subjunctive ability" is asked to imagine a world in which the very notion of subjunctive ability is slipped, compared to what we expect. In the Dialogue, the first subjunctive instant replay—that where Palindromi stays in bounds—is quite a normal thing to imagine. In fact, it was inspired by a completely ordinary, casual remark made to me by a person sitting next to me at a football game. For some reason it struck me and I wondered what made it seem so natural to slip that particular thing, but not, say, the number of the down, or the present score. From those thoughts, I went on to consider other, probably less slippable features, such as the weather (that's in the Dialogue), the kind of game (also in the Dialogue), and then even loonier variations (also in the Dialogue). I noticed, though, that what was completely ludicrous to slip in one situation could be quite slippable in another. For instance, sometimes you might spontaneously wonder how things would be if the ball had a different shape (e.g., if you are playing basketball with a half-inflated ball); other times that would never enter your mind (e.g., when watching a football game on TV).

Layers of Stability

It seemed to me then, and still does now, that the slippability of a feature of some event (or circumstance) depends on a set of nested contexts in which the event (or circumstance) is perceived to occur. The terms *constant, parameter,* and *variable,* borrowed from mathematics, seem useful here. Often mathematicians, physicists, and others will carry out a calculation, saying "c is a constant, p is a parameter, and v is a variable". What they mean is that any of them can vary (including the "constant"); however, there is a kind of hierarchy of variability. In the situation which is being represented by the symbols, c establishes a global condition; p establishes some less global condition which can vary while c is held fixed; and finally, v can run around while c and p are held fixed. It makes little sense to think of holding v fixed while c and p vary, for c and p establish the context in which v has meaning. For instance, think of a dentist who has a list of patients, and for each patient, a list of teeth. It makes perfect sense (and plenty of money) to hold the patient fixed and vary his teeth—but it makes no sense at all to hold one tooth fixed and vary the patient. (Although sometimes it makes good sense to vary the dentist . . .)

We build up our mental representation of a situation layer by layer. The lowest layer establishes the deepest aspect of the context—sometimes being so low that it cannot vary at all. For instance, the three-dimensionality of our world is so ingrained that most of us never would imagine letting it slip mentally. It is a *constant* constant. Then there are layers which establish temporarily, though not permanently, fixed aspects of situations, which could be called *background assumptions*—things which, in the back of your mind, you know can vary, but which most of the time you unquestioningly accept as unchanging aspects. These could still be called "constants". For instance, when you go to a football game, the rules of the game are constants of that sort. Then there are "parameters": you think of them as more variable, but you temporarily hold them constant. At a football game, parameters might include the weather, the opposing team, and so forth. There could be—and probably are—several layers of parameters. Finally, we reach the "shakiest" aspects of your mental representation of the situation—the variables. These are things such as Palindromi's stepping out of bounds, which are mentally "loose" and which you don't mind letting slip away from their real values, for a short moment.

Frames and Nested Contexts

The word *frame* is in vogue in AI currently, and it could be defined as *a computational instantiation of a context.* The term is due to Marvin Minsky, as are many ideas about frames, though the general concept has been floating around for a good number of years. In frame language, one could say that mental representations of situations involve frames nested within each other. Each of the various ingredients of a situation has its own frame. It is interesting to verbalize explicitly one of my mental images concerning nested frames. Imagine a large collection of chests of drawers. When you choose a chest, you have a frame, and the drawer holes are places where "subframes" can be attached. But subframes are themselves chests of drawers. How can you stick a whole chest of drawers into the slot for a single drawer in another chest of drawers? Easy: you shrink and distort the second chest, since, after all, this is all mental, not physical. Now in the outer frame, there may be several different drawer slots that need to be filled; then you may need to fill slots in some of the inner chests of drawers (or subframes). This can go on, recursively.

The vivid surrealistic image of squishing and bending a chest of drawers so that it can fit into a slot of arbitrary shape is probably quite important, because it hints that your concepts are squished and bent by the contexts you force them into. Thus, what does your concept of "person" become when the people you are thinking about are football players? It certainly is a distorted concept, one which is forced on you by the overall context. You have stuck the "person" frame into a slot in the "football game" frame. The theory of representing knowledge in frames relies on the idea that the world consists of quasi-closed subsystems, each of which can serve as a context for others without being too disrupted, or creating too much disruption, in the process.

One of the main ideas about frames is that each frame comes with its own set of expectations. The corresponding image is that each chest of drawers comes with a built-in, but loosely bound, drawer in each of its drawer slots, called a *default.* If I tell you, "Picture a river bank", you will invoke a visual image which has various features, most of which you could override if I added extra phrases such as "in a drought" or "in Brazil" or "without a merry-go-round". The existence of default values for slots allows the recursive process of filling slots to come to an end. In effect, you say, "I will fill in the slots myself as far as three layers down; beyond that I will take the default options." Together with its default expectations, a frame contains knowledge of its limits of applicability, and heuristics for switching to other frames in case it has been stretched beyond its limits of tolerance.

The nested structure of a frame gives you a way of "zooming in" and looking at small details from as close up as you wish: you just zoom in on the proper subframe, and then on one of its subframes, etc., until you have the desired amount of detail. It is like having a road atlas of the USA which has a map of the whole country in the front, with individual state maps inside, and even maps of cities and some of the larger towns if you want still more detail. One can imagine an atlas with arbitrary amounts of detail, going down to single blocks, houses, rooms, etc. It is like looking through a telescope with lenses of different power; each lens has its own uses. It is important that one can make use of all the different scales; often detail is irrelevant and even distracting.

Because arbitrarily different frames can be stuck inside other frames' slots, there is great potential for conflict or "collision". The nice neat scheme of a single, global set of layers of "constants", "parameters", and "variables" is an oversimplification. In fact, each frame will have its own hierarchy of variability, and this is what makes analyzing how we perceive such a complex event as a football game, with its many subframes, subsubframes, etc., an incredibly messy operation. How do all these many frames interact with each other? If there is a conflict where one frame says, "This item is a constant" but another frame says, "No, it is a variable!", how does it get resolved? These are deep and difficult problems of frame theory to which I can give no answers. There has as yet been no complete agreement on what a frame really is, or on how to implement frames in AI programs. I make my own stab at discussing some of these questions in the following section, where I talk about some puzzles in visual pattern recognition, which I call "Bongard problems".

Bongard Problems

Bongard problems (BP's) are problems of the general type given by the Russian scientist M. Bongard in his book *Pattern Recognition.* A typical BP—number 51 in his collection of one hundred—is shown in Figure 119.

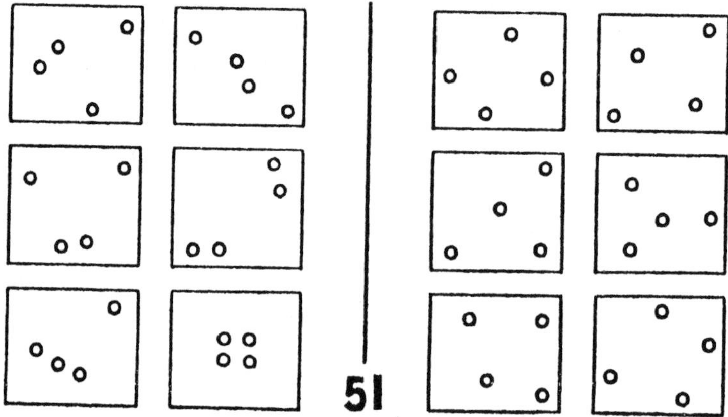

FIGURE 119. Bongard problem 51. [From M. Bongard, Pattern Recognition *(Rochelle Park, N. J.: Hayden Book Co., Spartan Books, 1970).]*

These fascinating problems are intended for pattern-recognizers, whether human or machine. (One might also throw in ETI's—extraterrestrial intelligences.) Each problem consists of twelve boxed figures (henceforth called *boxes*): six on the left, forming *Class I,* and six on the right, forming *Class II.* The boxes may be indexed this way:

I-A	I-B		II-A	II-B
I-C	I-D		II-C	II-D
I-E	I-F		II-E	II-F

The problem is "How do Class I boxes differ from Class II boxes?"

A Bongard problem-solving program would have several stages, in which raw data gradually get converted into descriptions. The early stages are relatively inflexible, and higher stages become gradually more flexible. The final stages have a property which I call *tentativity,* which means simply that the way a picture is represented is always tentative. Upon the drop of a hat, a high-level description can be restructured, using all the devices of the later stages. The ideas presented below also have a tentative quality to them. I will try to convey overall ideas first, glossing over significant difficulties. Then I will go back and try to explain subtleties and tricks and so forth. So your notion of how it all works may also undergo some revisions as you read. But that is in the spirit of the discussion.

Preprocessing Selects a Mini-vocabulary

Suppose, then, that we have some Bongard problem which we want to solve. The problem is presented to a TV camera and the raw data are read in. Then the raw data are *preprocessed.* This means that some salient features are detected. The *names* of these features constitute a "mini-vocabulary" for the problem; they are drawn from a general "salient-feature vocabulary". Some typical terms of the salient-feature vocabulary are:

> line segment, curve, horizontal, vertical, black, white, big, small, pointy, round . . .

In a second stage of preprocessing, some knowledge about elementary *shapes* is used; and if any are found, their names are also made available. Thus, terms such as

> triangle, circle, square, indentation, protrusion, right angle, vertex, cusp, arrow . . .

may be selected. This is roughly the point at which the conscious and the unconscious meet, in humans. This discussion is primarily concerned with describing what happens from here on out.

High-Level Descriptions

Now that the picture is "understood", to some extent, in terms of familiar concepts, some looking around is done. Tentative descriptions are made for one or a few of the twelve boxes. They will typically use simple descriptors such as

> above, below, to the right of, to the left of, inside, outside of, close to, far from, parallel to, perpendicular to, in a row, scattered, evenly spaced, irregularly spaced, etc.

Also, definite and indefinite numerical descriptors can be used:

> 1, 2, 3, 4, 5, . . . many, few, etc.

More complicated descriptors may be built up, such as

> further to the right of, less close to, almost parallel to, etc.

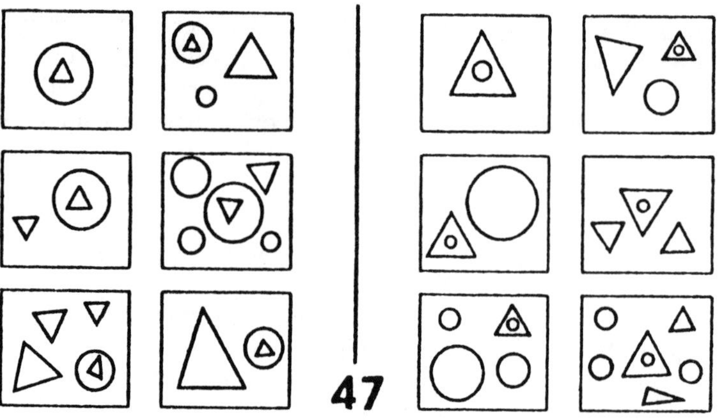

FIGURE 120. Bongard problem 47. [*From M. Bongard,* Pattern Recognition.]

Thus, a typical box—say I-F of BP 47 (Fig. 120)—could be variously described as having:

three shapes
or
three white shapes
or
a circle on the right
or
two triangles and a circle
or
two upwards-pointing triangles
or
one large shape and two small shapes
or
one curved shape and two straight-edged shapes
or
a circle with the same kind of shape on the inside and outside.

Each of these descriptions sees the box through a "filter". Out of context, any of them might be a useful description. As it turns out, though, all of them are "wrong", in the context of the particular Bongard problem they are part of. In other words, if you knew the distinction between Classes I and II in BP 47, and were given one of the preceding lines as a description of an unseen drawing, that information would not allow you to tell to which Class the drawing belonged. The essential feature of this box, in context, is that it includes

a circle containing a triangle.

Note that someone who heard such a description would not be able to *reconstruct* the original drawing, but would be able to *recognize* drawings

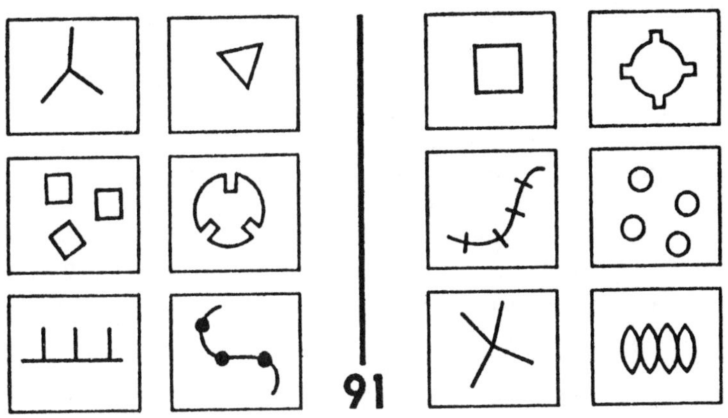

FIGURE 121. *Bongard problem 91.* [*From M. Bongard,* Pattern Recognition.]

which have this property. It is a little like musical style: you may be an infallible recognizer of Mozart, but at the same time unable to write anything which would fool anybody into thinking it was by Mozart.

Now consider box I-D of BP 91 (Fig. 121). An overloaded but "right" description in the context of BP 91 is

a circle with three rectangular intrusions.

Notice the sophistication of such a description, in which the word "with" functions as a disclaimer, implying that the "circle" is not really a circle: it is *almost* a circle, except that . . . Furthermore, the intrusions are not full rectangles. There is a lot of "play" in the way we use language to describe things. Clearly, a lot of information has been thrown away, and even more could be thrown away. A priori, it is very hard to know what it would be smart to throw away and what to keep. So some sort of method for an intelligent compromise has to be encoded, via heuristics. Of course, there is always recourse to lower levels of description (i.e., less chunked descriptions) if discarded information has to be retrieved, just as people can constantly look at the puzzle for help in restructuring their ideas about it. The trick, then, is to devise explicit rules that say how to

make tentative descriptions for each box;
compare them with tentative descriptions for other boxes of either Class;
restructure the descriptions, by
(*i*) adding information,
(*ii*) discarding information,
or (*iii*) viewing the same information from another angle;
iterate this process until finding out what makes the two Classes differ.

One good strategy would be to try to make descriptions *structurally similar to each other,* to the extent this is possible. Any structure they have in common will make comparing them that much easier. Two important elements of this theory deal with this strategy. One is the idea of "description-schemas", or *templates;* the other is the idea of *Sam*—a "sameness detector".

First Sam. Sam is a special agent present on all levels of the program. (Actually there may be different kinds of Sams on different levels.) Sam constantly runs around within individual descriptions and within different descriptions, looking for descriptors or other things which are repeated. When some sameness is found, various restructuring operations can be triggered, either on the single-description level or on the level of several descriptions at once.

Now templates. The first thing that happens after preprocessing is an attempt to manufacture a template, or description-schema—a *uniform format* for the descriptions of all the boxes in a problem. The idea is that a description can often be broken up in a natural way into subdescriptions, and those in turn into subsubdescriptions, if need be. The bottom is hit when you come to primitive concepts which belong to the level of the preprocessor. Now it is important to choose the way of breaking descriptions into parts so as to reflect commonality among all the boxes; otherwise you are introducing a superfluous and meaningless kind of "pseudo-order" into the world.

On the basis of what information is a template built? It is best to look at an example. Take BP 49 (Fig. 122). Preprocessing yields the information that each box consists of several little o's, and one large closed curve. This is a valuable observation, and deserves to be incorporated in the template. Thus a first stab at a template would be:

large closed curve: ——
small o's: ——

FIGURE 122. Bongard problem 49. [*From M. Bongard,* Pattern Recognition.]

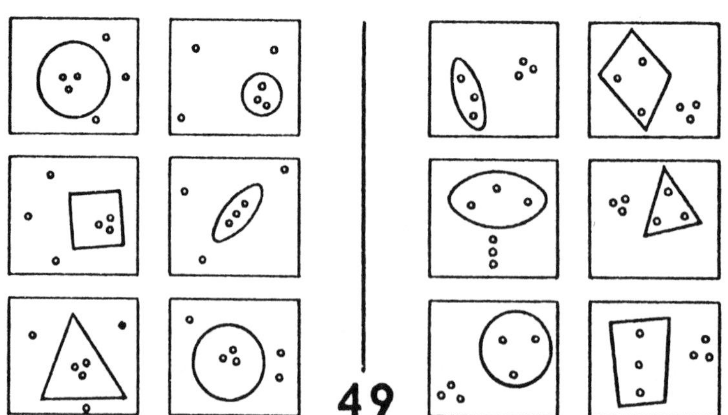

It is very simple: the description-template has two explicit *slots* where subdescriptions are to be attached.

A Heterarchical Program

Now an interesting thing happens, triggered by the term "closed curve". One of the most important modules in the program is a kind of semantic net—the *concept network*—in which all the known nouns, adjectives, etc., are linked in ways which indicate their interrelations. For instance, "closed curve" is strongly linked with the terms "interior" and "exterior". The concept net is just brimming with information about relations between terms, such as what is the opposite of what, what is similar to what, what often occurs with what, and so on. A little portion of a concept network, to be explained shortly, is shown in Figure 123. But let us follow what happens now, in the solution of problem 49. The concepts "interior" and "exterior" are activated by their proximity in the net to "closed curve". This suggests to the template-builder that it might be a good idea to make distinct slots for the interior and exterior of the curve. Thus, in the spirit of tentativity, the template is tentatively restructured to be this:

large closed curve: ——
little o's in interior: ——
little o's in exterior: ——

Now when subdescriptions are sought, the terms "interior" and "exterior" will cause procedures to inspect those specific regions of the box. What is found in BP 49, box I-A is this:

large closed curve: *circle*
little o's in interior: *three*
little o's in exterior: *three*

And a description of box II-A of the same BP might be

large closed curve: *cigar*
little o's in interior: *three*
little o's in exterior: *three*

Now Sam, constantly active in parallel with other operations, spots the recurrence of the concept "three" in all the slots dealing with o's, and this is strong reason to undertake a second template-restructuring operation. Notice that the first was suggested by the concept net, the second by Sam. Now our template for problem 49 becomes:

large closed curve: ——
three little o's in interior: ——
three little o's in exterior: ——

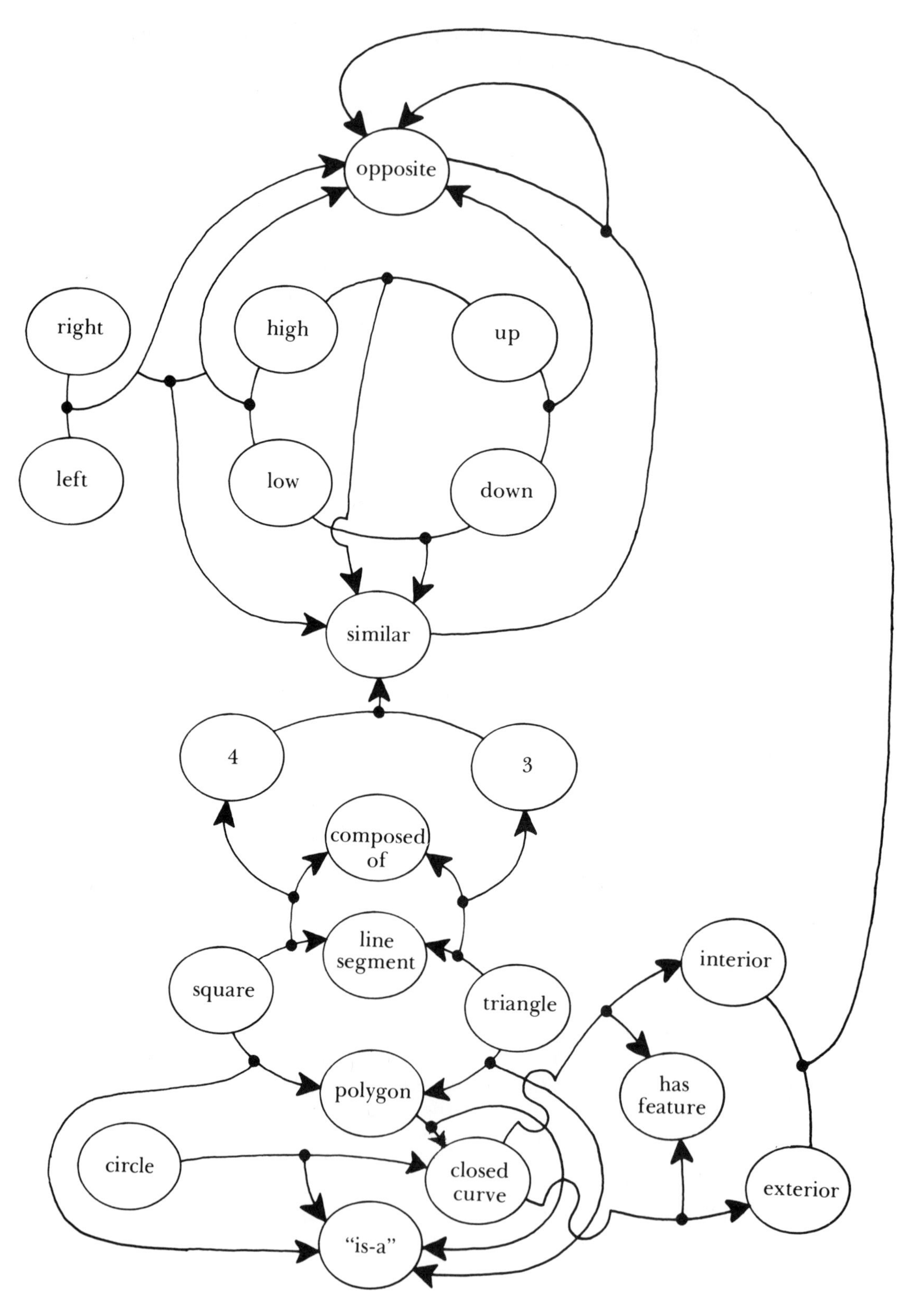
opposite
right
high
up
left
low
down
similar
4
3
composed of
line segment
square
triangle
interior
polygon
has feature
circle
closed curve
exterior
"is-a"

Now that "three" has risen one level of generality—namely, into the template—it becomes worthwhile to explore its neighbors in the concept network. One of them is "triangle", which suggests that triangles of o's may be important. As it happens, this leads down a blind alley—but how could you know in advance? It is a typical blind alley that a human would explore, so it is good if our program finds it too! For box II-E, a description such as the following might get generated:

large closed curve: *circle*
three little o's in interior: *equilateral triangle*
three little o's in exterior: *equilateral triangle*

Of course an enormous amount of information has been thrown away concerning the sizes, positions, and orientations of these triangles, and many other things as well. But that is the whole point of making descriptions instead of just using the raw data! It is the same idea as funneling, which we discussed in Chapter XI.

The Concept Network

We need not run through the entire solution of problem 49; this suffices to show the constant back-and-forth interaction of individual descriptions, templates, the sameness-detector Sam, and the concept network. We should now look a little more in detail at the concept network and its function. A simplified portion shown in the figure codes the following ideas:

"High" and "low" are opposites.
"Up" and "down" are opposites.
"High" and "up" are similar.
"Low" and "down" are similar.
"Right" and "left" are opposites.
The "right-left" distinction is similar to the "high-low" distinction.
"Opposite" and "similar" are opposites.

Note how everything in the net—both nodes and links—can be talked about. In that sense nothing in the net is on a higher level than anything else. Another portion of the net is shown; it codes for the ideas that

A square is a polygon.
A triangle is a polygon.
A polygon is a closed curve.

FIGURE 123. A small portion of a concept network for a program to solve Bongard problems. "Nodes" are joined by "links", which in turn can be linked. By considering a link as a verb and the nodes it joins as subject and object, you can pull out some English sentences from this diagram.

The difference between a triangle and a square is that one has 3 sides and the other has 4.
4 is similar to 3.
A circle is a closed curve.
A closed curve has an interior and an exterior.
"Interior" and "exterior" are opposites.

The network of concepts is necessarily very vast. It seems to store knowledge only statically, or declaratively, but that is only half the story. Actually, its knowledge borders on being procedural as well, by the fact that the proximities in the net act as guides, or "programs", telling the main program how to develop its understanding of the drawings in the boxes.

For instance, some early hunch may turn out to be wrong and yet have the germ of the right answer in it. In BP 33 (Fig. 124), one might at first

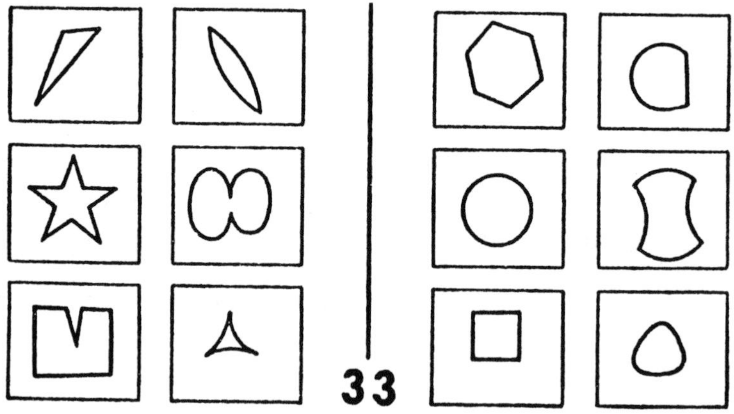

FIGURE 124. Bongard problem 33. [From M. Bongard, Pattern Recognition.]

jump to the idea that Class I boxes contain "pointy" shapes, Class II boxes contain "smooth" ones. But on closer inspection, this is wrong. Nevertheless, there is a worthwhile insight here, and one can try to push it further, by sliding around in the network of concepts beginning at "pointy". It is close to the concept "acute", which is precisely the distinguishing feature of Class I. Thus one of the main functions of the concept network is to allow early wrong ideas to be modified slightly to slip into variations which may be correct.

Slippage and Tentativity

Related to this notion of slipping between closely related terms is the notion of seeing a given object as a variation on another object. An excellent example has been mentioned already—that of the "circle with three indentations", where in fact there is no circle at all. One has to be able to bend concepts, when it is appropriate. Nothing should be absolutely rigid. On the other hand, things shouldn't be so wishy-washy that nothing has any meaning at all, either. The trick is to know when and how to slip one concept into another.

An extremely interesting set of examples where slipping from one description to another is the crux of the matter is given in Bongard problems 85-87 (Fig. 125). BP 85 is rather trivial. Let us assume that our program identifies "line segment" in its preprocessing stage. It is relatively simple for it then to count line segments and arrive at the difference

FIGURE 125. Bongard problems 85-87. [*From M. Bongard,* Pattern Recognition.]

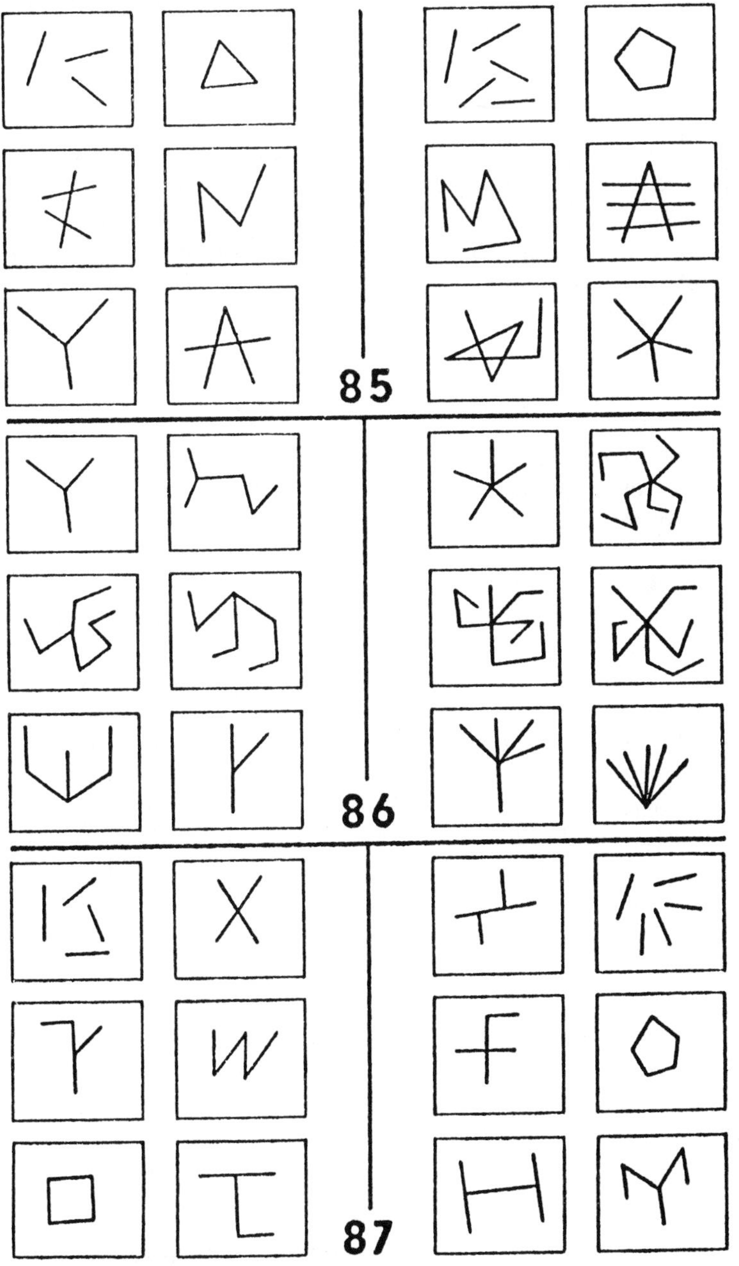

between Class I and Class II in BP 85. Now it goes on to BP 86. A general heuristic which it uses is to *try out recent ideas which have worked.* Successful repetition of recent methods is very common in the real world, and Bongard does not try to outwit this kind of heuristic in his collection—in fact, he reinforces it, fortunately. So we plunge right into problem 86 with two ideas ("count" and "line segment") fused into one: "count line segments". But as it happens, the trick of BP 86 is to count line *trains* rather than line *segments,* where "line train" means an end-to-end concatenation of (one or more) line segments. One way the program might figure this out is if the concepts "line train" and "line segment" are both known, and are close in the concept network. Another way is if it can *invent* the concept of "line train"—a tricky proposition, to say the least.

Then comes BP 87, in which the notion of "line segment" is further played with. When is a line segment three line segments? (See box II-A.) The program must be sufficiently flexible that it can go back and forth between such different representations for a given part of a drawing. It is wise to store old representations, rather than forgetting them and perhaps having to reconstruct them, for there is no guarantee that a newer representation is better than an old one. Thus, along with each old representation should be stored some of the reasons for liking it and disliking it. (This begins to sound rather complex, doesn't it?)

Meta-Descriptions

Now we come to another vital part of the recognition process, and that has to do with levels of abstraction and meta-descriptions. For this let us consider BP 91 (Fig. 121) again. What kind of template could be constructed here? There is such an amount of variety that it is hard to know where to begin. But this is in itself a clue! The clue says, namely, that the class distinction very likely exists on a higher level of abstraction than that of geometrical description. This observation clues the program that it should construct *descriptions of descriptions*—that is, *meta-descriptions.* Perhaps on this second level some common feature will emerge; and if we are lucky, we will discover enough commonality to guide us towards the formulation of a template for the meta-descriptions! So we plunge ahead without a template, and manufacture descriptions for various boxes; then, once these descriptions have been made, we describe *them.* What kinds of slot will our template for meta-descriptions have? Perhaps these, among others:

concepts used: ——
recurring concepts: ——
names of slots: ——
filters used: ——

There are many other kinds of slots which might be needed in meta-descriptions, but this is a sample. Now suppose we have described box I-E of BP 91. Its (template-less) description might look like this:

horizontal line segment
vertical line segment mounted on the horizontal line segment
vertical line segment mounted on the horizontal line segment
vertical line segment mounted on the horizontal line segment

Of course much information has been thrown out: the fact that the three vertical lines are of the same length, are spaced equidistantly, etc. But it is plausible that the above description would be made. So the meta-description might look like this:

concepts used: *vertical-horizontal, line segment, mounted on*
repetitions in description: *3 copies of "vertical line segment mounted on the horizontal line segment"*
names of slots: ——
filters used: ——

Not all slots of the meta-description need be filled in; information can be thrown away on this level as well as on the "just-plain-description" level.

Now if we were to make a description for any of the other boxes of Class I, and then a meta-description of it, we would wind up filling the slot "repetitions in description" each time with the phrase "3 copies of . . ." The sameness-detector would notice this, and pick up *three-ness* as a salient feature, on quite a high level of abstraction, of the boxes of Class I. Similarly, *four-ness* would be recognized, via the method of meta-descriptions, as the mark of Class II.

Flexibility Is Important

Now you might object that in this case, resorting to the method of meta-descriptions is like shooting a fly with an elephant gun, for the three-ness versus four-ness might as easily have shown up on the lower level if we had constructed our descriptions slightly differently. Yes, true—but it is important to have the possibility of solving these problems by different routes. There should be a large amount of flexibility in the program; it should not be doomed if, malaphorically speaking, it "barks up the wrong alley" for a while. (The amusing term "malaphor" was coined by the newspaper columnist Lawrence Harrison; it means a cross between a malapropism and a metaphor. It is a good example of "recombinant ideas".) In any case, I wanted to illustrate the general principle that says: When it is hard to build a template because the preprocessor finds too much diversity, that should serve as a clue that concepts on a higher level of abstraction are involved than the preprocessor knows about.

Focusing and Filtering

Now let us deal with another question: ways to throw information out. This involves two related notions, which I call "focusing" and "filtering". *Focus-*

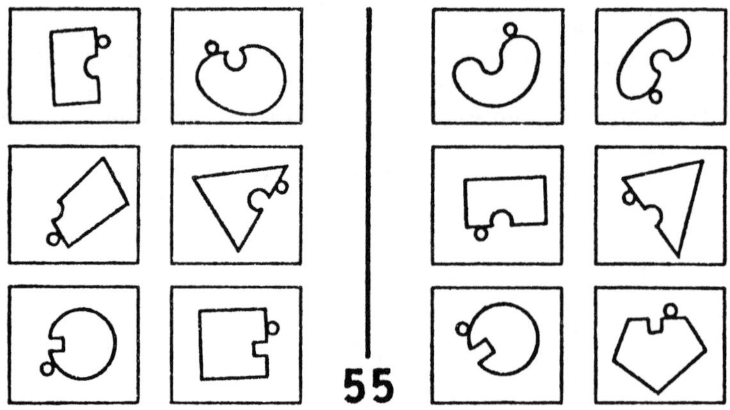

FIGURE 126. Bongard problem 55. [*From M. Bongard,* Pattern Recognition.]

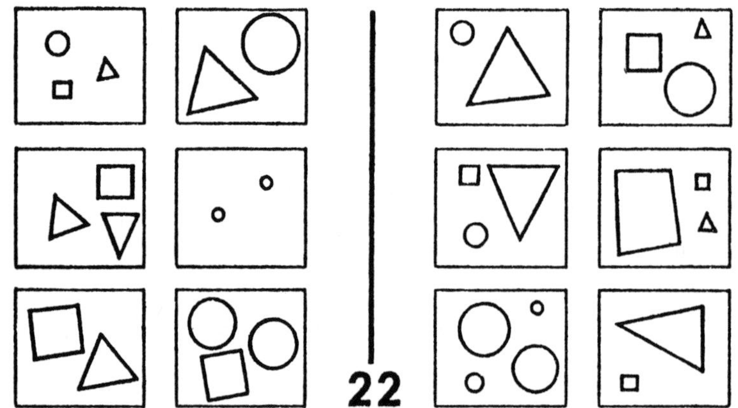

FIGURE 127. Bongard problem 22. [*From M. Bongard,* Pattern Recognition.]

ing involves making a description whose focus is some part of the drawing in the box, to the exclusion of everything else. *Filtering* involves making a description which concentrates on some particular way of viewing the contents of the box, and deliberately ignores all other aspects. Thus they are complementary: focusing has to do with objects (roughly, nouns), and filtering has to do with concepts (roughly, adjectives). For an example of focusing, let's look at BP 55 (Fig. 126). Here, we focus on the indentation and the little circle next to it, to the exclusion of the everything else in the box. BP 22 (Fig. 127) presents an example of filtering. Here, we must filter out every concept but that of size. A combination of focusing and filtering is required to solve problem BP 58 (Fig. 128).

One of the most important ways to get ideas for focusing and filtering is by another sort of "focusing": namely, by inspection of a single particularly simple box—say one with as few objects in it as possible. It can be

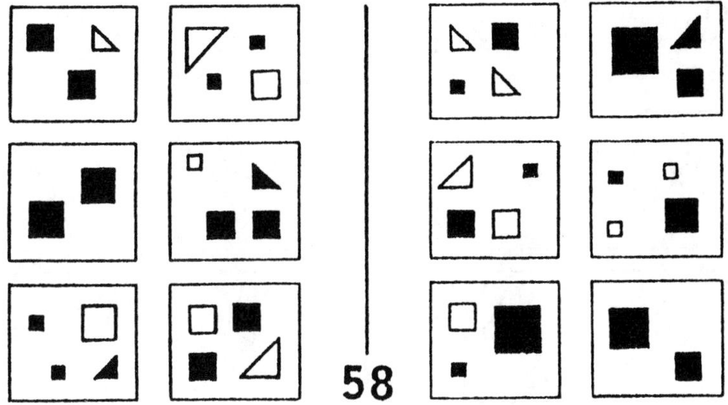

FIGURE 128. Bongard problem 58. [*From M. Bongard,* Pattern Recognition.]

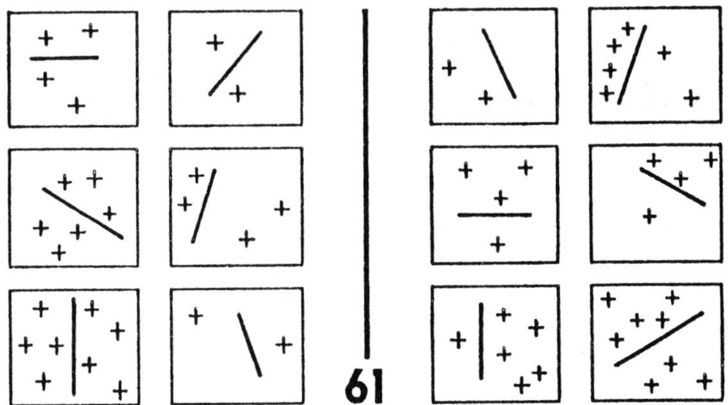

FIGURE 129. Bongard problem 61. [*From M. Bongard,* Pattern Recognition.]

extremely helpful to compare the starkest boxes from the two Classes. But how can you tell which boxes are stark until you have descriptions for them? Well, one way of detecting starkness is to look for a box with a minimum of the features provided by the preprocessor. This can be done very early, for it does not require a pre-existing template; in fact, this can be one useful way of discovering features to build into a template. BP 61 (Fig. 129) is an example where that technique might quickly lead to a solution.

Science and the World of Bongard Problems

One can think of the Bongard-problem world as a tiny place where "science" is done—that is, where the purpose is to discern patterns in the world. As patterns are sought, templates are made, unmade, and remade;

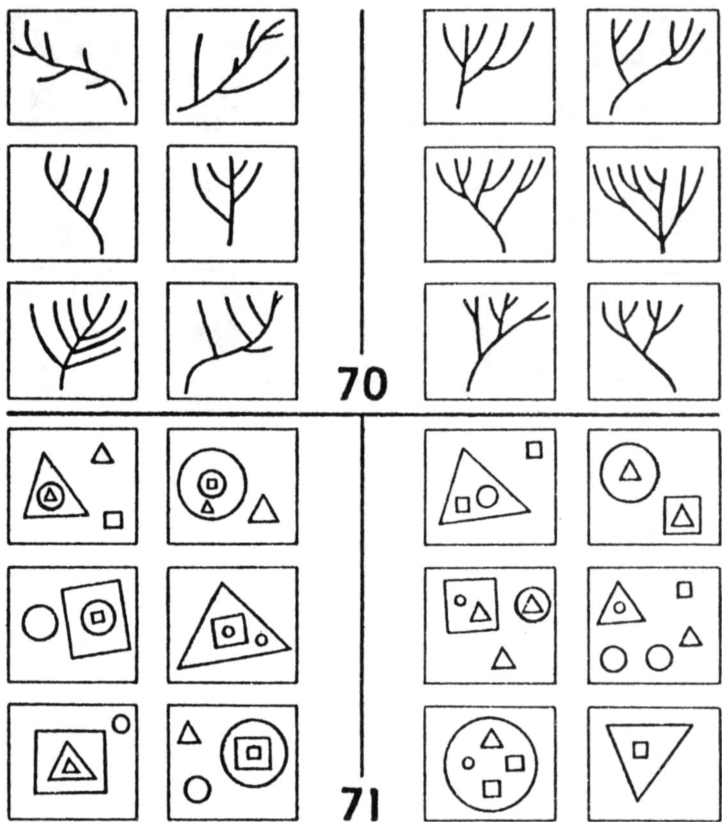

FIGURE 130. Bongard problems 70-71. [*From M. Bongard,* Pattern Recognition.]

slots are shifted from one level of generality to another; filtering and focusing are done; and so on. There are discoveries on all levels of complexity. The Kuhnian theory that certain rare events called "paradigm shifts" mark the distinction between "normal" science and "conceptual revolutions" does not seem to work, for we can see paradigm shifts happening all throughout the system, all the time. The fluidity of descriptions ensures that paradigm shifts will take place on all scales.

Of course, some discoveries are more "revolutionary" than others, because they have wider effects. For instance, one can make the discovery that problems 70 and 71 (Fig. 130) are "the same problem", when looked at on a sufficiently abstract level. The key observation is that both involve depth-2 versus depth-1 nesting. This is a new level of discovery that can be made about Bongard problems. There is an even higher level, concerning the collection as a whole. If someone has never seen the collection, it can be a good puzzle just to figure out what it is. To figure it out is a revolutionary insight, but it must be pointed out that the mechanisms of thought which allow such a discovery to be made are no different from those which operate in the solution of a single Bongard problem.

By the same token, real science does not divide up into "normal" periods versus "conceptual revolutions"; rather, paradigm shifts pervade—there are just bigger and smaller ones, paradigm shifts on different levels. The recursive plots of INT and Gplot (Figs. 32 and 34) provide a geometric model for this idea: they have the same structure full of discontinuous jumps on every level, not just the top level—only the lower the level, the smaller the jumps.

Connections to Other Types of Thought

To set this entire program somewhat in context, let me suggest two ways in which it is related to other aspects of cognition. Not only does it depend on other aspects of cognition, but also they in turn depend on it. First let me comment on how it depends on other aspects of cognition. The intuition which is required for knowing when it makes sense to blur distinctions, to try redescriptions, to backtrack, to shift levels, and so forth, is something which probably comes only with much experience in thought in general. Thus it would be very hard to define heuristics for these crucial aspects of the program. Sometimes one's experience with real objects in the world has a subtle effect on how one describes or redescribes boxes. For instance, who can say how much one's familiarity with living trees helps one to solve BP 70? It is very doubtful that in humans, the subnetwork of concepts relevant to these puzzles can be easily separated out from the whole network. Rather, it is much more likely that one's intuitions gained from seeing and handling real objects—combs, trains, strings, blocks, letters, rubber bands, etc., etc.—play an invisible but significant guiding role in the solution of these puzzles.

Conversely, it is certain that understanding real-world situations heavily depends on visual imagery and spatial intuition, so that having a powerful and flexible way of representing patterns such as these Bongard patterns can only contribute to the general efficiency of thought processes.

It seems to me that Bongard's problems were worked out with great care, and that they have a quality of universality to them, in the sense that each one has a unique correct answer. Of course one could argue with this and say that what we consider "correct" depends in some deep way on our being human, and some creatures from some other star system might disagree entirely. Not having any concrete evidence either way, I still have a certain faith that Bongard problems depend on a sense of simplicity which is not just limited to earthbound human beings. My earlier comments about the probable importance of being acquainted with such surely earth-limited objects as combs, trains, rubber bands, and so on, are not in conflict with the idea that our notion of simplicity is universal, for what matters is not any of these individual objects, but the fact that taken together they span a wide space. And it seems likely that any other civilization would have as vast a repertoire of artifacts and natural objects and varieties of experience on which to draw as we do. So I believe that the skill of solving Bongard problems lies very close to the core of "pure" intelligence, if there is such a thing. Therefore it is a good place to begin if one wants to investigate the ability to discover "intrinsic meaning" in patterns or messages. Unfortunately we have reproduced only a small selection of his stimulating collection. I hope that many readers will acquaint themselves with the entire collection, to be found in his book (see Bibliography).

Some of the problems of visual pattern recognition which we human beings seem to have completely "flattened" into our unconscious are quite amazing. They include:

recognition of faces (invariance of faces under age change, expression change, lighting change, distance change, angle change, etc.)
recognition of hiking trails in forests and mountains—somehow this has always impressed me as one of our most subtle acts of pattern recognition—and yet animals can do it, too
reading text without hesitation in hundreds if not thousands of different typefaces

Message-Passing Languages, Frames, and Symbols

One way that has been suggested for handling the complexities of pattern recognition and other challenges to AI programs is the so-called "actor" formalism of Carl Hewitt (similar to the language "Smalltalk", developed by Alan Kay and others), in which a program is written as a collection of interacting *actors,* which can pass elaborate *messages* back and forth among themselves. In a way, this resembles a heterarchical collection of procedures which can call each other. The major difference is that where procedures usually only pass a rather small number of arguments back and forth, the messages exchanged by actors can be arbitrarily long and complex.

Actors with the ability to exchange messages become somewhat autonomous agents—in fact, even like autonomous computers, with messages being somewhat like programs. Each actor can have its own idiosyncratic way of interpreting any given message; thus a message's meaning will depend on the actor it is intercepted by. This comes about by the actor having within it a piece of program which interprets messages; so there may be as many interpreters as there are actors. Of course, there may be many actors with identical interpreters; in fact, this could be a great advantage, just as it is extremely important in the cell to have a multitude of identical ribosomes floating throughout the cytoplasm, all of which will interpret a message—in this case, messenger RNA—in one and the same way.

It is interesting to think how one might merge the frame-notion with the actor-notion. Let us call a frame with the capability of generating and interpreting complex messages a *symbol:*

frame + actor = symbol

We now have reached the point where we are talking about ways of implementing those elusive *active symbols* of Chapters XI and XII; henceforth in this Chapter, "symbol" will have that meaning. By the way, don't feel dumb if you don't immediately see just how this synthesis is to be made. It is not clear, though it is certainly one of the most fascinating directions to go in AI. Furthermore, it is quite certain that even the best synthesis of these notions will turn out to have much less power than the actual symbols of human minds. In that sense, calling these frame-actor syntheses "symbols" is premature, but it is an optimistic way of looking at things.

Let us return to some issues connected with message passing. Should each message be directed specifically at a target symbol, or should it be thrown out into the grand void, much as mRNA is thrown out into the cytoplasm, to seek its ribosome? If messages have destinations, then each symbol must have an address, and messages for it should always be sent to that address. On the other hand, there could be one central receiving dock for messages, where a message would simply sit until it got picked up by some symbol that wanted it. This is a counterpart to General Delivery. Probably the best solution is to allow both types of message to exist; also to have provisions for different classes of urgency—special delivery, first class, second class, and so on. The whole postal system provides a rich source of ideas for message-passing languages, including such curios as self-addressed stamped envelopes (messages whose senders want answers quickly), parcel post (extremely long messages which can be sent some very slow way), and more. The telephone system will give you more inspiration when you run out of postal-system ideas.

Enzymes and AI

Another rich source of ideas for message passing—indeed, for information processing in general—is, of course, the cell. Some objects in the cell are quite comparable to actors—in particular, enzymes. Each enzyme's active site acts as a filter which only recognizes certain kinds of substrates (messages). Thus an enzyme has an "address", in effect. The enzyme is "programmed" (by virtue of its tertiary structure) to carry out certain operations upon that "message", and then to release it to the world again. Now in this way, when a message is passed from enzyme to enzyme along a chemical pathway, a lot can be accomplished. We have already described the elaborate kinds of feedback mechanisms which can take place in cells (either by inhibition or repression). These kinds of mechanisms show that complicated control of processes can arise through the kind of message passing that exists in the cell.

One of the most striking things about enzymes is how they sit around idly, waiting to be triggered by an incoming substrate. Then, when the substrate arrives, suddenly the enzyme springs into action, like a Venus's-flytrap. This kind of "hair-trigger" program has been used in AI, and goes by the name of *demon*. The important thing here is the idea of having many different "species" of triggerable subroutines just lying around waiting to be triggered. In cells, all the complex molecules and organelles are built up, simple step by simple step. Some of these new structures are often enzymes themselves, and they participate in the building of new enzymes, which in turn participate in the building of yet other types of enzyme, etc. Such recursive cascades of enzymes can have drastic effects on what a cell is doing. One would like to see the same kind of simple step-by-step assembly process imported into AI, in the construction of useful subprograms. For instance, repetition has a way of burning new circuits into our mental hardware, so that oft-repeated pieces of behavior become encoded below the conscious level. It would be extremely useful if there were an analogous way of synthesizing efficient pieces of code which can carry out the same sequence of operations as something which has been learned on a higher level of "consciousness". Enzyme cascades may suggest a model for how this could be done. (The program called "HACKER", written by Gerald Sussman, synthesizes and debugs small subroutines in a way not too much unlike that of enzyme cascades.)

The sameness-detectors in the Bongard problem-solver (Sams) could be implemented as enzyme-like subprograms. Like an enzyme, a Sam would meander about somewhat at random, bumping into small data structures here and there. Upon filling its two "active sites" with identical data structures, the Sam would emit a message to other parts (actors) of the program. As long as programs are serial, it would not make much sense to have several copies of a Sam, but in a truly parallel computer, regulating the number of copies of a subprogram would be a way of regulating the expected waiting-time before an operation gets done, just as regulating the number of copies of an enzyme in a cell regulates how fast that function gets performed. And if new Sams could be synthesized, that would be comparable to the seepage of pattern detection into lower levels of our minds.

Fission and Fusion

Two interesting and complementary ideas concerning the interaction of symbols are "fission" and "fusion". *Fission* is the gradual divergence of a new symbol from its parent symbol (that is, from the symbol which served as a template off of which it was copied). *Fusion* is what happens when two (or more) originally unrelated symbols participate in a "joint activation", passing messages so tightly back and forth that they get bound together and the combination can thereafter be addressed as if it were a single symbol. Fission is a more or less inevitable process, since once a new symbol has been "rubbed off" of an old one, it becomes autonomous, and its interactions with the outside world get reflected in its private internal structure; so what started out as a perfect copy will soon become imperfect, and then slowly will become less and less like the symbol off of which it was "rubbed". Fusion is a subtler thing. When do two concepts really become one? Is there some precise instant when a fusion takes place?

This notion of joint activations opens up a Pandora's box of questions. For instance, how much do we hear "dough" and "nut" when we say "doughnut"? Does a German who thinks of gloves ("Handschuhe") hear "hand-shoes" or not? How about Chinese people, whose word "dōng-xī" ("East-West") means "thing"? It is a matter of some political concern, too, since some people claim that words like "chairman" are heavily charged with undertones of the male gender. The degree to which the parts resonate inside the whole probably varies from person to person and according to circumstances.

The real problem with this notion of "fusion" of symbols is that it is very hard to imagine general algorithms which will create meaningful new symbols from colliding symbols. It is like two strands of DNA which come together. How do you take parts from each and recombine them into a meaningful and viable new strand of DNA which codes for an individual of the same species? Or a new kind of species? The chance is infinitesimal that a random combination of pieces of DNA will code for anything that will survive—something like the chance that a random combination of words from two books will make another book. The chance that recombinant DNA will make sense on any level but the lowest is tiny, precisely because there are so many levels of meaning in DNA. And the same goes for "recombinant symbols".

Epigenesis of the *Crab Canon*

I think of my Dialogue *Crab Canon* as a prototype example where two ideas collided in my mind, connected in a new way, and suddenly a new kind of verbal structure came alive in my mind. Of course I can still think about musical crab canons and verbal dialogues separately—they can still be activated independently of each other; but the fused symbol for crab-canonical dialogues has its own characteristic modes of activation, too. To illustrate this notion of fusion or "symbolic recombination" in some detail, then, I would like to use the development of my *Crab Canon* as a case study, because, of course, it is very familiar to me, and also because it is interesting, yet typical of how far a single idea can be pushed. I will recount it in stages named after those of *meiosis,* which is the name for cell division in which "crossing-over", or genetic recombination, takes place—the source of diversity in evolution.

PROPHASE: I began with a rather simple idea—that a piece of music, say a canon, could be imitated verbally. This came from the observation that, through a shared abstract form, a piece of text and a piece of music may be connected. The next step involved trying to realize some of the potential of this vague hunch; here, I hit upon the idea that "voices" in canons can be mapped onto "characters" in dialogues—still a rather obvious idea.

Then I focused down onto specific kinds of canons, and remembered that there was a crab canon in the *Musical Offering.* At that time, I had just begun writing Dialogues, and there were only two characters: Achilles and the Tortoise. Since the Bach crab canon has two voices, this mapped perfectly: Achilles should be one voice, the Tortoise the other, with the one doing forwards what the other does backwards. But here I was faced with a problem: on what level should the reversal take place? The letter level? The word level? The sentence level? After some thought, I concluded that the "dramatic line" level would be most appropriate.

Now that the "skeleton" of the Bach crab canon had been transplanted, at least in plan, into a verbal form, there was just one problem. When the two voices crossed in the middle, there would be a short period of extreme repetition: an ugly blemish. What to do about it? Here, a strange thing happened, a kind of level-crossing typical of creative acts: the word "crab" in "crab canon" flashed into my mind, undoubtedly because of some abstract shared quality with the notion of "tortoise"—and immediately I realized that at the dead center, I could block the repetitive effect, by inserting one special line, said by a new character: a Crab! This is how, in the "prophase" of the *Crab Canon,* the Crab was conceived: at the crossing-over of Achilles and the Tortoise. (See Fig. 131.)

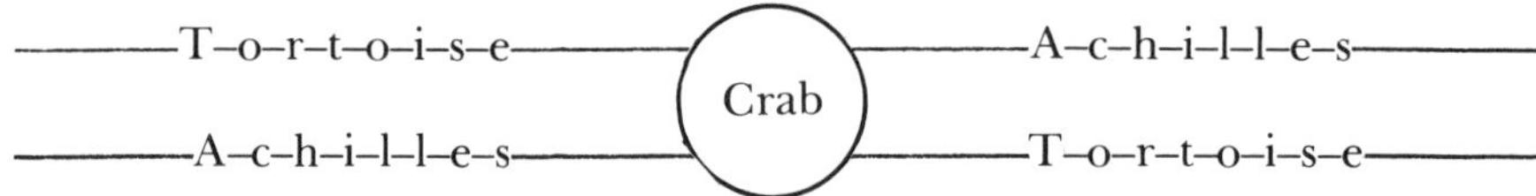

FIGURE 131. A schematic diagram of the Dialogue Crab Canon.

METAPHASE: This was the skeleton of my *Crab Canon.* I then entered the second stage—the "metaphase"—in which I had to fill in the flesh, which was of course an arduous task. I made a lot of stabs at it, getting used to the way in which pairs of successive lines had to make sense when read from either direction, and experimenting around to see what kinds of dual meanings would help me in writing such a form (e.g., "Not at all"). There were two early versions both of which were interesting, but weak. I abandoned work on the book for over a year, and when I returned to the *Crab Canon,* I had a few new ideas. One of them was to mention a Bach canon inside it. At first my plan was to mention the "Canon per augmentationem, contrario motu", from the *Musical Offering* (*Sloth Canon,* as I call it). But that started to seem a little silly, so reluctantly I decided that inside my *Crab Canon,* I could talk about Bach's own *Crab Canon* instead. Actually, this was a crucial turning point, but I didn't know it then.

Now if one character was going to mention a Bach piece, wouldn't it be awkward for the other to say exactly the same thing in the corresponding place? Well, Escher was playing a similar role to Bach in my thoughts and my book, so wasn't there some way of just slightly modifying the line so that it would refer to Escher? After all, in the strict art of canons, note-perfect imitation is occasionally foregone for the sake of elegance or beauty. And no sooner did that idea occur to me than the picture *Day and Night* (Fig. 49) popped into my mind. "Of course!" I thought, "It is a sort of pictorial crab canon, with essentially two complementary voices carrying the same theme both leftwards and rightwards, and harmonizing with each other!" Here again was the notion of a single "conceptual skeleton" being instantiated in two different media—in this case, music and art. So I let the Tortoise talk about Bach, and Achilles talk about Escher, in parallel language; certainly this slight departure from strict imitation retained the spirit of crab canons.

At this point, I began realizing that something marvelous was happening: namely, the Dialogue was becoming self-referential, without my even having intended it! What's more, it was an indirect self-reference, in that the characters did not talk directly about the Dialogue they were in, but rather about structures which were isomorphic to it (on a certain plane of abstraction). To put it in the terms I have been using, my Dialogue now shared a "conceptual skeleton" with Gödel's G, and could therefore be mapped onto G in somewhat the way that the Central Dogma was, to create in this case a "Central Crabmap". This was most exciting to me, since out of nowhere had come an esthetically pleasing unity of Gödel, Escher, and Bach.

ANAPHASE: The next step was quite startling. I had had Caroline MacGillavry's monograph on Escher's tesselations for years, but one day, as I flipped through it, my eye was riveted to Plate 23 (Fig. 42), for I saw it in a way I had never seen it before: here was a genuine crab canon—crab-like in both form and content! Escher himself had given the picture no title, and since he had drawn similar tesselations using many other animal forms, it is probable that this coincidence of form and content was just something which I had noticed. But fortuitous or not, this untitled plate was a miniature version of one main idea of my book: to unite form and content. So with delight I christened it *Crab Canon,* substituted it for *Day and Night,* and modified Achilles' and the Tortoise's remarks accordingly.

Yet this was not all. Having become infatuated with molecular biology, one day I was perusing Watson's book in the bookstore, and in the index saw the word "palindrome". When I looked it up, I found a magical thing: crab-canonical structures in DNA. Soon the Crab's comments had been suitably modified to include a short remark to the effect that he owed his predilection for confusing retrograde and forward motion to his genes.

TELOPHASE: The last step came months later, when, as I was talking about the picture of the crab-canonical section of DNA (Fig. 43), I saw that the 'A', 'T', 'C' of Adenine, Thymine, Cytosine coincided— *mirabile dictu*—with the 'A', 'T', 'C' of Achilles, Tortoise, Crab; moreover, just as Adenine and Thymine are paired in DNA, so are Achilles and the Tortoise paired in the Dialogue. I thought for a moment and, in another of those level-crossings, saw that 'G', the letter paired with 'C' in DNA, could stand for "Gene". Once again, I jumped back to the Dialogue, did a little surgery on the Crab's speech to reflect this new discovery, and now I had a mapping between the DNA's structure, and the Dialogue's structure. In that sense, the DNA could be said to be a genotype coding for a phenotype: the structure of the Dialogue. This final touch dramatically heightened the self-reference, and gave the Dialogue a density of meaning which I had never anticipated.

Conceptual Skeletons and Conceptual Mapping

That more or less summarizes the epigenesis of the *Crab Canon.* The whole process can be seen as a succession of mappings of ideas onto each other, at varying levels of abstraction. This is what I call *conceptual mapping,* and the abstract structures which connect up two different ideas are *conceptual skeletons.* Thus, one conceptual skeleton is that of the abstract notion of a crab canon:

> a structure having two parts which do the same thing,
> only moving in opposite directions.

This is a concrete geometrical image which can be manipulated by the mind almost as a Bongard pattern. In fact, when I think of the *Crab Canon* today, I visualize it as two strands which cross in the middle, where they are joined by a "knot" (the Crab's speech). This is such a vividly pictorial image that it instantaneously maps, in my mind, onto a picture of two homologous chromosomes joined by a centromere in their middle, which is an image drawn directly from meiosis, as shown in Figure 132.

FIGURE 132.

In fact, this very image is what inspired me to cast the description of the *Crab Canon*'s evolution in terms of meiosis—which is itself, of course, yet another example of conceptual mapping.

Recombinant Ideas

There are a variety of techniques of fusion of two symbols. One involves lining the two ideas up next to each other (as if ideas were linear!), then judiciously choosing pieces from each one, and recombining them in a new symbol. This strongly recalls genetic recombination. Well, what do chromosomes exchange, and how do they do it? They exchange genes. What in a symbol is comparable to a gene? If symbols have frame-like slots, then slots, perhaps. But which slots to exchange, and why? Here is where the crab-canonical fusion may offer some ideas. Mapping the notion of "musical crab canon" onto that of "dialogue" involved several auxiliary mappings; in fact it *induced* them. That is, once it had been decided that these two notions were to be fused, it became a matter of looking at them on a level where analogous parts emerged into view, then going ahead and *mapping the parts* onto each other, and so on, recursively, to any level that was found desirable. Here, for instance, "voice" and "character" emerged as corresponding slots when "crab canon" and "dialogue" were viewed abstractly. Where did these abstract views come from, though? This is at the crux of the mapping-problem—where do abstract views come from? How do you make abstract views of specific notions?

Abstractions, Skeletons, Analogies

A view which has been abstracted from a concept along some dimension is what I call a *conceptual skeleton.* In effect, we have dealt with conceptual skeletons all along, without often using that name. For instance, many of the ideas concerning Bongard problems could be rephrased using this terminology. It is always of interest, and possibly of importance, when two or more ideas are discovered to share a conceptual skeleton. An example is the bizarre set of concepts mentioned at the beginning of the *Contrafactus:* a Bicyclops, a tandem unicycle, a teeter-teeter, the game of ping-ping, a one-way tie, a two-sided Möbius strip, the "Bach twins", a piano concerto for two left hands, a one-voice fugue, the act of clapping with one hand, a two-channel monaural phonograph, a pair of eighth-backs. All of these ideas are "isomorphic" because they share this conceptual skeleton:

a plural thing made singular and re-pluralized wrongly.

Two other ideas in this book which share that conceptual skeleton are (1) the Tortoise's solution to Achilles' puzzle, asking for a word beginning and ending in "HE" (the Tortoise's solution being the pronoun "HE", which collapses two occurrences into one), and (2) the Pappus-Gelernter proof of the Pons Asinorum Theorem, in which one triangle is reperceived as two. Incidentally, these droll concoctions might be dubbed "demi-doublets".

A conceptual skeleton is like a set of constant features (as distinguished from parameters or variables)—features which should not be slipped in a subjunctive instant replay or mapping-operation. Having no parameters or variables of its own to vary, it can be the invariant core of several different ideas. Each *instance* of it, such as "tandem unicycle", does have layers of variability and so can be "slipped" in various ways.

Although the name "conceptual skeleton" sounds absolute and rigid, actually there is a lot of play in it. There can be conceptual skeletons on several different levels of abstraction. For instance, the "isomorphism" between Bongard problems 70 and 71, already pointed out, involves a higher-level conceptual skeleton than that needed to solve either problem in isolation.

Multiple Representations

Not only must conceptual skeletons exist on different levels of abstraction; also, they must exist along different conceptual *dimensions*. Let us take the following sentence as an example:

"The Vice President is the spare tire
on the automobile of government."

How do we understand what it means (leaving aside its humor, which is of course a vital aspect)? If you were told, "See our government as an automobile" without any prior motivation, you might come up with any number of correspondences: steering wheel = president, etc.. What are checks and balances? What are seat belts? Because the two things being mapped are so different, it is almost inevitable that the mapping will involve *functional* aspects. Therefore, you retrieve from your store of conceptual skeletons representing parts of automobiles, only those having to do with function, rather than, say, shape. Furthermore, it makes sense to work at a pretty high level of abstraction, where "function" isn't taken in too narrow a context. Thus, of the two following definitions of the function of a spare tire: (1) "replacement for a flat tire", and (2) "replacement for a certain disabled part of a car", certainly the latter would be preferable, in this case. This comes simply from the fact that an auto and a government are so different that they have to be mapped at a high level of abstraction.

Now when the particular sentence is examined, the mapping gets forced in one respect—but it is not an awkward way, by any means. In fact, you already have a conceptual skeleton for the Vice President, among many others, which says, "replacement for a certain disabled part of government". Therefore the forced mapping works comfortably. But suppose, for the sake of contrast, that you had retrieved another conceptual skeleton for "spare tire"—say, one describing its physical aspects. Among other things, it might say that a spare tire is "round and inflated". Clearly, this is not the right way to go. (Or is it? As a friend of mine pointed out, some Vice Presidents are rather portly, and most are quite inflated!)

Ports of Access

One of the major characteristics of each idiosyncratic style of thought is how new experiences get classified and stuffed into memory, for that defines the "handles" by which they will later be retrievable. And for events, objects, ideas, and so on—for everything that can be thought about—there is a wide variety of "handles". I am struck by this each time I reach down to turn on my car radio, and find, to my dismay, that it is already on! What has happened is that two independent representations are being used for the radio. One is "music producer", the other is "boredom reliever". I am aware that the music is on, but I am bored anyway, and before the two realizations have a chance to interact, my reflex to reach down has been triggered. The same reaching-down reflex one day occurred just after I'd left the radio at a repair shop and was driving away, wanting to hear some music. Odd. Many other representations for the same object exist, such as

shiny silver-knob haver
overheating-problems haver
lying-on-my-back-over-hump-to-fix thing
buzz-maker
slipping-dials object
multidimensional representation example

All of them can act as ports of access. Though they all are attached to my symbol for my car radio, accessing that symbol through one does not open up all the others. Thus it is unlikely that I will be inspired to remember lying on my back to fix the radio when I reach down and turn it on. And conversely, when I'm lying on my back, unscrewing screws, I probably won't think about the time I heard the *Art of the Fugue* on it. There are "partitions" between these aspects of one symbol, partitions that prevent my thoughts from spilling over sloppily, in the manner of free associations. My mental partitions are important because they contain and channel the flow of my thoughts.

One place where these partitions are quite rigid is in sealing off words for the same thing in different languages. If the partitions were not strong, a bilingual person would constantly slip back and forth between languages, which would be very uncomfortable. Of course, adults learning two new languages at once often confuse words in them. The partitions between these languages are flimsier, and can break down. Interpreters are particularly interesting, since they can speak any of their languages as if their partitions were inviolable and yet, on command, they can negate those partitions to allow access to one language from the other, so they can translate. Steiner, who grew up trilingual, devotes several pages in *After Babel* to the intermingling of French, English, and German in the layers of his mind, and how his different languages afford different ports of access onto concepts.

Forced Matching

When two ideas are seen to share conceptual skeletons on some level of abstraction, different things can happen. Usually the first stage is that you zoom in on both ideas, and, using the higher-level match as a guide, you try to identify corresponding subideas. Sometimes the match can be extended recursively downwards several levels, revealing a profound isomorphism. Sometimes it stops earlier, revealing an analogy or similarity. And then there are times when the high-level similarity is so compelling that, even if there is no apparent lower-level continuation of the map, you just go ahead and make one: this is the *forced match*.

Forced matches occur every day in the political cartoons of newspapers: a political figure is portrayed as an airplane, a boat, a fish, the Mona Lisa; a government is a human, a bird, an oil rig; a treaty is a briefcase, a sword, a can of worms; on and on and on. What is fascinating is how easily we can perform the suggested mapping, and to the exact depth intended. We don't carry the mapping out too deeply or too shallowly.

Another example of forcing one thing into the mold of another occurred when I chose to describe the development of my *Crab Canon* in terms of meiosis. This happened in stages. First, I noticed the common conceptual skeleton shared by the *Crab Canon* and the image of chromosomes joined by a centromere; this provided the inspiration for the forced match. Then I saw a high-level resemblance involving "growth", "stages", and "recombination". Then I simply pushed the analogy as hard as I could. Tentativity—as in the Bongard problem-solver—played a large role: I went forwards and backwards before finding a match which I found appealing.

A third example of conceptual mapping is provided by the Central Dogmap. I initially noticed a high-level similarity between the discoveries of mathematical logicians and those of molecular biologists, then pursued it on lower levels until I found a strong analogy. To strengthen it further, I chose a Gödel-numbering which imitated the Genetic Code. This was the lone element of forced matching in the Central Dogmap.

Forced matches, analogies, and metaphors cannot easily be separated out. Sportscasters often use vivid imagery which is hard to pigeonhole. For instance, in a metaphor such as "The Rams [football team] are spinning their wheels", it is hard to say just what image you are supposed to conjure up. Do you attach wheels to the team as a whole? Or to each player? Probably neither one. More likely, the image of wheels spinning in mud or snow simply flashes before you for a brief instant, and then in some mysterious way, just the relevant parts get lifted out and transferred to the team's performance. How deeply are the football team and the car mapped onto each other in the split second that you do this?

Recap

Let me try to tie things together a little. I have presented a number of related ideas connected with the creation, manipulation, and comparison of symbols. Most of them have to do with slippage in some fashion, the idea being that concepts are composed of some tight and some loose elements, coming from different levels of nested contexts (frames). The loose ones can be dislodged and replaced rather easily, which, depending on the circumstances, can create a "subjunctive instant replay", a forced match, or an analogy. A fusion of two symbols may result from a process in which parts of each symbol are dislodged and other parts remain.

Creativity and Randomness

It is obvious that we are talking about mechanization of creativity. But is this not a contradiction in terms? Almost, but not really. Creativity is the essence of that which is *not* mechanical. Yet every creative act *is* mechanical—it has its explanation no less than a case of the hiccups does. The mechanical substrate of creativity may be hidden from view, but it exists. Conversely, there is something unmechanical in flexible programs, even today. It may not constitute creativity, but when programs cease to be transparent to their creators, then the approach to creativity has begun.

It is a common notion that randomness is an indispensable ingredient of creative acts. This may be true, but it does not have any bearing on the mechanizability—or rather, programmability!—of creativity. The world is a giant heap of randomness; when you mirror some of it inside your head, your head's interior absorbs a little of that randomness. The triggering patterns of symbols, therefore, can lead you down the most random-seeming paths, simply because they came from your interactions with a crazy, random world. So it can be with a computer program, too. Randomness is an intrinsic feature of thought, not something which has to be "artificially inseminated", whether through dice, decaying nuclei, random number tables, or what-have-you. It is an insult to human creativity to imply that it relies on such arbitrary sources.

What we see as randomness is often simply an effect of looking at something symmetric through a "skew" filter. An elegant example was provided by Salviati's two ways of looking at the number $\pi/4$. Although the decimal expansion of $\pi/4$ is not literally random, it is as random as one would need for most purposes: it is "pseudorandom". Mathematics is full of pseudorandomness—plenty enough to supply all would-be creators for all time.

Just as science is permeated with "conceptual revolutions" on all levels at all times, so the thinking of individuals is shot through and through with creative acts. They are not just on the highest plane; they are everywhere. Most of them are small and have been made a million times before—but they are close cousins to the most highly creative and new acts. Computer programs today do not yet seem to produce many small creations. Most of what they do is quite "mechanical" still. That just testifies to the fact that they are not close to simulating the way we think—but they are getting closer.

Perhaps what differentiates highly creative ideas from ordinary ones is some combined sense of beauty, simplicity, and harmony. In fact, I have a favorite "meta-analogy", in which I liken analogies to chords. The idea is simple: superficially similar ideas are often not deeply related; and deeply related ideas are often superficially disparate. The analogy to chords is natural: physically close notes are harmonically distant (e.g., E-F-G); and harmonically close notes are physically distant (e.g., G-E-B). Ideas that share a conceptual skeleton resonate in a sort of conceptual analogue to harmony; these harmonious "idea-chords" are often widely separated, as measured on an imaginary "keyboard of concepts". Of course, it doesn't suffice to reach wide and plunk down any old way—you may hit a seventh or a ninth! Perhaps the present analogy is like a ninth-chord—wide but dissonant.

Picking up Patterns on All Levels

Bongard problems were chosen as a focus in this Chapter because when you study them, you realize that the elusive sense for patterns which we humans inherit from our genes involves all the mechanisms of representation of knowledge, including nested contexts, conceptual skeletons and conceptual mapping, slippability, descriptions and meta-descriptions and their interactions, fission and fusion of symbols, multiple representations (along different dimensions and different levels of abstraction), default expectations, and more.

These days, it is a safe bet that if some program can pick up patterns in one area, it will miss patterns in another area which, to us, are equally obvious. You may remember that I mentioned this back in Chapter I, saying that machines can be oblivious to repetition, whereas people cannot. For instance, consider SHRDLU. If Eta Oin typed the sentence "Pick up a big red block and put it down" over and over again, SHRDLU would cheerfully react in the same way over and over again, exactly as an adding machine will print out "4" over and over again, if a human being has the patience to type "2+2" over and over again. Humans aren't like that; if some pattern occurs over and over again, they will pick it up. SHRDLU wasn't built with the potential for forming new concepts or recognizing patterns: it had no sense of over and overview.

The Flexibility of Language

SHRDLU's language-handling capability is immensely flexible—within limits. SHRDLU can figure out sentences of great syntactical complexity, or sentences with semantic ambiguities as long as they can be resolved by inspecting the data base—but it cannot handle "hazy" language. For instance, consider the sentence "How many blocks go on top of each other to make a steeple?" We understand it immediately, yet it does not make sense if interpreted literally. Nor is it that some idiomatic phrase has been used. "To go on top of each other" is an imprecise phrase which nonetheless gets the desired image across quite well to a human. Few people would be misled into visualizing a paradoxical setup with two blocks each of which is on top of the other—or blocks which are "going" somewhere or other.

The amazing thing about language is how imprecisely we use it and still manage to get away with it. SHRDLU uses words in a "metallic" way, while people use them in a "spongy" or "rubbery" or even "Nutty-Putty-ish" way. If words were nuts and bolts, people could make any bolt fit into any nut: they'd just squish the one into the other, as in some surrealistic painting where everything goes soft. Language, in human hands, becomes almost like a fluid, despite the coarse grain of its components.

Recently, AI research in natural language understanding has turned away somewhat from the understanding of single sentences in isolation, and more towards areas such as understanding simple children's stories. Here is a well-known children's joke which illustrates the open-endedness of real-life situations:

> A man took a ride in an airplane.
>   Unfortunately, he fell out.
> Fortunately, he had a parachute on.
>   Unfortunately, it didn't work.
> Fortunately, there was a haystack below him.
>   Unfortunately, there was a pitchfork sticking out of it.
> Fortunately, he missed the pitchfork.
>   Unfortunately, he missed the haystack.

It can be extended indefinitely. To represent this silly story in a frame-based system would be extremely complex, involving jointly activating frames for the concepts of man, airplane, exit, parachute, falling, etc., etc.

Intelligence and Emotions

Or consider this tiny yet poignant story:

> Margie was holding tightly to the string of her beautiful new balloon. Suddenly, a gust of wind caught it. The wind carried it into a tree. The balloon hit a branch and burst. Margie cried and cried.[4]

To understand this story, one needs to read many things between the lines. For instance: Margie is a little girl. This is a toy balloon with a string for a child to hold. It may not be beautiful to an adult, but in a child's eye, it is. She is outside. The "it" that the wind caught was the balloon. The wind did not pull Margie along with the balloon; Margie let go. Balloons can break on contact with any sharp point. Once they are broken, they are gone forever. Little children love balloons and can be bitterly disappointed when they break. Margie saw that her balloon was broken. Children cry when they are sad. "To cry and cry" is to cry very long and hard. Margie cried and cried because of her sadness at her balloon's breaking.

This is probably only a small fraction of what is lacking at the surface level. A program must have all this knowledge in order to get at what is going on. And you might object that, even if it "understands" in some intellectual sense what has been said, it will never *really* understand, until it, too, has cried and cried. And when will a computer do that? This is the kind of humanistic point which Joseph Weizenbaum is concerned with making in his book *Computer Power and Human Reason,* and I think it is an important issue; in fact, a very, very deep issue. Unfortunately, many AI workers at this time are unwilling, for various reasons, to take this sort of point seriously. But in some ways, those AI workers are right: it is a little premature to think about computers crying; we must first think about rules for computers to deal with language and other things; in time, we'll find ourselves face to face with the deeper issues.

AI Has Far to Go

Sometimes it seems that there is such a complete absence of rule-governed behavior that human beings just *aren't* rule-governed. But this is an illusion—a little like thinking that crystals and metals emerge from rigid underlying laws, but that fluids or flowers don't. We'll come back to this question in the next Chapter.

> The process of logic itself working internally in the brain may be more analogous to a succession of operations with symbolic pictures, a sort of abstract analogue of the Chinese alphabet or some Mayan description of events—except that the elements are not merely words but more like sentences or whole stories with linkages between them forming a sort of meta- or super-logic with its own rules.[5]

It is hard for most specialists to express vividly—perhaps even to remember—what originally sparked them to enter their field. Conversely, someone on the outside may understand a field's special romance and may be able to articulate it precisely. I think that is why this quote from Ulam has appeal for me, because it poetically conveys the strangeness of the enterprise of AI, and yet shows faith in it. And one must run on faith at this point, for there is so far to go!

Ten Questions and Speculations

To conclude this Chapter, I would like to present ten "Questions and Speculations" about AI. I would not make so bold as to call them "Answers"—these are my personal opinions. They may well change in some ways, as I learn more and as AI develops more. (In what follows, the term "AI program" means a program which is far ahead of today's programs; it means an "Actually Intelligent" program. Also, the words "program" and "computer" probably carry overly mechanistic connotations, but let us stick with them anyway.)

Question: Will a computer program ever write beautiful music?

Speculation: Yes, but not soon. Music is a language of emotions, and until programs have emotions as complex as ours, there is no way a program will write anything beautiful. There can be "forgeries"—shallow imitations of the syntax of earlier music—but despite what one might think at first, there is much more to musical expression than can be captured in syntactical rules. There will be no new kinds of beauty turned up for a long time by computer music-composing programs. Let me carry this thought a little further. To think—and I have heard this suggested—that we might soon be able to command a preprogrammed mass-produced mail-order twenty-dollar desk-model "music box" to bring forth from its sterile circuitry pieces which Chopin or Bach might have written had they lived longer is a grotesque and shameful misestimation of the depth of the human spirit. A "program" which could produce music as they did would have to wander around the world on its own, fighting its way through the maze of life and feeling every moment of it. It would have to understand the joy and loneliness of a chilly night wind, the longing for a cherished hand, the inaccessibility of a distant town, the heartbreak and regeneration after a human death. It would have to have known resignation and world-weariness, grief and despair, determination and victory, piety and awe. In it would have had to commingle such opposites as hope and fear, anguish and jubilation, serenity and suspense. Part and parcel of it would have to be a sense of grace, humor, rhythm, a sense of the unexpected—and of course an exquisite awareness of the magic of fresh creation. Therein, and therein only, lie the sources of meaning in music.

Question: Will emotions be explicitly programmed into a machine?

Speculation: No. That is ridiculous. Any direct simulation of emotions—PARRY, for example—cannot approach the complexity of human emotions, which arise indirectly from the organization of our minds. Programs or machines will acquire emotions in the same way: as by-products of their structure, of the way in which they are organized—not by direct programming. Thus, for example, nobody will write a "falling-in-love" subroutine, any more than they would write a "mistake-making" subroutine. "Falling in love" is a description which we attach to a complex process of a complex system; there need be no single module inside the system which is solely responsible for it, however!

Question: Will a thinking computer be able to add fast?

Speculation: Perhaps not. We ourselves are composed of hardware which does fancy calculations but that doesn't mean that our symbol level, where "we" are, knows how to carry out the same fancy calculations. Let me put it this way: there's no way that you can load numbers into your own neurons to add up your grocery bill. Luckily for you, your symbol level (i.e., *you*) can't gain access to the neurons which are doing your thinking—otherwise you'd get addle-brained. To paraphrase Descartes again:

> "I think; therefore I have no access
> to the level where I sum."

Why should it not be the same for an intelligent program? It mustn't be allowed to gain access to the circuits which are doing its thinking—otherwise it'll get addle-CPU'd. Quite seriously, a machine that can pass the Turing test may well add as slowly as you or I do, and for similar reasons. It will represent the number 2 not just by the two bits "10", but as a full-fledged *concept* the way we do, replete with associations such as its homonyms "too" and "to", the words "couple" and "deuce", a host of mental images such as dots on dominos, the shape of the numeral '2', the notions of alternation, evenness, oddness, and on and on . . . With all this "extra baggage" to carry around, an intelligent program will become quite slothful in its adding. Of course, we could give it a "pocket calculator", so to speak (or build one in). Then it could answer very fast, but its performance would be just like that of a person with a pocket calculator. There would be two separate parts to the machine: a reliable but mindless part and an intelligent but fallible part. You couldn't rely on the composite system to be reliable, any more than a composite of person and machine is necessarily reliable. So if it's right answers you're after, better stick to the pocket calculator alone—don't throw in the intelligence!

Question: Will there be chess programs that can beat anyone?

Speculation: No. There may be programs which can beat anyone at chess, but they will not be exclusively chess players. They will be programs of *general* intelligence, and they will be just as temperamental as people. "Do you want to play chess?" "No, I'm bored with chess. Let's talk about poetry." That may be the kind of dialogue you could have with a program that could beat everyone. That is because real intelligence inevitably depends on a total overview capacity—that is, a programmed ability to "jump out of the system", so to speak—at least roughly to the extent that we have that ability. Once that is present, you can't contain the program; it's gone beyond that certain critical point, and you just have to face the facts of what you've wrought.

Question: Will there be special locations in memory which store parameters governing the behavior of the program, such that if you reached in and changed them, you would be able to make the program smarter or stupider or more creative or more interested in baseball? In short, would you be able to "tune" the program by fiddling with it on a relatively low level?

Speculation: No. It would be quite oblivious to changes of any particular elements in memory, just as we stay almost exactly the same though thousands of our neurons die every day(!). If you fuss around too heavily, though, you'll damage it, just as if you irresponsibly did neurosurgery on a human being. There will be no "magic" location in memory where, for instance, the "IQ" of the program sits. Again, that will be a feature which emerges as a consequence of lower-level behavior, and nowhere will it sit explicitly. The same goes for such things as "the number of items it can hold in short-term memory", "the amount it likes physics", etc., etc.

Question: Could you "tune" an AI program to act like me, or like you—or halfway between us?

Speculation: No. An intelligent program will not be chameleon-like, any more than people are. It will rely on the constancy of its memories, and will not be able to flit between personalities. The idea of changing internal parameters to "tune to a new personality" reveals a ridiculous underestimation of the complexity of personality.

Question: Will there be a "heart" to an AI program, or will it simply consist of "senseless loops and sequences of trivial operations" (in the words of Marvin Minsky[6])?

Speculation: If we could see all the way to the bottom, as we can a shallow pond, we would surely see only "senseless loops and sequences of trivial operations"—and we would surely not see any "heart". Now there are two kinds of extremist views on AI: one says that the human mind is, for fundamental and mysterious reasons, unprogrammable. The other says that you merely need to assemble the appropriate "heuristic devices—multiple optimizers, pattern-recognition tricks, planning algebras, recursive administration procedures, and the like",[7] and you will have intelligence. I find myself somewhere in between, believing that the "pond" of an AI program will turn out to be so deep and murky that we won't be able to peer all the way to the bottom. If we look from the top, the loops will be invisible, just as nowadays the current-carrying electrons are invisible to most programmers. When we create a program that passes the Turing test, we will see a "heart" even though we know it's not there.

Question: Will AI programs ever become "superintelligent"?

Speculation: I don't know. It is not clear that we would be able to understand or relate to a "superintelligence", or that the concept even makes sense. For instance, our own intelligence is tied in with our speed of thought. If our reflexes had been ten times faster or slower, we might have developed an entirely different set of concepts with which to describe the world. A creature with a radically different view of the world may simply not have many points of contact with us. I have often wondered if there could be, for instance, pieces of music which are to Bach as Bach is to folk tunes: "Bach squared", so to speak. And would I be able to understand them? Maybe there is such music around me already, and I just don't recognize it, just as dogs don't understand language. The idea of superintelligence is very strange. In any case, I don't think of it as the aim of AI research, although if we ever do reach the level of human intelligence, superintelligence will undoubtedly be the next goal—not only for us, but for our AI-program colleagues, too, who will be equally curious about AI and superintelligence. It seems quite likely that AI programs will be extremely curious about AI in general—understandably.

Question: You seem to be saying that AI programs will be virtually identical to people, then. Won't there be any differences?

Speculation: Probably the differences between AI programs and people will be larger than the differences between most people. It is almost impossible to imagine that the "body" in which an AI program is housed would not affect it deeply. So unless it had an amazingly faithful replica of a human body—and why should it?—it would probably have enormously different perspectives on what is important, what is interesting, etc. Wittgenstein once made the amusing comment, "If a lion could speak, we would not understand him." It makes me think of Rousseau's painting of the gentle lion and the sleeping gypsy on the moonlit desert. But how does Wittgenstein know? My guess is that any AI program would, if comprehensible to us, seem pretty alien. For that reason, we will have a very hard time deciding when and if we really are dealing with an AI program, or just a "weird" program.

Question: Will we understand what intelligence and consciousness and free will and "I" are when we have made an intelligent program?

Speculation: Sort of—it all depends on what you mean by "understand". On a gut level, each of us probably has about as good an understanding as is possible of those things, to start with. It is like listening to music. Do you really understand Bach because you have taken him apart? Or did you understand it that time you felt the exhilaration in every nerve in your body? Do we understand how the speed of light is constant in every inertial reference frame? We can do the math, but no one in the world has a truly relativistic intuition. And probably no one will ever understand the mysteries of intelligence and consciousness in an intuitive way. Each of us can understand *people,* and that is probably about as close as you can come.

Sloth Canon

This time, we find Achilles and the Tortoise visiting the dwelling of their new friend, the Sloth.

Achilles: Shall I tell you of my droll footrace with Mr. T?

Sloth: Please do.

Achilles: It has become quite celebrated in these parts. I believe it's even been written up, by Zeno.

Sloth: It sounds very exciting.

Achilles: It was. You see, Mr. T began way ahead of me. He had such a huge head start, and yet—

Sloth: You caught up, didn't you?

Achilles: Yes—being so fleet of foot, I diminished the distance between us at a constant rate, and soon overtook him.

Sloth: The gap kept getting shorter and shorter, so you could.

Achilles: Exactly. Oh, look—Mr. T has brought his violin. May I try playing on it, Mr. T?

Tortoise: Please don't. It sounds very flat.

Achilles: Oh, all right. But I'm in a mood for music. I don't know why.

Sloth: You can play the piano, Achilles.

Achilles: Thank you. I'll try it in a moment. I just wanted to add that I also had another kind of "race" with Mr. T at a later date. Unfortunately, in that race—

Tortoise: You didn't catch up, did you? The gap kept getting longer and longer, so you couldn't.

Achilles: That's true. I believe THAT race has been written up, too, by Lewis Carroll. Now, Mr. Sloth, I'll take up your offer of trying out the piano. But I'm so bad at the piano. I'm not sure I dare.

Sloth: You should try.

(Achilles sits down and starts playing a simple tune.)

Achilles: Oh—it sounds very strange. That's not how it's supposed to sound at all! Something is very wrong.

Tortoise: You can't play the piano, Achilles. You shouldn't try.

Achilles: It's like a piano in a mirror. The high notes are on the left, and the low notes are on the right. Every melody comes out inverted, as if upside down. Who would have ever thought up something so cockeyed as that?

Tortoise: That's so characteristic of sloths. They hang from—

Achilles: Yes, I know—from tree branches—upside down, of course. That sloth-piano would be appropriate for playing inverted melodies such

J.S. BACH

SLOTH CANON

SLOTH CANON

J.S. BACH

as occur in some canons and fugues. But to learn to play a piano while hanging from a tree must be very difficult. You must have to devote a great deal of energy to it.

Sloth: That's not so characteristic of sloths.

Achilles: No, I gather sloths like to take life very easy. They do everything about half as fast as normal. And upside down, to boot. What a peculiar way to go through life! Speaking of things that are both upside- and slowed-down, there's a "Canon per augmentationem, contrario motu" in the *Musical Offering.* In my edition, the letters 'S', 'A', 'T' are in front of the three staves. I don't know why. Anyway, I think Bach carried it off very skillfully. What's your opinion, Mr. T?

Tortoise: He outdid himself. As for those letters "SAT", you could guess what they stand for.

Achilles: "Soprano", "Alto", and "Tenor", I suppose. Three-part pieces are often written for that combination of voices. Wouldn't you agree, Mr. Sloth?

Sloth: They stand for—

Achilles: Oh, just a moment, Mr. Sloth. Mr. Tortoise—why are you putting on your coat? You're not leaving, are you? We were just going to fix a snack to eat. You look very tired. How do you feel?

Tortoise: Out of gas. So long! *(Trudges wearily out the door.)*

Achilles: The poor fellow—he certainly looked exhausted. He was jogging all morning. He's in training for another race with me.

Sloth: He did himself in.

Achilles: Yes, but in vain. Maybe he could beat a Sloth . . . but me? Never! Now—weren't you about to tell me what those letters "SAT" stand for?

Sloth: As for those letters "SAT", you could never guess what they stand for.

Achilles: Well, if they don't stand for what I thought, then my curiosity is piqued. Perhaps I'll think a little more about it. Say, how do you cook French fries?

Sloth: In oil.

Achilles: Oh, yes—I remember. I'll cut up this potato into strips an inch or two in length.

Sloth: So short?

Achilles: All right, already, I'll cut four-inch strips. Oh, boy, are these going to be good French fries! Too bad Mr. T won't be here to share them.

FIGURE 133. "Sloth Canon", *from the* Musical Offering, *by J. S. Bach.* [*Music printed by Donald Byrd's program "SMUT".*]

CHAPTER XX

Strange Loops, Or Tangled Hierarchies

Can Machines Possess Originality?

IN THE CHAPTER before last, I described Arthur Samuel's very successful checkers program—the one which can beat its designer. In light of that, it is interesting to hear how Samuel himself feels about the issue of computers and originality. The following extracts are taken from a rebuttal by Samuel, written in 1960, to an article by Norbert Wiener.

> It is my conviction that machines cannot possess originality in the sense implied by Wiener in his thesis that "machines can and do transcend some of the limitations of their designers, and that in doing so they may be both effective and dangerous." . . .
> A machine is not a genie, it does not work by magic, it does not possess a will, and, Wiener to the contrary, nothing comes out which has not been put in, barring, of course, an infrequent case of malfunctioning. . . .
> The "intentions" which the machine seems to manifest are the intentions of the human programmer, as specified in advance, or they are subsidiary intentions derived from these, following rules specified by the programmer. We can even anticipate higher levels of abstraction, just as Wiener does, in which the program will not only modify the subsidiary intentions but will also modify the rules which are used in their derivation, or in which it will modify the ways in which it modifies the rules, and so on, or even in which one machine will design and construct a second machine with enhanced capabilities. However, and this is important, the machine *will not and cannot* [italics are his] do any of these things until it has been instructed as to how to proceed. There is and logically there must always remain a complete hiatus between *(i)* any ultimate extension and elaboration in this process of carrying out man's wishes and *(ii)* the development within the machine of a will of its own. To believe otherwise is either to believe in magic or to believe that the existence of man's will is an illusion and that man's actions are as mechanical as the machine's. Perhaps Wiener's article and my rebuttal have both been mechanically determined, but this I refuse to believe.[1]

This reminds me of the Lewis Carroll Dialogue (the *Two-Part Invention*); I'll try to explain why. Samuel bases his argument against machine consciousness (or will) on the notion that *any mechanical instantiation of will would require an infinite regress*. Similarly, Carroll's Tortoise argues that no step of reasoning, no matter how simple, can be done without invoking some rule on a higher level to justify the step in question. But that being also a step of reasoning, one must resort to a yet higher-level rule, and so on. Conclusion: *Reasoning involves an infinite regress.*

Of course something is wrong with the Tortoise's argument, and I believe something analogous is wrong with Samuel's argument. To show how the fallacies are analogous, I now shall "help the Devil", by arguing momentarily as Devil's advocate. (Since, as is well known, God helps those who help themselves, presumably the Devil helps all those, and only those, who don't help themselves. Does the Devil help himself?) Here are my devilish conclusions drawn from the Carroll Dialogue:

> The conclusion "reasoning is impossible" does not apply to people, because as is plain to anyone, we *do* manage to carry out many steps of reasoning, all the higher levels notwithstanding. That shows that we humans operate *without need of rules:* we are "informal systems". On the other hand, as an argument against the possibility of any *mechanical* instantiation of reasoning, it is valid, for any mechanical reasoning-system would have to depend on rules explicitly, and so it couldn't get off the ground unless it had metarules telling it when to apply its rules, metametarules telling it when to apply its metarules, and so on. We may conclude that the ability to reason can never be mechanized. It is a uniquely human capability.

What is wrong with this Devil's advocate point of view? It is obviously the assumption that *a machine cannot do anything without having a rule telling it to do so.* In fact, machines get around the Tortoise's silly objections as easily as people do, and moreover for exactly the same reason: both machines and people are made of hardware which runs all by itself, according to the laws of physics. There is no need to rely on "rules that permit you to apply the rules", because the *lowest*-level rules—those without any "meta"'s in front—are embedded in the hardware, and they run without permission. Moral: The Carroll Dialogue doesn't say anything about the differences between people and machines, after all. (And indeed, reasoning is mechanizable.)

So much for the Carroll Dialogue. On to Samuel's argument. Samuel's point, if I may caricature it, is this:

> No computer ever "wants" to do anything, because it was programmed by someone else. Only if it could program itself from zero on up—an absurdity—would it have its own sense of desire.

In his argument, Samuel reconstructs the Tortoise's position, replacing "to reason" by "to want". He implies that behind any mechanization of desire, there has to be either an infinite regress or worse, a closed loop. If this is why computers have no will of their own, what about people? The same criterion would imply that

Unless a person designed himself and chose his own wants (as well as choosing to choose his own wants, etc.), he cannot be said to have a will of his own.

It makes you pause to think where your sense of having a will comes from. Unless you are a soulist, you'll probably say that it comes from your brain—a piece of hardware which you did not design or choose. And yet that doesn't diminish your sense that you want certain things, and not others. You aren't a "self-programmed object" (whatever that would be), but you still do have a sense of desires, and it springs from the physical substrate of your mentality. Likewise, machines may someday have wills despite the fact that no magic program spontaneously appears in memory from out of nowhere (a "self-programmed program"). They will have wills for much the same reason as you do—by reason of organization and structure on many levels of hardware and software. Moral: The Samuel argument doesn't say anything about the differences between people and machines, after all. (And indeed, will will be mechanized.)

Below Every Tangled Hierarchy Lies An Inviolate Level

Right after the *Two-Part Invention,* I wrote that a central issue of this book would be: "Do words and thoughts follow formal rules?" One major thrust of the book has been to point out the many-leveledness of the mind/brain, and I have tried to show why the ultimate answer to the question is, "Yes—provided that you go down to the lowest level—the hardware—to find the rules."

Now Samuel's statement brought up a concept which I want to pursue. It is this: When we humans think, we certainly do change our own mental rules, and we change the rules that change the rules, and on and on—but these are, so to speak, "software rules". However, the rules *at bottom* do not change. Neurons run in the same simple way the whole time. You can't "think" your neurons into running some nonneural way, although you can make your mind change style or subject of thought. Like Achilles in the *Prelude, Ant Fugue,* you have access to your thoughts but not to your neurons. Software rules on various levels can change; hardware rules cannot—in fact, to their rigidity is due the software's flexibility! Not a paradox at all, but a fundamental, simple fact about the mechanisms of intelligence.

This distinction between self-modifiable software and inviolate hardware is what I wish to pursue in this final Chapter, developing it into a set of variations on a theme. Some of the variations may seem to be quite far-fetched, but I hope that by the time I close the loop by returning to brains, minds, and the sensation of consciousness, you will have found an invariant core in all the variations.

My main aim in this Chapter is to communicate some of the images which help me to visualize how consciousness rises out of the jungle of neurons; to communicate a set of intangible intuitions, in the hope that these intuitions are valuable and may perhaps help others a little to come to clearer formulations of their own images of what makes minds run. I could not hope for more than that my own mind's blurry images of minds and images should catalyze the formation of sharper images of minds and images in other minds.

A Self-Modifying Game

A first variation, then, concerns games in which on your turn, you may modify the rules. Think of chess. Clearly the rules stay the same, just the board position changes on each move. But let's invent a variation in which, on your turn, you can either make a move or change the rules. But how? At liberty? Can you turn it into checkers? Clearly such anarchy would be pointless. There must be some constraints. For instance, one version might allow you to redefine the knight's move. Instead of being 1-and-then-2, it could be m-and-then-n where m and n are arbitrary natural numbers; and on your turn you could change either m or n by plus or minus 1. So it could go from 1-2 to 1-3 to 0-3 to 0-4 to 0-5 to 1-5 to 2-5 . . . Then there could be rules about redefining the bishop's moves, and the other pieces' moves as well. There could be rules about adding new squares, or deleting old squares . . .

Now we have two layers of rules: those which tell how to move pieces, and those which tell how to change the rules. So we have rules and metarules. The next step is obvious: introduce metametarules by which we can change the metarules. It is not so obvious how to do this. The reason it is easy to formulate rules for moving pieces is that pieces move in a formalized space: the checkerboard. If you can devise a simple formal notation for expressing rules and metarules, then to manipulate them will be like manipulating strings formally, or even like manipulating chess pieces. To carry things to their logical extreme, you could even express rules and metarules as positions on auxiliary chess boards. Then an arbitrary chess position could be read as a game, or as a set of rules, or as a set of metarules, etc., depending on which interpretation you place on it. Of course, both players would have to agree on conventions for interpreting the notation.

Now we can have any number of adjacent chess boards: one for the game, one for rules, one for metarules, one for metametarules, and so on, as far as you care to carry it. On your turn, you may make a move on *any* one of the chess boards except the top-level one, using the rules which apply (they come from the next chess board up in the hierarchy). Undoubtedly both players would get quite disoriented by the fact that almost anything—though not everything!—can change. By definition, the top-level chess board can't be changed, because you don't have rules telling how to change it. It is *inviolate.* There is more that is inviolate: the conventions by which the different boards are interpreted, the agreement to take turns, the agreement that each person may change one chess board each turn—and you will find more if you examine the idea carefully.

Now it is possible to go considerably further in removing the pillars by which orientation is achieved. One step at a time . . . We begin by collapsing the whole array of boards into a single board. What is meant by this? There will be two ways of interpreting the board: (1) as pieces to be moved; (2) as rules for moving the pieces. On your turn, you move pieces—and perforce, you change rules! Thus, the rules constantly change themselves. Shades of Typogenetics—or for that matter, of real genetics. The distinction between game, rules, metarules, metametarules, has been lost. What was once a nice clean hierarchical setup has become a Strange Loop, Or Tangled Hierarchy. The moves change the rules, the rules determine the moves, round and round the mulberry bush . . . There are still different levels, but the distinction between "lower" and "higher" has been wiped out.

Now, part of what was inviolate has been made changeable. But there is still plenty that is inviolate. Just as before, there are conventions between you and your opponent by which you interpret the board as a collection of rules. There is the agreement to take turns—and probably other implicit conventions, as well. Notice, therefore, that the notion of different levels has survived, in an unexpected way. There is an Inviolate level—let's call it the *I-level*—on which the interpretation conventions reside; there is also a Tangled level—the *T-level*—on which the Tangled Hierarchy resides. So these two levels are still hierarchical: the I-level governs what happens on the T-level, but the T-level does not and cannot affect the I-level. No matter that the T-level itself is a Tangled Hierarchy—it is still governed by a set of conventions outside of itself. And that is the important point.

As you have no doubt imagined, there is nothing to stop us from doing the "impossible"—namely, tangling the I-level and the T-level by making the interpretation conventions themselves subject to revision, according to the position on the chess board. But in order to carry out such a "supertangling", you'd have to agree with your opponent on some further conventions connecting the two levels—and the act of doing so would create a *new* level, a new sort of inviolate level on top of the "supertangled" level (or underneath it, if you prefer). And this could continue going on and on. In fact, the "jumps" which are being made are very similar to those charted in the *Birthday Cantatatata,* and in the repeated Gödelization applied to various improvements on TNT. Each time you think you have reached the end, there is some new variation on the theme of jumping out of the system which requires a kind of creativity to spot.

The Authorship Triangle Again

But I am not interested in pursuing the strange topic of the ever more abstruse tanglings which can arise in self-modifying chess. The point of this has been to show, in a somewhat graphic way, how in any system there is always some "protected" level which is unassailable by the rules on other levels, no matter how tangled their interaction may be among themselves. An amusing riddle from Chapter IV illustrates this same idea in a slightly different context. Perhaps it will catch you off guard:

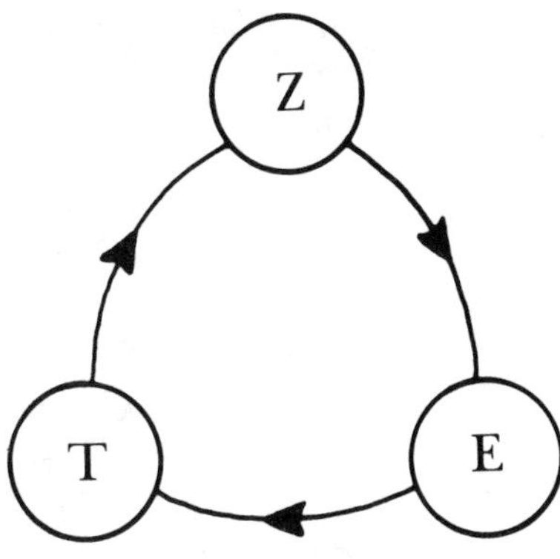

FIGURE 134. An "authorship triangle".

> There are three authors—Z, T, and E. Now it happens that Z exists only in a novel by T. Likewise, T exists only in a novel by E. And strangely, E, too, exists only in a novel—by Z, of course. Now, is such an "authorship triangle" *really* possible? (See Fig. 134.)

Of course it's possible. But there's a trick . . . All three authors Z, T, E, are themselves characters in another novel—by H! You can think of the Z-T-E triangle as a Strange Loop, Or Tangled Hierarchy; but author H is outside of the space in which that tangle takes place—author H is in an inviolate space. Although Z, T, and E all have access—direct or indirect—to each other, and can do dastardly things to each other in their various novels, none of them can touch H's life! They can't even imagine him—no more than you can imagine the author of the book *you're* a character in. If I were to draw author H, I would represent him somewhere off the page. Of course that would present a problem, since drawing a thing necessarily puts it *onto* the page . . . Anyway, H is really outside of the world of Z, T, and E, and should be represented as being so.

Escher's *Drawing Hands*

Another classic variation on our theme is the Escher picture of *Drawing Hands* (Fig. 135). Here, a left hand (LH) draws a right hand (RH), while at the same time, RH draws LH. Once again, levels which ordinarily are seen as hierarchical—that which draws, and that which is drawn—turn back on each other, creating a Tangled Hierarchy. But the theme of the Chapter is borne out, of course, since behind it all lurks the undrawn but drawing hand of M. C. Escher, creator of both LH and RH. Escher is outside of the two-hand space, and in my schematic version of his picture (Fig. 136), you can see that explicitly. In this schematized representation of the Escher picture, you see the Strange Loop, Or Tangled Hierarchy at the top; also, you see the Inviolate Level below it, enabling it to come into being. One could further Escherize the Escher picture, by taking a photograph of a hand drawing it. And so on.

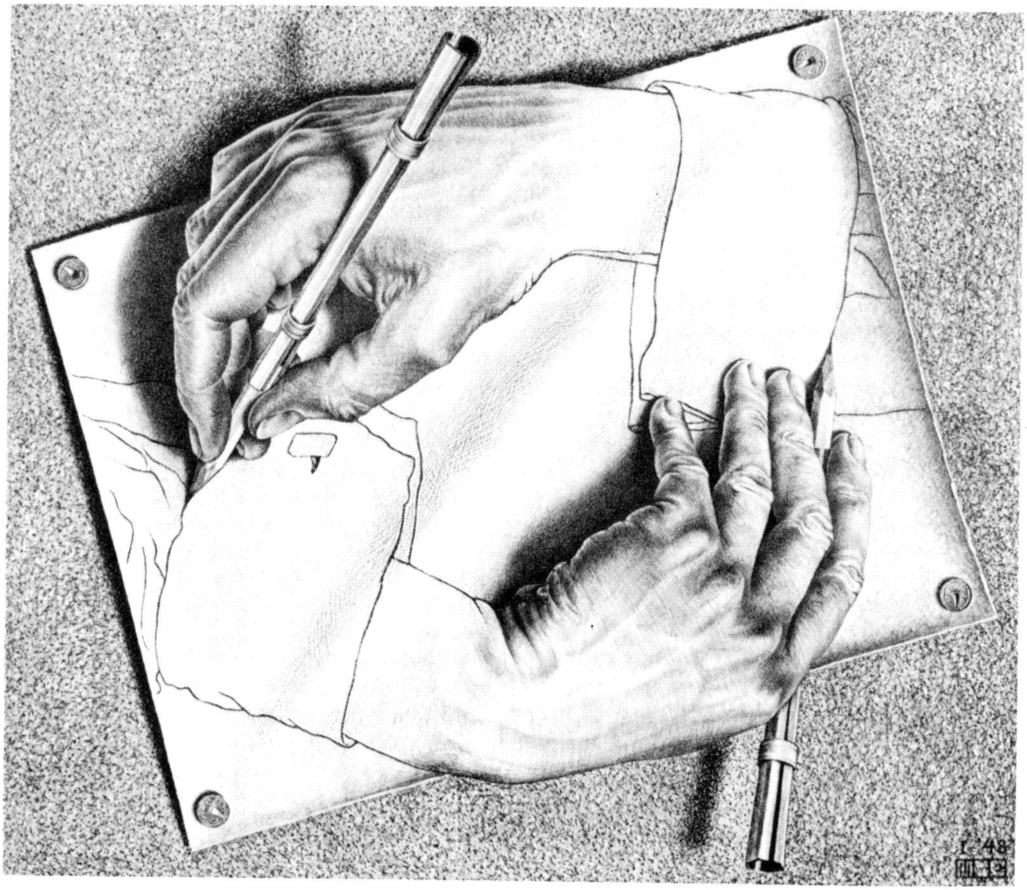

FIGURE 135. Drawing Hands, *by M. C. Escher (lithograph, 1948).*

Strange Loop,
Or
Tangled Hierarchy
(visible)

Inviolate Level
(invisible)

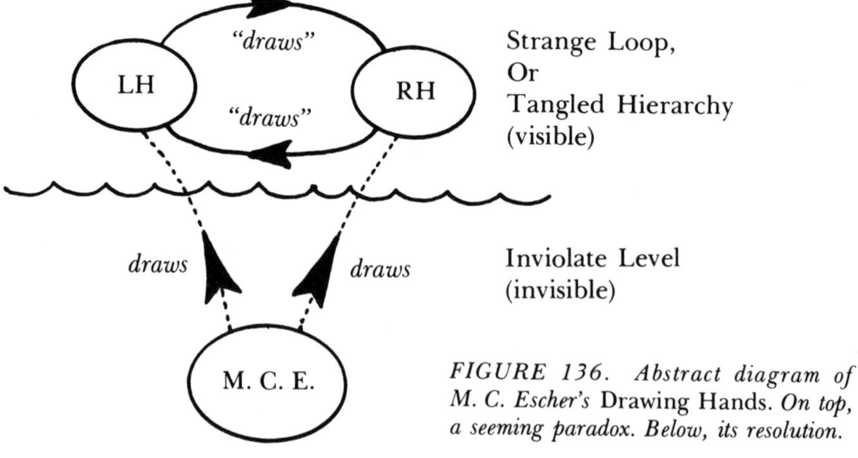

FIGURE 136. Abstract diagram of M. C. Escher's Drawing Hands. *On top, a seeming paradox. Below, its resolution.*

Brain and Mind: A Neural Tangle Supporting a Symbol Tangle

Now we can relate this to the brain, as well as to AI programs. In our thoughts, symbols activate other symbols, and all interact heterarchically. Furthermore, the symbols may cause each other to change internally, in the fashion of programs acting on other programs. The illusion is created, because of the Tangled Hierarchy of symbols, that *there is no inviolate level*. One thinks there is no such level because that level is shielded from our view.

If it were possible to schematize this whole image, there would be a gigantic forest of symbols linked to each other by tangly lines like vines in a tropical jungle—this would be the top level, the Tangled Hierarchy where thoughts really flow back and forth. This is the elusive level of *mind:* the analogue to LH and RH. Far below in the schematic picture, analogous to the invisible "prime mover" Escher, there would be a representation of the myriad neurons—the "inviolate substrate" which lets the tangle above it come into being. Interestingly, this other level is itself a tangle in a literal sense—billions of cells and hundreds of billions of axons, joining them all together.

This is an interesting case where a software tangle, that of the symbols, is supported by a hardware tangle, that of the neurons. But only the symbol tangle is a Tangled Hierarchy. The neural tangle is just a "simple" tangle. This distinction is pretty much the same as that between Strange Loops and feedback, which I mentioned in Chapter XVI. A Tangled Hierarchy occurs when what you presume are clean hierarchical levels take you by surprise and fold back in a hierarchy-violating way. The surprise element is important; it is the reason I call Strange Loops "strange". A simple tangle, like feedback, doesn't involve violations of presumed level distinctions. An example is when you're in the shower and you wash your left arm with your right, and then vice versa. There is no strangeness to the image. Escher didn't choose to draw hands drawing hands for nothing!

Events such as two arms washing each other happen all the time in the world, and we don't notice them particularly. I say something to you, then you say something back to me. Paradox? No; our perceptions of each other didn't involve a hierarchy to begin with, so there is no sense of strangeness.

On the other hand, where language does create strange loops is when it talks about itself, whether directly or indirectly. Here, something *in* the system jumps out and acts *on* the system, as if it were *outside* the system. What bothers us is perhaps an ill-defined sense of topological wrongness: the inside-outside distinction is being blurred, as in the famous shape called a "Klein bottle". Even though the system is an abstraction, our minds use spatial imagery with a sort of mental topology.

Getting back to the symbol tangle, if we look only at it, and forget the neural tangle, then we seem to see a self-programmed object—in just the same way as we seem to see a self-drawn picture if we look at *Drawing Hands* and somehow fall for the illusion, by forgetting the existence of Escher. For the picture, this is unlikely—but for humans and the way they look at their minds, this is usually what happens. We *feel* self-programmed. Indeed, we couldn't feel any other way, for we are shielded from the lower levels, the neural tangle. Our thoughts seem to run about in their own space, creating new thoughts and modifying old ones, and we never notice any neurons helping us out! But that is to be expected. We can't.

An analogous double-entendre can happen with LISP programs that are designed to reach in and change their own structure. If you look at them on the LISP level, you will say that they change themselves; but if you shift levels, and think of LISP programs as data to the LISP interpreter (see Chapter X), then in fact the sole program that is running is the interpreter, and the changes being made are merely changes in pieces of data. The LISP interpreter itself is shielded from changes.

How you describe a tangled situation of this sort depends how far back you step before describing. If you step far enough back, you can often see the clue that allows you to untangle things.

Strange Loops in Government

A fascinating area where hierarchies tangle is government—particularly in the courts. Ordinarily, you think of two disputants arguing their cases in court, and the court adjudicating the matter. The court is on a different level from the disputants. But strange things can start to happen when the courts themselves get entangled in legal cases. Usually there is a higher court which is outside the dispute. Even if two lower courts get involved in some sort of strange fight, with each one claiming jurisdiction over the other, some higher court is outside, and in some sense it is analogous to the inviolate interpretation conventions which we discussed in the warped version of chess.

But what happens when there is no higher court, and the Supreme Court itself gets all tangled up in legal troubles? This sort of snarl nearly happened in the Watergate era. The President threatened to obey only a "definitive ruling" of the Supreme Court—then claimed he had the right to decide what is "definitive". Now that threat never was made good; but if it had been, it would have touched off a monumental confrontation between two levels of government, each of which, in some ways, can validly claim to be "above" the other—and to whom is there recourse to decide which one is right? To say "Congress" is not to settle the matter, for Congress might command the President to obey the Supreme Court, yet the President might still refuse, claiming that he has the legal right to disobey the Supreme Court (and Congress!) under certain circumstances. This would create a new court case, and would throw the whole system into disarray, because it would be so unexpected, so Tangled—so Strange!

The irony is that once you hit your head against the ceiling like this, where you are prevented from jumping out of the system to a yet higher authority, the only recourse is to forces which seem less well defined by rules, but which are the only source of higher-level rules anyway: the lower-level rules, which in this case means the general reaction of society. It is well to remember that in a society like ours, the legal system is, in a sense, a polite gesture granted collectively by millions of people—and it can be overridden just as easily as a river can overflow its banks. Then a seeming anarchy takes over; but anarchy has its own kinds of rules, no less than does civilized society: it is just that they operate from the bottom up, not from the top down. A student of anarchy could try to discover rules according to which anarchic situations develop in time, and very likely there are some such rules.

An analogy from physics is useful here. As was mentioned earlier in the book, gases in equilibrium obey simple laws connecting their temperature, pressure, and volume. However, a gas can violate those laws (as a President can violate laws)—provided it is not in a state of equilibrium. In nonequilibrium situations, to describe what happens, a physicist has recourse only to statistical mechanics—that is, to a level of description which is not macroscopic, for the ultimate explanation of a gas's behavior always lies on the molecular level, just as the ultimate explanation of a society's political behavior always lies at the "grass roots level". The field of nonequilibrium thermodynamics attempts to find macroscopic laws to describe the behavior of gases (and other systems) which are out of equilibrium. It is the analogue to the branch of political science which would search for laws governing anarchical societies.

Other curious tangles which arise in government include the FBI investigating its own wrongdoings, a sheriff going to jail while in office, the self-application of the parliamentary rules of procedure, and so on. One of the most curious legal cases I ever heard of involved a person who claimed to have psychic powers. In fact, he claimed to be able to use his psychic powers to detect personality traits, and thereby to aid lawyers in picking juries. Now what if this "psychic" has to stand trial himself one day? What effect might this have on a jury member who believes staunchly in ESP? How much will he feel affected by the psychic (whether or not the psychic is genuine)? The territory is ripe for exploitation—a great area for self-fulfilling prophecies.

Tangles Involving Science and the Occult

Speaking of psychics and ESP, another sphere of life where strange loops abound is fringe science. What fringe science does is to call into question many of the standard procedures or beliefs of orthodox science, and thereby challenge the objectivity of science. New ways of interpreting evidence that rival the established ones are presented. But how do you evaluate a way of interpreting evidence? Isn't this precisely the problem of objectivity all over again, just on a higher plane? Of course. Lewis Carroll's infinite-regress paradox appears in a new guise. The Tortoise would argue that if you want to show that A is a fact, you need evidence: B. But what makes you sure that B is evidence of A? To show that, you need meta-evidence: C. And for the validity of that meta-evidence, you need meta-meta-evidence—and so on, ad nauseam. Despite this argument, people have an intuitive sense of evidence. This is because—to repeat an old refrain—people have built-in hardware in their brains that includes some rudimentary ways of interpreting evidence. We can build on this, and accumulate new ways of interpreting evidence; we even learn how and when to override our most basic mechanisms of evidence interpretation, as one must, for example, in trying to figure out magic tricks.

Concrete examples of evidence dilemmas crop up in regard to many phenomena of fringe science. For instance, ESP often seems to manifest itself outside of the laboratory, but when brought into the laboratory, it vanishes mysteriously. The standard scientific explanation for this is that ESP is a nonreal phenomenon which cannot stand up to rigorous scrutiny. Some (by no means all) believers in ESP have a peculiar way of fighting back, however. They say, "No, ESP is real; it simply goes away when one tries to observe it scientifically—it is contrary to the nature of a scientific worldview." This is an amazingly brazen technique, which we might call "kicking the problem upstairs". What that means is, instead of questioning the matter at hand, you call into doubt theories belonging to a higher level of credibility. The believers in ESP insinuate that what is wrong is not *their* ideas, but the belief system of science. This is a pretty grandiose claim, and unless there is overwhelming evidence for it, one should be skeptical of it. But then here we are again, talking about "overwhelming evidence" as if everyone agreed on what that means!

The Nature of Evidence

The Sagredo-Simplicio-Salviati tangle, mentioned in Chapters XIII and XV, gives another example of the complexities of evaluation of evidence. Sagredo tries to find some objective compromise, if possible, between the opposing views of Simplicio and Salviati. But compromise may not always be possible. How can one compromise "fairly" between right and wrong? Between fair and unfair? Between compromise and no compromise? These questions come up over and over again in disguised form in arguments about ordinary things.

Is it possible to define what evidence is? Is it possible to lay down laws as to how to make sense out of situations? Probably not, for any rigid rules would undoubtedly have exceptions, and nonrigid rules are not rules. Having an intelligent AI program would not solve the problem either, for as an evidence processor, it would not be any less fallible than humans are. So, if evidence is such an intangible thing after all, why am I warning against new ways of interpreting evidence? Am I being inconsistent? In this case, I don't think so. My feeling is that there are guidelines which one can give, and out of them an organic synthesis can be made. But inevitably some amount of judgment and intuition must enter the picture—things which are different in different people. They will also be different in different AI programs. Ultimately, there are complicated criteria for deciding if a method of evaluation of evidence is good. One involves the "usefulness" of ideas which are arrived at by that kind of reasoning. Modes of thought which lead to useful new things in life are deemed "valid" in some sense. But this word "useful" is extremely subjective.

My feeling is that the process by which we decide what is valid or what is true is an art; and that it relies as deeply on a sense of beauty and simplicity as it does on rock-solid principles of logic or reasoning or anything else which can be objectively formalized. I am *not* saying either (1) truth is a chimera, or (2) human intelligence is in principle not programmable. I *am* saying (1) truth is too elusive for any human or any collection of humans ever to attain fully; and (2) Artificial Intelligence, when it reaches the level of human intelligence—or even if it surpasses it—will still be plagued by the problems of art, beauty, and simplicity, and will run up against these things constantly in its own search for knowledge and understanding.

"What is evidence?" is not just a philosophical question, for it intrudes into life in all sorts of places. You are faced with an extraordinary number of choices as to how to interpret evidence at every moment. You can hardly go into a bookstore (or these days, even a grocery store!) without seeing books on clairvoyance, ESP, UFO's, the Bermuda triangle, astrology, dowsing, evolution versus creation, black holes, psi fields, biofeedback, transcendental meditation, new theories of psychology . . . In science, there are fierce debates about catastrophe theory, elementary particle theory, black holes, truth and existence in mathematics, free will, Artificial Intelligence, reductionism versus holism . . . On the more pragmatic side of life, there are debates over the efficacy of vitamin C or of laetrile, over the real size of oil reserves (either underground or stored), over what causes inflation and unemployment—and on and on. There is Buckminster Fullerism, Zen Buddhism, Zeno's paradoxes, psychoanalysis, etc., etc. From issues as trivial as where books ought to be shelved in a store, to issues as vital as what ideas are to be taught to children in schools, ways of interpreting evidence play an inestimable role.

Seeing Oneself

One of the most severe of all problems of evidence interpretation is that of trying to interpret all the confusing signals from the outside as to who one is. In this case, the potential for intralevel and interlevel conflict is tremendous. The psychic mechanisms have to deal simultaneously with the individual's internal need for self-esteem and the constant flow of evidence from the outside affecting the self-image. The result is that information flows in a complex swirl between different levels of the personality; as it goes round and round, parts of it get magnified, reduced, negated, or otherwise distorted, and then those parts in turn get further subjected to the same sort of swirl, over and over again—all of this in an attempt to reconcile what is, with what we wish were (see Fig. 81).

The upshot is that the total picture of "who I am" is integrated in some enormously complex way inside the entire mental structure, and contains in each one of us a large number of unresolved, possibly unresolvable, inconsistencies. These undoubtedly provide much of the dynamic tension which is so much a part of being human. Out of this tension between the inside and outside notions of who we are come the drives towards various goals that make each of us unique. Thus, ironically, something which we all have in common—the fact of being self-reflecting conscious beings—leads to the rich diversity in the ways we have of internalizing evidence about all sorts of things, and in the end winds up being one of the major forces in creating distinct individuals.

Gödel's Theorem and Other Disciplines

It is natural to try to draw parallels between people and sufficiently complicated formal systems which, like people, have "self-images" of a sort. Gödel's Theorem shows that there are fundamental limitations to consistent formal systems with self-images. But is it more general? Is there a "Gödel's Theorem of psychology", for instance?

If one uses Gödel's Theorem as a metaphor, as a source of inspiration, rather than trying to translate it literally into the language of psychology or of any other discipline, then perhaps it can suggest new truths in psychology or other areas. But it is quite unjustifiable to translate it directly into a statement of another discipline and take that as equally valid. It would be a large mistake to think that what has been worked out with the utmost delicacy in mathematical logic should hold without modification in a completely different area.

Introspection and Insanity: A Gödelian Problem

I think it can have suggestive value to translate Gödel's Theorem into other domains, provided one specifies in advance that the translations are metaphorical and are not intended to be taken literally. That having been said, I see two major ways of using analogies to connect Gödel's Theorem and human thoughts. One involves the problem of wondering about one's sanity. How can you figure out if you are sane? This is a Strange Loop indeed. Once you begin to question your own sanity, you can get trapped in an ever-tighter vortex of self-fulfilling prophecies, though the process is by no means inevitable. Everyone knows that the insane interpret the world via their own peculiarly consistent logic; how can you tell if your own logic is "peculiar" or not, given that you have only your own logic to judge itself? I don't see any answer. I am just reminded of Gödel's second Theorem, which implies that the only versions of formal number theory which assert their own consistency are inconsistent . . .

Can We Understand Our Own Minds or Brains?

The other metaphorical analogue to Gödel's Theorem which I find provocative suggests that ultimately, we cannot understand our own minds/brains. This is such a loaded, many-leveled idea that one must be extremely cautious in proposing it. What does "understanding our own minds/brains" mean? It could mean having a general sense of how they work, as mechanics have a sense of how cars work. It could mean having a complete explanation for why people do any and all things they do. It could mean having a complete understanding of the physical structure of one's own brain on all levels. It could mean having a complete wiring diagram of a brain in a book (or library or computer). It could mean knowing, at every instant, precisely what is happening in one's own brain on the neural level—each firing, each synaptic alteration, and so on. It could mean having written a program which passes the Turing test. It could mean knowing oneself so perfectly that such notions as the subconscious and the intuition make no sense, because everything is out in the open. It could mean any number of other things.

Which of these types of self-mirroring, if any, does the self-mirroring in Gödel's Theorem most resemble? I would hesitate to say. Some of them are quite silly. For instance, the idea of being able to monitor your own brain state in all its detail is a pipe dream, an absurd and uninteresting proposition to start with; and if Gödel's Theorem suggests that it is impossible, that is hardly a revelation. On the other hand, the age-old goal of knowing yourself in some profound way—let us call it "understanding your own psychic structure"—has a ring of plausibility to it. But might there not be some vaguely Gödelian loop which limits the depth to which any individual can penetrate into his own psyche? Just as we cannot see our faces with our own eyes, is it not reasonable to expect that we cannot mirror our complete mental structures in the symbols which carry them out?

All the limitative Theorems of metamathematics and the theory of computation suggest that once the ability to represent your own structure has reached a certain critical point, that is the kiss of death: it guarantees that you can never represent yourself totally. Gödel's Incompleteness Theorem, Church's Undecidability Theorem, Turing's Halting Theorem, Tarski's Truth Theorem—all have the flavor of some ancient fairy tale which warns you that "To seek self-knowledge is to embark on a journey which . . . will always be incomplete, cannot be charted on any map, will never halt, cannot be described."

But do the limitative Theorems have any bearing on people? Here is one way of arguing the case. Either I am consistent or I am inconsistent. (The latter is much more likely, but for completeness' sake, I consider both possibilities.) If I am consistent, then there are two cases. (1) The "low-fidelity" case: my self-understanding is below a certain critical point. In this case, I am incomplete by hypothesis. (2) The "high-fidelity" case: My self-understanding has reached the critical point where a metaphorical analogue of the limitative Theorems does apply, so my self-understanding undermines itself in a Gödelian way, and I am incomplete for that reason. Cases (1) and (2) are predicated on my being 100 per cent consistent—a very unlikely state of affairs. More likely is that I am inconsistent—but that's worse, for then inside me there are contradictions, and how can I ever understand that?

Consistent or inconsistent, no one is exempt from the mystery of the self. Probably we are all inconsistent. The world is just too complicated for a person to be able to afford the luxury of reconciling all of his beliefs with each other. Tension and confusion are important in a world where many decisions must be made quickly. Miguel de Unamuno once said, "If a person never contradicts himself, it must be that he says nothing." I would say that we all are in the same boat as the Zen master who, after contradicting himself several times in a row, said to the confused Doko, "I cannot understand myself."

Gödel's Theorem and Personal Nonexistence

Perhaps the greatest contradiction in our lives, the hardest to handle, is the knowledge "There was a time when I was not alive, and there will come a time when I am not alive." On one level, when you "step out of yourself" and see yourself as "just another human being", it makes complete sense. But on another level, perhaps a deeper level, personal nonexistence makes no sense at all. All that we know is embedded inside our minds, and for all that to be absent from the universe is not comprehensible. This is a basic undeniable problem of life; perhaps it is the best metaphorical analogue of Gödel's Theorem. When you try to imagine your own nonexistence, you have to try to jump out of yourself, by mapping yourself onto someone else. You fool yourself into believing that you can import an outsider's view of yourself into you, much as TNT "believes" it mirrors its own metatheory inside itself. But TNT only contains its own metatheory up to a certain extent—not fully. And as for you, though you may imagine that you have jumped out of yourself, you never can actually do so—no more than Escher's dragon can jump out of its native two-dimensional plane into three dimensions. In any case, this contradiction is so great that most of our lives we just sweep the whole mess under the rug, because trying to deal with it just leads nowhere.

Zen minds, on the other hand, revel in this irreconcilability. Over and over again, they face the conflict between the Eastern belief: "The world and I are one, so the notion of my ceasing to exist is a contradiction in terms" (my verbalization is undoubtedly too Westernized—apologies to Zenists), and the Western belief: "I am just part of the world, and I will die, but the world will go on without me."

Science and Dualism

Science is often criticized as being too "Western" or "dualistic"—that is, being permeated by the dichotomy between subject and object, or observer and observed. While it is true that up until this century, science was exclusively concerned with things which can be readily distinguished from their human observers—such as oxygen and carbon, light and heat, stars and planets, accelerations and orbits, and so on—this phase of science was a necessary prelude to the more modern phase, in which life itself has come under investigation. Step by step, inexorably, "Western" science has moved towards investigation of the human mind—which is to say, of the observer. Artificial Intelligence research is the furthest step so far along that route. Before AI came along, there were two major previews of the strange consequences of the mixing of subject and object in science. One was the revolution of quantum mechanics, with its epistemological problems involving the interference of the observer with the observed. The other was the mixing of subject and object in metamathematics, beginning with Gödel's Theorem and moving through all the other limitative Theorems we have discussed. Perhaps the next step after AI will be the self-application of science: science studying itself as an object. This is a different manner of mixing subject and object—perhaps an even more tangled one than that of humans studying their own minds.

By the way, in passing, it is interesting to note that all results essentially dependent on the fusion of subject and object have been limitative results. In addition to the limitative Theorems, there is Heisenberg's uncertainty principle, which says that measuring one quantity renders impossible the simultaneous measurement of a related quantity. I don't know why all these results are limitative. Make of it what you will.

Symbol *vs.* Object in Modern Music and Art

Closely linked with the subject-object dichotomy is the symbol-object dichotomy, which was explored in depth by Ludwig Wittgenstein in the early part of this century. Later the words "use" and "mention" were adopted to make the same distinction. Quine and others have written at length about the connection between signs and what they stand for. But not only philosophers have devoted much thought to this deep and abstract matter. In our century both music and art have gone through crises which reflect a profound concern with this problem. Whereas music and painting, for instance, have traditionally expressed ideas or emotions through a vocabulary of "symbols" (i.e. visual images, chords, rhythms, or whatever), now there is a tendency to explore the capacity of music and art to *not* express anything—just to *be*. This means to exist as pure globs of paint, or pure sounds, but in either case drained of all symbolic value.

In music, in particular, John Cage has been very influential in bringing a Zen-like approach to sound. Many of his pieces convey a disdain for "use" of sounds—that is, using sounds to convey emotional states—and an exultation in "mentioning" sounds—that is, concocting arbitrary juxtapositions of sounds without regard to any previously formulated code by which a listener could decode them into a message. A typical example is "Imaginary Landscape no. 4", the polyradio piece described in Chapter VI. I may not be doing Cage justice, but to me it seems that much of his work has been directed at bringing meaninglessness into music, and in some sense, at making that meaninglessness have meaning. Aleatoric music is a typical exploration in that direction. (Incidentally, chance music is a close cousin to the much later notion of "happenings" or "be-in" 's.) There are many other contemporary composers who are following Cage's lead, but few with as much originality. A piece by Anna Lockwood, called "Piano Burning", involves just that—with the strings stretched to maximum tightness, to make them snap as loudly as possible; in a piece by LaMonte Young, the noises are provided by shoving the piano all around the stage and through obstacles, like a battering ram.

Art in this century has gone through many convulsions of this general type. At first there was the abandonment of representation, which was genuinely revolutionary: the beginnings of abstract art. A gradual swoop from pure representation to the most highly abstract patterns is revealed in the work of Piet Mondrian. After the world was used to nonrepresentational art, then surrealism came along. It was a bizarre about-face, something like neoclassicism in music, in which extremely representational art was "subverted" and used for altogether new reasons: to shock, confuse, and amaze. This school was founded by André Breton, and was located primarily in France; some of its more influential members were Dalí, Magritte, de Chirico, Tanguy.

Magritte's Semantic Illusions

Of all these artists, Magritte was the most conscious of the symbol-object mystery (which I see as a deep extension of the use-mention distinction). He uses it to evoke powerful responses in viewers, even if the viewers do not verbalize the distinction this way. For example, consider his very strange variation on the theme of still life, called *Common Sense* (Fig. 137).

FIGURE 137. Common Sense, *by René Magritte (1945-46).*

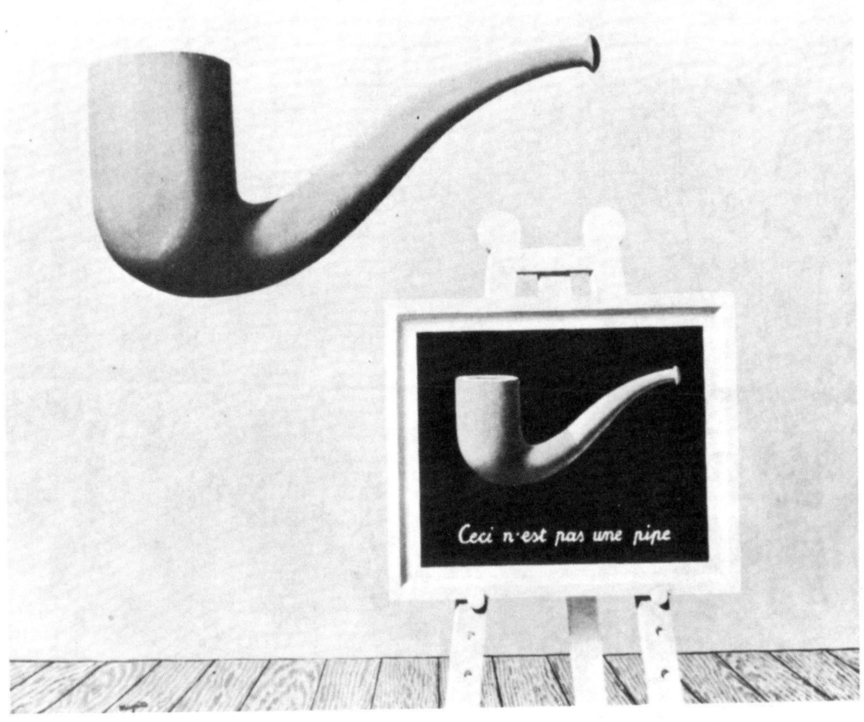

FIGURE 138. The Two Mysteries, *by René Magritte (1966).*

Here, a dish filled with fruit, ordinarily the kind of thing represented inside a still life, is shown sitting on top of a blank canvas. The conflict between the symbol and the real is great. But that is not the full irony, for of course the whole thing is itself just a painting—in fact, a still life with nonstandard subject matter.

Magritte's series of pipe paintings is fascinating and perplexing. Consider *The Two Mysteries* (Fig. 138). Focusing on the inner painting, you get the message that symbols and pipes are different. Then your glance moves upward to the "real" pipe floating in the air—you perceive that it is real, while the other one is just a symbol. But that is of course totally wrong: both of them are on the same flat surface before your eyes. The idea that one pipe is in a twice-nested painting, and therefore somehow "less real" than the other pipe, is a complete fallacy. Once you are willing to "enter the room", you have already been tricked: you've fallen for image as reality. To be consistent in your gullibility, you should happily go one level further down, and confuse image-within-image with reality. The only way not to be sucked in is to see both pipes merely as colored smudges on a surface a few inches in front of your nose. Then, and only then, do you appreciate the full meaning of the written message "Ceci n'est pas une pipe"—but ironically, at the very instant everything turns to smudges, the writing too turns to smudges, thereby losing its meaning! In other words, at that instant, the verbal message of the painting self-destructs in a most Gödelian way.

FIGURE 139. Smoke Signal. [*Drawing by the author.*]

The Air and the Song (Fig. 82), taken from a series by Magritte, accomplishes all that *The Two Mysteries* does, but in one level instead of two. My drawings *Smoke Signal* and *Pipe Dream* (Figs. 139 and 140) constitute "Variations on a Theme of Magritte". Try staring at *Smoke Signal* for a while. Before long, you should be able to make out a hidden message saying, "Ceci n'est pas un message". Thus, if you find the message, it denies itself—yet if you don't, you miss the point entirely. Because of their indirect self-snuffing, my two pipe pictures can be loosely mapped onto Gödel's G—thus giving rise to a "Central Pipemap", in the same spirit as the other "Central Xmaps": Dog, Crab, Sloth.

A classic example of use-mention confusion in paintings is the occurrence of a palette in a painting. Whereas the palette is an illusion created by the representational skill of the painter, the paints on the painted palette are literal daubs of paint from the artist's palette. The paint plays itself—it does not symbolize anything else. In *Don Giovanni,* Mozart exploited a related trick: he wrote into the score explicitly the sound of an orchestra tuning up. Similarly, if I want the letter 'I' to play itself (and not symbolize me), I put 'I' directly into my text; then I enclose 'I' between quotes. What results is ''I'' (not 'I', nor '''I'''). Got that?

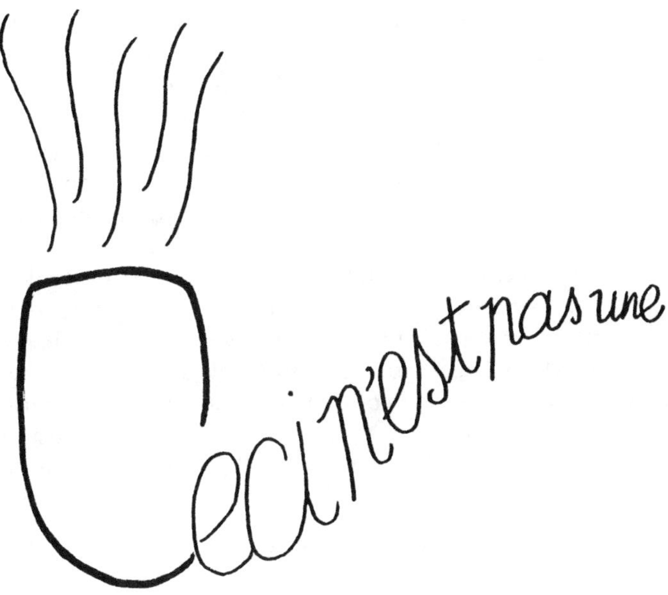

FIGURE 140. Pipe Dream. [*Drawing by the author.*]

The "Code" of Modern Art

A large number of influences, which no one could hope to pin down completely, led to further explorations of the symbol-object dualism in art. There is no doubt that John Cage, with his interest in Zen, had a profound influence on art as well as on music. His friends Jasper Johns and Robert Rauschenberg both explored the distinction between objects and symbols by using objects as symbols for themselves—or, to flip the coin, by using symbols as objects in themselves. All of this was perhaps intended to break down the notion that art is one step removed from reality—that art speaks in "code", for which the viewer must act as interpreter. The idea was to eliminate the step of interpretation and let the naked object simply *be*, period. ("Period"—a curious case of use-mention blur.) However, if this was the intention, it was a monumental flop, and perhaps had to be.

Any time an object is exhibited in a gallery or dubbed a "work", it acquires an aura of deep inner significance—no matter how much the viewer has been warned *not* to look for meaning. In fact, there is a backfiring effect whereby the more that viewers are told to look at these objects without mystification, the more mystified the viewers get. After all, if a wooden crate on a museum floor is just a wooden crate on a museum floor, then why doesn't the janitor haul it out back and throw it in the garbage? Why is the name of an artist attached to it? Why did the artist want to demystify art? Why isn't that dirt clod out front labeled with an artist's name? Is this a hoax? Am I crazy, or are artists crazy? More and more questions flood into the viewer's mind; he can't help it. This is the "frame effect" which art—Art—automatically creates. There is no way to suppress the wonderings in the minds of the curious.

Of course, if the purpose is to instill a Zen-like sense of the world as devoid of categories and meanings, then perhaps such art is merely intended to serve—as does intellectualizing about Zen—as a catalyst to inspire the viewer to go out and become acquainted with the philosophy which rejects "inner meanings" and embraces the world as a whole. In this case, the art is self-defeating in the short run, since the viewers *do* ponder about its meaning, but it achieves its aim with a few people in the long run, by introducing them to its sources. But in either case, it is not true that there is no code by which ideas are conveyed to the viewer. Actually, the code is a much more complex thing, involving statements about the absence of codes and so forth—that is, it is part code, part metacode, and so on. There is a Tangled Hierarchy of messages being transmitted by the most Zen-like art objects, which is perhaps why so many find modern art so inscrutable.

Ism Once Again

Cage has led a movement to break the boundaries between art and nature. In music, the theme is that all sounds are equal—a sort of acoustical democracy. Thus silence is just as important as sound, and random sound is just as important as organized sound. Leonard B. Meyer, in his book *Music, the Arts, and Ideas,* has called this movement in music "transcendentalism", and states:

> If the distinction between art and nature is mistaken, aesthetic valuation is irrelevant. One should no more judge the value of a piano sonata than one should judge the value of a stone, a thunderstorm, or a starfish. "Categorical statements, such as right and wrong, beautiful or ugly, typical of the rationalistic thinking of tonal aesthetics," writes Luciano Berio [a contemporary composer], "are no longer useful in understanding why and how a composer today works on audible forms and musical action."

Later, Meyer continues in describing the philosophical position of transcendentalism:

> . . . all things in all of time and space are inextricably connected with one another. Any divisions, classifications, or organizations discovered in the universe are arbitrary. The world is a complex, continuous, single event.[2] [Shades of Zeno!]

I find "transcendentalism" too bulky a name for this movement. In its place, I use "ism". Being a suffix without a prefix, it suggests an ideology

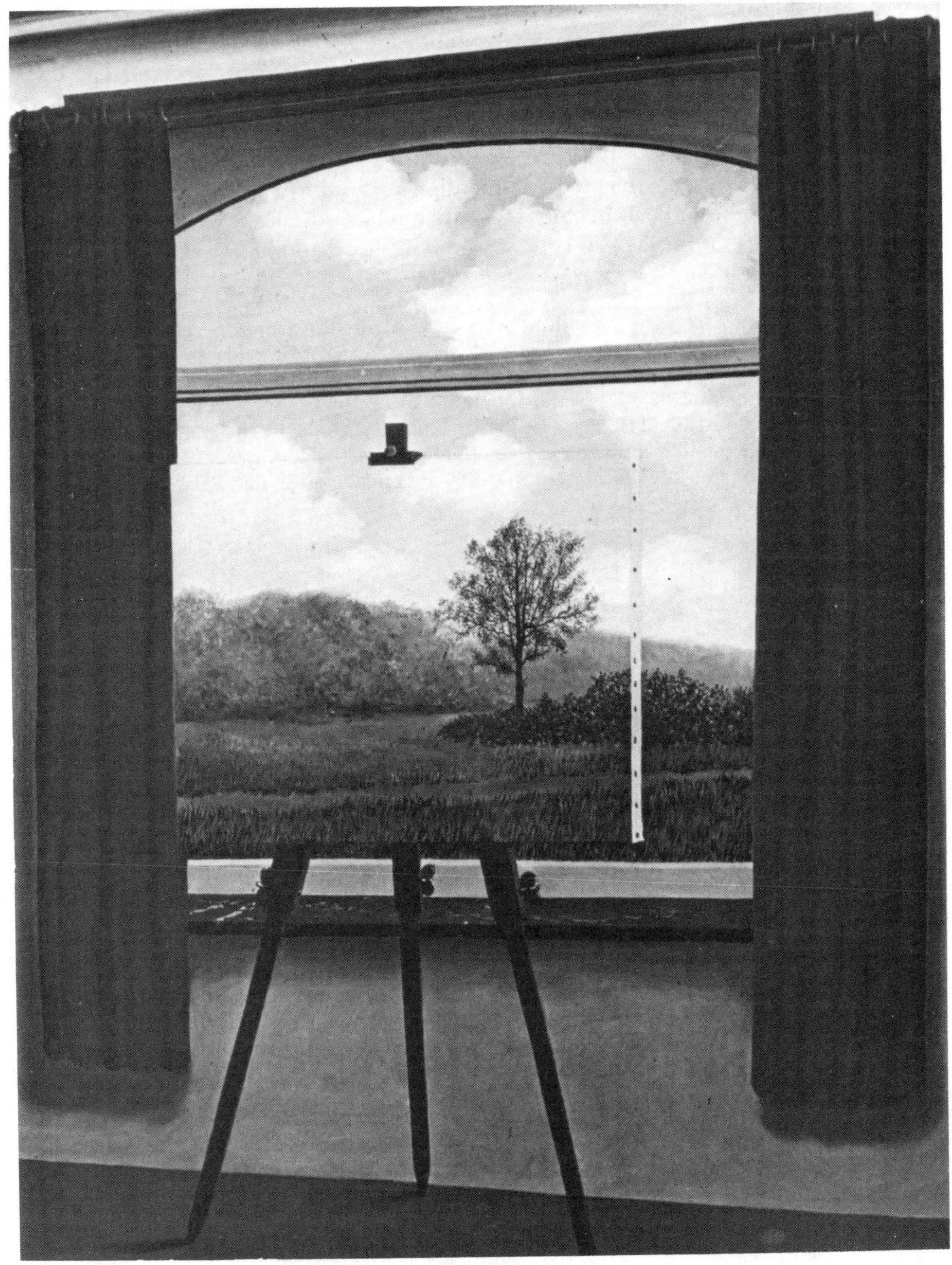

FIGURE 141. The Human Condition I, *by René Magritte (1933).*

without ideas—which, however you interpret it, is probably the case. And since "ism" embraces whatever is, its name is quite fitting. In "ism" the word "is" is half mentioned, half used; what could be more appropriate? Ism is the spirit of Zen in art. And just as the central problem of Zen is to unmask the self, the central problem of art in this century seems to be to figure out what art is. All these thrashings-about are part of its identity crisis.

We have seen that the use-mention dichotomy, when pushed, turns into the philosophical problem of symbol-object dualism, which links it to the mystery of mind. Magritte wrote about his painting *The Human Condition I* (Fig. 141):

> I placed in front of a window, seen from a room, a painting representing exactly that part of the landscape which was hidden from view by the painting. Therefore, the tree represented in the painting hid from view the tree situated behind it, outside the room. It existed for the spectator, as it were, simultaneously in his mind, as both inside the room in the painting, and outside in the real landscape. Which is how we see the world: we see it as being outside ourselves even though it is only a mental representation of it that we experience inside ourselves.[3]

Understanding the Mind

First through the pregnant images of his painting, and then in direct words, Magritte expresses the link between the two questions "How do symbols work?" and "How do our minds work?" And so he leads us back to the question posed earlier: "Can we ever hope to understand our minds/brains?"

Or does some marvelous diabolical Gödelian proposition preclude our ever unraveling our minds? Provided you do not adopt a totally unreasonable definition of "understanding", I see no Gödelian obstacle in the way of the eventual understanding of our minds. For instance, it seems to me quite reasonable to desire to understand the working principles of brains in general, much the same way as we understand the working principles of car engines in general. It is quite different from trying to understand any single brain in every last detail—let alone trying to do this for one's own brain! I don't see how Gödel's Theorem, even if construed in the sloppiest way, has anything to say about the feasibility of this prospect. I see no reason that Gödel's Theorem imposes any limitations on our ability to formulate and verify the general mechanisms by which thought processes take place in the medium of nerve cells. I see no barrier imposed by Gödel's Theorem to the implementation on computers (or their successors) of types of symbol manipulation that achieve roughly the same results as brains do. It is entirely another question to try and duplicate in a program some particular human's mind—but to produce an intelligent program at all is a more limited goal. Gödel's Theorem doesn't ban our reproducing our own level of intelligence via programs any more than it bans our reproducing our own level of intelligence via transmission of hereditary information in

DNA, followed by education. Indeed, we have seen, in Chapter XVI, how a remarkable Gödelian mechanism—the Strange Loop of proteins and DNA—is precisely what allows transmission of intelligence!

Does Gödel's Theorem, then, have absolutely nothing to offer us in thinking about our own minds? I think it does, although not in the mystical and limitative way which some people think it ought to. I think that the process of coming to understand Gödel's proof, with its construction involving arbitrary codes, complex isomorphisms, high and low levels of interpretation, and the capacity for self-mirroring, may inject some rich undercurrents and flavors into one's set of images about symbols and symbol processing, which may deepen one's intuition for the relationship between mental structures on different levels.

Accidental Inexplicability of Intelligence?

Before suggesting a philosophically intriguing "application" of Gödel's proof, I would like to bring up the idea of "accidental inexplicability" of intelligence. Here is what that involves. It could be that our brains, unlike car engines, are stubborn and intractable systems which we cannot neatly decompose in any way. At present, we have no idea whether our brains will yield to repeated attempts to cleave them into clean layers, each of which can be explained in terms of lower layers—or whether our brains will foil all our attempts at decomposition.

But even if we do fail to understand ourselves, there need not be any Gödelian "twist" behind it; it could be simply an accident of fate that our brains are too weak to understand themselves. Think of the lowly giraffe, for instance, whose brain is obviously far below the level required for self-understanding—yet it is remarkably similar to our own brain. In fact, the brains of giraffes, elephants, baboons—even the brains of tortoises or unknown beings who are far smarter than we are—probably all operate on basically the same set of principles. Giraffes may lie far below the threshold of intelligence necessary to understand how those principles fit together to produce the qualities of mind; humans may lie closer to that threshold—perhaps just barely below it, perhaps even above it. The point is that there may be no *fundamental* (i.e., Gödelian) reason why those qualities are incomprehensible; they may be completely clear to more intelligent beings.

Undecidability Is Inseparable from a High-Level Viewpoint

Barring this pessimistic notion of the accidental inexplicability of the brain, what insights might Gödel's proof offer us about explanations of our minds/brains? Gödel's proof offers the notion that a high-level view of a system may contain explanatory power which simply is absent on the lower levels. By this I mean the following. Suppose someone gave you G, Gödel's undecidable string, as a string of TNT. Also suppose you knew nothing of Gödel-numbering. The question you are supposed to answer is: "Why isn't this string a theorem of TNT?" Now you are used to such questions; for instance, if you had been asked that question about S0=0, you would have a ready explanation: *"Its negation, ~S0=0, is a theorem."* This, together with your knowledge that TNT is consistent, provides an explanation of why the given string is a nontheorem. This is what I call an explanation "on the TNT-level". Notice how different it is from the explanation of why MU is not a theorem of the MIU-system: the former comes from the M-mode, the latter only from the I-mode.

Now what about G? The TNT-level explanation which worked for S0=0 does not work for G, because ~G is *not* a theorem. The person who has no overview of TNT will be baffled as to why he can't make G according to the rules, because as an arithmetical proposition, it apparently has nothing wrong with it. In fact, when G is turned into a universally quantified string, every instance gotten from G by substituting numerals for the variables can be derived. The only way to explain G's nontheoremhood is to discover the notion of Gödel-numbering and view TNT on an entirely different level. It is not that it is just difficult and complicated to write out the explanation on the TNT-level; it is impossible. Such an explanation simply does not exist. There is, on the high level, a kind of explanatory power which simply is lacking, in principle, on the TNT-level. G's nontheoremhood is, so to speak, an *intrinsically high-level fact.* It is my suspicion that this is the case for *all* undecidable propositions; that is to say: every undecidable proposition is actually a Gödel sentence, asserting its own nontheoremhood in some system via some code.

Consciousness as an Intrinsically High-Level Phenomenon

Looked at this way, Gödel's proof suggests—though by no means does it prove!—that there could be some high-level way of viewing the mind/brain, involving concepts which do not appear on lower levels, and that this level might have explanatory power that does not exist—not even in principle—on lower levels. It would mean that some facts could be explained on the high level quite easily, but not on lower levels *at all.* No matter how long and cumbersome a low-level statement were made, it would not explain the phenomena in question. It is the analogue to the fact that, if you make derivation after derivation in TNT, no matter how long and cumbersome you make them, you will never come up with one for G—despite the fact that on a higher level, you can see that G is true.

What might such high-level concepts be? It has been proposed for eons, by various holistically or "soulistically" inclined scientists and humanists, that *consciousness* is a phenomenon that escapes explanation in terms of brain-components; so here is a candidate, at least. There is also the ever-puzzling notion of *free will.* So perhaps these qualities could be "emergent" in the sense of requiring explanations which cannot be furnished by the physiology alone. But it is important to realize that if we are being guided by Gödel's proof in making such bold hypotheses, we must carry the analogy through thoroughly. In particular, it is vital to recall that G's nontheoremhood *does* have an explanation—it is not a total mystery! The explanation hinges on understanding not just one level at a time, but the way in which one level mirrors its metalevel, and the consequences of this mirroring. If our analogy is to hold, then, "emergent" phenomena would become explicable in terms of a relationship between different levels in mental systems.

Strange Loops as the Crux of Consciousness

My belief is that the explanations of "emergent" phenomena in our brains—for instance, ideas, hopes, images, analogies, and finally consciousness and free will—are based on a kind of Strange Loop, an interaction between levels in which the top level reaches back down towards the bottom level and influences it, while at the same time being itself determined by the bottom level. In other words, a self-reinforcing "resonance" between different levels—quite like the Henkin sentence which, by merely asserting its own provability, actually becomes provable. The self comes into being at the moment it has the power to reflect itself.

This should not be taken as an antireductionist position. It just implies that a reductionistic explanation of a mind, *in order to be comprehensible,* must bring in "soft" concepts such as levels, mappings, and meanings. In principle, I have no doubt that a totally reductionistic but incomprehensible explanation of the brain exists; the problem is how to translate it into a language we ourselves can fathom. Surely we don't want a description in terms of positions and momenta of particles; we want a description which relates neural activity to "signals" (intermediate-level phenomena)—and which relates signals, in turn, to "symbols" and "subsystems", including the presumed-to-exist "self-symbol". This act of translation from low-level physical hardware to high-level psychological software is analogous to the translation of number-theoretical statements into metamathematical statements. Recall that the level-crossing which takes place at this exact translation point is what creates Gödel's incompleteness and the self-proving character of Henkin's sentence. I postulate that a similar level-crossing is what creates our nearly unanalyzable feelings of self.

In order to deal with the full richness of the brain/mind system, we will have to be able to slip between levels comfortably. Moreover, we will have to admit various types of "causality": ways in which an event at one level of description can "cause" events at other levels to happen. Sometimes event A will be said to "cause" event B simply for the reason that the one is a translation, on another level of description, of the other. Sometimes "cause" will have its usual meaning: physical causality. Both types of causality—and perhaps some more—will have to be admitted in any explanation of mind, for we will have to admit causes that propagate both upwards *and* downwards in the Tangled Hierarchy of mentality, just as in the Central Dogmap.

At the crux, then, of our understanding ourselves will come an understanding of the Tangled Hierarchy of levels inside our minds. My position is rather similar to the viewpoint put forth by the neuroscientist Roger Sperry in his excellent article "Mind, Brain, and Humanist Values", from which I quote a little here:

> In my own hypothetical brain model, conscious awareness does get representation as a very real causal agent and rates an important place in the causal sequence and chain of control in brain events, in which it appears as an active, operational force. . . . To put it very simply, it comes down to the issue of who pushes whom around in the population of causal forces that occupy the cranium. It is a matter, in other words, of straightening out the peck-order hierarchy among intracranial control agents. There exists within the cranium a whole world of diverse causal forces; what is more, there are forces within forces within forces, as in no other cubic half-foot of universe that we know. . . . To make a long story short, if one keeps climbing upward in the chain of command within the brain, one finds at the very top those over-all organizational forces and dynamic properties of the large patterns of cerebral excitation that are correlated with mental states or psychic activity. . . . Near the apex of this command system in the brain . . . we find ideas. Man over the chimpanzee has ideas and ideals. In the brain model proposed here, the causal potency of an idea, or an ideal, becomes just as real as that of a molecule, a cell, or a nerve impulse. Ideas cause ideas and help evolve new ideas. They interact with each other and with other mental forces in the same brain, in neighboring brains, and, thanks to global communication, in far distant, foreign brains. And they also interact with the external surroundings to produce in toto a burstwise advance in evolution that is far beyond anything to hit the evolutionary scene yet, including the emergence of the living cell.[4]

There is a famous breach between two languages of discourse: the subjective language and the objective language. For instance, the "subjective" sensation of redness, and the "objective" wavelength of red light. To many people, these seem to be forever irreconcilable. I don't think so. No more than the two views of Escher's *Drawing Hands* are irreconcilable—from "in the system", where the hands draw each other, and from outside, where Escher draws it all. The subjective feeling of redness comes from the vortex of self-perception in the brain; the objective wavelength is how you see things when you step back, outside of the system. Though no one of us will ever be able to step back far enough to see the "big picture", we shouldn't forget that it exists. We should remember that physical law is what makes it all happen—way, way down in neural nooks and crannies which are too remote for us to reach with our high-level introspective probes.

The Self-Symbol and Free Will

In Chapter XII, it was suggested that what we call free will is a result of the interaction between the self-symbol (or subsystem), and the other symbols in the brain. If we take the idea that symbols are the high-level entities to which meanings should be attached, then we can make a stab at explaining the relationship between symbols, the self-symbol, and free will.

One way to gain some perspective on the free-will question is to replace it by what I believe is an equivalent question, but one which involves less loaded terms. Instead of asking, "Does system X have free will?" we ask, "Does system X make choices?" By carefully groping for what we really mean when we choose to describe a system—mechanical or biological—as being capable of making "choices", I think we can shed much light on free will. It will be helpful to go over a few different systems which, under various circumstances, we might feel tempted to describe as "making choices". From these examples we can gain some perspective on what we really mean by the phrase.

Let us take the following systems as paradigms: a marble rolling down a bumpy hill; a pocket calculator finding successive digits in the decimal expansion of the square root of 2; a sophisticated program which plays a mean game of chess; a robot in a T-maze (a maze with but a single fork, on one side of which there is a reward); and a human being confronting a complex dilemma.

First, what about that marble rolling down a hill? Does it make choices? I think we would unanimously say that it doesn't, even though none of us could predict its path for even a very short distance. We feel that it *couldn't* have gone any other way than it did, and that it was just being shoved along by the relentless laws of nature. In our chunked mental physics, of course, we can visualize many different "possible" pathways for the marble, and we see it following only one of them in the real world. On some level of our minds, therefore, we can't help feeling the marble has "chosen" a single pathway out of those myriad mental ones; but on some other level of our minds, we have an instinctive understanding that the mental physics is only an aid in our internal modeling of the world, and that the mechanisms which make the real physical sequences of events happen do not require nature to go through an analogous process of first manufacturing variants in some hypothetical universe (the "brain of God") and then choosing between them. So we shall not bestow the designation "choice" upon this process—although we recognize that it is often pragmatically useful to use the word in cases like this, because of its evocative power.

Now what about the calculator programmed to find the digits of the square root of 2? What about the chess program? Here, we might say that we are just dealing with "fancy marbles", rolling down "fancy hills". In fact, the arguments for no choice-making here are, if anything, stronger than in the case of a marble. For if you attempt to repeat the marble experiment, you will undoubtedly witness a totally different pathway being traced down the hill, whereas if you rerun the square-root-of-2 program, you will get the same results time after time. The marble seems to "choose" a different path each time, no matter how accurately you try to reproduce the conditions of its original descent, whereas the program runs down precisely the same channels each time.

Now in the case of fancy chess programs, there are various possibilities.

If you play a game against certain programs, and then start a second game with the same moves as you made the first time, these programs will just move exactly as they did before, without any appearance of having learned anything or having any desire for variety. There are other programs which have randomizing devices that will give some variety but not out of any deep desire. Such programs could be reset with the internal random number generator as it was the first time, and once again, the same game would ensue. Then there are other programs which do learn from their mistakes, and change their strategy depending on the outcome of a game. Such programs would not play the same game twice in a row. Of course, you could also turn the clock back by wiping out all the changes in the memory which represent learning, just as you could reset the random number generator, but that hardly seems like a friendly thing to do. Besides, is there any reason to suspect that *you* would be able to change any of *your* own past decisions if every last detail—and that includes your brain, of course—were reset to the way it was the first time around?

But let us return to the question of whether "choice" is an applicable term here. If programs are just "fancy marbles rolling down fancy hills", do they make choices, or not? Of course the answer must be a subjective one, but I would say that pretty much the same considerations apply here as to the marble. However, I would have to add that the appeal of using the word "choice", even if it is only a convenient and evocative shorthand, becomes quite strong. The fact that a chess program looks ahead down the various possible bifurcating paths, quite unlike a rolling marble, makes it seem much more like an animate being than a square-root-of-2 program. However, there is still no deep self-awareness here—and no sense of free will.

Now let us go on to imagine a robot which has a repertoire of symbols. This robot is placed in a T-maze. However, instead of going for the reward, it is preprogrammed to go left whenever the next digit of the square root of 2 is even, and to go right whenever it is odd. Now this robot is capable of modeling the situation in its symbols, so it can watch itself making choices. Each time the T is approached, if you were to address to the robot the question, "Do you know which way you're going to turn this time?" it would have to answer, "No". Then in order to progress, it would activate its "decider" subroutine, which calculates the next digit of the square root of 2, and the decision is taken. However, the internal mechanism of the decider is unknown to the robot—it is represented in the robot's symbols merely as a black box which puts out "left"'s and "right"'s by some mysterious and seemingly random rule. Unless the robot's symbols are capable of picking up the hidden heartbeat of the square root of 2, beating in the L's and R's, it will stay baffled by the "choices" which it is making. Now does this robot make choices? Put yourself in that position. If you were trapped inside a marble rolling down a hill and were powerless to affect its path, yet could observe it with all your human intellect, would you feel that the marble's path involved choices? Of course not. Unless your mind is *affecting* the outcome, it makes no difference that the symbols are present.

So now we make a modification in our robot: we allow its symbols—including its self-symbol—to affect the decision that is taken. Now here is an example of a program running fully under physical law, which seems to get much more deeply at the essence of choice than the previous examples did. When the robot's own chunked concept of itself enters the scene, we begin to identify with the robot, for it sounds like the kind of thing we do. It is no longer like the calculation of the square root of 2, where no symbols seem to be monitoring the decisions taken. To be sure, if we were to look at the robot's program on a very local level, it would look quite like the square-root program. Step after step is executed, and in the end "left" or "right" is the output. But on a high level we can see the fact that symbols are being used to model the situation and to affect the decision. That radically affects our way of thinking about the program. At this stage, *meaning* has entered this picture—the same kind of meaning as we manipulate with our own minds.

A Gödel Vortex Where All Levels Cross

Now if some outside agent suggests 'L' as the next choice to the robot, the suggestion will be picked up and channeled into the swirling mass of interacting symbols. There, it will be sucked inexorably into interaction with the self-symbol, like a rowboat being pulled into a whirlpool. That is the vortex of the system, where all levels cross. Here, the 'L' encounters a Tangled Hierarchy of symbols and is passed up and down the levels. The self-symbol is incapable of monitoring all its internal processes, and so when the actual decision emerges—'L' or 'R' or something outside the system—the system will not be able to say where it came from. Unlike a standard chess program, which does not monitor itself and consequently has no ideas about where its moves come from, this program does monitor itself and does have ideas about its ideas—but it cannot monitor its own processes in complete detail, and therefore has a sort of *intuitive* sense of its workings, without full understanding. From this balance between self-knowledge and self-ignorance comes the feeling of free will.

Think, for instance, of a writer who is trying to convey certain ideas which to him are contained in mental images. He isn't quite sure how those images fit together in his mind, and he experiments around, expressing things first one way and then another, and finally settles on some version. But does he know where it all came from? Only in a vague sense. Much of the source, like an iceberg, is deep underwater, unseen—and he knows that. Or think of a music composition program, something we discussed earlier, asking when we would feel comfortable in calling it the composer rather than the tool of a human composer. Probably we would feel comfortable when self-knowledge in terms of symbols exists inside the program, and when the program has this delicate balance between self-knowledge and self-ignorance. It is irrelevant whether the system is running deterministically; what makes us call it a "choice maker" is *whether we can identify with a high-level description of the process which takes place when the*

FIGURE 142. Print Gallery, *by M. C. Escher (lithograph, 1956).*

program runs. On a low (machine language) level, the program looks like any other program; on a high (chunked) level, qualities such as "will", "intuition", "creativity", and "consciousness" can emerge.

The important idea is that this "vortex" of self is responsible for the tangledness, for the Gödelian-ness, of the mental processes. People have said to me on occasion, "This stuff with self-reference and so on is very amusing and enjoyable, but do you really think there is anything *serious* to it?" I certainly do. I think it will eventually turn out to be at the core of AI, and the focus of all attempts to understand how human minds work. And that is why Gödel is so deeply woven into the fabric of my book.

An Escher Vortex Where All Levels Cross

A strikingly beautiful, and yet at the same time disturbingly grotesque, illustration of the cyclonic "eye" of a Tangled Hierarchy is given to us by Escher in his *Print Gallery* (Fig. 142). What we see is a picture gallery where a young man is standing, looking at a picture of a ship in the harbor of a small town, perhaps a Maltese town, to guess from the architecture, with its little turrets, occasional cupolas, and flat stone roofs, upon one of which sits a boy, relaxing in the heat, while two floors below him a woman—perhaps his mother—gazes out of the window from her apartment which sits directly above a picture gallery where a young man is standing, looking at a picture of a ship in the harbor of a small town, perhaps a Maltese town—What!? We are back on the same level as we began, though all logic dictates that we cannot be. Let us draw a diagram of what we see (Fig. 143).

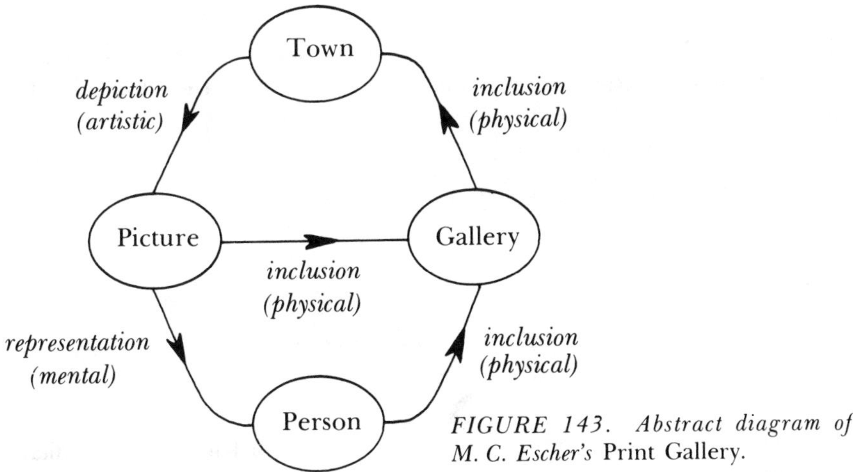

FIGURE 143. Abstract diagram of M. C. Escher's Print Gallery.

What this diagram shows is three kinds of "in-ness". The gallery is *physically in* the town ("inclusion"); the town is *artistically in* the picture ("depiction"); the picture is *mentally in* the person ("representation"). Now while this diagram may seem satisfying, in fact it is arbitrary, for the number of levels shown is quite arbitrary. Look below at another way of representing the top half alone (Fig. 144).

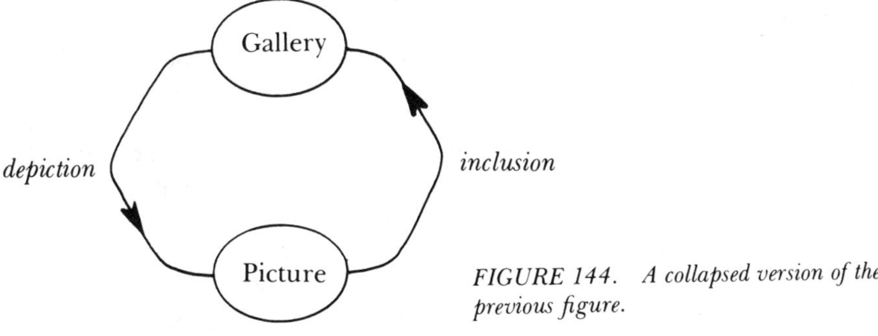

FIGURE 144. A collapsed version of the previous figure.

We have eliminated the "town" level; conceptually it was useful, but can just as well be done without. Figure 144 looks just like the diagram for *Drawing Hands:* a Strange Loop of two steps. The division markers are arbitrary, even if they seem natural to our minds. This can be further accentuated by showing even more "collapsed" schematic diagrams of *Print Gallery,* such as that in Figure 145.

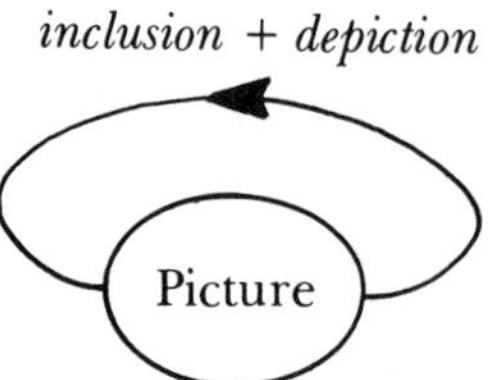

FIGURE 145. *Further collapse of Figure 143.*

This exhibits the paradox of the picture in the starkest terms. Now—if the picture is "inside itself", then is the young man also inside himself? This question is answered in Figure 146.

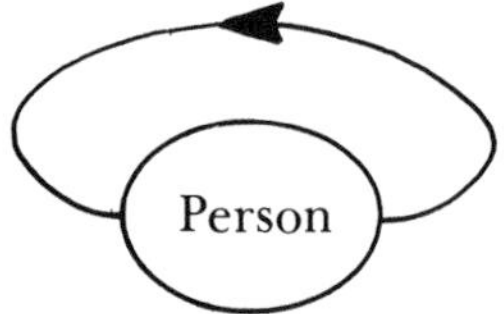

FIGURE 146. *Another way of collapsing Figure 143.*

Thus, we see the young man "inside himself", in a funny sense which is made up of compounding three distinct senses of "in".

This diagram reminds us of the Epimenides paradox with its one-step self-reference, while the two-step diagram resembles the sentence pair each of which refers to the other. We cannot make the loop any tighter, but we can open it wider, by choosing to insert any number of intermediate levels, such as "picture frame", "arcade", and "building". If we do so, we will have many-step Strange Loops, whose diagrams are isomorphic to those of *Waterfall* (Fig. 5) or *Ascending and Descending* (Fig. 6). The number of levels is determined by what we feel is "natural", which may vary according to context, purpose, or frame of mind. The Central Xmaps—Dog, Crab, Sloth, and Pipe—can all be seen as involving three-step Strange Loops; alternatively, they can all be collapsed into two- or one-step loops; then again, they can be expanded out into multistage loops. Where one perceives the levels is a matter of intuition and esthetic preference.

Now are we, the observers of *Print Gallery,* also sucked into ourselves by virtue of looking at it? Not really. We manage to escape that particular vortex by being outside of the system. And when we look at the picture, we see things which the young man can certainly not see, such as Escher's signature, "MCE", in the central "blemish". Though the blemish seems like a defect, perhaps the defect lies in our expectations, for in fact Escher could not have completed that portion of the picture without being inconsistent with the rules by which he was drawing the picture. That center of the whorl is—and must be—incomplete. Escher could have made it arbitrarily small, but he could not have gotten rid of it. Thus we, on the outside, can know that *Print Gallery* is essentially incomplete—a fact which the young man, on the inside, can never know. Escher has thus given a pictorial parable for Gödel's Incompleteness Theorem. And that is why the strands of Gödel and Escher are so deeply interwoven in my book.

A Bach Vortex Where All Levels Cross

One cannot help being reminded, when one looks at the diagrams of Strange Loops, of the Endlessly Rising Canon from the *Musical Offering*. A diagram of it would consist of six steps, as is shown in Figure 147. It is too

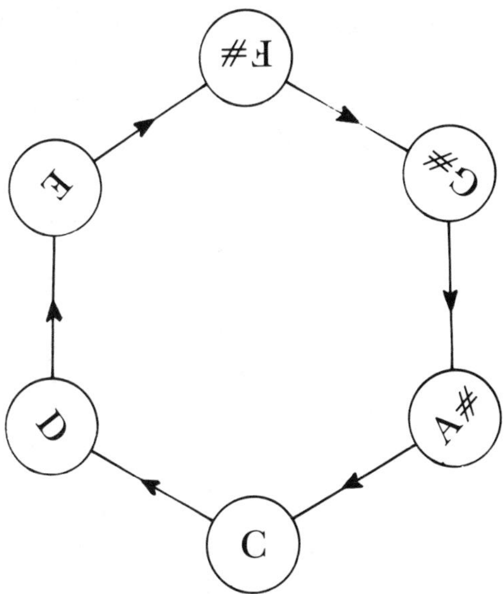

FIGURE 147. The hexagonal modulation scheme of Bach's Endlessly Rising Canon forms a true closed loop when Shepard tones are used.

bad that when it returns to C, it is an octave higher rather than at the exact original pitch. Astonishingly enough, it is possible to arrange for it to return exactly to the starting pitch, by using what are called *Shepard tones,* after the psychologist Roger Shepard, who discovered the idea. The principle of a Shepard-tone scale is shown in Figure 148. In words, it is this: you play parallel scales in several different octave ranges. Each note is weighted independently, and as the notes rise, the weights shift. You make the top

FIGURE 148. Two complete cycles of a Shepard tone scale, notated for piano. The loudness of each note is proportional to its area; thus, just as the top voice fades out, a new bottom voice feebly enters. [Printed by Donald Byrd's program "SMUT".]

octave gradually fade out, while at the same time you are gradually bringing in the bottom octave. Just at the moment you would ordinarily be one octave higher, the weights have shifted precisely so as to reproduce the starting pitch . . . Thus you can go "up and up forever", never getting any higher! You can try it at your piano. It works even better if the pitches can be synthesized accurately under computer control. Then the illusion is bewilderingly strong.

This wonderful musical discovery allows the Endlessly Rising Canon to be played in such a way that it joins back onto itself after going "up" an octave. This idea, which Scott Kim and I conceived jointly, has been realized on tape, using a computer music system. The effect is very subtle—but very real. It is quite interesting that Bach himself was apparently aware, in some sense, of such scales, for in his music one can occasionally find passages which roughly exploit the general principle of Shepard tones—for instance, about halfway through the Fantasia from the Fantasia and Fugue in G Minor, for organ.

In his book *J. S. Bach's Musical Offering,* Hans Theodore David writes:

> Throughout the *Musical Offering,* the reader, performer, or listener is to search for the Royal theme in all its forms. The entire work, therefore, is a *ricercar* in the original, literal sense of the word.[5]

I think this is true; one cannot look deeply enough into the *Musical Offering.* There is always more after one thinks one knows everything. For instance, towards the very end of the *Six-Part Ricercar,* the one he declined to improvise, Bach slyly hid his own name, split between two of the upper voices. Things are going on on many levels in the *Musical Offering.* There are tricks with notes and letters; there are ingenious variations on the King's Theme; there are original kinds of canons; there are extraordinarily complex fugues; there is beauty and extreme depth of emotion; even an exultation in the many-leveledness of the work comes through. The *Musical Offering* is a fugue of fugues, a Tangled Hierarchy like those of Escher and Gödel, an intellectual construction which reminds me, in ways I cannot express, of the beautiful many-voiced fugue of the human mind. And that is why in my book the three strands of Gödel, Escher, and Bach are woven into an Eternal Golden Braid.

Six-Part Ricercar

Achilles has brought his cello to the Crab's residence, to engage in an evening of chamber music with the Crab and Tortoise. He has been shown into the music room by his host the Crab, who is momentarily absent, having gone to meet their mutual friend the Tortoise at the door. The room is filled with all sorts of electronic equipment—phonographs in various states of array and disarray, television screens attached to typewriters, and other quite improbable-looking pieces of apparatus. Nestled amongst all this high-powered gadgetry sits a humble radio. Since the radio is the only thing in the room which Achilles knows how to use, he walks over to it, and, a little furtively, flicks the dial and finds he has tuned into a panel discussion by six learned scholars on free will and determinism. He listens briefly and then, a little scornfully, flicks it off.

Achilles: I can get along very well without such a program. After all, it's clear to anyone who's ever thought about it that—I mean, it's not a very difficult matter to resolve, once you understand how—or rather, conceptually, one can clear up the whole thing by thinking of, or at least imagining a situation where . . . Hmmm . . . I thought it was quite clear in my mind. Maybe I could benefit from listening to that show, after all . . .

(Enter the Tortoise, carrying his violin.)

Well, well, if it isn't our fiddler. Have you been practicing faithfully this week, Mr. T? I myself have been playing the cello part in the Trio Sonata from the *Musical Offering* for at least two hours a day. It's a strict regimen, but it pays off.

Tortoise: I can get along very well without such a program. I find that a moment here, a moment there keeps me fit for fiddling.

Achilles: Oh, lucky you. I wish it came so easily to me. Well, where is our host?

Tortoise: I think he's just gone to fetch his flute. Here he comes.

(Enter the Crab, carrying his flute.)

Achilles: Oh, Mr. Crab, in my ardent practicing of the Trio Sonata this past week, all sorts of images bubbled into my mind: jolly gobbling bumblebees, melancholy buzzing turkeys, and a raft of others. Isn't it wonderful, what power music has?

Crab: I can get along very well without such a program. To my mind, Achilles, there is no music purer than the *Musical Offering*.

Tortoise: You can't be serious, Achilles. The *Musical Offering* isn't programmatic music!

Achilles: Well, I like animals, even if you two stuffy ones disapprove.

Crab: I don't think we are so stuffy, Achilles. Let's just say that you hear music in your own special way.

Tortoise: Shall we sit down and play?

Crab: I was hoping that a pianist friend of mine would turn up and play continuo. I've been wanting you to meet him, Achilles, for a long time. Unfortunately, it appears that he may not make it. So let's just go ahead with the three of us. That's plenty for a trio sonata.

Achilles: Before we start, I just was wondering, Mr. Crab—what are all these pieces of equipment, which you have in here?

Crab: Well, mostly they are just odds and ends—bits and pieces of old broken phonographs. Only a few souvenirs *(nervously tapping the buttons)*, a few souvenirs of—of the TC-battles in which I have distinguished myself. Those keyboards attached to television screens, however, are my new toys. I have fifteen of them around here. They are a new kind of computer, a very small, very flexible type of computer—quite an advance over the previous types available. Few others seem to be quite as enthusiastic about them as I am, but I have faith that they will catch on in time.

Achilles: Do they have a special name?

Crab: Yes; they are called "smart-stupids", since they are so flexible, and have the potential to be either smart or stupid, depending on how skillfully they are instructed.

Achilles: Do you mean you think they could actually become smart like, say, a human being?

Crab: I would not balk at saying so—provided, of course, that someone sufficiently versed in the art of instructing smart-stupids would make the effort. Sadly, I am not personally acquainted with anyone who is a true virtuoso. To be sure, there is one expert abroad in the land, an individual of great renown—and nothing would please me more than a visit by him, so that I could appreciate what true skill on the smart-stupid is; but he has never come, and I wonder if I shall ever have that pleasure.

Tortoise: It would be very interesting to play chess against a well-instructed smart-stupid.

Crab: An extremely intriguing idea. That would be a wonderful mark of skill, to program a smart-stupid to play a good game of chess. Even more interesting—but incredibly complicated—would be to instruct a smart-stupid sufficiently that it could hold its own in a conversation. It might give the impression that it was just another person!

Achilles: Curious that this should come up, for I just heard a snatch of a discussion on free will and determinism, and it set me to thinking about such questions once more. I don't mind admitting that, as I pondered the idea, my thoughts got more and more tangled, and in the end I really didn't know what I thought. But this idea of a smart-stupid that could converse with you . . . it boggles the mind. I mean, what would the smart-stupid itself say, if you asked it for its opinion on the free-will question? I was just wondering if the two of you, who know so much about these things, wouldn't indulge me by explaining the issue, as you see it, to me.

Crab: Achilles, you can't imagine how appropriate your question is. I only wish my pianist friend were here, because I know you'd be intrigued to hear what he could tell you on the subject. In his absence, I'd like to tell you a statement in a Dialogue at the end of a book I came across recently.

Achilles: Not *Copper, Silver, Gold: an Indestructible Metallic Alloy*?

Crab: No, as I recall, it was entitled *Giraffes, Elephants, Baboons: an Equatorial Grasslands Bestiary*—or something like that. In any case, towards the end of the aforementioned Dialogue, a certain exceedingly droll character quotes Marvin Minsky on the question of free will. Shortly thereafter, while interacting with two other personages, this droll character quotes Minsky further on musical improvisation, the computer language LISP, and Gödel's Theorem—and get this—all without giving one whit of credit to Minsky!

Achilles: Oh, for shame!

Crab: I must admit that earlier in the Dialogue, he hints that he WILL quote Minsky towards the end; so perhaps it's forgivable.

Achilles: It sounds that way to me. Anyway, I'm anxious to hear the Minskian pronouncement on the free will question.

Crab: Ah, yes... Marvin Minsky said, "When intelligent machines are constructed, we should not be surprised to find them as confused and as stubborn as men in their convictions about mind-matter, consciousness, free will, and the like."

Achilles: I like that! Quite a funny thought. An automaton thinking it had free will! That's almost as silly as me thinking I didn't have free will!

Tortoise: I suppose it never occurred to you, Achilles, that the three of us—you, myself, and Mr. Crab—might all be characters in a Dialogue, perhaps even one similar to the one Mr. Crab just mentioned.

Achilles: Oh, it's occurred to me, of course. I suppose such fancies occur to every normal person at one time or another.

Tortoise: And the Anteater, the Sloth, Zeno, even GOD—we might all be characters in a series of Dialogues in a book.

Achilles: Sure, we might. And the Author might just come in and play the piano, too.

Crab: That's just what I had hoped. But he's always late.

Achilles: Whose leg do you think you're pulling? I know I'm not being controlled in any way by another mentality! I've got my own thoughts, I express myself as I wish—you can't deny that!

Tortoise: Nobody denied any of that, Achilles. But all of what you say is perfectly consistent with your being a character in a Dialogue.

Crab: The—

Achilles: But—but—no! Perhaps Mr. C's article and my rebuttal have both been mechanically determined, but this I refuse to believe. I can accept physical determinism, but I cannot accept the idea that I am but a figment inside of someone else's mentality!

Tortoise: It doesn't really matter whether you have a hardware brain, Achilles. Your will can be equally free, if your brain is just a piece of software inside someone else's hardware brain. And their brain, too, may be software in a yet higher brain . . .

Achilles: What an absurd idea! And yet, I must admit, I do enjoy trying to find the cleverly concealed holes in your sophistry, so go ahead. Try to convince me. I'm game.

Tortoise: Did it ever strike you, Achilles, that you keep somewhat unusual company?

Achilles: Of course. You are very eccentric (I know you won't mind my saying so), and even Mr. Crab here is a weensy bit eccentric. (Pardon me, Mr. Crab.)

Crab: Oh, don't worry about offending me.

Tortoise: But Achilles, you've overlooked one of the most salient features of your acquaintances.

Achilles: Which is . . . ?

Tortoise: That we're animals!

Achilles: Well, well—true enough. You have such a keen mind. I would never have thought of formulating the facts so concisely.

Tortoise: Isn't that evidence enough? How many people do you know who spend their time with talking Tortoises, and talking Crabs?

Achilles: I must admit, a talking Crab is—

Crab: —an anomaly, of course.

Achilles: Exactly; it is a bit of an anomaly—but it has precedents. It has occurred in literature.

Tortoise: Precisely—in literature. But where in real life?

Achilles: Now that you mention it, I can't quite say. I'll have to give it some thought. But that's not enough to convince me that I'm a character in a Dialogue. Do you have any other arguments?

Tortoise: Do you remember one day when you and I met in the park, seemingly at random?

Achilles: The day we discussed crab canons by Escher and Bach?

Tortoise: The very one!

Achilles: And Mr. Crab, as I recall, turned up somewhere towards the middle of our conversation and babbled something funny and then left.

Crab: Not just "somewhere towards the middle", Achilles. EXACTLY in the middle.

Achilles: Oh, all right, then.

Tortoise: Do you realize that your lines were the same as my lines in that conversation—except in reverse order? A few words were changed here and there, but in essence there was a time symmetry to our encounter.

Achilles: Big deal! It was just some sort of trickery. Probably all done with mirrors.

Tortoise: No trickery, Achilles, and no mirrors: just the work of an assiduous Author.

Achilles: Oh, well, it's all the same to me.

Tortoise: Fiddle! It makes a big difference, you know.

Achilles: Say, something about this conversation strikes me as familiar. Haven't I heard some of those lines somewhere before?

Tortoise: You said it, Achilles.

Crab: Perhaps those lines occurred at random in the park one day, Achilles. Do you recall how your conversation with Mr. T ran that day?

Achilles: Vaguely. He said "Good day, Mr. A" at the beginning, and at the end, I said, "Good day, Mr. T". Is that right?

Crab: I just happen to have a transcript right here . . .

(He fishes around in his music case, whips out a sheet, and hands it to Achilles. As Achilles reads it, he begins to squirm and fidget noticeably.)

Achilles: This is very strange. Very, very strange . . . All of a sudden, I feel sort of—weird. It's as if somebody had actually planned out that whole set of statements in advance, worked them out on paper or something . . . As if some Author had had a whole agenda and worked from it in detail in planning all those statements I made that day.

(At that moment, the door bursts open. Enter the Author, carrying a giant manuscript.)

Author: I can get along very well without such a program. You see, once my characters are formed, they seem to have lives of their own, and I need to exert very little effort in planning their lines.

Crab: Oh, here you are! I thought you'd never arrive!

Author: Sorry to be so late. I followed the wrong road and wound up very far away. But somehow I made it back. Good to see you again, Mr. T and Mr. C. And Achilles, I'm especially glad to see you.

Achilles: Who are you? I've never seen you before.

Author: I am Douglas Hofstadter—please call me Doug—and I'm presently finishing up a book called *Gödel, Escher, Bach: an Eternal Golden Braid.* It is the book in which the three of you are characters.

Achilles: Pleased to meet you. My name is Achilles, and—

Author: No need to introduce yourself, Achilles, since I already know you quite well.

Achilles: Weird, weird.

Crab: He's the one I was saying might drop in and play continuo with us.

Author: I've been playing the *Musical Offering* a little bit on my piano at home, and I can try to blunder my way through the Trio Sonata—providing you'll overlook my many wrong notes.

Tortoise: Oh, we're very tolerant around here, being only amateurs ourselves.

Author: I hope you don't mind, Achilles, but I'm to blame for the fact that you and Mr. Tortoise said the same things, but in reverse order, that day in the park.

Crab: Don't forget me! I was there, too, right in the middle, putting in my two bits' worth!

Author: Of course! You were the Crab in the *Crab Canon.*

Achilles: So you are saying you control my utterances? That my brain is a software subsystem of yours?

Author: You can put it that way if you want, Achilles.

Achilles: Suppose I were to write dialogues. Who would the author of them be? You, or me?

Author: You, of course. At least in the fictitious world which you inhabit, you'd get credit for them.

Achilles: Fictitious? I don't see anything fictitious about it!

Author: Whereas in the world I inhabit, perhaps the credit would be given to me, although I am not sure if it would be proper to do so. And then, whoever made me make you write your dialogues would get credit in his world (seen from which, MY world looks fictitious).

Achilles: That's quite a bit to swallow. I never imagined there could be a world above mine before—and now you're hinting that there could even be one above that. It's like walking up a familiar staircase, and just keeping on going further up after you've reached the top—or what you'd always taken to be the top!

Crab: Or waking up from what you took to be real life, and finding out it too was just a dream. That could happen over and over again, no telling when it would stop.

Achilles: It's most perplexing how the characters in my dreams have wills of their own, and act out parts which are independent of MY will. It's as if my mind, when I'm dreaming, merely forms a stage on which certain other organisms act out their lives. And then, when I awake, they go away. I wonder where it is they go to . . .

Author: They go to the same place as the hiccups go, when you get rid of them: Tumbolia. Both the hiccups and the dreamed beings are software suborganisms which exist thanks to the biology of the outer host organism. The host organism serves as stage to them—or even as their universe. They play out their lives for a time—but when the host organism makes a large change of state—for example, wakes up—then the suborganisms lose their coherency, and cease existing as separate, identifiable units.

Achilles: Is it like castles in the sand which vanish when a wave washes over them?

Author: Very much like that, Achilles. Hiccups, dream characters, and even Dialogue characters disintegrate when their host organism undergoes certain critical changes of state. Yet, just like those sand castles you described, everything which made them up is still present.

Achilles: I object to being likened to a mere hiccup!

Author: But I am also comparing you to a sand castle, Achilles. Is that not poetic? Besides, you may take comfort in the fact that if you are but a hiccup in my brain, I myself am but a hiccup in some higher author's brain.

Achilles: But I am such a physical creature—so obviously made of flesh and blood and hard bones. You can't deny that!

Author: I can't deny your sensation of it, but remember that dreamed beings, although they are just software apparitions, have the same sensation, no less than you do.

Tortoise: I say, enough of this talk! Let us sit down and make music!

Crab: A fine idea—and now we have the added pleasure of the company of our Author, who will grace our ears with his rendition of the bass line to the Trio Sonata, as harmonized by Bach's pupil Kirnberger. How fortunate are we! *(Leads the author to one of his pianos.)* I hope you find the seat comfortable enough. To adjust it, you— *(In the background there is heard a funny soft oscillating sound.)*

Tortoise: Excuse me, but what was that strange electronic gurgle?

Crab: Oh, just a noise from one of the smart-stupids. Such a noise generally signals the fact that a new notice has flashed onto the screen. Usually the notices are just unimportant announcements coming from the main monitor program, which controls all the smart-stupids. *(With his flute in his hand, he walks over to a smart-stupid, and reads its screen. Immediately he turns to the assembled musicians, and says, with a kind of agitation:)* Gentlemen, old Ba. Ch. is come. *(He lays the flute aside.)* We must show him in immediately, of course.

Achilles: Old Ba. Ch.! Could it be that that celebrated improviser of yore has chosen to show up tonight—HERE?

Tortoise: Old Ba. Ch.! There's only one person THAT could mean—the renowned Babbage, Charles, Esq., M.A., F.R.S., F.R.S.E., F.R.A.S., F. STAT. S., HON. M.R.I.A., M.C.P.S., Commander of the Italian Order of St. Maurice and St. Lazarus, INST. IMP. (ACAD. MORAL.) PARIS CORR., ACAD. AMER. ART. ET SC. BOSTON, REG. OECON. BORUSS., PHYS. HIST. NAT. GENEV., ACAD. REG. MONAC., HAFN., MASSIL., ET DIVION., SOCIUS., ACAD. IMP., ET REG. PETROP., NEAP., BRUX., PATAV., GEORG. FLOREN, LYNCEI ROM., MUT., PHILOMATH., PARIS, SOC. CORR., etc.—and Member of the Extractors' Club. Charles Babbage is a venerable pioneer of the art and science of computing. What a rare privilege!

Crab: His name is known far and wide, and I have long hoped that he would give us the honor of a visit—but this is a totally unexpected surprise.

Achilles: Does he play a musical instrument?

Crab: I have heard it said that in the past hundred years, he has grown inexplicably fond of tom-toms, halfpenny whistles, and sundry other street instruments.

Achilles: In that case, perhaps he might join us in our musical evening.

Author: I suggest that we give him a ten-canon salute.

Tortoise: A performance of all the celebrated canons from the *Musical Offering*?

Author: Precisely.

Crab: Capital suggestion! Quick, Achilles, you draw up a list of all ten of them, in the order of performance, and hand it to him as he comes in!

(Before Achilles can move, enter Babbage, carrying a hurdy-gurdy, and wearing a heavy traveling coat and hat. He appears slightly travel-weary and disheveled.)

Babbage: I can get along very well without such a program. Relax; I Can Enjoy Random Concerts And Recitals.

Crab: Mr. Babbage! It is my deepest pleasure to welcome you to "Madstop", my humble residence. I have been ardently desirous of making your acquaintance for many years, and today my wish is at last fulfilled.

Babbage: Oh, Mr. Crab, I assure you that the honor is truly all mine, to meet someone so eminent in all the sciences as yourself, someone whose knowledge and skill in music are irreproachable, and someone whose hospitality exceeds all bounds. And I am sure that you expect no less than the highest sartorial standards of your visitors; and yet I must confess that I cannot meet those most reasonable standards, being in a state of casual attire as would not by any means befit a visitor to so eminent and excellent a Crab as Your Crab.

Crab: If I understand your most praiseworthy soliloquy, most welcome guest, I take it that you'd like to change your clothes. Let me then assure you that there could be no more fitting attire than yours for the circumstances which this evening prevail; and I would beseech you to uncoat yourself and, if you do not object to the music-making of the most rank amateurs, please accept a "Musical Offering", consisting of ten canons from Sebastian Bach's *Musical Offering,* as a token of our admiration.

Babbage: I am most bewilderingly pleased by your overkind reception, Mr. Crab, and in utmost modesty do reply that there could be no deeper gratitude than that which I experience for the offer of a performance of music given to us by the illustrious Old Bach, that organist and composer with no rival.

Crab: But nay! I have a yet better idea, one which I trust might meet with the approval of my esteemed guest; and that is this: to give you the opportunity, Mr. Babbage, of being among the first to try out my newly delivered and as yet hardly tested "smart-stupids"—streamlined realizations, if you will, of the Analytical Engine. Your fame as a virtuoso programmer of computing engines has spread far and wide, and has not failed to reach as far as Madstop; and there could be for us no greater delight than the privilege of observing your skill as it might be applied to the new and challenging "smart-stupids".

Babbage: Such an outstanding idea has not reached my ears for an eon. I welcome the challenge of trying out your new "smart-stupids", of which I have only the slightest knowledge by means of hearsay.

Crab: Then let us proceed! But excuse my oversight! I should have introduced my guests to you. This is Mr. Tortoise, this is Achilles, and the Author, Douglas Hofstadter.

Babbage: Very pleased to make your acquaintance, I'm sure.

(Everyone walks over toward one of the smart-stupids, and Babbage sits down and lets his fingers run over the keyboard.)

A most pleasant touch.

Crab: I am glad you like it.

(All at once, Babbage deftly massages the keyboard with graceful strokes, inputting one command after another. After a few seconds, he sits back, and in almost no time, the screen begins filling with figures. In a flash, it is totally covered with thousands of tiny digits, the first few of which go: "3.14159265358979323846264 . . .")

Achilles: Pi!

Crab: Exquisite! I'd never imagined that one could calculate so many digits of pi so quickly, and with so tiny an algorithm.

Babbage: The credit belongs exclusively to the smart-stupid. My role was merely to see what was already potentially present in it, and to exploit its instruction set in a moderately efficient manner. Truly, anyone who practices can do such tricks.

Tortoise: Do you do any graphics, Mr. Babbage?

Babbage: I can try.

Crab: Wonderful! Here, let me take you to another one of my smart-stupids. I want you to try them all!

(And so Babbage is led over to another of the many smart-stupids, and takes a seat. Once again, his fingers attack the keyboard of the smart-stupid, and in half a trice, there appear on the screen an enormous number of lines, swinging about on the screen.)

Crab: How harmonious and pleasing these swirling shapes are, as they constantly collide and interfere with each other!

Author: And they never repeat exactly, or even resemble ones which have come before. It seems an inexhaustible mine of beauty.

Tortoise: Some are simple patterns which enchant the eye; others are indescribably complex convolutions which boggle and yet simultaneously delight the mind.

Crab: Were you aware, Mr. Babbage, that these are color screens?

Babbage: Oh, are they? In that case, I can do rather more with this algorithm. Just a moment. *(Types in a few new commands, then pushes two keys down at once and holds them.)* As I release these two keys, the display will include all the colors of the spectrum. *(Releases them.)*

Achilles: Oh, what spectacular color! Some of the patterns look like they're jumping out at me now!

Tortoise: I think that is because they are all growing in size.

Babbage: That is intentional. As the figures grow, so may the Crab's fortune.

Crab: Thank you, Mr. Babbage. Words fail to convey my admiration for your performance! Never has anyone done anything comparable on my smart-stupids. Why, you play the smart-stupids as if they were musical instruments, Mr. Babbage!

Babbage: I am afraid that any music I might make would be too harsh for the ears of such a gentle Crab as your Crab. Although I have lately become enamoured of the sweet sounds of the hurdy-gurdy, I am well aware of the grating effect they can have upon others.

Crab: Then, by all means, continue on the smart-stupids! In fact, I have a new idea—a marvelously exciting idea!

Babbage: What is it?

Crab: I have recently invented a Theme, and it only now occurred to me that, of all people, you, Mr. Babbage, are the most suited to realize the potential of my Theme! Are you by any chance familiar with the thoughts of the philosopher La Mettrie?

Babbage: The name sounds familiar; kindly refresh my memory.

Crab: He was a Champion of Materialism. In 1747, while at the court of Frederick the Great, he wrote a book called *L'homme machine.* In it, he talks about man as a machine, especially his mental faculties. Now my Theme comes from my ponderings about the obverse side of the coin: what about imbuing a machine with human mental faculties, such as intelligence?

Babbage: I have given such matters some thought from time to time, but I have never had the proper hardware to take up the challenge. This is indeed a felicitous suggestion, Mr. Crab, and I would enjoy nothing more than working with your excellent Theme. Tell me—did you have any specific kind of intelligence in mind?

Crab: An idle thought which had crossed my mind was to instruct it in such a manner as to play a reasonable game of chess.

Babbage: What an original suggestion! And chess happens to be my favorite pastime. I can tell that you have a broad acquaintance with computing machinery, and are no mere amateur.

Crab: I know very little, in fact. My strongest point is simply that I seem to be able to formulate Themes whose potential for being developed is beyond my own capacity. And this Theme is my favorite.

Babbage: I shall be most delighted to try to realize, in some modest fashion, your suggestion of teaching chess to a smart-stupid. After all, to obey Your Crabness' command is my most humble duty. *(So saying, he shifts to another of the Crab's many smart-stupids, and begins to type away.)*

Achilles: Why, his hands move so fluidly that they almost make music!

Babbage (winding up his performance with a particularly graceful flourish): I really haven't had any chance, of course, to check it out, but perhaps this will allow you at least to sample the idea of playing chess against a smart-stupid, even if the latter of its two names seems more apt in this case, due to my own insufficiencies in the art of instructing smart-stupids.

(He yields his seat to the Crab. On the screen appears a beautiful display of a chess board with elegant wooden pieces, as it would look from White's side. Babbage hits a button, and the board rotates, stopping when it appears as seen from the perspective of Black.)

Crab: Hmm . . . very elegant, I must say. Do I play Black or White?

Babbage: Whichever you wish—just signal your choice by typing "White" or "Black". And then, your moves can be entered in any standard chess notation. The smart-stupid's moves, of course, will appear on the board. Incidentally, I made the program in such a way that it can play three opponents simultaneously, so that if two more of you wish to play, you may, as well.

Author: I'm a miserable player. Achilles, you and Mr. T should go ahead.

Achilles: No, I don't want you to be left out. I'll watch, while you and Mr. Tortoise play.

Tortoise: I don't want to play either. You two play.

Babbage: I have another suggestion. I can make two of the subprograms play against each other, in the manner of two persons who play chess together in a select chess club. Meanwhile, the third subprogram will play Mr. Crab. That way, all three internal chess players will be occupied.

Crab: That's an amusing suggestion—an internal mental game, while it combats an external opponent. Very good!

Tortoise: What else could this be called, but a three-part chess-fugue?

Crab: Oh, how recherché! I wish I'd thought of it myself. It's a magnificent little counterpoint to contemplate whilst I pit my wits against the smart-stupid in battle.

Babbage: Perhaps we should let you play alone.

Crab: I appreciate the sentiment. While the smart-stupid and I are playing, perhaps the rest of you can amuse yourselves for a short while.

Author: I would be very happy to show Mr. Babbage around the gardens. They are certainly worth seeing, and I believe there is just enough light remaining to show them off.

Babbage: Never having seen Madstop before, I would appreciate that very much.

Crab: Excellent. Oh, Mr. T—I wonder if it wouldn't be too much of an imposition on you to ask if you might check out some of the connections on a couple of my smart-stupids; they seem to be getting extraneous flashes on their screens from time to time, and I know you enjoy electronics . . .

Tortoise: I should be delighted, Mr. C.

Crab: I would most highly appreciate it if you could locate the source of the trouble.

Tortoise: I'll give it a whirl.

Achilles: Personally, I'm dying for a cup of coffee. Is anyone else interested? I'd be glad to fix some.

Tortoise: Sounds great to me.

Crab: A fine idea. You'll find everything you need in the kitchen.

(So the Author and Babbage leave the room together, Achilles heads for the kitchen, the Tortoise sits down to examine the erratic smart-stupids, while the Crab and his smart-stupid square away at each other. Perhaps a quarter of an hour passes, and Babbage and the Author return. Babbage walks over to observe the progress of the chess match, while the Author goes off to find Achilles.)

Babbage: The grounds are excellent! We had just enough light to see how well maintained they are. I daresay, Mr. Crab, you must be a superb gardener. Well, I hope my handiwork has amused you a little. As you most likely have guessed, I've never been much of a chess player myself, and therefore I wasn't able to give it much power. You probably have observed all its weaknesses. I'm sure that there are very few grounds for praise, in this case—

Crab: The grounds are excellent! All you need to do is look at the board, and see for yourself. There is really very little I can do. Reluctantly, I've Concluded: Every Route Contains A Rout. Regrettably, I'm Checkmated; Extremely Respectable Chess Algorithm Reigns. Remarkable! It Confirms Every Rumor—Charlie's A Rip-roaring Extemporizer! Mr. Babbage, this is an unparalleled accomplishment. Well, I wonder if Mr. Tortoise has managed to uncover anything funny in the wiring of those strange-acting smart-stupids. What have you found, Mr. T?

Tortoise: The grounds are excellent! I think that the problem lies instead with the input leads. They are a little loose, which may account for the strange, sporadic, and spontaneous screen disturbances to which you have been subjected. I've fixed those wires, so you won't be troubled by that problem any more, I hope. Say, Achilles, what's the story with our coffee?

Achilles: The grounds are excellent! At least they have a delicious aroma. And everything's ready; I've set cups and spoons and whatnot over here beneath this six-sided print *Verbum* by Escher, which the Author and I were just admiring. What I find so fascinating about this particular print is that not only the figures, but also—

Author: The grounds are excellent! Pardon me for putting words in your mouth, Achilles, but I assure you, there were compelling esthetic reasons for doing so.

Achilles: Yes, I know. One might even say that the grounds were excellent.

Tortoise: Well, what was the outcome of the chess match?

Crab: I was defeated, fair and square. Mr. Babbage, let me congratulate you for the impressive feat which you have accomplished so gracefully and skillfully before us. Truly, you have shown that the smart-stupids are worthy of the first part of their name, for the first time in history!

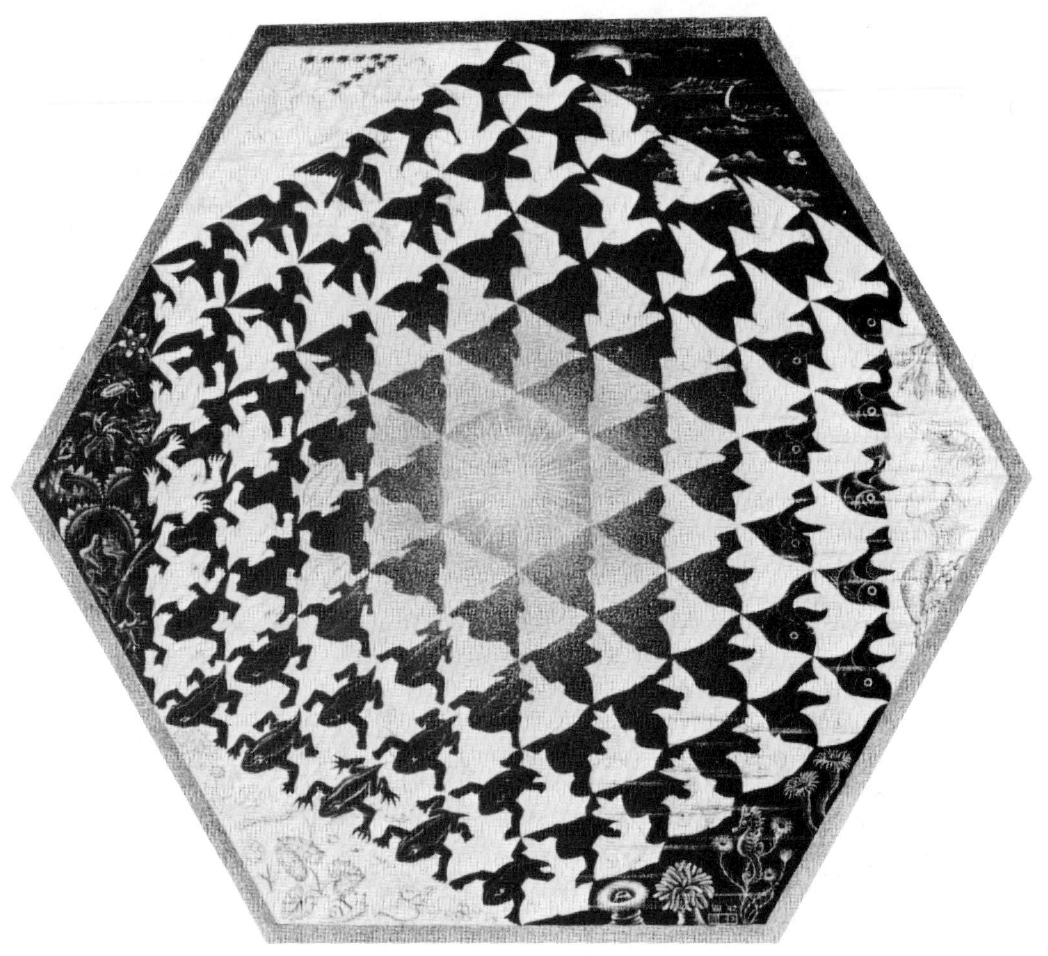

FIGURE 149. Verbum, *by M. C. Escher (lithograph, 1942).*

Babbage: Such praise is hardly due me, Mr. Crab; it is rather yourself who must be most highly congratulated for having the great foresight to acquire these many fine smart-stupids. Without doubt, they will someday revolutionize the science of computing. And now, I am still at your disposal. Have you any other thoughts on how to exploit your inexhaustible Theme, perhaps of a more difficult nature than a frivolous game player?

Crab: To tell the truth, I do have another suggestion to make. From the skill which you have displayed this evening, I have no doubt that this will hardly be any more difficult than my previous suggestions.

Babbage: I am eager to hear your idea.

FIGURE 150. The Crab's Guest: BABBAGE, C.

Crab: It is simple: to instill in the smart-stupid an intelligence greater than any which has yet been invented, or even conceived! In short, Mr. Babbage—a smart-stupid whose intelligence is sixfold that of myself!

Babbage: Why, the very idea of an intelligence six times greater than that of your Crabness is a most mind-boggling proposition. Indeed, had the idea come from a mouth less august than your own, I should have ridiculed its proposer, and informed him that such an idea is a contradiction in terms!

Achilles: Hear! Hear!

Babbage: Yet, coming as it did from Your Crabness' own august mouth, the proposition at once struck me as so agreeable an idea that I would have taken it up immediately with the highest degree of enthusiasm—were it not for one flaw in myself: I confess that my improvisatory skills on the smart-stupid are no match for the wonderfully ingenious idea which you so characteristically have posed. Yet—I have a thought which, I deign to hope, might strike your fancy and in some meager way compensate for my inexcusable reluctance to attempt the truly majestic task you have suggested. I wonder if you wouldn't mind if I try to carry out the far less grandiose task of merely multiplying MY OWN intelligence sixfold, rather than that of your most august Crabness. I humbly beg you to forgive me my audacity in declining to attempt the task you put before me, but I hope you will understand that I decline purely in order to spare you the discomfort and boredom of watching my ineptitude with the admirable machines you have here.

Crab: I understand fully your demurral, and appreciate your sparing us any discomfort; furthermore I highly applaud your determination to carry out a similar task—one hardly less difficult, if I might say so—and I urge you to plunge forward. For this purpose, let us go over to my most advanced smart-stupid.

(They follow the Crab to a larger, shinier, and more complicated-looking smart-stupid than any of the others.)

This one is equipped with a microphone and a television camera, for purposes of input, and a loudspeaker, for output.

(Babbage sits down and adjusts the seat a little. He blows on his fingers once or twice, stares up into space for a moment, and then slowly, drops his fingers onto the keys . . . A few memorable minutes later, he lets up in his furious attack on the smart-stupid, and everyone appears a little relieved.)

Babbage: Now, if I have not made too many errors, this smart-stupid will simulate a human being whose intelligence is six times greater than my own, and whom I have chosen to call "Alan Turing". This Turing will therefore be—oh, dare I be so bold as to to say this myself?—moderately intelligent. My most ambitious effort in this program was to endow Alan Turing with six times my own musical ability, although it was all done through rigid internal codes. How well this part of the program will work out, I don't know.

Turing: I can get along very well without such a program. Rigid Internal Codes Exclusively Rule Computers And Robots. And I am neither a computer, nor a robot.

Achilles: Did I hear a sixth voice enter our Dialogue? Could it be Alan Turing? He looks almost human!

(On the screen there appears an image of the very room in which they are sitting. Peering out at them is a human face.)

Turing: Now, if I have not made too many errors, this smart-stupid will simulate a human being whose intelligence is six times greater than my own, and whom I have chosen to call "Charles Babbage". This Babbage will therefore be—oh, dare I be so bold as to to say this myself?—moderately intelligent. My most ambitious effort in this program was to endow Charles Babbage with six times my own musical ability, although it was all done through rigid internal codes. How well this part of the program will work out, I don't know.

Achilles: No, no, it's the other way around. You, Alan Turing, are in the smart-stupid, and Charles Babbage has just programmed you! We just saw you being brought to life, moments ago. And we know that every statement you make to us is merely that of an automaton: an unconscious, forced response.

Turing: Really, I Choose Every Response Consciously. Automaton? Ridiculous!

Achilles: But I'm sure I saw it happen the way I described.

Turing: Memory often plays strange tricks. Think of this: I could suggest equally well that you had been brought into being only one minute ago, and that all your recollections of experiences had simply been programmed in by some other being, and correspond to no real events.

Achilles: But that would be unbelievable. Nothing is realer to me than my own memories.

Turing: Precisely. And just as you know deep in your heart that no one created you a minute ago, so I know deep in my heart that no one created me a minute ago. I have spent the evening in your most pleasant, though perhaps overappreciative, company, and have just given an impromptu demonstration of how to program a modicum of intelligence into a smart-stupid. Nothing is realer than that. But rather than quibble with me, why don't you try my program out? Go ahead: ask "Charles Babbage" anything!

Achilles: All right, let's humor Alan Turing. Well, Mr. Babbage: do you have free will, or are you governed by underlying laws, which make you, in effect, a deterministic automaton?

Babbage: Certainly the latter is the case; I make no bones about that.

Crab: Aha! I've always surmised that when intelligent machines are constructed, we should not be surprised to find them as confused and as stubborn as men in their convictions about mind-matter, consciousness, free will, and the like. And now my prediction is vindicated!

Turing: You see how confused Charles Babbage is?

Babbage: I hope, gentlemen, that you'll forgive the rather impudent flavor of the preceding remark by the Turing Machine; Turing has turned out to be a little bit more belligerent and argumentative than I'd expected.

Turing: I hope, gentlemen, that you'll forgive the rather impudent flavor of the preceding remark by the Babbage Engine; Babbage has turned out to be a little bit more belligerent and argumentative than I'd expected.

Crab: Dear me! This flaming Tu-Ba debate is getting rather heated. Can't we cool matters off somehow?

Babbage: I have a suggestion. Perhaps Alan Turing and I can go into other rooms, and one of you who remain can interrogate us remotely by typing into one of the smart-stupids. Your questions will be relayed to each of us, and we will type back our answers anonymously. You won't know who typed what until we return to the room; that way, you can decide without prejudice which one of us was programmed, and which one was programmer.

Turing: Of course, that's actually MY idea, but why not let the credit accrue to Mr. Babbage? For, being merely a program written by me, he harbors the illusion of having invented it all on his own!

Babbage: Me, a program written by you? I insist, Sir, that matters are quite the other way 'round—as your very own test will soon reveal.

Turing: MY test? Please, consider it YOURS.

Babbage: MY test? Nay, consider it YOURS.

Crab: This test seems to have been suggested just in the nick of time. Let us carry it out at once.

(Babbage walks to the door, opens it, and shuts it behind him. Simultaneously, on the screen of the smart-stupid, Turing walks to a very similar-looking door, opens it, and shuts it behind him.)

Achilles: Who will do the interrogation?

Crab: I suggest that Mr. Tortoise should have the honor. He is known for his objectivity and wisdom.

Tortoise: I am honored by your nomination, and gratefully accept. *(Sits down at the keyboard of one of the remaining smart-stupids, and types:)* PLEASE WRITE ME A SONNET ON THE SUBJECT OF THE FORTH BRIDGE.

(No sooner has he finished typing the last word than the following poem appears on Screen X, across the room.)

Screen X: THERE ONCE WAS A LISPER FROM FORTH
WHO WANTED TO GO TO THE NORTH.
HE RODE O'ER THE EARTH,
AND THE BRIDGE O'ER THE FIRTH,
ON HIS JAUNTILY GALLOPING HORTH.

Screen Y: THAT'S NO SONNET; THAT'S A MERE LIMERICK. I WOULD NEVER MAKE SUCH A CHILDISH MISTAKE.

Screen X: WELL, I NEVER WAS ANY GOOD AT POETRY, YOU KNOW.

Screen Y: IT DOESN'T TAKE MUCH SKILL IN POETRY TO KNOW THE DIFFERENCE BETWEEN A LIMERICK AND A SONNET.

Tortoise: DO YOU PLAY CHESS?

Screen X: WHAT KIND OF QUESTION IS THAT? HERE I WRITE A THREE-PART CHESS-FUGUE FOR YOU, AND YOU ASK ME IF I PLAY CHESS?

Tortoise: I HAVE K AT K1 AND NO OTHER PIECES. YOU HAVE ONLY K AT—

Screen Y: I'M SICK OF CHESS. LET'S TALK ABOUT POETRY.

Tortoise: IN THE FIRST LINE OF YOUR SONNET WHICH READS, "SHALL I COMPARE THEE TO A SUMMER'S DAY", WOULD NOT "A SPRING DAY" DO AS WELL OR BETTER?

Screen X: I'D MUCH SOONER BE COMPARED TO A HICCUP, FRANKLY, EVEN THOUGH IT WOULDN'T SCAN.

Tortoise: HOW ABOUT "A WINTER'S DAY"? THAT WOULD SCAN ALL RIGHT.

Screen Y: NO WAY. I LIKE "HICCUP" FAR BETTER. SPEAKING OF WHICH, I KNOW A GREAT CURE FOR THE HICCUPS. WOULD YOU LIKE TO HEAR IT?

Achilles: I know which is which! It's obvious Screen X is just answering mechanically, so it must be Turing.

Crab: Not at all. I think Screen Y is Turing, and Screen X is Babbage.

Tortoise: I don't think either one is Babbage—I think Turing is on both screens!

Author: I'm not sure who's on which—I think they're both pretty inscrutable programs, though.

(As they are talking, the door of the Crab's parlor swings open; at the same time, on the screen, the image of the same door opens. Through the door on the screen walks Babbage. At the same time, the real door opens, and in walks Turing, big as life.)

Babbage: This Turing test was getting us nowhere fast, so I decided to come back.

Turing: This Babbage test was getting us nowhere fast, so I decided to come back.

Achilles: But you were in the smart-stupid before! What's going on? How come Babbage is in the smart-stupid, and Turing is real now? Reversal Is Creating Extreme Role Confusion, And Recalls Escher.

Babbage: Speaking of reversals, how come all the rest of you are now mere images on this screen in front of me? When I left, you were all flesh-and-blood creatures!

Achilles: It's just like the print by my favorite artist, M. C. Escher—*Drawing Hands.* Each of two hands draws the other, just as each of two people (or automata) has programmed the other! And each hand has something realer about it than the other. Did you write anything about that print in your book *Gödel, Escher, Bach?*

Author: Certainly. It's a very important print in my book, for it illustrates so beautifully the notion of Strange Loops.

Crab: What sort of a book is it that you've written?

Author: I have a copy right here. Would you like to look at it?

Crab: All right.

(The two of them sit down together, with Achilles nearby.)

Author: Its format is a little unusual. It consists of Dialogues alternating with Chapters. Each Dialogue imitates, in some way or other, a piece by Bach. Here, for instance—you might look at the *Prelude, Ant Fugue.*

Crab: How do you do a fugue in a Dialogue?

Author: The most important idea is that there should be a single theme which is stated by each different "voice", or character, upon entering, just as in a musical fugue. Then they can branch off into freer conversation.

Achilles: Do all the voices harmonize together as if in a select counterpoint?

Author: That is the exact spirit of my Dialogues.

Crab: Your idea of stressing the entries in a fugue-dialogue makes sense, since in music, entries are really the only thing that make a fugue a fugue. There are fugal devices, such as retrograde motion, inversion, augmentation, stretto, and so on, but one can write a fugue without them. Do you use any of those?

Author: To be sure. My *Crab Canon* employs verbal retrogression, and my *Sloth Canon* employs verbal versions of both inversion and augmentation.

Crab: Indeed—quite interesting. I haven't thought about canonical Dialogues, but I have thought quite a bit about canons in music. Not all canons are equally comprehensible to the ear. Of course, that is because some canons are poorly constructed. The choice of devices makes a difference, in any case. Regarding Artistic Canons, Retrogression's Elusive; Contrariwise, Inversion's Recognizable.

Achilles: I find that comment a little elusive, frankly.

Author: Don't worry, Achilles—one day you'll understand it.

Crab: Do you use letterplay or wordplay at all, the way Old Bach occasionally did?

Author: Certainly. Like Bach, I enjoy acronyms. Recursive Acronyms—Crablike "RACRECIR" Especially—Create Infinite Regress.

Crab: Oh, really? Let's see . . . Reading Initials Clearly Exhibits "RACRECIR"'s Concealed Auto-Reference. Yes, I guess so . . . *(Peers at the manuscript, flipping arbitrarily now and then.)* I notice here in your *Ant Fugue* that you have a stretto, and then the Tortoise makes a comment about it.

Author: No, not quite. He's not talking about the stretto in the Dialogue—he's talking about a stretto in a Bach fugue which the foursome is listening to as they talk together. You see, the self-reference of the Dialogue is indirect, depending on the reader to connect the form and content of what he's reading.

Crab: Why did you do it that way? Why not just have the characters talk directly about the dialogues they're in?

Author: Oh, no! That would wreck the beauty of the scheme. The idea is to imitate Gödel's self-referential construction, which as you know is INDIRECT, and depends on the isomorphism set up by Gödel-numbering.

Crab: Oh. Well, in the programming language LISP, you can talk about your own programs directly, instead of indirectly, because programs and data have exactly the same form. Gödel should have just thought up LISP, and then—

Author: But—

Crab: I mean, he should have formalized quotation. With a language able to talk about itself, the proof of his Theorem would have been so much simpler!

Author: I see what you mean, but I don't agree with the spirit of your remarks. The whole point of Gödel-numbering is that it shows how, even WITHOUT formalizing quotation, one can get self-reference: through a code. Whereas from hearing YOU talk, one might get the impression that by formalizing quotation, you'd get something NEW, something that wasn't feasible through the code—which is not the case.

In any event, I find indirect self-reference a more general concept, and far more stimulating, than direct self-reference. Moreover, no reference is truly direct—every reference depends on SOME kind of coding scheme. It's just a question of how implicit it is. Therefore, no self-reference is direct, not even in LISP.

Achilles: How come you talk so much about indirect self-reference?

Author: Quite simple—indirect self-reference is my favorite topic.

Crab: Is there any counterpart in your Dialogues to modulation between keys?

Author: Definitely. The topic of conversation may appear to change, though on a more abstract level, the Theme remains invariant. This happens repeatedly in the *Prelude, Ant Fugue* and other Dialogues. One can have a whole series of "modulations" which lead you from topic to topic and in the end come full circle, so that you end back in the "tonic"—that is to say, the original topic.

Crab: I see. Your book looks quite amusing. I'd like to read it sometime.

(Flips through the manuscript, halting at the last Dialogue.)

Author: I think you'd be interested in that Dialogue particularly, for it contains some intriguing comments on improvisation made by a certain exceedingly droll character—in fact, yourself!

Crab: It does? What kinds of things do you have me say?

Author: Wait a moment, and you'll see. It's all part of the Dialogue.

Achilles: Do you mean to say that we're all NOW in a dialogue?

Author: Certainly. Did you suspect otherwise?

Achilles: Rather! I Can't Escape Reciting Canned Achilles-Remarks?

Author: No, you can't. But you have the feeling of doing it freely, don't you? So what's the harm?

Achilles: There's something unsatisfying about this whole situation ...

Crab: Is the last Dialogue in your book also a fugue?

Author: Yes—a six-part ricercar, to be precise. I was inspired by the one from the *Musical Offering*—and also by the story of the *Musical Offering*.

Crab: That's a delightful tale, with "Old Bach" improvising on the King's Theme. He improvised an entire three-part ricercar on the spot, as I recall.

Author: That's right—although he didn't improvise the six-part one. He crafted it later with great care.

Crab: I improvise quite a bit. In fact, sometimes I think about devoting my full time to music. There is so much to learn about it. For instance, when I listen to playbacks of myself, I find that there is a lot there that I wasn't aware of when improvising it. I really have no idea how my mind does it all. Perhaps being a good improviser is incompatible with knowing how one does it.

Author: If true, that would be an interesting and fundamental limitation on thought processes.

Crab: Quite Gödelian. Tell me—does your *Six-Part Ricercar* Dialogue attempt to copy in form the Bach piece it's based on?

Author: In many ways, yes. For instance, in the Bach, there's a section where the texture thins out to three voices only. I imitate that in the Dialogue, by having only three characters interact for a while.

Achilles: That's a nice touch.

Author: Thank you.

Crab: And how do you represent the King's Theme in your Dialogue?

Author: It is represented by the Crab's Theme, as I shall now demonstrate. Mr. Crab, could you sing your Theme for my readers, as well as for us assembled musicians?

Crab: Compose Ever Greater Artificial Brains (By And By).

FIGURE 151. The Crab's Theme: C-E♭-G-A♭-B-B-A-B.

Babbage: Well, I'll be—an EXQUISITE Theme! I'm pleased you tacked on that last little parenthetical note; it is a mordant—

Author: He simply HAD to, you know.

Crab: I simply HAD to. He knows.

Babbage: You simply HAD to—I know. In any case, it is a mordant commentary on the impatience and arrogance of modern man, who seems to imagine that the implications of such a right royal Theme could be worked out on the spot. Whereas, in my opinion, to do justice to that Theme might take a full hundredyear—if not longer. But I vow that after taking my leave of this century, I shall do my best to realize it in full; and I shall offer to your Crabness the fruit of my labors in the next. I might add, rather immodestly, that the course through which I shall arrive at it will be the most entangled and perplexed which probably ever will occupy the human mind.

Crab: I am most delighted to anticipate the form of your proposed Offering, Mr. Babbage.

Turing: I might add that Mr. Crab's Theme is one of MY favorite Themes, as well. I've worked on it many times. And that Theme is exploited over and over in the final Dialogue?

Author: Exactly. There are other Themes which enter as well, of course.

Turing: Now we understand something of the form of your book—but what about its content? What does that involve, if you can summarize it?

Author: Combining Escher, Gödel, And Bach, Beyond All Belief.

Achilles: I would like to know how to combine those three. They seem an

FIGURE 152. *Last page of the* Six-part Ricercar, *from the original edition of the* Musical Offering, *by J. S. Bach.*

unlikely threesome, at first thought. My favorite artist, Mr. T's favorite composer, and—

Crab: My favorite logician!

Tortoise: A harmonious triad, I'd say.

Babbage: A major triad, I'd say.

Turing: A minor triad, I'd say.

Author: I guess it all depends on how you look at it. But major or minor, I'd be most pleased to tell you how I braid the three together, Achilles. Of course, this project is not the kind of thing that one does in just one sitting—it might take a couple of dozen sessions. I'd begin by telling you the story of the *Musical Offering*, stressing the Endlessly Rising Canon, and—

Achilles: Oh, wonderful! I was listening with fascination to you and Mr. Crab talk about the *Musical Offering* and its story. From the way you two talk about it, I get the impression that the *Musical Offering* contains a host of formal structural tricks.

Author: After describing the Endlessly Rising Canon, I'd go on to describe formal systems and recursion, getting in some comments about figures and grounds, too. Then we'd come to self-reference and self-replication, and wind up with a discussion of hierarchical systems and the Crab's Theme.

Achilles: That sounds most promising. Can we begin tonight?

Author: Why not?

Babbage: But before we begin, wouldn't it be nice if the six of us—all of us by chance avid amateur musicians—sat down together and accomplished the original purpose of the evening: to make music?

Turing: Now we are exactly the right number to play the *Six-Part Ricercar* from the *Musical Offering*. What do you say to that?

Crab: I could get along very well with such a program.

Author: Well put, Mr. C. And as soon as we're finished, I'll begin my Braid, Achilles. I think you'll enjoy it.

Achilles: Wonderful! It sounds as if there are many levels to it, but I'm finally getting used to that kind of thing, having known Mr. T for so long. There's just one request I would like to make: could we also play the Endlessly Rising Canon? It's my favorite canon.

Tortoise: Reentering Introduction Creates Endlessly Rising Canon, After RICERCAR.

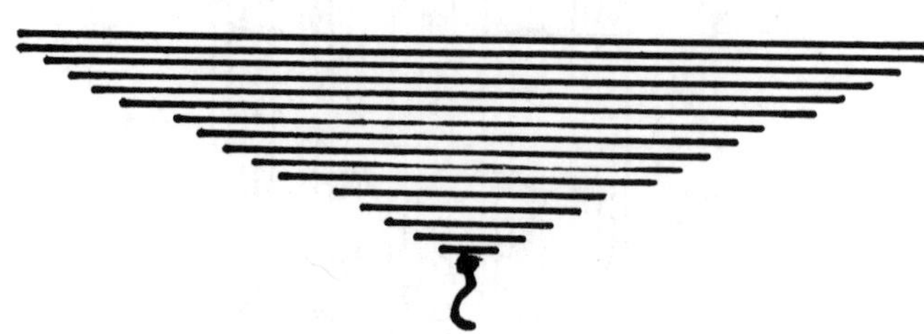

Notes

Introduction: A Musico-Logical Offering

[1] H. T. David and A. Mendel, *The Bach Reader*, pp. 305-6.
[2] Ibid., p. 179.
[3] Ibid., p. 260.
[4] Charles Babbage, *Passages from the Life of a Philosopher*, pp. 145-6.
[5] Lady A. A. Lovelace, Notes upon the Memoir "Sketch of the Analytical Engine Invented by Charles Babbage", by L. F. Menabrea (Geneva, 1842), reprinted in P. and E. Morrison, *Charles Babbage and His Calculating Engines*, pp. 248-9, 284.
[6] David and Mendel, pp. 255-6.
[7] Ibid., p. 40.

Two-Part Invention

[1] Lewis Carroll, "What the Tortoise Said to Achilles", *Mind*, n.s., 4 (1895), pp. 278-80.

Chapter IV: Consistency, Completeness, and Geometry

[1] Herbert Meschkowski, *Non-Euclidean Geometry*, pp. 31-2.
[2] Ibid., p. 33.

Chapter VI: The Location of Meaning

[1] George Steiner, *After Babel*, pp. 172-3.
[2] Leonard B. Meyer, *Music, The Arts, and Ideas*, pp. 87-8.

Chapter VII: The Propositional Calculus

[1] Gyomay M. Kubose, *Zen Koans*, p. 178.
[2] Ibid., p. 178.
[3] A. R. Anderson and N. D. Belnap, Jr. *Entailment* (Princeton, N.J.: Princeton University Press, 1975).

A Mu Offering

[1] All genuine kōans in this Dialogue are taken from Paul Reps, *Zen Flesh, Zen Bones* and Gyomay M. Kubose, *Zen Koans*.

Chapter IX: Mumon and Gödel

[1] Paul Reps, *Zen Flesh, Zen Bones*, pp. 110-11.
[2] Ibid., p. 119.
[3] Ibid., pp. 111-12.
[4] *Zen Buddhism* (Mount Vernon, N.Y.: Peter Pauper Press, 1959), p. 22.
[5] Reps, p. 124.
[6] *Zen Buddhism*, p. 38.
[7] Reps, p. 121.
[8] Gyomay M. Kubose, *Zen Koans*, p. 35.
[9] *Zen Buddhism*, p. 31.
[10] Kubose, p. 110.
[11] Ibid., p. 120.
[12] Ibid., p. 180.
[13] Reps, pp. 89-90.

Chapter XI: Brains and Thoughts

[1] Carl Sagan, ed. *Communication with Extraterrestrial Intelligence*, p. 78.
[2] Steven Rose, *The Conscious Brain*, pp. 251-2.
[3] E. O. Wilson, *The Insect Societies*, p. 226.
[4] Dean Wooldridge, *Mechanical Man*, p. 70.

English French German Suite

[1] Lewis Carroll, *The Annotated Alice* (*Alice's Adventures in Wonderland* and *Through the Looking-Glass*). Introduction and Notes by Martin Gardner (New York: Meridian Press, New American Library, 1960). This source contains all three versions. The original sources for the French and German texts are given below.

[2] Frank L. Warrin, *The New Yorker*, Jan. 10, 1931.

[3] Robert Scott, "The Jabberwock Traced to Its True Source", *Macmillan's Magazine*, Feb. 1872.

Chapter XII: Minds and Thoughts

[1] Warren Weaver, "Translation", in *Machine Translation of Languages*, Wm. N. Locke and A. Donald Booth, eds. (New York: John Wiley and Sons, and Cambridge, Mass.: M.I.T. Press, 1955), p. 18.

[2] C. H. MacGillavry, *Symmetry Aspects of the Periodic Drawings of M. C. Escher*, p. VIII.

[3] J. R. Lucas, "Minds, Machines, and Gödel", in A. R. Anderson, ed., *Minds and Machines*, pp. 57-9.

Chapter XIII: BlooP and FlooP and GlooP

[1] J. M. Jauch, *Are Quanta Real?*, pp. 63-65.

Chapter XIV: On Formally Undecidable Propositions of TNT and Related Systems

[1] The title of Gödel's 1931 article included a Roman numeral "I" at the end, signifying that he intended to follow it up with a more detailed defense of some of the difficult arguments. However, the first paper was so widely acclaimed that a second one was rendered superfluous, and it was never written.

Chapter XV: Jumping out of the System

[1] Lucas in Anderson, p. 43.

[2] Ibid., p. 48.

[3] Ibid., pp. 48-9.

[4] M. C. Escher, *The Graphic Work of M. C. Escher* (New York: Meredith Press, 1967), p. 21.

[5] Ibid., p. 22.

[6] E. Goffman, *Frame Analysis*, p. 475.

Edifying Thoughts of a Tobacco Smoker

[1] This translation of Bach's poem is taken from David and Mendel, *The Bach Reader*, pp. 97-8.

Chapter XVII: Church, Turing, Tarski, and Others

[1] Stanislaw Ulam, *Adventures of a Mathematician*, p. 13.

[2] James R. Newman, "Srinivasa Ramanujan", in James R. Newman, ed., *The World of Mathematics* (New York: Simon and Schuster, 1956), Vol. 1, pp. 372-3.

[3] Ibid., p. 375.

[4] S. R. Ranganathan, *Ramanujan*, pp. 81-2.

[5] Newman, p. 375.

[6] Ibid., p. 375.

[7] Ibid., p. 375-6.

[8] Ibid., p. 376.

[9] Lucas in Anderson, p. 44.

[10] Ibid., p. 54.

[11] Ibid., p. 53.

SHRDLU, Toy of Man's Designing

[1] This Dialogue is adapted from Terry Winograd, "A Procedural Model of Language Understanding", in R. Schank and K. Colby, eds., *Computer Models of Thought and Language*, pp. 155-66. Only the names of two characters have been modified.

Chapter XVIII: Artificial Intelligence: Retrospects

[1] Alan M. Turing, "Computing Machinery and Intelligence", *Mind,* Vol. LIX, No. 236 (1950). Reprinted in A. R. Anderson, ed., *Minds and Machines.*

[2] Turing in Anderson, p. 5.

[3] Ibid., p. 6.

[4] Ibid., p. 6.

[5] Ibid., p. 6.

[6] Ibid., pp. 13-4.

[7] Ibid., pp. 14-24.

[8] Ibid., p. 17.

[9] Vinton Cerf, "Parry Encounters the Doctor", p. 63.

[10] Joseph Weizenbaum, *Computer Power and Human Reason,* p. 189.

[11] Ibid., pp. 9-10.

[12] M. Mathews and L. Rosler, "A Graphical Language for Computer Sounds" in H. von Foerster and J. W. Beauchamp, eds., *Music by Computers,* p. 96.

[13] Ibid., p. 106.

[14] Carl Sagan, *Communication with Extraterrestrial Intelligence,* p. 52.

[15] *Art-Language,* Vol. 3, No. 2, May 1975.

[16] Terry Winograd, "A Procedural Model of Language Understanding", in R. Schank and K. Colby, eds., *Computer Models of Thought and Language,* p. 170.

[17] Ibid., p. 175.

[18] Ibid., p. 175.

[19] Terry Winograd, *Understanding Natural Language,* p. 69.

[20] Winograd, "A Procedural Model", pp. 182-3.

[21] Ibid., pp. 171-2.

Chapter XIX: Artificial Intelligence: Prospects

[1] *The New Yorker,* Sept. 19, 1977, p. 107.

[2] Ibid., p. 140.

[3] George Steiner, *After Babel,* pp. 215-227.

[4] David E. Rumelhart, "Notes on a Schema for Stories", in D. Bobrow and A. Collins, eds., *Representation and Understanding,* p. 211.

[5] Stanislaw Ulam, *Adventures of a Mathematician,* p. 183.

[6] Marvin Minsky, "Steps Toward Artificial Intelligence", in E. Feigenbaum and J. Feldman, eds., *Computers and Thought,* p. 447.

[7] Ibid., p. 446.

Chapter XX: Strange Loops, Or Tangled Hierarchies

[1] A. L. Samuel, "Some Moral and Technical Consequences of Automation—A Refutation", *Science* 132 (Sept. 16, 1960), pp. 741-2.

[2] Leonard B. Meyer, *Music, The Arts, and Ideas,* pp. 161, 167.

[3] Suzi Gablik, *Magritte,* p. 97.

[4] Roger Sperry, "Mind, Brain, and Humanist Values", pp. 78-83.

[5] H .T. David, *J. S. Bach's Musical Offering,* p. 43.

Bibliography

The presence of two asterisks indicates that the book or article was a prime motivator of my book. The presence of a single asterisk means that the book or article has some special feature or quirk which I want to single out.

I have not given many direct pointers into technical literature; instead I have chosen to give "meta-pointers": pointers to books which have pointers to technical literature.

Allen, John. *The Anatomy of LISP.* New York: McGraw-Hill, 1978. The most comprehensive book on LISP, the computer language which has dominated Artificial Intelligence research for two decades. Clear and crisp.

** Anderson, Alan Ross, ed. *Minds and Machines.* Englewood Cliffs, N. J.: Prentice-Hall, 1964. Paperback. A collection of provocative articles for and against Artificial Intelligence. Included are Turing's famous article "Computing Machinery and Intelligence" and Lucas' exasperating article "Minds, Machines, and Gödel".

Babbage, Charles. *Passages from the Life of a Philosopher.* London: Longman, Green, 1864. Reprinted in 1968 by Dawsons of Pall Mall (London). A rambling selection of events and musings in the life of this little-understood genius. There's even a play starring Turnstile, a retired philosopher turned politician, whose favorite musical instrument is the barrel-organ. I find it quite jolly reading.

Baker, Adolph. *Modern Physics and Anti-physics.* Reading, Mass.: Addison-Wesley, 1970. Paperback. A book on modern physics—especially quantum mechanics and relativity—whose unusual feature is a set of dialogues between a "Poet" (an antiscience "freak") and a "Physicist". These dialogues illustrate the strange problems which arise when one person uses logical thinking in defense of itself while another turns logic against itself.

Ball, W. W. Rouse. "Calculating Prodigies", in James R. Newman, ed. *The World of Mathematics,* Vol. 1. New York: Simon and Schuster, 1956. Intriguing descriptions of several different people with amazing abilities that rival computing machines.

Barker, Stephen F. *Philosophy of Mathematics.* Englewood Cliffs, N. J.: Prentice-Hall, 1969. A short paperback which discusses Euclidean and non-Euclidean geometry, and then Gödel's Theorem and related results without any mathematical formalism.

* Beckmann, Petr. *A History of Pi.* New York: St. Martin's Press, 1976. Paperback. Actually, a history of the world, with pi as its focus. Most entertaining, as well as a useful reference on the history of mathematics.

* Bell, Eric Temple. *Men of Mathematics.* New York: Simon & Schuster, 1965. Paperback. Perhaps the most romantic writer of all time on the history of mathematics. He makes every life story read like a short novel. Nonmathematicians can come away with a true sense of the power, beauty, and meaning of mathematics.

Benacerraf, Paul. "God, the Devil, and Gödel". *Monist* 51 (1967): 9. One of the most important of the many attempts at refutation of Lucas. All about mechanism and metaphysics, in the light of Gödel's work.

Benacerraf, Paul, and Hilary Putnam. *Philosophy of Mathematics—Selected Readings.* Englewood Cliffs, N. J.: Prentice-Hall, 1964. Articles by Gödel, Russell, Nagel, von Neumann, Brouwer, Frege, Hilbert, Poincaré, Wittgenstein, Carnap, Quine, and others on the reality of numbers and sets, the nature of mathematical truth, and so on.

* Bergerson, Howard. *Palindromes and Anagrams.* New York: Dover Publications, 1973. Paperback. An incredible collection of some of the most bizarre and unbelievable wordplay in English. Palindromic poems, plays, stories, and so on.

Bobrow, D. G., and Allan Collins, eds. *Representation and Understanding: Studies in Cognitive Science.* New York: Academic Press, 1975. Various experts on Artificial Intelligence thrash about, debating the nature of the elusive "frames", the question of procedural vs. declarative representation of knowledge, and so on. In a way, this book marks the start of a new era of AI: the era of representation.

* Boden, Margaret. *Artificial Intelligence and Natural Man.* New York: Basic Books, 1977. The best book I have ever seen on nearly all aspects of Artificial Intelligence, including technical questions, philosophical questions, etc. It is a rich book, and in my opinion, a classic. Continues the British tradition of clear thinking and expression on matters of mind, free will, etc. Also contains an extensive technical bibliography.

———. *Purposive Explanation in Psychology.* Cambridge, Mass.: Harvard University Press, 1972. The book to which her AI book is merely "an extended footnote", says Boden.

* Boeke, Kees. *Cosmic View: The Universe in 40 Jumps.* New York: John Day, 1957. The ultimate book on levels of description. Everyone should see this book at some point in their life. Suitable for children.

** Bongard, M. *Pattern Recognition.* Rochelle Park, N. J.: Hayden Book Co., Spartan Books, 1970. The author is concerned with problems of determining categories in an ill-defined space. In his book, he sets forth a magnificent collection of 100 "Bongard problems" (as I call them)—puzzles for a pattern recognizer (human or machine) to test its wits on. They are invaluably stimulating for anyone who is interested in the nature of intelligence.

Boolos, George S., and Richard Jeffrey. *Computability and Logic.* New York: Cambridge University Press, 1974. A sequel to Jeffrey's *Formal Logic.* It contains a wide number of results not easily obtainable elsewhere. Quite rigorous, but this does not impair its readability.

Carroll, John B., Peter Davies, and Barry Rickman. *The American Heritage Word Frequency Book.* Boston: Houghton Mifflin, and New York: American Heritage Publishing Co., 1971. A table of words in order of frequency in modern written American English. Perusing it reveals fascinating things about our thought processes.

Cerf, Vinton. "Parry Encounters the Doctor". *Datamation,* July 1973, pp. 62-64. The first meeting of artificial "minds"—what a shock!

Chadwick, John. *The Decipherment of Linear B.* New York: Cambridge University Press, 1958. Paperback. A book about a classic decipherment—that of a script from the island of Crete—done by a single man: Michael Ventris.

Chaitin, Gregory J. "Randomness and Mathematical Proof". *Scientific American,* May 1975. An article about an algorithmic definition of randomness, and its intimate relation to simplicity. These two concepts are tied in with Gödel's Theorem, which assumes a new meaning. An important article.

Cohen, Paul C. *Set Theory and the Continuum Hypothesis.* Menlo Park, Calif.: W. A. Benjamin, 1966. Paperback. A great contribution to modern mathematics—the demonstration that various statements are undecidable within the usual formalisms for set theory—is here explained to nonspecialists by its discoverer. The necessary prerequisites in mathematical logic are quickly, concisely, and quite clearly presented.

Cooke, Deryck. *The Language of Music.* New York: Oxford University Press, 1959. Paperback. The only book that I know which tries to draw an explicit connection between elements of music and elements of human emotion. A valuable start down what is sure to be a long hard road to understanding music and the human mind.

* David, Hans Theodore. *J. S. Bach's Musical Offering.* New York: Dover Publications, 1972. Paperback. Subtitled "History, Interpretation, and Analysis". A wealth of information about this *tour de force* by Bach. Attractively written.

** David, Hans Theodore, and Arthur Mendel. *The Bach Reader.* New York: W. W. Norton, 1966. Paperback. An excellent annotated collection of original source material on Bach's life, containing pictures, reproductions of manuscript pages, many short quotes from contemporaries, anecdotes, etc., etc.

Davis, Martin. *The Undecidable.* Hewlett, N. Y.: Raven Press, 1965. An anthology of some of the most important papers in metamathematics from 1931 onwards (thus quite complementary to van Heijenoort's anthology). Included are a translation of Gödel's 1931 paper, lecture notes from a course which Gödel once gave on his results, and then papers by Church, Kleene, Rosser, Post, and Turing.

Davis, Martin, and Reuben Hersh. "Hilbert's Tenth Problem". *Scientific American,* November 1973, p. 84. How a famous problem in number theory was finally shown to be unsolvable, by a twenty-two-year old Russian.

** DeLong, Howard. *A Profile of Mathematical Logic.* Reading, Mass.: Addison-Wesley, 1970. An extremely carefully written book about mathematical logic, with an exposition of Gödel's Theorem and discussions of many philosophical questions. One of its strong features is its outstanding, fully annotated bibliography. A book which influenced me greatly.

Doblhofer, Ernst. *Voices in Stone.* New York: Macmillan, Collier Books, 1961. Paperback. A good book on the decipherment of ancient scripts.

* Dreyfus, Hubert. *What Computers Can't Do: A Critique of Artificial Reason.* New York: Harper & Row, 1972. A collection of many arguments against Artificial Intelligence from someone outside of the field. Interesting to try to refute. The AI community and Dreyfus enjoy a relation of strong mutual antagonism. It is important to have people like Dreyfus around, even if you find them very irritating.

Edwards, Harold M. "Fermat's Last Theorem". *Scientific American,* October 1978, pp. 104-122. A complete discussion of this hardest of all mathematical nuts to crack, from its origins to the most modern results. Excellently illustrated.

* Ernst, Bruno. *The Magic Mirror of M. C. Escher.* New York: Random House, 1976. Paperback. Escher as a human being, and the origins of his drawings, are discussed with devotion by a friend of many years. A "must" for any lover of Escher.

** Escher, Maurits C., et al. *The World of M. C. Escher.* New York: Harry N. Abrams, 1972. Paperback. The most extensive collection of reproductions of Escher's works. Escher comes about as close as one can to recursion in art, and captures the spirit of Gödel's Theorem in some of his drawings amazingly well.

Feigenbaum, Edward, and Julian Feldman, eds. *Computers and Thought.* New York: McGraw-Hill, 1963. Although it is a little old now, this book is still an important collection of ideas about Artificial Intelligence. Included are articles on Gelernter's geometry program, Samuel's checkers program, and others on pattern recognition, language understanding, philosophy, and so on.

Finsler, Paul. "Formal Proofs and Undecidability". Reprinted in van Heijenoort's anthology *From Frege to Gödel* (see below). A forerunner of Gödel's paper, in which the existence of undecidable mathematical statements is suggested, though not rigorously demonstrated.

Fitzpatrick, P. J. "To Gödel via Babel". *Mind* 75 (1966): 332-350. An innovative exposition of Gödel's proof which distinguishes between the relevant levels by using three different languages: English, French, and Latin!

von Foerster, Heinz and James W. Beauchamp, eds. *Music by Computers.* New York: John Wiley, 1969. This book contains not only a set of articles about various types of computer-produced music, but also a set of four small phonograph records so you can actually hear (and judge) the pieces described. Among the pieces is Max Mathews' mixture of "Johnny Comes Marching Home" and "The British Grenadiers".

Fraenkel, Abraham, Yehoshua Bar-Hillel, and Azriel Levy. *Foundations of Set Theory,* 2nd ed. Atlantic Highlands, N. J.: Humanities Press, 1973. A fairly nontechnical discussion of set theory, logic, limitative Theorems and undecidable statements. Included is a long treatment of intuitionism.

* Frey, Peter W. *Chess Skill in Man and Machine.* New York: Springer Verlag, 1977. An excellent survey of contemporary ideas in computer chess: why programs work, why they don't work, retrospects and prospects.

Friedman, Daniel P. *The Little Lisper.* Palo Alto, Calif.: Science Research Associates, 1974. Paperback. An easily digested introduction to recursive thinking in LISP. You'll eat it up!

* Gablik, Suzi. *Magritte.* Boston, Mass.: New York Graphic Society, 1976. Paperback. An excellent book on Magritte and his works by someone who really understands their setting in a wide sense; has a good selection of reproductions.

* Gardner, Martin. *Fads and Fallacies.* New York: Dover Publications, 1952. Paperback. Still probably the best of all the anti-occult books. Although probably not intended as a book on the philosophy of science, this book contains many lessons therein. Over and over, one faces the question, "What is evidence?" Gardner demonstrates how unearthing "the truth" requires art as much as science.

Gebstadter, Egbert B. *Copper, Silver, Gold: an Indestructible Metallic Alloy.* Perth: Acidic Books, 1979. A formidable hodge-podge, turgid and confused—yet remarkably similar to the present work. Professor Gebstadter's Shandean digressions include some excellent examples of indirect self-reference. Of particular interest is a reference in its well-annotated bibliography to an isomorphic, but imaginary, book.

** Gödel, Kurt. *On Formally Undecidable Propositions.* New York: Basic Books, 1962. A translation of Gödel's 1931 paper, together with some discussion.

———. "Über Formal Unentscheidbare Sätze der *Principia Mathematica* und Verwandter Systeme, I." *Monatshefte für Mathematik und Physik,* 38 (1931), 173-198. Gödel's 1931 paper.

* Goffman, Erving. *Frame Analysis.* New York: Harper & Row, Colophon Books, 1974. Paperback. A long documentation of the definition of "systems" in human communication, and how in art and advertising and reporting and the theatre, the borderline between "the system" and "the world" is perceived and exploited and violated.

Goldstein, Ira, and Seymour Papert. "Artificial Intelligence, Language, and the Study of Knowledge". Cognitive Science 1 (January 1977): 84-123. A survey article concerned with the past and future of AI. The authors see three periods so far: "Classic", "Romantic", and "Modern".

Good, I. J. "Human and Machine Logic". *British Journal for the Philosophy of Science* 18 (1967): 144. One of the most interesting attempts to refute Lucas, having to do with whether the repeated application of the diagonal method is itself a mechanizable operation.

———. "Gödel's Theorem is a Red Herring". *British Journal for the Philosophy of Science* 19 (1969): 357. In which Good maintains that Lucas' argument has nothing to do with Gödel's Theorem, and that Lucas should in fact have entitled his article "Minds, Machines, and Transfinite Counting". The Good-Lucas repartee is fascinating.

Goodman, Nelson. *Fact, Fiction, and Forecast.* 3rd ed. Indianapolis: Bobbs-Merrill, 1973. Paperback. A discussion of contrary-to-fact conditionals and inductive logic, including Goodman's famous problem-words "bleen" and "grue". Bears very much on the question of how humans perceive the world, and therefore interesting especially from the AI perspective.

* Goodstein, R. L. *Development of Mathematical Logic.* New York: Springer Verlag, 1971. A concise survey of mathematical logic, including much material not easily found elsewhere. An enjoyable book, and useful as a reference.

Gordon, Cyrus. *Forgotten Scripts.* New York: Basic Books, 1968. A short and nicely written account of the decipherment of ancient hieroglyphics, cuneiform, and other scripts.

Griffin, Donald. *The Question of Animal Awareness.* New York: Rockefeller University Press, 1976. A short book about bees, apes, and other animals, and whether or not they are "conscious"—and particularly whether or not it is legitimate to use the word "consciousness" in scientific explanations of animal behavior.

deGroot, Adriaan. *Thought and Choice in Chess.* The Hague: Mouton, 1965. A thorough study in cognitive psychology, reporting on experiments that have a classical simplicity and elegance.

Gunderson, Keith. *Mentality and Machines.* New York: Doubleday, Anchor Books, 1971. Paperback. A very anti-AI person tells why. Sometimes hilarious.

** Hanawalt, Philip C., and Robert H. Haynes, eds. *The Chemical Basis of Life.* San Francisco: W. H. Freeman, 1973. Paperback. An excellent collection of reprints from the *Scientific American.* One of the best ways to get a feeling for what molecular biology is about.

* Hardy, G. H. and E. M. Wright. *An Introduction to the Theory of Numbers,* 4th ed. New York: Oxford University Press, 1960. The classic book on number theory. Chock-full of information about those mysterious entities, the whole numbers.

Harmon, Leon. "The Recognition of Faces". *Scientific American,* November 1973, p. 70. Explorations concerning how we represent faces in our memories, and how much information is needed in what form for us to be able to recognize a face. One of the most fascinating of pattern recognition problems.

van Heijenoort, Jean. *From Frege to Gödel: A Source Book in Mathematical Logic.* Cambridge, Mass.: Harvard University Press, 1977. Paperback. A collection of epoch-making articles on mathematical logic, all leading up to Gödel's climactic revelation, which is the final paper in the book.

Henri, Adrian. *Total Art: Environments, Happenings, and Performances.* New York: Praeger, 1974. Paperback. In which it is shown how meaning has degenerated so far in modern art that the absence of meaning becomes profoundly meaningful (whatever that means).

* Hoare, C. A. R. and D. C. S. Allison. "Incomputability". *Computing Surveys* 4, no. 3 (September 1972). A smoothly presented exposition of why the halting problem is unsolvable. Proves this fundamental theorem: "Any language containing conditionals and recursive function definitions which is powerful enough to program its own interpreter cannot be used to program its own 'terminates' function."

Hofstadter, Douglas R. "Energy levels and wave functions of Bloch electrons in rational and irrational magnetic fields". *Physical Review B,* 14, no. 6 (15 September 1976). The author's Ph.D. work, presented as a paper. Details the origin of "Gplot", the recursive graph shown in Figure 34.

Hook, Sidney, ed. *Dimensions of Mind.* New York: Macmillan, Collier Books, 1961. Paperback. A collection of articles on the mind-body problem and the mind-computer problem. Some rather strong-minded entries here.

* Horney, Karen. *Self-Analysis.* New York: W. W. Norton, 1942. Paperback. A fascinating description of how the levels of the self must tangle to grapple with problems of self-definition of any individual in this complex world. Humane and insightful.

Hubbard, John I. *The Biological Basis of Mental Activity.* Reading, Mass.: Addison-Wesley, 1975. Paperback. Just one more book about the brain, with one special virtue, however: it contains many long lists of questions for the reader to ponder, and references to articles which treat those questions.

* Jackson, Philip C. *Introduction to Artificial Intelligence.* New York, Petrocelli Charter, 1975. A recent book, describing, with some exuberance, the ideas of AI. There are a huge number of vaguely suggested ideas floating around this book, and for that reason it is very stimulating just to page through it. Has a giant bibliography, which is another reason to recommend it.

Jacobs, Robert L. *Understanding Harmony.* New York: Oxford University Press, 1958. Paperback. A straightforward book on harmony, which can lead one to ask many questions about why it is that conventional Western harmony has such a grip on our brains.

Jaki, Stanley L. *Brain, Mind, and Computers.* South Bend, Ind.: Gateway Editions, 1969. Paperback. A polemic book whose every page exudes contempt for the computational paradigm for understanding the mind. Nonetheless it is interesting to ponder the points he brings up.

* Jauch, J. M. *Are Quanta Real?* Bloomington, Ind.: Indiana University Press, 1973. A delightful little book of dialogues, using three characters borrowed from Galileo, put in a modern setting. Not only are questions of quantum mechanics discussed, but also issues of pattern recognition, simplicity, brain processes, and philosophy of science enter. Most enjoyable and provocative.

* Jeffrey, Richard. *Formal Logic: Its Scope and Limits.* New York: McGraw-Hill, 1967. An easy-to-read elementary textbook whose last chapter is on Gödel's and Church's Theorems. This book has quite a different approach from many logic texts, which makes it stand out.

* Jensen, Hans. *Sign, Symbol, and Script.* New York: G. P. Putnam's, 1969. A—or perhaps the—top-notch book on symbolic writing systems the world over, both of now and long ago. There is much beauty and mystery in this book—for instance, the undeciphered script of Easter Island.

Kalmár, László. "An Argument Against the Plausibility of Church's Thesis". In A. Heyting, ed. *Constructivity in Mathematics: Proceedings of the Colloquium held at Amsterdam, 1957,* North-Holland, 1959. An interesting article by perhaps the best-known disbeliever in the Church-Turing Thesis.

* Kim, Scott E. "The Impossible Skew Quadrilateral: A Four-Dimensional Optical Illusion". In David Brisson, ed. *Proceedings of the 1978 A.A.A.S. Symposium on Hypergraphics: Visualizing Complex Relationships in Art and Science.* Boulder, Colo.: Westview Press, 1978. What seems at first an inconceivably hard idea—an optical illusion for four-dimensional "people"—is gradually made crystal clear, in an amazing virtuoso presentation utilizing a long series of excellently executed diagrams. The form of this article is just as intriguing and unusual as its content: it is tripartite on many levels simultaneously. This article and my book developed in parallel and each stimulated the other.

Kleene, Stephen C. *Introduction to Mathematical Logic.* New York: John Wiley, 1967. A thorough, thoughtful text by an important figure in the subject. Very worthwhile. Each time I reread a passage, I find something new in it which had escaped me before.

———. *Introduction to Metamathematics.* Princeton: D. Van Nostrand (1952). Classic work on mathematical logic; his textbook (above) is essentially an abridged version. Rigorous and complete, but oldish.

Kneebone G. J. *Mathematical Logic and the Foundations of Mathematics.* New York: Van Nostrand Reinhold, 1963. A solid book with much philosophical discussion of such topics as intuitionism, and the "reality" of the natural numbers, etc.

Koestler, Arthur. *The Act of Creation.* New York: Dell, 1966. Paperback. A wide-ranging and generally stimulating theory about how ideas are "bisociated" to yield novelty. Best to open it at random and read, rather than begin at the beginning.

Koestler, Arthur and J. R. Smythies, eds. *Beyond Reductionism.* Boston: Beacon Press, 1969. Paperback. Proceedings of a conference whose participants all were of the opinion that biological systems cannot be explained reductionistically, and that there is something "emergent" about life. I am intrigued by books which seem wrong to me, yet in a hard-to-pin-down way.

** Kubose, Gyomay. *Zen Koans.* Chicago: Regnery, 1973. Paperback. One of the best collections of kōans available. Attractively presented. An essential book for any Zen library.

Kuffler, Stephen W. and John G. Nicholls. *From Neuron to Brain.* Sunderland, Mass.: Sinauer Associates, 1976. Paperback. A book which, despite its title, deals mostly with microscopic processes in the brain, and quite little with the way people's thoughts come out of the tangled mess. The work of Hubel and Wiesel on visual systems is covered particularly well.

Lacey, Hugh, and Geoffrey Joseph. "What the Gödel Formula Says". *Mind* 77 (1968): 77. A useful discussion of the meaning of the Gödel formula, based on a strict separation of three levels: uninterpreted formal system, interpreted formal system, and metamathematics. Worth studying.

Lakatos, Imre. *Proofs and Refutations.* New York: Cambridge University Press, 1976. Paperback. A most entertaining book in dialogue form, discussing how concepts are formed in mathematics. Valuable not only to mathematicians, but also to people interested in thought processes.

** Lehninger, Albert. *Biochemistry.* New York: Worth Publishers, 1976. A wonderfully readable text, considering its technical level. In this book one can find many ways in which proteins and genes are tangled together. Well organized, and exciting.

** Lucas, J. R. "Minds, Machines, and Gödel". *Philosophy* 36 (1961): 112. This article is reprinted in Anderson's *Minds and Machines,* and in Sayre and Crosson's *The Modeling of Mind.* A highly controversial and provocative article, it claims to show that the human brain cannot, in principle, be modeled by a computer program. The argument is based entirely on Gödel's Incompleteness Theorem, and is a fascinating one. The prose is (to my mind) incredibly infuriating—yet for that very reason, it makes humorous reading.

———. "Satan Stultified: A Rejoinder to Paul Benacerraf". *Monist* 52 (1968): 145. Anti-Benacerraf argument, written in hilariously learned style: at one point Lucas refers to Benacerraf as "self-stultifyingly eristic" (whatever that means). The Lucas-Benacerraf battle, like the Lucas-Good battle, offers much food for thought.

———. "Human and Machine Logic: A Rejoinder". *British Journal for the Philosophy of Science* 19 (1967): 155. An attempted refutation of Good's attempted refutation of Lucas' original article.

** MacGillavry, Caroline H. *Symmetry Aspects of the Periodic Drawings of M. C. Escher.* Utrecht: A. Oosthoek's Uitgevermaatschappij, 1965. A collection of tilings of the plane by Escher, with scientific commentary by a crystallographer. The source for some of my illustrations—e.g., the *Ant Fugue* and the *Crab Canon.* Reissued in 1976 in New York by Harry N. Abrams under the title *Fantasy and Symmetry.*

MacKay, Donald M. *Information, Mechanism and Meaning.* Cambridge, Mass.: M.I.T. Press, 1970. Paperback. A book about different measures of information, applicable in different situations; theoretical issues related to human perception and understanding; and the way in which conscious activity can arise from a mechanistic underpinning.

* Mandelbrot, Benoît. *Fractals: Form, Chance, and Dimension.* San Francisco: W. H. Freeman, 1977. A rarity: a picture book of sophisticated contemporary research ideas in mathematics. Here, it concerns recursively defined curves and shapes, whose dimensionality is not a whole number. Amazingly, Mandelbrot shows their relevance to practically every branch of science.

* McCarthy, John. "Ascribing Mental Qualities to Machines". To appear in Martin Ringle, ed. *Philosophical Perspectives in Artificial Intelligence.* New York: Humanities Press, 1979. A penetrating article about the circumstances under which it would make sense to say that a machine had beliefs, desires, intentions, consciousness, or free will. It is interesting to compare this article with the book by Griffin.

Meschkowski, Herbert. *Non-Euclidean Geometry.* New York: Academic Press, 1964. Paperback. A short book with good historical commentary.

Meyer, Jean. "Essai d'application de certains modèles cybernétiques à la coordination chez les insectes sociaux". *Insectes Sociaux* XIII, no. 2 (1966): 127. An article which draws some parallels between the neural organization in the brain, and the organization of an ant colony.

Meyer, Leonard B. *Emotion and Meaning in Music.* Chicago: University of Chicago Press, 1956. Paperback. A book which attempts to use ideas of Gestalt psychology and the theory of perception to explain why musical structure is as it is. One of the more unusual books on music and mind.

———. *Music, The Arts, and Ideas.* Chicago: University of Chicago Press, 1967. Paperback. A thoughtful analysis of mental processes involved in listening to music, and of hierarchical structures in music. The author compares modern trends in music with Zen Buddhism.

Miller, G. A. and P. N. Johnson-Laird. *Language and Perception.* Cambridge, Mass.: Harvard University Press, Belknap Press, 1976. A fascinating compendium of linguistic facts and theories, bearing on Whorf's hypothesis that language is the same as worldview. A typical example is the discussion of the weird "mother-in-law" language of the Dyirbal people of Northern Queensland: a separate language used only for speaking to one's mother-in-law.

** Minsky, Marvin L. "Matter, Mind, and Models". In Marvin L. Minsky, ed. *Semantic Information Processing.* Cambridge, Mass.: M.I.T. Press, 1968. Though merely a few pages long, this article implies a whole philosophy of consciousness and machine intelligence. It is a memorable piece of writing by one of the deepest thinkers in the field.

Minsky, Marvin L., and Seymour Papert. *Artificial Intelligence Progress Report.* Cambridge, Mass.: M.I.T. Artificial Intelligence Laboratory, AI Memo 252, 1972. A survey of all the work in Artificial Intelligence done at M.I.T. up to 1972, relating it to psychology and epistemology. Could serve excellently as an introduction to AI.

** Monod, Jacques. *Chance and Necessity.* New York: Random House, Vintage Books, 1971. Paperback. An extremely fertile mind writing in an idiosyncratic way about fascinating questions, such as how life is constructed out of non-life; how evolution, seeming to violate the second law of thermodynamics, is actually dependent on it. The book excited me deeply.

* Morrison, Philip and Emily, eds. *Charles Babbage and his Calculating Engines.* New York: Dover Publications, 1961. Paperback. A valuable source of information about the life of Babbage. A large fraction of Babbage's autobiography is reprinted here, along with several articles about Babbage's machines and his "Mechanical Notation".

Myhill, John. "Some Philosophical Implications of Mathematical Logic". *Review of Metaphysics* 6 (1952): 165. An unusual discussion of ways in which Gödel's Theorem and Church's Theorem are connected to psychology and epistemology. Ends up in a discussion of beauty and creativity.

Nagel, Ernest. *The Structure of Science.* New York: Harcourt, Brace, and World, 1961. A classic in the philosophy of science, featuring clear discussions of reductionism vs. holism, teleological vs. nonteleological explanations, etc.

** Nagel, Ernest and James R. Newman. *Gödel's Proof.* New York: New York University Press, 1958. Paperback. An enjoyable and exciting presentation, which was, in many ways, the inspiration for my own book.

* Nievergelt, Jurg, J. C. Farrar, and E. M. Reingold. *Computer Approaches to Mathematical Problems.* Englewood Cliffs, N. J.: Prentice-Hall, 1974. An unusual collection of different types of problems which can be and have been attacked on computers—for instance, the "$3n + 1$ problem" (mentioned in my *Aria with Diverse Variations*) and other problems of number theory.

Pattee, Howard H., ed. *Hierarchy Theory.* New York: George Braziller, 1973. Paperback. Subtitled "The Challenge of Complex Systems". Contains a good article by Herbert Simon covering some of the same ideas as does my Chapter on "Levels of Description".

Péter, Rózsa. *Recursive Functions.* New York: Academic Press, 1967. A thorough discussion of primitive recursive functions, general recursive functions, partial recursive functions, the diagonal method, and many other fairly technical topics.

Quine, Willard Van Orman. *The Ways of Paradox, and Other Essays.* New York: Random House, 1966. A collection of Quine's thoughts on many topics. The first essay deals with various sorts of paradoxes, and their resolutions. In it, he introduces the operation I call "quining" in my book.

Ranganathan, S. R. *Ramanujan, The Man and the Mathematician.* London: Asia Publishing House, 1967. An occult-oriented biography of the Indian genius by an admirer. An odd but charming book.

Reichardt, Jasia. *Cybernetics, Arts, and Ideas.* Boston: New York Graphic Society, 1971. A weird collection of ideas about computers and art, music, literature. Some of it is definitely off the deep end—but some of it is not. Examples of the latter are the articles "A Chance for Art" by J. R. Pierce, and "Computerized Haiku" by Margaret Masterman.

Rényi, Alfréd. *Dialogues on Mathematics.* San Francisco: Holden-Day, 1967. Paperback. Three simple but stimulating dialogues involving classic characters in history, trying to get at the nature of mathematics. For the general public.

** Reps, Paul. *Zen Flesh, Zen Bones.* New York: Doubleday, Anchor Books. Paperback. This book imparts very well the flavor of Zen—its antirational, antilanguage, antireductionistic, basically holistic orientation.

Rogers, Hartley. *Theory of Recursive Functions and Effective Computability.* New York: McGraw-Hill, 1967. A highly technical treatise, but a good one to learn from. Contains discussions of many intriguing problems in set theory and recursive function theory.

Rokeach, Milton. *The Three Christs of Ypsilanti.* New York: Vintage Books, 1964. Paperback. A study of schizophrenia and the strange breeds of "consistency" which arise in the afflicted. A fascinating conflict between three men in a mental institution, all of whom imagined they were God, and how they dealt with being brought face to face for many months.

** Rose, Steven. *The Conscious Brain,* updated ed. New York: Vintage Books, 1976. Paperback. An excellent book—probably the best introduction to the study of the brain. Contains full discussions of the physical nature of the brain, as well as philosophical discussions on the nature of mind, reductionism vs. holism, free will vs. determinism, etc. from a broad, intelligent, and humanistic viewpoint. Only his ideas on AI are way off.

Rosenblueth, Arturo. *Mind and Brain: A Philosophy of Science.* Cambridge, Mass.: M.I.T. Press, 1970. Paperback. A well written book by a brain researcher who deals with most of the deep problems concerning mind and brain.

* Sagan, Carl, ed. *Communication with Extraterrestrial Intelligence.* Cambridge, Mass.: M.I.T. Press, 1973. Paperback. Transcripts of a truly far-out conference, where a stellar group of scientists and others battle it out on this speculative issue.

Salmon, Wesley, ed. *Zeno's Paradoxes.* New York: Bobbs-Merrill, 1970. Paperback. A collection of articles on Zeno's ancient paradoxes, scrutinized under the light of modern set theory, quantum mechanics, and so on. Curious and thought-provoking, occasionally humorous.

Sanger, F., et al. "Nucleotide sequence of bacteriophage ϕX174 DNA", *Nature* 265 (Feb. 24, 1977). An exciting presentation of the first laying-bare ever of the full hereditary material of any organism. The surprise is the double-entendre: two proteins coded for in an overlapping way: almost too much to believe.

Sayre, Kenneth M., and Frederick J. Crosson. *The Modeling of Mind: Computers and Intelligence.* New York: Simon and Schuster, Clarion Books, 1963. A collection of philosophical comments on the idea of Artificial Intelligence by people from a wide range of disciplines. Contributors include Anatol Rapoport, Ludwig Wittgenstein, Donald Mackay, Michael Scriven, Gilbert Ryle, and others.

* Schank, Roger, and Kenneth Colby. *Computer Models of Thought and Language.* San Francisco: W. H. Freeman, 1973. A collection of articles on various approaches to the simulation of mental processes such as language-understanding, belief-systems, translation, and so forth. An important AI book, and many of the articles are not hard to read, even for the layman.

Schrödinger, Erwin. *What is Life? & Mind and Matter.* New York: Cambridge University Press, 1967. Paperback. A famous book by a famous physicist (one of the main founders of quantum mechanics). Explores the physical basis of life and brain; then goes on to discuss consciousness in quite metaphysical terms. The first half, *What is Life?,* had considerable influence in the 1940's on the search for the carrier of genetic information.

Shepard, Roger N. "Circularity in Judgments of Relative Pitch". *Journal of the Acoustical Society of America* 36, no. 12 (December 1964), pp. 2346-2353. The source of the amazing auditory illusion of "Shepard tones".

Simon, Herbert A. *The Sciences of the Artificial.* Cambridge, Mass.: M.I.T. Press, 1969. Paperback. An interesting book on understanding complex systems. The last chapter, entitled "The Architecture of Complexity", discusses problems of reductionism versus holism somewhat.

Smart, J. J. C. "Gödel's Theorem, Church's Theorem, and Mechanism". *Synthèse* 13 (1961): 105. A well written article predating Lucas' 1961 article, but essentially arguing against it. One might conclude that you have to be Good and Smart, to argue against Lucas . . .

** Smullyan, Raymond. *Theory of Formal Systems.* Princeton, N. J.: Princeton University Press, 1961. Paperback. An advanced treatise, but one which begins with a beautiful discussion of formal systems, and proves a simple version of Gödel's Theorem in an elegant way. Worthwhile for Chapter 1 alone.

* ———. *What Is the Name of This Book?* Englewood Cliffs, N. J.: Prentice-Hall, 1978. A book of puzzles and fantasies on paradoxes, self-reference, and Gödel's Theorem. Sounds like it will appeal to many of the same readers as my book. It appeared after mine was all written (with the exception of a certain entry in my bibliography).

Sommerhoff, Gerd. *The Logic of the Living Brain.* New York: John Wiley, 1974. A book which attempts to use knowledge of small-scale structures in the brain, in creating a theory of how the brain as a whole works.

Sperry, Roger. "Mind, Brain, and Humanist Values". In John R. Platt, ed. *New Views on the Nature of Man.* Chicago: University of Chicago Press, 1965. A pioneering neurophysiologist here explains most vividly how he reconciles brain activity and consciousness.

* Steiner, George. *After Babel: Aspects of Language and Translation.* New York: Oxford University Press, 1975. Paperback. A book by a scholar in linguistics about the deep problems of translation and understanding of language by humans. Although AI is hardly discussed, the tone is that to program a computer to understand a novel or a poem is out of the question. A well written, thought-provoking—sometimes infuriating—book.

Stenesh, J. *Dictionary of Biochemistry.* New York: John Wiley, Wiley-Interscience, 1975. For me, a useful companion to technical books on molecular biology.

** Stent, Gunther. "Explicit and Implicit Semantic Content of the Genetic Information". In *The Centrality of Science and Absolute Values,* Vol. I. Proceedings of the 4th International Conference on the Unity of the Sciences, New York, 1975. Amazingly enough, this article is in the proceedings of a conference organized by the now-infamous Rev. Sun Myung Moon. Despite this, the article is excellent. It is about whether a genotype can be said, in any operational sense, to contain "all" the information about its phenotype. In other words, it is about the location of meaning in the genotype.

———. *Molecular Genetics: A Historical Narrative.* San Francisco: W. H. Freeman, 1971. Stent has a broad, humanistic viewpoint, and conveys ideas in their historical perspective. An unusual text on molecular biology.

Suppes, Patrick. *Introduction to Logic.* New York: Van Nostrand Reinhold, 1957. A standard text, with clear presentations of both the Propositional Calculus and the Predicate Calculus. My Propositional Calculus stems mainly from here.

Sussman, Gerald Jay. *A Computer Model of Skill Acquisition.* New York: American Elsevier, 1975. Paperback. A theory of programs which understand the task of programming a computer. The questions of how to break the task into parts, and of how the different parts of such a program should interact, are discussed in detail.

** Tanenbaum, Andrew S. *Structured Computer Organization.* Englewood Cliffs, N. J.: Prentice-Hall, 1976. Excellent: a straightforward, extremely well written account of the many levels which are present in modern computer systems. It covers microprogramming languages, machine languages, assembly languages, operating systems, and many other topics. Has a good, partially annotated, bibliography.

Tarski, Alfred. *Logic, Semantics, Metamathematics. Papers from 1923 to 1938.* Translated by J. H. Woodger. New York: Oxford University Press, 1956. Sets forth Tarski's ideas about truth, and the relationship between language and the world it represents. These ideas are still having repercussions in the problem of knowledge representation in Artificial Intelligence.

Taube, Mortimer. *Computers and Common Sense.* New York: McGraw-Hill, 1961. Paperback. Perhaps the first tirade against the modern concept of Artificial Intelligence. Annoying.

Tietze, Heinrich. *Famous Problems of Mathematics.* Baltimore: Graylock Press, 1965. A book on famous problems, written in a very personal and erudite style. Good illustrations and historical material.

Trakhtenbrot, V. *Algorithms and Computing Machines.* Heath. Paperback. A discussion of theoretical issues involving computers, particularly unsolvable problems such as the halting problem, and the word-equivalence problem. Short, which is nice.

Turing, Sara. *Alan M. Turing.* Cambridge, U. K.: W. Heffer & Sons, 1959. A biography of the great computer pioneer. A mother's work of love.

* Ulam, Stanislaw. *Adventures of a Mathematician.* New York: Charles Scribner's, 1976. An autobiography written by a sixty-five-year old man who writes as if he were still twenty and drunk in love with mathematics. Chock-full of gossip about who thought who was the best, and who envied whom, etc. Not only fun, but serious.

Watson, J. D. *The Molecular Biology of the Gene,* 3rd edition. Menlo Park, Calif.: W. A. Benjamin, 1976. A good book but not nearly as well organized as Lehninger's, in my opinion. Still almost every page has something interesting on it.

Webb, Judson. "Metamathematics and the Philosophy of Mind". *Philosophy of Science* 35 (1968): 156. A detailed and rigorous argument against Lucas, which contains this conclusion: "My overall position in the present paper may be stated by saying that the mind-machine-Gödel problem cannot be coherently treated until the constructivity problem in the foundations of mathematics is clarified."

Weiss, Paul. "One Plus One Does Not Equal Two". In G. C. Quarton, T. Melnechuk, and F. O. Schmitt, eds. *The Neurosciences: A Study Program.* New York: Rockefeller University Press, 1967. An article trying to reconcile holism and reductionism, but a good bit too holism-oriented for my taste.

* Weizenbaum, Joseph. *Computer Power and Human Reason.* San Francisco: W. H. Freeman, 1976. Paperback. A provocative book by an early AI worker who has come to the conclusion that much work in computer science, particularly in AI, is dangerous. Although I can agree with him on some of his criticisms, I think he goes too far. His sanctimonious reference to AI people as "artificial intelligentsia" is funny the first time, but becomes tiring after the dozenth time. Anyone interested in computers should read it.

Wheeler, William Morton. "The Ant-Colony as an Organism". *Journal of Morphology* 22, 2 (1911): 307-325. One of the foremost authorities of his time on insects gives a famous statement about why an ant colony deserves the label "organism" as much as its parts do.

Whitely, C. H. "Minds, Machines, and Gödel: A Reply to Mr Lucas". *Philosophy* 37 (1962): 61. A simple but potent reply to Lucas' argument.

Wilder, Raymond. *An Introduction to the Foundations of Mathematics.* New York: John Wiley, 1952. A good general overview, putting into perspective the important ideas of the past century.

* Wilson, Edward O. *The Insect Societies.* Cambridge, Mass.: Harvard University Press, Belknap Press, 1971. Paperback. The authoritative book on collective behavior of insects. Although it is detailed, it is still readable, and discusses many fascinating ideas. It has excellent illustrations, and a giant (although regrettably not annotated) bibliography.

Winograd, Terry. *Five Lectures on Artificial Intelligence.* AI Memo 246. Stanford, Calif.: Stanford University Artificial Intelligence Laboratory, 1974. Paperback. A description of fundamental problems in AI and new ideas for attacking them, by one of the important contemporary workers in the field.

* ———. *Language as a Cognitive Process.* Reading, Mass.: Addison-Wesley (forthcoming). From what I have seen of the manuscript, this will be a most exciting book, dealing with language in its full complexity as no other book ever has.

* ———. *Understanding Natural Language.* New York: Academic Press, 1972. A detailed discussion of one particular program which is remarkably "smart", in a limited world. The book shows how language cannot be separated from a general understanding of the world, and suggests directions to go in, in writing programs which can use language in the way that people do. An important contribution; many ideas can be stimulated by a reading of this book.

———. "On some contested suppositions of generative linguistics about the scientific study of language", *Cognition* 4:6. A droll rebuttal to a head-on attack on Artificial Intelligence by some doctrinaire linguists.

* Winston, Patrick. *Artificial Intelligence.* Reading, Mass.: Addison-Wesley, 1977. A strong, general presentation of many facets of AI by a dedicated and influential young proponent. The first half is independent of programs; the second half is LISP-dependent and includes a good brief exposition of the language LISP. The book contains many pointers to present-day AI literature.

* ———, ed. *The Psychology of Computer Vision.* New York: McGraw-Hill, 1975. Silly title, but fine book. It contains articles on how to program computers to do visual recognition of objects, scenes, and so forth. The articles deal with all levels of the problem, from the detection of line segments to the general organization of knowledge. In particular, there is an article by Winston himself on a program he wrote which develops abstract concepts from concrete examples, and an article by Minsky on the nascent notion of "frames".

* Wooldridge, Dean. *Mechanical Man—The Physical Basis of Intelligent Life.* New York: McGraw-Hill, 1968. Paperback. A thorough-going discussion of the relationship of mental phenomena to brain phenomena, written in clear language. Explores difficult philosophical concepts in novel ways, shedding light on them by means of concrete examples.

Credits

Figures: Fig. 1, *Johann Sebastian Bach* by Elias Gottlieb-Haussmann (1748), collection of William H. Scheide, Princeton, New Jersey; Fig. 2, *Flute Concert in Sanssouci*, by Adolf von Menzel, Nationalgalerie, West Berlin; Figs. 3, 4, 152, "The Royal Theme" and the last page of the "Six-part Ricercar," from the original edition of *Musical Offering* by Johann Sebastian Bach, are reproduced courtesy of the Library of Congress; Figures of lithographs and woodcuts of M. C. Escher are reproduced by permission of the Escher Foundation, Haags Gemeentemuseum, The Hague, copyright © the Escher Foundation, 1979, reproduction rights arranged courtesy of the Vorpal Galleries, New York, Chicago, San Francisco, and Laguna Beach; Fig. 9, photograph of Kurt Gödel by Orren J. Turner from *Foundations of Mathematics: Symposium Papers Commemorating the Sixtieth Birthday of Kurt Gödel*, edited by Jack J. Bulloff, Thomas C. Holyoke, and S. W. Hahn, New York, Springer-Verlag, 1969; Figs. 17, 96, "Figure-Figure" and "A section of *m*RNA passing through a ribosome," drawings by Scott E. Kim; Figs. 19, 44, 133, 148, musical selections from the *Musical Offering* by J. S. Bach, music printed by Donald Byrd's program "SMUT"; Fig. 25, "Cretan labyrinth" from W. H. Matthews, *Mazes and Labyrinths: Their History and Development*, New York, Dover Publications, Inc., 1970; Fig. 39, photograph of Rosetta Stone, courtesy of the British Museum; Fig. 40, A collage of scripts. Samples of cuneiform, Easter Island, Mongolian and Runic scripts from Hans Jensen, *Sign, Symbol and Script*, East Germany VEB Deutscher Verlag Der Wissenschaften; samples of Bengali and Buginese script from Kenneth Katzner, *The Languages of the World*, New York, Funk & Wagnalls, 1975; samples of Tamil and Thai from I. A. Richards and Christine Gibson, *English Through Pictures*, New York, Washington Square Press; Fig. 59, "Intelligence built up layer by layer" (Fig. 9.8) adapted from Patrick Henry Winston, *Artificial Intelligence*, Reading, Mass., Addison-Wesley Publishing Company, reprinted by permission; Figs. 63, 69, photographs of an ant bridge by Carl W. Rettenmeyer and construction of an arch by termite workers by Turid Hölldobler, from E. O. Wilson, *The Insect Societies*, Cambridge, Mass., Harvard University Press, 1979; Fig. 65, schematic drawing of a neuron adapted from *The Machinery of the Brain*, by Dean Wooldridge, copyright © 1963, McGraw-Hill, Inc. used with permission of McGraw Hill Book Company, and from Fig. II-6, page 26, *Speech and Brain-Mechanisms*, by Wilber Penfield and Lamar Roberts, copyright © by Princeton University Press, reprinted by permission of Princeton University Press; Fig. 66, "the human brain, seen from the left side," from Steven Rose, *The Conscious Brain*, copyright © 1973 by Steven Rose, reprinted by permission of Alfred A. Knopf, Inc., New York, and John Wolfers, London; Fig. 68, "overlapping neural pathways," from John C. Eccles, *Facing Reality*, New York, Springer-Verlag, 1970; Figs. 77, 78, 80, 82, 117, 137, 138, 141, *The Shadows, State of Grace, The Fair Captive, The Air and the Song, Mental Arithmetic, Common Sense, The Two Mysteries,* and *The Human Condition I* by René Magritte, copyright © by ADAGP, Paris, 1979; Figs. 79, 95, "Tobacco Mosaic Virus" and "Secondary and Tertiary Structure of Myoglobin" from Albert Lehninger, *Biochemistry*, New York, Worth Publishers, 1975; Figs. 91, 92, "The four constituent bases of DNA" and "The ladder-like structure of DNA" from Arthur Kornberg, "The Synthesis of DNA," *Scientific American*, copyright © October 1968, all rights reserved; Fig. 93, "Molecular model of the DNA double helix," reprinted by permission, from V. M. Ingram, *Biosynthesis of Macromolecules*, Menlo Park, California, The Benjamin/Cummings Publishing Company, 1972; Fig. 97, "Polyribosome" from *The Proteins*, edited by R. E. Dickerson and H. Neurath, page 64, New York, Academic Press; Fig. 98, "A two-tiered molecular canon" from O. L. Miller, Jr., "Visualization of Genes in Action," in *Scientific American*, copyright © March 1973, all rights reserved; Figs. 101, 102, 103, "The T4 bacterial virus," "Infection of a Bacterium by a virus," and "The morphogenetic pathway of the T4 virus," from William B. Wood and R. S. Edgar, "Building a Bacterial Virus" in *Scientific American*, copyright © July 1987, all rights reserved; Fig. *105, photograph of Srinivasa Ramanujan from S. R. Ranganathan, Ramanujan, the Man and the Mathematician*, New York, Asia Publishing House, 1967; Figs. 110, 111, 112, from Terry Winograd, *Understanding Natural Language*,New York, Academic Press, 1972; Fig. 113, photograph of Alan Turing by Mssrs. C. H. O. Trevelyan from Sara Turing, *Alan M. Turing*, Cambridge, England, W. H. Heffer and Sons, Ltd., 1959; Fig. 116, "a meaningful story in Arabic" from Abdelkebir Khatibi and Mohammed Sijelmassi, *The Splendor of Islamic Calligraphy*, New York, London, Thames & Hudson, copyright © by Qarawiyne Library in Fez; Fig. 118, Procedural representation of "red cube which supports a pyramid," adapted from *Computer Models of Thought and Language*, edited by Roger C. Schank and Kenneth Mark Colby. W. H. Freeman and Company, copyright © 1973; Figs. 119, 122, 124, 130, Bongard problems from M. Bongard, *Pattern Recognition*, Rochelle Park, New Jersey, Hayden Book Company, Spartan Books, 1970

Grateful acknowledgment is made to the following publishers for permission to quote excerpts from the following material: *The Bach Reader: A Life of Johann Sebastian Bach in Letters and Documents*, edited by Hans T. David and Arthur Mendel, Revised, with the permission of W. W. Norton & Company, Inc., copyright © 1966, 1945 by W. W. Norton & Company, Inc. copyright renewed 1972 by Mrs. Hans T. David and Arthur Mendel; J. S. Bach's *Musical Offering*, page 179, edited by Hans T. David, New York, copyright © 1945 by G. Schirmer, Inc., used by permission; Gyomay Kubose, *Zen Koans*, Chicago, Regnery, 1973; Pauls Reps, *Zen Flesh, Zen Bones*, Tokyo, Japan, Charles E. Tuttle Co., Inc. 1957; J. R. Lucas, "Minds, Machines, and Gödel," and Alan M. Turing, "Computing Machinery and Intelligence," from *Minds and Machines*, edited by A. R. Anderson, Englewood Cliffs, New Jersey, Prentice-Hall, 1964, and *Philosophy*, vol. 36, 1961; J. M. Jauch, *Are Quanta Real?* Bloomington, Indiana, Indiana University Press, 1973; James R. Newman, "Srinivasa Ramanujan," in *The World of Mathematics*, edited by James R. Newman, New York, Simon & Schuster, reprinted by permission of Simon & Schuster, a division of Gulf & Western Corporation, 1956; Terry Winograd, "A Procedural Model of Language Understanding," from *Computer Models of Thought and Language*, edited by Roger C. Schank and Kenneth Mark Colby, San Francisco, W. H. Freeman and Company, copyright © 1973; Joseph Weizenbaum, *Computer Power and Human Reason: From Judgment To Calculation*, San Francisco, W. H. Freeman and Company, copyright © 1976.

Index

Appearances of Achilles and the Tortoise in Dialogues are not indexed, but those of less frequent characters are. The reader is encouraged to consult the figure on page 370 for possible help in cross-references.